Microbial Surfactants in Pharmaceuticals and Cosmetics

Editors

R.Z. Sayyed
Department of Biological Sciences and Chemistry
University of Nizwa, Sultanate of Oman

Shilpa Mujumdar
Department of Microbiology
Modern College of Engineering
Pune, India

CRC Press
Taylor & Francis Group
Boca Raton London New York

CRC Press is an imprint of the
Taylor & Francis Group, an **informa** business

First edition published 2025
by CRC Press
2385 NW Executive Center Drive, Suite 320, Boca Raton FL 33431

and by CRC Press
4 Park Square, Milton Park, Abingdon, Oxon, OX14 4RN

CRC Press is an imprint of Taylor & Francis Group, LLC

Library of Congress Cataloging-in-Publication Data (applied for)

ISBN: 978-1-032-54438-0 (hbk)
ISBN: 978-1-032-54439-7 (pbk)
ISBN: 978-1-003-42486-4 (ebk)

DOI: 10.1201/9781003424864

Typeset in Times New Roman
by Prime Publishing Services

Foreword

Biosurfactants (BS) and bioemulsifiers (BE) derived from microbes are today's green compounds. Many different microorganisms produce these chemicals. BS and BE come from biological sources and have qualities comparable to commercial surfactants. They are created by changing the structure of the substrate and the growing conditions. They have been employed as an environmentally benign alternative to synthetic chemical surfactants and emulsifiers. They have specific benefits over their chemical equivalents, especially in the medical, pharmaceutical, and cosmetics industries. Their unique properties such as low toxicity, stability at wider pH and temperature range, high surface activity, biodegradability, and excellent emulsifying and demulsifying abilities make them ideal molecules for commercial applications.

BS have received importance in medicine due to their antimicrobial, antiadhesive activity, and anticancer actions. They have also proven effective in gene finding and intracellular product recovery. In cosmetics, BS act as wetting agents, solubilizers, dispersants, foaming agents, cleansers, detergents, and emulsion-forming agents. They also have an important function in formulations. Emerging roles of biosurfactants and bioemulsifiers in different industries will replace hazardous chemical surfactants and emulsifiers and direct these industries toward green synthesis. Furthermore, the growing consumer preference for biological goods over chemicals drives up commercial demand for BS and BE.

The book Microbial Surfactants in Pharmaceutical and Cosmetics provides a solid balance of fundamental and advanced principles on the manufacture, synthesis, and use of biosurfactants and bioemulsifiers, as well as their applications in pharmaceuticals, drug design, medicine, and cosmetics. The writers presented the book in a straightforward and understandable manner, with well-illustrated illustrations, procedures, figures, and fresh field data. I am confident that this book will benefit students, researchers, instructors, and entrepreneurs in the pharmaceutical and cosmetics industries.

Professor Ts. Dr. Wong Ling Shing
Pro Vice-Chancellor, INTI International University, Malaysia

Preface

Biosurfactants and bioemulsifers are green and sustainable molecules of microbial origin. Biosurfactants produced by different microorganisms are of different structures produced naturally, unlike surfactants. These particles are mainly prepared by altering the substrate's structure or growth settings. Hence, these biosurfactants have a unique advantage over their chemical counterparts, particularly in medicine, drugs, and cosmetics.

They are suitable for commercial applications because of their low toxicity, wide pH and temperature range, high surface activity, biodegradability, and excellent emulsifying and demulsifying abilities. They gained particular importance in medicine due to their antimicrobial, antiadhesive activity, and anticancer actions. They are also used successfully in gene discovery and the recovery of intracellular products. In cosmetics, biosurfactants act as wetting agents, solubilizers, dispersants, foaming agents, cleansers, detergents, and emulsion-forming agents and thus play a significant role in formulations. Emerging roles of biosurfactants and bioemulsifiers in different industries will replace chemical surfactants and direct these industries toward green synthesis.

The increased commercial demand for biosurfactants is due to their green properties, which raised their use in industries such as personal care, soaps, detergents, and industrial cleaning. Furthermore, the consumer's awareness of biological products over chemicals boosts the industry demand in the current timeline.

The prime goal of this book is to provide a mixture of basic and advanced concepts on the production, synthesis and application of biosurfactants and bioemulsifiers in the field of pharmacy, drug designing, medicine, and cosmetics. This volume is presented in an easy-to-understand manner, with well-illustrated diagrams, protocols, figures, and recent data. As such, this book will be helpful for students, researchers, teachers, and entrepreneurs in medicine, pharmacy, and cosmetics.

R.Z. Sayyed
Shilpa Mujumdar

List of Abbreviations

α-MES	-	α-methyl ether sulfonates
ABC	-	ATP-binding cassette
ADBAC	-	Alkyl dimethyl benzyl ammonium chloride
ADEBAC	-	Alkyl dimethyl ethylbenzyl ammonium chloride
AFM	-	Atomic force microscopy
AgNPs	-	Silver nanoparticles
AI	-	Artificial Intelligence
AI-2	-	Autoinducer-2
AMPs	-	Anti-microbial peptides
AMR	-	Anti-microbial resistance
ANN	-	Artificial Neural Networks
ANN-GA	-	Artificial Neural Intelligence coupled with Genetic Algorithm
APDT	-	Antibacterial photodynamic therapies
ARDS	-	Acute respiratory distress syndrome
ARTP	-	Atmospheric and room-temperature plasma
ATP	-	Adenosine tri phosphate
ATR	-	Attenuated total reflectance
AuNPs	-	Gold nanoparticles
BB	-	Bacterial biosurfactants
BE	-	Bioemulsifier
BE/Bes	-	Bioemulsifiers
BS	-	Biosurfactant
BSA	-	Bovine serum albumin
BSs	-	Biosurfactants
CAGR	-	Compounded annual growth rate
CF	-	Cystic fibrosis
CFU	-	Colony forming unit
CGLs	-	Complex glycolipids
CIR	-	Cosmetic ingredient review
CLRs	-	C-type lectin receptors
CLSI	-	Clinical & Laboratory Standards Institute
CMC	-	Critical micelle concentration
CNS	-	Central nervous system
COMs	-	Complex organic molecules
CoV	-	Coronavirus

CP	-	Cerato-platanin
CPNP	-	Cosmetic products notification portal
CPP	-	Critical packing parameter
CRISPR	-	Clustered regularly interspaced short palindromic repeats
DAG	-	Diacylglycerol
DBT	-	Dibenzothiophene
DDAC	-	Didecylammonium chloride
DDS	-	Drug delivery system
diRL	-	di-rhamnolipids
DIS	-	Draft international standard
DLS	-	Dynamic light scattering
DO	-	Dissolved oxygen
DPPH	-	1-diphenyl-2-picrylhydrazyl radical scavenging assays
DSS	-	Dextran sulfate sodium
EEB	-	Endophytic bacterial bio-surfactants
ELISA	-	Enzyme linked immonosorbent assay
ELSD	-	Evaporative light scattering detector
EM	-	Electron microscopy
EOR	-	Enhanced oil recovery
EPR	-	Enhanced permeability and retention
EPS	-	Extracellular polymeric substance
ESBLs	-	Extended-spectrum beta-lactamases
ESI-MS	-	Electrospray ion-MS
EU	-	European Union
EU/ml	-	Emulsification units per milliliter
FA	-	Fatty acid
FD & C Act	-	Federal Food, Drug, and Cosmetic Act
FDA	-	Food and Drug Administration (US)
FIC	-	Fractional inhibitory concentration
FMCG	-	Fast-moving consumer goods
FT-IR	-	Transform-infrared spectroscopy
G6P	-	Glucose-6-phosphate
GLs	-	Glycolipids
GMPs	-	Good manufacturing practices
GRAS	-	Generally recognized as safe
GSH	-	Glutathione
GSLs	-	Glycosphingolipids
HAI	-	Healthcare-associated infections
HLB	-	Hydrophilic-lipophilic balance
HMP	-	Human microbiome project
HMW	-	High molecular weight
HPLC	-	High-performance liquid chromatography
HPTLC	-	High-performance thin-layer chromatography
HSV	-	Herpes simplex virus
HSV-1	-	Herpes simplex virus 1
HUVEC	-	Human umbilical vein endothelial cells
IL-6	-	Interleukin-6
INCI	-	International nomenclature cosmetic ingredient

ISO	-	International Organization for Standardization
ISPR	-	In-situ product removal
LAB	-	Lactic acid bacteria
LAS	-	Linear alkylbenzene sulfonate
LHB	-	Lipophilic-hydrophilic balance
LMW	-	Low molecular weight
LPMOs	-	Lytic polysaccharide monooxygenase
MALDI-TOF	-	Matrix-assisted laser desorption/ionization-time of flight
MAPK	-	Mitogen-activated protein kinase
MDDS	-	Microemulsion drug delivery systems
MDR	-	Multi-drug resistant
MDROs	-	Multi-drug resistant organisms
MEL-A	-	Mannosylerythritol lipid A
MEL-B	-	Mannosylerythritol lipid B
MEL-C	-	Mannosylerythritol lipid C
MELs	-	Mannoselerythritol lipids
MEOR	-	Microbial enhanced oil recovery
MERS-CoV	-	Middle East Respiratory Syndrome Coronavirus
MEVs	-	Microbial-derived extracellular vesicles
MGFA	-	Methylated guanidino fatty acid
MIC	-	Minimal inhibitory concentrations
MMP	-	Mitochondrial membrane potential
monoRL	-	mono-rhamnolipids
MRI	-	Magnetic resonance imaging
MRSA	-	Methicillin-resistant Staphylococcus aureus
MS	-	Mass spectrometer
MSM	-	Minimal Salt Medium
MTS	-	3-(4,5-dimethylthiazol-2-yl)-5-(3-carboxymethoxyphenyl)-2-(4-sulfophenyl)-2H-tetrazolium
MTT	-	3-[4,5-dimethylthiazol-2-yl]-2,5 diphenyl tetrazolium bromide
MWCO	-	Molecular weight cutoff
NAC	-	Non-albicans candida
N-AHLs	-	N-acyl homoserine lactones
NF-KB	-	Nuclear factor kappa B
NHEJ	-	Non-homologous end joining
NLCs	-	Nanostructured lipid carriers
NLRs	-	NOD-like receptors
NMR	-	Nuclear magnetic resonance
nPMMA	-	Novel polymethyl methacrylate
nPMMA	-	Polymethyl methacrylate nanoparticles
NPs	-	Nanoparticles
NRLPs	-	Non-ribosomal lipid peptides
NRPs	-	Non-ribosomal peptides
NRPSs	-	Non-ribosomal peptide synthetases
O/W	-	Oil-in-water
OlsB	-	Ornithine acyl-ACP N-acyltransferase
OMP	-	Outer membrane protein

ODD	-	Oral drug delivery
ORF	-	Open reading frame
PAMPs	-	Pathogen-associated molecular patterns
PBD	-	Plackett-Burman Design
PEL	-	Pectate lyase
PFU	-	Plaque-forming units
PG	-	Polygalacturonase
PGP	-	Plant growth promoting
PGPR	-	Plant growth promoting rhizobacteria
PHA	-	Polyhydroxyalkanoate
PHB	-	Polyhydroxybutyrate
PKB	-	Protein kinase B
PLGA	-	Poly lactic-co-glycolic acid
PNL	-	Pectin lyase
PAH	-	Polycyclic aromatic hydrocarbon
PRRs	-	Pattern recognition receptors
RCI	-	Renewable carbon index
RCNMV	-	Red clover necrotic mosaic virus
RIG	-	Retinoic acid-inducible gene
RiLP	-	Ribosomally synthesized lipopeptides
RIPT	-	Repeat insult patch testing
RLs	-	Rhamnolipids
ROS	-	Reactive oxygen species
RP-HPLC	-	Reverse phase high pressure liquid chromatography
RSM	-	Response Surface Methodology
SARS-CoV	-	Severe Acute Respiratory Syndrome Coronavirus
SDS	-	Sodium dodecyl sulfate
SEM	-	Scanning electron microscopy
SGLs	-	Simple glycolipids
SIMS	-	Secondary ion-MS
SLA	-	Sophorolipid A
SLNs	-	Solid lipid nanoparticles
SLPs	-	Soporolipids
SLS	-	Sodium lauryl sulphate
SOPs	-	Standard operating procedures
SRB	-	Sulfate-reducing bacteria
SrfA	-	Surfactin A
SrfB	-	Surfactin B
SrfC	-	Surfactin C
SrfD	-	Surfactin D
SrfTEII	-	Surfactin external thioesterase
SSF	-	Solid-state fermentation
STAT	-	Signal transducer and activation of transcription
TA	-	Toxin-antitoxin
TB	-	Toluidine blue
TDM	-	Trehalose dimycolate
TE-I	-	Thiol esterase
TEM	-	Transmission electron microscopy

TGA	-	Thermogravimetric analysis
TL-1	-	Trehalose monocorynomycolate
TL-2		Trehalose dicorynomycolate
TLC	-	Thin-layer chromatography
TLRs	-	Toll-like receptors
TLs	-	Trehalolipids
TMV	-	Tobacco mosaic virus
TNF	-	Tumor necrosis factor
VNPs	-	Virus-based nanoparticles
W/O	-	Water-in-oil
XRD	-	X-ray diffraction
ZnONPs	-	Zin oxide nanoparticles

Contents

CHAPTER

1

Application of Microbial Biosurfactants and Factors Needed for their Production: An Overview

Umesh Pravin Dhuldhaj[1*], R.Z. Sayyed[2], and Chandan R. Bora[3]

[1] School of Life Sciences, Swami Ramanand Teerth Marathwada University, Nanded, Maharashtra - 431606, India

[2] Department of Microbiology, PSGVP Mandal's SI Patil Arts, GB Patel Science and STKVS Commerce College, Shahada, India

[3] NOVO Cellular Medicine Institute, Fidelity Healthcare Limited, Port of Spain, Trinidad, Spain

1. Introduction

Biosurfactant are well known compounds for its beneficial activities, including the reduction of surface tension, interface tension, acting as a good emulsifier and thickener, serving as a detergent and contributing to the pseudo solubility of immiscible solvents. Pseudo solubilization makes hydrophobic substances available to microbes for consumption as substrates.

Microbes, such as bacteria, fungi and yeast, can extracellularly synthesize biosurfactants as secondary metabolites (Tripathy et al., 2018; Dhuldhaj and Bora, 2021). These microbes have the capability to survive in extreme conditions through the production of protective compounds, which are not observed in microbes belonging to normal environments (Fenical, 1993). Biosurfactants are formed by two groups, with the head groups being amphiphilic in nature, while end groups are hydrophobic. The head regions include several functional groups such mono-, oligo-, or polysaccharides, peptides or proteins, and the end groups are formed by saturated, unsaturated, and hydroxylated fatty acids or fatty alcohols (Pacwa-Plociniczak et al., 2011). On the basis of molecular weight, these compounds can be grouped into low molecular weight biosurfactants and high molecular weight biosurfactants. Low molecular weight biosurfactants (e.g. Glycolipids, lipopeptides and phospholipids) are involved in the reduction of surface and interface tension while high molecular weight biosurfactants

* Corresponding author: umeshpd12@gmail.com

(e.g. particulate and polymeric) are involved in emulsion formations and act as stabilizing agents (Ron and Rosenberg, 2001; Shekhar et al., 2015). Biosurfactants, with respect to chemical compositions can be grouped into glycolipids, lipopeptides, fatty acids/phospholipids/neutral lipids, polymeric biosurfactants and particulate biosurfactants (Stancu, 2015; Muthusamy et al., 2008).

These compounds are highly specific in function, required in trace amounts, and have wide significance over synthetic surfactants, being less toxic and highly active even in extreme conditions (Joshi et al., 2013; Pirog et al., 2015). It covers economic applications such as industrial purposes like oil recovery and mining, and it also contributes to environmental protection through bioremediation, aiding in the removal of hydrocarbons and heavy metals from contaminated sites (Desai and Banat, 1997; Banat et al., 2000; Shah et al., 2022; Saranraj et al., 2022b). Heavy metals are resistant to degradation, but microbial systems can transform them by altering their oxidation states through biochemical reactions, such as oxidation-reduction reactions and alkylation. This transformation aids their removal by forming complexes with such secondary metabolites (Gao et al., 2012). Biosurfactants also play a role in microbial biofilm formations and quorum sensing (Rodrigues et al., 2006), influencing adhesion and anti-adhesion to the substratum (Neu, 1996; Ron and Rosenberg, 2001). The production of biosurfactants can be initiated at the site of requirement by introducing microbes capable of producing them. In this chapter, we focus on the factors necessary for biosurfactant production and its applications.

2. Biosynthesis of Surfactants

Microbes can biosynthesize biosurfactants under regulated conditions, such as sufficient availability of hydrophobic nutrients, appropriate cell density, and growth limiting conditions (Hausmann and Syldatk, 2014; Gautam and Tyagi, 2006; Satpute et al., 2010; Das et al., 2008b; Roongsawang et al., 2010; Kubicki et al., 2019). Pathways for biosurfactant synthesis and the putative genes required for the marine microbes can be revealed by using related non-marine microbes with the help of bioinformatic tools (Kubicki et al., 2019) e.g. Glucose lipids from Alcanivorax (Schneiker et al., 2006) and trehalose lipids from Rhodococcus (Sambles and White, 2015).

Biosurfactant fatty acids and hydroxyl fatty acids are synthesized through primary carbon metabolism, while the synthesis of lipiamino acid biosurfactant is probably done through the activation of fatty acids catalyzed by enzymes like acyltransferase or N-acyl amino acid synthase, similar to the bacterial synthesis of N-acyl amino acids (Brady et al., 2004, Brady and Clardy, 2005, Van Wagoner et al., 2006). The biosynthesis of surfactin biosurfactant occurs through the complex of four enzymatic subunits known as the surfactin synthetase complex. The enzymatic subunits involved are SrfA, SrfB, SrfC and SrfD, with molecular weights of 402 kDa, 401 kDa, 144 kDa and 40 kDa, respectively (Steller et al., 2004). The first three subunits are enzymes, while the latter one is a protein. The first three subunits are involved in forming seven modules comprising 24 catalytic domains. Each module contributes substrates to the newly synthesizing heptapeptide chain, and the latter subunit is involved in the initiation of surfactin biosynthesis (Peypoux et al., 1999, Seydlová and Svobodová, 2008). SrfA and SrfB are involved in the thioester bond cleavage and the transpeptidation reactions result in the elongation of the initiation product. During the biosynthesis of surfactin, the primary target substrates are identified by domain A (i.e. adenylation domain) which is analogous to aminoacyl-tRNA synthetase. It further catalyzes the activation of the substrate as aminoacyladenylate and is hydrolyzed by using Mg^{2+}-dependent ATP, resulting in the release of pyrophosphate (Dieckmann et al., 1995). Furthermore,the aminoacyladenylate intermediate formed binds to the free thiol group of phosphopantetheine (ppan) cofactors which were linked to domain T (thiolation domain), composed of around 80 amino acids and also known as the peptidyl carrier protein. Domain T is located downstream of domain A, while domain C is present between domain A and T of the respective module. Domain C is a condensation domain formed by around 450 amino acids (Keating et al., 2002). The aminoacyladenylate intermediate bound to domain T, having activated thioester group, faces nucleophilic attack catalyzed by domain C with its free α-amino group (Belshaw et al., 1999). Surfactin can be biosynthesized by three subunits only, namely SrfA, SrfB and SrC, with the process initiated by the subunit SrfD. SrfD induces the formation of surfactin through the formation of β-hydroxyacyl-glutamate by the transfer of fatty acid substrates to the Glu-module (Steller et al., 2004). At the peptide synthetase reaction center, the surfactin synthetase with acyltransferase enzyme possesses SrfTEII (external thioesterase), which proofreads unnecessary charging of the phosphopantetheine cofactor (Schwarzer et al., 2002). The release of the lipoheptapeptidyl intermediate takes place through hydrolysis or intramolecular reactions aided by TE-I (thioesterase domain fused with SrfC) and SrfC which condenses the last amino acid (Peypoux et al., 1999, Bruner et al., 2002, Tseng et al., 2002). The condensation domain adds only D-amino acids to the elongating chain, hence, there is an optional domain present i.e. Domain E (Epimerization domain) (Linne and Marahiel, 2000). Domain E racemizes L-leu bound to domain T; hence, the presence of L and D-amino acids gives a unique peculiarity to surfactin to interact with cellular targets (Seydlová and Svobodová, 2008).

2.1 Production of Surfactants

In a study involving the supplementation of whey as substrate with Lactobacillus species, it was found that several strains (including *Lactobacillus casei*, *Lactobacillus rhamnosus*, *Lactobacillus pentosus*, and *Lactobacillus coryniformis* torquens) are capable of producing biosurfactant, with *Lactobacillus pentosus* identified as the dominant surfactant producing strain (Rodrigues et al., 2006a). Another study by Saisa-ard et al. (2014) was the first to report that 16 microbial species are able to produce biosurfactants. These species include *Caryophanon*; *Castellaniella*; *Filibacter*; *Geminicoccus*; *Georgenia*; *Luteimonas*; *Mesorhizobium*; *Mucilaginibacter*; *Nubsella*; *Paracoccus*; *Pedobacter*; *Psychrobacter*; *Rahnella*; *Sphingobium*; *Sphingopyxis* and *Sporosarcina*. Microbes can grow in the vegetable oil by using it as the carbon source (Tummler et al., 2003, Thaniyavarn et al., 2006). Microbes growing on vegetable oils along with the hydrocarbons, produce biosurfactants that assist the microbes in dispersing oil through emulsification (More et al., 2017, Ron and Rosenberg, 2002, Kumar et al., 2007). Microbes thriving in oil fields, such as *Alcanivorax dieselolei* B-5 and *Halomonas* sp. TG39, naturally produces surfactants, providing them with a convenient way to utilize the oil as a substrate for their survival (Qiao and Shao, 2010, Gutierrez et al., 2013). It is also reported that the bacterial species *Bacillus subtilis* spreads multicellular colonies and forms aerial structures by maintaining a low surface tension in the surrounding fluid medium through the production of surfactin biosurfactant (See details in Table 1) (Angelini et al., 2009). Surfactants are often derived from petroleum products, however, microbial biosurfactants are gaining more attention, as compared to synthetic surfactants, due to their eco-friendly, environmentally protective properties. They have several applications in crude oil recovery, health care and the food industries (Desai and Banat, 1997, Banat, 1995a & b, Fiechter, 1992, Klekner and Kosaric, 1993, Muller-Hurtig et al., 1993, Velikonja and Kosaric, 1993; Zaman et al., 2022).

Table 1: List of some of microbes producing biosurfactant

Producing organisms	Biosurfactant	References
Pseudomonas aeruginosa strain BN10	Rhamnolipid	Christova et al., 2013, Jarvis et al., 1949
Pseudozyma Antarctica	Mannosylerythriol lipids	Kitamoto et al., 1990
Rhodococcus sp., *Nocardia* sp., *Arthrobacter* sp., and *Mycobacterium* sp.	Trehalose lipids	Lang and Wagner, 1987
Ustilago maydis	Cellobiolipids	Teichmann et al., 2007b
Candida sp.	Sophorolipids	Cooper and Paddock, 1984
Bacillus licheniformis and *Bacillus subtilis*	Peptide-lipid and lichenysin	Yakimov et al., 1997, Begley et al., 2009
Pseudomonas fluorescens	Viscosin	Banat et al., 2010
Serratia marcescens	Serrawettin	Lai et al., 2009
Bacillus sp.,	Fengycin	Vanittanakom et al., 1986
Arthrobacter sp.,	Arthrofactine	Morikawa et al., 1993
Bacillus brevis, Brevibacterium brevis	Gramicidins	Krauss and Chan, 1983
Bacillus polymyxa, Brevibacterium polymyxa	Polymyxins	Suzuki et al., 1965
Myroides sp., *Pseudomonas* sp., *Thiobacillus* sp. *Agrobacterium* sp., *Gluconobacter* sp.	Ornithine lipids	Desai and Banat, 1997
Nocardia erythropolis, Thiobacillus thiooxidans, Candida lepus, Acinetobacter sp., *Pseudomonas* sp., *Micrococcus* sp., *Mycococcus* sp., *Candida* sp., *Penicillium* sp., and *Aspergillus* sp.	Phospholipids, fatty acids or neutral lipids	Kappeli and Finnerty, 1979
Acinetobacter sp., *Rhodococcus erythropolis*	Phosphatidylethanolamine	Kappeli and Finnerty, 1979, Kretschmer et al., 1982
Candida lipolytica	Liposan	
Candida tropicalis	Mannan	Cirigliano and Carman, 1984, Pandya et al., 2018
Pseudomonas aeruginosa	Protein PA	
Acinetobacter calcoaceticus	Emulsan and biodispersan	Rosenberg et al., 1988
Pseudomonas strain	Polymeric biosurfactants	Shekhar et al., 2015
Pseudomonas nautica	Extracellular biosurfactants	Husain et al., 1997

Contd...

Contd...

Producing organisms	Biosurfactant	References
Pseudomonas fluorescens	Trehalose lipid-o-dialkyl monoglycerides-protein	Desai et al., 1988
Bacillus sp.	Lipopeptides, Lichenysin, surfactin, lipid protein complex and subtilisin	Shekhara et al., 2014
Acinetobacter calcoaceticus RAG-1	Emulsan	Nerurkar et al., 2009
Rhodococcus erythropolis	Glycolipid, polysaccharides,	Shulga et al., 1990
Rhodococcus sp.	free fatty acids and Trehalose dicorynomycolate	Neu et al. 1990, Singer and Finnerty, 1990; Neu et al., 1992
Halomonas sp.	Glycoprotein, glycolipids	Pepi et al., 2005; Gutiérrez et al., 2007a
Candida bombicola	sophorolipids	Casas and Garcia-Ochoa, 1999
Yarrowia lipolytica	complex of lipid carbohydrate protein	Zinjarde and Pant, 2002
Yarrowia lipolytica, IMUFRJ 50682	Yansan	
Torulopis petrophilum	sophorolipids	Cooper and Paddock, 1983
Kurtzmanomyces sp. I-11	Mannosylerythritol lipids (MEL)	Kakugawa et al., 2002

2.2 Factors Affecting Biosurfactant Production

Several environmental and nutrient factors contribute to the production of Biosurfactant (Saharan et al., 2011). The important characteristics of biosurfactant i.e. can be used to testify the quality of produced biosurfactant (Ikhwani et al., 2017). The characterization and identifications of biosurfactants can be done with the important environmental factor i.e. pH.

2.2.1 Nutrient Factors and Salt Concentration

For the mass production of biosurfactants, using fermentation methods in bioreactors, an adequate supply of nutrients is necessary. Nutrients play a crucial role in optimizing biosurfactant production, and various factors, including carbon, nitrogen, phosphate, etc., influence the production of surfactants. as mentioned below:

2.2.1.1 Carbon

Microbes primarily utilize hydrocarbons, carbohydrates, fats, vegetable oils, petroleum oils, and their derivatives as major carbon sources for biosurfactant production. Additionally, a cost-effective fatty acids mixture derived from soap stock oil can serve as a good carbon source, especially when combined with nitrogen sources (Shabtai, 1990). Aerophilic microbes including bacteria such as species belonging to the genera *Pseudomonas*, *Bacillus* and *Acinetobacter*), yeast (e.g. yeast species belongs to the genera *Candida* and *Pseudozyma*) and fungi (such as species belonging to the genera *Aspergillus* and *Fusarium*), utilize these carbon sources in aqueous media for biosurfactant production (e Silva et al., 2014).

2.2.1.2 Nitrogen

Nitrogen, sourced as NH_4^+, NO_3^-, urea or amino acids, plays a crucial role as a nutrient affecting biosurfactant production in microbes (Duvnjak et al., 1983, Robert et al., 1989, Haba et al., 2000). Some of the microbes produce or overproduce biosurfactant in nitrogen limiting conditions (Suzuki et al., 1974, Guerra-Santos et al., 1984, Christofi and Ivshina, 2002).

2.2.1.3 Phosphate

Phosphates play a crucial role in biosurfactant production, serving as an essential component in various biochemical reactions and part of the energy currency, ATP. In the medium phosphates are added in the form of potassium and dipotassium salts to regulate the desired pH. The phosphate present in the salts is solely responsible for the maintenance of buffer pH, while potassium plays a role in intracellular cations present in the bacteria as the source of energy (Epstein, 2003). These phosphates also act as the necessary intermediate for the biosynthesis of biosurfactants. The glucose present in the medium is broken down in the glycolytic pathway and forms glucose-6-phosphate (G6P), which is the important precursor for the carbohydrate which is a hydrophilic group in the biosurfactant (Nurfarahin et al., 2018).

2.2.2 Environmental Factors

Surfactants have significant environmental applications in oil recovery, bioremediation, and the mitigation of heavy metals, among others (Desai and Banat, 1997, Banat et al. 2000). Environmental factors also impact the production of biosurfactants, including temperature, incubation period, aeration, agitation and pH (Fig. 1) (Joshi et al., 2013).

2.2.2.1 Temperature

The physical environmental factor, temperature, has a significant effect on biosurfactant production, as observed in the case of *Bacillus subtilis* CN2 when exposed to temperature range 25 to 125 °C (Bezza and Chirwa, 2017). Similar results were also obtained in a study with the bacterial species *B. subtilis* and *B. tequilensis*. These species were found to tolerate an extreme temperature range (Marajan et al., 2020) and the biosurfactant produced by these species had a negligible effect up to 100 °C (Khopade et al., 2012). However, in some investigations contrary to the above observations, variations in the hydrophilicity of surfactants have been shown with an increase in temperature, especially in anionic types of surfactants (Minana-Peŕez et al., 1995, Velasquez et al., 2010; Saranraj et al., 2022a). Increase in temperature also has a negative effect on the lipophilicity of the surfactants (Goethals et al., 2001).

2.2.2.2 Incubation Period

The production of biosurfactant from bacterial species *B. subtilis* and *B. tequilensis* was observed to increase with a longer incubation period, resulting in higher productivity of biosurfactant and biomass, along with a reduction in surface tension. Reduction in surface tension was consistently observed throughout the growth of *B. subtilis,* from the exponential phase to the stationary phase (Marajan et al., 2020).

2.2.2.3 Aeration and Agitation Factors

Agitation and aeration have a positive effect on the growth and enrichments of biosurfactants, along with the foam formations (Yao et al., 2015). The optimum rates of agitation and aeration have significantly increased the production of biosurfactants (Yeh et al., 2006). Agitation and aeration maintain the optimum flow of oxygen in the medium, leading to an increase in the production of lipopeptide biosurfactants (Davis et al., 1999, Davis et al., 2001, Kim et al., 1997). Uncontrolled agitation negatively affects the bacterial growth and biosurfactant production due to extreme foaming (Davis et al., 1999, Davis et al., 2001, Joshi et al., 2013).

2.2.2.4 pH

Higher pH values generally have no significant impact on biosurfactant activities (Khopade et al., 2012). In the pH range 2-12, there is negligible or no significant effect on surface activities such as surface tensions (Marajan et al., 2020). In some cases, an increase in pH has been observed to reduce the surface tension of the medium (Abouseoud et al., 2008).

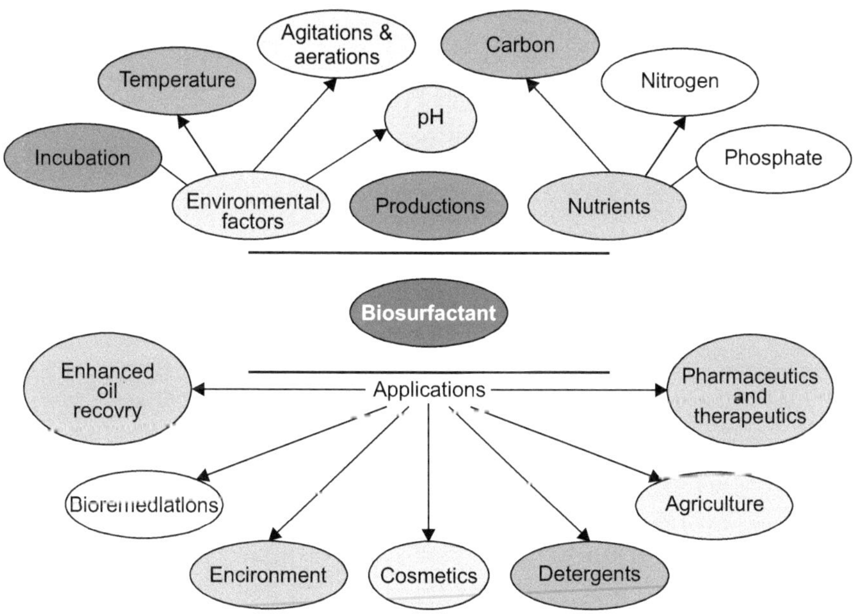

Fig. 1: Factors affecting production and applications of biosurfactants.

3. Regulation of Biosurfactants

Similar to the production system, the microbial system also has a regulating system that controls the production and characterization of biosurfactants. Two gene systems, GacA/GacS are reported in *Pseudomonas* species, functioning as a master switch. Suppression of these genes, due to mutations or other factors can result in the loss of lipopolysachharide production (Kitten et al., 1998, Koch et al., 2002, Dubern et al., 2005, De Bruijn et al., 2007, 2008). This two gene system is subject to further study to reveal more functions, as there is limited information about the signal that stimulates lipopolysachharides synthesis (Heeb and Haas, 2001, De Souza et al., 2003, Haas and Defago, 2005). The biosurfactant collectins and lipopolysachharides can bind to the amphiphilic lipid A moiety, oligosaccharides and polysaccharides domain of variable O-antigens, which are components of Gram-Negative bacterial cell walls (Kalina et al., 1995; van Iwaarden et al., 1994).

The well-known system of communication and regulation in biofilm forming and Lipopolysaccharide producing microbes is quorum sensing (Ali et al., 2022a; Ali et al., 2022b). The luxI type gene in *Pseudomonas* species expresses N-acyl homoserine lactones (N-AHLs), which directs the quorum sensing (Raaijmakers et al., 2010). Another regulatory system involved in the biosynthesis of surfactin is the Operon SrfA which codes for the surfactin synthetase complex and also regulates sporulation and development of competence (Hamoen et al., 2003). The operon SrfA is composed of four other modular open reading frames : SrfA-A, SrfA-B, SrfA-C and SrfA-D. Enzymes expressed by these open reading frames include L-leucine epimerase (coded by the gene present at the 3' region of SrfA-A and SrfA-B), homologue of fatty acid thioesterase type I (coded by the genes present at the end of SrfA-C) and mammalian thioesterases type II (coded by the SrfA-D) (Peypoux et al., 1999). One more important gene necessary for the production of surfactin is sfp, which is present downstream of the srfA operon (Nakano et al., 1992). This gene encodes enzymes priming non-ribosomal peptide synthesis and belongs to the 4'-phosphopantetheinases superfamily. Additionally, the sfp converts the inactive synthetase apoform to the holoform with cofactor (Lambalot et al., 1996, Seydlová and Svobodová, 2008).

4. Applications of Biosurfactants

Biosurfactants play a crucial role in the oil industry and have various valuable industrial applications with lower purity specifications (Fig. 1) (Desai et al., 1994, Fiechter, 1992, Velikonja and Kosaric, 1993, Banat, 1995a, Chakrabarty, 1985, Sarkar et al., 1989, Shennan and Levi, 1987). Biosurfactant needs a minute amount in comparison to synthetic surfactant as they are specific and very selective and also efficient under broad range oil and reservoir conditions (Pines et al., 1983, Ramsay, 1989, Rosenberg, 1986). Biosurfactants have a wide range of industrial applications, including emulsification, foaming, detergency, wetting, dispersing and solubilization (Gautam and Tyagi, 2006). In addition to industrial applications, biosurfactants find uses in pharmacology, pharmaceutics, food processing, etc. (Chetan et al., 2018). Their significance over synthetic surfactants is notable in terms of environmental protection, bioremediation and industrial oil recovery (Poremba et al., 1991). The biosurfactant trehalolipids produced by Nocardia rhodochrous have been reported to recover around 30% oil from the underground (Rapp et al., 1977).

4.1 Enhance Oil Recovery in Industries

The microbial system producing surfactants in situ can effectively remove crude oil from contaminated soil and employ a strategic mechanism to recover a significant amount of residual oil from (Gudina et al., 2013, Almeida et al., 2004, Bordoloi and Konwar, 2008, Banat et al., 2010). Microbial enhancement of oil recovery from residual oil involves the utilization of microbial metabolites such as biomass, biopolymers, gases, acids, solvents, enzymes, and biosurfactants, especially in marginal and depleted reservoirs (Lazar et al., 2007, Brown, 2010).

4.2 Bioremediation

Biosurfactants have the capacity to desorb pollutants from their particulate state.

When these pollutants are exposed to microbes with the ability to degrade them, the presence of biosurfactants can increase the rate of biodegradation of these pollutants (Deschenes et al., 1995, Thibault et al., 1996, Christofi et al., 1998, Ivshina et al., 1998, Christofi and Ivshina, 2002). Biosurfactants play a crucial role in the bioremediation of marine systems affected by oil spills. With their excellent emulsification properties, biosurfactants facilitate the migration of organic hydrophobic compounds from soil to the water phase. Moreover, biosurfactants contribute to the removal of crude oil from contaminated soil and the biodegradation of pollutants (Aronstein et al., 1991, Laha and Luthy, 1991, Laha and Luthy, 1992, Scheibenbogen et al., 1994,Van Dyke et al., 1993). Hydrocarbons and polycyclic hydrocarbons, which are pollutants, often have low solubility in water and tend to easily adsorb to soil particles. This reduces their bioavailability to microbes and hinders their biodegradation. These hydrophobic pollutants are also known to be toxic to humans and animals (Makkar and Rockne, 2003). However, the secretion of microbial metabolites and the application of biosurfactants can facilitate the desorption of particulate pollutants and increase their solubility in water. This, in turn, makes the pollutants more accessible to microbes and enhances the process of biodegradation (Rockne et al., 2002).

4.3 Uses in Environment

For environmental management and protection from oil spills, various biological methods are employed. One such approach is the direct introduction of microbes to the contaminated sites, known as bioaugmentation (e.g. applying petro oxidizing microorganisms). Another method involves introducing certain compounds (e.g. biosurfactants) that stimulate the enhancement of the natural microbial community (i.e. biostimulation) influencing the rate of biodegradation (Pirog et al., 2015, Tyagi et al., 2011, Ławniczak et al., 2013). Oil spills can be managed with the help of microbes through biodegradation (Das and Chandran, 2011). The production of biosurfactants is often associated with microbes that grow on biofuels and has significant industrial and biotechnological applications (Rodríguez-Rodríguez et al., 2012). Organisms producing biosurfactants thrive in environments contaminated with hydrophobic hydrocarbon compounds as these biosurfactant possess high surface and emulsifying properties (Rodríguez-Rodríguez et al., 2012, Menezes Bento et al., 2005, Correa Bicca et al., 1999, Desai and Banat, 1997). Biosurfactants are naturally synthesized compounds used in the bioremediation, aiding in the degradation of pollutants, detoxification of industrial effluents, and addressing oil spills. It has been reported that biosurfactants, along with the biodegradation of hydrophobic pollutants, also contribute to the removal of contaminants such as Cd, BP and Zn from contaminated soil, for example, through the use of Rhamnolipids. Mechanisms involved in the reduction of metal toxicity by biosurfactants include:

1. **Interaction with Microbial Surfaces:** Rhamnolipids interact with microbial surfaces, leading to alterations in the uptake of cadmium (Cd) by the microbes.
2. **Complexation of Cd with Rhamnolipids:** Rhamnolipids can form complexes with cadmium, reducing its toxicity and potentially facilitating its removal from the environment.
3. **Binding with Heavy Metals:** High molecular weight surfactants, such as emulsan, can interact with heavy metals like uranium, binding with them. This has been observed in the case of *A. calcoaceticus*.

Mechanisms involved in the reduction of metal toxicity by biosurfactants include:

1. Rhamnolipids interact with microbial surfaces leading to alterations in the uptake of cadmium (Cd) by the microbes.
2. Complexation of Cd with Rhamnolipids: Rhamnolpids can form complexes with cadmium, reducing its toxicity and potentially facilitating its removal from the environment.
3. Binding with Heavy Metals: High molecular weight surfactants, such as emulsan, can interact with heavy metals like Uranium, binding with them. This has been observed in the case of *A. calcoaceticus* (Ron and Rosenberg, 2001, Gautam and Tyagi, 2006). Pollutants in the solid phase often bind with soil particles, leading to a gradual release into the environment from the solid phase to the aqueous phase. This process results in reduced availability of these pollutants to microbes, consequently limiting biodegradation. The application of biosurfactant increases desorption of these pollutants and also increases the solubility in aqueous phase as it has both hydrophilic and hydrophobic components (Mihelcic et al., 1993). Rhamnolipids produced by *Pseudomonas aeruginosa* have antimicrobial properties; at lower concentrations they inhibit the growth of pathogenic algae (e.g. *Heterosigma akashiwo, Prorocentrum dentatum* and *Gymnodinium* sp.) through the disruptions of cell membranes. Biosurfactants, by disrupting the cell membranes, specifically the plasma membranes of these microbes, enter the cells and cause deformation of the inner structure. This disturbance in proper functioning and metabolism ultimately leads to cell lyses (Wang et al., 2005).

4.4 Pharmaceuticals and Therapeutics

Surfactin, a well known biosurfactant produced by the genus Bacillus, plays a significant role in enzyme inhibition, immunomodulation, antimicrobial, antitumor, antiviral activities,etc. (Cameotra and Makkar, 2004, Fernandes et al., 2007). Surfactin exhibits low toxicity in the environment and can form ion channels in the lipid bilayer of microbes, making it an effective antimicrobial agent, specifically against Gram negative bacteria. Other than its antimicrobial activity, surfactin also possesses several pharmacological properties such as prevention of hemolysis, inhibition of fibrin clot formation (Cameotra and Makkar, 2004) and adenosine 3,5-monophosphate phosphodiesterase inhibition (Hosono and Suzuki, 1983, Fernandes et al., 2007).

Drug delivery is a crucial aspect of therapy, and the use of microemulsion drug delivery systems has proven to be significant. These systems typically consist of lipids, surfactant, co-surfactant and co-solvents (Wu et al., 2017). This improved drug delivery mehtod enhances the availability of hydrophobic drugs can be made available for oral, nasal, ocular, topical, and intravenous delivery, allowing them to reach their target point (Rodrigues, 2015, Tang et al., 2008, Ohadi et al., 2020). Self-emulsifying formulations within drug delivery systems incorporate oils and surfactants. This formulation aids in dissolving a large amount of drugs and increases permeability, contributing to more effective drug delivery (Patel and Sawant, 2009). The optimization of drug delivery system formulations involves various parameters including, the efficiency of emulsification, the capacity for drug solubilization, acceptable intestinal permeability, and drug stability over an extended period (Agrawal et al., 2012). Microemulsions prepared from biosurfactants are particularly well-suited for drug delivery formulations that meet these criteria. Biosurfactant-based microemulsions are thermodynamically stable for extended periods, exhibit a high rate of solubilization and emulsification, and can enhance drug permeability (Date and Nagarsenker, 2008). Emulgels represent an interesting formulation for drug delivery systems, incorporating a surfactant. Emulgels consist of emulsions of water in oil or oil in water mixed with a gelling agent. This formulation offers enhanced stability and efficiency compared to other drug delivery systems (Ajazuddin et al., 2013).

Surfactants have been recognized for their wound healing properties (Custer et al., 1971). For example, a formulation for mastitis, which involves a mixture of commonly used amphoteric surfactants, specifically tertiary alkylamine oxides and alkyl betaine, known as C13G has been reported to possess lesser toxicity, antimicrobial and anti-inflammatory activity, and wound healing properties (Michaels, 1983, Stieritz, 1982, Amin et al., 1984).

4.4.1 Gene Delivery

Surfactants, including Gemini surfactants, have been explored for gene delivery applications (Gharaei-Fathabad, 2011). In gene delivery, a complex is formed between nucleic acids and surfactants,which is then delivered to the targeted site. The release of nucleic acids from the complex to the target site can be controlled by altering pH and providing reducing conditions (Cardoso et al., 2016, Sadiq et al., 2022).

4.4.2 Antimicrobial Activity

Surfactants, including biosurfactants, exhibit not only antimicrobial potential (Volkering et al., 1995, Prasad et al., 2015) but also play a role in the defense mechanisms against infections and inflammations. In the human body, surfactants are synthesized by the epithelial cells in the lungs. These lipoprotein-based surfactants are secreted into the extracellular space, where they reduce surface tension at the air/liquid interface (Wright, 2003). Biosurfactants that are FDA-approved and commercialized, such as Cubicin®, are lipopeptides with antimicrobial properties. Cubicin is used for the treatment of skin infections (Giuliani et al., 2007). In the context of drug resistant microbes, daptomycin, which interacts with pulmonary surfactant, has shown significant antimicrobial activity against Staphylococcus aureus, including methicillin-resistant and multidrug-resistant strains produced by *Strepotomyces roseosporus* (McHenney et al., 1996, Tally and De Bruin, 2000). This highlights the potential of surfactants, including biosurfactants, in innovative and effective therapeutic approaches. Surfactin, a biosurfactant produced by Bacillus subtilis, has demonstrated significant antiviral activity against various viruses. Studies have shown that surfactin exhibits antiviral effects against Semliki Forest virus, HSV-1, HSV-2 (Herpes simplex virus), feline calicivirusactive, simian immunodeficiency virus, murine encephalomyocarditis, and others (Kracht et al., 1999, Seydlová and Svobodová, 2008).

4.4.3 Anti-human Immunodeficiency Virus and Sperm-immobilizing Activity

The chances of the occurrence of the human immunodeficiency virus (HIV)/AIDS is high in the age range of 15-49 years. It can be controlled through the application of effective and safe microbicides with the help of surfactants. It has been reported that Biosurfactant (i.e. sophorolipid) produced by *Candida bombicola* and its structural analogue Sophorolipid diacetate ethyl ester exhibit effective antiviral activity against the HIV (Shah et al., 2005). Sophorolipids also showed positive activity as agents for the virucidal and spermicidal activities similar to those of typical surfactants, such as nonoxynol-9 (Seydlová and Svobodová, 2008).

4.4.4 Anticancer Activity

Surfactin has demonstrated effectiveness in controlling cancer cells. In a study it was found that the biosurfactant Surfactin can block proliferations of cancer cells, specifically in the human colon carcinoma cell line. Additionally it has been reported that Surfactin exhibits potent antitumor activity against Ehrlich's ascites carcinoma cells (Kameda et al., 1974). Surfactin has the ability to arrest the cell cycle and induce apoptosis by suppressing ERK and PI3K/Akt signals,which are crucial for cell survival and regulation (Kim et al., 2007, Chen et al., 2006).

4.4.5 Anti-adhesive Agents

Surfactants are effective against the bacteria which form biofilms. Surfactants exhibit anti-adhesive activities through which they prevent biofilm formation. Additionally, they are very effective against the Gram-negative bacteria such as *Salmonella typhimurium*, *Salmonella enteric*, *Escherichia coli* and *Proteus mirabilis* and can be utilized as medical appliances in the form of antimicrobial agents (Rodrigues et al., 2006b).

4.5 Cosmetics

Biosurfactants find applications in the cosmetics industry, particularly in the treatment of skin conditions such as dandruff and acne. They play a role in the development of various cosmetic products, including those for bathing, baby care, and as special ingredients in mascara, cleansers, lipstick, and toothpaste (Lahiry and Sinha, 2017). The unique characteristics of surfactants, including emulsification, de-emulsification, foaming, wetting, solubilization, antimicrobial activity, stability and binding capacity make them suitable compounds for the development of cleansers, skin care products and wound healing applications (Lahiry and Sinha, 2017, Tugrul and Cansunar, 2005, Tuleva et al., 2002). The biosurfactant glycolipid, namely Mannosylerythritol lipids are produced by species of *Pseudozyma* and *Ustilago* (Arutchelvi et al., 2008). It has several derivatives such as A, B and C. This glycolipid has demonstrated cell protection during oxidative stress and free radical scavenging activity specifically against superoxide anion. These properties make it a suitable compound for the development of anti-aging skin care products (Takahashi et al., 2012). Animals and human beings are naturally protected by their hair, serving as a kind of exoskeleton that shields against heat and UV rays. In one the study, it was discovered that the biosurfactant oligomers are effective in forming a hair mask andproviding conditioning (Owen and Fan, 2013a & b). Biosurfactant produced by *Pseudomonas antarctica* find applications in cosmetics as skin nourishment products (Masaru et al., 2007, Bhattacharya et al., 2017). Additionally, biosurfactants produced by Pseudomonas aeruginosa, like rhamnolipid, possess antimicrobial properties and serve as important ingredients (2%) in shampoo formulations. This contributes to making the scalp odor free and provides a soft glow to the hair (Desanto, 2008). Biosurfactants can also be effectively utilized as a key component in shower gels. They constitute a significant portion of the formulation, typically comprising sophorolipid (1-20%), anionic surfactant (1-20%) and also foaming surfactant (0-10%), along with other components such as electrolytes, detergent additives, and water (Cox et al., 2013). Furthermore, biosurfactants like rhamnolipids and sophorolipids, in combination with oleic oil (10%) are major ingredients in the formulations of various moisturizing skin cleansers and body cleansers (Allef et al., 2014).

4.6 Detergents and Cleaners

Biosurfactants find numerous applications in the detergent industry due to their emulsification properties, surface activity, ability to reduce surface tension, stability, and low toxicity. Effective detergent formulations typically include one or more types of surfactants, constituting 10-20% of the total content, which are responsible for the cleaning properties. Surfactants exhibit lower sensitivity to hard water compared to soaps and soap solutions (Osadebe et al., 2018, Okpokwasili and Nwabuzor, 1988). Biosurfactants produced by the bacterial genus Nocardiopsis possess greater stability and lower toxicity, positioning them as a significant alternative to synthetic Sodium Dodecyl Sulfate (SDS) in toothpaste formulations (Das et al., 2013). Biosurfactants produced by the bacterial genus Nocardiopsis possess greater stability and lower toxicity, positioning them as a significant alternative to synthetic Sodium dodecyl sulfate (SDS) in toothpaste formulations (Das et al., 2013).

4.7 Agriculture

Biosurfactants play a more effective role in agriculture by remediating soil from contaminations, improving soil quality, and removing plant pathogens (Rahman 2014, Bee et al., 2019, Ravinder et al., 2022). Rhamnolipids, a type of biosurfactant when exposed to plants, enhance immunity similar to mammals by inducing signaling pathways also known as microbe-associated molecular patterns (Vatsa et al., 2010). Glycolipid biosurfactants, such as sophorolipids, cellobiose lipids, mannosylerythritol lipids, and rhamnolipids, exhibit antimicrobial activities against phytopathogens (e.g. *Sclerotinia sclerotiorum*, *Phomopsis helianthi, Botrytis cinerea, Alternaria tomatophila, Alternaria solani*, etc.), thereby protecting plants and enhancing plant immunity (Cortés-Sánchez 2020). In a study, the biosurfactant rhamnolipid produced by Pseudomonas species demonstrated pesticidal effects on plant pests (Mnif and Ghribi, 2016).

4.8 Surfactants Used in Food Industry

Biosurfactants such as lecithin from yolk sac or plant derived sources, have been utilized as natural emulsifiers in the food industry for a long time..They are used in the preparation of several food products including mayonnaise, salad creams, dressings, deserts, etc. (Kralova and Sjöblom, 2009, Xu et al., 1998). In the food industry biosurfactants are not only used for the purpose of emulsification but also as the thickeners, foaming agents, humectants and for stabilizing emulsions and reduction of interface and surface tensions (Nitschke and Costa, 2007, Shoeb et al., 2013, Sourav et al., 2015, Vijayakumar and Saravanan, 2015, Santos et al., 2016). In the food industry biosurfactants have two major applications: they are utilized in the treatment and cleaning of surfaces of containers, and they serve as significant additives in food products (Nitschke and Silva, 2018).

5. Conclusion

With the increasing recognition of the beneficial aspects of biosurfactants, there is a growing demand in the modern era. Microbes can synthesize specific biosurfactants needed, when provided with the necessary substrates, nutrients, and environmental conditions. These microbial-synthesized compounds are advantageous over synthetic ones due to their lower toxicity and enhanced degradability. Biosurfactants find applications as wetting agents, foaming agents, and emulsifiers in the cosmetic and pharmaceutical industries. Biosurfactants find applications as wetting agents, foaming agents, and emulsifiers in the cosmetic and pharmaceutical industries. They also play a crucial role in mining and environmental protection, including bioremediation, removal of heavy metals and contaminants, oil spill control, detoxification, and biodegradation. They also play a crucial role in mining and environmental protection including bioremediation, removal of heavy metals and contaminants, oil spill control, detoxification, and biodegradation. Therefore, to advance the development and production of novel biosurfactants, exploring new regions such as the marine environment or extreme environments becomes essential.

References

Abouseoud, M., Yataghene, A., Amrane, A. & Maachi, R. (2008). Biosurfactant production by free and alginate entrapped cells of Pseudomonas fluorescens. *J. Ind. Microbiol. Biotechnol.*, 35(11): 1303-1308.

Agrawal, S., Giri, T.K., Tripathi, D.K. & Alexander, A. (2012). A review on novel therapeutic strategies for the enhancement of solubility for hydrophobic drugs through lipid and surfactant based self micro emulsifying drug delivery system: A novel approach. *Am. J. Drug Disc. Dev.*, 2(4): 143-183.

Ajazuddin, Alexander A., Khichariya, A., Gupta, S., Patel, R.J., Giri, T.K. & Tripathi, D.K. (2013). Recent expansions in an emergent novel drug delivery technology: Emulgel. *Cont. Release*, 171(2): 122-132.

Allef, P., Hartung, C. & Schilling, M. (2014). Aqueous hair and skin cleaning compositions comprising biosurfactants. Patent US 20140349902 A1.

Ali, S.A.M., Sayyed, R.Z., Mir, M.I., Hameeda, B., Khan, Y., Alkhanani, M.F. et al. (2022a). Induction of systemic resistance in maize and antibiofilm activity of surfactin from *Bacillus velezensis* MS20. *Front. Microbiol.*, 13, 879739. https://doi.org/10.3389/fmicb.2022.879739.

Ali, S.A.M., Sayyed, R.Z., Reddy, M.S., Enshasy, H.E. & Hameeda, B. (2022b). Delving through quorum sensing and CRISPRi strategies for enhanced surfactin production. *In:* Sayyed, R.Z. (Eds), *Biosurfatnats: Production and Applications in Bioremediation/Reclaimation*. 59-79. CRC Press: Taylor & Francis Group, USA.

Almeida, P.F., Moreira, R.S., Almeida, R.C.C., Guimarães, A.K., Carvalho, A.S., Quintella, C., et al. (2004). Selection and application of microorganisms to improve oil recovery. *Engineering in Life Sciences*, 4: 319-325.

Amin, M.M., Smith, A.R., Anderson, K.L., Hahn, E.C. & Gustafsson, B.K. (1984). Evaluation of a surfactant mixture C31G as a teat dip by a modified excised teat model. *J. Dairy Sci.*, 67: 421-426.

Angelini, T.E., Roper, M., Kolter, R., Weitz, D.A., Brenner. (2009). *Bacillus subtilis* spreads by surfing on waves of surfactant. *PNAS*, 106(43): 18109-18113.

Aronstein, B., Cavillo, Y. & Alexender, M. (1991). Effect of surfactants at low concentrations on the desorption and biodegradation of sorbed aromatic compounds in soil. *Environ Sci Technol*, 25: 1728-1731.

Arutchelvi, J.I., Bhaduri, S., Uppara, P.V. & Doble, M. (2008). Mannosylerythritol lipids: A review. *J Ind Microbiol Biotechnol*, 35: 1559-1570.

Banat, I.M., Makkar, R.S. & Cameotra, S.S. (2000). Potential commercial applications of microbial surfactants. *Appl Microbiol Biot*, 53: 495-508.

Banat, I.M. (1995a). Biosurfactants production and possible uses in microbial enhanced oil recovery and oil pollution remediation: A review. *Bioresource Technol.*, 51: 1-12.

Banat, I.M. (1995b). Characterization of biosurfactants and their use in pollution removal—State of the art. *Acta Biotechnol.*, 15: 251-267.

Banat, I.M., Franzetti, A., Gandolfi, I., Bestetti, G., Martinotti, M.G., Fracchia, L. et al. (2010). Microbial biosurfactants production, applications and future potential. *Applied Microbiology and Biotechnology*, 87: 427-444.

Begley, M., Cotter, P.D., Hill, C. & Ross, R.P. (2009). Identification of a novel two-peptide lantibiotic, lichenicidin, following rational genome mining for Lan M proteins. *Applied Environmental Microbiology*, 75: 5451-5460.

Bee, H., Khan, M.Y. & Sayyed, R.Z. (2019). Microbial surfactants and their significance in agriculture. *In:* Sayyed Reddy Antonious (Eds), *PGPR: Prospects for Sustainable Agriculture.* 205-216. Springer-Nature, Singapore.

Belshaw, P.J., Walsh, C.T. & Stachelhaus, T. (1999). Aminoacyl-CoAs as probes of condensation domain selectivity in nonribosomal peptide synthesis. *Science*, 284: 486-489.

Bezza, F.A. & Chirwa, E.M.N. (2017). The role of lipopeptide biosurfactant on microbial remediation of aged polycyclic aromatic hydrocarbon (PAHs)-contaminated soil. *Chemical Eng. Journal*, 309: 563-576.

Bhattacharya, B., Ghosh, T.K. & Das, N. (2017). Application of bio-surfactants in cosmetics and pharmaceutical industry. *Sch. Acad. J. Pharm.*, 6(7): 320-329.

Bordoloi, N.K. & Konwar, B.K. (2008). Microbial surfactant-enhanced mineral oil recovery under laboratory conditions. *Colloids and Surfaces B: Biointerfaces*, 63: 73-82.

Brady, S.F. & Clardy, J. (2005). N-acyl derivatives of arginine and tryptophan isolated from environmental DNA expressed in Escherichia coli. *Org. Lett.*, 7: 3613-3616.

Brady, S.F., Chao, C.J. & Clardy, J. (2004). Long-chain N-acyltyrosine synthases from environmental DNA. *Appl. Environ. Microbiol.*, 70: 6865-6870.

Brown, L.R. (2010). Microbial enhanced oil recovery (MEOR). *Current Opinion in Microbiology*, 13: 316-320.

Bruner, S.D., Weber, T., Kohli, R.M., Schwarzer, D., Marahiel, M.A., Walsh, C.T., Stubbs, M.T. (2002). Structural basis for the cyclization of the lipopeptide antibiotic surfactin by the thioesterase domain SrfTE. *Structure*, 10: 301-310.

Cameotra, S.S. & Makkar, R.S. (2004). Recent applications of biosurfactants as biological and immunological molecules. *Current Opinion in Microbiology*, 7: 262-266.

Cardoso, A.M.S., Silva, S.G., Luisa do Vale, M., Marques, E.F., Pedroso de Lima, M.C. & Jurado, A.S. (2016). Chapter 7 - Gene delivery mediated by gemini surfactants: Structure–activity relationships. *In:* Grumezescu, A.M. (Ed.), *Engineering of Nanobiomaterials*. 227-256. William Andrew Publishing.

Casas, J.A. & Garcia-Ochoa, F. (1999). Sophorolipid production by *Candida bombicola* medium composition and culture methods. *Journal of Bioscience and Bioengineering*, 88: 488-494.

Chakrabarty, A.M. (1985). Genetically-manipulated microorganisms and their products in the oil service industries. *Trends Biotechnol.*, 3: 32-38.

Chen, J., Song, X., Zhang, H. & Qu, Y. (2006). Production, structure elucidation and anticancer properties of sophorolipid from *Wickerhamiella domercqiae*. *Enzyme and Microbial Technology*, 39(3): 501-506.

Chetan, D.M., Keerthana, Prabhu, A., Manjula, Ranjit S., Bhat, R.S. (2018). Biosurfactants: An alternative to the synthetic surfactants and their production by bacteria isolated from solid waste. *Journal of Pure and Applied Microbiology*, 12(3): 1561-1567.

Christofi, N. & Ivshina, I.B. (2002). Microbial surfactants and their use in field studies of soil remediation. *Journal of Applied Microbiology*, 93: 915-929.

Christofi, N., Ivshina, I.B., Kuyukina, M.S. & Philp, J.C. (1998). Biological treatment of crude oil contaminated soil in Russia. *In:* Lerner, D.N. and Walton, N.R.G. (Eds), *Contaminated Land and Groundwater: Future Directions*. 45–51. Engineering Geology Special Publication, 14. London: Geological Society.

Christova, N., Petrov, P. & Kabaivanova, L. (2013). Biosurfactant production by *Pseudomonas aeruginosa* BN10 cells entrapped in cryogels. *Z. Naturforsch.*, 68c: 47-52.

Cirigliano, M. & Carman, G. (1984). Isolation of a bioemulsifier from *Candida lipodytica*. *Appl. Environ. Microbiol.*, 48: 747-750.

Cooper, D.G. & Paddock, D.A. (1983). Torulopsis petrophilum and surface activity. *Appl. Environ. Microbiol.*, 46: 1426-1429.

Cooper, D.G. & Paddock, D.A. (1984). Production of a biosurfactant from Torulopsis bombicola. *Appl. Environ. Microbiol.*, 47: 173-176.

Correa, Bicca F., Colombo, Fleck L. & Záchia Ayub, M.A. (1999). Production of biosurfactant by hydrocarbon degrading Rhodococcus rubber and Rhodococcus erythropolis. *Rev Microbiol.*, 30: 231-236.

Cortés-Sánchez, A.J. (2020). Surfactants of microbial origin and its application in foods. *Scientific Research and Essays*, 15(1): 11-17.

Cox, T.F., Crawford, R.J., Gregory, L.G., Hosking, S.L., Kotsakis, P. (2013). Inventors; Conopco, Inc., assignee. Mild to the skin, foaming detergent composition. United States patent US 8,563,490. Oct 22.

Custer, J., Edlich, R.F., Prusek, M., Madden, J., Panky, P. & Wagensteen, D.H. (1971). Studies in the management of the contaminated wound. V. An assessment of the effectiveness of phisohex and betadine surgical scrub solutions. *Am. J. Surg.*, 121: 572.

Das, I., Roy, S. & Chandni, S. (2013). Biosurfactant from marine actinobacteria and its application in cosmetic formulation of toothpaste. *Pharm Lett.*, 5: 1–6.

Das, N. & Chandran, P. (2011). Microbial degradation of petroleum hydrocarbon contaminants: An overview. *Biotechnol. Res. Int.* (Article ID 941810): 1-13.

Das, P., Mukherjee, S. & Sen, R. (2008b). Genetic regulations of the biosynthesis of microbial surfactants: An overview. *Biotechnol. Genet. Eng. Rev.*, 25: 165-186.

Date, A.A. & Nagarsenker, M.S. (2008). Parenteral microemulsions: An overview. *Int. J. Pharm.*, 355(1-2): 19-30.

Davis, D.A., Lynch, H.C. & Varley, J. (1999). The production of Surfactin in batch culture by Bacillus subtilis ATCC 21332 is strongly influenced by the conditions of nitrogen metabolism. *Enzyme Microbial Technology*, 25: 322-329.

Davis, D.A., Lynch, H.C. & Varley, J. (2001). The application of foaming for the recovery of surfactin from B. subtilis ATCC 21332 cultures. *Enzyme Microbial Technology*, 28: 346-354.

De Bruijn, I., de Kock, M.J., deWaard, P., van Beek, T.A. & Raaijmakers, J.M. (2008). Massetolide: A biosynthesis in Pseudomonas fluorescens. *J. Bacteriol.*, 190: 2777–2789.

De Bruijn, I., de Kock, M.J.D., Yang, M., de Waard, P., van Beek, T.A. & Raaijmakers, J.M. (2007). Genome-based discovery, structure prediction and functional analysis of cyclic lipopeptide antibiotics in Pseudomonas species. *Mol. Microbiol.*, 63: 417-428.

De Souza, J.T., De Boer, M., De Waard, P., van Beek, T.A. & Raaijmakers, J.M. (2003). Biochemical, genetic, and zoosporicidal properties of cyclic lipopeptide surfactants produced by Pseudomonas fluorescens. *Appl. Environ. Microb.*, 69: 7161-7172.

Desai, J.D. & Banat, I.M. (1997). Microbial production of surfactants and their commercial potential. *Microbiol. Mol. Biol. Rev.*, 61(1): 47.

Desai, A.J., Patel, K.M. & Desai, J.D. (1988). Emulsifier production by *Pseudomonas fluorescens* during the growth of hydrocarbons. *Curr. Sci.*, 57(9): 500-501.

Desai, A.J., Patel, K.M. & Desai, J.D. (1994). Advances in production of biosurfactants and their commercial applications. *J. Sci. Ind. Res.*, 53: 619-629.

Desanto, K. (2008). Rhamnolipid-based formulations. Patent WO 2008013899 A2.

Deschenes, L., Lafrance, P. & Villeneuve, J.P. (1995). The effect of anionic surfactant on the mobilisation and biodegradation of PAHs in creosote-contaminated soil. *Hydrological Sciences*, 40: 471-484.

Dhuldhaj, U.P. & Bora, C.R. (2021). Microbial surfactants: An overview. *In:* Sayyad, R. and Hameeda Bee (Eds), *Microbial Surfactants: Production and Applications*. Volume 1: 1-26. CRC Press, USA.

Dieckmann, R., Lee, Y.O., van Liempt, H., von Dohren, H. & Kleinkauf, H. (1995). Expression of an active adenylate-forming domain of peptide synthesis corresponding to acyl-CoAsynthetases. *FEBS Lett.*, 357: 212-216.

Dubern, J.F., Lagendijk, E.L., Lugtenberg, B.J.J. & Bloemberg, G.V. (2005). The heat shock genes dnaK, dnaJ, and grpE are involved in regulation of putisolvin biosynthesis in Pseudomonas putida PCL1445. *J Bacteriol*, 187: 5967-5976.

Duvnjak, Z., Cooper, D.G. & Kosaric, N. (1983). Effect of nitrogen source on surfactant production by Arthrobacter paraffineus ATCC 19558. *In:* Zadic, J.E., Cooper, D.G., Jack, T.R. and Kosaric, N. (Eds), *Microbial Enhanced Oil Recovery*. 66–72. Tulsa, OK: Pennwell.

e Silva, N.M.P.R., Rufino, R.D., Luna, J.M., Santos, V.A. & Sarubbo, L.A. (2014). Screening of *Pseudomonas* species for biosurfactant production using low-cost substrates. *Biocatalysis and Agricultural Biotechnology*, 3(2): 132-139.

Epstein, W. (2003). The roles and regulation of potassium in bacteria. *Prog. Nucleic Acid Res. Mol. Biol.*, 75: 293-320.

Fenical, W. (1993). Chemical studies of marine bacteria: Developing a new resource. *Chem Rev*, 93: 1673-1683.

Fernandes, P.A.C., de Arruda, I.R., dos Santos, A.F., de Araújo, A.A., Maior, A.M.S. & Ximenes, E.A. (2007). Antimicrobial activity of surfactants produced by Bacillus subtilis R14 against multidrug-resistant bacteria. *Brazilian Journal of Microbiology*, 38: 704-709.

Fiechter, A. (1992). Biosurfactants: Moving towards industrial application. *Trends Biotechnol.*, 10: 208-217.

Gao, L., Kano, N., Sato, Y., Li, C., Zhang, S. & Imaizumi, H. (2012). Behavior and distribution of heavy metals including rare earth elements, thorium, and uranium in sludge from industry water treatment plant and recovery method of metals by biosurfactants application. *Bioinorg. Chem. Appl.*, 2012: 173819. doi: 10.1155/2012/173819.

Gautam, K.K. & Tyagi, V.K. (2006). Microbial surfactants: A review. *Journal of Oleo Science*, 55(4): 155-166.

Gharaei-Fathabad, E. (2011). Biosurfactants in pharmaceutical industry: A minireview. *American Journal of Drug Discovery and Development*, 1(1): 58-69.

Giuliani, A., Pirri, G. & Fabiole Nicoletto, S. (2007). Antimicrobial peptides: An overview of a promising class of therapeutics. *Central European Journal of Biology*, 2: 1-33.

Goethals, G., Fernandez, A., Martin, P., Miñana-Perez, M., Scorzza, C., Villa, P. & Gode, P. (2001). Spacer arm influence on glucidoamphiphile compound properties. *Carbohydr. Polym.*, 45: 147-154.

Guerra-Santos, L.H., Kappeli, O. & Fiechter, A. (1984). Pseudomonas aeruginosa biosurfactant production in continuous culture with glucose as carbon source. *Applied and Environmental Microbiology*, 48: 301-305.

Gudiña, E.J., Pereira, J.F.B., Costa, R., Coutinho, J.A.P., Teixeira, T.A. & Rodrigues, L.R. (2013). Biosurfactant-producing and oil-degrading Bacillus subtilis strains enhance oil recovery in laboratory sand-pack columns. *Journal of Hazardous Materials*, 261: 106- 113.

Gutierrez, T., Mulloy, B., Black, K. & Green, D.H. (2007a). Glycoprotein emulsifiers from two marine *Halomonas* species: Chemical and physical characterization. *J. Appl. Microbiol.*, 103: 1716-1727.

Gutierrez, T., Berry, D., Yang, T., Mishamandani, S., McKay, L. Teske, A. & Aitken, M.D. (2013). Role of bacterial exopolysaccharides (EPS) in the fate of the oil released during the Deepwater Horizon oil spill. *PloS One*, 8(6): e67717.

Haas, D. & Defago, G. (2005). Biological control of soil-borne pathogens by fluorescent pseudomonads. *Nat Rev Microbiol*, 3: 307-319.

Haba, E., Espuny, M.J., Busquets, M. & Manresa, A. (2000). Screening and production of rhamnolipids by Pseudomonas aeruginosa 47T2 NCIB 40044 from waste frying oils. *Journal of Applied Microbiology*, 88: 379-387.

Hamoen, L.W., Venema, G. & Kuipers, O.P. (2003). Controlling competence in Bacillus subtilis: Shared use of regulators. *Microbiology*, 149: 9-17.

Hausmann, R. & Syldatk, C. (2014). Types and classification of microbial surfactants. *In:* Kosaric, N., Varder-Sukan, F. (Eds). *Biosurfactants: Production and Utilization—Processes, Technologies, and Economics*. 3-18. CRC Press, Taylor & Francis Group: Boca Raton, FL, USA.

Heeb, S. & Haas, D. (2001). Regulatory roles of the GacS/GacA two component system in plant associated and other Gram negative bacteria. *Mol Plant Microbe In*, 14: 1351-1363.

Hosono, K. & Suzuki, H. (1983). Acylpeptides, the inhibitors of cyclic adenosine 3′,5′-monophosphodiesterase. III. Inhibition of cyclic AMP phosphodiesterase. *J. Antibiot. Japan*, 36: 679-683.

Husain, D.R., Goutx, M., Acquaviva, M., Gilewicz, M. & Bertrand, J.C. (1997). The effect of temperature on eicosane substrate uptake modes by a marine bacterium *Pseudomonas nautica* strain 617: Relationship with the biochemical content of cells and supernatants. *World J. Microbiol. Biotechnol.*, 13: 587-590.

Ikhwani, A.Z.N., Nurlaila, H.S., Ferdinand, F.D.K., Fachria, R., Hasan, A.E.Z., Yani, M. & Suryani, S.I. (2017). Preliminary study: Optimization of pH and salinity for biosurfactant production from Pseudomonas aeruginosa in diesel fuel and crude oil medium. *IOP Conf. Series: Earth and Environmental Science*, 58: 1-7.

Ivshina, I.B., Kuyukina, M.S., Philp, J.C. & Christofi, N. (1998). Oil desorption from mineral and organic materials using biosurfactant complexes produced by Rhodococcus species. *World Journal of Microbiology and Biotechnology*, 14: 711-717.

Jarvis, F.G. & Johnson, M.J. (1949). A glyco-lipid produced by Pseudomonas aeruginosa. *Am Chem Soc*, 71: 4124-4126.

Joshi, S.J., Geetha, S.J., Yadav, S. & Desaia, A.J. (2013). Optimization of bench-scale production of biosurfactant by Bacillus licheniformis R2, ICESD 2013: January 19-20, Dubai, UAE, *APCBEE Procedia*, 5: 232-236.

Kakugawa, K., Tamai, M., Imamura, K., Miyamoto, K. & Miyoshi, S. (2002). Isolation of yeast *Kurtzmanomyces* sp. I-11, novel producer of mannosylerythriotol lipid. *Biosci. Biotechnol. Biochem.*, 66: 188-191.

Kalina, M., Blau, H., Riklis, S. & Kravtsov, V. (1995). Interaction of surfactant protein A with bacterial lipopolysaccharide. *Am. J. Physiol.*, 268: L144–L151.

Kameda, Y., Oira, S., Matsui, K., Kanatomo, S. & Hase, T. (1974). Antitumor activity of Bacillus natto. V. Isolation and characterization of surfactin in the culture medium of Bacillus natto KMD 2311. *Chem. Pharm. Bull.*, 22: 938-944.

Kappeli, O. & Finnerty, W.R. (1979). Partition of alkane by an extracellular vesicle derived from hexadecane grown Acinetobacter. *J. Bacteriol.*, 140: 707-712.

Keating, T.A., Marshall, C.G., Walsh, C.T. & Keating, A.E. (2002).The structure of VibH represents nonribosomal peptide synthetase condensation, cyclization and epimerization domains. *Nat. Struct. Biol.*, 9: 522-526.

Khopade, A., Ren, B., Liu, X.Y., Mahadik, K., Zhang, L. & Kokare, C. (2012). Production and characterization of biosurfactants from marine Streptomyces species B3J. *Colloid Interface Sci.*, 367(1): 311-318.

Kim, H.S., Yoon, B.D., Lee, C.H., Suh, H.H., Oh, H.M., Katsuragi, T. & Tani, Y. (1997). Production and properties of a lipopeptide biosurfactant from Bacillus subtilis C9. *J Ferment Bioeng*, 84: 41-46.

Kim, S.Y., Kim, J.Y., Kim, S.H., Bae, H.J., Yi, H., Yoon, S.H. et al. (2007). Surfactin from Bacillus subtilis displays anti-proliferative effect via apoptosis induction, cell cycle arrest and survival signaling suppression. *FEBS Lett.*, 581: 865-871.

Kitamoto, D., Haneishi, K., Nakahara, T. & Tabuchi, T. (1990). Production of mannosylerythritol lipids by Candida antarctica from vegetable-oils. *Agric Biol Chem*, 54: 37-40.

Kitten, T., Kinscherf, T.G., McEvoy, J.L. & Willis, D.K. (1998). A newly identified regulator is required for virulence and toxin production in Pseudomonas syringae. *Mol Microbiol*, 28: 917-929.

Klekner, V. & Kosaric, N. (1993). Biosurfactants for cosmetics. *In:* N. Kosaric (Ed.), *Biosurfactants: Production, Properties, Applications*. 329–372. Marcel Dekker, Inc., New York, N.Y.

Koch, B., Nielsen, T.H., Sorensen, D., Andersen, J.B., Christophersen, C., Molin, S. et al. (2002). Lipopeptide production in Pseudomonas sp. strain DSS73 is regulated by components of sugar beet seed exudate via the Gac two component regulatory system. *Appl Environ Microb*, 68: 4509-4516.

Kracht, M.A., Rokos, H., Ozel, M., Kowal, M., Pauli, G. Vater, J. et al. (1999). Antiviral and hemolytic activities of surfactin isoforms and their methyl ester derivatives. *The Journal of Antibiotics*, 52(7): 613-619.

Kralova, I. & Sjöblom, J. (2009). Surfactants used in food industry: A review. *Journal of Dispersion Science and Technology*, 30: 1363-1383.

Krauss, E.M. & Chan, S.I. (1983). Complexation and phase transfer of nucleotides by gramicidine S. *Biochemistry*, 22: 4280-4285.

Kretschmer, A., Bock, H. & Wagner, F. (1982). Chemical and physical characterization of interfacial-active lipids from *Rhodococcus erythropolis* grown on n-alkanes. *Appl. Environ. Microbiol.*, 44: 864-870.

Kubicki, S., Bollinger, A., Katzke, N., Jaeger, K.E., Loeschcke, A. & Thies, S. (2019). Marine biosurfactants: Biosynthesis, structural diversity and biotechnological applications. *Marine Drugs*, 17(408): 1-30.

Kumar, A.S., Mody, K. & Jha, B. (2007). Evaluation of biosurfactant/bioemulsifier production by a marine bacterium. *Bull Environ Contam Toxicol*, 79: 617-621.

Laha, S. & Luthy, R. (1991). Inhibition of phenanthrene mineralization by nonionic surfactants in soil water systems. *Environ Sci Technol*, 25: 1920-1930.

Laha, S. & Luthy, R. (1992). Effects of nonionic surfactants on the solubilization and mineralization of phenanthrene in soil water systems. *Biotechnology and Bioengineering*, 40: 1367-1380.

Lahiry, S. & Sinha, R. (2017). Biosurfactant: Pharmaceutical perspective. *Journal of Analytical & Pharmaceutical Research*, 4(3): 1-3.

Lai, C.C., Huang, Y.C., Wei, Y.H. & Chang, J.S. (2009). Biosurfactant-enhanced removal of total petroleum hydrocarbons from contaminated soil. *J. Hazard Mater*, 167: 609-614.

Lambalot, R.H., Gehring, A.M., Flugel, R.S., Zuber, P., LaCelle, M., Marahiel, M.A. et al. (1996). A new enzyme superfamily – The phosphopantetheinyl transferases. *Chem Biol.*, 3: 923-936.

Lang, S. & Wagner, G. (1987). Structure and properties of biosurfactants. *In:* Kosaric, N., Cairns, W.L., Gray, N.C.C. (Eds), *Biosurfactants and Biotechnology*. 21-47. Marcel Dekker, Inc., New York.

Ławniczak, Ł., Marecik, R. & Chrzanowski (2013). Contributions of biosurfactants to natural or induced bioremediation. *Appl. Microbiol. Biotechnol.*, 97(6): 2327-2339.

Lazar, I., Petrisor, I.G. & Yen, T.F. (2007). Microbial enhanced oil recovery (MEOR). *Petroleum Science and Technology*, 25: 1353-1366.

Linne, U. & Marahiel, M.A. (2000). Control of directionality in nonribosomal peptide synthesis: Role of the condensation domain in preventing misinitiation and timing of epimerization. *Biochemistry*, 39: 10439-10447.

Makkar, R.S. & Rockne, K.J. (2003). Comparison of synthetic surfactants and biosurfactants in enhancing biodegradation of polycyclic aromatic hydrocarbons. *Environmental Toxicology and Chemistry*, 22(10): 2280-2292.

Marajan, C., Alias, S., Ramasamy, K. & Abdul-Talib, S. (2020). The effect of incubation time, temperature and pH variations on the surface tension of biosurfactant produced by Bacillus spp. AIP Conference Proceedings 2020, 0200471-7 (2018); https://doi.org/10.1063/1.5062673.

Masaru, K., Michiko, S. & Shuhei, Y. (2007). Skin care cosmetic and skin and agent for preventing skin roughness containing biosurfactants (World Patent 2007/060956). Toyo Boseki Kabu Shiki Kaisha and National Industrial Science and Technology, Osaka, Japan.

McHenney, M.A. & Baltz, R.H. (1996). Gene transfer and transposition mutagenesis in Streptomyces roseosporus: Mapping of insertions that influence daptomycin or pigment production. *Microbiology*, 142: 2363-2373.

Menezes, Bento F., de Oliveira Camargo, F.A., Okeke, B.C., Frankenberger Jr., W.T. (2005). Diversity of biosurfactant producing microorganisms isolated from soils contaminated with diesel oil. *Microbiol Res.*, 160: 249-255.

Michaels, E.B., Hahn, E.C. & Kenyon, A.J. (1983). Effect of C31G, an ant±microbial surfactant, on healing of incised guinea pig wounds. *Am. J. Vet. Res.*, 44: 1378.

Mihelcic, J.R., Lueking, D.R., Mitzell, R.J. & Stapleton, J.M. (1993). Bioavailability of sorbed and separate-phase chemicals. *Biodegradation*, 4: 141-153.

Miñana-Pérez, M., Graciaa, A., Lachaise, J. & Salager, J.L. (1995). Solubilization of polar oils with extended surfactants. *Colloids Surf A*, 100: 217-224.

Mnif, I. & Ghribi, D. (2016). Glycolipid biosurfactants: Main properties and potential applications in agriculture and food industry. *Journal of the Science of Food and Agriculture*, 96(13): 4310-4320.

More, P.A., Jagtap, C.B., Joshi, P.A., Tiwari, O.K. & Kumar, P. (2017). Evaluation of bioemulsifier production by Comamonas sp. CBJ1 isolated from soil contaminated soil. *Indian Journal of Geo Marine Sciences*, 46(7): 1365-1370.

Morikawa, M., Daido, H., Takao, T., Murata, S., Shimonishi, Y. & Imanaka, T. (1993). A new lipopeptide biosurfactant produced by *Arthrobacter* sp. strain MIS38. *J. Bacteriol.*, 175(20): 6459-6466.

Muller-Hurtig, R., Wagner, F., Blaszczyk, R. & Kosaric, N. (1993). Biosurfactants for environmental control. *In:* N. Kosaric (Ed.), *Biosurfactants: Production, Properties, Applications.* 447–469. Marcel Dekker, Inc., New York, N.Y.

Muthusamy, K., Gopalakrishnan, S., Sivachidambaram, P. & Ravi, T.K. (2008). Biosurfactants: Properties, commercial production and application. *Current Science*, 94(6): 736-747.

Nakano, M.M., Corbell, N., Besson, J. & Zuber, P. (1992). Isolation and characterization of sfp: A gene that functions in the production of the lipopeptide biosurfactant, surfactin, in Bacillus subtilis. *J. Bacteriol.*, 182: 3274-3277.

Nerurkar, A.S., Hingurao, K.S. & Suthar, H.G. (2009). Bioemulsfiers from marine microorganisms. *Journal of Science Industrial Research*, 68: 273-277.

Neu, T.R. (1996). Significance of bacterial surface active compounds in interaction of bacteria with interfaces. *Microbiol Rev*, 60(1): 151-166.

Neu, T.R., Vandermei, H.C. & Busscher, B.H.J. (1992). Biofilm association with health. 21-34. *In-Biofilm Science and Technology*, NATO ASI Series, Kluwer Academic Publishers, Dordrecht.

Neu, T.R., Härtner, T. & Poralla, K. (1990). Surface active properties of viscosin: A peptidolipid antibiotic. *Appl. Microbiol. Biotechnol.*, 32: 518-520.

Nitschke, M. & Costa, S.G.V.A.O. (2007). Biosurfactants in the food industry. *Trends in Food Science and Technology*, 18: 252-259.

Nitschke, M. & Silva, S.S.E. (2018). Recent food applications of microbial surfactants. *Critical Reviews in Food Science and Nutrition*, 58(4): 631-638.

Nurfarahin, A.H., Mohamed, M.S. & Phang, L.Y. (2018). Culture medium development for microbial-derived surfactants production—An overview. *Molecules*, 23: 1049-1075.

Ohadi, M., Shahravan, A., Dehghannoudeh, N., Eslaminejad, T., Banat, I.M., Dehghannoud, G. et al. (2020). Potential use of microbial surfactant in microemulsion drug delivery system: A systematic review. *Drug Design, Development and Therapy*, 14: 541-550.

Okpokwasili, G.C. & Nwabuzor, C.N. (1988). Primary biodegradation of anionic surfactants in laundry detergents. *Chemosphere*, 17: 2175-2182.

Osadebe, A.U., Onyiliogwu, C.A., Suleiman, B.M. & Okpokwasili, G.C. (2018). Microbial degradation of anionic surfactants from laundry detergents commonly discharged into a riverine ecosystem. *Journal of Applied Life Sciences International*, Article no. JALSI.40131, 16(4): 1-11.

Owen, D. & Fan, L. (2013a). Oligomeric biosurfactants in dermatocosmetic compositions. Patent US 8431523 B2.

Owen, D. & Fan, L. (2013b). Polymeric biosurfactants. Patent US 8586541 B2.

Pacwa-Plociniczak, M., Plaza, G.A., Piotrowska-Seget, Z. & Cameotra, S.S. (2011). Environmental applications of biosurfactants: Recent advances. *International Journal Molecular Science*, 12: 633-654.

Pandya, U., Dhuldhaj, U. & Sahay, N. (2018). Bioactive mushroom polysaccharides as antitumor: An overview. *Natural Product Research*, 33(18): 2668–2680 (IF: 2.488). DOI: 10.1080/14786419.2018.1466129.

Patel, D. & Sawant, K.K. (2009). Self microemulsifying drug delivery system: Formulation development and biopharmaceutical evaluation of lipophilic drugs. *Curr. Drug Deliv.*, 6(4): 419-424.

Pepi, M., Cesaro, A., Liut, G. & Baldi, F. (2005). An Antarctic psychrotrophic bacterium Halomonas sp. ANT-3b, growing on n-hexadecane, produces a new emulsifying glycolipid. *FEMS Microbiol Ecol.*, 53: 157–166.

Peypoux, F., Bonmatin, J.M. & Wallach, J. (1999). Recent trends in the biochemistry of surfactin. *Appl. Microbiol. Biotechnol.*, 51: 553-563.

Pines, O., Bayer, E.A. & Gutnick, D.L. (1983). Localization of emulsanlike polymers associated with the cell surface of Acinetobacter calcoaceticus. *J. Bacteriol.*, 154: 893-905.

Pirog, T.P., Konon, A.D. & Savenko, I.V. (2015). Microbial surfactants in environmental technologies. *Biotechnologia Acta*, 8(4): 21-39.

Poremba, K., Wilfried, G., Siegmund, L. & Wagner, F. (1991). Marine biosurfactants, III. Toxicity testing with marine microorganisms and comparison with synthetic surfactants. *Zeitschrift für Naturforschung C*, 46.3-4 (1991): 210-216.

Prasad, B., Kaur, H.P. & Kaur, S. (2015). Potential biomedical and pharmaceutical applications of microbial surfactants. *WJPPS*, 4(4): 1557-1575.

Qiao, N. & Shao, Z. (2010). Isolation and characterization of a novel biosurfactant produced by hydrocarbon-degrading bacterium Alcanivorax dieselolei B-5. *Journal of Applied Microbiology*, 108(4): 1207-1216.

Raaijmakers, J.M., de Bruijin, I., Nybroe, O. & Ongena, M. (2010). Natural functions of lipopeptides from Bacillus and Pseudomonas: More than surfactants and antibiotics. *FEMS Microbiol Rev*, 34: 1037-1062.

Rahman, S.R. (2014). Rhamnolipid biosurfactants—Past, present, and future scenario of global market. *Frontiers in Microbiology*, 5: 454.

Ramsay, J.A., Cooper, D.G. & Neufeld, R.J. (1989). Effect of oil reservoir conditions on production of water insoluble levan by Bacillus licheniformis. *Geomicrobiol. J.*, 7: 155-165.

Rapp, P., Bock, H., Urban, E., Wagner, F., Gebetsberger, W. & Schulz, W. (1977). Use of trehalose lipids in enhanced oil recovery. *DESCHEMA Monogr. Biotechnol.*, 81: 177-185.

Ravinder, R., Manasa, M., Roopa, D., Bukhari, N.A., Hatamleh, A.A., Khan, M.Y. et al. (2022). Biosurfactant producing multifarious Streptomyces puniceus RHPR9 of Coscinium fenestratum rhizosphere promotes plant growth in chilli. *Plos One*, 17(3): e0264975. https://doi.org/10.1371/ journal.pone.0264975.

Robert, M., Mercade, M.E., Bosch, M.P., Parra, J.L., Espuny, M.J., Manresa, M.J. et al. (1989). Effect of the carbon source on biosurfactant production by Pseudomonas aeruginosa 44T. *Biotechnology Letters*, 11: 871-874.

Rockne, K.J., Shor, L.M., Young, L.Y., Taghon, G.L. & Kosson, D.S. (2002). Distributed sequestration and release of PAHs in weathered sediment: The role of sediment structure and organic carbon properties. *Environ Sci Technol*, 36: 2636-2644.

Rodrigues, L., Moldes, A., Teixeira, J. & Oliveira, R. (2006a). Kinetic study of fermentative biosurfactant production by lactobacillus strains. *Biochem. Eng. J.*, 28: 109-116.

Rodrigues, L., Banat, I.M., Teixeira, J. & Oliveira, R. (2006b). Biosurfactants: Potential applications in medicine. *Journal of Antimicrobial Chemotherapy*, 57(4): 609-618.

Rodrigues, L.R. (2015). Microbial surfactants: Fundamentals and applicability in the formulation of nano-sized drug delivery vectors. *Journal of Colloid and Interface Science*, 449: 304-316.

Rodríguez-Rodríguez, C.E., Zúñiga-Chacón, C. & Barboza-Solano, C. (2012). Evaluation of growth in diesel fuel and surfactants production ability by bacteria isolated from fuels in Costa Rica. *Revista de la Sociedad Venezolana de Microbiología*, 32: 116-120.

Ron, E.Z. & Rosenberg, E. (2001). Natural roles of biosurfactants. *Env Microbiol*, 3(4): 229-236.

Ron, E.Z. & Rosenberg, E. (2002). Biosurfactant and oil bioremediation. *Curr Opin Biotechnol*, 13: 249-252.

Roongsawang, N., Washio, K. & Morikawa, M. (2010). Diversity of nonribosomal peptide synthetases involved in the biosynthesis of lipopeptide biosurfactants. *Int. J. Mol. Sci.*, 12: 141-172.

Rosenberg, E. (1986). Microbial surfactants. *Crit. Rev. Biotechnol.*, 3: 109-132.

Rosenberg, E., Rubinovitz, C., Legmann, R. & Ron, E.Z. (1988). Purification and chemical properties of Acinetobacter calcoaceticus A2 biodispersan. *Appl Environ Microbiol*, 54: 323-326.

Sadiq, M.B., Khan, M.R. & Sayyed, R.Z. (2022). Biosurfactant mediated synthesis and stabilization of nanoparticles. *In:* Sayyed, R.Z. (Eds), *Biosurfactants: Production and Applications in Bioremediation/Reclamation.* 158-168. CRC Press, Taylor & Francis Group, USA.

Saharan, B.S., Sahu, R.K. & Sharma, D. (2011). A review on biosurfactants: Fermentation, current developments and perspectives. *Genet Engin Biotechnol J*, 2011: 1-14.

Saisa-ard, K., Saimmai, A. & Maneerat, S. (2014). Characterization and phylogenetic analysis of biosurfactant-producing bacteria isolated from palm oil contaminated soils. *Songklanakarin J. Sci. Technol.*, 36(2): 163-175.

Sambles, C.M. & White, D.A. (2015). Genome sequence of Rhodococcus sp. strain PML026, atrehalolipid biosurfactant producer and biodegrader of oil and alkanes. *Genome Announc.*, 3: e00433-15.

Santos, D.K.F., Rufino, R.D., Luna, J.M., Santos, V.A. & Sarubbo, L.A. (2016). Biosurfactants: Multifunctional biomolecules of the 21st century. *International Journal of Molecular Sciences*, 17(3): 401.

Saranraj, P., Sayyed, R.Z., Sivasakthivelan, P., Hasan, M.S., Al-Tawaha, A.R.M.A. & Amala, K. (2022a). Microbial biosurfactants: Methods of investigation, characterization, current market value and applications. *In:* Sayyed, R.Z. (Eds), *Biosurfactants: Production and Applications in Bioremediation/Reclaimation.* 19-34. CRC Press, Taylor & Francis Group, USA.

Saranraj, P., Sayyed, R.Z., Hamzah, K.J., Asokan, N., Sivasakthivelan, P. & Al-Tawaha, A.R.M.A. (2022b). Efficient substrates for microbial synthesis of biosurfactants. *In:* Sayyed, R.Z. (Eds), *Biosurfactants: Production and Applications in Bioremediation/ Reclaimation.* 1-18. CRC Press, Taylor & Francis Group, USA.

Sarkar, A.K., Sharma, M.M., Goursaud, J.C. & Georgiou, G. (1989). A critical evaluation of MEOR processes. United States: N. p., 1989. Web.

Satpute, S.K., Bhuyan, S.S., Pardesi, K.R., Mujumdar, S.S., Dhakephalkar, P.K., Shete, A.M. et al. (2010). Molecular genetics of biosurfactant synthesis in microorganisms. *In:* Sen, R. (Ed.), *Biosurfactants. Advances in Experimental Medicine and Biology*. 672: 14-41. Springer: New York, NY, USA.

Scheibenbogen, K., Zytner, R., Lee, H. & Trevors, J. (1994). Enhanced removal of selected hydrocarbon from soil by *Pseudomonas aeruginosa* UG2 biosurfactants and some chemical surfactants. *J Chem Tech Biotechnol*, 59: 53-59.

Schneiker, S., Martins dos Santos, V.A.P., Bartels, D., Bekel, T., Brecht, M., Buhrmester, J. et al. (2006). Genome sequence of the ubiquitous hydrocarbon-degrading marine bacterium Alcanivorax borkumensis. *Nat. Biotechnol.*, 24: 997-1004.

Schwarzer, D., Mootz, H.D., Linne, U. & Marahiel, M.A. (2002). Regeneration of misprimed nonribosomal peptide synthetases by type II thioesterases. *Proc. Natl. Acad. Sci.* U.S.A., 99: 14083-14088.

Seydlová, G. & Svobodová, J. (2008). Review of surfactin chemical properties and the potential biomedical applications. *Cent. Eur. J. Med.*, 3(2): 123-133.

Shabtai, Y. (1990). Production of exopolysaccharides by Acinetobacter strains in a controlled fed batch fermentation process using soap stock oil (SSO) as carbon source. *Int. J. Biol. Macromol.*, 12: 145-152.

Shah, V., Doncel, G.F., Seyoum, T., Eaton, M., Zalenskaya, I., Hagver, R. et al. (2005). Sophorolipids, microbial glycolipids with anti-human immunodeficiency virus and sperm-immobilizing activities. *Antimicrobial Agents and Chemotherapy*, 49(10): 4093-4100.

Shah, I, Hamid, B., Zaman, M., Fatima, S., Farooq, S., Datta, R. et al. (2022). Microbial biosurfactants: An eco-friendly approach for bioremediation of contaminated environments. *In:* Sayyed, R.Z. (Eds), *Biosurfactants: Production and Applications in Bioremediation/ Reclamation.* 197-207. CRC Press, Taylor & Francis Group, USA.

Shekhar, S., Sundaramanickam, A. & Balasubramanian, T. (2015). Biosurfactant producing microbes and their potential applications: A review. *Critical Reviews in Environmental Science and Technology*, 45(14): 1522-1554.

Shennan, J.L. & Levi, J.D. (1987). In situ microbial enhanced oil recovery. 163-181. *In:* N. Kosaric, W.L. Cairns, N.C.C. Gray (Eds), *Biosurfactants and Biotechnology.* Marcel Dekker, Inc., New York, N.Y.

Shoeb, E., Akhlaq, F., Badar, U., Akhter, J. & Imtiaz, S. (2013). Classification and industrial applications of biosurfactants. *Academic Research International*, 4(3): 243-252.

Shulga, A.N., Karpenko, E.V., Eliseev, S.A., Turovsky, A.A. & Koronelli, T.V. (1990). Extracellular lipids and surface-active properties of the bacterium *Rhodococcus erythropolis* depending on the source of carbon nutrition. *Mikrobiologya*, 59: 443-447.

Singer, M.E. & Finnerty, W.R. (1990). Physiology of biosurfactant synthesis by Rhodococcus species H13A. *Can. J. Microbiol.*, 36: 741-745.

Sourav, D., Malik, S., Ghosh, A., Saha, R. & Saha, B. (2015). A review on natural surfactants. *RSC Advances*, 5(81): 65757-65767.

Stancu, M.M. (2015). Response of Rhodococcus erythropolis strain IBBPo1 to toxic organic solvents. *Brazilian Journal of Microbiology*, 46(4): 1009-1018.

Steller, S., Sokoll, A., Wilde, C., Bernhard, F., Franke, P., Vater, J. et al. (2004). Initiation of surfactin biosynthesis and the role of the SrfD-thioesterase protein. *Biochemistry*, 43: 11331-11343.

Stieritz, D.D., Bondi, A., McDermott, D. & Michaels, E.B. (1982). A burned mouse model to evaluate anti-pseudomonas activity of topical agents. *J. Antimicrob. Chemother.*, 9: 133.

Suzuki, T., Tanaka, H. & Itoh, S. (1974). Sucrose lipids of Arthrobacter, Corynebacteria and Nocardia grown on sucrose. *Agricultural and Biological Chemistry*, 38: 557-563.

Suzuki, T., Hayashi, K., Fujikawa, K. & Tsukamoto, K. (1965). Elucidation of the structure of polymyxin B1. *J. Biol. Chem.*, 57: 226.

Takahashi, M., Morita, T., Fukuoka, T., Imura, T. & Kitamoto, D. (2012). Glycolipid biosurfactants, mannosylerythritol lipids, show antioxidant and protective effects against H_2O_2-induced oxidative stress in cultured human skin fibroblasts. *J. Oleo. Sci.*, 61: 457-464.

Tally, F.P. & De Bruin, M.F. (2000). Development of daptomycin for gram-positive infections. *J. Antimicrob. Chemother.*, 46: 523-526.

Tang, B., Cheng, G., Gu, J.C. & Xu, C.H. (2008). Development of solid self-emulsifying drug delivery systems: Preparation techniques and dosage forms. *Drug Discov Today*, 13(13-14): 606-612.

Teichmann, B., Linne, U., Hewald, S., Marahiel, M.A. & Bolker, M. (2007b). A biosynthetic gene cluster for a secreted cellobiose lipid with antifungal activity from Ustilago maydis. *Molecular Microbiology*, 66: 525-533.

Thaniyavarn, J., Chongchin, A., Wanitsuksombut, N., Thaniyavarn, S., Pinpharnichakarn, P., Leepipatpibbon, N. et al. (2006). Biosurfactant production by Pseudomonas aeruginosa A41 using palm oil as carbon source. *J. Gen. Appl. Microbiol*, 52: 215-222.

Thibault, S.L., Anderson, M. & Frankenberger, W.T. (1996). Influence of surfactants on pyrene desorption and degradation in soils. *Applied and Environmental Microbiology*, 62: 283-287.

Tripathy, D.B., Mishra, A., Clark, J. & Farmer, T. (2018). Synthesis, chemistry, physicochemical properties and industrial applications of amino acid surfactants: A review. *C.R. Chimie*, 21: 112-130.

Tseng, C.C., Bruner, S.D., Kohli, R.M., Marahiel, M.A., Walsh, C.T., Siber, S.A. et al. (2002). Characterization of the surfactin synthetase C-terminal thioesterase domain as a cyclic depsipeptide synthese. *Biochemistry*, 41: 13350-13359.

Tugrul, T. & Cansunar, E. (2005). Detecting surfactant-producing microorganisms by the drop collapse test. *World Journal of Microbiology and Biotechnology*, 21(6): 851-853.

Tuleva, B.K., Ivanov, G.R. & Christova, N.E. (2002). Biosurfactant production by a new *Pseudomonas putida* strain. *Z Naturforsch C*, 57(3-4): 356-360.

Tummler, K., Effenberger, F. & Syldark, C. (2003). An integrated microbial/enzymatic process for production of rhamnolipids and L-(+)-rhamnose from rapeseed oil with Pseudomonas sp. DSM 2874. *Eur. J. Lipid Sci. Tech.*, 105: 563-571.

Tyagi, M., da Fonseca, M.M. & Carvalho, C.C. (2011). Bioaugmentation and biostimulation strategies to improve the effectiveness of bioremediation processes. *Biodegradation*, 22(2): 231-241.

Van Dyke, M., Gulley, S., Lee, H. & Trevors, J. (1993). Evaluation of microbial surfactants for recovery of hydrophobic compounds in soil. *J. Ind. Microbiol. Biotechnol.*, 11: 163-170.

van Iwaarden, J.F., Pikaar, J.C., Storm, J., Brouwer, E., Verhoef, J., Oosting, R.S. et al. (1994). Binding of surfactant protein A to the lipid A moiety of bacterial lipopolysaccharides. *Biochem. J.*, 303: 407–411.

Van Wagoner, R.M., Clardy, J. & Fee, M. (2006). An N-acyl amino acid synthase from an uncultured soil microbe: Structure, mechanism, and acyl carrier protein binding. *Structure*, 14: 1425-1435.

Vanittanakom, N., Loeffler, W., Koch, U. & Jung, G. (1986). Fengycin – A novel antifungal lipopeptide antibiotic produced by *Bacillus subtilis* F-29-3. *J. Antibiotics*, 39(7): 888-901.

Vatsa, P., Sanchez, L., Clement, C., Baillieul, F. & Dorey, S. (2010). Rhamnolipid biosurfactants as new players in animal and plant defense against microbes. *International Journal of Molecular Sciences*, 11(12): 5095-5108.

Velasquez, J., Scorzza, C., Vejar, F., Forgiarini, A.M., Antón, R.E. & Salager, J.L. (2010). Effect of temperature and other variables on the optimum formulation of anionic extended surfactant–alkane–brine systems. *J Surfact Deterg*, 13: 69-73.

Velikonja, J. & Kosaric, N. (1993). Biosurfactant in food applications. *In:* N. Kosaric (Ed.), *Biosurfactants: Production, Properties, Applications*. 419–446. Marcel Dekker Inc., New York, N.Y.

Vijayakumar, S. & Saravanan, V. (2015). Biosurfactants – Types, sources and applications. *Research Journal of Microbiology*, 10(5): 181.

Volkering, F., Breure, A.M., van Andel, J.G. & Rulkens, W.H. (1995). Influence of nonionic surfactants on bioavailability and biodegradation of polycyclic aromatic hydrocarbons. *Applied and Environmental Microbiology*, 61(5): 1699-1705.

Wang, X., Gong, L., Liang, S., Han, X., Zhu, C. & Li, Y. (2005). Algicidal activity of rhamnolipid biosurfactants produced by Pseudomonas aeruginosa. *Harmful Algae*, 4: 433-443.

Wright, J.R. (2003). Pulmonary surfactant: A front line of lung host defense. *J. Clin. Invest.*, 111: 1453-1455.

Wu, Y.S., Ngai, S.C., Goh, B.H., Chan, K.G., Lee, L.H. & Chuah, L.H. (2017). Anticancer activities of surfactin and potential application of nanotechnology assisted surfactin delivery. *Front Pharmacol.*, 8: 761.

Xu, W., Nikolov, A., Wasan, D.T., Gonsalves, A. & Borwankar, R.P. (1998). Fat particle structure and stability of food emulsions. *J. Food Sci.*, 63(2): 183-188.

Yakimov, M., Amro, M. & Bock, M. (1997). The potential of *Bacillus licheniformis* strains for in situ enhanced oil recovery. *Journal of Petroleum Science Engineering*, 18: 147-160.

Yao, S., Zhao, S., Lu, Z., Gao, Y., Lv, F. & Bie, X. (2015). Control of agitation and aeration rates in the production of surfactin in foam overflowing fed-batch culture with industrial fermentation. *Rev Argent Microbiol.*, 47(4): 344-349.

Yeh, M.S., Wei, Y.H. & Chang, J.S. (2006). Bioreactor design for enhanced carrier-assisted surfactin production with Bacillus subtilis. *Process Biochem.*, 41: 1799-1805.

Zaman, M., Hassan, S., Fatima, S., Hamid, B., Farooq, S. et al. (2022). Biosurfactants production and applications in food. *In:* Sayyed, R.Z. and Enshasy, H.E. (Eds), *Biosurfactants: Production and Applications in Food and Agriculture*, Vol II, 225-241. CRC Press, Taylor & Francis Group, USA.

Zinjarde, S. & Pant, A. (2002). Emulsifier from tropical marine yeast, *Yarrowia lipolytica* NCIM 3589. *J. Basic Microbiol.*, 42: 67-73.

CHAPTER

2

Biosurfactant Production: Limitations and New Approaches

Suman Ganger[1,2*], Deepak Sonawane[2], Ninad Mhatre[2], Fatema Saiger[2], and Amit Pratap[2]

[1] Department of Microbiology, Vivekanand Education Society's College of Arts, Science and Commerce (Autonomous), Chembur, Mumbai, India

[2] Department of Oils, Oleochemicals and Surfactants Technology, Institute of Chemical Technology, Nathalal Parekh Marg, Matunga, Mumbai, India

1. Introduction

Surface-active agents or surfactants play a pivotal role in various aspects of our daily lives, constituting a significant component in many commonly used products such as laundry detergents, surface cleaners, fabric softeners, cosmetics, body care, pharmaceuticals, agriculture, food processing and the petroleum industry. Surfactant molecules, characterized by their amphiphilic nature with a hydrophilic head and hydrophobic tail, interact at interfaces between aqueous and non-aqueous systems, influencing surface tension, emulsification, wetting and foaming. These surfactants, mostly chemicals, are predominated by alkyl sulfates or sulfonates with straight or branched chains and are sourced from either petrochemical or oleochemical origins, as outlined by Desai and Banat (1997). Over the past few decades, the global utilization of surfactants has witnessed substantial growth. The surfactants market size is projected to increase from USD 43.5 billion in 2022 to USD 57.8 billion by 2028, at a compound annual growth rate (CAGR) of 4.9% (*Surfactants Market*, 2023). Companies incorporating surfactants in their products are now exploring the substitution of chemical surfactants due to their high levels of toxicity and low biodegradability with sustainable bio-based surfactants (Johnson et al., 2021). According to a prior study, the European Union's CO_2 emissions might be reduced by 37% if surfactant synthesis were done with renewable resources rather than petrochemicals (M. Patel, 2003). Holiferm, a UK based company had appointed Clarke, a life-cycle analysis team which predicts that 1.5 tons of CO_2 could be eliminated from greenhouse

* Corresponding author: sumanganger@gmail.com

gas emissions by substituting 1 metric ton of sophorolipids for 1 metric ton of a conventional ethoxylated surfactant (Bettenhausen, 2022).

Bio-based surfactants are the promising and greener sustainable solutions to the conventional petrochemical based surfactants due to their biodegradable properties and low (or no) toxicity (Agger & Zeuner, 2022) and more environmentally friendly manufacturing process (Liyana et al., 2022). These comprise any surfactant manufactured from renewable biological material, such as carbohydrates, fats or proteins which can be substantiated using C-14 analysis. The International Organization for Standardization (ISO) / Draft International Standard (DIS) 21680 defines a bio-based surfactant as "a surfactant wholly or partly derived from biomass (based on biogenic carbon)" (*ISO/DIS 21680(En), Surface Active Agents-Bio-Based Surfactants-Requirements and Test Methods*, 2020). Bio-based or 'green' surfactants also called oleosurfactants are produced chemically, enzymatically or in cellular factories; the latter termed as biosurfactants (Agger & Zeuner, 2022). Despite being safer and environmentally friendly, bio-based surfactants currently constitute only 4% of the market share in the surfactant production sector, with the majority being dominated by chemically produced bio-based surfactants (Hausmann & Henkel, 2022). The primary reason for such a low market share is the economic non-viability of their production as compared to chemical counterparts. In comparison to synthetic surfactants, which cost US $4 per kg, bio-based surfactants cost approximately US $34 per kg (*Research and Market. Global Biosurfactants Market Size, Segments, Outlook, and Revenue Forecast 2022-2028 by Product, Application, and Region*, 2022). Degradation of the environment, dwindling petrochemical stocks, consumer demand for safer and cleaner products and a global initiative for fight towards climate change have spurred research into new renewable bioresources for the efficient synthesis of bio-based products including surfactants. The Renewable Carbon Initiative seeks to expedite and facilitate the shift of all organic materials and chemicals from fossil carbon to renewable carbon. A carbon circular economy is created by the exchange of renewable carbon between the atmosphere and the biosphere. The Renewable Carbon Index (RCI) calculates the renewable material content in a product. It's calculated by dividing the carbons from renewable sources by the total carbons in an active ingredient. A high RCI surfactant, for example, contains a high percentage of carbon atoms from natural resources. TegraSurf™, a bio-based surfactant produced by Integrity Bio-Chemicals boasts of a RCI of at least 58%, and as high as 94%, indicating a significant presence of carbon atoms from natural sources (*TegraSurf by Integrity Bio-Chemicals*, 2021).

Bio-based surfactants can be classified based on their production process. Figure 1 summarizes the same along with representative examples. Soap is an example of a bio-based surfactant, representing one of the earliest man-made surfactants. It is produced chemically through the saponification of vegetable oils or animal fats and an alkali such as sodium or potassium hydroxide resulting in the formation of soap and glycerol. Rahayu et al. (2021) introduced an environmentally friendly technology for transforming spent cooking oil into soap. Other specialized chemically produced bio-based surfactants include α-Methyl Ether Sulfonates (α-MES), which are manufactured using palm oil through transesterification (Low et al., 2021) and Alkyl Polyglucosides produced from bio-ethylene oxide (Croda, 2016). Typically, chemical biosurfactants are produced using refined substrates such as carbohydrates (glucose, lactose, sucrose etc.), fatty alcohols, fatty acids and oils. However, this

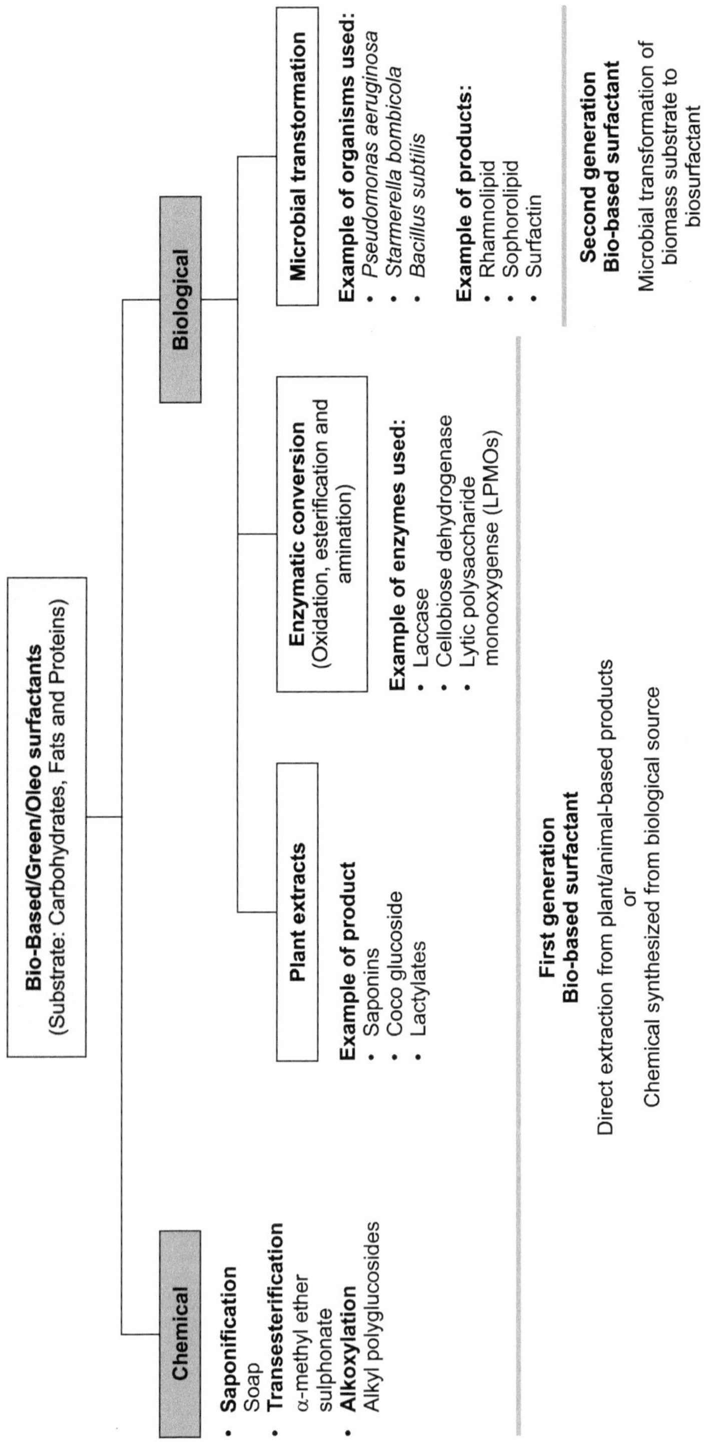

Fig. 1: Classification of bio-based surfactants according to production process.

production method can lead to increased costs, the use of chemicals, and a shift from fossil oil to tropical oils, which may not be environmentally beneficial (Hausmann & Henkel, 2022). The second category of bio-based surfactants is derived from organic production methods and is further subdivided into three types: plant-extracted surfactants, biocatalyzed surfactants using enzymes, and microbe-derived surfactants called biosurfactants. Saponins extracted from plant materials such as leaves, roots, flowers, etc. via solvent-based approaches are one of the best examples of plant-extracted surfactants (Ribeiro et al., 2013). Plant derived surfactants depend on CO_2 as the main substrate sequestered from the atmosphere by the plants. However, growing of plants needs cultivable land, fertilizers and water as vital resources which necessitates allocation of natural resources for their production. On the other hand, biocatalysis approach for surfactant synthesis is an influential and multipurpose tool to replace and/or complement chemical routes. Linking the surfactant hydrophilic head and hydrophobic tail enzymatically involves the use of traditional enzymes like lipases and glycoside hydrolases. Recently, proteases have been used for creation of amino-acid based surfactants. Other enzymes employed include laccase, cellobiose dehydrogenase and lytic polysaccharide monooxygenase (LPMOs). Enzymes have been highly effective in providing chemical functionality specific to a given site, hence facilitating the synthesis of surfactants through click chemistry linkages (Agger & Zeuner, 2022). Until today, the enzyme synthesis approach has not led to large scale production volumes of surfactants. However, several biotech businesses are exploring the potential while optimizing their biocatalytic platform technology, making it an ideal direction to advance (Hausmann & Henkel, 2022). Surfactants which are directly mined from animal-based or plant-based raw materials or directly synthesized chemically from biological sources are called first-generation biosurfactants (Nagtode et al., 2023).

In congruence with the plant-extracted and enzyme catalyzed surfactants, the third type of biologically produced surfactants are microbial biosurfactants. Diverse types and species of microorganisms (yeast, bacteria and fungi) can transform a broad variety of biomass substrates into biosurfactants via the intricate biocatalytic pathways contained in their genomes. They are referred to as second-generation biosurfactants (Nagtode et al., 2023). Since the middle of the 20th century, biosurfactants have been documented in literature. The rhamnolipid produced by the bacteria *Pseudomonas aeruginosa* is known since 1949 (Jarvis & Johnson, 1949), the fungus *Ustilago maydis* secretes cellobioselipids, also referred to as ustilagic acid was discovered in 1951 (Haskins & Thorn, 1951), Tulloch originally described sophorolipids, which are derived from the yeast *Starmerella bombicola* (previously called *Candida bombicola*), in 1968 (Tulloch et al., 1968) and mannosylerythritol lipids, produced by *Candida antarctica* in 1990 (Kitamoto et al., 1990). The above classes belong to the group of surfactants called glycolipids. Another important group called lipopeptides contains surfactin (Arima & Tamura, 1968) and iturin (Maget-Dana et al., 1992) produced by *Bacillus subtilis.* Sarubbo et al., 2022, gives a list of biosurfactant classes and their microbial source (Sarubbo et al., 2022). Biosurfactants derived from microbes have a high potential for commercialization since they may be generated on a large scale, with a shorter process cycle, less space usage, no competition with food, independence from seasonal and climatic conditions and can be tweaked for higher outputs using recombinant technologies (Rodrigues et al., 2017). Microbe based biosurfactants show diverse structures that can have profound effects on their potential industrial application ranging from oil recovery and bioremediation to biomedical, agriculture and food applications (Thistleton, 2022). A review by Kashif et al. and Santos et al. gives exhaustive applications of different classes of biosurfactants (Kashif et al., 2022; Santos et al., 2016). Table 1 provides a list of some companies that have successfully commercialized and are using biosurfactants in their products (*Biosurfactants Market, By Product Type (Glycolipids, Lipopeptides & Lipoproteins, Surfactin, Others), By Application (Personal Care, Food Processing, Others), By Region Forecast to 2030*, 2022).

Table 1: List of companies with commercialization of biosurfactants in their products

Company	Biosurfactant	Product Name	Applications	Reference
Kaneka	Surfactin	Kaneka Surfactin	Cosmetics	(*Biosurfactant - KANEKA Surfactin*, 2023)
Holiferm	Sophorolipid	HoliSurf LF, HoliSurf HF, HoneySurf AG	Hair Conditioners, Shampoo and Rinses	(*Holiferm Products - Biosurfactants from Renewable Feedstocks*, 2023)
DOW	Acid sophorolipid	EcoSense™ GL-60 HA	Shampoo and Body Wash	(Dow, 2022)
Unilever & Evonik	Rhamnolipid	Quix	Hand dishwashing liquid	(Krauter, 2019)
Givaudan Active Beauty	Sophorolipid	Sopholiance® S	Fragrance and Beauty	(*Sopholiance™ S - Givaudan*, 2023)

Biosurfactants are further divided into low molecular mass and high molecular mass groups based on their molecular mass, with typical mass ranges between 500 to 1500 Daltons (Naughton et al., 2019). Low molecular mass biosurfactants such as lipopeptides, glycolipids, and phospholipids are superior at lowering surface or interfacial tension. High molecular mass biosurfactants lipoproteins and lipopolysaccharide are better at stabilizing emulsions and referred to as bioemulsifiers (Nagtode et al., 2023). The structures of the biosurfactants determine their classification and most importantly their unique physicochemical properties which further decides their potential use in varied applications. Diverse microorganisms produce a variety of biosurfactants; yet they share a common amphiphilic nature, typically being anionic, neutral or cationic. Long-chain fatty acids or saturated and unsaturated hydrocarbons are frequently the hydrophobic moiety's constituents, whereas organic acids, alcohols or other carbohydrate functional groups make up the hydrophilic moiety (Sarubbo et al., 2022).

Biosurfactants have a long way to go before they can rival high-volume, low-value surfactants. The projected growth of the biosurfactants market anticipates an increase of USD 3.51 billion between 2022 and 2027, with a CAGR of 5.6%. (*Biosurfactants Market Analysis Europe, North America, APAC, Middle East and Africa, South America US, Canada, China, Germany, France - Size and Forecast 2023-2027*, 2023). The recent initiatives of corporations to introduce new biosurfactant products having direct commercial applications have fueled research and innovations in this field. Evonik Industry is the first company to mass produce rhamnolipid from butane using recombinant *P. aeruginosa*. German biotech company Biotensidon GmbH is the first to create economically viable methods for producing rhamnolipids on an industrial scale (5,000 tonnes per year) (Eslami et al., 2020). AGAE Technologies produces different purities of rhamnolipids, such as R90, R95 (*AGAE Produces Record Yields of Rhamnolipids at Its Lowest Cost by Fermentation*, 2022). Ismail et al. in their chapter on 'Overview of Bio-Based Surfactant – Recent Development, Industrial Challenge, and Future Outlook' have detailed a table containing commercially available biosurfactants, their manufacturers, and their applications (Liyana Ismail et al., 2022).

In this chapter, we have discussed the key factors which determine biosurfactant production including their limitations. The newer approaches to make biosurfactant production feasible and economically viable are discussed.

2. Biosurfactant Production

The production of amphiphilic biosurfactants is based on the genetic recipe present within the microorganism, either naturally occurring or engineered, and the ingredients i.e. media components provided in a suitable fermentative environment. It is generally acknowledged that there are four primary possibilities for the synthesis of such amphiphilic biosurfactants – (i) Hydrophilic and hydrophobic molecules of the biosurfactant structure are synthesized separately from the de novo route, (ii) the hydrophilic molecules are synthesized using de novo and the hydrophobic molecules on the basis of the available substrate inducer, (iii) the hydrophobic molecules are synthesized using de novo and the hydrophilic molecules on the basis of the available substrate inducer and (iv) both the hydrophilic and hydrophobic molecules rely on the available substrates (P. Patel et al., 2023). Whatever be the mechanism employed, the synthesis of biosurfactants involves many metabolic processes which are regulated. The media components, especially the complex waste feed stocks greatly influence the type and yield of biosurfactants. An overview of biosurfactant production is represented in Fig. 2. Industrially relevant microbial strain selection is the most crucial step, focusing on those with a high capacity to produce biosurfactants. In addition to screening microbial strains, genetic engineering may be taken into account to improve biosurfactant production through the manipulation of metabolic pathways. Media formulation is equally important for a successful production process. A nutrient-rich medium is carefully formulated to promote the growth of microorganisms and increasing biosurfactant production. It includes the selection of favorable and economic sources of carbon, nitrogen and trace elements. Optimization of the process parameters such as pH, temperature, aeration and agitation levels of the culture are meticulously regulated to establish a favorable atmosphere for the development of microorganisms and the synthesis of biosurfactants. Throughout the fermentation process, constant monitoring and modification is necessary to maintain ideal conditions. Throughout the production process, growth kinetics (cell growth) and biosurfactant production kinetics are closely monitored to understand their interplay. The process of harvesting involves analyzing growth and production curves to determine the best time to collect biosurfactants. The individual microorganisms and the properties of the biosurfactant are taken into consideration while selecting the harvesting procedure, which may involve filtration, centrifugation, or other methods. Following harvesting, the biosurfactant is extracted from the culture broth using extraction techniques. The specific characteristics of the biosurfactant determine which extraction technique

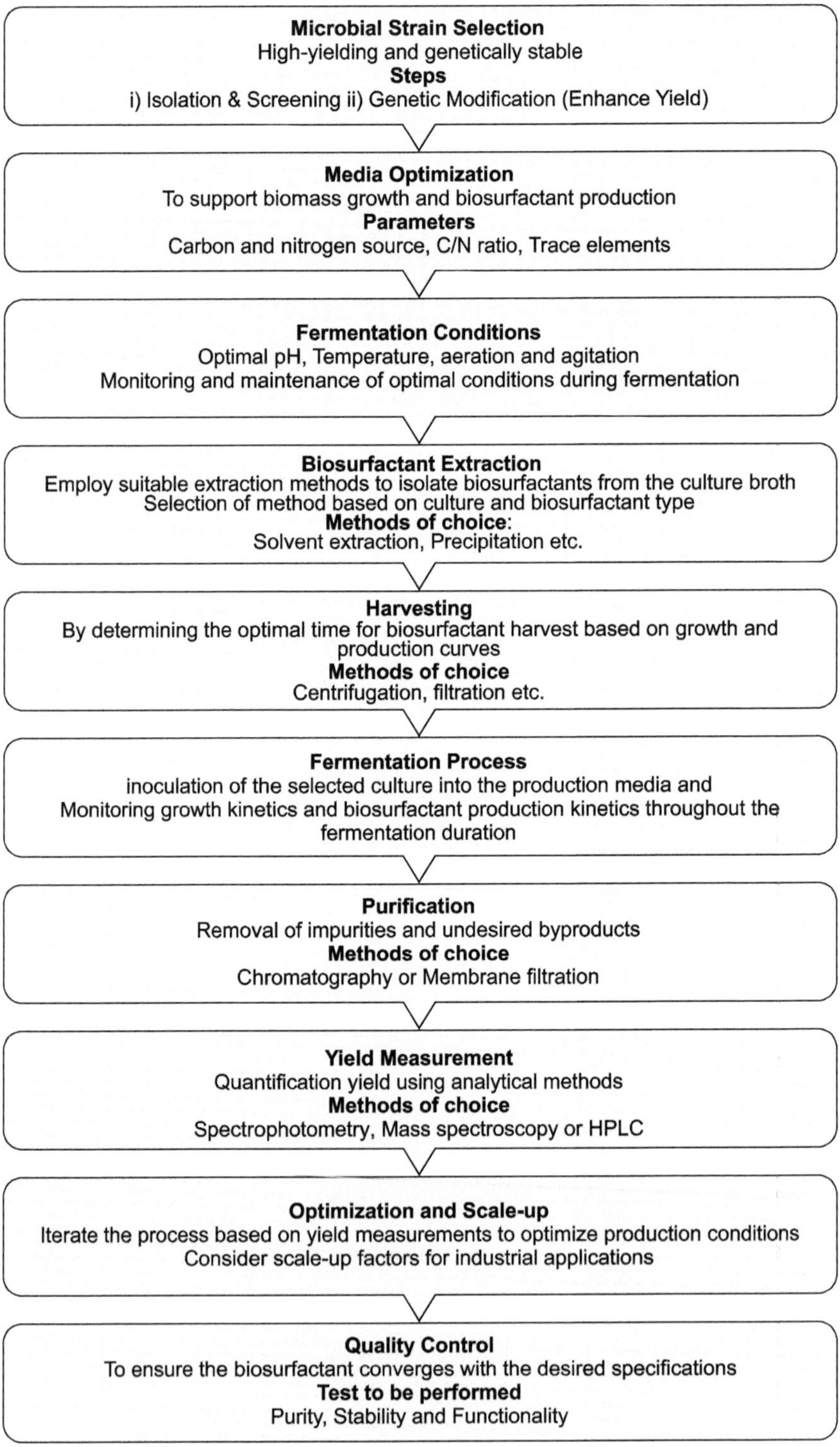

Fig. 2: An overview of biosurfactant production.

solvent extraction or precipitation—to use. The recovered biosurfactant is then purified to remove contaminants and byproducts. Effective purification is achieved by applying methods like solvent extraction, chromatography, or membrane filtration. Subsequently, the biosurfactant yield is measured using analytical techniques such as mass spectrometry, spectrophotometry, and HPLC. By comparing the amount of biosurfactant obtained to the starting biomass or substrate, the yield is calculated. Finally, optimization and scale-up is attained, which involve iterative processes based on yield measurements to improve production conditions for maximum efficiency. Considerations for reactor scale-up are explored, especially in the context of industrial applications. Strict quality control procedures are used to make sure the biosurfactant satisfies the desired specifications. Purity, stability and functionality tests are performed to ensure that the biosurfactant will work as intended. This comprehensive approach ensures a systematic and optimized biosurfactant production process, starting with the selection of microorganisms and ending with quality control evaluations.

2.1 Microbial Strain Selection

A large number of varied microbial species possess the capability to produce surface-active compounds. Biosurfactants have a wide range of physiologically significant roles for microorganisms, including adhesion to the substrate, cell-to-cell communication, differentiation, motility, high ionic strength, defense against noxious substances, energy and carbon storage (Ron & Rosenberg, 2001). In aqueous environments, aerobic bacteria produce biosurfactants to facilitate the movement of insoluble substrates through cell membranes and promote their growth (Patel et al., 2023). Hence many microbes require and produce minute quantities of biosurfactants and on screening are positive for the same, but their production potential relies on various environmental and physiological conditions for e.g. Quorum sensing coordination plays a fundamental role in the synthesis of rhamnolipids and surfactin (Satpute et al., 2010). The type of microbe and the type of nutrients are the main factors that determine how much biosurfactant is produced. The biosurfactant production capacity of a microorganism is also limited to its genetic capacity and the role played by the biosurfactant in the growth and maintenance of the producer organism. The majority of microbes capable of producing biosurfactants have been filtered from a variety of industrial waste sites such as contaminated soils, effluents, and wastewater discharge points (Kiran et al., 2015). Consequently, by using industrial waste as a carbon source and optimizing other production parameters, these organisms can be explored for enhanced biosurfactant production. However, the synthesis of rhamnolipids by the opportunistic, Gram-negative bacterium *P. aeruginosa* is a great example of how some of these species are also opportunistic to humans (Jensen et al., 2007). Hence, it forms a regulatory perspective for the use of a microbe and/or their products in the Fast-Moving Consumer Goods (FMCG) and pharmaceutical companies, organisms must have the status of GRAS (Generally Recognized as Safe). Bacteria like *B. subtilis,* Yeast like *Yarrowia lipolytica, Saccharomyces cerevisiae,* and *Kluyveromyces lactis* fall into the GRAS category and have been explored for their biosurfactant and/or bioemulsifier production (Sharma et al., 2023). Some of the hydrocarbon-degrading bacteria have been found to be excellent producers of biosurfactants, as they need to solubilize hydrocarbons as a source of carbon for their growth. *Acinetobacter* spp. are known to be hydrocarbon degraders and have been employed for bioremediation purposes. By employing waste from the oil industry, Choi et al. (1996) identified *Acinetobacter calcoaceticus RAG-1*, which generates a commercially significant biosurfactant dubbed "emulsan". Hence the inherent genetic limitations, media conditions supporting the growth of organisms and pathogenic features of the producer organisms act as great limiting factors. This chapter explores multiple approaches for selection and improvement of existing and novel strains for biosurfactant production from the environment.

2.2 Media Optimization

Media formulation is a critical step in metabolite production as it directly impacts biomass production, product quality, yield and the efficiency of the purification process. It largely depends on the requirement of the cell biomass and the metabolite. Effectively formulated media contribute to stable and consistent output and significantly influence the production cost (Sharon et al., 2023). For a longtime, the production medium has been the subject of research to optimize biosurfactant production. This not only lowers the costs (by using waste feed stocks) but also modifies the amount of nutrients accessible to the organism, thereby affecting the biosurfactant yield. The principal challenge for commercializing biosurfactants is associated with media formulation and recovery, accounting for 30% and 60-80% of the total manufacturing cost, respectively (Sharon et al., 2023). This also depends on the type of biosurfactant that can increase to even 80%. The most researched class of biosurfactants, rhamnolipids, continues to face difficulties with high manufacturing costs and low yields. Several of the synthetic processes needed or involved in the synthesis of biosurfactants are harsh environment processes (Liyana et al., 2022) including hazardous solvents and toxic acid/base catalysts (Nagtode et al., 2023). Enzymes have been working as an alternative to chemical catalysts, but their cost and slow reaction rates act as a hurdle in their utilization. Therefore, having appropriate reaction optimizations for solvents or catalysts is a significant problem in the synthesis or purification of biosurfactants. The majority of raw material expenses for high-volume, low-value products like rhamnolipids and sophorolipids are attributed to the cost of fermentation media (Chong & Li, 2017; Dhanarajan & Sen, 2014). The primary cost factors for high-value, low-volume items like surfactin are the materials needed for recycling them and the need to keep an aseptic medium until the very end of the packaging. The cost of the glucose and oleic acid needed to generate 90.7 million kg of sophorolipid was estimated to be $150 million in the sophorolipid manufacturing model. This amount accounts for almost 75% of the overall operating cost. This can be reduced by replacing expensive substrates with industrial by-products and cheap agricultural raw materials (Dhanarajan & Sen, 2014). Another study found that using cheap substrates for biosurfactant production increases yields by 1.5 times and reduces average costs by 60-80%. The cost of a fermentation medium based on cheese curds and molasses was found to be €2.6-3.8/liter. The use of complex substrates and the use of antifoam agents during the biosurfactant synthesis process necessitate additional purification stages, which raises the cost of further processing. As a result, different cultivation techniques have been used to lessen the production of foam (Dhanarajan & Sen, 2014; Rangarajan & Sen, 2013). Therefore, the use of affordable and environmentally suitable substrates and process optimization to increase yield would be crucial for cost-effective biosurfactant production (Sharon et al., 2023; Thio et al., 2022) which is influenced by the requirement of the cell biomass and the metabolite production. Abalos et al. (2002) used response surface methodology to optimize the culture media so that *P. aeruginosa* AT10 could produce rhamnolipids. They came to the conclusion that the carbon and nitrogen supply along with phosphate and iron source are influential for the production (Abalos et al., 2002). To minimize the raw material cost, extensive studies are carried out to understand and evaluate use of substitute sources such as industrial waste material. Use of these waste materials has proven to be beneficial for production with respective cost as well as yield. Comprehensive evaluation is critical to understand the exact role of nitrogen in biosurfactant production. Supplementing culture medium with an inducer that initiates biosurfactant metabolism is a simple way to increase biosurfactant productivity. They also affect the chemical structure of the biosurfactant. Mostly they act as supplementary carbon sources (Nurfarahin et al., 2018), however, they have been poorly investigated. Hydrophobic inducers like vegetable oil helps in synthesis of the hydrophobic part of biosurfactants directly, altering their chemical structure and, ultimately, their potential applications. Hydrophilic inducers like metal ions function as cofactors for microbial growth, leading to improved biosurfactant production. Therefore, using alternate culture media coupled with inducers is an intriguing possibility to lower the cost of producing biosurfactants. However, further research is required with respect to their types, purity, concentrations etc (De Oliveira Schmidt et al., 2021). Studies have proven that a combination of different industrial waste can enhance production and lower the production cost. Some of the novel waste which have been employed in the biosurfactant production includes frying oils (Haba et al., 2000), as well as other oil wastes from soybean and corn (Nitschke et al., 2005), palm kernel fatty acid distillate (Thio et al., 2022). A review by Wongsirichot et al. (2021) elucidates the prospective operation of unconventional feed stocks as sources of sugars, fatty acids and nitrogen for sophorolipid production. In order to do away with the necessity for purified substrates, the review also makes the case for combining alternate feed supplies (Wongsirichot et al., 2021). Researchers should, however, consider early on the variability and availability of such alternate substrates at the relevant scale. Nevertheless, for practical reasons, first-generation refined substrates like refined sugar, oils, and/or fatty acids are still utilized in industrial production today. As sophorolipids have shown (Baccile et al., 2017), this has a major negative impact on the sustainability of microbial biosurfactants.

2.2.1 Carbon Source

Biosurfactants are bio-based amphipathic molecules. Chemically, they comprise a hydrophobic moiety (saturated, unsaturated, hydroxylated, or branching - hydroxyl fatty acid, long-chain fatty acid) and a hydrophilic moiety (ester, carboxylate, hydroxyl group, phosphate, peptide, amino acid, or carbohydrate) much like in any surfactant (Nagtode et al., 2023; Sharon et al., 2023). Therefore, both water-soluble and water-insoluble carbon source substrates are employed in the synthesis of biosurfactants. Widely used water-soluble substrates are glucose, ethanol, glycerol, and mannitol, and water-insoluble substrates are sunflower oil, coconut oil, soybean oil, olive oil, and n-alkanes (Thio et al., 2022). Several research studies have indicated that water-insoluble substrates used in the manufacture of biosurfactants showed increased output in comparison to substrates soluble in water. It is contributed by the hydrophobic nature of the substrate which increases the availability of the substrate for consumption by the organism (Thio et al., 2022). Additionally, they are rich in carbon content and can be hydrolyzed into long-chain fatty acids. The high carbon content and free fatty acid content of vegetable oil and its derivatives promotes the production of biosurfactants. However, given the competition in the food market and the high cost of refined vegetable oils, it is more practical to use vegetable oil industry by-products, such as used cooking oil and soap stock, as a substrate for the synthesis of biosurfactants (Thio et al., 2022).

It is widely acknowledged that the use of purified carbon sources increases the yield and decreases the pressure on downstream processing, but it also increases the raw material cost. Therefore, a conscious decision for the selection of carbon source is required to strike a balance between the cost incurred by raw material, the yield obtained, and the optimization of downstream processing. Multiple attempts are made to use industrial waste water-soluble and insoluble carbon sources (Chong & Li, 2017).

2.2.2 Nitrogen Source

Nitrogen plays a vital part in biosurfactant production by influencing microbial growth and metabolism. Rhamnolipids are a class of secondary metabolites produced primarily by *P. aeruginosa* when cell growth reaches a stable state and the nitrogen supply runs out. Hence earlier concentration of nitrogen was closely monitored to achieve steady state due to exhaustion of nitrogen. It was also observed that rhamnolipids are not strictly secondary metabolites. They are produced during the exponential phase as well. Several studies also reported that high concentration of $NaNO_3$ resulted in a higher rhamnolipid yield. Whether nitrogen deficiency or excess is better for increasing rhamnolipid synthesis is yet to be determined but it is widely accepted that the choice of nitrogen source can also influence the total cost and environmental impact of biosurfactant production. Researchers often seek to identify cost-effective and sustainable nitrogen sources to optimize biosurfactant production processes (Wu et al., 2019). Saikia et al. received positive results during evaluating the effect of nitrogen sources such as ammonium nitrate, ammonium chloride, ammonium sulphate, sodium nitrate and potassium nitrate, on biosurfactant production (Saikia et al., 2012). Table 2 summarizes the effect of different carbon and nitrogen source on the yields of different classes of biosurfactants (Chong & Li, 2017; Dhanarajan & Sen, 2014; Onwosi & Odibo, 2012; Wongsirichot et al., 2021).

Apart from carbon and nitrogen source, C/N ratio in the production media is also crucial (Abalos et al., 2002; Onwosi & Odibo, 2012) along with optimization of various macro and micro nutrients, specifically phosphates and iron ratio should be carried out as it directly and indirectly affects the type, quality and quantity of biosurfactant (Abalos et al., 2002).

2.3 Downstream Processing

The aim of downstream processing is to isolate, purify and concentrate the product from the complex bulk matrix. The selection of techniques to be employed for the downstream processing is based on the purity intended, final product intended application, operational economics, scalability and the possibility of process integration with other downstream stages. The economy of bio-production is heavily dependent on the careful and effective planning of downstream purification processes, which can make up to 70-80% of the entire cost of production (Rangarajan & Clarke, 2016; Santos et al., 2016). This is also true in case of biosurfactants where the production of low-cost biosurfactants is difficult due to the complicated recovery processes (Santos

Table 2: Effect of carbon and nitrogen source on the yields of different classes of biosurfactants

Surfactant	Organism	Carbon source	Nitrogen source	Yield (g/L)
Rhamnolipid	*Pseudomonas nitroreducens*	Glucose	Sodium nitrate	4.4
	Pseudomonas aeruginosa Z41	Waste frying oil	-	9.0
	Pseudomonas putida B12	Molasses	-	0.52
	Enterobacter asburiae NRRL B-59189	Sodium citrate	-	0.51
	Enterobacter asburiae NRRL B-59189	Glycerol and Sodium citrate	-	2.0
Sophorolipid	*Starmerella bombicola* ATCC 22214	Cheese whey (10%) and Olive oil (10.5%)	Yeast extract (0.25–0.3%)	6.23
		Deproteinized whey concentrate (100 g/L lactose) and Rapeseed oil (Initial: 105 g/L) (Fed: 3 kg)	Yeast extract (4 g/L)	280.0
		Sunflower oil cake (50 g/L) and Crude soybean oil (50 g/L)	Ammonium nitrate (1 g/ L)	10.0
		Soy molasses and Oleic acid	Soy molasses	53.0
Mannosylerythritol lipid + cellobiose lipid	*Ustilago maydis*	Glucose	-	23.0
Trehaloselipid	*Rhodococcus wratislaviensis* BN38	Hexadecane	-	3.1
	Rhodococcus erythropolis DSM 43215	Mihagol L	-	32.0
Lipopeptide	*Bacillus subtilis* C9	Glucose	-	7.0
Surfactin	*Bacillus subtilis* ATCC 21332	Glucose	-	6.45
Surfactin + Fengycin	*Bacillus circulans*	Glucose	-	6.98

et al., 2016). Most biosurfactant production procedures require the use of spent whole-cell culture broths or other crude preparations due to budgetary constraints (Marchant et al., 2014). Precipitation at very high pH levels and extraction using organic solvents are the most popular methods for recovering and purifying crude biosurfactants (Lin

& Jiang, 1997), though the same cannot be used at commercial application purposes. Conventional downstream processes like precipitation, solvent extraction, membrane ultrafiltration, adsorption chromatography techniques have their limitations of removal of certain impurities such as proteins, salts and other media components, leading to partially purified or purified mixtures like in case of lipopeptide mixtures. For further purification, superior techniques like reverse phase high pressure liquid chromatography (RP-HPLC) are employed (Rangarajan & Clarke, 2016) which would escalate the final product price. Each downstream approach has pros and cons of its own and hence more than one technique is applied to achieve a better purification and higher yield. Employing organic solvents for biosurfactant recovery increases the overall production cost due to use of large amounts of expensive solvents. Though solvent recovery can facilitate reuse, production is batch mode and hence time consuming and crude biosurfactant is obtained which narrows down commercial exploitation. Some solvents e.g. chloroform are toxic for human health and the environment. Hence search for inexpensive solvents with low toxicity is of paramount importance for biosurfactant extractions at industrial scale (Marchant & Banat, 2012a; Wisjnuprapto et al., 2011). Owing to the problems associated with traditional techniques, In situ product removal (ISPR) was developed by Mulligan and Gibbs as a solution which targets continuous separation of end products from fermentation broth (Mulligan & Gibbs, 1990) and provides partial relief to the end-product inhibitory effects of biosurfactants. Solvent-free methods like foam fractionation can be favored to overcome the disadvantages of solvent recovery. Foam formation is a natural phenomenon in any biosurfactant production, hence biosurfactant molecules due to high surface activity can get adsorbed to air bubbles in the culture medium and stable foams can assist continuous removal of the product there by accomplishing production and extraction in a single stage. However, foam adversely affects the productivity by interfering with the mass and heat transfer process, bioreactor clogging, nutrients and cells getting adsorbed on to the foam (Najmi et al., 2018; Winterburn et al., 2011). The severe foam generation during the fermentation process, reduces the bioreactor's working efficiency and causes significant product loss in the foam overflow creating a choke point of productivity in bioreactors as seen with rhamnolipid production (Long et al., 2017). Ultrafiltration is another solvent-free extraction and isolation recovery technique used in biosurfactant purification. It is possible to separate surfactant micelles formed above their critical micelle concentration (CMC) from cell-free supernatants using filters with a certain molecular weight cutoff (MWCO) (Mohamad Pauzi et al., 2019). For example, the molecular size of surfactin monomers is about 994–1050 Da whereas after generating micelles the molecular size increases to 30–100 kDa (Lin & Jiang, 1997). One of the drawbacks of ultrafiltration is simultaneous concentration of other macromolecules like proteins and carbohydrates along with targeted biosurfactants. This can be overcome by using a selected solvent at appropriate concentration which can break the surfactant micelles without breaking other macromolecule structures. Sung-Chyr et al., used methanol at 50% to break surfactin micelles purified initially using ultrafiltration of 30 kDa MWCO and achieved a yield of 95% (Lin & Jiang, 1997). Mubarak et al. used a cross-flow ultrafiltration with a 10 kDa hydrosart membrane to recover and purify surfactin from the fermentation broth (Mubarak et al., 2014). Parameters such as transmembrane pressure, pH, and initial concentration of surfactant may affect the ultrafiltration process (Sen & Swaminathan, 2005). While process optimization has greatly increased the yield and productivity of lipopeptide production, much less attention has been paid to the downstream processing options that can be used to create customized lipopeptide products with varying levels of purity. Furthermore, the utilization of lipopeptides for high-end applications is severely limited by the lack of developed methods for fractionating lipopeptide mixtures into families or specific isoforms. (Rangarajan & Clarke, 2016). Table 3 lists some of the common traditional methods of downstream processing with their advantages and disadvantages (Santos et al., 2016). The methods can be classified on the mode of operation as batch (acid precipitation, organic solvent extraction and ammonium sulphate precipitation) and continuous (adsorption, centrifugation, ion-exchange chromatography, foam fractionation, ultrafiltration). To make current recovery methods more competitive and financially sustainable, more research and development is needed to improve them.

Table 3: Common traditional methods of downstream processing for biosurfactant isolation

Method	Principle	Advantages	Disadvantages
Precipitation at low pH	At low pH, biosurfactant becomes insoluble	Low cost, crude biosurfactant easily obtained	Low purity and less yield, not suitable for commercial application, batch process
Solvent Extraction	The hydrophobic end makes them soluble in organic solvents	Effective recovery of crude biosurfactant, partial purification, reuse of solvents	Partial purification, product degradation, loss of activity, expensive solvents, environmentally unsustainable, batch process
Ammonium sulphate precipitation	Salting-out of polymeric or protein-rich biosurfactants	Effective for certain type of polymeric biosurfactant	Narrow usability due to selectivity, product degradation, batch process
Adsorption Chromatography	Adsorption to activated carbon or polystyrene resins and desorption using organic solvents	Cheaper, highly pure biosurfactant, reusability, continuous recovery	Lack of selectivity, needs greater optimization, high cost of adsorbents
Ion-exchange chromatography	Ion-exchange resins with charged biosurfactants bonded to them are eluted with buffer	High purity, high capacity, reusability, fast recovery	Cost, affected by temperature, pH conditions, non-selective under certain conditions
Foam Fractionation	Biosurfactant generate foam and separate by adsorbing to it	High purity, continuous recovery	Clogging of bioreactor, heat and mass transfer are affected
Ultrafiltration	Surfactants form micelles or vesicles above their critical micelle concentration (CMC) resulting in larger molecular diameters than that of the single unassociated molecules which helps in their separation	Sustainable, fast, one-step recovery, reusability, economical, minimal denaturation, higher recovery yield, carried out at room temperature	Simultaneous concentration of other macromolecules like proteins and carbohydrates

2.4 Product Yield and Cost

The selection of cheaper raw materials that the microorganisms consume is a crucial step for improving the economic feasibility of biosurfactant production from them. This is done to maximize the yield of the biosurfactant (above other metabolites) and minimize the cost of that material, for example by using waste products (Bognolo, 1998). Substrates with lower prices and less impact can be highly alluring, but if process efficiencies drastically decline, this will undoubtedly have a detrimental effect on the biosurfactants total costs and sustainability. When compared to first-generation biosurfactants, this may have even higher costs and an even greater impact per kilogram of generated biosurfactant; yet, researchers investigating substitute substrates frequently overlook this (Hausmann & Henkel, 2022). When the organism produces a significant amount of lipopeptides, it is possible to use these partially purified lipopeptides directly in environmental applications and microbial enhanced oil recovery through the use of a straightforward precipitation procedure. On the other hand, downstream processing of lipopeptides will usually require a series of successive processes which include concentration, purification, and fractionation techniques in order to attain the level of purity that is required for high end medicinal and biomedical applications (Rangarajan & Clarke, 2016). To accurately weigh the relative costs of biosurfactants in contrast to their advantages, a thorough analysis of the lifespan costs of biosurfactants in relation to their synthetic equivalents is also required (Johnson et al., 2021).

The goals of earlier biosurfactant research were to increase yields and lower production costs. This led in neglecting other crucial research, including demonstrating the use of biosurfactants in industrial applications and comparing their foamability, cleanliness and wettability with those of synthetic surfactants (Johnson et al., 2021). Even though there is lot of fundamental and advanced research focused towards biosurfactant production, their exist limitations compared to the chemical counterpart which includes search for hyper producing strains, low production yield (Mukherjee et al., 2006) specific functionality (Wisjnuprapto et al., 2011), difficulties in process optimization (Marchant & Banat, 2012b; Nurfarahin et al., 2018), inefficient purification methodologies (Najmi et al., 2018) and most importantly the overall cost (Bognolo, 1999; Dhanarajan & Sen, 2014). All these factors create an economical bottleneck of the biosurfactant production and impede its percolation in a market predominantly controlled by surfactants of synthetic origin (Hausmann & Henkel, 2022). For this reason, innovative and novel approaches are crucial in positioning biosurfactants as a commercially viable product.

3. Recent Approaches in Biosurfactant Production

3.1 Simultaneous Production of High Value Products along with Biosurfactants

In addition to being exclusively synthesized by some microbes, biosurfactants are co-produced by a diverse range of microbes with a number of other valuables side products, such as microbial lipids, polyhydroxyalkanoates, and ethanol etc. (Shah et al., 2021). Biosynthesis of pectinase, such as pectate lyase (PEL), polygalacturonase (PG) and pectin lyase (PNL) are co-produced in addition to biosurfactants by *B. subtilis* BKDS1 (Kavuthodi et al., 2015). Similarly, amylase enzyme production along with lipopeptide biosurfactant from *Bacillus methylotrophicus* DCS1 was studied (Hmidet et al., 2019). A solid-state fermentation on sugarcane bagasse co-produced of 9.1g/L rhamnolipid of 84% emulsion index and 8.4 g/L ethanol using strain of *Saccharomyces cerevisiae* and *P. aeruginosa* (Guzmán-López et al., 2021). Extracellular rhamnolipid and polyhydroxyalkanoate (PHA) accumulated in cells was found to be co-produced simultaneously by *Enterobacter aerogenes*. β-hydroxyalkanoic acid which further gets transformed into rhamnolipid or PHA is produced during the stationary phase of growth which helps in co-production of these molecules in single fermentation (Shekhar et al., 2015). *Thermus thermophilus* HB8 also co-produced rhamnolipid and PHA using dextrose and sodium gluconate as carbon source (Chong & Li, 2017). It was found that initial phosphate (PO_4^{3-}) concentration governed the production of 3-hydroxyoctanoate, a monomer of PHA co-polymer to be accumulated in the cell and production of rhamnolipid. At initial phosphate concentration of 25 mM production of rhamnolipid was 200 mg/L whereas accumulation PHA was >300 mg/L after 72 hr. of fermentation cycle. LCMS analysis of rhamnolipid revealed a wide range of molecules of mono, di-rhamnolipid with varying lipid chain length along with several saturated as well as unsaturated fatty moieties (Pantazaki et al., 2011; Varjani & Upasani, 2019). 400 mg/L rhamnolipid and 36% PHA is co-produced from hydrolyzed palm oil by *P. aeruginosa* IFO3924 (Hori et al., 2011). Use of waste cooking oil as carbon source by *Burkholderia thailandensis* was studied yielding to 2200 mg/L of di-rhamnolipid $(Rha)_2$-$(C14)_2$ along with intracellular accumulation of biopolymers like polyhydroxybutyrate (PHB) and PHA (Kourmentza et al., 2017). Amplified production of rhamnolipid up to 3.78 gm/L from mutated *B. thailandensis* (ΔphbB1) with elevated mono-rhamnolipid content was reported compared to wild type strain producing 1.28 g/L of rhamnolipid (Funston et al., 2017). Co-production could be a valuable technique to reduce production cost. As biosurfactant is released into extracellular media and if secondary product is accumulated intracellularly may aid to ease of separation or isolation of both the molecules. However, in some cases product recovery procedure is the main constraint on this since it is expensive and time-consuming to purify goods with the same polarity against organic solvents. To overcome this problem, current methods have been altered such that one product can be expressed while the other is inhibited by substrate alteration. As demonstrated by cultivation modified Candida species in the presence of elevated glucose levels, increasing the synthesis of sophorolipid biosurfactants while inhibiting the β-oxidation pathway, which reduced and/or blocked the synthesis of polyhydroxyalkanoates (Roelants et al., 2013).

3.2 Genetic Engineering Strategies for Enhancing Biosurfactant Production

Significant progress in biosurfactant production has proven difficult to achieve using conventional methods such as statistical optimization of media and process control and fermentation (Jimoh & Lin, 2019a). Therefore, as the development of microbial strains presents a huge opportunity to lower production costs, the existing genetic engineering technologies can be utilized to meet the demand for novel, competitive and environmentally friendly biosurfactant molecules (Bouassida et al., 2018; Kuhn et al., 2010). In comparison to parent strains, the engineered strains produce a higher yield of desired products while using the same amount of ingredients (Bouassida et al., 2018). Microbial genetic makeup plays a crucial role in determining the yield of all biotechnological products, as organism genes are responsible for the ability to generate metabolites (Mukherjee et al., 2006). The process known as genetic engineering involves the use of biotechnological techniques to facilitates building biosynthetic pathways, modifying the sequence of already-existing biosurfactant synthetic genes, and manipulating the genes or operons of a single organism, including heritable and non-heritable DNA. Consequently, the genetic techniques applied to enhance biosurfactant production are of utmost relevance.

3.2.1 Recombinant DNA Technology

rDNA technology involves the genetic modification of host microbe by inserting desired gene of interest to form recombinant microbial strain. Selection of host microbes is based on compatibility of the foreign gene in the host cell ecosystem and its mechanism to produce desired biosurfactant. Recombinant DNA technology has been effective in helping scientists boost biosurfactant yield. For instance, rhlAB gene expressed in *Escherichia coli* cell, exhibited excess rhamnolipid by means of genetic engineering (Cabrera-Valladares et al., 2006). Genetically modified *E. coli* by RhlA expression caused 1.3-times rise in biosynthesis of fatty acid demonstrating that RhlA is both sufficient and necessary for rhamnolipid biosynthesis (Zhu & Rock, 2008). A gene for surfactin obtained from *Bacillus licheniformis* NIOT06 was expressed heterologous in *E. coli* M15 (Anburajan et al., 2015). Study on Sfp phosphopantetheinyl transferase revealed its responsibility for polyketide activation, fatty acid, and non-ribosomal peptide synthetase enzymes suggesting its vital role in lipopeptide synthesis (Jimoh & Lin, 2019b). The arthrofactin synthetase gene cluster was engineered (cloned and sequenced) to produce lipopeptides with better characteristics, thereby taking advantage of its industrial application (Roongsawang et al., 2003). Genome reduction has become emerging powerful tool for designing metabolic pathways for production of biosurfactant as studied on *Bacillus amyloliquefaciens* LL3 (BAM LL3) where in 4.8% unwanted genome was removed to obtain strain GR167 and was found to better in terms of surfactin production. Similarly, strain GR167ID was derived from the genome of GR167 by slicing biosynthetic gene clusters of fengycin and iturin and adding PR_{suc} promoters of BAM LL3 strain. GR167IDS was observed to have higher transcription of srfA operon by 678-times and surfactant production was elevated by 10.4 times (Zhang et al., 2021). Mannosylerythritol lipids (MELs) produced merging segments of gene responsible for acyl transferase which were extracted and merged form various fungal species was found to have explicit antimicrobial activity. Also changes in the gene of Mac enzymes can modify the final molecule of MEL's fatty-acid profile as per application demand (Becker et al., 2021).

3.2.2 Mutagenesis

Mutagenesis means modifying or changing the genome of a microbial strain (Bouassida et al., 2018). This process is also widely employed in the creation of hyperproducing microbial strains engineered for enhanced biosurfactant synthesis. Numerous mutation methods, including site-directed mutagenesis, DNA ligase treatment, substitutions, deletions and insertions, have been suggested by scientists to the exiting traditional methods of physical (e.g. UV irradiation) or chemical mutagenesis (e.g. methyl-N'-nitro-N-nitrosoguanidine) (Bouassida et al., 2018; J. Xu & Zhang, 2016). Developing hyper-producing mutants is one way to deal with the financial limitations related to biosurfactant manufacturing. Based on the literature search, the genera of *Acinetobacter, Pseudomonas* and *Bacillus*—which generate emulsan, rhamnolipid and surfactin respectively—have been the only recipients of the hyperproducing mutants for enhanced biosurfactant biosynthesis (Mukherjee et al., 2006). Use of atmospheric and room-temperature plasma (ARTP) mutation of *Pseudomonas* sp. L01 elevated rhamnolipid production by 2.7 fold (Yang et al., 2023). A study reported an increase in biosurfactant when an ion beam was used to produce a *B. subtilis* E8 mutant (Gong et al., 2009). Also, a *P. aeruginosa* MR01 mutant obtained by exposing *P. aeruginosa* to gamma radiation was found to have higher i.e. 1.5-times more biosurfactant production ability (Lotfabad et al., 2010). *B. subtilis* MS1 a chemically mutated strain caused a 2-fold increase in yield of biosurfactant. An improvement in emulsion index was also observed (Asgher et al., 2020).

3.2.3 Overexpression of Extracellular Peptides

Feasibility of improving surfactin production by modification of gene regulatory elements to direct the flow of the two peptides ComX pheromone and CSF was demonstrated (Jung et al., 2012). Increased production of surfactants during low-density cell growth can be achieved by overexpressing extracellular peptides, ComX, and PhrC. In another study, 48 hr. of cultivation of *B. subtilis* (pHT43-comX-phrC) strain, surfactant production was elevated by 6.4 times. Both external signaling inputs and unique response pathways in *B. subtilis* contributed to the enhancement of surfactin synthesis through the establishment of genetic competence and quorum response triggers at low cell density (Jung et al., 2012). Overexpression of certain enzymes like BKD, FabHB, and LipALM were done to obtain *B. subtilis* 168S16 which produced 5.6 g/l of surfactin, a 1.5-fold higher than the background strain 168S12. Also significantly enhanced surfactin cellular tolerance by overexpression of putative self-resistance-associated proteins, resulting in an additional 8.5-fold rise in surfactin titer (Wu et al., 2019).

3.2.4 CRISPR - Gene Editing

The acronym "CRISPR" means Clustered Regularly Interspaced Short Palindromic Repeats, a family of DNA sequences found in the genome of bacteria and archaea (Ganger et al., 2023). Several novel Cas nuclease proteins associated with CRISPR like Cas14/Cas12f nucleases, Cas12b/C2c1 nuclease, Cas12a/Cpf1 nuclease, Cas13/C2c2 nuclease and CasMINI have been developed for the purpose of modifying the genome other than the traditional Cas9 nuclease (Biswas et al., 2021). Every CRISPR-Cas system has its unique characteristics, such as variations in size, cleavage sites and PAM regions. For example, Cas12a, as opposed to Cas9, provides additional advantages in knock-in techniques and enhances the effectiveness of Non-homologous end joining (NHEJ) based gene insertion. Compared between Cas9 and Cas12a, Cas12b is smaller and provides greater versatility in gRNA engineering (Liu et al., 2020). Scientists from Stanford University developed an efficient, as functional as Cas12a and compact CasMINI protein from Cas12f. This technique expedites the range of uses that CRISPR-Cas systems provide across several biotechnology domains, such as genetic engineering (Xu et al., 2021). Utilizing CRISPR for modifying microbes for higher yield of biosurfactants is one of the recently developed approaches. A research reported that employing CRISPRi technology to inhibit the genes on the branch metabolic pathways of amino acid biosynthesis increased surfactin production in *B. subtilis* by 4.69 times (Wang et al., 2019). In a different research, the srfA promoter was replaced with eight endogenous strong auto-inducible phase-dependent promoters utilizing the CRISPR-Cas9 technology, resulting in strains that may find value in industry (Liu et al., 2021). Modified strains utilizing the CRISPR-Cas9 gene editing method to produce acidic or lactonic sophorolipid have been developed. This modified strain yielded 53.64 g/L of acidic sophorolipid form and 45.32 g/L of lactonic sophorolipid. Also at fermenter scale 129.7 g/L of acidic form sophorolipid utilizing cheap raw materials such as rapeseed oil and glucose was produced (Xia et al., 2023).

3.2.5 Substitution, Replacement and Modifications of Amino Acids

When residual amino acid having a side chain is replaced with a different amino acid residue of similar properties, it is referred to as amino acid modification. This procedure depends on a variety of side chains, which include polar side chains with no charge (such as cysteine, glycine, serine, etc.) and side chains with aromaticity (such as tyrosine, histidine etc.), as well as side chains with acidity (e.g. glutamic acid), side chains with basicity (such as arginine, lysine, histidine) and side chains with non-polar moiety (e.g. alanine, proline, valine, isoleucine, tryptophan) (Jimoh et al., 2021). A target gene's transcription level is directly impacted by the promoter's intensity and this in turn influences how this intensity is expressed. Additionally, promoter replacement is one of the simplest and most efficient strategies to control how important genes are expressed (Jiao et al., 2017). In srfA operon presence, surfactin yields following promoter swap were examined in two previous investigations. However, in these investigations, sequences of promoter to be substituted and microbial strain of surfactin were different, which led to inconsistent results (Coutte et al., 2010; Sun et al., 2009). Insertion of a functional sfp gene in *B. subtilis* 168 resulted in the constitutive promoter PrepU replacing the native srfA promoter. Lipopeptide synthetase expression was inhibited when the surfactin operon promoter was changed to a constitutive one. This resulted in a little increase in surfactin production, indicating that the precursor supply may be the cause of surfactin overproduction (Coutte et al., 2010). A study revealed that substituting Pspac, a hybrid promoter based on the *E. coli* lac operon and the *B. subtilis* bacteriophage SP01, for PsrfA resulted in ten times higher surfactin production (Sun et al., 2009). Similar results were observed in another investigation, where substituting the constitutive Pveg for the native srfA promoter may dramatically enhance surfactin production (Willenbacher et al., 2016).

3.3 Omics: An Advanced Method for Identifying Producers of Biosurfactants

3.3.1 Metabolomics

The entire metabolic makeup of a microbial cell is studied by a field called metabolomics. It reflects physiological condition and metabolic activity by taking an overview of all metabolites and molecules in the microbial cell, yielding a remarkably high level of information (Worley & Powers, 2012). In a study by use of LC-MS/MS to examine the chemistry of different marine creatures, several molecular families and commercially significant compounds such as valinomycin, actinomycin D and desferrioxamine E have been identified. It's interesting to note that in addition to these additional metabolites, surfactin biosurfactant was also discovered in this investigation. This implied that the identification of specialized chemicals can be accomplished well with untargeted metabolic profiling (Floros et al., 2016; Gaur et al., 2022; Sirohi, 2022) and particularly the recognition and identification of biosurfactant producers among several non-producers (Adetunji & Olaniran, 2021). A metabolomics investigation of *Bacillus velezensis* FZB42's bactericidal action on the phytopathogen *Xanthomonas campestris* revealed that the killing mechanism involved the siderophore, lipopeptides, and bacillibactin. The mono- and co-cultures of *B. velezensis* included bacillibactin and the lipopeptides bacillomycin, surfactin and fengycin. Among the 24 distinct forms found in the co-culture supernatant, mass spectrophotometric analysis indicated that these four chemicals were important. Moreover, a discrepancy in the side chain of fengycin and surfactin was discovered (Mácha et al., 2021). The metabolomics investigation by mass spectrometry of 250 ecologically diverse *Pseudomonas* strains revealed structural links in the molecular networking. It was discovered that cyclic lipopeptides belonging to peptide families, such as putisolvins, tolaasins, orfamides, xantholysins and massetolides, exhibited biosurfactant characteristics (Nguyen et al., 2016). In addition, Serratia marcescens' molecular networking has revealed lipopeptides with antibacterial properties. Glucosamine derivatives with butyric acid, fatty acid chain, valine and glucose were among the lipopeptides that took the form of serratamolide, which has two L-serine residues connected to two fatty acid chains (lengths ranging from C10 to C12) (Clements et al., 2021). Using *Rhodotorula mucilaginosa* 50-3-19/20B yeast, media-dependent metabolomic analysis showed the existence of antibiotic and anticancer chemicals in various solvent extracts. It was discovered that glycolipids have the largest molecular family inside the molecular network. These molecules are classified as fatty acid polyesters with varying degrees of acetylation (Buedenbender et al., 2020). These findings indicated that using metabolomics techniques biosurfactant producing microbes could be identified.

3.3.2 Sequencing Based

The most prevalent, varied, adaptable life form on earth is microorganisms that have been fostered by their surroundings. They are necessary to keep our ecosystem functioning and using these bacteria for human benefit has become a crucial tool (S.J. Varjani et al., 2015). Conventionally, the well-known and established culture and identification procedures were used, which resulted in the vast majority of microbial strains or species remaining unidentified (Dhanjal & Sharma, 2018). Therefore, to detect and obtain comprehensive insights into the biosurfactant generating capability of microorganisms, omics approach which combines metagenomics (whole DNA) and metatranscriptomics (total RNA) and metaproteomics (total protein) was established (Malik et al., 2022).

3.3.2.1 Metagenomics

Metagenomics is the omics method used to identify the unidentified microorganisms. In this technique, prior enrichment of culture is not done; instead, the genetic material is extracted straight from environmental samples. DNA isolation, metagenome library building, 16S rRNA sequencing and screening or data analysis are all part of metagenomics (Datta et al., 2020; Dhanjal & Sharma, 2018). There are two methods used in metagenomics screening: while function-based screening is not dependent on an existing metagenomics library, sequence-based screening relies on already determined sequences, i.e., primers are generated based on known coding genes. Using this technique, DNA was cloned to produce sequences with desired functionality that could potentially lead to the identification of new sequences (Datta et al., 2020; Ngara & Zhang, 2018). It facilitates the processing and comprehension of the genetic data of uncultivated microorganisms, as well as their interactions, metabolic pathways and roles. This leads to the discovery and identification of novel microorganisms and their products (Datta et al., 2020; Malik et al., 2021). One such benefit of the metagenomics technique is the identification of new strains that produce biosurfactants. Metagenomic libraries provide researchers with valuable information about the functions, interactions and identification of microorganisms within and between diverse microbial communities (Dhanjal & Sharma, 2018).

However, this method has certain drawbacks. For example, sequence-based metagenomics cannot identify rare microorganisms, nor can it distinguish between expressed and unexpressed genes. This means genes responsible for desired functionality cannot be found by metagenomics. (Hazen et al., 2013; Malik et al., 2021). Therefore, to identify gene expression, metagenomics is primarily used in conjunction with metatranscriptomics.

3.3.2.2 Metatranscriptomics

Understanding gene activity through the analysis of microbiome mRNA transcripts is known as metatranscriptomics. This method explains which genes, as a result of changes in the environment, are overexpressed or underexpressed (Malik et al., 2022). Metatranscriptomics involves three primary steps: (1) direct RNA collection from ambient samples and enrichment of mRNA, (2) mRNA conversion to cDNA and (3) next-generation sequencing (Araújo et al., 2020; Ranjan et al., 2015). Through functional screening, a metagenome library was created and a clone known as 3C6 was found. This clone contained the novel protein MBSPI with biosurfactant activity. According to da Silva Araújo et al. the MBSP1 sequence had the greatest resemblance to *Haloferax lucentense*, a recognised hydrocarbon degrader. In a study, a metagenomic library was built using *P. putida*, *Streptomyces lividans* fosmid and *E. coli*. By using the paraffin spray test, the *P. putida* clone was found to have biosurfactant activity, and ornithine acyl-ACP N-acyltransferase (olsB) was found to be the enzyme responsible for this activity. Even though the fosmid was inactive in *E. coli*, overexpression of the olsB gene was seen under the T7 promoter, which led to the synthesis of lyso-ornithine lipid with biosurfactant activity (Williams et al., 2019). A report suggested a transcriptome method to investigate the impact of L and D-leucine on *B. velezensis* BS-37, a bacteria producing 1 g/L of surfactin biosurfactant when glycerol was present. When 10 mM D-Leu was added, the output dropped to 2.5 g/L, but it surged to 2 g/L upon the adding 10 mM L-Leu. The strain BS-37 transcriptome's RPKM analysis (reads per kilobase per million mapped reads) showed that L-Leu was used as a precursor while D-Leu acted as a competitive inhibitor. These findings were consistent with higher transcription levels of the genes encoding surfactin synthetase, the branched chain amino acid synthesis pathway and the glycerol utilization pathway. Therefore, understanding and development of high surfactin generating strains was aided by metatranscriptomics (Zhou et al., 2018). Using a transcriptome analysis technique, the surfactin synthase efficiency of *Bacillus amyloliquefaciens* MT45 and *B. amyloliquefaciens* strain DSM7T was examined. According to the research, strain MT45 exhibited improved carbon metabolism, fatty acid biosynthesis, and surfactin production pathway enrichment. Additionally, strain MT45 was discovered to have higher levels of surfactin production than strain DSM7T due to an upregulation of surfactin synthase, which was reported to be 9–49 fold higher (Zhi et al., 2017). Although the technique of metatranscriptomics is useful for analyzing the activities of genes and the regulation of genes in the microbial community, its usefulness is limited by the short half-life, instability and susceptibility to degradation of mRNA (Ranjan et al., 2015; Zhang et al., 2021). Consequently, the alternative omics strategy known as metaproteomics is employed to get over these constraints (Malik et al., 2022).

3.3.2.3 Metaproteomics

The study of proteins expressed by microorganisms found in environmental sample material is known as metaproteomics. It examines and presents data regarding the microbial community's functional genes (Malik et al., 2022; Maron et al., 2007). Metaproteomics analysis creates a connection between the functional and genomic information of microbial communities (Gaur et al., 2022; Maron et al., 2007). To examine novel biosurfactant proteins, two fungal strains were obtained from an oil-contaminated marine site: *Aspergillus terreus* MUT271 and *Trichoderma harzianum* MUT290. Proteomics allowed for the detection of a single common protein that was a member of the cerato-platanin (CP) family. They noticed that the CPs from both strains that were examined have biosurfactant and bioemulsifier characteristics. On the other hand, *T. harzianum*'s CP demonstrated superior surfactant activity and may be used in industry (Pitocchi et al., 2020). According to a paper, adding glycerol to *Burkholderia* sp. C3 improved the way that Dibenzothiophene (DBT) was broken down and raised the production of rhamnolipids. The rate of degradation rose by 25–30% while the rate of rhamnolipids biosynthesis rose at the same time. To find and examine the overexpression of the enzymes involved in rhamnolipids biosynthesis and DBT breakdown, proteomics was utilized (Ortega Ramirez et al., 2020). Researchers can discover new functional genes and biochemical pathways using the metaproteomics approach, as well as determine which proteins are expressed in response to changes in the substrate or environmental factors (Maron et al., 2007). Despite these drawbacks, metaproteomics offers more information than both metagenomics and metatranscriptomic. Although the omics approach has numerous advantages, there are still a lot of issues that need to be resolved. However, the multi-metaomics strategy, which combines two or more approaches, needs to be taken into consideration in order to obtain more accurate data and profound insights into genes (Malik et al., 2022).

3.4 Statistical Optimization

Cell-based bioprocesses, being interconnected, pose difficulties in monitoring when individual parameters are optimized. To address the simultaneous handling of multiple data, techniques such as Response Surface Methodology (RSM), Taguchi and Plackett-Burman Design (PBD) have been frequently employed for media optimization in biosurfactant production processes (Eswari et al., 2016; Santos et al., 2016; Syed-Hassan & Zaini, 2016). While traditional statistical designs are often suitable for biosurfactant optimization, there are cases where complex non-linear biological interactions are not fully exploitable. In such instances, advanced computational methods based on artificial intelligence, such as Artificial Neural Networks (ANN), can be employed to predict biosurfactant production based on dependent variables. For instance, ANN was used to simulate complex system performance in predicting biosurfactant yield, surface tension reduction and emulsion by *Klebsiella* sp. *FKOD36*

(Ahmad et al., 2016). Additionally, Artificial Intelligence (AI) methods, like Artificial Neural Intelligence coupled with Genetic Algorithm (ANN-GA) have been explored for media optimization (Ahmad et al., 2016; Sivapathasekaran & Sen, 2013). Comparisons between ANN and statistical designs have indicated that ANN tends to be more accurate in predicting optimum conditions for biosurfactant production, as demonstrated in studies, where biosurfactant production yield was significantly increased through the combined use of ANN and RSM. Sivapathasekaran & Sen also used the ANN strategy with *Bacillus circulans* MTCC 8281. After determining the most important factors for biosurfactant production (glucose, urea, $SrCl_2$, and $MgSO_4$), ANN was used to enhance biosurfactant production by 70% (Sivapathasekaran & Sen, 2013). Pal et al. improved the biosurfactant (glycolipid containing trehalose as the major carbohydrate) production in *Rhodococcus erythropolis* MTCC 2794 and increased its production yield by 3.5 fold by using ANN coupled with genetic algorithm (Pal et al., 2009).

3.5 Fed Batch Process

The major disadvantage of variable volume fed-batch systems is that much of the fermenter volume is not utilized until the end of the batch and consequently, the duration of the batch is limited by the fermenter volume. New studies have gained some success to overcome these limitations. For instance, a high rhamnolipid titer of 150g/l was obtained using sequential fed batch strategy with a fill and draw strategy. It showed a 163% and 102% increase over the traditional batch and fed-batch processes, respectively (He et al., 2017). In another study yield of 70.56g/l using pH stat controlled fed batch fermentation for production of rhamnolipids. Overproduction of rhamnolipids with yield 240g/l was achieved using fed batch operation under tight DO control in a lab scale fermenter (Bazsefidpar et al., 2019). Jingjing Jiang et al., demonstrated a new approach of using sequential fed batch strategy with a high cell density resulting in a yield of 665g/l with enhanced mass transfer efficiency (Jiang et al., 2021). Fed batch production of MELs resulted in a titer of 61g/l using foam control (Yang et al., 2023). Hence a variation of the traditional fed batch process can be explored for higher biosurfactant yield.

4. Conclusion

Biosurfactants exhibit great promise in supplanting synthetic petroleum-derived surfactants and have remarkable environmental applications; yet, their high manufacturing costs hinder their market penetration. Non availability of hyper producers, substrate cost, high downstream processing and lower biosurfactant yield are some of the reasons to limit biosurfactant production. One tactic that might be used to improve the economics of the production process is co-production, or the simultaneous manufacture of additional value-added molecules. In addition, genetic engineering using recombinant DNA technology can be exploited further to create a robust and versatile microbe for biosurfactant production. The omics techniques have the potential to identify novel biosurfactants by circumventing the tedious and time-consuming biosurfactant screening procedures. All of these methods might help increase in the structural diversity, which has been identified as a significant flaw in the majority of glycolipid biosurfactants. Future studies may concentrate on using novel, inexpensive, renewable substrates to produce biosurfactants using genetic modified strains that will produce molecules that are specifically designed for desired application.

References

Abalos, A., Maximo, F., Manresa, M.A. & Bastida, J. (2002). Utilization of response surface methodology to optimize the culture media for the production of rhamnolipids by *Pseudomonas aeruginosa* AT10. *Journal of Chemical Technology & Biotechnology*, 77(7): 777–784. https://doi.org/10.1002/jctb.637

Adetunji, A.I. & Olaniran, A.O. (2021). Production and potential biotechnological applications of microbial surfactants: An overview. *Saudi Journal of Biological Sciences*, 28(1), 669–679. https://doi.org/10.1016/j.sjbs.2020.10.058

AGAE Produces Record Yields of Rhamnolipids at Its Lowest Cost by Fermentation. (2022). [News Report]. https://www.prnewswire.com/news-releases/agae-produces-record-yields-of-rhamnolipids-at-its-lowest-cost-by-fermentation-301621340.html

Agger, J.W. & Zeuner, B. (2022). Bio-based surfactants: Enzymatic functionalization and production from renewable resources. *Current Opinion in Biotechnology*, 78: 102842. https://doi.org/10.1016/j.copbio.2022.102842

Ahmad, Z., Crowley, D., Marina, N. & Jha, S. Kr. (2016). Estimation of biosurfactant yield produced by Klebseilla sp. FKOD36 bacteria using artificial neural network approach. *Measurement*, 81: 163–173. https://doi.org/10.1016/j.measurement.2015.12.019

Anburajan, L., Meena, B., Raghavan, R.V., Shridhar, D., Joseph, T.C., Vinithkumar, N.V. & et al. (2015). Heterologous expression, purification, and phylogenetic analysis of oil-degrading biosurfactant biosynthesis genes from the marine sponge-associated Bacillus licheniformis NIOT-06. *Bioprocess and Biosystems Engineering*, 38(6): 1009–1018. https://doi.org/10.1007/s00449-015-1359-x

Araújo, S.C.D.S., Silva-Portela, R.C.B., De Lima, D.C., Da Fonsêca, M.M.B., Araújo, W.J., Da Silva, U.B. et al. (2020). MBSP1: A biosurfactant protein derived from a metagenomic library with activity in oil degradation. *Scientific Reports*, 10(1): 1340. https://doi.org/10.1038/s41598-020-58330-x

Asgher, M., Afzal, M., Qamar, S.A. & Khalid, N. (2020). Optimization of biosurfactant production from chemically mutated strain of Bacillus subtilis using waste automobile oil as low-cost substrate. *Environmental Sustainability*, 3(4): 405–413. https://doi.org/10.1007/s42398-020-00127-9

Bazsefidpar, S., Mokhtarani, B., Panahi, R. & Hajfarajollah, H. (2019). Overproduction of rhamnolipid by fed-batch cultivation of Pseudomonas aeruginosa in a lab-scale fermenter under tight DO control. *Biodegradation*, 30(1): 59–69. https://doi.org/10.1007/s10532-018-09866-3

Becker, F., Stehlik, T., Linne, U., Bölker, M., Freitag, J. & Sandrock, B. (2021). Engineering Ustilago maydis for production of tailor-made mannosylerythritol lipids. *Metabolic Engineering Communications*, 12: e00165. https://doi.org/10.1016/j.mec.2021.e00165

Bettenhausen, C. (2022, May). C&EN, As the personal care industry races to drop petroleum-based ingredients, chemical companies are rolling out a dizzying array of biobased surfactants. Choosing from among them isn't easy. https://pubs.acs.org/doi/10.1021/cen-10015-cover#

Biosurfactant—KANEKA Surfactin. (2023). https://www.kaneka.co.jp/en/business/qualityoflife/nbd_002.html

Biosurfactants Market Analysis Europe, North America, APAC, Middle East and Africa, South America US, Canada, China, Germany, France—Size and Forecast 2023-2027 (IRTNTR44099; p. 174). (2023). https://www.technavio.com/report/biosurfactants-market-industry-analysis

Biosurfactants Market, By Product Type (Glycolipids, Lipopeptides & Lipoproteins, Surfactin, Others), By Application (Personal care, Food processing, Others), By Region Forecast to 2030. (2022). Emergen Research. https://www.emergenresearch.com/industry-report/biosurfactants-market

Biswas, S., Zhang, D. & Shi, J. (2021). CRISPR/Cas systems: Opportunities and challenges for crop breeding. *Plant Cell Reports*, 40(6): 979–998. https://doi.org/10.1007/s00299-021-02708-2

Bognolo, G. (1998). Biosurfactants as emulsifying agents for hydrocarbons. *Colloids and Surfaces A: Physicochemical and Engineering Aspects*, 152(1–2): 41–52. https://doi.org/10.1016/S0927-7757(98)00684-0

Bouassida, M., Ghazala, I., Ellouze-Chaabouni, S. & Ghribi, D. (2018). Improved biosurfactant production by Bacillus subtilis SPB1 mutant obtained by random mutagenesis and its application in enhanced oil recovery in a sand system. *Journal of Microbiology and Biotechnology*, 28(1): 95–104. https://doi.org/10.4014/jmb.1701.01033

Buedenbender, L., Kumar, A., Blümel, M., Kempken, F. & Tasdemir, D. (2020). Genomics- and metabolomics-based investigation of the deep-sea sediment-derived yeast, Rhodotorula mucilaginosa 50-3-19/20B. *Marine Drugs*, 19(1): 14. https://doi.org/10.3390/md19010014

Cabrera-Valladares, N., Richardson, A.-P., Olvera, C., Treviño, L.G., Déziel, E., Lépine, F. et al.& (2006). Monorhamnolipids and 3-(3-hydroxyalkanoyloxy) alkanoic acids (HAAs) production using *Escherichia coli* as a heterologous host. *Applied Microbiology and Biotechnology*, 73(1): 187–194. https://doi.org/10.1007/s00253-006-0468-5

Choi, J.-W., Choi, H.-G. & Lee, W.-H. (1996). Effects of ethanol and phosphate on emulsan production by Acinetobacter calcoaceticus RAG-1. *Journal of Biotechnology*, 45(3): 217–225. https://doi.org/10.1016/0168-1656(95)00175-1

Chong, H. & Li, Q. (2017). Microbial production of rhamnolipids: Opportunities, challenges and strategies. *Microbial Cell Factories*, 16(1): 137. https://doi.org/10.1186/s12934-017-0753-2

Clements, T., Rautenbach, M., Ndlovu, T., Khan, S. & Khan, W. (2021). A metabolomics and molecular networking approach to elucidate the structures of secondary metabolites produced by Serratia marcescens strains. *Frontiers in Chemistry*, 9: 633870. https://doi.org/10.3389/fchem.2021.633870

Coutte, F., Leclère, V., Béchet, M., Guez, J.-S., Lecouturier, D., Chollet-Imbert, M. et al. & (2010). Effect of *pps* disruption and constitutive expression of *srfA* on surfactin productivity, spreading and antagonistic properties of *Bacillus subtilis* 168 derivatives. *Journal of Applied Microbiology*, 109(2): 480–491. https://doi.org/10.1111/j.1365-2672.2010.04683.x

Croda. (2016). Turning the world's products green. https://www.crodaindustrialspecialties.com/en-gb/sustainability/sustainable-manufacturing

De Oliveira Schmidt, V.K., De Souza Carvalho, J., De Oliveira, D. & De Andrade, C.J. (2021). Biosurfactant inducers for enhanced production of surfactin and rhamnolipids: An overview. *World Journal of Microbiology and Biotechnology*, 37(2): 21. https://doi.org/10.1007/s11274-020-02970-8

Desai, J.D. & Banat, I.M. (1997). Microbial production of surfactants and their commercial potential. *Microbiol. Mol. Biol. Rev.*, 61. https://doi: 10.1128/mmbr.61.1.47-64.1997

Dhanarajan, G. & Sen, R. (2014). Cost analysis of biosurfactant production from a scientist's perspective. *In:* N. Kosaric & F.V. Sukan (Eds.), *Biosurfactants*. 164–175. CRC Press. https://doi.org/10.1201/b17599-12

Dhanjal, D.S. & Sharma, D. (2018). Microbial metagenomics for industrial and environmental bioprospecting: The unknown envoy. *In:* J. Singh, D. Sharma, G. Kumar & N.R. Sharma (Eds.), *Microbial Bioprospecting for Sustainable Development*. 327–352. Springer Singapore. https://doi.org/10.1007/978-981-13-0053-0_18

Dow. (2022). EcoSense™ GL-60 HA Surfactant.pdf.

Eslami, P., Hajfarajollah, H. & Bazsefidpar, S. (2020). Recent advancements in the production of rhamnolipid biosurfactants by *Pseudomonas aeruginosa*. *RSC Advances*, 10(56): 34014–34032. https://doi.org/10.1039/D0RA04953K

Eswari, J.S., Anand, M. & Venkateswarlu, C. (2016). Optimum culture medium composition for lipopeptide production by Bacillus subtilis using response surface model-based ant colony optimization. *Sadhana*, 41(1): 55–65. https://doi.org/10.1007/s12046-015-0451-x

Floros, D.J., Jensen, P.R., Dorrestein, P.C. & Koyama, N. (2016). A metabolomics guided exploration of marine natural product chemical space. *Metabolomics*, 12(9): 145. https://doi.org/10.1007/s11306-016-1087-5

Funston, S.J., Tsaousi, K., Smyth, T.J., Twigg, M.S., Marchant, R. & Banat, I.M. (2017). Enhanced rhamnolipid production in Burkholderia thailandensis transposon knockout strains deficient in polyhydroxyalkanoate (PHA) synthesis. *Applied Microbiology and Biotechnology*, 101(23–24): 8443–8454. https://doi.org/10.1007/s00253-017-8540-x

Ganger, S., Harale, G. & Majumdar, P. (2023). Clustered regularly interspaced short palindromic Repeats/CRISPR-Associated (CRISPR/Cas) systems: Discovery, structure, classification, and general mechanism. *In:* D. Pandita & A. Pandita (Eds.), *CRISPR/Cas-mediated Genome Editing in Plants* (First edition). 66–97. Apple Academic Press.

Gaur, V.K., Sharma, P., Gupta, S., Varjani, S., Srivastava, J.K., Wong, J.W.C. et al. (2022). Opportunities and challenges in omics approaches for biosurfactant production and feasibility of site remediation: Strategies and advancements. *Environmental Technology & Innovation*, 25: 102132. https://doi.org/10.1016/j.eti.2021.102132

Gong, G., Zheng, Z., Chen, H., Yuan, C., Wang, P., Yao, L. & Yu, Z. (2009). Enhanced production of surfactin by Bacillus subtilis E8 mutant obtained by ion beam implantation. *Food Technology and Biotechnology*, 47(1): 27–31.

Guzmán-López, O., Cuevas-Díaz, M.D.C., Martínez Toledo, A., Contreras-Morales, M.E., Ruiz-Reyes, C.I. & Ortega Martínez, A.D.C. (2021). Fenton-biostimulation sequential treatment of a petroleum-contaminated soil amended with oil palm bagasse (*Elaeis guineensis*). *Chemistry and Ecology*, 37(6): 573–588. https://doi.org/10.1080/02757540.2021.1909003

Haba, E., Espuny, M.J., Busquets, M. & Manresa, A. (2000). Screening and production of rhamnolipids by Pseudomonas aeruginosa 47T2 NCIB 40044 from waste frying oils. *Journal of Applied Microbiology*, 88(3): 379–387. https://doi.org/10.1046/j.1365-2672.2000.00961.x

Hausmann, R. & Henkel, M. (Eds.). (2022). *Biosurfactants for the Biobased Economy* (Vol. 181). Springer International Publishing. https://doi.org/10.1007/978-3-031-07337-3

He, N., Wu, T., Jiang, J., Long, X., Shao, B. & Meng, Q. (2017). Toward high-efficiency production of biosurfactant rhamnolipids using sequential fed-batch fermentation based on a fill-and-draw strategy. *Colloids and Surfaces B: Biointerfaces*, 157, 317–324. https://doi.org/10.1016/j.colsurfb.2017.06.007

Hmidet, N., Jemil, N. & Nasri, M. (2019). Simultaneous production of alkaline amylase and biosurfactant by Bacillus methylotrophicus DCS1: Application as detergent additive. *Biodegradation*, 30(4): 247–258. https://doi.org/10.1007/s10532-018-9847-8

Holiferm Products—Biosurfactants from renewable feedstocks. (2023). https://holiferm.com/products/

Hori, K., Ichinohe, R., Unno, H. & Marsudi, S. (2011). Simultaneous syntheses of polyhydroxyalkanoates and rhamnolipids by Pseudomonas aeruginosa IFO3924 at various temperatures and from various fatty acids. *Biochemical Engineering Journal*, 53(2): 196–202. https://doi.org/10.1016/j.bej.2010.10.011

ISO/DIS 21680(en), *Surface Active Agents-bio-based Surfactants – Requirements and Test Methods* (21680). (2020). https://www.iso.org/obp/ui#iso:std:iso:21680:dis:ed-1:v1:en

Jarvis F.G. & Johnson, M.J. (1949). *A Glyco-lipide Produced by Pseudomonas Aeruginosa.pdf*. 71(12): 4124–4126. https://doi.org/doi.org/10.1021/ja01180a073

Jiang, J., Zhang, D., Niu, J., Jin, M. & Long, X. (2021). Extremely high-performance production of rhamnolipids by advanced sequential fed-batch fermentation with high cell density. *Journal of Cleaner Production*, *326*, 129382. https://doi.org/10.1016/j.jclepro.2021.129382

Jiao, S., Li, X., Yu, H., Yang, H., Li, X. & Shen, Z. (2017). In situ enhancement of surfactin biosynthesis in *Bacillus subtilis* using novel artificial inducible promoters. *Biotechnology and Bioengineering*, 114(4): 832–842. https://doi.org/10.1002/bit.26197

Jimoh, A.A. & Lin, J. (2019a). Biosurfactant: A new frontier for greener technology and environmental sustainability. *Ecotoxicology and Environmental Safety*, 184: 109607. https://doi.org/10.1016/j.ecoenv.2019.109607

Jimoh, A.A. & Lin, J. (2019b). Heterologous Expression of Sfp-Type Phosphopantetheinyl transferase is indispensable in the biosynthesis of lipopeptide biosurfactant. *Molecular Biotechnology*, 61(11): 836–851. https://doi.org/10.1007/s12033-019-00209-y

Jimoh, A.A., Senbadejo, T.Y., Adeleke, R. & Lin, J. (2021). Development and genetic engineering of hyper-producing microbial strains for improved synthesis of biosurfactants. *Molecular Biotechnology*, 63(4): 267–288. https://doi.org/10.1007/s12033-021-00302-1

Johnson, P., Trybala, A., Starov, V. & Pinfield, V.J. (2021). Effect of synthetic surfactants on the environment and the potential for substitution by biosurfactants. *Advances in Colloid and Interface Science*, 288: 102340. https://doi.org/10.1016/j.cis.2020.102340

Jung, J., Yu, K.O., Ramzi, A.B., Choe, S.H., Kim, S.W. & Han, S.O. (2012). Improvement of surfactin production in *Bacillus subtilis* using synthetic wastewater by overexpression of specific extracellular signaling peptides, *comX* and *phrC*. *Biotechnology and Bioengineering*, 109(9): 2349–2356. https://doi.org/10.1002/bit.24524

Kashif, A., Rehman, R., Fuwad, A., Shahid, M.K., Dayarathne, H.N.P., Jamal, A. et al. & (2022). Current advances in the classification, production, properties and applications of microbial biosurfactants – A critical review. *Advances in Colloid and Interface Science*, 306: 102718. https://doi.org/10.1016/j.cis.2022.102718

Kavuthodi, B., Thomas, S. & Sebastian, D. (2015). Co-production of pectinase and biosurfactant by the newly isolated strain Bacillus subtilis BKDS1. *British Microbiology Research Journal*, 10(2): 1–12. https://doi.org/10.9734/BMRJ/2015/19627

Kiran, G.S., Ninawe, A.S., Lipton, A.N., Pandian, V. & Selvin, J. (2015). Rhamnolipid biosurfactants: Evolutionary implications, applications and future prospects from untapped marine resources. *Critical Reviews in Biotechnology*, 1–17. https://doi.org/10.3109/07388551.2014.979758

Kourmentza, C., Freitas, F., Alves, V. & Reis, M.A.M. (2017). Microbial conversion of waste and surplus materials into high-value added products: The case of biosurfactants. *In:* V.C. Kalia & P. Kumar (Eds.), *Microbial Applications* (Vol. 1). 29–77. Springer International Publishing. https://doi.org/10.1007/978-3-319-52666-9_2

Krauter, J. (2019, December). Unilever and Evonik partner to launch green cleaning ingredien. https://www.unilever.com/news/press-and-media/press-releases/2019/unilever-and-evonik-partner-to-launch-green-cleaning-ingredient

Kuhn, D., Blank, L.M., Schmid, A. & Bühler, B. (2010). Systems biotechnology – Rational whole-cell biocatalyst and bioprocess design. *Engineering in Life Sciences*, 10(5): 384–397. https://doi.org/10.1002/elsc.201000009

Lin, S.-C. & Jiang, H.-J. (1997). Recovery and purification of the lipopeptide biosurfactant of Bacillus subtilis by ultrafiltration. *Biotechnology Techniques*, 11(6): 413–416. https://doi.org/10.1023/A:1018468723132

Liu, F.-F., Liu, Y.-F., Qiao, Y.-W., Guo, Y.-Z., Kuang, F.-Y., Lin, X.-Q. et al. (2021). Improvement surfactin production by substitution of promoters in Bacillus subtilis TD7. *Applied Environmental Biotechnology*, 6(1): 31–41. https://doi.org/10.26789/AEB.2021.01.004

Liu, K., Sun, Y., Cao, M., Wang, J., Lu, J.R. & Xu, H. (2020). Rational design, properties, and applications of biosurfactants: A short review of recent advances. *Current Opinion in Colloid & Interface Science*, 45: 57–67. https://doi.org/10.1016/j.cocis.2019.12.005

Liyana Ismail, N., Shahruddin, S. & Othman, J. (2022). Perspective Chapter: Overview of bio-based surfactant – recent development, industrial challenge, and future outlook. *In:* A. Kumar Dutta (Ed.), *Surfactants and Detergents—Updates and New Insights*. IntechOpen. https://doi.org/10.5772/intechopen.100542

Long, X., Shen, C., He, N., Zhang, G. & Meng, Q. (2017). Enhanced rhamnolipids production via efficient foam-control using a stop valve as a foam breaker. *Bioresource Technology*, 224: 536–543. https://doi.org/10.1016/j.biortech.2016.10.072

Lotfabad, T.B., Abassi, H., Ahmadkhaniha, R., Roostaazad, R., Masoomi, F., Zahiri, H. et al. (2010). Structural characterization of a rhamnolipid-type biosurfactant produced by Pseudomonas aeruginosa MR01: Enhancement of di-rhamnolipid proportion using gamma irradiation. *Colloids and Surfaces B: Biointerfaces*, 81(2): 397–405. https://doi.org/10.1016/j.colsurfb.2010.06.026

Low, S.Y., Tan, J.Y., Ban, Z.H. & Siwayanan, P. (2021). Performance of green surfactants in the formulation of heavy-duty laundry liquid detergents (HDLD) with special emphasis on palm based alpha methyl ester sulfonates (α-MES). *Journal of Oleo Science*, 70(8): 1027–1037. https://doi.org/10.5650/jos.ess21078

Mácha, H., Marešová, H., Juříková, T., Švecová, M., Benada, O., Škríba, A. et al., & (2021). Killing effect of Bacillus velezensis FZB42 on a Xanthomonas campestris pv. Campestris (Xcc) strain newly isolated from cabbage Brassica oleracea Convar. Capitata (L.): A metabolomic study. *Microorganisms*, 9(7): 1410. https://doi.org/10.3390/microorganisms9071410

Malik, G., Arora, R., Chaturvedi, R. & Paul, M.S. (2022). Implementation of genetic engineering and novel omics approaches to enhance bioremediation: A focused review. *Bulletin of Environmental Contamination and Toxicology*, 108(3): 443–450. https://doi.org/10.1007/s00128-021-03218-3

Marchant, R. & Banat, I.M. (2012a). Biosurfactants: A sustainable replacement for chemical surfactants? *Biotechnology Letters*, 34(9): 1597–1605. https://doi.org/10.1007/s10529-012-0956-x

Marchant, R. & Banat, I.M. (2012b). Microbial biosurfactants: Challenges and opportunities for future exploitation. *Trends in Biotechnology*, 30(11): 558–565. https://doi.org/10.1016/j.tibtech.2012.07.003

Marchant, R., Funston, S., Uzoigwe, C., Rahman, P.K.S.M. & Banat, I.M. (2014). Production of biosurfactants from nonpathogenic bacteria. *In: Biosurfactants*. 84–93. CRC Press. https://doi.org/10.1201/b17599-7

Maron, P.-A., Ranjard, L., Mougel, C. & Lemanceau, P. (2007). Metaproteomics: A new approach for studying functional microbial ecology. *Microbial Ecology*, 53(3): 486–493. DOI: 10.1007/s00248-006-9196-8

Mohamad Pauzi, S., Anak Halbert, D., Azizi, S., Ahmad, N.A.A., Ahmad, N. & Marpani, F. (2019). Effect of organic antifoam's concentrations on filtration performance. *Journal of Physics: Conference Series*, 1349(1): 012141. https://doi.org/10.1088/1742-6596/1349/1/012141

Mubarak, M.Q.E., Mohd Isa, M.H. & Hassan, A.R. (2014, September 17). Single-step cross-flow ultrafiltration for recovery and purification of surfactin produced by *Bacillus subtilis* ATCC 21332. International Conference on Chemical, Environment & Biological Sciences (CEBS-2014) Sept. 17-18, 2014 Kuala Lumpur (Malaysia). https://doi.org/10.15242/IICBE.C914034

Mukherjee, S., Das, P. & Sen, R. (2006). Towards commercial production of microbial surfactants. *Trends in Biotechnology*, 24(11): 509–515. https://doi.org/10.1016/j.tibtech.2006.09.005

Mulligan, C.N. & Gibbs, B.F. (1990). Recovery of biosurfactants by ultrafiltration. *Journal of Chemical Technology & Biotechnology*, 47(1): 23–29. https://doi.org/10.1002/jctb.280470104

Nagtode, V.S., Cardoza, C., Yasin, H.K.A., Mali, S.N., Tambe, S.M., Roy, P. et al. (2023). Green surfactants (biosurfactants): A petroleum-free substitute for sustainability – Comparison, applications, market, and future prospects. *ACS Omega*, 8(13): 11674–11699. https://doi.org/10.1021/acsomega.3c00591

Najmi, Z., Ebrahimipour, G., Franzetti, A. & Banat, I.M. (2018). In situ downstream strategies for cost-effective bio/surfactant recovery. *Biotechnology and Applied Biochemistry*, 65(4): 523–532. https://doi.org/10.1002/bab.1641

Naughton, P.J., Marchant, R., Naughton, V. & Banat, I.M. (2019). Microbial biosurfactants: Current trends and applications in agricultural and biomedical industries. *Journal of Applied Microbiology*, 127(1): 12–28. https://doi.org/10.1111/jam.14243

Nguyen, D.D., Melnik, A.V., Koyama, N., Lu, X., Schorn, M., Fang, J. et al. (2016). Indexing the Pseudomonas specialized metabolome enabled the discovery of poaeamide B and the bananamides. *Nature Microbiology*, 2(1): 16197. https://doi.org/10.1038/nmicrobiol.2016.197

Nitschke, M., Costa, S.G.V.A.O., Haddad, R.G., Gonçalves, L.A., Eberlin, M.N. & Contiero, J. (2005). Oil wastes as unconventional substrates for rhamnolipid biosurfactant production by *Pseudomonas aeruginosa* LBI. *Biotechnology Progress*, 21(5): 1562–1566. https://doi.org/10.1021/bp050198x

Nurfarahin, A., Mohamed, M. & Phang, L. (2018). Culture medium development for microbial-derived surfactants production—An overview. *Molecules*, 23(5): 1049. https://doi.org/10.3390/molecules23051049

Onwosi, C.O. & Odibo, F.J.C. (2012). Effects of carbon and nitrogen sources on rhamnolipid biosurfactant production by Pseudomonas nitroreducens isolated from soil. *World Journal of Microbiology and Biotechnology*, 28(3): 937–942. https://doi.org/10.1007/s11274-011-0891-3

Ortega Ramirez, C.A., Kwan, A. & Li, Q.X. (2020). Rhamnolipids induced by glycerol enhance dibenzothiophene biodegradation in Burkholderia sp. C3. *Engineering*, 6(5): 533–540. https://doi.org/10.1016/j.eng.2020.01.006

Pal, M.P., Vaidya, B.K., Desai, K.M., Joshi, R.M., Nene, S.N. & Kulkarni, B.D. (2009). Media optimization for biosurfactant production by Rhodococcus erythropolis MTCC 2794: Artificial intelligence versus a statistical approach. *Journal of Industrial Microbiology & Biotechnology*, 36(5): 747–756. https://doi.org/10.1007/s10295-009-0547-6

Pantazaki, A.A., Papaneophytou, C.P. & Lambropoulou, D.A. (2011). Simultaneous polyhydroxyalkanoates and rhamnolipids production by Thermus thermophilus HB8. *AMB Express*, 1(1): 17. https://doi.org/10.1186/2191-0855-1-17

Patel, M. (2003). Surfactants based on renewable raw materials: carbon dioxide reduction potential and policies and measures for the European Union. *Journal of Industrial Ecology*, 7(3–4): 47–62. https://doi.org/10.1162/108819803323059398

Patel, P., Patel, H., Sharma, J., Shrimali, S., Sharma, S. & Saraf, M. (2023). Microbial biosurfactants: An overview of their uses, classification, types, properties, and biosynthesis. *Acta Scientific Microbiology*, 61–71. https://doi.org/10.31080/ASMI.2023.06.1260

Pitocchi, R., Cicatiello, P., Birolo, L., Piscitelli, A., Bovio, E., Varese, G.C. et al. (2020). Cerato-platanins from marine fungi as effective protein biosurfactants and bioemulsifiers. *International Journal of Molecular Sciences*, 21(8): 2913. https://doi.org/10.3390/ijms21082913

Rahayu, S., Pambudi, K.A., Afifah, A., Fitriani, S.R., Tasyari, S., Zaki, M. & Djamahar, R. (2021). Environmentally safe technology with the conversion of used cooking oil into soap. *Journal of Physics: Conference Series*, 1869(1): 012044. https://doi.org/10.1088/1742-6596/1869/1/012044

Rangarajan, V. & Clarke, K.G. (2016). Towards bacterial lipopeptide products for specific applications—A review of appropriate downstream processing schemes. *Process Biochemistry*, 51(12): 2176–2185. https://doi.org/10.1016/j.procbio.2016.08.026

Rangarajan, V. & Sen, R. (2013). An inexpensive strategy for facilitated recovery of metals and fermentation products by foam fractionation process. *Colloids and Surfaces B: Biointerfaces*, 104: 99–106. https://doi.org/10.1016/j.colsurfb.2012.12.007

Ranjan, R., Rani, A. & Kumar, R. (2015). Exploration of microbial cells: the storehouse of bio-wealth through metagenomics and metatranscriptomics. *In:* V.C. Kalia (Ed.), *Microbial Factories*. 7–27. Springer India. https://doi.org/10.1007/978-81-322-2598-0_2

Research and Market. Global Biosurfactants Market Size, Segments, Outlook, and Revenue Forecast 2022-2028 by Product, Application, and Region. (2022, November). Research and Markets. https://www.researchandmarkets.com/report/biosurfactant

Ribeiro, B.D., Alviano, D.S., Barreto, D.W. & Coelho, M.A.Z. (2013). Functional properties of saponins from sisal (Agave sisalana) and juá (Ziziphus joazeiro): Critical micellar concentration, antioxidant and antimicrobial activities. *Colloids and Surfaces A: Physicochemical and Engineering Aspects*, 436: 736–743. https://doi.org/10.1016/j.colsurfa.2013.08.007

Rodrigues, M.S., Moreira, F.S., Cardoso, V.L. & De Resende, M.M. (2017). Soy molasses as a fermentation substrate for the production of biosurfactants using Pseudomonas aeruginosa ATCC 10145. *Environmental Science and Pollution Research*, 24(22): 18699–18709. https://doi.org/10.1007/s11356-017-9492-5

Roelants, S.L.K.W., Saerens, K.M.J., Derycke, T., Li, B., Lin, Y., Van De Peer, Y. et al. (2013). *Candida bombicola* as a platform organism for the production of tailor-made biomolecules. *Biotechnology and Bioengineering*, 110(9): 2494–2503. https://doi.org/10.1002/bit.24895

Ron, E.Z. & Rosenberg, E. (2001). Natural roles of biosurfactants: Minireview. *Environmental Microbiology*, 3(4): 229–236. https://doi.org/10.1046/j.1462-2920.2001.00190.x

Roongsawang, N., Hase, K., Haruki, M., Imanaka, T., Morikawa, M. & Kanaya, S. (2003). Cloning and characterization of the gene cluster encoding arthrofactin Synthetase from Pseudomonas sp. MIS38. *Chemistry & Biology*, 10(9): 869–880. https://doi.org/10.1016/j.chembiol.2003.09.004

Saikia, R.R., Deka, S., Deka, M. & Banat, I.M. (2012). Isolation of biosurfactant-producing Pseudomonas aeruginosa RS29 from oil-contaminated soil and evaluation of different nitrogen sources in biosurfactant production. *Annals of Microbiology*, 62(2): 753–763. https://doi.org/10.1007/s13213-011-0315-5

Santos, D., Rufino, R., Luna, J., Santos, V. & Sarubbo, L. (2016). Biosurfactants: Multifunctional biomolecules of the 21st century. *International Journal of Molecular Sciences*, 17(3): 401. https://doi.org/10.3390/ijms17030401

Sarubbo, L.A., Silva, M.D.G.C., Durval, I.J.B., Bezerra, K.G.O., Ribeiro, B.G., Silva, I.A. et al. (2022). Biosurfactants: Production, properties, applications, trends, and general perspectives. *Biochemical Engineering Journal*, 181: 108377. https://doi.org/10.1016/j.bej.2022.108377

Satpute, S.K., Bhuyan, S.S., Pardesi, K.R., Mujumdar, S.S., Dhakephalkar, P.K., Shete, A.M. & Chopade, B.A. (2010). Molecular genetics of biosurfactant synthesis in microorganisms. *In:* R. Sen (Ed.), *Biosurfactants* (Vol. 672). 14–41. Springer New York. https://doi.org/10.1007/978-1-4419-5979-9_2

Sen, R. & Swaminathan, T. (2005). Characterization of concentration and purification parameters and operating conditions for the small-scale recovery of surfactin. *Process Biochemistry*, 40(9): 2953–2958. https://doi.org/10.1016/j.procbio.2005.01.014

Shah, A.V., Srivastava, V.K., Mohanty, S.S. & Varjani, S. (2021). Municipal solid waste as a sustainable resource for energy production: State-of-the-art review. *Journal of Environmental Chemical Engineering*, 9(4): 105717. https://doi.org/10.1016/j.jece.2021.105717

Sharma, D., Singh, D., Sukhbir-Singh, G.M., Karamchandani, B.M., Aseri, G.K., Banat, I.M. & Satpute, S.K. (2023). Biosurfactants: Forthcomings and regulatory affairs in food-based industries. *Molecules*, 28(6): 2823. https://doi.org/10.3390/molecules28062823

Sharon, A.B., Ahuekwe, E.F., Nzubechi, E.G., Oziegbe, O. & Oniha, M. (2023). Statistical optimization strategies on waste substrates for solving high-cost challenges in biosurfactants production: A review. *IOP Conference Series: Earth and Environmental Science*, 1197(1): 012004. https://doi.org/10.1088/1755-1315/1197/1/012004

Shekhar, S., Sundaramanickam, A. & Balasubramanian, T. (2015). Biosurfactant producing microbes and its potential applications: A review. *Critical Reviews in Environmental Science and Technology*, 1522–1554. https://doi.org/10.1016/1352-2310(94)90055-8

Sirohi, R. (2022). Shearography and its applications – A chronological review. *Light: Advanced Manufacturing*, 3(1): 1. https://doi.org/10.37188/lam.2022.001

Sivapathasekaran, C. & Sen, R. (2013). Performance evaluation of an ANN–GA aided experimental modeling and optimization procedure for enhanced synthesis of marine biosurfactant in a stirred tank reactor. *Journal of Chemical Technology & Biotechnology*, 88(5): 794–799. https://doi.org/10.1002/jctb.3900

Sopholiance™ S - Givaudan. (2023). https://cosmetics.specialchem.com/product/i-givaudan-sopholiance-s

Sun, H., Bie, X., Lu, F., Lu, Y., Wu, Y. & Lu, Z. (2009). Enhancement of surfactin production of *Bacillus subtilis* fmbR by replacement of the native promoter with the Pspac promoter. *Canadian Journal of Microbiology*, 55(8): 1003–1006. https://doi.org/10.1139/W09-044

Surfactants Market (CH 3464). (2023). Markets and Markets. https://www.marketsandmarkets.com/Market-Reports/biosurfactants-market-493.html

Syed-Hassan, S.S.A. & Zaini, M.S.M. (2016). Optimization of the preparation of activated carbon from palm kernel shell for methane adsorption using Taguchi orthogonal array design. *Korean Journal of Chemical Engineering*, 33(8): 2502–2512. https://doi.org/10.1007/s11814-016-0072-z

TegraSurf by Integrity Bio-Chemicals. (2021). TegraSurf. https://www.integritybiochem.com/industries/specialty-chemicals/tegrasurf/

Thio, C.W., Lim, W.H., Md. Shah, U.K. & Phang, L.-Y. (2022). Palm kernel fatty acid distillate as substrate for rhamnolipids production using *Pseudomonas* sp. LM19. *Green Chemistry Letters and Reviews*, 15(1): 83–92. https://doi.org/10.1080/17518253.2021.2023223

Thistleton, S. (2022). Biosurfactants: The Lowdown. https://x-cellr8.com/2022/07/22/biosurfactants-the-lowdown/

Varjani, S.J., Rana, D.P., Jain, A.K., Bateja, S. & Upasani, V.N. (2015). Synergistic ex-situ biodegradation of crude oil by halotolerant bacterial consortium of indigenous strains isolated from on shore sites of Gujarat, India. *International Biodeterioration & Biodegradation*, 103: 116–124. https://doi.org/10.1016/j.ibiod.2015.03.030

Varjani, S. & Upasani, V.N. (2019). Evaluation of rhamnolipid production by a halotolerant novel strain of Pseudomonas aeruginosa. *Bioresource Technology*, 288: 121577. https://doi.org/10.1016/j.biortech.2019.121577

Wang, C., Cao, Y., Wang, Y., Sun, L. & Song, H. (2019). Enhancing surfactin production by using systematic CRISPRi repression to screen amino acid biosynthesis genes in Bacillus subtilis. *Microbial Cell Factories*, 18(1): 90. https://doi.org/10.1186/s12934-019-1139-4

Willenbacher, J., Mohr, T., Henkel, M., Gebhard, S., Mascher, T., Syldatk, C. & Hausmann, R. (2016). Substitution of the native srfA promoter by constitutive P in two *B. subtilis* strains and evaluation of the effect on Surfactin production. *Journal of Biotechnology*, 224: 14–17. https://doi.org/10.1016/j.jbiotec.2016.03.002

Williams, W., Kunorozva, L., Klaiber, I., Henkel, M., Pfannstiel, J., Van Zyl, L.J. et al. (2019). Novel metagenome-derived ornithine lipids identified by functional screening for biosurfactants. *Applied Microbiology and Biotechnology*, 103(11): 4429–4441. https://doi.org/10.1007/s00253-019-09768-1

Winterburn, J.B., Russell, A.B. & Martin, P.J. (2011). Integrated recirculating foam fractionation for the continuous recovery of biosurfactant from fermenters. *Biochemical Engineering Journal*, 54(2): 132–139. https://doi.org/10.1016/j.bej.2011.02.011

Wisjnuprapto, N.A., Helmy, Q., Kardena, E. & Funamizu, N. (2011). Strategies toward commercial scale of biosurfactant production as potential substitute for its chemically counterparts. *International Journal of Biotechnology*, 12(1/2): 66. https://doi.org/10.1504/IJBT.2011.042682

Wongsirichot, P., Ingham, B. & Winterburn, J. (2021). A review of sophorolipid production from alternative feedstocks for the development of a localized selection strategy. *Journal of Cleaner Production*, 319: 128727. https://doi.org/10.1016/j.jclepro.2021.128727

Worley, B. & Powers, R. (2012). Multivariate analysis in metabolomics. *Current Metabolomics*, 1(1): 92–107. https://doi.org/10.2174/2213235X11301010092

Wu, T., Jiang, J., He, N., Jin, M., Ma, K. & Long, X. (2019). High-performance production of biosurfactant rhamnolipid with nitrogen feeding. *Journal of Surfactants and Detergents*, 22(2): 395–402. https://doi.org/DOI 10.1002/jsde.12256

Xia, Y., Shi, Y., Chu, J., Zhu, S., Luo, X., Shen, W. & Chen, X. (2023). Efficient Biosynthesis of acidic/lactonic sophorolipids and their application in the remediation of cyanobacterial harmful algal blooms. *International Journal of Molecular Sciences*, 24(15): 12389. https://doi.org/10.3390/ijms241512389

Xu, J. & Zhang, W. (2016). Strategies used for genetically modifying bacterial genome: Site-directed mutagenesis, gene inactivation, and gene over-expression. *Journal of Zhejiang University - SCIENCE B*, 17(2): 83–99. https://doi.org/10.1631/jzus.B1500187

Xu, X., Chemparathy, A., Zeng, L., Kempton, H.R., Shang, S., Nakamura, M. & Qi, L.S. (2021). Engineered miniature CRISPR-Cas system for mammalian genome regulation and editing. *Molecular Cell*, 81(20): 4333–4345.e4. https://doi.org/10.1016/j.molcel.2021.08.008

Yang, Q., Shen, L., Yu, F., Zhao, M., Jin, M., Deng, S. & Long, X. (2023). Enhanced fermentation of biosurfactant mannosylerythritol lipids on the pilot scale under efficient foam control with addition of soybean oil. *Food and Bioproducts Processing*, 138: 60–69. https://doi.org/10.1016/j.fbp.2023.01.002

Zhang, Y., Thompson, K.N., Branck, T., Yan Yan, Nguyen, L.H., Franzosa, E.A. & Huttenhower, C. (2021). Metatranscriptomics for the human microbiome and microbial community functional profiling. *Annual Review of Biomedical Data Science*, 4(1): 279–311. https://doi.org/10.1146/annurev-biodatasci-031121-103035

Zhi, Y., Wu, Q. & Xu, Y. (2017). Genome and transcriptome analysis of surfactin biosynthesis in Bacillus amyloliquefaciens MT45. *Scientific Reports*, 7(1): 40976. https://doi.org/10.1038/srep40976

Zhou, D., Hu, F., Lin, J., Wang, W. & Li, S. (2018). Genome and transcriptome analysis of Bacillus velezensis BS-37, an efficient surfactin. *Microbiology Open*, 1–14. https://doi.org/10.1002/mbo3.794

Zhu, K. & Rock, C.O. (2008). RhlA converts β-hydroxyacyl-acyl carrier protein intermediates in fatty acid synthesis to the β-hydroxydecanoyl-β-hydroxydecanoate component of rhamnolipids in *Pseudomonas aeruginosa*. *Journal of Bacteriology*, 190(9): 3147–3154. https://doi.org/10.1128/JB.00080-08

CHAPTER

3

Methods of Biosurfactants Production

Nileema S. Gore[1*], Priyanka S. Patil[1], and Archana S. Harke[2]

[1] Institute of Biosciences and Technology, MGM University, Chhatrapati Sambhajinagar, India

[2] Novo Nordisk Foundation Center for Biosustainability Reconstruction DTU Microbes Initiative, DTU – Technical University of Denmark, Copenhagen, Capital Region, Denmark

1. Introduction

Due to their exceptional qualities and numerous uses in fields ranging from bioremediation to medicines, biosurfactants, a broad set of surface-active chemicals generated by microbes, have attracted considerable attention They have established themselves as potential alternatives to synthetic surfactants due to their eco-friendliness and biodegradability (Banat et al., 2010). Nevertheless, despite their potential, a number of industrial restrictions ingrained in conventional manufacturing techniques have prevented biosurfactants from being widely used (Valiathan and Radhakrishnan, 2023).

This analysis explores the crucial elements of biosurfactant production, highlighting the roadblocks to scalability and financial feasibility (Sarubbo et al., 2022). This review highlights the difficulties caused by low yields, lengthy fermentation periods, and expensive extraction methods by analyzing the current procedures. It also examines the effects of conventional manufacturing techniques on the environment, highlighting the urgent need for sustainable alternatives.

A growing body of study has evolved in reaction to these obstacles with the goal of developing cutting-edge methods that get beyond these restrictions. This study seeks to present a thorough overview of these novel approaches, covering process intensification, substrate optimization, and genetic and enzyme engineering. It also explores how synthetic biology and microbial consortia could revolutionize the manufacturing of biosurfactants.

* Corresponding author: goreneel@gmail.com

To demonstrate how these innovative techniques have started to transform the biosurfactant production environment through a thorough review of case studies and success stories. By taking this action can hope to provide a glimpse into this field's bright future and its potential to alter a variety of industries.

In the subsequent sections, we embark on an exploration of the intricacies involved in biosurfactant production, scrutinizing the challenges that have hindered its advancement and delineating a pathway towards a more efficient and sustainable future. Through this systematic approach, tried to initiate a new epoch wherein biosurfactants become pivotal to both industrial innovation and environmental stewardship.

2. Biosurfactants

A remarkable family of surface-active chemicals known as biosurfactants stands out due to its natural origins and variety of chemical configurations. Biosurfactants are made by a variety of microorganisms, including bacteria, fungi, and yeast, in contrast to their synthetic equivalents, which come from petrochemical sources. These substances are able to play crucial roles in a variety of biological and industrial processes because of their special capacity to lower surface tension at fluid interfaces (Nagtode et al., 2023). The different advantages, applications and challenges related to biosurfactants are added in Fig. 1.

The amphipathic nature of biosurfactants is one of their distinguishing features. As a result, their molecular structure contains both hydrophilic (water-loving) and hydrophobic (water-repelling) components. They are able to interact with interfaces between immiscible substances, such as oil and water, because of their dual personality (Markande et al., 2021). As a result, biosurfactants are used in a wide range of sectors, from food and agriculture to pharmaceuticals and environmental cleanup (Kumar et al., 2021).

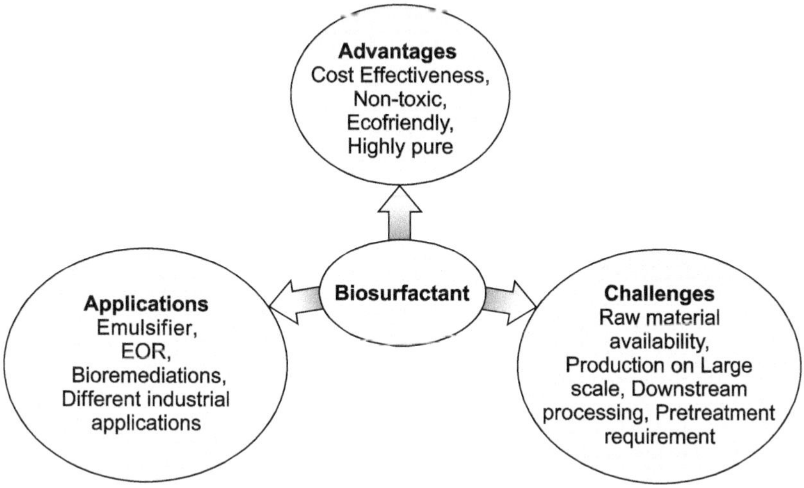

Fig. 1: Biosurfactants: Advantages, applications and challenges.

Biosurfactants have a wide range of positive characteristics in addition to their ability to lower surface tension. Because of their inherent biodegradability, they are environmentally beneficial and protect natural ecosystems from the long-lasting effects of synthetic surfactants. Additionally, biosurfactants have shown significant antibacterial, emulsifying, and even medicinal characteristics, boosting their applicability in numerous biotechnology domains (Kumari et al., 2023).

The potential of biosurfactants has drawn more attention as society moves closer to sustainable and environmentally friendly practices. Researchers and businesses are working together to explore new manufacturing methods, identify new uses, and exploit the full potential of these natural substances (Gayathiri et al., 2022).

This introduction lays the groundwork for a thorough investigation of biosurfactants, including information on their history, characteristics, uses, and most recent developments in their manufacture and application. This is the set up of road to a more sustainable and environmentally responsible future by exploring the intricate details of these remarkable molecules.

3. Current Production Methods

The cultivation of microorganisms that naturally create these surface-active substances is necessary for the creation of biosurfactants. There are various existing techniques for producing biosurfactants, each with a unique set of benefits and restrictions,

3.1 Fermentation

This process is the most popular for making biosurfactants. In a controlled bioreactor environment, microorganisms (such as bacteria or yeast) are grown, and during their metabolic and development processes, they produce biosurfactants. The benefits include the possibility of mass production and optimization of high yields (Saharan et al., 2011). Specific biosurfactant types may require adjustments to controlled circumstances. These limitations may call for specialized tools and knowledge. Depending on the microorganism and circumstances, fermentation processes can take a long time (Gurkok, 2021, Saranraj et al., 2022).

3.2 Cultivation by Batch and Fed-batch

In fed-batch cultivation, nutrients are provided progressively over time to prolong the growth period as opposed to batch cultivation, when all nutrients are added at the beginning. This type of cultivation improves control over the availability of nutrients. It can result in increased production and yields. It requires exact monitoring and management of nutritional inputs (Kronemberger et al., 2010).

3.3 Submerged Fermentation

In bioreactors, microorganisms are grown in a liquid medium where they make biosurfactants. It is suitable for bacteria that grow in submerged environments, enabling effective mixing and aeration. But due to the need for aeration, it may be energy-intensive (Velioğlu and Ürek, 2015).

3.4 Fermentation in a Solid State

In this technique, microorganisms develop on a solid substrate (such as bran or agricultural waste) with a low moisture content. This can be economical and makes use of agricultural byproducts. This process is adapted to some actinomycetes and fungi. In the process there is toughness in managing and keeping track of microbial development (Barbosa et al., 2021).

3.5 Extraction and Purification

Following fermentation, biosurfactants are often isolated from the microbial culture by centrifugation, filtration, or solvent extraction. To separate the biosurfactant from other biological components, purification techniques might come next. The ability to concentrate and separate biosurfactants from the fermentation broth is advantageous. The associated limitations are that some extraction techniques can be wasteful and resource-intensive (Singh, 2012).

3.6 Optimization of Growth Conditions

In order to increase biosurfactant synthesis, this strategy concentrates on adjusting variables including pH, temperature, substrate concentration, and aeration. This may result in increased biosurfactant quality and production. This process requires a detailed grasp of the needs of the particular microorganism (Najafi et al., 2010).

Each of these approaches has advantages and disadvantages of its own. The microorganism employed, the kind of biosurfactant being generated, and the intended application all have a role in the production process selection. Additionally, ongoing research intends to create methods for producing biosurfactants that are more effective and sustainable.

4. Production Limitations

Although promising, the manufacture of biosurfactants has a number of drawbacks that could prevent their broad use in a variety of industries. The following are some major biosurfactant production restrictions which are pointed out in Fig. 2 also.

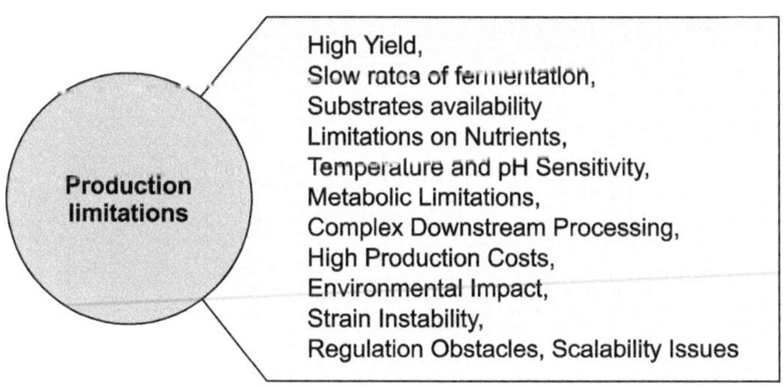

Fig. 2: Production limitations of biosurfactants.

Getting high yields in the manufacturing of biosurfactants is one of the main challenges. Because many microorganisms create biosurfactants in comparatively tiny amounts naturally, yield optimization measures are required (Reis, 2013). Microorganisms that produce biosurfactants frequently develop slowly, which might cause fermentation times to be prolonged. Due to the length of the lead time, production costs may rise and commercial viability may be hampered. A limiting factor for the synthesis of biosurfactants and microbial growth can be the availability and choice of suitable substrates. Specific carbon sources that may not be affordable or easily accessible are needed by some microbes (Banat et al., 2014). The formation of biosurfactants may be impacted by critical nutrients (such as nitrogen and phosphorus) that are present in the fermentation medium. Subpar yields can result from insufficient fertilizer availability (Vieira et al., 2021). Microorganisms that produce biosurfactants frequently have particular pH and temperature requirements for development and biosurfactant production. Production efficiency can suffer from deviations from these conditions (Ostendorf et al., 2019). Lower yields may result from microorganisms prioritizing alternative metabolic pathways over biosurfactant synthesis. Approaches in genetic and metabolic engineering can be used to solve this (Moutinho et al., 2021). Biosurfactant extraction and purification from the fermentation broth can be difficult and time-consuming. To obtain high-purity biosurfactants, effective downstream processing methods are essential (Venkataraman et al., 2022). Particularly for small-scale operations, the expense of production tools, media, and downstream processing might pose serious obstacles. To be competitive, a company must use economical production techniques (Venkataraman et al., 2022).

Some conventional biosurfactant production techniques could produce waste or have an impact on the environment. This could negate the environmental benefits of biosurfactants (Makkar and Cameotra, 2002). Over time, certain biosurfactant-producing strains may experience genetic mutations or lose their capacity to create biosurfactants, which could compromise production reliability and consistency (Desai and Banat, 1997). Due to the constantly changing nature of biotechnological processes, it might be difficult to meet regulatory requirements and secure appropriate permissions for the manufacturing of biosurfactants. It might be challenging to increase biosurfactant production from a lab to a commercial scale. Process optimization must be done carefully if consistent quality and yield are to be achieved at higher quantities (Sarubbo et al., 2022).

Continuous research and development, such as genetic engineering, process improvement, and the creation of new substrates, are required to overcome these constraints. Overcoming these obstacles will help biosurfactants become more widely used in a variety of industries.

5. Environmental and Sustainability Concerns

While the manufacturing of biosurfactants has a number of advantages over the production of synthetic surfactants in terms of the environment, it is not without its own set of environmental and sustainability issues. Some crucial things to remember in this context are shown in Fig. 3.

Some biosurfactant production methods may require substantial amounts of resources, such as water, energy, and raw materials. Depending on the specific production process, this can contribute to resource depletion and environmental impact (Olasanmi and Thring, 2018). If biosurfactant production relies on agricultural feedstocks or crops, it may compete with food production or lead to land-use change, potentially resulting in deforestation or habitat loss (Ruiz-Gonzalez and Vicente, 2023). Some of the substrates utilized in the manufacturing of biosurfactants could not be long-term renewable or sustainable. Concerns about resource depletion may arise, for instance, from dependency on particular crops or nonrenewable carbon sources. Some production processes could produce waste by-products or effluents that need to be properly controlled. This also contains leftovers from extraction or fermentation procedures. If not handled properly, the use of chemicals for procedures like substrate

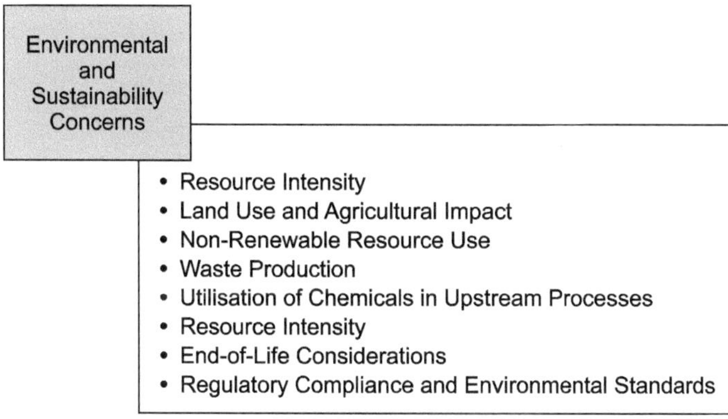

Fig. 3: Environmental and sustainability concerns in biosurfactant production.

preparation, sterilization, and pH adjustment might have negative environmental effects. Pollution or the release of dangerous substances could result (Reis et al., 2013). The disposal or treatment of waste generated during the production process, as well as the fate of used fermentation media, can be important aspects to consider from an environmental standpoint. Ensuring compliance with environmental regulations and standards is crucial to minimizing negative impacts associated with biosurfactant production (Schonhoff et al., 2022).

Addressing these concerns involves adopting best practices, conducting life cycle assessments, and implementing sustainable production strategies. Additionally, ongoing research focuses on developing eco-friendlier and resource-efficient methods for biosurfactant production to mitigate environmental impacts.

6. Emerging Approaches

Emerging approaches towards biosurfactant production are as shown in Fig. 4. These are paving the way for more efficient, sustainable, and scalable methods. The noteworthy strategies regarding these are,

Enzyme engineering is a strategy that entails improving the activity and specificity of the biosurfactant synthesis-related enzymes. Researchers want to boost biosurfactant yields and boost production effectiveness by modifying enzymes (Helmy et al., 2011). Pathways for the synthesis of biosurfactants can be made more efficient through genetic alteration of microorganisms. Scientists can modify microbes to create more biosurfactants with desired qualities by inserting or boosting particular genes (Das et al., 2008). Metabolic engineering methods include rerouting cellular energy towards the synthesis of biosurfactants by modifying the metabolic pathways of bacteria. Researchers can improve biosurfactant output by fine-tuning metabolic fluxes (Gaur et al., 2022). Biosurfactant production is one of the specialized biological systems that may be designed and built using synthetic biology approaches. Scientists can engineer microbes with genetic circuits and pathways that are best for synthesizing biosurfactants (Rodrigues and Rodrigues, 2023). A crucial field of research involves investigating alternate and sustainable substrates for the generation of biosurfactants. The goal of this strategy is to find affordable, easily accessible feedstocks that

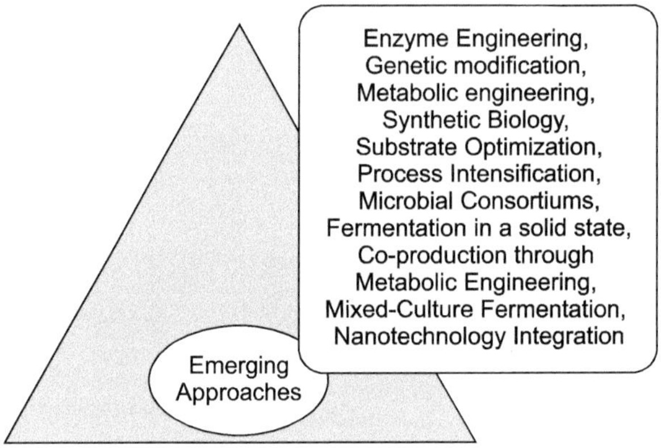

Fig. 4: Emerging approaches towards biosurfactant production.

can enable efficient biosurfactant manufacturing (Asgher et al., 2020). Process intensification involves streamlining the fermentation procedure to increase output and decrease turnaround times. High cell density fermentation, continuous fermentation, and fed-batch processes are among methods that can be used to accomplish this (Tiso et al., 2020). Synergistic effects in the synthesis of biosurfactants might result from the use of a consortium of microorganisms with complementary metabolic capacities. In order to increase total yields, this strategy makes use of the cooperative interactions between several species (Darvishi et al., 2011). Solid substrates with little water content are used for solid-state fermentation. This method is gaining popularity for the synthesis of biosurfactants because, in comparison to submerged fermentation, it may be more resource-efficient and environmentally benign (Rubio-Ribeaux et al., 2023). Researchers are looking into the idea of engineering microbes to co-produce useful substances like biofuels or bio-based chemicals in addition to biosurfactants (Yadav et al., 2021). Using mixed cultures of microorganisms can increase the generation of biosurfactants. This method takes advantage of the various metabolic capacities of several species, which could lead to higher yields. Nanotechnology and biosurfactant integration enable the creation of sophisticated formulations with enhanced stability, distribution, and application characteristics. This creates new opportunities for specialized applications possible (Ingle et al., 2023).

These novel methods demonstrate how biosurfactant production research is dynamic and creative. Researchers are well-positioned to overcome historical constraints and realize the full potential of biosurfactants across a variety of industries by leveraging developments in biotechnology and metabolic engineering.

7. Biotechnological Advancements

Production of biosurfactants has advanced significantly thanks to biotech. Bacteria, fungi, and yeasts all produce biosurfactants, which are molecules with surface activity. They are widely used in a variety of industries, including food, pharmaceuticals, agriculture, and the environment and petroleum. Some significant developments in the creation of biosurfactants are exhibited in Fig. 5 and explained.

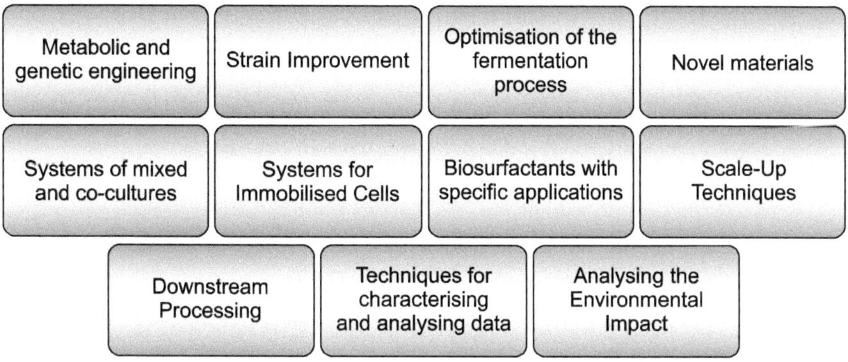

Fig. 5: Biotechnological advancements in the production of biosurfactants.

Techniques from genetic engineering have been used to increase the capacity of microorganisms to produce biosurfactants. To improve the microorganism's traits and increase yield, this entails changing its genetic makeup (Jimoh et al., 2021). Researchers have been able to recognize and isolate microorganism strains with high production rates through repetitive screening and selection techniques. The foundation for additional advancements is then established using these strains (Adetunji et al., 2021). Conditions for the generation of biosurfactants have improved as a result of developments in bioprocessing and fermentation technologies. This entails optimizing variables including pH, temperature, aeration, and nutrient availability (Ghribi et al., 2012). For the synthesis of biosurfactants, researchers are looking into alternative and waste-derived substrates. This lowers the cost of production while also increasing the process' sustainability (Hu et al., 2021). A co-cultivation or mixed culture system that combines several microorganisms has shown potential in boosting biosurfactant production. This strategy utilizes the synergistic interactions between many species (Alves et al., 2019). A matrix or other support material can be used to immobilize microorganisms, which can boost the synthesis of biosurfactants. The microbes can be used again using this method, which also makes downstream processing easier (Khondee et al., 2015). In order to create biosurfactants specifically suited for uses like increased oil recovery, bioremediation, and medicinal formulations, scientists are now engineering microorganisms (Muthusamy et al., 2008). For procedures to be commercially viable, scaling up biosurfactant synthesis from lab-scale to industrial-scale processes is essential (Amani et al., 2010). Higher yields and better purity of biosurfactants have been made possible by improvements in downstream processing procedures, such as extraction and purification techniques (Venkataraman et al., 2022). For biosurfactants to be successfully applied in a variety of industries, more accurate analytical techniques provide a deeper comprehension of their properties (Sarubbo et al., 2022). In order to compare biosurfactants to conventional surfactants and evaluate their environmental sustainability, researchers are undertaking thorough analyses of the production processes for biosurfactants (Schonhoff et al., 2022).

8. Future Prospects and Industry Impact of Biosurfactants

Because they have the potential to replace synthetic surfactants in a variety of industries, biosurfactants are receiving more and more attention. Here are some biosurfactant prospects for the future and their business effects.

8.1 Sustainable and Environmentally Friendly Alternatives

Biosurfactants are biodegradable and can be produced from renewable resources. This makes them a more sustainable and environmentally friendly alternative to synthetic surfactants derived from petrochemicals (Thavasi et al., 2011).

8.2 Reduced Environmental Impact

The use of biosurfactants can lead to reduced environmental pollution and toxicity compared to conventional surfactants. This is particularly important in applications like bioremediation, where the goal is to clean up contaminated environments (Johnson et al., 2021).

8.3 Enhanced Oil Recovery (EOR)

Biosurfactants have shown promise in improving the efficiency of oil recovery from reservoirs. Their ability to reduce interfacial tension between oil and water can lead to increased oil production (McInerney et al., 2005).

8.4 Bioremediation and Environmental Cleanup

Biosurfactants can aid in the breakdown and removal of pollutants from soil and water, making them valuable in environmental cleanup efforts (Sarubbo et al., 2015).

8.5 Industry for Personal Care and Cosmetics

The use of biosurfactants in formulations for personal care items like shampoos, soaps, and skincare products is being investigated. They are less abrasive on the skin and can offer mild washing (Vecino et al., 2017).

8.6 Food and Beverages Industry

In food and beverage products, biosurfactants can be employed as emulsifiers and stabilizers, improving texture and stability (Campos et al., 2013).

8.7 Drug Delivery and Pharmaceuticals

Pharmaceutical formulations may use biosurfactants, especially in drug delivery systems where they might improve solubility and bioavailability (Liao et al., 2021).

8.8 Agrochemicals and Agriculture

Biosurfactants have a variety of uses in agriculture, including aiding nutrient intake, enhancing plant growth, and enhancing the efficacy of insecticides Patel et al., 2023).

8.9 Smelting ore and mining

In mining operations, biosurfactants can be employed to help separate valuable minerals from ores. They may lessen the surface tension that exists between water and minerals, allowing for easier flotation (Khoshdast et al., 2011).

8.10 Treatment of wastewater

By promoting the decomposition of organic matter and improving the removal of pollutants, biosurfactants can contribute to improving the efficacy of wastewater treatment systems (Malkapuram et al., 2021).

8.11 Market Development and Economic Effects

The market for biosurfactants is anticipated to rise as people become more conscious of sustainability and environmental effects. For businesses engaged in their manufacturing and application, this can result in economic prospects (Mohanty et al., 2021).

8.12 Regulations to be Considered

Biosurfactants may increasingly be encouraged or required for use by regulatory authorities, which could accelerate their uptake (Effendi et al., 2018).

Although biosurfactants have numerous advantages, yet there are concerns that need to be resolved, including cost effectiveness, scaling-up issues, and rivalry with long-established synthetic surfactants. Realizing the full potential of biosurfactants across diverse industries will require continued research and development efforts.

9. Conclusion

In conclusion, in many industries, biosurfactants hold enormous promise as sustainable substitutes for synthetic surfactants. However, some production restrictions make their wider adoption difficult. Among them are difficulties with scaling, high production costs, and variation in yield from various microbial strains. Innovative strategies have been investigated to address these obstacles.

The ability to produce biosurfactants has been improved by using genetic and metabolic engineering techniques. These strategies entail changing the genetic make-up of microorganisms to boost productivity and enhance traits. To lower production costs and improve sustainability, innovative substrates made from waste and renewable resources have also been researched.

Additionally, strategies using immobilized cells, mixed culture systems, and co-cultivation have shown promise in boosting biosurfactant synthesis. These methods promote repeated use by utilizing synergistic interactions between many microorganisms.

Biosurfactants are positioned to play a significant role in a variety of industries, including agriculture, pharmaceuticals, bioremediation, and more, as the demand for sustainable and environmentally friendly solutions keeps growing. The manufacturing constraints of biosurfactants are likely to be gradually addressed with continued research and development, opening the door for their wider adoption in standard industrial processes. These developments have the potential to drastically lessen the environmental effect of many industries while fostering a more sustainable and environmentally friendly future.

References

Adetunji, C.O., Jeevanandam, J., Anani, O.A., Inobeme, A., Thangadurai, D. et al. (2021). Strain improvement methodology and genetic engineering that could lead to an increase in the production of biosurfactants. *In: Green Sustainable Process for Chemical and Environmental Engineering and Science*, 299-315. Elsevier.

Alves, A.R., Sequeira, A.M. & Cunha, Â. (2019). Increase in bacterial biosurfactant production by co-cultivation with biofilm-forming bacteria. *Lett. Appl. Microbiol.*, 69(1): 79-86.

Amani, H., Mehrnia, M.R., Sarrafzadeh, M.H., Haghighi, M. & Soudi, M.R. (2010). Scale up and application of biosurfactant from Bacillus subtilis in enhanced oil recovery. *Appl. Biochem. Biotechnol.*, 162: 510-523.

Asgher, M., Afzal, M., Qamar, S.A. & Khalid, N. (2020). Optimization of biosurfactant production from chemically mutated strain of *Bacillus subtilis* using waste automobile oil as low-cost substrate. *J. Environ. Sustain.*, 3: 405-413.

Banat, I.M., Franzetti, A., Gandolfi, I., Bestetti, G., Martinotti, M.G., Fracchia, L. & Marchant, R. (2010). Microbial biosurfactants production, applications and future potential. *Appl. Microbiol. Biotechnol.*, 87: 427-444.

Banat, I.M., Satpute, S.K., Cameotra, S.S., Patil, R. & Nyayanit, N.V. (2014). Cost effective technologies and renewable substrates for biosurfactants' production. *Front. Microbiol.*, 5: 697.

Barbosa, J.R., da Silva, S.B. & de Carvalho Junior, R.N. (2021). Biosurfactant production by solid-state fermentation, submerged fermentation, and biphasic fermentation. *In: Green Sustainable Process for Chemical and Environmental Engineering and Science*, 155-171. Elsevier.

Campos, J.M., Montenegro Stamford, T.L., Sarubbo, L.A., de Luna, J.M., Rufino, R.D. & Banat, I.M. (2013). Microbial biosurfactants as additives for food industries. *Biotechnol. Prog.*, 29(5): 1097-1108.

Darvishi, P., Ayatollahi, S., Mowla, D. & Niazi, A. (2011). Biosurfactant production under extreme environmental conditions by an efficient microbial consortium, ERCPPI-2. *Colloids Surf. B.*, 84(2): 292-300.

Das, P., Mukherjee, S. & Sen, R. (2008). Genetic regulations of the biosynthesis of microbial surfactants: An overview. *Biotechnol. Genet. Eng. Rev.*, 25(1): 165-186.

Desai, J.D. & Banat, I.M. (1997). Microbial production of surfactants and their commercial potential. *Microbiol. Mol. Biol.*, 61(1): 47-64.

Effendi, A.J., Kardena, E. & Helmy, Q. (2018). Biosurfactant-enhanced petroleum oil bioremediation. *Microbial Action on Hydrocarbons*, 143-179.

Gaur, V.K., Sharma, P., Gupta, S., Varjani, S., Srivastava, J.K., Wong, J.W. & Ngo, H.H. (2022). Opportunities and challenges in omics approaches for biosurfactant production and feasibility of site remediation: Strategies and advancements. *Environ. Technol. Innov.*, 25: 102132.

Gayathiri, E., Prakash, P., Karmegam, N., Varjani, S., Awasthi, M.K. & Ravindran, B. (2022). Biosurfactants: Potential and eco-friendly material for sustainable agriculture and environmental safety—A review. *Agron.*, 12(3): 662.

Ghribi, D., Abdelkefi-Mesrati, L., Mnif, I., Kammoun, R., Ayadi, I., Saadaoui, I. et al. (2012). Investigation of antimicrobial activity and statistical optimization of Bacillus subtilis SPB1 biosurfactant production in solid-state fermentation. *Biomed Res. Int.*, 2012.

Gurkok, S. (2021). Important parameters necessary in the bioreactor for the mass production of biosurfactants. *In:* Green Sustainable Process for Chemical and Environmental Engineering and Science. 347-365. Elsevier.

Helmy, Q., Kardena, E., Funamizu, N. & Wisjnuprapto. (2011). Strategies toward commercial scale of biosurfactant production as potential substitute for its chemically counterparts. *Int. J. Biotechnol.*, 12(1-2): 66-86.

Hu, X., Subramanian, K., Wang, H., Roelants, S.L., To, M.H., Soetaert, W. et al. (2021). Guiding environmental sustainability of emerging bioconversion technology for waste-derived sophorolipid production by adopting a dynamic life cycle assessment (dLCA) approach. *Environ. Pollut.*, 269: 116101.

Ingle, A.P., Saxena, S., Moharil, M., Rai, M. & Da Silva, S.S. (2023). Biosurfactants in nanotechnology: Recent advances and applications. *Biosurfactants and Sustainability: From Biorefineries Production to Versatile Applications*, 173-194.

Jimoh, A.A., Senbadejo, T.Y., Adelekc, R. & Lin, J. (2021). Development and genetic engineering of hyper-producing microbial strains for improved synthesis of biosurfactants. *Mol. Biotechnol.*, 63: 267-288.

Johnson, P., Trybala, A., Starov, V. & Pinfield, V.J. (2021). Effect of synthetic surfactants on the environment and the potential for substitution by biosurfactants. *Adv. Colloid Interface Sci.*, 288: 102340.

Khondee, N., Tathong, S., Pinyakong, O., Müller, R., Soonglerdsongpha, S., Ruangchainikom, C. et al. (2015). Lipopeptide biosurfactant production by chitosan-immobilized Bacillus sp. GY19 and their recovery by foam fractionation. *Biochem. Eng. J.*, 93: 47-54.

Khoshdast, H., Sam, A., Vali, H. & Noghabi, K.A. (2011). Effect of rhamnolipid biosurfactants on performance of coal and mineral flotation. *Int. Biodeterior. Biodegradation.*, 65(8): 1238-1243.

Kronemberger, F.A., Borges, C.P. & Freire, D.M. (2010). Fed-batch biosurfactant production in a bioreactor. *Int. Rev. Chem. Eng.*, 2(4): 513-518.

Kumar, A., Singh, S.K., Kant, C., Verma, H., Kumar, D., Singh, P.P. et al. (2021). Microbial biosurfactant: A new frontier for sustainable agriculture and pharmaceutical industries. *Antioxid.*, 10(9): 1472.

Kumari, R., Singha, L.P. & Shukla, P. (2023). Biotechnological potential of microbial bio-surfactants, their significance and diverse applications. *FEMS Microbes.*, xtad015.

Liao, Y., Li, Z., Zhou, Q., Sheng, M., Qu, Q., Shi, Y. et al. (2021). Saponin surfactants used in drug delivery systems: A new application for natural medicine components. *Int. J. Pharm.*, 603: 120709.

Makkar, R. & Cameotra, S. (2002). An update on the use of unconventional substrates for biosurfactant production and their new applications. *Appl. Microbiol. Biotechnol.*, 58: 428-434.

Malkapuram, S.T., Sharma, V., Gumfekar, S.P., Sonawane, S., Sonawane, S., Boczkaj, G. & Seepana, M.M. (2021). A review on recent advances in the application of biosurfactants in wastewater treatment. *Sustain. Energy Technol. Assess.*, 48: 101576.

Markande, A.R., Patel, D. & Varjani, S. (2021). A review on biosurfactants: Properties, applications and current developments. *Bioresour. Technol.*, 330: 124963.

McInerney, M.J., Nagle, D.P. & Knapp, R.M. (2005). Microbially enhanced oil recovery: Past, present, and future. *Petroleum Microbiology*, 215-237.

Mohanty, S.S., Koul, Y., Varjani, S., Pandey, A., Ngo, H.H., Chang, J.S. et al. (2021). A critical review on various feedstocks as sustainable substrates for biosurfactants production: A way towards cleaner production. *Microb. Cell Factories*, 20(1): 1-13.

Moutinho, L.F., Moura, F.R., Silvestre, R.C. & Romão-Dumaresq, A.S. (2021). Microbial biosurfactants: A broad analysis of properties, applications, biosynthesis, and techno-economical assessment of rhamnolipid production. *Biotechnol. Prog.*, 37(2): e3093.

Muthusamy, K., Gopalakrishnan, S., Ravi, T.K. & Sivachidambaram, P. (2008). Biosurfactants: Properties, commercial production and application. *Curr. Sci.*, 736-747.

Nagtode, V.S., Cardoza, C., Yasin, H.K.A., Mali, S.N., Tambe, S.M., Roy, P. et al. (2023). Green surfactants (biosurfactants): A petroleum-free substitute for sustainability – Comparison, applications, market, and future prospects. *ACS Omega*, 8(13): 11674-11699.

Najafi, A.R., Rahimpour, M.R., Jahanmiri, A.H., Roostaazad, R., Arabian, D. & Ghobadi, Z. (2010). Enhancing biosurfactant production from an indigenous strain of *Bacillus mycoides* by optimizing the growth conditions using a response surface methodology. *J. Chem. Eng.*, 163(3): 188-194.

Olasanmi, I.O. & Thring, R.W. (2018). The role of biosurfactants in the continued drive for environmental sustainability. *Sustainability*, 10(12): 4817.

Ostendorf, T.A., Silva, I.A., Converti, A. & Sarubbo, L.A. (2019). Production and formulation of a new low-cost biosurfactant to remediate oil-contaminated seawater. *J. Biotechnol.*, 295: 71-79.

Patel, P., Patel, R., Mukherjee, A. & Munshi, N.S. (2023). Microbial biosurfactants for green agricultural technology. *In: Sustainable Agriculture Reviews 60: Microbial Processes in Agriculture*. 389-413. Cham: Springer Nature Switzerland.

Reis, R.S., Pacheco, G.J., Pereira, A.G. & Freire, D.M.G. (2013). Biosurfactants: Production and applications. *Biodegradation - Life of Science*, 2: 31-61.

Rodrigues, J.L. & Rodrigues, L.R. (2023). Synthetic biology approaches for biosurfactants production by lactic acid bacteria. *In: Lactic Acid Bacteria as Cell Factories*, 311-334. Woodhead Publishing.

Rubio-Ribeaux, D., Mata da Costa, R.A., Montero Rodríguez, D., Marques, N.S.A.D.A., Silva, G.M. & da Silva, S.S. (2023). Biosurfactant production by solid-state fermentation in biorefineries. *Biosurfactants and Sustainability: From Biorefineries Production to Versatile Applications*. 95-115.

Ruiz-Gonzalex, M.X. & Vicente, O. (2023). Agrobiodiversity: Conservation, threats, challenges, and strategies for the 21st century. *AgroLife Sci. J.*, 12(1): 174-185.

Saharan, B.S., Sahu, R.K. and Sharma, D. (2011). A review on biosurfactants: Fermentation, current developments and perspectives. *J. Genet. Eng. Biotechnol.*, 2011(1): 1-14.

Saranraj, P., Sivasakthivelan, P., Hamzah, K.J., Hasan, M.S., Sayyed, R.Z. & Tawaha, A.R.M.A. (2022). Microbial fermentation technology for biosurfactants production. *Biosurfactants: Production and Applications in Food and Agriculture*. 2.

Sarubbo, L.A., Maria da Gloria, C.S., Durval, I.J.B., Bezerra, K.G.O., Ribeiro, B.G., Silva, I.A. et al. (2022). Biosurfactants: Production, properties, applications, trends, and general perspectives. *Biochem. Eng. J.*, 181: 108377.

Sarubbo, L.A., Rocha Jr, R.B., Luna, J.M., Rufino, R.D., Santos, V.A. & Banat, I.M. (2015). Some aspects of heavy metals contamination remediation and role of biosurfactants. *Chem. Ecol.*, 31(8): 707-723.

Schonhoff, A., Ihling, N., Schreiber, A. & Zapp, P. (2022). Environmental impacts of biosurfactant production based on substrates from the sugar industry. *ACS Sustain. Chem. Eng.*, 9345-9358.

Singh, V. (2012). Biosurfactant – Isolation, production, purification & significance. *Int. J. Sci. Res. Publ.*, 2(7).

Thavasi, R., Subramanyam Nambaru, V.R.M., Jayalakshmi, S., Balasubramanian, T. & Banat, I.M. (2011). Biosurfactant production by *Pseudomonas aeruginosa* from renewable resources. *Indian J. Microbiol.*, 51: 30-36.

Tiso, T., Ihling, N., Kubicki, S., Biselli, A., Schonhoff, A., Bator, I. et al. (2020). Integration of genetic and process engineering for optimized rhamnolipid production using *Pseudomonas putida. Front. Bioeng. Biotechnol.*, 8: 976.

Valiathan, S. & Radhakrishnan, P. (2023). Applications of biotechnological approaches in the food industry. *Novel and Alternative Methods in Food Processing: Biotechnological, Physicochemical, and Mathematical Approaches*. 177.

Vecino, X., Cruz, J.M., Moldes, A.B. & Rodrigues, L.R. (2017). Biosurfactants in cosmetic formulations: Trends and challenges. *Crit. Rev. Biotechnol.*, 37(7): 911-923.

Velioğlu, Z. & Ürek, R.Ö. (2015). Biosurfactant production by Pleurotus ostreatus in submerged and solid-state fermentation systems. *Turk. J. Biol.*, 39(1): 160-166.

Venkataraman, S., Rajendran, D.S., Kumar, P.S., Vo, D.V.N. & Vaidyanathan, V.K. (2022). Extraction, purification and applications of biosurfactants based on microbial-derived glycolipids and lipopeptides: A review. *Environ. Chem. Lett.*, 1-22.

Vieira, I.M.M., Santos, B.L.P., Ruzene, D.S. & Silva, D.P. (2021). An overview of current research and developments in biosurfactants. *J. Ind. Eng. Chem.*, 100: 1-18.

Yadav, B., Talan, A., Tyagi, R.D. & Drogui, P. 2021. Concomitant production of value-added products with polyhydroxyalkanoate (PHA) synthesis: A review. *Bioresour. Technol.*, 337: 125419.

CHAPTER

4

Biosurfactants from Extremophilic Microorganisms as a Sustainable Alternative

Priyanka Patel[1]*, Shreyas Bhatt[1], Peter Poczai[2], Waleed Hassan Almalki[3], and R.Z. Sayyed[4]

[1] Department of Life Sciences, Hemchandracharya North Gujarat University, Patan, India

[2] Finnish Museum of Natural History, University of Helsinki, FI-00014 Helsinki, Finland

[3] Department of Pharmacology, College of Pharmacy, Umm Al-Qura University, Makkah, Saudi Arabia

[4] Department of Microbiology, PSGVP Mandal's, SI Patil Arts, GB Patel Science and STKV Sangh Commerce College, Shahada, India

1. Introduction

Extremophiles are microorganisms that can survive in extreme environments (Kochhar et al., 2022). Under extreme conditions such temperature and pH fluctuations and high salinity, among others traditional techniques are ineffective. Not only are extremophiles well adapted to survive in these environments, they are cheaper and excellent sustainable alternatives to the traditional techniques. They adapt to these environments with biochemical and physiological changes and produce products like extremolytes, extremozymes and biosurfactants, which are found to be useful in a wide range of industries like sustainable agriculture, food, cosmetics, and pharmaceuticals. These products also play a crucial role in bioremediation, production of biofuels, biorefining, and astrobiology (Kochhar et al., 2022). Extremophiles that can survive in more than one type of extreme environment are called polyextremophiles (Gupta et al., 2014). Extremolytes are small organic molecules, either synthesized or taken up by extremophilic bacteria that accumulate inside the cells. They protect macromolecules and cell structures of extremophiles (Becker and Wittmann, 2020, Kochhar et al., 2022). Extremozymes, the enzymes isolated from extremophiles, are highly stable

* Corresponding author: pu2989@gmail.com

at extreme temperatures and pH and are resistant to denaturing agents (Gupta et al., 2014, Kochhar et al., 2022). Various types of extremophiles are mentioned in Table 1 (Gayathiri et al., 2022, Schultz and Rosado, 2020).

Table 1: Biosurfactants produced by various Extremophiles (Gayathiri et al., 2022, Schultz and Rosado, 2020)

Types of extremophiles	An organism with optimal growth at	Examples	Biosurfactant
Thermophiles	Grow at temperatures above 45 °C.	*Synechococcus lividus*, *Sulfolobus* sp.	Rhamno lipids
Hyperthermophiles	Grow at temperatures above 80 °C.	*Pyrolobus fumarii*	
Psychrophiles (or) Cryophiles	Grow at temperatures of 15 °C or lower.	*Psychrobacter*, *Methanogenium* sp.	Trehalose lipid
Halophiles	Grow at a concentration of dissolved salts of 50 g/L (= 5% m/v) or above.	*Halobacteriaceae*, *Dunaliella salina*, *Halanaerobacter* sp.	Glycolipids
Osmophiles	Grow in environments with a high sugar concentration.	*Saccharomyces rouxii*, *Saccharomyces bailii*	
Acidophiles	Grow at pH levels of 3.0 or below.	*Cyanidium caldarium*, *Ferroplasma* sp., *Picrophilus oshimae*	Lipoprotein
Alkaliphiles	Grow at pH levels of 9.0 or above.	*Natronobacterium*, *Bacillus firmus*, *Spirulina* spp.	Lipopeptides
Piezophiles (or) Barophiles	Grow in hydrostatic pressures above 10 MPa (= 99 atm = 1,450 psi).	*Pyrococcus* sp., *Methanosarcina Kandleri*	
Xerophiles	Grow at water activity below 0.8. They are "xerotolerant", implying tolerant to dry conditions.	*Deinococcus* sp., *lichens*, *Methanosarcina barkeri*	

Biosurfactants are defined as biologically derived microbial surface-active materials (Patel et al., 2022). Biosurfactants are amphiphilic compounds which are produced by plants and organisms. An alternative environment friendly technique of remediation of pollutants involves the role of biosurfactants and biosurfactant-producing microbes. Production of biosurfactants depends on physico-chemical factors, moreover it is directly proportional to the growth of microbes (Patel et al., 2022).

2. Isolation of Extremophilic Biosurfactant Producers

Presently, its a challenge to produce biosurfactants from extremophilic organisms for their use in different biotechnological fields (Schultz and Rosado, 2020). Currently, to overcome the challenges for biosurfactant production by cultivated and uncultivated extremophilic microbial species from extreme environments are shown in Fig. 1 (Schultz and Rosado, 2020).

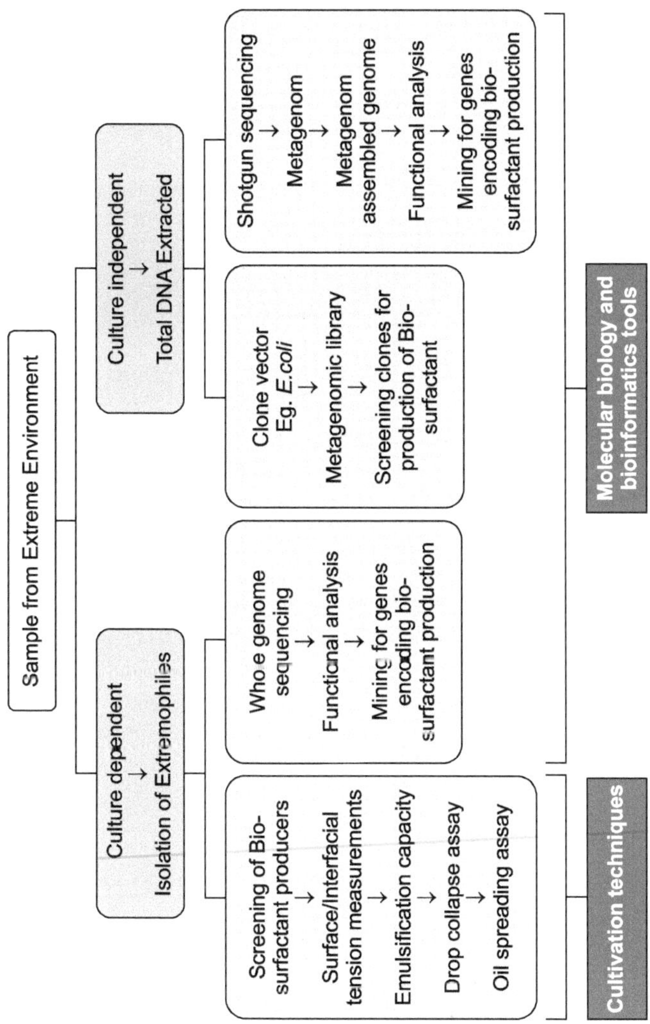

Fig. 1: Method for screening extremophiles through culture dependent and independent approaches (Schultz and Rosado, 2020).

There are two techniques for isolation of biosurfactant producing. culture-dependent and culture-independent extremophilic bacteria. In the culture-dependent method, isolation of extremophiles is done by either screening for biosurfactant producers or whole genome sequencing followed by mining for genes encoding biosurfactant production. In the culture-independent method the total DNA extraction is done directly by either cloning through vector or shotgun sequencing followed by metagenomically assembled genes and mining for genes encoding biosurfactant production Fig. 1 (Schultz and Rosado, 2020). Metagenomics may lead to the discovery of unknown biosurfactants and is especially suitable for extreme environments (Cowan et al., 2014, Jackson et al., 2015). Biosurfactants can be characterized by TLC, FTIR, NMR, etc. Another substitute way is to use genetic engineering techniques, modifying microorganisms to increase the secretion of biosurfactants and minimizing costs, as is performed with mesophiles (Assadi and Tabatabaee, 2010, Bachmann et al., 2014). This option is mostly used for mesophiles to produce enzymes and could be used for extremophiles to enhance biosurfactant production.

3. Biosurfactants from Extremophilic Organisms

3.1 Biosurfactants from Thermophilic Microorganisms

Thermophilic environments include natural environments, viz., hot springs, volcanoes and deserts (LopezVazquez et al., 2014). Most of the biosurfactant production is sourced from mesophiles (Sharaf et al., 2014), however, few studies have been reported for biosurfactant production under thermophilic environments also, viz., microorganisms isolated from oil reservoirs at 60°-160 °C (Tabatabaee et al., 2005). *Bacillus* species have been isolated from deep oil reservoirs in southern Iran, at 65°-110 °C (Daryasafar et al., 2016). *Bacillus licheniformis* isolated from the oil reservoir of SouthWest Iran at 50 °C also showed biosurfactant production. Under optimum conditions, biosurfactants produced by *Bacillus licheniformis* can reduce the surface tension from 72 to 23.8 mN m^{-1} and the tension interface from 36.8 to 0.93 mN m^{-1} (Schultz and Rosado, 2020). Another study reported, the same sp. isolated from the Western Desert of Egypt, showing a maximum emulsification index of 96% and a reduction of surface tension to 36 mN m^{-1} after 72 h at 45 °C (El-Sheshtawy et al., 2015).

3.2 Biosurfactants from Psychrophilic Microorganisms

Microorganisms from cold environments, such as glaciers, alpine regions, snow, deep ocean environments, polar oceans and ice caps (Feller, 2013), are considered as reservoirs of biomolecules viz., biosurfactants, enzymes and antibiotics (Schultz and Rosado, 2020). Biosurfactants from cold environment organisms can interact with water, ice and hydrophobic compounds (Perfumo et al., 2018). *Pseudomonas, Pseudoalteromonas, Halomonas, Marinomonas, Burkholderia, Rhodococcus* and *Bacillus* are psychrotolerants with a capacity to produce biosurfactants in cold environments viz., deep and polar marine sediments (Gesheva et al., 2010, Janek et al., 2013, Konishi et al., 2014). Studies have reported that *Rhodococcus* sp. is predominant followed by *Bacillus* sp. having the capacity to produce biosurfactants at 4 °C, with a highest emulsification index of 62.5% from marine environments of the Canadian Arctic region (Cai et al., 2014).

3.3 Biosurfactants from Halophilic Microorganisms

Biosurfactant producing halophilic microorganisms have been isolated from salterns from Argentina (Nercessian et al., 2015). Studies reported that the archaea viz., *Haloarcula argentinensis, Haloarcula japonica, Haloarcula vallismortis, Halorubrum tebenquichense, Halobacterium salinarum, Halobacterium* sp., *Halobacterium piscisalsi* and the bacteria viz., *Salinibacter ruber Salicola* sp. are capable of producing biosurfactants at high salt concentrations.

3.4 Biosurfactants from Acidophilic/Alkaliphilic Microorganisms

The best known biosurfactant producers are neutrophilic bacteria. There is limited data available on biosurfactant production with optimal functions occurring at extreme pH values (Ivanova et al., 2016, Elazzazy et al., 2015). The maximum stable biosurfactant production occurred in the pH range of 5 to 9, and in the temperature range of 30°-100 °C. The lipopeptidic nature of a biosurfactant was reported at pH values of 4 to 12 and salinity of 1 to 10% of NaCl. Moreover, it has been extensively used in the pharmaceutical, cosmetics and food industries as well as for bioremediation in marine environments. Studies have reported that *Acidithiobacillus thiooxidans* produced 7 to 11 mg/L of biosurfactant in the pH range of 2-3 (Karwowska et al., 2015). Another study involving biosurfactant production by acidophilic bacteria was developed by Arulazhagan et al. (2017), who isolated the species *Stenotrophomonas maltophilia* from a mining site in Saudi Arabia that can produce biosurfactants capable of reducing surface tension and effectively degrading Polycyclic Aromatic Hydrocarbon (PAH) under acidic extreme conditions (pH 2) at 30 °C (Schultz and Rosado, 2020).

4. Properties of Biosurfactants

There are various properties of Biosurfactants as mentioned below (Gayathiri et al., 2022):

- Surface active Biomolecules - Produced by microorganisms with a wide range of applications
- Specificity - Increase the pseudo-solubility
- Temperature and pH tolerance - Biosurfactants and their surface activity are resistant towards environmental factors
- Relative ease of preparation - The producer microbes naturally grow on different substrates
- Hydrophilic and Hydrophobic regions - Increase the bioavailability of hydrophobic substrates through solubilization/desorption
- Decrease interfacial surface tension - Critical Micelle Concentration (CMC) is about several times lower than that of chemical surfactants
- Biodegradability, Biocompatibility and Digestibility - Suitable for environmental applications such as bioremediation/biosorption

5. Food and Agricultural Wastes as Sources for Biosurfactant Production

Agricultural industry waste is considered as a significant carbon source. Sugar beets, banana and citrus fruits, have been studied for their potential to produce biosurfactants

(Panadare and Rathod, 2015). The type of biosurfactant produced is mainly dependent on the substrates utilized (Gayathiri et al., 2022). Glycolipids are extracted from fruit and vegetable by products (Singh et al., 2019). To enhance dough rheology, water retention, ingredient mixing and handling, emulsifiers could be used in baking industries. Biosurfactants are good additives and preservatives in food manufacturing industries (Zaman et al., 2022, Manga et al., 2019). The production of biosurfactants through *Lactobacillus* using sugar beet molasses and glycerol as raw materials resulted in the formation of glycolipids and glycolipopetides as multicomponent mixtures. (Kaur et al., 2015).

6. Factors affecting Biosurfactant Production

There are various factors affecting the production of biosurfactants as mentioned below (Pattanathu et al., 2008).

- Carbon source
- Nitrogen source
- Salt concentration
- Ionic strength
- Aeration and agitation
- Environmental factors
- Fermentation conditions
- Nutrient bioavailability such as carbon and nitrogen sources

7. Applications of Biosurfactants

There are various applications of extremophilic biosurfactants in different fields including Agriculture, Cosmetic industries, Environmental and Bioremediation Industries, Pharmaceuticals and composting (Patel et al., 2022, Saranraj et al., 2022a, 2022b). Some of them are mentioned in Table 2.

7.1 Biosurfactant Used in Agriculture

Extremophiles play an important role in maintaining plant growth and productivity in areas with extreme conditions viz., low temperature, high salinity, drought conditions, etc. as biofertilizers, bioinoculants, biofortifiers and biocontrol agents (Kochhar et al., 2022, Yadav and Saxena 2018). Biosurfactants are used in the bioremediation of soils to improve soil quality, antimicrobial activity, promote plant defense and are alternatives for chemical surfactants (Gayathiri et al., 2022). They decrease water infiltration in arid soils by increasing hydrophilization, which indicate an expansion of sustainable agriculture in desert regions (Ravindara et al., 2022, Bee et al., 2019, Shekhar et al., 2015, Markande et al., 2021). Green technology can be used in agriculture to ensure long term feasibility. Biosurfactants have the potential to increase the nutrients supply to microorganisms which surround plants along with improving soil health. Biosurfactants derived from *Pseudomonas* sp. and *Burkholderia* sp. might be used as safe biopesticides as well as efficient antimicrobial agents against plant pathogens for biocontrol in sustainable agriculture practices. Agronomists utilize biosurfactants to improve antimicrobial activity of microorganisms (Mnif, 2013). Biosurfactants could enhance PGPR activities, soil quality and degrade pesticides.

Table 2: Applications of biosurfactants in various fields (Gayathiri et al., 2022)

Medicine	Bioremediation	Industrial	Pharmaceutical	Food	Agriculture	Cosmetic
Antimicrobial properties	Oil recovery	Textile		Additives	Inhibition of toxicants	Moisturizing factor
Therapeutic values Multi drug resistance Coating agents	Diesel degradation	Uranium extraction		Emulsifier in various foods	Wettability	Skin treatments
Anti-cancer properties	MEOR			Increase the self life	Distribution of fertilizers	Formulations of cosmetic products
					Bioavailability of nutrients	Anti-aging property
					Pesticides	
					Soil reclamation	

Pseudomonas sp. has been shown to exhibit Plant Growth Promoting (PGP) attributes (Gayathiri et al., 2022). Biosurfactants, such as rhamnolipids and lipopeptides produced by *Pseudomonas* sp. H have the potential to act as biocontrol agents against pathogens, biopesticides, fungicides, anti-zoospores, degrade hydrophobic herbicides like 2,4,5-trichlorophenoxyacetic acid, possess hydrocarbon degrading capacity and furthermore, enhance plant tolerance by boosting plant immune systems (Sachdev and Cameotra, 2013, Garcia-Reyes, 2018, Rawat et al., 2020) Moreover, glycolipid biosurfactants produced by *Pseudomonas* sp. could increase methyl parathion and endosulfan solubilization (Nair et al., 2015). Biosurfactants could be used to enhance micronutrients accessibility to growing plants.

7.2 Cosmetic Industries

Sophorolipid, a biosurfactant has been reported as a natural moisturizer which is synthesized by *Torulopsis bombicola*. The sophorolipids, rhamnolipids, and mannosylerythritol lipids have maximum application in cosmetic industries (Kitamoto, 2002). Due to anti-microbial properties the Sophorolipids play a vital role in the treatment of acne, dandruff and body odors. Rhamnolipids also play a crucial role in the production of deodorants, nail care products, toothpaste and as anti-wrinkle agents (Muhammad and Mahsa, 2014). Studies have reported that Mannosylerythritol lipids are used as anti-aging skincare products and lipopeptides are also used as anti-wrinkle agents (Morita et al., 2013, Meena and Kanwar, 2015). Due to non-toxic and non-irritant properties of Biosurfactants as well as their eco-friendly and safe nature they are extensively used by cosmetic industries (Urum et al., 2006).

7.3 Pharmaceuticals

Biosurfactants have been frequently used as antibacterial, antifungal, antiviral, adhesive agents, immunomodulatory molecules and vaccines as well as in gene therapy. The application of cyclic lipopeptide biosurfactants has also been reported in nanoparticle formation (Huang et al., 2018). In the Microemulsions Drug Delivery Systems (MDDS) biosurfactants could be competitively preferred to synthetic surfactants because of their non-toxic and safe properties. Studies have reported that among all the biosurfactants, lipopeptide and glycolipids are highly preferred (Ohadi et al., 2020). During gene transfection, biosurfactant based liposomes have great significance as compared to synthetic liposomes (Kitamoto, 2002). The problems regarding the drug delivery system are effectively solved by the microspheres, nanoparticles, micelles and liposomes (Rienzo et al., 2015, Eslaminejad et al., 2016a, Eslaminejad et al., 2016b, Sharma and Sharma, 2020). The use of biosurfactants along with antibiotic or chemotherapeutic agents can be an available viable approach to combat diseases, biofilm development as well as microbial growth (Gayathiri et al., 2022).

7.4 Environment and Bioremediation

The biosurfactants produced by *Bacillus licheniformis* have been reported to reduce the viscosity of crude oil and also removing crude oil from oil tanks (Banat et al., 2010, Banat et al., 1991). Biosurfactants can also be used in the pulp and paper, textiles and ceramics industries. Moreover, Biosurfactants minimize surface tension between water and hydrocarbon mixtures. Studies have reported that rhamnolipids produced by *Pseudomonas aeruginosa* have high potential to process micelle formation. The rhamnolipids from *Pseudomonas aeruginosa* J4 and surfactin from *Bacillus subtilis* showed high bioremediation potential to biodegrade contaminants in water and soil (Sadiq et al., 2022, Shah et al., 2022, Ali et al., 2022a, 2022b, Gayathiri et al., 2022).

7.5 Industries

Better biosurfactant production could be achieved by *Pseudomonas aeruginosa* when it is grown on nitrate and protease peptone media (Mulligan et al., 1989). The strain *Pseudomonas aeruginosa* J16 has been reported to synthesize better di-rhamnolipid production than mono-rhamnolipid (Wei and Chu, 1998). Moreover, Biosurfactants have been reported to remove heavy metals from soil and sludge (Guan et al., 2017). Studies have shown that the heavy molecular weight biosurfactants have been recognized to possess high emulsifying properties, whereas the low molecular biosurfactants have reduced surface tension properties (Gayathiri et al., 2022). However, the combined effect of the biosurfactants produced by *Acidithiobacillus* sp. and *Meyerozyma quilliermondii* has been reported to have remarkably high bioleaching properties against heavy metals viz., copper, cadmium (Camargo et al., 2018).

7.6 Composting

Rhamnolipids along with tween 80 have been successfully used in composting as they promote microbial growth significantly. Synergistic activity of *Bacillus* sp. and *Streptomyces* sp. has been shown to result in a high breakdown rate of organic materials during the composting process (Gayathiri et al., 2022). Microorganisms along with rhamnolipids may hasten the decomposition of organic matter into humic substances, which result in high humic acid concentration (Zhang et al., 2008). Rhamnolipids produced from organic matter decomposition and the enhanced activity of diazotrophs may increase the overall soil nitrogen status (Molina et al., 2013).

8. Conclusion

The study concluded that Extremophilic microorganisms and their products, such as biosurfactants have been beneficial compared to conventional methods. Moreover, they have massive potential to extend their uses into bioremediation, medicine, pharmaceutical products, cosmetic industries, food industries, agricultural fields, as well as composting. Further, in the future, researchers should focus on genetic modification of biofuel producing extremophilic organisms that can help to overcome the confines of the efficiency of the conventional methods. Biosurfactants from extremophilic microorganisms have garnered attention for future applications due to their safe and environmentally friendly properties.

Conflict of Interest

Authors have been declared that they have no conflict of Interest.

Acknowledgements

Authors are thankful to the Department of Life Sciences, HNGU, Patan.

References

Ali, S.A.M., Sayyed, R.Z., Mir, M.I., Hameeda, B., Khan, Y. Alkhanani, M.F. et al. (2022a). Induction of systemic resistance in maize and antibiofilm activity of surfactin from *Bacillus velezensis* MS20. *Front. Microbiol.*, 13: 879739. https://doi.org/10.3389/fmicb.2022.879739

Ali, S.A.M., Sayyed, R.Z., Reddy, M.S., Enshasy, H.E. & Hameeda, B. (2022b). Delving through quorum sensing and CRISPRi strategies for enhanced surfactin production. *In:* Sayyed, R.Z. (Ed.), *Biosurfactants: Production and Applications in Bioremediation/Reclaimation.* 59-79. CRC Press, Taylor & Francis Group, USA.

Arulazhagan, P., Al-Shekri, K., Huda, Q., Godon, J.J., Basahi, J.M. & Jeyakumar, D. (2017). Biodegradation of polycyclic aromatic hydrocarbons by an acidophilic Stenotrophomonas maltophilia strain AJH1 isolated from a mineral mining site in Saudi Arabia. *Extremoph.*, 21: 163-174.

Assadi, M. & Tabatabaee, M.S. (2010). Biosurfactants and their use in upgrading petroleum vacuum distillation residue: A review. *Int. J. Environ. Res.*, 4: 549-572.

Bachmann, R.T., Johnson, A.C. & Edyvean, R.G.J. (2014). Biotechnology in the petroleum industry: An overview. *Int. Biodeterior. Biodegrad.*, 86: 225-237.

Banat, I.M., Franzetti, A., Gandolfi, I., Bestetti, G., Martinotti, M.G. et al. (2010). Microbial biosurfactants production, applications. *Appl. Microbiol. Biotechnol.*, 87: 427-444.

Banat, I.M., Samarah, N., Murad, M., Horne, R. & Banerjee, S. (1991). Biosurfactant production and use in oil tank clean-up. *World J. Microbiol. Biotechnol.*, 7: 80-88.

Becker, J. & Wittmann, C. (2020). Microbial production of extremolytes – High-value active ingredients for nutrition, health care, and well-being. *Curr. Opin. Biotechnol.*, 65: 118-128.

Bee, H., Khan, M.Y. & Sayyed, R.Z. (2019). Microbial surfactants and their significance in agriculture. *In:* Sayyed Reddy Antonious (Ed.), *PGPR: Prospects for Sustainable Agriculture,* 205-216. Springer-Nature, Singapore.

Cai, Q., Zhang, B., Chen, B., Zhu, Z., Lin, W. & Cao, T. (2014). Screening of biosurfactant producers from petroleum hydrocarbon contaminated sources. *Mar. Pollut. Bull.*, 86: 402-410.

Camargo, F.P., doPrado, P.F., Tonello, P.S., Santos, A.C.A.D. & Duarte, I.C.S. (2018). Bioleaching of toxic metals from sewage sludge by co-inoculation of Acidithiobacillus and the biosurfactant-producing yeast Meyerozyma guilliermondii. *J. Environ. Manag.*, 211: 28-35.

Cowan, D.A., Makhalanyane, T.P., Dennis, P.G. & Hopkins, D.W. (2014). Microbial ecology and biogeochemistry of continental Antarctic soils. *Front. Microbiol.*, 5: 154.

Daryasafar, A., Jamialahmadi, M., Moghaddam, M.B. & Moslemi, B. (2016). Using biosurfactant producing bacteria isolated from an Iranian oil field for application in microbial enhanced oil recovery. *Pet. Sci. Technol.*, 34: 739-746.

Elazzazy, A.M., Abdelmoneim, T.S. & Almaghrabi, A.O. (2015). Isolation and characterization of biosurfactant production under extreme environmental conditions by alkali-halo-thermophilic bacteria from Saudi Arabia. *Saudi. J. Biol. Sci.*, 22: 466-475.

El-Sheshtawy, H.S., Aiad, I., Osman, M.E., Abo-ELnasr, A.A. & Kobisy, A.S. (2015). Production of biosurfactant from Bacillus licheniformis for microbial enhanced oil recovery and inhibition of the growth of sulphate reducing bacteria. *Egypt. J. Pet.*, 24: 155-162.

Eslaminejad, T., Nematollahi-Mahani, S.N. & Ansari, M. (2016). Cationic β-cyclodextrin–Chitosan conjugates as potential carrier for pmcherry-c1 gene delivery. *Mol. Biotechnol.*, 58: 287-298.

Eslaminejad, T., Nematollahi-Mahani, S.N. & Ansari, M. (2016). Synthesis, characterization, and cytotoxicity of the plasmid EGFP-p53 loaded on pullulan–spermine magnetic nanoparticles. *J. Magn. Magn. Mater.*, 402: 34-43.

Schultz, J. & Alexandre, S.R. (2020). Extreme environments: A source of biosurfactants for biotechnological applications. *Extremoph.*, 1-18.

Feller, G. (2013). Psychrophilic enzymes: From folding to function and biotechnology. *Scient.*, 512840: 1-28.

Garcia-Reyes, S., Yanez-Ocampo, G., Wong-Villarreal, A., Rajaretinam, R.K., Thavasimuthu, C. & Patino, R. (2018). Partial characterization of a biosurfactant extracted from Pseudomonas sp. B0406 that enhances the solubility of pesticides. *Environ. Technol.*, 39: 2622-2631.

Gayathiri, E., Prakash, P., Karmegam, N., Varjani, S., Awasthi, M.K. & Ravindran, B. (2022). Biosurfactants: Potential and eco-friendly material for sustainable agriculture and environmental safety—A review. *Agron.*, 12(662): 1-35.

Gesheva, V., Stackebrandt, E. & Vasileva-Tonkova, E. (2010). Biosurfactant production by halotolerant Rhodococcus fascians from Casey Station, Wilkes Land, Antarctica. *Curr. Microbiol.*, 61: 112-117.

Guan, R., Yuan, X., Wu, Z., Wang, H., Jiang, L. Li, Y. & Zeng, G. (2017). Functionality of surfactants in waste-activated sludge treatment: A review. *Sci. Total Environ.*, 609: 1433-1442.

Gupta, G.N., Srivastava, S., Khare, S.K. & Prakash, V. (2014). Extremophiles: An overview of microorganism from extreme environments. *Int. J. Agric. Environ. Biotechnol.*, 7(2): 371-380.

Huang, W., Lang, Y., Hakeem, A., Lei, Y., Gan, L. & Yang, X. (2018). Surfactin-based nanoparticles loaded with doxorubicin to overcome multidrug resistance in cancers. *Int. J. Nanomed.*, 13: 1723-1736.

Ivanova, A.E., Sokolovaa, D.S. & Yu, A. (2016). Hydrocarbon biodegradation and surfactant production by acidophilic mycobacteria. *Microbiol.*, 85: 317-324.

Jackson, S.A., Borchert, E., O'Gara, F. & Dobson, A.D. (2015). Metagenomics for the discovery of novel biosurfactants of environmental interest from marine eco-systems. *Curr. Opin. Biotechnol.*, 33: 176-182.

Janek, T., Lukaszewicz, M. & Krasowska, A. (2013). Identification and characterization of biosurfactants produced by the Arctic bacterium *Pseudomonas putida* BD2. *Colloids Surf. B: Biointerfaces*, 110: 379-386.

Karwowska, W.A., Wojtkowska, M. & Andrzejewska, D. (2015). The influence of metal speciation in combustion waste on the efficiency of Cu, Pb, Zn, Cd, Ni and Cr bioleaching in a mixed culture of sulfur-oxidizing and biosurfactant-producing bacteria. *J. Hazard Mater.*, 299: 35-41.

Kaur, H.P., Prasad, B. & Kaur, S. (2015). A review on applications of biosurfactants produced from unconventional inexpensive wastes in the food and agriculture industry. *World J. Pharm. Res.*, 4: 827-842.

Kitamoto, D. (2002). Functions and potential applications of glycolipid biosurfactants – From energy-saving materials to gene delivery carriers. *J. Biosci. Bioeng.*, 94: 187-201.

Kochhar, N., Kavya, I.K., Shrashti, S., Anshika, G., Rawat, V.S. et al. (2022). Perspectives on the microorganism of extreme environments and their applications. *Curr. Res. Microbial Sci.*, 3(2022): 100134.

Konishi, M., Nishi, S., Fukuoka, T., Kitamoto, D., Watsuji, T.O. et al. (2014). Deep sea *Rhodococcus* sp. BS-15, lacking the phytopathogenic fas genes, produces a novel gluco-triose lipid biosurfactant. *Mar. Biotechnol.*, 6: 484-493.

Lopez-Vazquez, C.M., Kubare, M., Saroj, D.P., Chikamba, C., Schwarz, J. et al. (2014). Thermophilic biological nitrogen removal in industrial wastewater treatment. *Appl. Microbiol. Biotechnol.*, 98: 945-956.

Manga, E.B., Celik, P.A., Cabuk, A. & Banat, I.M. (2021). Biosurfactants: Opportunities for the development of a sustainable future. *Curr. Opin. Colloid Interface Sci.*, 56: 101514.

Markande, A.R., Patel, D. & Varjani, S. (2021). A review on biosurfactants: Properties, applications and current developments. *Bioresour. Technol.*, 330: 124963.

Meena, K.R. & Kanwar, S.S. (2015). Lipopeptides as antifungal and antibacterial agents: Applications in food safety and therapeutics. *Biomed. Res. Int.*, 2015: 473050.

Mnif, I. (2013). Improvement of bread dough quality by *Bacillus subtilis* SPB1 biosurfactant addition: Optimized extraction using response surface methodology. *J. Sci. Food Agric.*, 93(12): 3055-3064.

Molina, M.J., Soriano, M.D., Ingelmo, F. & Llinares, J. (2013). Stabilisation of sewage sludge and vinasse bio-wastes by vermicomposting with rabbit manure using *Eisenia fetida. Bioresour. Technol.*, 137: 88-97.

Morita, T., Fukuoka, T., Imura, T. & Kitamoto, D. (2013). Production of mannosylerythritol lipids and their application in cosmetics. *Appl. Microbiol. Biotechnol.*, 97: 4691-4700.

Muhammad, I.M. & Mahsa, S.S. (2014). Rhamnolipids: Well-characterized glycolipids with potential broad applicability as biosurfactants. *Ind. Biotechnol.*, 10: 285-291.

Mulligan, C.N., Mahmourides, G. & Gibbs, B.F. (1989). The influence of phosphate metabolism on biosurfactant production by *Pseudomonas aeruginosa. J. Biotechnol.*, 12: 199-209.

Nair, A.M., Rebello, S., Rishad, K.S., Asok, A.K. & Jisha, M.S. (2015). Biosurfactant facilitated biodegradation of Quinalphos at high concentrations by *Pseudomonas aeruginosa* Q10. *Soil Sediment Contam. Int. J.*, 24: 542-553.

Nercessian, D., Di Meglio, L., De Castro, R. & Paggi, R. (2015). Exploring the multiple biotechnological potential of halophilic microorganisms isolated from two Argentinean salterns. *Extremoph.*, 19: 1133-1143.

Ohadi, M., Shahravan, A., Dehghannoudeh, N., Eslaminejad, T., Banat, I.M. & Dehghannoudeh, G. (2020). Potential use of microbial surfactant in microemulsion drug delivery system: A systematic review. *Drug Des. Devel. Ther.*, 14: 541-550.

Panadare, D.C. & Rathod, V.K. (2015). Applications of waste cooking oil other than biodiesel: A review. *Iran. J. Chem. Eng.*, 12: 55-76.

Pattanathu, K.S.M. Rahman & Gakpe, E. (2008). Production, characterisation and applications of biosurfactants – Review. *Biotechnol.*, 7: 360-370.

Perfumo, A., Banat, I.M. & Marchant, R. (2018). Going green and cold: Biosurfactants from low-temperature environments to biotechnology applications. *Trends Biotechnol.*, 36: 277-289.

Patel, P., Bhatt, S., Patel, H., Marcelino, L.A. & Sayyed, R.Z. (2022). Biosurfactant – A biomolecule and its potential applications. *In:* R.Z. Sayyed, H.A. El Enshasy (Eds.), *Biosurfactant: Production and Applications in Food and Agriculture.* Vol. II. 133-149. CRC Press, Taylor & Francis Group, USA.

Ravinder, R., Manasa, M., Roopa, D., Bukhari, N.A., Hatamleh, A.A., Khan, M.Y. et al.,(2022). Biosurfactant producing multifarious Streptomyces puniceus RHPR9 of Coscinium fenestratum rhizosphere promotes plant growth in chilli. *Plos One* 17, 3: e0264975. https://doi.org/10.1371/journal.pone.0264975

Rawat, G., Dhasmana, A. & Kumar, V. (2020). Biosurfactants: The next generation biomolecules for diverse applications. *Environ. Sustain.*, 3: 353-369.

Rienzo, M.A.D.D., Banat, I.M., Dolman, B., Winterburn, J. & Martin, P.J. (2015). Sophorolipid biosurfactants: Possible uses as antibacterial and antibiofilm agent. New Biotechnol. 32: 720-726.

Sadiq, M.B., Khan, M.R. & Sayyed, R.Z. (2022). Biosurfactant mediated synthesis and stabilization of nanoparticles. *In:* Sayyed, R.Z. (Eds), *Biosurfactants: Production and Applications in Bioremediation/Reclamation.* 158-168. CRC Press, Taylor & Francis Group, USA.

Shah, I., Hamid, B., Zaman, M., Fatima, S., Farooq, S., Datta, R. et al. (2022). Microbial biosurfactants: An eco-friendly approach for bioremediation of contaminated environments. *In:* Sayyed, R.Z. (Eds), *Biosurfactants: Production and Applications in Bioremediation/ Reclamation.* 197-207. CRC Press, Taylor & Francis Group, USA.

Sachdev, D.P. & Cameotra, S.S. (2013). Biosurfactants in agriculture. *Appl. Microbiol. Biotechnol.*, 97: 1005-1016.

Saranraj, P., Sayyed, R.Z., Sivasakthivelan, P., Hasan, M.S., Al-Tawaha, A.R.M.A. & Amala, K. (2022a). Microbial biosurfactants: Methods of investigation, characterization, current market value and applications. *In:* Sayyed, R.Z. (Eds), *Biosurfactants: Production and*

Applications in Bioremediation/Reclamation. 19-34. CRC Press, Taylor & Francis Group, USA.

Saranraj, P., Sayyed, R.Z., Hamzah, K.J., Asokan, N., Sivasakthivelan, P. & Al-Tawaha, A.R.M.A. (2022b). *In:* Sayyed, R.Z. (Eds), *Biosurfactants: Production and Applications in Bioremediation/Reclamation.* 1-18. CRC Press, Taylor & Francis Group, USA.

Sharaf, H., Abdoli, M., Hajfarajollah, H., Samie, N., Alidoust, L., Abbasi, H. et al. (2014). First report of a lipopeptide biosurfactant from thermophilic bacterium Aneurinibacillus thermoaerophilus MK01 newly isolated from municipal landfll site. *Appl Biochem Biotechnol*, 173: 1236-1249

Sharma, P. & Sharma. N. (2020). Microbial Biosurfactants – An ecofriendly boon to industries for green revolution. *Recent Pat. Biotechnol.*, 14: 169-183.

Shekhar, S., Sundaramanickam, A. & Balasubramanian, T. (2015). Biosurfactant producing microbes and their potential applications: A review. *Crit. Rev. Environ. Sci. Technol.*, 45(14): 1522-1554.

Singh, P., Patil, Y. & Rale, V. (2019). Biosurfactant production: Emerging trends and promising strategies. *J. Appl. Microbiol.*, 126: 2-13.

Tabatabaee, A., Mazaheri-Assadi, M., Noohi, A.A. & Sajadian, V.A. (2005). Isolation of biosurfactants producing bacteria from oil reservoirs. *Iran J. Environ. Health Sci. Eng.*, 2: 6-12.

Urum, K., Grigson, S., Pekdemir, T. & McMenamy, S. (2006). A comparison of the efficiency of different surfactants for removal of crude oil from contaminated soils. *Chemosph.*, 62: 1403-1410.

Wei, Y.H. & Chu, I.M. (1998). Enhancement of surfactin production in iron-enriched media by Bacillus subtilis ATCC 21332. *Enzym. Microb. Technol.*, 22: 724-728.

Yadav, A.N. & Saxena, A.K. (2018). Biodiversity and biotechnological applications of halophilic microbes for sustainable agriculture. *J. Appl. Biol. Biotechnol.*, 6(1): 48-55.

Zhang, Q., Weimin, C.A.I. & Juan, W. (2008). Stimulatory effects of biosurfactant produced by *Pseudomonas aeruginosa* BSZ-07 on rice straw decomposing. *J. Environ. Sci.*, 20: 975-980.

Zaman, M., Hassan, S., Fatima, S., Hamid, B., Farooq, S., Qayoom, I. et al. (2022). Biosurfactants production and applications in food. *In:* Sayyed, R.Z. and Enshasy, H.E. (Eds), *Biosurfactants: Production and Applications In Food and Agriculture.* Vol II. 225-241. CRC Press, Taylor & Francis Group, USA.

CHAPTER

5

Biosurfactants and Bioemulsifiers: Classification

A. García Romero[1], and E. Dantán-González[2*]

[1] Laboratorio de Biotecnología Ambiental, Centro de Investigación en Biotecnología, Universidad Autónoma del Estado de Morelos. Av. Universidad 1001, Colonia Chamilpa, Cuernavaca, Morelos, México

[2] Laboratorio de Estudios Ecogenómicos, Centro de Investigación en Biotecnología, Universidad Autónoma del Estado de Morelos, Cuernavaca, Morelos, Mexico

1. Introduction

Biosurfactants represent a structurally diverse group of secondary metabolites produced mainly by microorganisms. They possess significant surface and interfacial activity resulting from the presence of both hydrophobic and hydrophilic moieties within a single molecule, leading to a decrease in both surface and interfacial tension. The hydrophilic group can include mono-, oligo-, or polysaccharides, peptides, or proteins. They are classified based on the nature of the head group in solution, which can be anionic, cationic, amphoteric (zwitterionic), or non-ionic. This classification is based on their charge and polarity (Pan et al., 2020, Atta et al., 2020). On the other hand, the hydrophobic group usually includes saturated, unsaturated, and hydroxylated fatty acids or fatty alcohols (Lang, 2002). Biosurfactants, with their amphiphilic nature, can effectively increase the surface area of hydrophobic water-insoluble substances, enhancing their bioavailability. Furthermore, these compounds modify the properties of bacterial cell surfaces, allowing them to be utilized in numerous processes such as emulsification, foaming, detergency, welting, dispersing, or solubilizing (Desai and Banat, 1997, Banat et al., 2010, Mulligan and Gibbs et al., 2004).

Biosurfactants are classified based on their biochemical composition as either low molecular weight or high molecular weight (Campos et al., 2013, Jahan et al., 2020). Low molecular weight biosurfactants, such as lipopeptides, glycolipids, and phospholipids, effectively lower surface, and interfacial tension, while high molecular weight biosurfactants, including polymeric and particulate biosurfactants, act as emulsion stabilizers (Santos et al., 2016, Varjani and Upasani et al., 2017, Henkel and

* Corresponding author: edantan@uaem.mx

Hausmann, 2019). They can be categorized by their chemical composition into several groups including glycolipids (such as rhamnolipids, sophorolipids, trehalolipids, and mannosylitrol lipids), lipopeptides (like surfactin, lichenin, iturin, fengycin, and serrwettin), and fatty acids/phospholipids/neutral lipids (Antonioli-Júnior et al., 2022). Examples of polymeric biosurfactants consist of emulsan, alasan, biodispesan, and liposan, while particulate biosurfactants include vesicles and whole cells.

2. Low Molecular Weight Surfactant Biomolecules

Low molecular weight biosurfactants reduce interfacial and surface tension (Fenibo et al., 2019). Physiologically, these substances are known to support microbial mobility in a complex environment. They reduce the interfacial tension between the cell and the external environment, thereby facilitating the growth, reproduction, and colonization of microbial communities. Furthermore, the complex carbohydrate-carbohydrate interactions within the sugar head groups confer unique properties to these glycol conjugates. This maintains the hydrophilic-hydrophobic balance and allows these molecules to achieve diverse self-assembly patterns. These compounds include glycolipids, lipopeptides and phospholipids (Hommel and Ratledge, 1993).

2.1 Glycolipids

Glycolipids are an interesting and diverse group of biosurfactants. They contain carbohydrates linked to long chain aliphatic acids or hydroxyl fatty acids; the hydrophobic portion is either long chain aliphatic or hydroxyl aliphatic acids attached to the mono-, di-, tri- and tetra-saccharide hydrophilic sugar through an ether or ester linkage. They are found in the cell membranes of bacteria, fungi, plants, and animals in the form of sterylglycosides, glycosylceramides, and diacylglycerol glycosides (Kitamato et al., 2002). In addition, these molecules reduce the interfacial tension between the cell and the external environment thereby facilitating the promotion of growth, reproduction and colonization of microbial communities and can dissolve contaminating hydrocarbons (tetradecane, aliphatic and aromatic) during fermentation and make it available as a substrate for microbes (Inés and Dhouha, 2015). This family can be further subdivided into rhamnolipids, sophorolipids, trehalolipids, and mannosylerythritol lipids. Several microbial glycolipids are used as energy sources and store extracellular carbon (Bauer et al., 2006, Warnecke and Heinz, 2010, Dhasayan et al., 2014, Parthipan et al., 2018). Moreover, the complex carbohydrate-carbohydrate interactions within the sugar head groups give these glycol conjugates unique properties. This maintains the hydrophilic-hydrophobic balance and allows these molecules to achieve diverse self-assembly patterns.

2.1.1 Rhamnolipids

Rhamnolipids are a group of glycolipids that have a hydrophilic head composed of one or two rhamnose molecules, known as mono-rhamnolipids (monoRL) and di-rhamnolipids (diRL), respectively, while the hydrophobic tail is composed of one or two fatty acid chains of varying length and complexity. They are considered anionic surfactants because these compounds are retained less by soil than nonionic or neutral surfactants due to the negative surface charge of the soil (Bordas et al., 2007). *Pseudomonas* species are dominant in the production of rhamnolipids containing

L-rhamnose and 3-hydroxy fatty acids (Cameotra and Singh 2009). These compounds can effectively reduce the surface tension of water from 72 mN/m to 27 mN/m with a varying CMC of 5-200 mg/L. Under acidic conditions, the extracellularly released rhamnolipid aggregates protect the bacterial cell membrane by adapting vesicle or lamellar assembly, whereas in neutral or alkaline environments, these surfactant compounds perform intermediate cell-associated physiological functions (Abdel-Mawgoud et al., 2010).

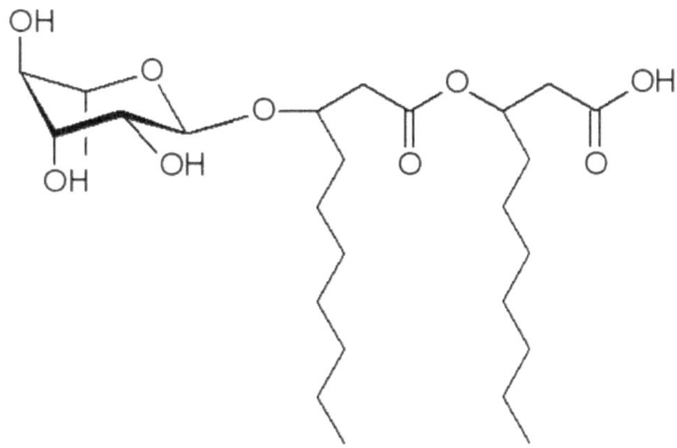

Fig. 1: Chemical structure of mono-rhamnolipid or rhamnolipid 1.

2.1.2 *Sophorolipids*

Sophorolipids are dimeric carbohydrates of sophorose linked by a β-glycosidic linkage to a long hydroxy fatty acid chain. These molecules exist in either an acidic form, in which the fatty acid tail is free, or a lactonic form, in which the carboxyl group of the fatty acid chain is linked to the hydroxyl group of the sophorose sugar by an intramolecular ester linkage. The acidic form is known to play a role in improving solubility and foamability, while the lactonic form significantly influences antimicrobial activity (Konishi et al., 2008). At least six to nine different hydrophobic sophorolipids are found in a mixture (Hommel and Ratledge, 1993, Lo and Ju, 2009). Sophorolipids are not effective as emulsifiers, although they can lower surface and interfacial tension (Cavalero and Cooper, 2003, Daverey and Pakshirajan, 2009).

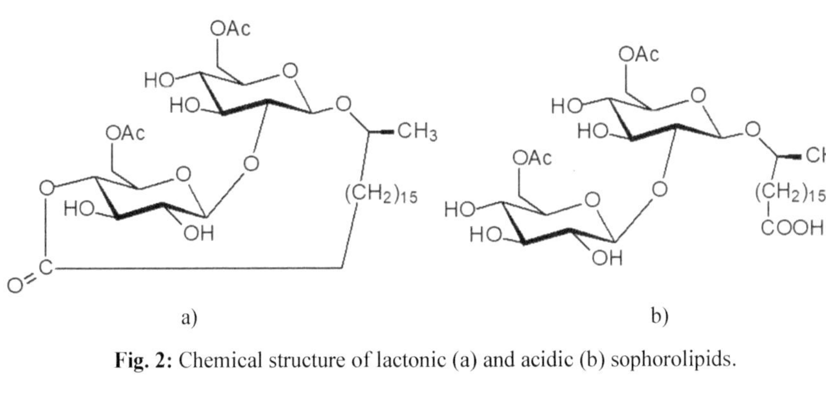

Fig. 2: Chemical structure of lactonic (a) and acidic (b) sophorolipids.

2.1.3 Trehalolipids

Trehalolipids (TLs), or trehalose lipids, represent a significant proportion of glycolipids containing the disaccharide trehalose attached to the mycolic acid (α-branched β-hydroxy fatty acid chains) at the C-6 and C-69 positions (Philp et al., 2002). They are one of the best known biosurfactants. However, they differ from rhamnolipids and sophorolipids in both composition and activity. They are active in reducing water surface and interfacial tension. Mostly gram-positive bacteria with high GC content such as *Nocardia*, *Mycobacteria* and *Corynebacteria* species are considered to be the main producers of trehalose biosurfactants. These molecules exhibit varying degrees of saturation, along with different shapes and sizes, and express promising physiological activities under extreme conditions. The trehalolipids produced by *Rhodococcus* form emulsions and are known to be stable under extreme conditions, including temperature (20-100 °C), pH (2-10), and salinity (5-25% w/v) (Zaragoza et al., 2009). The two molecular forms of TLs, known as trehalose monocorynomycolate (TL-1) and dicorynomycolate (TL-2), reduce the surface tension of water from 72 to 32 and 36 mN/m and the interfacial tension of the water/n-hexadecane system from 43 to 14 and 17 mN/m, respectively. The physiological role of trehalolipids is elucidated in terms of their bioremediation potentials, as these molecules increase the bioavailability of contaminants. Moreover, the significant antimicrobial activity against gram-positive bacteria and some pathogenic fungal species are also some of the key attributes of TLs (Hunter et al., 2006, Peng et al., 2007, Ortiz et al., 2009).

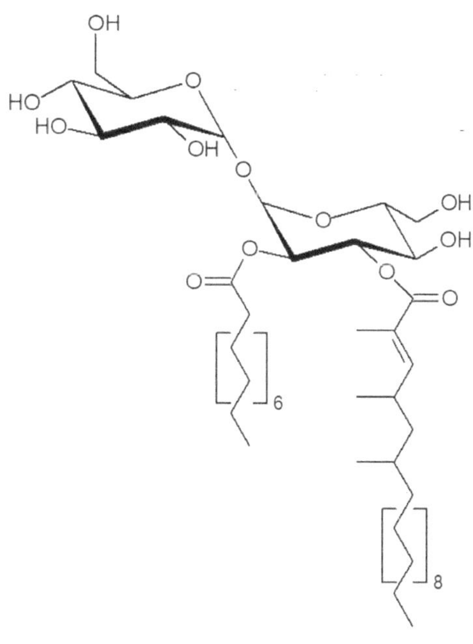

Fig. 3: Chemical structure of Trehalose lipids from *M. tuberculosis*.

2.2 Lipopeptides and Lipoproteins

While there is a relationship between peptides and proteins, similarly there is one between lipopeptides and lipoproteins. We will explain the characteristics of each of these relationships. The main difference between a peptide and a protein is their size and structure. Both are polymers formed by linking amino acids, however, they differ in chain lengths and structural complexities. Lipopeptides are compounds consisting of a lipid and peptide combination. Because of their surfactant properties, these materials can reduce the surface tension that exists between two phases, such as water and oil. Lipopeptides have applications in molecular biology, pharmacology, and nanotechnology, among others (Hamley, 2021). Lipopeptides may be essential to the structure and functioning of cell membranes in the biological realm. Certain lipopeptides have occasionally been shown to have antimicrobial properties, meaning that they have the ability to prevent the growth of microorganisms such as bacteria and fungi. Because of these antimicrobial properties, lipopeptides are of interest in the development of novel antibacterial drugs (Koutsopoulos et al., 2012). In the field of nanotechnology, lipopeptides are being further explored as elements of drug delivery systems and as emulsifiers in the construction of nanomaterials. The composition of lipopeptides can vary, but in general they usually have a combination of lipid and amino acid residues. The characterization of lipopeptides involves the study of these elements and the identification of specific chemical bonds in the molecule. Here are some key points related to the composition of lipopeptides: Peptide bonds: Lipopeptides contain peptide bonds between amino acid residues. These bonds are covalent and are formed by condensation between the amino group of one amino acid and the amino group of another amino acid.

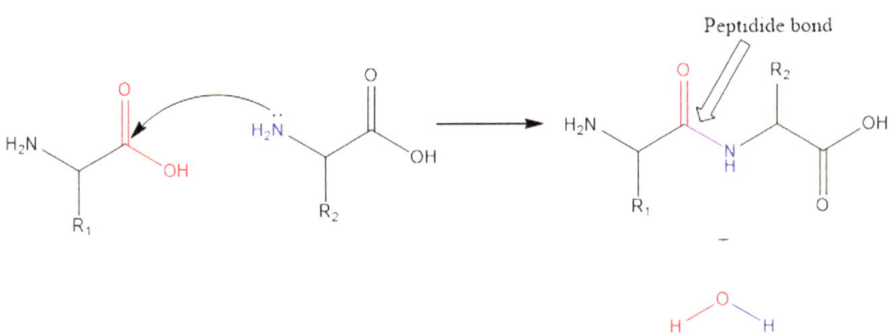

Fig. 4: Chemical reaction between two amino acids to generate a peptide bond.

Ester and amide bonds: Lipid residues are usually linked to the peptide quantity through ester or amide bonds, depending on the type of lipid and how it is linked to the peptide composition. As an example, some lipopeptides hold lipids together through ester bonds, while others can have amide bonds.

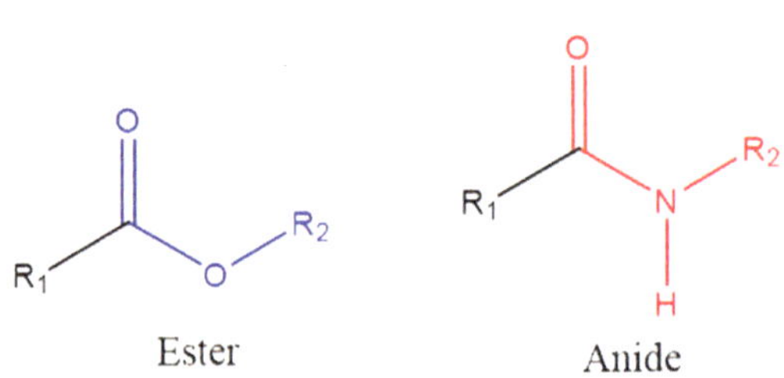

Fig. 5: Two functional groups that can be obtained through amino acid reactions.

Cyclic constructions: Certain lipopeptides have cyclic constructions, where the peptide chain forms a closed ring. This can significantly impair its features and functionality.

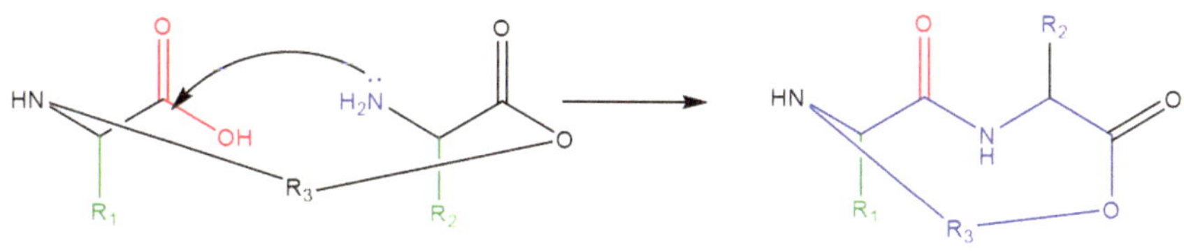

Fig. 6: Intramolecular cyclization reaction of amino acids for the formation of cyclic structures.

Disulfide bonds: Sometimes, lipopeptides may contain disulfide bonds between cysteine residues in the peptide chain. These bonds contribute to the three-dimensional balance of the molecule.

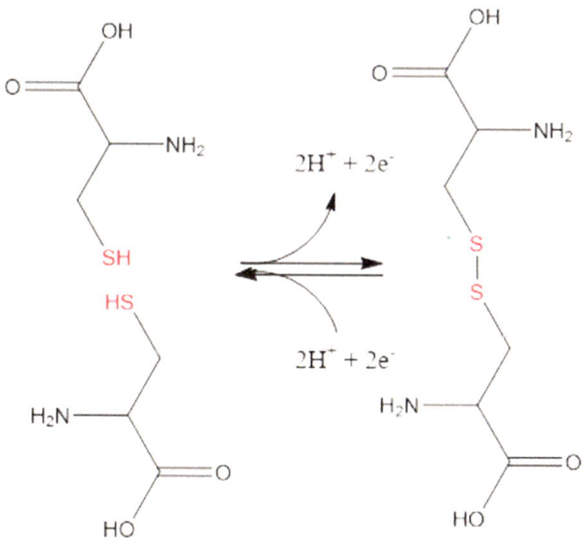

Fig. 7: Formation of disulfide bridges in two cysteines.

Each specific lipopeptide may have a unique composition, and the strict identification of its structural elements and chemical bonds is carried out by a combination of these analytical techniques. There are several examples of lipopeptides that have been studied in various environments. Some notable examples are Cyclic polypeptides: One example is colistin, an antibiotic used to treat bacterial infections. Colistin has a repetitive lipopeptide composition and has been of clinical interest because of its antimicrobial activity, especially against certain drug-resistant bacteria.

Daptomycin: This is another example of a lipopeptide used as an antibiotic. Daptomycin is used to treat infections caused by gram-positive bacteria, and its mechanism of action involves binding to the bacterial cell membrane.

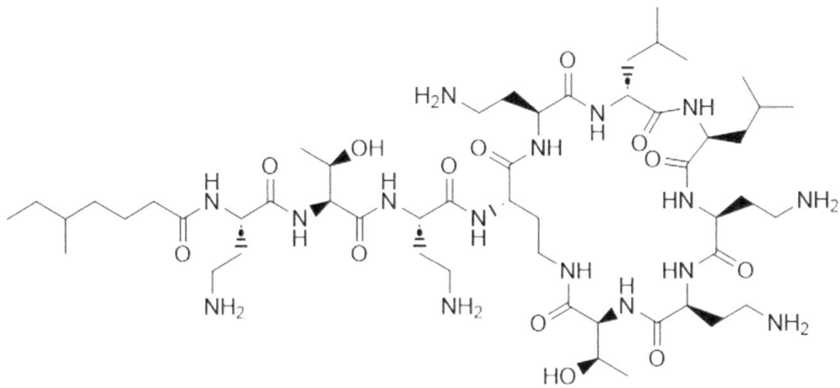

Fig. 8: Colistin structure.

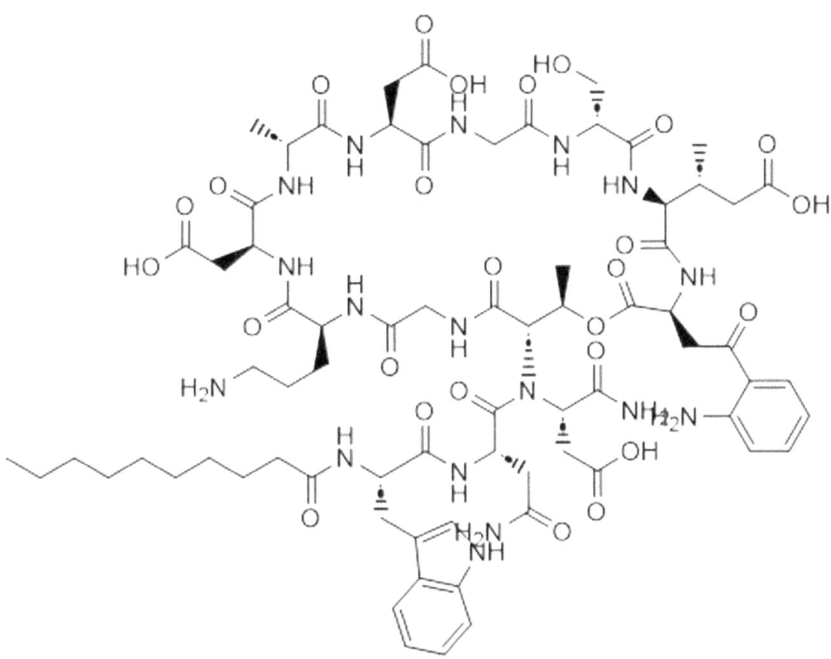

Fig. 9: Daptomycin structure.

Iturin: It is another lipopeptide produced by certain *Bacillus* species. Iturin also has antifungal properties and has been studied for its potential in the biological control of plant pathologies caused by fungi.

These are just a few examples, and there are many other lipopeptides with different functionalities and properties. It should be noted that research on lipopeptides continues, and new compounds and applications are being discovered. Bacteria of the genus *Pseudomonas* are ubiquitous in nature. Pseudomonas present a fascinating

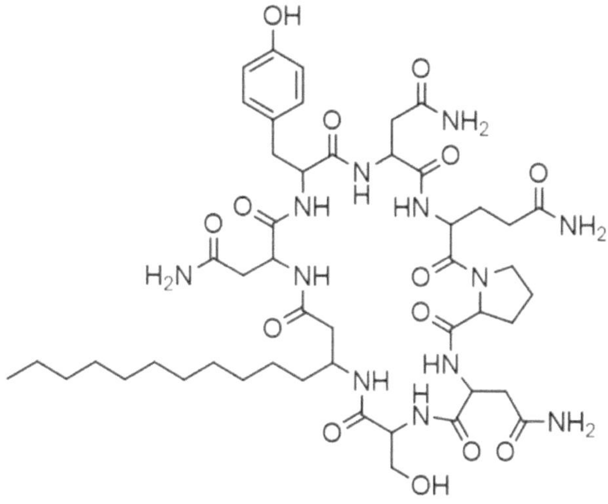

Fig. 10: Iturin D structure.

metabolic diversity that correlates with their ability to colonize a radically wide range of ecological niches. As a result, these bacteria are a prolific source of natural products. The biosynthesis of the latter is constantly orchestrated by sets of chemical signals resulting from intraspecific or interspecific communication. (Götze and Stallforth, 2020). In particular, non-ribosomal lipopeptides with different biological functions play relevant roles in the lifestyle of Pseudomonads. In this review, we will focus on the molecular constructs, characteristics, biosynthetic pathways, and biological processes, the importance of lipoproteins in *Streptococcus pneumoniae* and *Staphylococcus aureus*. Lipoproteins are a family of surface-exposed proteins in bacteria involved in diverse but critical cellular functions ranging from fitness to virulence (Bartual et al., 2018).

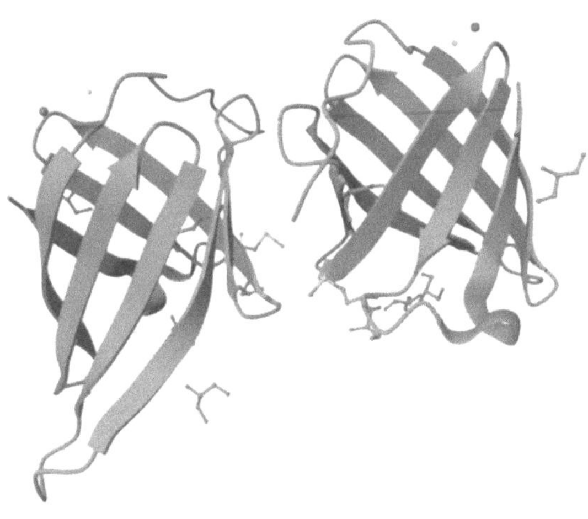

Fig. 11: Crystal structure of putative lipoprotein SP_0198 from *Streptococcus pneumoniae* (3GE2).

Functionalities of *Pseudomonas* lipopeptides. This review is not only aimed at biochemists, but also serves as a complete guide for all scientists (micro/biologists, ecologists, and environmental scientists) working in this multidisciplinary field.

2.3 Fatty Acids (Fatty Acid Ester and Sugar Fatty Acid Esters)

Fatty acids are a diverse and essential group of molecules that play a critical role in the functioning of living organisms. These molecules serve as energy stores, structural elements, and important participants in various biological processes. Fatty acids are organic molecules consisting of a long hydrocarbon chain with a carboxyl group (–COOH) at one end.

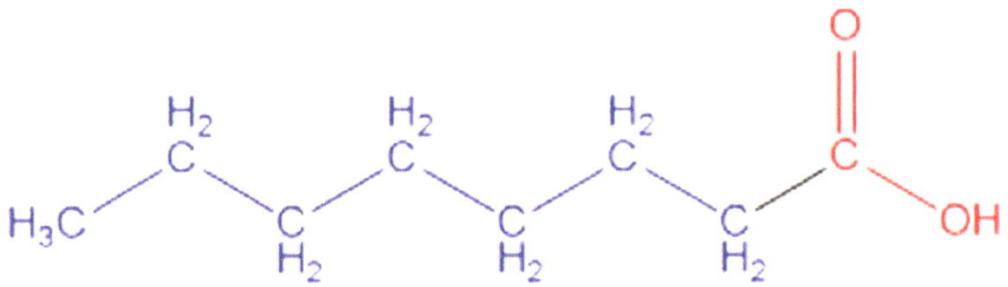

Fig. 12: An example of a saturated fatty acid is octanoic acid. The coloring of the aliphatic chains is blue, indicating the nonpolar part of the molecule, while the section in red represents the polar functional group due to the presence of the carboxylic acid.

These hydrocarbon chains can vary in length and can have single (C-C) and double (C=C) bonds between the carbon atoms in the chain and can be divided into two main groups: saturated and unsaturated.

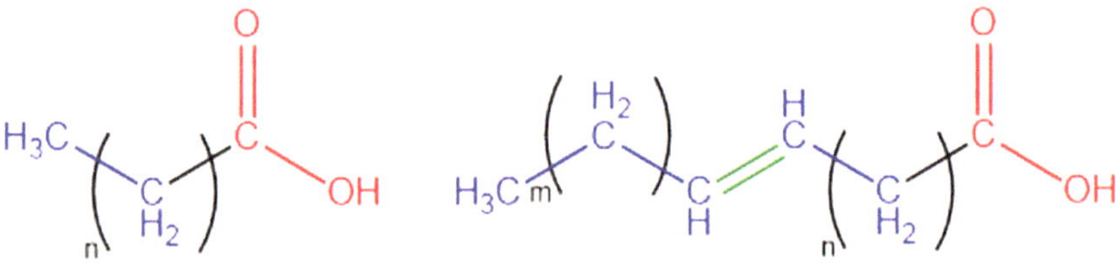

Fig. 13: Examples of fatty acids where the aliphatic chains are simplified with parentheses with subscripts using letters n and m, where n > 2 and m > 1, a) saturated, and b) unsaturated.

Saturated fatty acids have no double bonds between the carbon atoms in their hydrocarbon chains, making them relatively straight and rigid. In contrast, unsaturated fatty acids have one or more double bonds that cause kinks in their chains. The length of the chain of carbon atoms in a fatty acid can vary considerably. Short-chain fatty acids have fewer than 6 carbon atoms, medium-chain fatty acids have between 6 and 12, and long-chain fatty acids have more than 12. Chain length also affects physical and biological properties (Nelson and Cox, 2017).

Fatty acids are named according to the number of carbon atoms in their chain and the location of the double bonds. For example, oleic acid is an unsaturated fatty acid with 18 carbon atoms and a double bond at the ninth carbon.

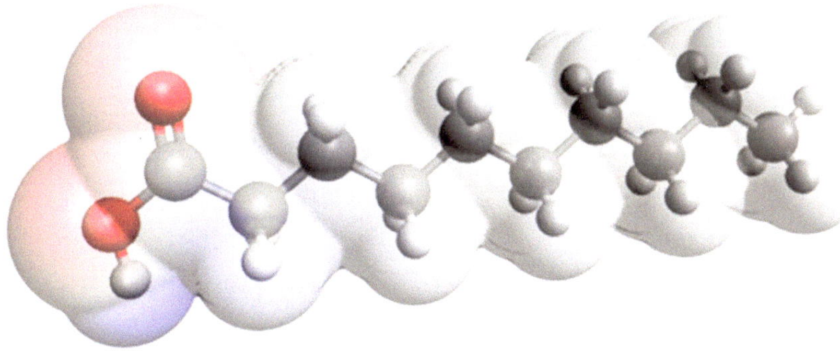

Fig. 14: Electron density map of the decanoic acid molecule, where red coloring indicates the presence of electron density (δ^-), while blue coloring represents absence (δ^+). Meanwhile the absence of color signifies neutrality in terms of electron density.

Table 1: Some saturated fatty acids and their nomenclature

IUPAC-preferred name	Structure
Butyric acid	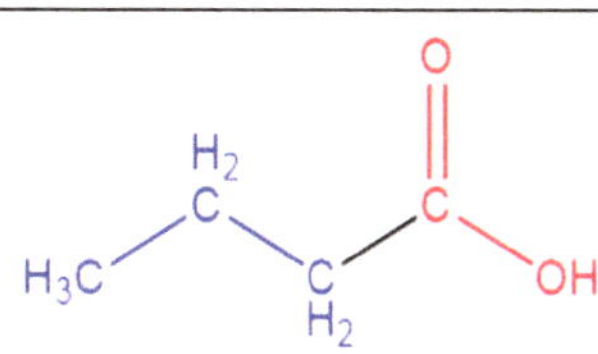
Heptanoic acid	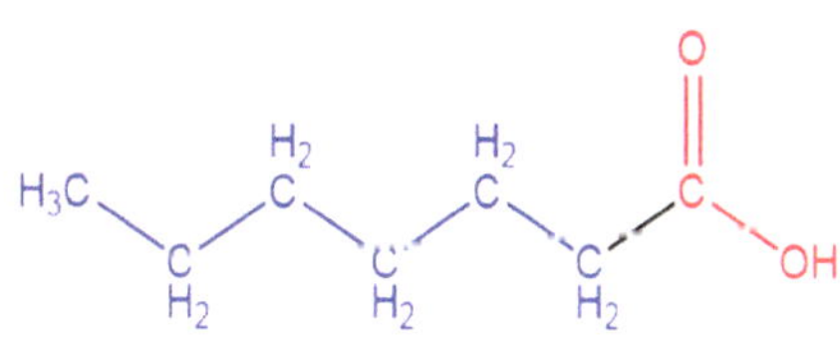
Decanoic acid	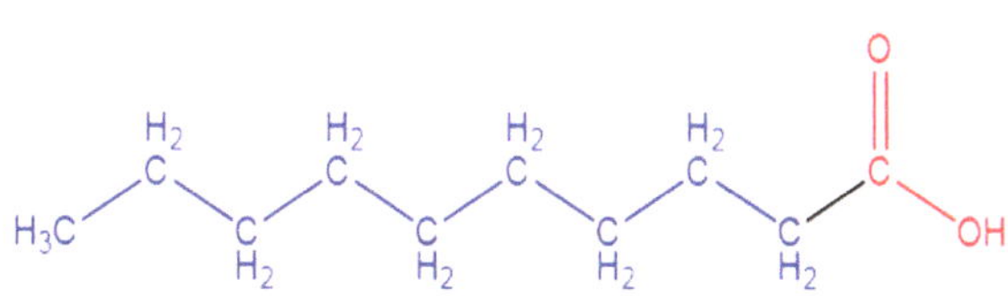

Saturation refers to the presence of single bonds between carbon atoms in the fatty acid chain, while unsaturation refers to the presence of double bonds. The number and location of double bonds in an unsaturated chain can affect its physical and chemical properties (Bender et al., 2013). In unsaturated fatty acids, the spatial configuration of the double bonds can be cis or trans. This can affect their biological behavior and physical properties.

Saturated fatty acids have carbon chains without double bonds between carbon atoms, making them straight and compact. This allows saturated fatty acid molecules to pack tightly into triglycerides, increasing their density and melting point. As a result, triglycerides with saturated fatty acids are typically solid at room temperature and tend to form crystalline structures. Meanwhile, unsaturated fatty acids have one or more double bonds in their carbon chains, introducing a curve or twist into the chain.

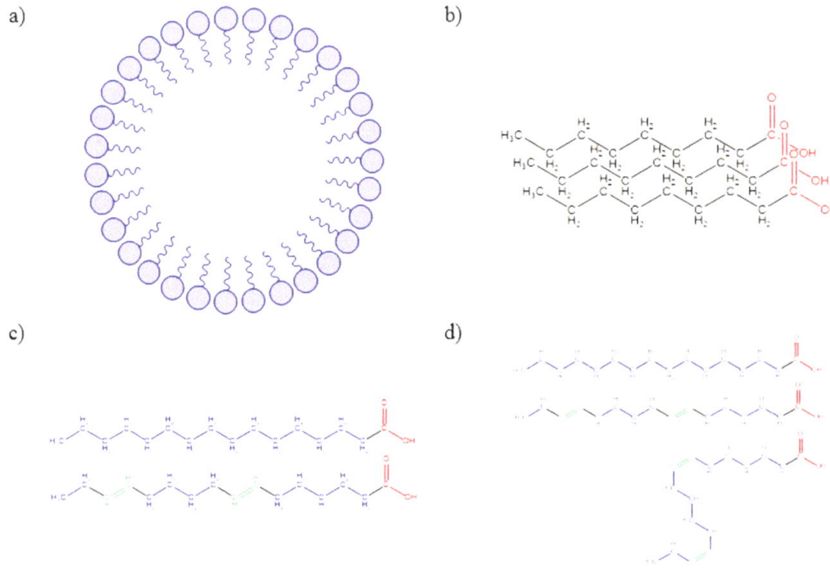

Fig. 15: Schemes of possible intermolecular interactions in fatty acids according to their physicochemical properties: a) micelle, b) interaction between saturated fatty acids, c) interaction between saturated and trans-unsaturated fatty acids (E configuration), d) interaction between saturated, trans-unsaturated (E configuration), and cis-unsaturated (Z configuration) fatty acids.

This hinders the packing of unsaturated fatty acid molecules into triglycerides. As a result, triglycerides with unsaturated fatty acids are more fluid and tend to be liquid at room temperature. In addition, the double bonds in unsaturated fatty acids can lead to weaker interactions, such as London dispersion forces, which are not as strong as the van der Waals interactions in saturated fatty acids. This can affect the stability and shelf life of unsaturated triglycerides. The intermolecular interactions of triglycerides with saturated fatty acids are characterized by compact packing and solid structures at room temperature due to strong van der Waals forces. Conversely, triglycerides with unsaturated fatty acids tend to be more fluid and liquid due to packing difficulties and weaker interactions between unsaturated fatty acid molecules. These differences in interactions affect the physical and chemical properties of triglycerides and have implications for their use in the food industry and on human health (Devlin, 2015).

2.3.1 Fatty Acid Esters

Fatty acid esters are chemical compounds formed by the esterification of fatty acids with an alcohol. They are commonly found in nature as components of lipids and triglycerides, which serve as energy storage molecules in organisms. Fatty acid esters have a wide range of applications, including their use as emulsifiers, solvents, and surfactants in various industries. These compounds play a critical role in the food industry, where they are used to improve the texture and stability of products such as margarine, chocolate, and salad dressings.

Formation: Fatty acid esters are formed by a chemical reaction known as esterification, in which a fatty acid reacts with an alcohol (or other compound containing a hydroxyl group) in the presence of an acid catalyst. This reaction results in the removal of water molecules, forming the ester bond.

Structure: Fatty acid esters consist of a fatty acid molecule and an alcohol molecule linked by an ester linkage (–COO–). The length and saturation of the fatty acid chain and the type of alcohol used can vary, resulting in a wide range of fatty acid esters with different properties.

Table 2: Examples of some fatty acid esters where the black section belongs to the alcohols

IUPAC-preferred name	Structure
Methyl butyrate	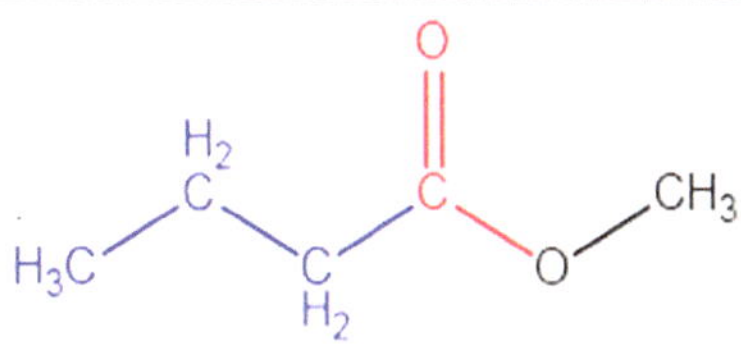
Methyl heptanoate	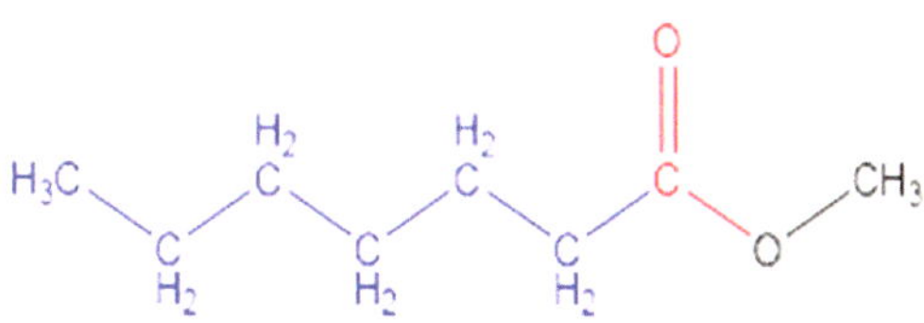
Ethyl decanoate	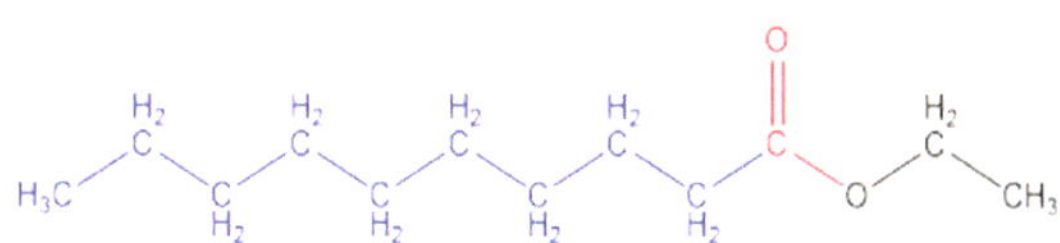

Natural occurrence: Fatty acid esters are abundant in nature. They are essential components of lipids, including triglycerides (fats and oils), phospholipids (important in cell membranes), and waxes (found in plants and animals). These natural esters serve various biological functions, such as energy storage and structural support.

Industrial applications: Fatty acid esters have many industrial applications. They are commonly used as emulsifiers, which help mix substances that would normally separate, such as oil and water. Emulsifiers are essential in the production of foods such as mayonnaise, salad dressings, and chocolate, as well as in cosmetics and pharmaceuticals.

Surfactant properties: Fatty acid esters also act as surfactants, which means they can lower the surface tension of liquids and improve the wetting and spreading properties of substances. This makes them valuable in detergents, cleaning products and personal care products such as shampoos and soaps.

Biodegradability: Fatty acid esters are often preferred in various industries because they are biodegradable and environmentally friendly. They can be used in environmentally friendly formulations and products, contributing to sustainability efforts.

Health considerations: Some fatty acid esters are used as food additives and their safety for consumption has been rigorously evaluated. However, it is important to consider the source and quality of these esters in processed foods, as well as potential health concerns associated with their consumption in excessive amounts.

Synthesis: Fatty acid esters can be synthesized in the laboratory or obtained from natural sources, depending on their intended use. They can be derived from vegetable oils, animal fats or synthesized by chemical processes. Fatty acid esters are versatile compounds with a wide range of applications in the food, cosmetic, pharmaceutical and chemical industries. Their natural occurrence, ability to act as emulsifiers and surfactants, biodegradability and potential health benefits contribute to their widespread use in various products and processes. Fatty acid esters are ester compounds formed from fatty acids and alcohols with a wide range of industrial applications. Sugar fatty acid esters, a specific subset of these compounds, are used in the food and cosmetic industries for their emulsifying and stabilizing properties, particularly in natural and environmentally friendly product formulations.

2.3.2 Sugar Fatty Acid Esters

Sugar fatty acid esters, also known as sugar esters, a category of ester compounds formed by the reaction of fatty acids with sugar molecules such as glucose or sucrose. They are widely used in the food and cosmetic industries as emulsifiers and stabilizers. Sugar fatty acid esters are known for their biocompatibility and environmentally friendly properties, making them a preferred choice for formulating natural and eco-friendly products. They are often used to create stable emulsions in food products, to improve the texture of baked goods, and to improve the consistency of creams and lotions in cosmetics.

Formation: Sugar fatty acids, or sugar esters, are ester compounds formed by the reaction of fatty acids with sugar molecules. This reaction involves esterification, where the hydroxyl (OH) group of the sugar molecule reacts with the carboxyl (COOH) group of the fatty acid, resulting in the formation of an ester bond (–COO–) between them.

Types of sugars: Common sugars used in the synthesis of sugar fatty acids include glucose, sucrose, and maltose. The specific sugar used can affect the properties and applications of the resulting sugar ester.

Emulsification: Sugar fatty acids are well known for their emulsifying properties. They can help stabilize and maintain the even distribution of immiscible substances, such as oil and water, in various products. This makes them valuable in the food industry, particularly for creating stable emulsions in products such as salad dressings and mayonnaise.

Natural origin: Sugar Fatty Acids can be derived from natural sources such as sugar beet or sugar cane, making them a preferred choice for natural and organic product formulations. They are considered biocompatible and environmentally friendly.

Food industry: In the food industry, sugar fatty acids are used as food additives, mainly as emulsifiers and stabilizers. They can improve the texture and consistency of food products while extending their shelf life. In addition, they are often used in baking to improve dough properties and enhance the texture of baked goods (Kurtzman et al., 2010).

Cosmetic industry: Sugar esters are also used in the cosmetics and personal care industry. They can be found in various skin and hair care products, including creams, lotions, and shampoos, where they act as emulsifiers and contribute to the overall stability and texture of the product.

Biodegradability: Like other esters, sugar fatty acids are biodegradable, which is in line with environmental sustainability goals. Their biocompatibility and minimal environmental impact make them attractive for environmentally friendly product formulations.

Regulation: Sugar esters used in food and cosmetics are typically subject to regulatory oversight to ensure their safety for consumption or topical application. Regulatory agencies set guidelines and allowable levels for their use in various products. Sugar fatty acids, or sugar esters, are ester compounds formed by the combination of fatty acids and sugar molecules. They are valued for their emulsifying properties, natural origin, biodegradability, and versatility, making them essential components in the food, cosmetic and personal care industries. Their ability to stabilize and improve the properties of various products has contributed to their widespread use in formulations.

2.3.3 Phospholipids

Phospholipids are versatile molecules with a wide range of applications in various industries, including food, pharmaceuticals, cosmetics, biotechnology and more. Their ability to form the structural basis of cell membranes, to act as emulsifiers, and serve as delivery systems for bioactive compounds make them indispensable in modern industry and research.

Cell membrane structure: Phospholipids play a fundamental role in biology by forming the lipid bilayer structure of cell membranes. This bilayer consists of two layers of phospholipid molecules with hydrophilic (water-attracting) heads facing outward and hydrophobic (water-repelling) tails facing inward. This arrangement provides a semipermeable barrier that regulates the passage of molecules in and out of cells.

Emulsifiers and stabilizers: Phospholipids are excellent emulsifiers and stabilizers in the food and beverage industry. They can help mix and stabilize immiscible substances such as oil and water in products such as salad dressings, mayonnaise, and creamy sauces. Phospholipids contribute to the uniform texture and consistency of these products.

Pharmaceuticals and drug delivery: In the pharmaceutical industry, phospholipids are used in drug delivery systems and liposomal formulations. Liposomes are spherical vesicles composed of phospholipid bilayers that can encapsulate both hydrophobic and hydrophilic drugs. This technology enables targeted drug delivery and improved drug bioavailability (Devlin, 2015).

Nutraceuticals: Phospholipids are used in the manufacture of nutraceuticals and dietary supplements. For example, phosphatidylcholine, a type of phospholipid, is widely used in dietary supplements for its potential health benefits, particularly in liver health and cognitive function.

Cosmetics and personal care: Phospholipids are incorporated into various cosmetic and personal care products. They can improve the stability, texture and moisturizing properties of creams, lotions, and skin care formulations. Phospholipids are also valued for their ability to create liposomal delivery systems for cosmetic actives.

Biotechnology and research: Phospholipids are essential in biotechnology for the development of liposomal drug carriers, cell culture media and membrane-related studies. They are valuable tools in research for the study of cell membranes, lipid bilayers and lipid-protein interactions.

Industrial processes: Phospholipids are used in various industrial processes, such as the production of biodiesel. Phospholipids present in vegetable oils can lead to unwanted foaming and soap formation during biodiesel production. Understanding and managing phospholipids is critical to the efficiency of this industry.

Health supplements: Phospholipid supplements are marketed for their potential health benefits, including support of brain health, liver function, and cardiovascular health. Phospholipids such as phosphatidylserine and phosphatidylcholine are among the most popular dietary supplements (Devlin et al., 2007).

3. High Molecular Weight Surfactant Biomolecules (Bioemulsifier)

High molecular weight bioemulsifiers include amphipathic polysaccharides, proteins, lipopolysaccharides, lipoproteins, or complex combinations of these biopolymers synthesized by a wide variety of microorganisms (Perfumo et al., 2010, Inés and Dhouha, 2015). They exhibit remarkable structural diversity and possess the ability to reduce surface and interfacial tension, either at the surface or at the interface, and/or to emulsify hydrophobic compounds. Examples of high molecular weight bioemulsifiers are emulsifiers, fatty acids, phospholipids, neutral lipids, exopolysaccharides, vesicles, and fimbriae. In addition to low toxicity and biodegradability, these molecules exhibit remarkable physicochemical properties, such as resistance to extreme conditions of pH, temperature, and salinity. Due to their ability to form and break emulsions, solubilize substances, mobilize, and disperse, and reduce viscosity, they hold great promise for environmental applications. These include serving as enhancers of hydrocarbon biodegradation and facilitating microbial enhanced oil recovery. They also have biomedical applications due to their antimicrobial and anti-adhesive activities and their involvement in immune responses (Inés and Dhouha, 2015).

3.1 Polymeric Microbial Surfactants (Emulsans, Alasans, Liposans)

3.1.1 Emulsan

Emulsan, a polyanionic emulsifier, is an extracellular acylated polysaccharide biosurfactant with an average molecular weight of approximately 1000 kDa (Martinez-Checa et al., 2007, Panilaitis et al., 2007, Inés and Dhouha, 2015). The structure of emulsan is composed of two main components, the polysaccharide backbone, and the fatty acid side chains. The polysaccharide backbone consists of three amino sugars, D-galactosamine, D-galactosaminouronic acid, and dideoxydiaminohexose in a 1:1:1 ratio (Zhang et al., 1997). The fatty acid side chains range in length from 10 to 22 carbon atoms and constitute 5-23% (w/w) of the polymer (Panilaitis et al., 2007, Arli et al., 2011). The combination of hydrophilic anionic sugar backbone repeating units together with the hydrophobic side groups results in the amphipathic behavior of Emulsan and therefore its ability to form stable oil-in-water emulsions (Panilaitis et al., 2007). On the other hand, Mercaldi et al. (2008) had discovered that emulsan is a complex of two polymers containing approximately 80% lipopolysaccharide and 20% high molecular weight exopolysaccharide.

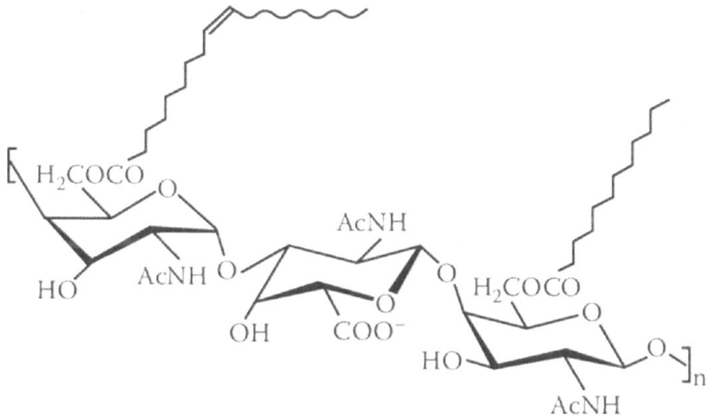

Fig. 16: Structure chemical of emulsan.

3.1.2 Alasans

Alasan biosurfactants are chemically a high molecular weight complex structure of polysaccharide and protein. They are involved in emulsifying hydrocarbons (paraffins, crude oils, long chain aliphatic and polyaromatic hydrocarbons) and assist in solubilizing polyaromatic hydrocarbons (Bhattacharya et al., 2017). These molecules are produced by *Acinetobacter radioresistens*, where the polysaccharide component is unusual in that it contains covalently bound alanine. The protein component appears to play an important role in both the structure and activity of the complex (Toren et al., 2001). Proteins of the complex are necessary for the activity and structure of the complex, whereas polysaccharides (apo-alasan) do not have emulsifying properties and therefore do not exhibit the properties of alasan itself, such as change in hydrodynamic shape due to temperature induction (Toren et al., 2001).

3.1.3 Liposans

Another example of high molecular weight bioemulsifiers is liposan, an extracellular water-soluble emulsifier (Cirigliano and Carman, 1985). This type of biosurfactant was first reported in the yeast *Candida lipolytic* and is also produced by *Acinetobacter* radioresistant. Chemically, liposan contains 83% carbohydrates and 17% proteins, with a carbohydrate content similar to that of emulsans. Other physiological functions associated with these molecules are their role as bioflocculants and growth stimulators. In addition, due to the substrate specificity and unique properties of surfactant activity, emulsans is usually considered for advanced biotechnological applications, even in dilute concentrations. Among these polymers, lipomannan is one of the best studied polymeric biosurfactants, which are chemically polysaccharide-protein complexes (Santos et al., 2016) produced by *Candida tropicalis* (Galabova et al., 2014). These have a similar structure to the microbial whole cell lipid and consist of approximately

50-60 residues that have glycosidic linkages with diglyceride type fatty acids. Lipomann is mostly found on the cell surface of mycobacteria and plays a role in virulence and survival in host cells (Sawettanai et al., 2019).

3.2 Particulate Biosurfactants (Microemulsions, Whole Cells, Vesicles)

Particulate biosurfactants play an important role in microbial uptake and, chemically, are complexes of proteins, phospholipids, and lipopolysaccharides that form extracellular membrane vesicles for the compartmentalization of microemulsion-forming hydrocarbons, which play an important role in alkane uptake by microbial cells. It is related to surfactants associated with particulate structures such as microemulsions, whole cells or vesicles. Microemulsions are colloidal systems characterized by the dispersion of nanoscale droplets of an immiscible liquid (usually oil) in another immiscible liquid (usually water), stabilized by surfactant molecules. These systems have an exclusive composition where the surfactant molecules form a layer near the dispersed droplets, preventing them from fusing or separating.

3.2.1 Microemulsions

Microemulsions consist primarily of three essential components: water, surfactant, and oil, with the occasional addition of co-surfactants for improved balance. These microemulsions exhibit various content and structural variations, including bicontinuous microemulsions, water-in-oil (W/O), and oil-in-water (O/W) formulations. The central role of surfactants in the formation and stability of microemulsions cannot be overemphasized. Surfactants play a critical role in facilitating the formation of tiny, stable droplets by reducing the interfacial tension between water and oil. The resulting microemulsions have a transparent and clear appearance that is indicative of their thermodynamic stability. The prevention of coalescence of the droplets and the maintenance of the equilibrium of the interfacial tension are fundamental factors that contribute to the safety profile of the microemulsions (Santos et al., 2016). Characteristically, the droplets in microemulsions exhibit a size range of 20 to 200 nm, significantly smaller than that observed in conventional emulsions. This reduced size contributes significantly to improved visual clarity and safety. Their spontaneous formation allows for easy preparation without significant energy input. The microemulsion zone, characterized as a thermodynamic safety zone, represents a distinctive phase behavior. Factors such as surfactant concentration, structure and temperature influence this zone. Ongoing research and development efforts continue, driven by the diverse applications and unique properties of microemulsions (Hayes and Smith, 2019).

3.2.2 Whole Cells

On the other hand, "whole cells" refer to intact and undamaged microorganisms or biological entities that have not undergone significant alteration or cell fractionation. In contrast to cell extracts or isolated cellular components, whole cells retain their original cellular composition, including organelles and cell membranes. This preservation of structural integrity extends to the cytoplasm, all organelles, or cellular components, and, if present, the cell membrane, and the cell wall. The functionality and inherent interrelationships of a cell depend on its intact composition. Whole cells maintain their biological activity, allowing them to perform their natural functions such as growth, metabolism, and reproduction. Consequently, the study of whole cells is essential to understanding the behavior of microorganisms in their native environment, as it provides insight into their unaltered physiological processes and interactions (Simpson et al., 2021).

Several scientific disciplines, including molecular biology, biotechnology, and microbiology, make extensive use of whole cells in their investigations. These cells serve as valuable model systems that allow experiments to be conducted in physiologically relevant environments, facilitate the study of the effects of environmental factors, and enhance the understanding of intricate cellular processes. Whole cells are particularly useful in industrial bioprocessing for the production of enzymes, antibiotics, and biofuels. In fermentation processes, whole cells are cultured to harness their metabolic activity for the efficient synthesis of desired products (Hayes and Smith, 2019). In addition, whole cells play a central role in bioremediation processes that aim to remove contaminants from the surrounding environment. The use of whole cells ensures the optimization of the potential of specific microbes to effectively degrade or modify contaminants. This comprehensive approach highlights the diverse applications of whole cells in various scientific and industrial fields (Simpson et al., 2021).

In the field of vaccine development, there are instances where the construction of vaccines involves the use of whole cells. Inactivated or attenuated whole cells can induce an immune response that provides protection against infectious diseases. While working with whole cells preserves native cellular capabilities, the associated processing can be challenging compared to working with cell extracts. Working with whole cells may require additional steps to separate and purify specific cellular components. Cell fractionation techniques facilitate the isolation of specific components, such as proteins or nucleic acids, from the whole cell, allowing targeted studies of these components (Saranraj et al., 2022). In summary, whole cells contribute significantly to a comprehensive understanding of the behaviors and functions of microbes in their native state. This characteristic makes whole cells indispensable entities in a wide range of scientific and industrial applications.

3.2.3 Vesicles

Vesicles, defined as tiny bubble-like structures, are formed by a lipid bilayer. Within the biosurfactant environment, vesicles can either encapsulate or mimic surfactant molecules, offering potential applications in drug delivery due to their ability to encapsulate and transport molecules. These tiny, membrane-enclosed sacs play key roles in various cellular functions, including communication, storage, and transport. Analogous to cell membranes, vesicular sacs are composed of lipid bilayers and can contain various materials such as lipids and proteins. Vesicles are essential for material exchange between cells, intracellular transport, and secretion processes (Bnyan et al., 2018). Consisting of a lipid bilayer similar to cell membranes, vesicles have an ellipsoidal or spherical structure that provides a confined space for material storage. Vesicle initiation can occur by membrane sprouting, where a distinct vesicle is formed by pinching off a portion of the membrane. Depending on the environmental context, different methods of vesicle production, such as endocytosis, exocytosis or intracellular trafficking, manifest themselves (Kreuter, 2014).

Different types of vesicles play different roles within cellular processes. Endosomes, for example, are vesicles generated by the endocytosis process when chemicals enter a cell. Exosomes, on the other hand, are small vesicles that are consistently implicated in intercellular communication and are actively released by cells into the extracellular environment. Synaptic vesicles, located in nerve terminals, serve as repositories for neurotransmitters that are subsequently released during synaptic transmission. Lysosomes, as membrane-bound organelles, contain digestive enzymes for intracellular use (Kreuter, 2014).

Intracellular transport is a fundamental function of vesicles, facilitating the movement of materials between different cellular compartments. For example, vesicles play a crucial role in the transfer of proteins from the endoplasmic reticulum to the Golgi apparatus, which subsequently facilitates their transport to other cellular locations. This orchestration of vesicular activities underscores their indispensable contributions to intricate cellular processes and compartmentalized functions (Hayes et al., 2019).

Endocytosis and exocytosis are fundamental cellular processes involving vesicular dynamics. Exocytosis is characterized by vesicles attaching to the cell membrane and releasing their contents into the extracellular space. Conversely, endocytosis is the cellular process by which materials from the environment are engulfed by cells, incorporated into vesicles, and subsequently transported within the cell. In cell signaling, extracellular vesicles such as exosomes participate in intercellular communication by transporting materials such as proteins and nucleic acids across cells, affecting a range of physiological functions.

On the other hand, in the area of drug delivery within nanotechnology, artificial vesicles, commonly referred to as liposomes, have been developed for drug delivery purposes. By encapsulating drugs and targeting them to specific tissues or cells, these vesicles offer the potential to mitigate adverse effects associated with drug administration. The study of vesicles is an integral part of both cell and molecular biology due to their importance in biological functions. Ongoing research on vesicle formation and function provides valuable insights into various cellular processes, contributing to a deeper understanding of cellular functions and molecular mechanisms. In summary, vesicles are versatile structures with a wide range of functions that are integral to cellular transport, communication, and compartmentalization, thereby maintaining the overall functionality and integrity of cells. These particulate biosurfactants find application in various contexts of biotechnology and bioprocessing (Bhadra et al., 2023).

For example, in the field of enhanced oil recovery (EOR) within the petroleum industry, microemulsions and microorganisms capable of biosurfactant production can be strategically employed to increase the recovery of oil from reservoirs. In addition, in bioremediation efforts, the use of whole cells capable of biosurfactant production holds promise for increasing the accessibility of oil to microbial degradation, thereby potentially improving the bioremediation of oil-contaminated environments. These applications take advantage of the unique properties of biosurfactants by exploiting their ability to interact with surfaces and enhance the solubility or dispersion of hydrophobic substances (Bhadra et al., 2023).

4. Conclusions

Biosurfactants and bioemulsifiers are the amphiphilic secondary metabolites, unlike their chemical counterparts, biosurfactants have been found successful in achieving an environmentally friendly process without any toxicity. They are readily degradable and therefore have application in waste management, bioremediation and manage the bioavailability of toxicants in the environment. These molecules possess significant surface and interfacial activity as these molecules have both hydrophobic and hydrophilic moieties in their structure. In this study, we detail the nature, types, and characteristic chemicals of biosurfactants and bioemulsifiers. Furthermore, the composition of different classes of biosurfactants and their applications in various fields were briefly discussed. In this way, understanding the chemical structure allows us to understand the unique properties that make these compounds effective in biotechnological applications. In addition, consideration of factors such as hydrophobic and hydrophilic moieties, molecular weight, and stereochemistry contribute to a comprehensive understanding of their structure-function relationships.

References

Abdel-Mawgoud, A.M., Lepine, F. & Deziel, E. (2010). Rhamnolipids: Diversity of structures, microbial origins, and roles. *Appl. Microbiol. Biotechnol.*, 86: 1323-1336.

Antonioli-Júnior, R., Poloni. J.F., Pinto, É.S.M. & Dorn, M. (2022). Interdisciplinary overview of lipopeptide and protein-containing biosurfactants. *Genes*, 14: 76.

Arli, S.D., Trivedi, U.B. & Patel, K.C. (2011). Curdlan-like exopolysaccharide production by *Cellulomonas flavigena* UNP3 during growth on hydrocarbon substrates. *World J. Microbiol. Biotechnol.*, 27: 1415-1422.

Atta, D.Y., Negash, B.M., Yekeen, N. & Habte, A.D. (2021). A state-of-the-art review on the application of natural surfactants in enhanced oil recovery. *J. Mol. Liq.*, 321: 114888.

Banat, I.M., Franzetti, A., Gandolfi, I., Bestetti, G., Martinotti, G., Fracchia, L. et al. (2010). Microbial biosurfactants production, applications, and future potential. *Appl. Microbiol. Biotechnol.*, 87: 427-444.

Bartual, S., Alcorlo, M., Martínez-Caballero, S., Molina, R. & Hermoso, J. (2018). Three-dimensional structures of lipoproteins from *Streptococcus pneumoniae* and *Staphylococcus aureus*. *International Journal of Medical Microbiology*, 308(6): 692-704.

Bauer, J., Brandenburg, K., Zahringer, U. & Rademann, J. (2006). Chemical synthesis of a glycolipid library by a solid-phase strategy allows elucidation of the structural specificity of immunostimulation by rhamnolipids. *Chem. Eur. J.*, 12: 7116-712.

Bender, D., Botham, K., Weil, P., Kennelly, P., Murray, R. & Rodwell, V. (2013). HARPER. Bioquimica ilustrada (29th ed.). McGraw-Hill.

Bhadra, S., Chettri, D. & Kumar Verma, A. (2023). Biosurfactants: Secondary metabolites involved in the process of bioremediation and biofilm removal. *Applied Biochemistry and Biotechnology*, 195(9): 5541-5567.

Bhattacharya, B., Ghosh, T.K. & Das, N. (2017). Application of bio-surfactants in cosmetics and pharmaceutical industry. *Sch. Acad. J. Pharm.*, 6(7): 320-329.

Bnyan, R., Khan, I., Ehtezazi, T., Saleem, I., Gordon, S., O'Neill, F. et al (2018). Surfactant effects on lipid-based vesicles properties. *Journal of Pharmaceutical Sciences*, 107(5): 1237-1246.

Bordas, F., Lafrance, P. & Villemur, R. (2007). Conditions for effective removal of pyrene from an artificially contaminated soil using *Pseudomonas aeruginosa* 57SJ rhamnolipids. *Environmental Pollution*, 138: 69-76.

Cameotra, S.S. & Singh, P. (2009). Synthesis of rhamnolipid biosurfactant and mode of hexadecane uptake by *Pseudomonas* species. *Microb Cell Factories*, 8: 16.

Campos, J.M., Montenegro Stamford, T.L., Sarubbo, L.A., de Luna, J.M., Rufino, R.D. & Banat, I.M. (2013). Microbial biosurfactants as additives for food industries. *Biotechnol. Prog.*, 29: 1097-1108.

Cavalero, D.A. & Cooper, D.G. (2003). The effect of medium composition on the structure and physical state of sophorolipids produced by *Candida bombicola* ATCC 22214. *J. Biotechnol.,* 103: 31-41.

Chakrabarti, S. (2012). Bacterial Biosurfactant: Characterization, Antimicrobial and Metal Remediation Properties. National Institute of Technology, Surat, India.

Cirigliano, M.C. & Carman, G.M. (1985). Purification and characterization of liposan, a bioemulsifier from *Candida lipolytica*. *Appl. Environ. Microbiol.,* 50: 846-850.

Daverey, A. & Pakshirajan, K. (2009). Production, characterization, and properties of sophorolipids from the yeast *Candida bombicola* using a low-cost fermentative medium. *Appl. Biochem. Biotechnol.*, 158(3): 663-674.

Desai, J.D. & Banat, I.M. (1997). Microbial production of surfactants and their commercial potential. *Microbiol. Mol. Biol. R.,* 61: 47-64.

Devlin, A.M., Singh, R., Wade, R.E., Innis, S.M., Bottiglieri, T. & Lentz, S.R. (2007). Hypermethylation of Fads2 and altered hepatic fatty acid and phospholipid metabolism in mice with hyperhomocysteinemia. *Journal of Biological Chemistry*, 282: 37082-37090.

Devlin, T.M. (2015). Bioquímica. con aplicaciones clínicas. *Reverte*, 1215 pp.

Dhasayan, A., Kiran, G.S. & Selvin, J. (2014). Production and characterization of glycolipid biosurfactant by *Halomonas* sp. MB-30 for potential application in enhanced oil recovery. *Appl. Biochem. Biotechnol.*, 174: 2571-2584.

Fenibo, E.O., Douglas, S.I. & Stanley, H.O. (2019). A review on microbial surfactants: Production, classifications, properties and characterization. *J. Adv. Microbiol.,* 18: 1-22.

Galabova, D.A., Sotirova, E., Karpenkoy, & Karpenk, O. (2014). Role of microbial surface-active compounds in environmental protection. *In:* Fanum, M. (Ed.), *The Role of Colloidal Systems in Environmental Protection*. 41-83. Elsevier B.V.

Götze, S. & Stallforth, P. (2020). Structure, properties, and biological functions of nonribosomal lipopeptides from pseudomonads. *Natural Product Reports*, 37(1): 29-54.

Hamley, I.W. (2021). Lipopeptides for vaccine development. *Bioconjugate Chemistry*, 32(8): 1472-1490.

Hayes, D.G. & Smith, G.A. (2019). Biobased surfactants: Overview and industrial state of the art. *Biobased Surfactants*, 3-38.

Hayes, D.G., Solaiman, D.K. & Ashby, R.D. (2019). Biobased Surfactants: Synthesis, Properties, and Applications. Elsevier.

Henkel, M. & Hausmann, R. (2019). Chapter 2—Diversity and classification of microbial surfactants. *In:* Hayes, D.G., Solaiman, D.K., Ashby, R.D. (Eds), *Biobased Surfactants*. 3rd ed. 41-63. AOCS Press, London, UK.

Hommel, R.K. & Ratledge, C. (1993). Biosynthetic mechanisms of low molecular weight surfactants and their precursor molecules. *In:* Kosaric, N. (Ed.), *Biosurfactants: Production, properties, application*. 3-63. Marcel Dekker, Inc.: New York, NY, USA.

Hunter, R.L., Olsen, M., Jagannath, C. & Actor, J.K. (2006). Trehalose 6, 60-dimycolate and lipid in the pathogenesis of caseating granulomas of tuberculosis in mice. *Am. J. Pathol.,* 168: 1249-1261.

Inés, M. & Dhouha, G. (2015). Glycolipid biosurfactants: Potential related biomedical and biotechnological applications. *Carbohydr Res.*, 416: 59-69.

Jahan, R., Bodratti, A.M., Tsianou, M. & Alexandridis, P. (2020). Biosurfactants, natural alternatives to synthetic surfactants: Physicochemical properties and applications. *Adv. Colloid Interface Sci.,* 275: 102061.

Kitamato, D. (2002). Functions and potential applications of glycolipid biosurfactants from energy-saving materials to gene delivery carriers. *J. Bio. Sci.Bio. Eng.,* 94: 187-201.

Konishi, M., Fukuoka, T., Morita, T., Imura, T. & Kitamoto, D. (2008). Production of new types of sophorolipids by *Candida batistae*. *J. Oleo. Sci.*, 57: 359-369.

Koutsopoulos, S., Kaiser, L., Eriksson, H. & Zhang, S. (2012). Designer peptidesurfactants stabilize diverse functional membrane proteins. *Chemical Society Reviews*, 41(5): 1721-1728.

Kreuter, J. (2014). Colloidal Drug Delivery Systems. CRC Press eBooks.

Kurtzman, C.P., Price, N.P.J., Ray, K. & Kuo, T.M. (2010). Production of sophorolipid biosurfactants by multiple species of the *Starmerella* (Candida) *bombicola* yeast clade. *Fems Microbiology Letters*, 311(2): 140-146.

Lang, S. (2002). Biological amphiphiles (microbial biosurfactants). *Curr. Opin. Colloid Inter. Sci.*, 7: 12-20.

Lo, C.M. & Ju, K.L. (2009). Sophorolipids-induced cellulase production in cocultures of Hypocrea jecorina Rut C30 and *Candida bombicola*. *Enzyme Microb. Technol.*, 44: 107-111.

Martinez-Checa, F., Toledo, F.L., El Mabrouki, K., Quesada, E. & Calvo, C. (2007). Characteristics of bioemulsifier V2-7 synthesized in culture media added of hydrocarbons: Chemical composition, emulsifying activity, and rheological properties. *Bioresour. Technol.*, 98: 3130-3135.

Mercaldi, M.P., Dams-Kozlowska, H., Panilaitis, B., Joyce, A.P. & Kaplan, D.L. (2008). Discovery of the dual polysaccharide composition of emulsan and the isolation of the emulsion stabilizing component. *Biomacromolecules*, 9: 1988-1996.

Mulligan, C.N. & Gibbs, B.F. (2004). Types, production, and applications of biosurfactants. *Proc. Indian Nat. Sci. Acad.*, 1: 31-55.

Nelson, D.L. & Cox, M.M. (2017) Lehninger Principles of Biochemistry. 7th Edition, W.H. Freeman, New York, 1328.

Ortiz, A., Teruel, J.A., Espuny, M.J., Marques, A., Manresa, A. & Aranda, F.J. (2009). Interactions of a bacterial biosurfactant trehalose lipid with phosphatidylserine membranes. *Chem. Phys. Lipids*, 158: 46-53.

Pan, F., Zhang, Z., Zhang, X. & Davarpanah, A. (2020). Impact of anionic and cationic surfactants interfacial tension on the oil recovery enhancement. *Powder Technol.*, 373: 93-98.

Panilaitis, B., Castro, G.R., Solaiman, D. & Kaplan, D.L. (2007). Biosynthesis of emulsan biopolymers from agro-based feedstocks. *J. Appl. Microbiol.*, 102: 531-537.

Parthipan, P., Sabarinathan, D., Angaiah, S. & Rajasekar, A. (2018). Glycolipid biosurfactant as an eco-friendly microbial inhibitor for the corrosion of carbon steel in vulnerable corrosive bacterial strains. *J. Mol. Liq.*, 261: 473-479.

Peng, F., Liu, Z., Wang, L. & Shao, Z. (2007). An oil-degrading bacterium: Rhodococcus erythropolis strain 3C-9 and its biosurfactants. *J. Appl. Microbiol.*, 102: 1603-1611.

Perfumo, A.S., Smyth, T.J.P., Marchant, R. & Banat, I. (2010). Production and roles of biosurfactants and bioemulsifiers in accessing hydrophobic substrates. Chap. 47. *In:* Timmis, K.N. (Ed.), *Handbook of Hydrocarbon and Lipid Microbiology*. 2(7): 1501-1512.

Philp, J.C., Kuyukina, M.S., Ivshina, I.B., Dunbar, S.A., Christofi, N., Lang, S. & Wray, V. (2002). Alkanotrophic *Rhodococcus ruber* as a biosurfactant producer. *Appl. Microbiol. Biotechnol.*, 59: 318-324.

Santos, D.K.F., Rufino, R.D., Luna, J.M., Santos, V.A. & Sarubbo, L.A. (2016). Biosurfactants: Multifunctional Biomolecules of the 21st Century. *Int. J. Mol. Sci.*, 17: 401.

Saranraj, P., Zayyed, R.Z., Sivasakthivelan, P., Devi, M.D., Al-Tawaha, A.R.M. & Sivasakthi, S. (2022). Microbial biosurfactants sources, classification, properties and mechanism of interaction. *Microbial Surfactants*. 1-24. CRC Press.

Sawettanai, N.H., Leelayuwapan, N., Karoonuthaisiri, S., Ruchirawat & Boonyarattanakalin, S. (2019). Synthetic lipomannan glycan microarray reveals the importance of α(1,2) mannose branching in DC-SIGN binding. *J. Org. Chem.*, 84(12): 7606-7617.

Simpson, M.L., Sayler, G.S., Fleming, J.T. & Applegate, B. (2001). Whole-cell biocomputing. *Trends in Biotechnology*, 19(8): 317-323.

Toren, A.S., Navon-Venezia, E.Z. & Rosenberg, E. (2001). Emulsifying activities of purified alasan proteins from *Acinetobacter radioresistens* KA53. *Applied and Environmental Microbiology*, 67(3): 1102-1106.

Varjani, S.J. & Upasani, V.N. (2017). Critical review on biosurfactant analysis, purification and characterization using rhamnolipid as a model biosurfactant. *Bioresour. Technol.*, 232: 389-397.

Warnecke, D. & Heinz, E. (2010). Glycolipid headgroup replacement: A new approach for the analysis of specific functions of glycolipids in vivo. *Eur. J. Cell Biol.*, 89(1): 53-61.

Zaragoza, A., Aranda, F.J., Espuny, M.J., Teruel, J.A., Marqués, A., Manresa, Á. & Ortiz, A. (2009). Mechanism of membrane permeabilization by a bacterial trehalose lipid biosurfactant produced by Rhodococcus sp. *Langmuir*, 25: 7892-7898.

Zhang, J.A., Gorkovenko, R.A., Gross, A.L. & Kaplan, D.L. (1997). Incorporation of 2-hydroxyl fatty acids by *Acinetobacter calcoaceticus* RAG-1 to tailor emulsan structure. *Int. J. Biol. Macromol.*, 20: 9-21.

CHAPTER

6

Bioemulsifier (BE) and Biosurfactant (BS): Green Solutions for Pharmaceutical and Cosmetic Industries

Lavina A. Pinto[1], Smita S. Bhuyan[2], Dimple O. Parate[1], Dhanashree A. Kulkarni[1], Neelima J. Kulkarni[3], and Shilpa S. Mujumdar*[1]

[1] Department of Microbiology, Modern College of Arts, Science and Commerce (Autonomous), Shivajinagar, Pune - 411005, Maharashtra, India

[2] CRI - The Clinical Research Institute GmbH, Arnulfstrasse 19, 80335 Munich, Germany

[3] Department of Microbiology, P.E.S. Modern College of Arts, Science and Commerce (Autonomous), Ganeshkhind, Pune - 411016, Maharashtra, India

1. Introduction

Emulsions are used to enhance the effectiveness of pharmaceutical and cosmetic preparations targeted for internal as well as external usage (Khan et al., 2011). External pharmaceutical and cosmetic emulsions are used to prevent evaporation of moisturizers, to remove oil soluble dirt from skin, for prolonged release of drugs (Singh et al., 2019). Internal pharmaceutical and cosmetic emulsions are used to mask taste of unpleasant ingredients in the drugs, as immunological adjuvants in vaccine preparations, to prolong release of drugs, to improve solubility and bioavailability of drugs, and to pass metabolic effects in laxative preparations (Singh et al., 2013, Bhattacharya et al., 2017). However, care should be taken that these emulsions do not affect the odour, colour and appearance of the pharmaceutical and cosmetic preparations. The phenomenon of coalescence of dispersed phase, and creaming should be prevented in emulsion preparations (Mophosa and Jideani, 2018).

Emulsions are thermodynamically unstable and thus emulsifiers or surfactants should be added in the emulsion preparations to stabilize them. The major function of bioemulsifiers (BE) and biosurfactants (BS) in pharmaceutical processing is to

* Corresponding author: hodmicro@moderncollegepune.edu.in

Table 1: Applications of bioemulsifiers (BE) and biosurfactants (BS) in pharmaceutical and cosmetic industries

Organism	Name of the BE/BS	BE/BS		Function	Reference
Bacteria					
Acinetobacter calcoaceticus RAG 1	Emulsan	BE	Glycoprotein	Immuno-adjuvant, Drug delivery agent	Morita et al., 2013, Banat et al., 2019
Acinetobacter radioresistens	Alasan	BE	Glycoprotein	Stabiliser of micro-emulsions	Banat et al., 2019, Mouaziz et al., 2018
Bacillus brevis	Gramicidin	BS	Lipopeptide	Antibiotic, Anti-microbial agent	Santos et al., 2018
Bacillus siamensis	Surfactin	BS	Lipopeptide	Anti-inflammatory activity, Anti-cancer activity, Anti-bacterial activity	Ohadi et al., 2020, Fernandes et al., 2007, Seydlová and Svobodová 2008
Bacillus subtilis	Fengycin	BS	Lipopeptide	Anti-fungal activity, Anti-bacterial activity, Haemolytic	Banat et al., 2019, Fernandes et al., 2007
Bacillus subtilis K1	Surfactin	BS	Lipopeptide	Immunomodulator siRNA delivery agent, Anti-adhesion agent, Inhibits fibrin clot formation	Ohadi et al., 2020, Shim et al., 2009, Fernandes et al., 2007, Sharrel et al., 2014
Bacillus subtilis K1	Iturin	BS	Lipopeptide	Immuno-adjuvant	Santos et al., 2018
Lactobacillus sp.	-	BS	Glycoprotein	Antimicrobial agent, Anti-adhesive activity	Madhu and Prapulla, 2013, Fracchia et al., 2010
Planococcus maritimus	-	BS	Glycolipid	Antimicrobial activity, Anti-cancer activity	Waghmode et al., 2020
Pseudomonas aeruginosa DS10-129	Rhamnolipid	BS	Glycolipid	Anti-cancer activity Antimicrobial activity Anti-adhesion activity Immunomodulator, Drug delivery agent	Thanomsub et al., 2006, Lang and Wullbrandt, 1999

Contd...

Contd...

Organism	Name of the BE/BS	BE/BS		Function	Reference
Pseudozyma spp.	MEL	BS	Glycolipid	Drug delivery, Anti-ageing, Anti-oxidant	Morita et al., 2013, Takahashi et al., 2012, Ueno et al., 2007
Rhodococcus erythropolis	Trehalose dimycolate (TDM)	-	Glycolipid	Immunomodulator	Chereshnev et al., 2010
Fungi					
Candida bombicola *Candida tropicalis*	Sophorolipids	BS	Glycolipid	Immunomodulator, Anti-adhesive activity, Anti-cancer activity, Antiviral activity	Lang and Wagner, 1993, Lourith and Kanlayavattanakul, 2009, Shah et al., 2005, Hagler et al., 2007
Candida glabrata UCP0995	Iunasan	BS	Glycolipid	Antimicrobial activity, Anti-adhesive activity	Luna et al., 2011
Candida lipolytica *Yarrowia lipolytica*	Liposan	BE	Glycoprotein	Stabiliser of microemulsions	Cirigliano Carman, 1985, Amaral et al., 2006
Saccharomyces cerevisiae *Kluyveromyces marxianus*	Mannoproteins	BE	Glycoprotein	Stabiliser of microemulsions	Banat et al., 2019
Ustilago maydis	Cellobiose lipids, MEL	BS	Glycolipid	Antimicrobial agent	Morita et al., 2013

Note: Data not available.

enhance the solubility of water-insoluble drugs. BE and BS also act as solubilizing and wetting agents to enable drug incorporation into the delivery vehicles. They promote emulsification at the time of manufacture by stabilizing the dispersed phase and the encapsulated drugs. They also control emulsion stability during the shelf-life of the pharmaceutical and cosmetic formulations. BE and BS control the rheological properties of formulation by their interaction with continuous and dispersed phases. They enable the penetration of drugs across cell walls and membranes, skin, and other biological interfaces. BE and BS interact with the stratum corneum (outermost layer of the skin) in order to impart permeability to the cosmetic and therapeutic agents (Fracchia et al., 2010). These interactions should be such that they promote penetration of the therapeutic agents into the skin and allow the cosmetic agents to remain on the skin surface. They reduce evaporation by forming a thin occlusive layer when applied on the skin. This gives a smooth appearance to the skin (Alizadeh-Sani 2018).

Bioemulsifers and biosurfactants are the amphiphilic molecules produced by microorganisms that exhibit surface activity and emulsifying properties. Although the terms biosurfactant and bioemulsifier are often used interchangeably, they have marked differences in their physicochemical properties and physiological roles (Uzoigwe et al., 2015, Bee et al., 2029, Patel et al., 2022). BE and BS can solubilize both hydrophobic and hydrophilic therapeutic agents. Synthetic surfactants show toxicity and hence are not approved for manufacturing pharmaceutical formulations. On the other hand, BE and BS are safe and exhibit no or low toxicity, they are biodegradable, biocompatible, and effective even at low concentrations hence can be used in pharmaceutical and cosmetic formulations. BE and BS can adapt to a wide range of pH, temperature and salinity. BE is anionic in higher pH ranges and non-ionic lower pH ranges. The addition of electrolytes can trigger transition from micellar to lamellar structure which is valuable for tailoring a drug delivery system for a specific therapeutic agent. BE and BS possess antioxidant, antibacterial and antifungal properties and hence the formulations need not include preservatives (Bhattacharya et al., 2017, Shah et al., 2022, Ali et al., 2022a, Saranraj et al., 2022b). Owing to their high nutrient quality and medicinal properties they can also enhance nutritive value of the products (Zaman et al., 2022). Apart from this, they are easily degraded by microorganisms. In addition, they can be produced from cheap and easily available raw materials as well as industrial wastes and by-products depending on the application (Singh et al., 2013, Bhattacharya et al., 2017). A number of microorganisms including bacteria, fungi and yeast produce BE and BS which show tremendous potential applications in pharmaceutical and cosmetic industries (Table 1).

2. Properties of BE and BS Essential for Pharmaceutical and Cosmetic Industries

2.1 Stability

Pharmaceutical and cosmetic emulsions should be stable. The stability of an emulsion is its ability to retain its properties over time. Instabilities generally occur in an emulsion through flocculation, creaming, coalescence or ostwald ripening. Factors that contribute to the instability of the emulsions include the force of attraction between the dispersed phases, influence of buoyancy or centripetal force and change of an inhomogeneous structure over time (Bhattacharya et al., 2017).

2.2 Compatibility

Emulsions should be compatible with other ingredients, that is, no marked changes should be observed in the structure or function of the encapsulated active agents under the influence of the emulsifiers and surfactants. They should not interfere with the stability or efficacy of the active agent (Bhattacharya et al., 2017).

2.3 Safety and Non-toxicity

The emulsions used for pharmaceutical and cosmetic formulations should be safe and non-toxic. The emulsifying agent or the surfactant should possess little or no odor, taste, or color (Vecino et al., 2017).

2.4 Applicable Routes of Administration for Different Types of Emulsions

Emulsions are now well recognized drug delivery systems for water insoluble drugs. Emulsions need to be carefully designed with respect to the target and route of administration. Emulsions need to be administered either by the oral or intravenous, intramuscular and subcutaneous routes. The emulsions mentioned for this use are generally O/W or W/O/W formulations. The formulations need to be sterilized and free of endotoxin. Moreover, the formulation should be physico-chemically stable and isotonic with blood (pH 7.4). These formulations are extensively used for parenteral nutrition and as drug carrier systems (Marti-Mestres and Nielloud, 2002). Topical and transdermal routes include, percutaneous absorption that involves the transfer of drug from the skin into the microcirculation through a series of diffusion barriers. The main advantages of topical formulations include- effectiveness in low doses and drug delivery at a specific site. Challenges in topical application include maintaining a long contact between the drug and the skin which may lead to individual sensitization. The term ointment is used for semisolid formulations and can make use of both O/W and W/O emulsions. The term cream refers to semisolid formulations with relatively low viscosity and consisting mostly of O/W emulsions and is widely used in pharmaceutical and cosmetic formulation. Creams can penetrate the skin and are used for wiping lesions, whereas ointments do not penetrate the skin surface and are used on dry lesions (Murthy and Shivakumar, 2010).

Oral drug delivery (ODD) is the convenient route of drug administration because it is cost effective, requires least sterility constraints, flexible in dose design and ease of production. Emulsions enhance the bioavailability of poorly absorbed drugs and also protect the active agents from gastric media. Laxative preparations or enteric nutrition are administered through oral routes. The active agents have to get dissolved in the aqueous fluid prior to absorption and be sufficiently lipophilic to move across the lipid membrane of the gut. The metabolism of an emulsion vehicle used as a drug delivery system itself plays a crucial role in drug release and absorption (Kalepu et al., 2013).

3. Application of BS and BE Used in the Pharmaceutical Industry

Biosurfactants offer astonishing benefits to many industries such as oil, food, agriculture and have gained importance especially in the pharmaceutical and cosmetics industrial segments. Some of them have been reported as reliable multi-functional molecules because of their excellent properties such as biocompatibility, low toxicity, emulsifying ability, foaming, skin hydrating and hair repairing properties. In addition to these, they have antimicrobial activity and antiadhesive properties which enable them to compete with chemically synthesised antibiotics and the recovery of intracellular products respectively (Varvaresou and Iakovou, 2015). In pharmaceutical industries they are mainly used in drug delivery, in the recovery of intracellular products, as an anticancer agent, antimicrobial agent and immunological adjuvants and immunomodulatory agents.

3.1 Drug Delivery

Many drugs are normally unable to diffuse through the membrane; by encapsulating them in liposomes they can be delivered through the lipid bilayer. A liposome is a spherical vesicle with a phospholipid bilayer membrane used to deliver an antigen, an antibiotic, an allergen, a drug substance or a gene drug into a cell without triggering an immune rejection reaction (Faivre and Rosilio, 2010). As shown in Fig. 1, Liposomes interact with cells by endocytosis by phagocytic cells; adsorption to the cell surface by interacting with cell-surface components; adsorption to the cell surface by weak hydrophobic or electrostatic forces; fusion with the plasma cell membrane with simultaneous release of liposomal content into the cytoplasm or transfer of liposomes to cellular or sub-cellular membranes without release of the liposome contents (Ohadi et al., 2020). The efficiency of liposome formation and delivery of active compounds depend on the charge properties of the active compounds and their interaction with vesicle forming molecules. Furthermore, it depends on the composition of the molecules used for the formation of the vesicular structure along with its interaction with serum proteins, lipoproteins, opsonin and phagocytic systems, target cells and other components of the body. For the treatment of diseases like cancers, tumours and HIV, the drug can be easily and effectively delivered by means of the liposomes. Multiple emulsions and microemulsions show enhanced drug bioavailability by increasing the drug solubility and controlling the drug release. They also protect the active agent from chemical or biological attack (Faivre and Rosilio, 2010).

Glycolipids like rhamnolipids can be used in liposome formulations for drug delivery (Faivre and Rosilio, 2010). MEL-A-containing vesicles have shown dramatically increased transfection efficacy of cationic liposomes, exhibiting significantly higher levels of gene expression and thus can be used for gene delivery. Exact mechanism for the increased transfection efficiency is not known but is thought to be the acceleration of the fusion between DNA-liposome complexes and the plasma membrane or the destabilizing effect on the endosome membrane (Sharma et al., 2009, Ueno et al., 2007). The cells must be able to receive signals from the extracellular medium in the normal and pathological processes such as cell recognition, cell adhesion, signal transduction and protein sorting. Glycolipids are recognized by receptors or complimentary carbohydrates in the extracellular medium. Binding specificity depends on positioning, accessibility and the sequence of the carbohydrate and lipid moieties present in glycolipids (Ohadi et al., 2020). There can be allosteric mechanisms where changes in the structure in a distal part of the molecule results in alteration of the orientation or conformation of the carbohydrate moiety at the interface, allowing or restricting ligand binding (Faivre and Rosilio, 2010).

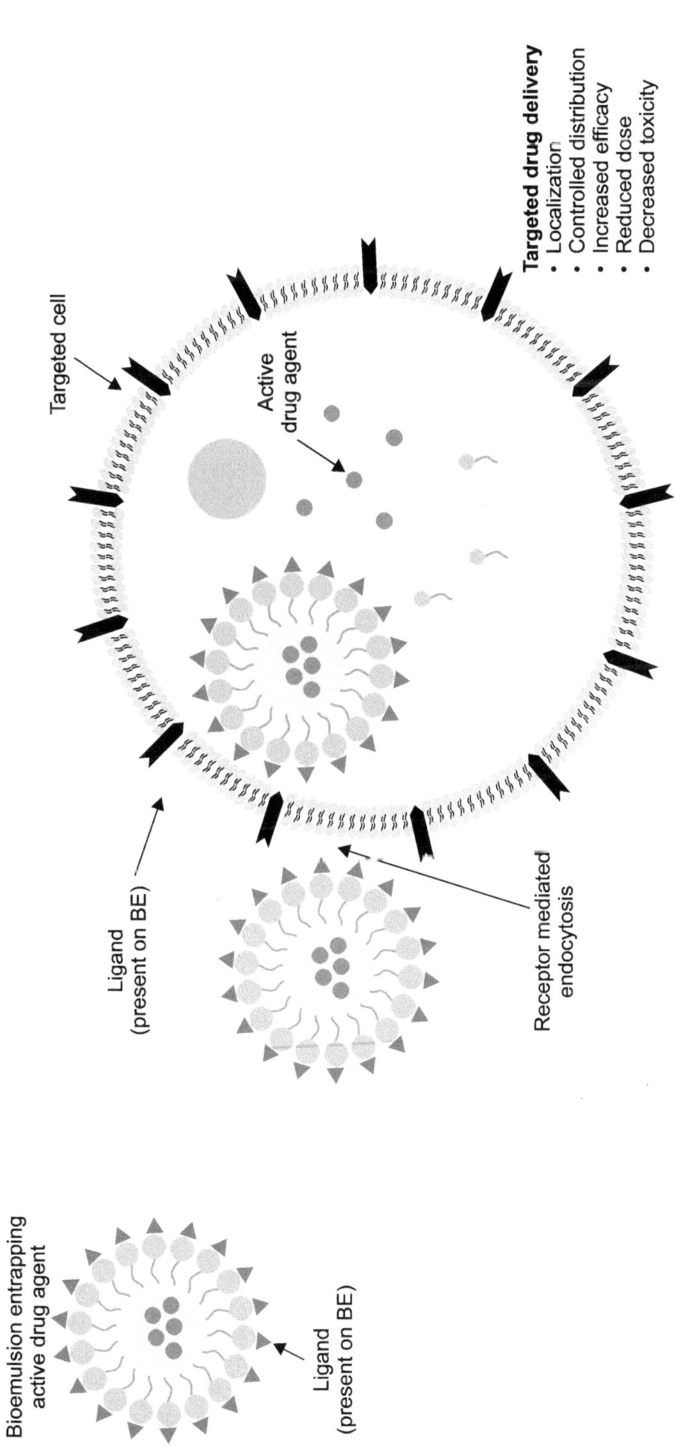

Fig. 1: Targeted drug delivery system by using micro-emulsions.

3.2 Immunological Adjuvants and Immune Modulatory Agents

Immunological adjuvants are added to a vaccine to modify or boost the immune response. Use of immunological adjuvants supports higher amounts of antibody production thereby imparting long lasting protection and minimizing the amount of injected foreign material. These adjuvants only stimulate the immune system's response against the target antigen and do not provide immunity by themselves (Rodrigues et al., 2006). When some potent non-toxic, non-pyrogenic bacterial lipopeptides such as Iturin AL and herbicolin A coupled to poly-l-lysine (MLR-PLL) are mixed with the antigen of interest, a marked improvement in the humoral immune response is observed in rabbits and chickens. Emulsan produced by *Acinetobacter calcoaceticus*, when administered with antigens showed significant adjuvant properties (Sajid et al., 2020).

Immunomodulators are agents used to modify the activity of the immune system. They decrease the inflammatory response and are often used in organ transplantation to prevent rejection of the new organ. Biosurfactants demonstrate immunomodulatory activities either by suppression or activation of the immune system (Sajid et al., 2020). Glycolipids are found to modulate both humoral and cellular immune systems e.g., sophorolipids have demonstrated decreased sepsis related mortality in vivo in a rat model of septic peritonitis. They can be used against autoimmune diseases such as rheumatoid arthritis and for the treatment of multiple myeloma (Hardin et al., 2007). Sophorolipids have also decreased IgE production in vitro in U266 cells. Sophorolipids act as anti-inflammatory agents hence can be used to cure diseases which are associated with altered IgE production such as asthma, rhinosinusitis, atopic eczema and hay fever (Hardin et al., 2007). Rhamnolipids also act as immunomodulators, they exhibit chemotactic recruitment of neutrophils and increased oxidative responses, histamine release from mast cells and increased production of serotonin (Sajid et al., 2020). Rhamnolipids inhibit the phagocytic ability of macrophages either by interfering with the internalization process or by inhibiting the fusion of particle encapsulated phagosomes to lysosomes inside the macrophages. Surfactin is shown to promote suppression of cell survival signalling, inhibition of platelets' aggregation properties and anti-inflammatory properties. In addition to that, Surfactin down-regulates expression of several surface molecules like CD40, CD54, CD80 and MHC class-II from lipopolysaccharide-activated macrophages. Trehalose lipids BS produced by *Rhodococcus* also express immuno-modulatory activities (Shah et al., 2005, Sajid et al., 2020, Hagler et al., 2007).

3.3 Antimicrobial Activity

Rapidly increasing antimicrobial resistance compels the discovery of new and safe antibiotics. Biosurfactants are remarkably placed as antimicrobial agents because of their noteworthy properties such as bactericidal, bacteriostatic, antiadhesive, and antibiofilm activities. In addition, they also show synergistic and adjuvant effects with antibiotics. Extensive research is carried out on biosurfactant producing microbes and their application as antimicrobial agents. They gain importance because they can be used as alternative medicines (Rufino et al., 2011). Table 2 includes examples of BE and BS and their antimicrobial activity against different microorganisms. Biosurfactants mainly act as antimicrobial agents by acting on the cell wall of the bacteria and fungi. Biosurfactants alter the structure of the cell membrane and thus

Table 2: Antimicrobial Activity of Bioemulsifiers (BE) and Biosurfactants (BS)

BS/BE producing organism	Source	Type of BE/BS	Antimicrobial activity against	References
B. stratosphericus *Pseudomonas aeruginosa* C2	Soil (oil contaminated)	Lipopeptide and rhamnolipid	*Escherichia coli, Staphylococcus aureus*	Sana et al., 2018
Bacillus amyloliquefaciens	WWTP (Wastewater treatment plant)	Gycolipid (Surfactin)	*Salmonella typhimurium, Escherichia coli, Klebsiella pneumoniae, Acinetobacter* sp., *Enterobacter* sp., *Salmonella enterica* sp. *Staphylococcus aureus, Bacillus cereus, Micrococcus* sp, *Enterococcus* sp. *Cryptococcus neoformans, Candida albicans*	Ndlovu et al., 2017
Bacillus cereus NK1	Soil	Lipopeptide	*Escherichia fergusonii, Klebsiella varicolla, Escherichia coli, Bacillus subtilis, Bacillus licheniformis, Bacillus megaterium, Saccharomyces cerevisiae*	Sriram et al., 2011
Bacillus halotolerans	-	Lipopeptide	*Staphylococcus aureus, Pseudomonas aeruginosa, Escherichia coli, Saccharomyces cerevisiae*	Etemadzadeh et al., 2023
Bacillus subtilis	Marine	Lipopeptide	*Candida albicans*	Yuliani et al., 2018
Bacillus subtilis	Intestine of Apis mellifera and honey	Gycolipid (Surfactin)	*Listeria monocytogenes*	Sabate and Audisio 2013
Bacillus thuringiensis	Soil	Lipopeptide (Fengycin)	*Escherichia coli, Staphylococcus epidermidis, Candida albicans, Aspergillus niger*	Roy et al., 2013
Lacobacillus spp.	ATCC 29544 (American Type Culture Collection)	-	*Cronobacter sakazakii*	Campana et al., 2019

Contd...

Contd...

BS/BE producing organism	Source	Type of BE/BS	Antimicrobial activity against	References
Lactobacillus paracasei	Dairy plant	-	*Eescherichia coli, Lactobacillus casei, Staphylococcus aureus, Staphylococcus epidermidis, Streptococcus agalactiae, Candida albicans*	Gudina et al., 2010
Lactobacillus plantarum	Kanjika (rice based fermented product)	Glycoprotein	*Escherichia coli, Salmonella typhi, Yersinia enterocolitica, Staphylococcus aureus*	Madhu and Prapulla, 2013
Pseudomonas aeruginosa	WWTP	Rhamnolipid	*Salmonella typhimurium, Escherichia coli, Klebsiella pneumoniae, Acinetobacter* sp., *Enterobacter* sp., *Salmonella enterica* sp. *Staphylococcus aureus, Bacillus cereus, Micrococcus* sp, *Enterococcus* sp. *Cryptococcus neoformans, Candida albicans*	Ndlovu et al., 2017
Staphylococcus saprophyticus SBPS 15	Coastal sediments (petroleum hydrocarbon contaminated)	Glycolipid	*Klebsiella pneumonia, Escheria coli, Staphylococcus aureus, Vibrio cholera, Bacillus subtilis*	Naughton et al., 2019, Mani et al., 2016
Starmerella bombicola MTCC1910	MTCC (Microbial Type Culture Collection)	Glycolipid (Sophorolipid)	*Candida albicans, Candida husitaniae, Candida tropicalis, Candida glabrata*	Haque et al., 2016

Note: '-' Data not available.

Table 3: Anti-adhesion Activity of Bioemulsifiers (BE) and Biosurfactants (BS)

Producer organism	Natural source/Manufacturer	Type	Antiadhesive activity against	Reference
Bacillus cereus NK1	Soil	Lipopeptide	*Pseudomonas aeruginosa, Staphylococcus epidermidis*	Sriram et al., 2011
B. licheniformis	Organic ammendant ENZYVEBA® Nucleobase 2, Italy	Lipopeptide	*Escherichia coli, Staphylococcus aureus,*	Rivardo et al., 2009, 2011
L. plantarum CFR2194	-	Glycoprotein	*Staphylococcus epidermidis*	Gundina et al., 2015
Lactobacillus paracasei	Dairy		*Escherichia coli, Pseudomonas aeruginosa, Lactobacillus casei, Lactobacillus reuteri, Staphylococcus aureus, Staphylococcus epidermidis, Streptococcus agalactiae, Streptococcus mutans*	Gudina et al., 2010
Pseudomonas aeruginosa	Marine soil	Rhamnolipid	*Bacillus pumilus, Yarrowia lipolytica*	Dusane et al., 2012
Pseudomonas aeruginosa AP02-1	Hot spring	Rhamnolipid	*Bacillus subtilis, Staphylococcus aureus, Micrococcus luteus*	Perfumo et al., 2006, Quinn et al., 2013

Note: '-' Data not available

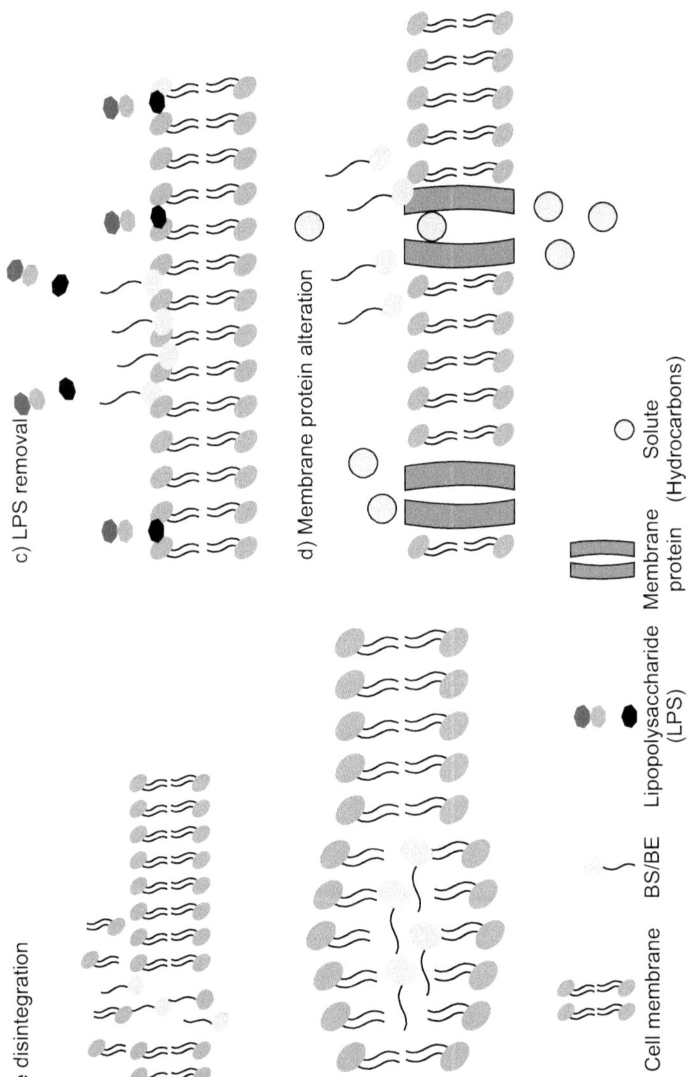

Fig. 2: Mechanism of antimicrobial activity of BS and BE.

affect the release of metabolites (Fig 2). The alteration of the cell surface property is achieved by changing hydrophobicity, biomorphology, electrokinetic potential and functional groups present on the cell surface (Kaczorek et al., 2018). Research papers suggest that rhamnolipid biosurfactants are more effective against Gram positive than Gram negative bacteria as the cell wall in them may exclude these molecules (Sotirova, 2009). It was also reported that rhamnolipids damage the cell membrane by inserting acyl tails and causing leakage of cytoplasmic components and changing the configuration of the cell. It was observed by Sana et al. (2018) that in case of *E. coli* and *S. aureus,* hydrophilic and hydrophobic parts of rhamnolipids interacted with nonpolar parts of the cell membrane. This leads to pore formation in the cell wall, plasma membrane and subsequently leakage of cytoplasmic components which results in cell death (Ortiz et al., 2006).

Surfactin is reported to cause forceful membrane-distortion which depends on the size of the peptide ring and the peptide moiety piercing the cell membrane. This creates an alteration in the charge on the membrane surface (Heerklotz and Seelig, 2001). A similar mechanism is reported by lipopeptide biosurfactants (Sana et al., 2018). Fengycin, a cyclic peptide, has been reported to exhibit strong antifungal activity by inhibiting a few enzymes (Steller and Vater, 2000). The ampiphilic nature of fengycin showed an affinity for lipid bilayers. Roy et al. (2013) reported antifungal action of fengycin like peptides against *C. albicans* by exhibiting membrane blebs which suggests loss of contact between the cell membrane and the cell wall.

It has been observed that the multidrug resistance (MDR) efflux pump present in some Gram-negative bacteria make them resistant to numerous antibiotics (Girish and Smith, 2008). This problem can be solved by lipopeptides and rhamnolipids as they specifically act on cell surfaces only. Lipopetides showed antimicrobial, antifungal and antiviral properties which are associated with their membrane lytic properties (Borsanyiova et al., 2016). Saphorolipids also exhibited antiviral, antibacterial and antibiotic adjuvant characteristics with similar mechanisms like lipopeptides (Shah et al., 2005, Borsanyiova et al., 2016).

Ayed et al. (2015) studied the wound healing property of Lipopeptides produced by *B. mojavensis* A21. The lipopeptides exhibited strong antioxidant activity and were shown to be promising wound healing agents *in vitro* in experimental rats.

3.4 Anti-adhesive Activity

A biofilm is an irreversible accumulation of surface accompanying microbial cells that are bounded in an extracellular polymeric substance matrix. Many microbiologically sensitive surfaces such as potable water system piping, living tissues, indwelling medical devices, industrial or natural aquatic systems provide favourable conditions for biofilm formation (Donlan, 2002). Biofilms are microbial colonies that remain attached to a surface and are composed of exopolysaccharides or slime secreted by the adherent bacteria. In single-species biofilms form in multiple steps and require intercellular signalling. Bacteria that have formed adherent biofilms exist either as a tightly packed unit or as columns of loosely associated cells, some fixed, and others motile. Water channels between pillars of cells in such biofilms allow nutrients to disperse. Biofilms are medically and industrially important because they can accumulate on a wide variety of substrates and are resistant to antimicrobial agents and detergents (Donlan, 2002, Satpute et al., 2016). Biofilms generated on materials such as catheters and prosthesis remain a significant challenge in the biomedical field,

Table 4: Summary of anticancer activities by different BS/BE (last 10 years)

Cancer type	Type of BS or BE	Cell line	Producer microorganism	Activity	Assay/Detection technique	Treatment dose	Activity time (h)	Reference
Breast	Surfactin	MCF-7	*B. subtilis* CSY 191	Growth inhibition	MTT	IC50 = 9.65μM	24	Wu et al., 2017
Breast	Surfactin	-	*B. subtilis*	Inhibition of invasion, migration, and colony formation	Wound healing, Matrigel invasion, gelatin zymography, ChIP	10 μM	24	Wu et al., 2017
Breast	?	?	*B. subtilis* 573	Growth inhibition	MTS	C50 = 93μM	48	Duarte et al., 2014, Wu et al., 2017
Breast	Surfactin	Bcap-37	*B. subtilis* Hs0121	Growth inhibition	MTT	IC50 = 29 ± 2.4μM	24	Liu, X et al., 2010
Breast	Rhamnolipid	MCF-7	*Pseudomonas aeruginosa* MR01	Induce apoptosis	MTT	IC50 = 0.0017 μg/ml	24	Rahimi et al., 2019
Breast		MDA MB 231	*Pseudomonas* sp. strain ICTB	Growth inhibition	Cytotoxicity assay	IC50 in μg/ml 86,58	24	Kumar et al., 2018
Cervical	Surfactin	HeLa	*B. subtilis* HSO121	Growth inhibition	MTT	IC50 = 37 ± 4.5μM	24	Liu, X et al., 2010 Wu et al., 2017
Cervical	Monoolein	HeLa, MCF-7, HepG2	*Exophiala dermatitidis*	Growth inhibition	MTT	IC50 =10 μg/ml	48	Chiewpattanakul et al., 2010
Cervical	Rhamnolipid -1 Rhamno-lipid-2	HeLa	*Pseudomonas* sp. strain ICTB	Growth inhibition	Cytotoxicity assay	IC50 in μg/ml 123,88	-	Kumar et al., 2018
Cervix adeno-carcinoma	e-poly-L-lysine	Hela S3,	*Bacillus subtilis* SDNS	Growth inhibition	MTT	-		El-Sersy et al., 2012
Colon	Surfactin	HCT15	*B. circulans* DMS-2	Growth inhibition	MTT	IC50 = 77μM	24	Sivapathasekaran et al., 2010

Contd...

Contd...

Cancer type	Type of BS or BE	Cell line	Producer microorganism	Activity	Assay/Detection technique	Treatment dose	Activity time (h)	Reference
Esophageal	trehalose lipid tetraester (THL)	BV-173 and SKW-3	*Nocardia farcinica* strain BN26	Concentration-dependent antiproliferative activity	MTT	IC 50 18.7 (17.4-20.2) µM 28.9 (26.8-31.2) µM		Christov et al., 2015
Hepatocellular	Surfactin	BEL7402	*B. subtilis* HSO121	Growth inhibition	MTT	IC50 = 35 ± 12µM	24	Liu, X et al., 2010
Hepatocellular	Surfactin	HepG2	*B. natto* TK-1	Apoptosis induction	DCFH-DA and analysis of [Ca2+]i	41µM	-	Wang et al., 2013
Hepatocellular liver cancer	Rhamnolipid -1 Rhamno-lipid-2	HepG2	*Pseudomonas* sp. strain ICTB	Growth inhibition	Cytotoxicity assay	IC50 in µg/ml 140,79	-	Kumar et al., 2018
Hepatocel-lular liver carcinoma	e-poly-L-lysine	HepG2	*Bacillus subtilis* SDNS	Growth inhibition	MTT	-	-	El-Sersy et al., 2012
Leukemia	Monoolein	U937	*Exophiala dermatitidis*	Growth inhibition	MTT	IC50 =10 µg/ml	48	Chiewpattanakul et al., 2010
Lung cancer	Rhamnolipid -1 Rhamno-lipid-2	A549	*Pseudomonas* sp. strain ICTB	Growth inhibition	Cytotoxicity assay	IC50 in µg/ml 54,98	-	Kumar et al., 2018
Oral epidermoid	-	KB-3-1	*B. subtilis* HSO121	Growth inhibition	MTT	IC50 = 57 ± 2.6µM	24	Liu, X et al., 2010, Wu et al., 2017
Pancreatic	-	SW1990	*B. subtilis* HSO121	Growth inhibition	MTT	IC50 = 58 ± 1.6µM	24	Liu, X et al., 2010 Wu et al., 2017
Rat melanoma	-	B16	*B. subtilis* HSO121	Growth inhibition	MTT	IC50 = 20 ± 1.5µM	24	Liu, X et al., 2010 Wu et al., 2017

Note: '-' Data not available.

as they tend to be more tolerant to antimicrobial treatments (Mireles et al., 2001). Biofilm infection can be limited by preventing microbial adhesion to the surfaces of medical devices. Another remarkable property of biosurfactants is its inhibitory activity against adhesion and colonization of bacterial and fungal cells on surfaces, particularly those of biomedical interest. They do so by affecting the development of flagella, and bringing about changes in the attachment capability of bacteria (Splendiani et al., 2006). Biosurfactants are able to modify bacterial surface hydrophobicity and consequently, microbial adhesion to solid surfaces.

Biosurfactants produced by *Lactococcus lactis* and *Streptococcus thermophilus* are able to decrease the number of bacteria in a multi-species biofilm on voice prosthesis (Rodrigues et al., 2004). Mireles et al. (2001) observed that Surfactin could promote dispersal of a preformed biofilm and inhibits *Salmonella enterica* adhesion to PVC. They also found that surfactin and rhamnolipids are able to prevent the *Escherichia coli* and *Proteus mirabilis* biofilm attachment. According to Meylheuc et al. (2001) rhamnolipid preconditioned PTFE surface was able to reduce *Listeria monocytogenes* attachment. Table 3 enlists the Anti-adhesion activity of the BS and BE produced by different organisms.

3.5 Anti-Cancer Activity

Cancer is still the highest cause of death worldwide, which needs to be addressed soon. In cancer treatment chemotherapy remains inevitable though chemotherapeutic drugs are cytotoxic and have side effects on human health (Sak, 2012, Dey et al., 2015). In this regard, there is an urge to search and develop a new anticancer agent that can act specifically and selectively on target cancer cells (Gudiña et al., 2016, Wu et al., 2017). Biosurfactants, mainly lipopeptides and glycolipids, have been studied extensively for their potential as anti-cancer agents. These compounds have been involved in numerous intercellular molecular recognition stages comprising signal transduction, cell immune response and cell differentiation. In addition to this, BS has advantages such as low toxicity, biodegradability and high efficacy, which are significant features required for any drug to act as an anti-cancer agent. Different mechanisms which are essential as anti-cancer agents as anticipated by biosurfactants include, cell cycle inhibition, inhibition of invasion, inhibition of crucial signalling pathways mainly Atk pathway, MAPK signalling pathway, activator of transcription (JAK/STAT), Janus kinase/ signal transducer, activation of natural killer T (NKT) cells and induction of apoptosis in cancer cells (Gudiña et al., 2016) as shown in Fig 3. Advantage of Biosurfactant over the other anti cancer drugs is that it increases cell permiability, which leads to metabolite leakage thus can contribute as potential anticancer agent. Although BS has great potential and diversity in their chemical structure, very few known BS have been studied for anticancer activity. Hence, there is a need to find novel BS or BE exhibiting good anticancer activity which may offer different mechanisms of action on different targets.

Recently, surfactin, has shown cytotoxic effects against different cancer types, such as colon cancer, breast cancer, leukemia and hepatoma (Gudiña et al., 2016). Surfactin, shows anticancer activity by interaction of fatty acid moiety, membrane-bound phospholipids and its peptide moiety (Liu et al., 2010). Surfactin exhibits different mechanisms for its anticancer activity according to the cancer cell type. These activities include, inhibition of matrix metalloproteinases, signalling pathways

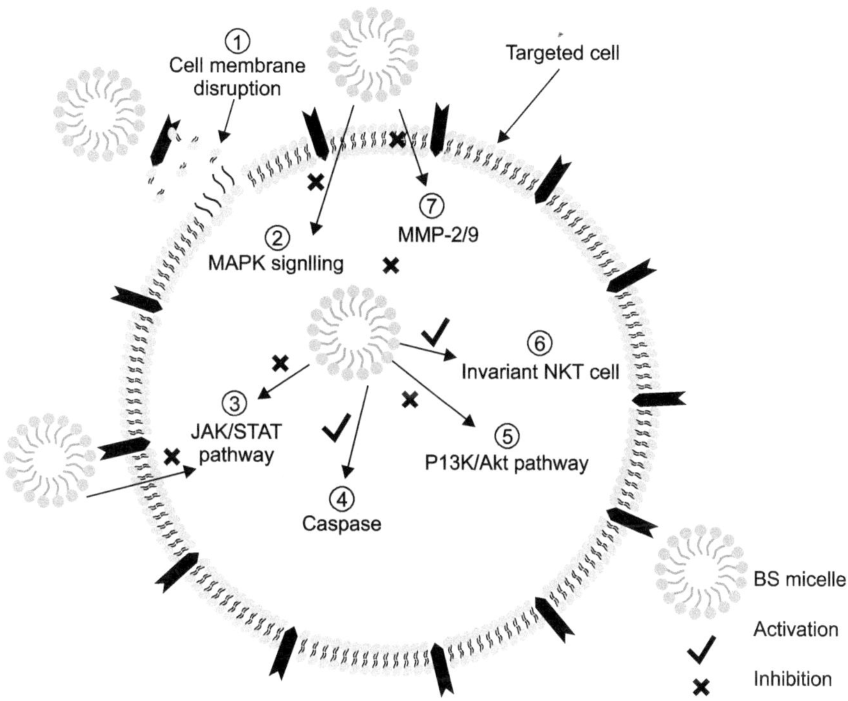

Fig. 3: Anticancer activity exhibited by BS in cancer therapy.

and induction of apoptosis (Park et al., 2013, Gudiña et al., 2016). Iturins contain cyclic peptide chains conjugated with beta-amino fatty acids (13 to 17 carbons). They hinder proliferation and inhibit the Akt signalling network which leads to apoptosis induction in breast cancer cells. Another lipopeptide FW523-3, blocks the ERK and p38 signalling pathways. Apratoxins, cyclic lipopeptides, exhibit induced apoptosis and inhibition of the IL-6 signalling pathway in human bone osteosarcoma U2OS cells. Other derivatives of apratoxin also showed cytotoxicity on human NCI-H460 lung cancer cells, colon cancer cells and mouse models (Thornburg et al., 2013).

Glycolipids such as rhamnolipids exhibit anti-cancer potential against cervical cancer, colon cancer and breast cancer. It does so by interacting with tubulin via several hydrophobic, vander-waal interactions and hydrogen bonding with certain amino acids (Arginine, Tyrosine, Leucine, Asparagine, Alanine) present on tubulin. Many glycolipids inhibit growth of HeLa cells and also induce apoptosis. They also may penetrate into the mitochondria and generate oxidative stress which activates p53 mediated intrinsic apoptosis pathway (Ramalingam et al., 2016). Table 4 summarizes main potential anticancer activity exhibited by different BS in the last 10 years.

4. Role of Biosurfactants and Bioemulsifers in Combating COVID-19 Pandemic

Recently, action of BE and BS on the coronavirus was studied by Smith et al. (2020). According to their study, owing to their amphiphilic and biodegradable nature and low cytotoxicity biosurfactants are better suited in dealing with the current pandemic. As BE and BS can be used very effectively in sanitizer and soap preparations, their importance was proved not only in the cosmetic and pharmaceutical industry but also in biomedical applications. Biosurfactants can penetrate the coronavirus membrane, destroy it and encapsulate the fragments of the coronavirus by forming micelles and washing it away (Smith et al., 2020).

5. Applications of BE and BS in the Cosmetic Industry

Chemical surfactants and emulsifiers in cosmetic formulations have the potential to cause skin irritations and allergic reactions. They interact with proteins and remove lipids from the epidermal surface. Surfactants cause disorganization of the intercellular structure of lipids (Vecino et al., 2017). Therefore, BE and BS prove to be better alternatives to chemical surfactants in cosmetic formulations. BE and BS can be used in formulating cosmetic products like hair colour and hair care products, skin care and foot care products, creams, lotions, body massage products, soap and bathing products, lipsticks, eye shadows, mascaras, shaving creams, acne pads, antiseptics, antiperspirants and deodorants, contact lens solution, lubricated condoms and other health and beauty products (Vecino et al., 2017, Ahmadi-Ashtiani et al., 2020). Table 5 summarizes the applications of BE and BS in the cosmetic industry.

Table 5: Applications of Bioemulsifiers (BE) and Biosurfactants (BS) in the cosmetic industry

Relevant cosmetic product/branch	BS/BE	Function	Reference
Anti-aging skin care products	MEL	Protects cell against oxidative stress	Battacharya et al., 2017, Takahashi et al., 2012
Conditioning hair mask	Oligomeric BS	Moisturises the hair	Battacharya et al., 2017, Owen and Fan, 2013
Moisturizing skin cleanser	Rhamnolipids and Sophorolipids	Cleansing, moisturising and antimicrobial effect	Battacharya et al., 2017, Allef et al., 2014
Shampoo	Rhamnolipids	Antimicrobial effect	Battacharya et al., 2017, Desanto et al., 2008
Shower gel	Rhamnolipids and Sophorolipids	Cleansing and antimicrobial effect	Battacharya et al., 2017, Cox et al., 2013
Skin nourishing	Rhamnolipids	Protects against harsh climate and UV	Battacharya et al., 2017, Suzuki et al., 2011

The emulsifier function is probably the most important property of BS and BE in the formulation of cosmetic emulsions as these emulsions are easy to apply and are relatively inexpensive. BS and BE also assist the permeation of active agents in the skin (Marchant and Banat, 2012). Along with emulsifying properties, other properties such as anti-adhesive, anti-fungal, anti-bacterial and anti-viral activities make BS and BE highly beneficial for cosmetic and personal care applications. Fatty acid chains of BS and BE act as natural antioxidants by preventing generation of free radicals by the UV light. Properties such as the critical micelle concentration (CMC), hydrophilic–lipophilic balance (HLB), and the ionic performance determine the choice of BE and BS in cosmetic formulations. The concentration of surfactants at which the surface becomes saturated with surfactant molecules, and micelles start to form is known as CMC. Lower the CMC of a surfactant the greater is its surface-active efficiency (Rufino et al., 2011). The balance between the hydrophobic and hydrophilic moiety of the BE is also known as HLB. Greater the lipophilic character of BS and BE, lower are their HLB values. Based on the HLB values a BE or BS can act as an emulsifier, wetting agent or antifoaming agent. Since the lipid film on the skin favours oil-soluble active agents, W/O emulsions with HLB values between 1 and 4, are preferable for dermatological applications. However, O/W emulsions with HLB values between 8 and 16 have a semi-solid or liquid consistency and are less greasy and are therefore appreciated more by the consumers. Biosurfactants and bioemulsifiers can be classified as anionic, cationic, non-ionic or amphoteric based on the charge on the polar head. As compared with cationic BE and BS, the anionic BE and BS have greater wetting, foaming and emulsifying properties. However anionic BE and BS cause high skin and eye irritation. Cationic BS and BE show excellent anti-bacterial activity (Vecino et al., 2017, Bhattacharya et al., 2017). However, before introducing BS and BE in cosmetic formulations, it is necessary to evaluate the toxicity of these compounds (Vecino et al., 2017).

The skin contains natural resident microflora that include bacteria, viruses and many types of fungi. Around 90% of skin microflora is *Staphylococcus epidermidis*, other organisms include *Propionibacterium, Staphylococcus, Micrococcus, Corynebacterium* and *Acinetobacter* (Percival et al., 2012). These organisms protect the human skin from infections and other environmental pollutants. Pathogens such as *Staphylococcus aureus* are also present on the human skin. Some BS have been shown to exhibit inhibitory effects against different pathogenic microorganisms like *S. aureus, Streptococcus mutans, E. coli* and *P. aeruginosa* (Gudiña et al., 2010). Moreover, in the presence of BE, skin cells increase the oxygen consumption as a defence mechanism. Substances with the ability to rebalance the skin microflora are known as prebiotic ingredients. BS and BE obtained from GRAS (Generally Regarded as Safe) organisms such as organisms belonging to genus *Lactobacillus, Bacillus, Kluyveromyces, Candida* and *Saccharomyces*, could be included as potential prebiotics in cosmetic formulations (Bhattacharya et al., 2017). The microflora of skin with acne could be rebalanced by promoting the growth of beneficial bacteria like *S. epidermidis*, and inhibiting the growth of harmful bacteria such as *P. acnes.* The ingredients used in cosmetic formulations should not destroy or negatively affect the natural microflora of the skin. It is important to identify and eliminate all substances that can generate irritations or allergic reactions on the skin (Vecino et al., 2017).

Glycolipids such as rhamnolipids containing cosmetics have been patented for their use as anti-wrinkle and anti-ageing products (Lourith and Kanlayavattanakul, 2009). Rhamnolipids produced by *P. aeruginosa* can be used to formulate a shampoo as they show antimicrobial activity. The formulation also maintains lustre and leaves the scalp free from odour up to three days. Oligomeric BS can be used in topical formulations like anti-aging facial gels, anti-aging creams and conditioning hair masks. Mannosylerythritol lipids (MEL) can be used as active agents in skin care cosmetic products to prevent skin roughness (Varvaresou and Iakovou, 2015). MELs have potential as anti-aging skin care ingredients based on the studies carried out by Takahashi et al. (2012). They evaluated the antioxidant capacity of MEL derivatives using fibroblasts NB1RGB cells. MEL-C showed the highest antioxidant activity and also presented good protective effects in cells against oxidative stress. BS obtained from *Nocardiopsis* can prove to be an alternative to SDS in cosmetic toothpaste formulations as they are more effective and less toxic (Vecino et al., 2017, Varvaresou and Iakovou, 2015, Bhattacharya et al., 2017).

6. Conclusion

Biosurfactants show tremendous potential in the pharmaceutical and cosmetic industry. Owing to their biodegradability, ability to be produced from renewable resources and exhibition of a spectrum of activities like anticancer, antimicrobial, antiadhesive and immunomodulator, not to forget the drug delivery ability make them way superior to their synthetic counterparts Although their use has been limited because they are economically uncompetitive, their lucrative properties make them excellent candidates for translation research to replace the synthetic surfactants in day to day usage. With concerted efforts to upscale to industry standards, they could well be the eco-friendly alternative in the pharma and cosmetic industry for a cleaner and greener tomorrow.

Chemically synthesized surfactants and emulsifiers are extensively used in cosmetic, pharmaceutical, agriculture, food and petroleum industries. However, according to the current scenario, biosurfactants can easily replace chemical surfactants as they possess many green properties compared to them. To achieve this, there is a need to understand physiology, biochemistry and genetics of the biosurfactant producing strains and to expand the process technology to reduce production costs.

References

Ahmadi-Ashtiani, H.R., Baldisserotto, A., Cesa, E., Manfredini, S., Zadeh, H.S., Gorab, M.G. et al. (2020). Microbial biosurfactants as key multifunctional ingredients for sustainable cosmetics. *Cosmetics*, 7(2): 1-46.

Ali, S.A.M., Sayyed, R.Z., Mir, M.I., Hameeda, B., Khan, Y., Alkhanani, M.F. et al. (2022a). Induction of systemic resistance in maize and antibiofilm activity of surfactin from Bacillus velezensis MS20. *Front. Microbiol.*, 13: 879739. https://doi.org/10.3389/fmicb.2022.879739

Alizadeh-Sani, M., Hamishehkar, H., Khezerlou, A., Azizi-Lalabadi, M., Azadi, Y., Nattagh-Eshtivani, E. et al. (2018). Bioemulsifiers derived from microorganisms: Applications in the drug and food industry. *Advanced Pharmaceutical Bulletin*, 8(2): 1-191.

Allef, P., Hartung, C. & Schilling, M. (2014). Aqueous hair and skin cleaning compositions comprising biosurfactants. Patent US 20140349902 A1

Amaral, P.F.F., Da Silva, J.M., Lehocky, B.M., Barros-Timmons, A.M.V., Coelho, M.A.Z., Marrucho, I.M. et al. (2006). Production and characterization of a bioemulsifier from *Yarrowia lipolytica*. *Process Biochemistry*, 41(8): 1894-1898.

Banat, I.M. and Thavasi, R. (2019). Microbial biosurfactants and their environmental and industrial applications. CRC Press, Taylor and Francis Group. UK.

Bee, H., Khan, M.Y. & Sayyed, R.Z. (2019). Microbial surfactants and their significance in agriculture. *In:* Sayyed Reddy Antonious (Eds), *PGPR: Prospects for Sustainable Agriculture*. 205-216. Springer-Nature, Singapore.

Bhadoriya, S.S., Madoriya, N., Shukla, K. & Parihar, M.S. (2013). Biosurfactants: A new pharmaceutical additive for solubility enhancement and pharmaceutical development. *Biochem. Pharmacol. Open Access.*, 2: 1-113.

Bhattacharya, B., Ghosh, T.K. & Das N. (2017). Application of biosurfactants in cosmetics and pharmaceutical industry. *Scholars Academic Journal of Pharmacy*, 6: 320-329.

Borsanyiova, M., Patil, A., Mukherji, R., Prabhune, A. & Bopegamage, S. (2016). Biological activity of sophorolipids and their possible use as antiviral agents. *Folia Microbiologica*, 61(1): 85-89.

Borzeix, F., Hillion G., Marchal, C. & Stoltz, C. (1995). Use of sophorolipids in cosmetic and dermatological compositions. European Patent No. EP0820273B1

Brown, M.J. (1991). Biosurfactants for cosmetic applications. *International Journal of Cosmetic Science*, 13(2): 61-64.

Brown, P., Butts, C.P. & Eastoe, J. (2013). Stimuli-responsive surfactants. *Soft Matter.*, 9(8): 2365-2374.

Blenkinsopp, A. & Paxton, P. (1998). Symptoms in the Pharmacy: A Guide to the Management of Common Illness, 3rd ed. Oxford: Blackwell Science.

Cameotra, S.S. & Makkar, R.S. (2004). Recent applications of biosurfactants as biological and immunological molecules. *Current Opinion in Microbiology*, 7(3): 262-266.

Campana, R., Federici, S., Ciandrini, E., Manti, A. & Baffone, W. (2019). *Lactobacillus* spp. inhibit the growth of *Cronobacter sakazakii* ATCC 29544 by altering its membrane integrity. *Journal of Food Science and Technology*, 56(8): 3962-3967.

Chereshnev, V.A., Gein, S.V., Baeva, T.A., Galkina, T.V., Kuyukina, M.S. & Ivshina, I.B. (2010). Modulation of cytokine secretion and oxidative metabolism of innate immune effectors by *Rhodococcus* biosurfactant. *Bulletin of Experimental Biology and Medicine*, 149(6): 1- 734.

Chiewpattanakul, P., Phonnok, S., Durand, A., Marie, E. & Thanomsub, B.W. (2010). Bioproduction and anticancer activity of biosurfactant produced by the dematiaceous fungus *Exophiala dermatitidis* SK80. *J. Microbiol. Biotechnol.*, 20(12): 1664-1671.

Christova, N., Lang, S., Wray, V., Kaloyanov, K., Konstantinov, S. & Stoineva, I. (2015). Production, structural elucidation and in vitro antitumor activity of trehalose lipid biosurfactant from *Nocardia farcinica* strain. *J. Microbiol. Biotechnol.*, 25(4): 439-447.

Cirigliano, M.C. & Carman, G.M. (1985). Purification and characterization of liposan, a bioemulsifier from *Candida lipolytica. Applied and Environmental Microbiology*, 50(4): 846-850.

Cox, T.F., Crawford, R.J., Gregory, L.G., Hosking, S.L. & Kotsakis, P. (2013). U.S. Patent No. 8,563,490. Washington, DC: U.S. Patent and Trademark Office.

De Rienzo, M.A.D., Banat, I.M., Dolman, B., Winterburn, J. & Martin, P.J. (2015). Sophorolipid biosurfactants: Possible uses as antibacterial and anti-biofilm agent. *New Biotechnology*, 32(6): 720-726.

Desanto, K. (2008). Rhamnolipid-based formulations. World Patent WO 2008013899 A3.

Dey, G., Bharti, R., Sen, R. & Mandal M. (2015). Microbial amphiphiles: A class of promising new-generation anticancer agents. *Drug Discovery Today*. 20(1): 136-146.

Diniz Rufino, R., Moura de Luna, J., de Campos Takaki, G.M. & Asfora Sarubbo, L. (2014). Characterization and properties of the biosurfactant produced by *Candida lipolytica* UCP 0988. *Electronic Journal of Biotechnology*, 17(1): 34-38.

Donlan, R.M. (2002). Biofilms: Microbial life on surfaces. *Emerging Infectious Diseases*. 8(9): 881.

Dusane, D.H., Dam, S., Nancharaiah, Y.V., Kumar, A.R., Venugopalan, V.P. & Zinjarde, S.S. (2012). Disruption of *Yarrowia lipolytica* biofilms by rhamnolipid biosurfactant. *Aquatic Biosystems*, 8(1): 1-7.

Doe, J. (15 Jan 1999). The dictionary of Substances and Their Effects. Royal Society of Chemistry. http://www.rsc.org/dose/title of subordinate document.

Eastoe, J., Hatzopoulos, H.M. & Tabor, R. Micellar systems, UK. Tharwat Tadros et al. (2013). Encyclopedia of colloid and interface science. Wokingham, Berkshire, UK. 685-731.

El-Sersy, N.A., Abdelwahab, A.E., Abouelkhiir, S.S., Abou-Zeid, D.M. & Sabry, S.A. (2012). Antibacterial and anticancer activity of ε-poly-L-lysine (ε-PL) produced by a marine *Bacillus subtilis* sp. *Journal of Basic Microbiology*, 52(5): 513-522.

Etemadzadeh, S.S., Emtiazi, G. & Soltanian, S. (2023). Production of biosurfactant by salt-resistant *Bacillus* in lead-supplemented media: Application and toxicity. *Int. Microbiol.*, 26(4): 869-880. doi: 10.1007/s10123-023-00334-4. Epub 2023 Feb 22.

Faivre, V. & Rosilio, V. (2010). Interest of glycolipids in drug delivery: From physicochemical properties to drug targeting. *Expert Opinion on Drug Delivery*, 7(9): 1031-1048.

Fernandes, A.P.V., Arruda, I.R., Fernando, A., Santos, B.A., Araújo, A.A.A., Maior, A.M.A. et al. (2007). Antimicrobial activity of surfactants produced by *Bacillus subtilis* R14 against multidrug-resistant bacteria. *Brazilian Journal of Microbiology*, 38: 704-709.

Fracchia, L., Cavallo, M., Allegrone, G. & Martinotti, M.G. (2010). A lactobacillus-derived biosurfactant inhibits biofilm formation of human pathogenic *Candida albicans* biofilm producers. *Appl. Microbiol. Biotechnol.*, 2: 827-837.

Frelichowska, J., Bolzinger, M.A., Valour, J.P., Mouaziz, H., Pelletier, J. & Chevalier, Y. (2009). Pickering w/o emulsions: Drug release and topical delivery. *International Journal of Pharmaceutics*, 368(1-2): 7-15.

Gharaei-Fathabad, E. (2010). Biosurfactants in pharmaceutical industry: A mini review. *American Journal of Drug Discovery and Development*, 1: 58-69.

Girish, C.K. & Smith T.K. (2008). Effects of feeding blends of grains naturally contaminated with *Fusarium* mycotoxins on small intestinal morphology of turkeys. *Poultry Science*, 87(6): 1075-1082.

Gołek, P., Bednarski, W., Brzozowski, B. & Dziuba, B. (2009). The obtaining and properties of biosurfactants synthesized by bacteria of the genus *Lactobacillus*. *Annals of Microbiology*. 59(1): 119-126.

Gudiña, E.J., Rocha, V., Teixeira, J.A. & Rodrigues, L.R. (2010). Antimicrobial and antiadhesive properties of a biosurfactant isolated from *Lactobacillus paracasei* ssp. *paracasei* A20. *Letters in Applied Microbiology*, 50(4): 419-424.

Gudiña, E.J., Rangarajan, V., Sen, R. & Rodrigues, L.R. (2013). Potential therapeutic applications of biosurfactants. *Trends in Pharmacological Sciences*, 34(12): 667-675.

Gudiña, E.J., Teixeira, J.A. & Rodrigues, L.R. (2016). Biosurfactants produced by marine microorganisms with therapeutic applications. *Marine Drugs*, 14(2): 38.

Hagler, M., Smith-Norowitz, T.A., Chice, S., Wallner, S.R., Viterbo, D., Mueller, C. et al. (2007). Sophorolipids decrease IgE production in U266 cells by downregulation of BSAP (Pax5), TLR-2, STAT3 and IL-6. *Journal of Allergy and Clinical Immunology*, 119(1): 263.

Hanen Ben Ayed, Sana Bardaa, Dorsaf Moalla, Mourad Jridi, Hana Maalej, Zouheir Sahnoun et al. (2015). Wound healing and in vitro antioxidant activities of lipopeptides mixture produced by *Bacillus mojavensis* A21. *Process Biochemistry*, 50(6): 1023-1030.

Haque, F., Alfatah, M., Ganesan, K. & Bhattacharyya, M.S. (2016). Inhibitory effect of sophorolipid on *Candida albicans* biofilm formation and hyphal growth. *Scientific Reports*, 6: 23575.

Hardin, R., Pierre, J., Schulze, R., Mueller, C.M., Fu, S.L., Wallner, S.R. et al. (2007). Sophorolipids improve sepsis survival: Effects of dosing and derivatives. *Journal of Surgical Research*, 142(2): 314-319.

Heerklotz, H. & Seelig, J. (2001). Detergent-like action of the antibiotic peptide surfactin on lipid membranes. *Biophysical Journal*, 81(3): 1547-1554.

Inès, M. & Dhouha, G. (2015). Glycolipid biosurfactants: Potential related biomedical and biotechnological applications. *Carbohydrate Research*, 416: 59-69.

Irwin, M. and Suzanne, R. Rosenthal. (2018). Immunomodulators. Crohn's and Colitis Foundation. New York.

Kaczorek, E., Pacholak, A., Zdarta, A. & Smułek, W. (2018. The impact of biosurfactants on microbial cell properties leading to hydrocarbon bioavailability increase. *Colloids and Interfaces*, 2(3): 35.

Kalepu, S., Manthina, M. & Padavala, V. (2013). Oral lipid-based drug delivery systems – An overview. *Acta Pharmaceutica Sinica B*, 3(6): 361-372.

Kanlayavattanakul, M. & Lourith, N. (2010). Lipopeptides in cosmetics. *International Journal of Cosmetic Science*, 32(1): 1-8.

Khan, B.A., Akhtar, N., Khan, H.M.S., Waseem, K., Mahmood, T., Rasul, A. & Khan, H. (2011). Basics of pharmaceutical emulsions: A review. *African Journal of Pharmacy and Pharmacology*, 5(25): 2715-2725.

Kumar, R. & Das, A.J. (2018). Application of rhamnolipids in medical sciences. *Rhamnolipid Biosurfactant*, 79-87. Springer.

Lang, S. & Wullbrandt, D. (1999). Rhamnose lipids – biosynthesis, microbial production and application potential. *Applied Microbiology and Biotechnology*, 51(1): 22-32.

Liu, X., Tao, X., Zou, A., Yang, S., Zhang, L. & Mu, B. (2010). Effect of the microbial lipopeptide on tumour cell lines: Apoptosis induced by disturbing the fatty acid composition of cell membrane. *Protein Cell*, 1: 584-594.

Lourith, N. & Kanlayavattanakul, M. (2009). Natural surfactants used in cosmetics: Glycolipids. *International Journal of Cosmetic Science*, 31(4): 255-261.

Luna, J.M., Rufino, R.D., Sarubbo, L.A., Rodrigues, L.R., Teixeira, J.A. & de Campos-Takaki, G.M. (2011). Evaluation antimicrobial and antiadhesive properties of the biosurfactant Lunasan produced by *Candida sphaerica* UCP 0995. *Current Microbiology*, 62(5): 1527-1534.

Madhu, A.N. & Prapulla, S.G. (2014). Evaluation and functional characterization of a biosurfactant produced by *Lactobacillus plantarum* CFR 2194. *Applied Biochemistry and Biotechnology*, 172(4): 1777-1789.

Mani, P., Dineshkumar, G., Jayaseelan, T., Deepalakshmi, K., Kumar, C.G. & Balan, S.S. (2016). Antimicrobial activities of a promising glycolipid biosurfactant from a novel marine *Staphylococcus saprophyticus* SBPS 15. 3 *Biotech.*, 6(2): 163.

Maphosa, Y. & Jideani, V.A. (2018). Factors affecting the stability of emulsions stabilised by biopolymers. *In: Science and Technology Behind Nanoemulsions*. IntechOpen.

Marchant, R. & Banat, I.M. (2012). Biosurfactants: A sustainable replacement for chemical surfactants. *Biotechnology Letters*, 34(9): 1597-1605.

Marti-Mestres, G. & Nielloud, F. (2002). Emulsions in health care applications – An overview. *Journal of Dispersion Science and Technology*, 23(1-3): 419-439.

Mireles, J.R., Toguchi, A. & Harshey, R.M. (2001). *Salmonella enterica serovar Typhimurium* swarming mutants with altered biofilm-forming abilities: Surfactin inhibits biofilm formation. *Journal of Bacteriology*, 183(20): 5848-5854.

Mnif, I. & Ghribi, D. (2015). Review lipopeptides biosurfactants: Mean classes and new insights for industrial, biomedical, and environmental applications. *Peptide Science*, 104(3): 129-147.

Morita, T., Fukuoka, T., Imura, T. & Kitamoto, D. (2013). Production of mannosylerythritol lipids and their application in cosmetics. *Applied Microbiology and Biotechnology*, 97(11): 4691-4700.

Morita, T., Kawamura, D., Morita, N., Fukuoka, T., Imura, T., Sakai, H. et al. (2013). Characterization of mannosylerythritol lipids containing hexadecatetraenoic acid produced from cuttlefish oil by *Pseudozyma churashimaensis* OK96. *Journal of Oleo Science*, 62(5): 319-327.

Mujumdar, S., Bhandari, T. & Atre, C. (2017). Industrial and biological applications of bioemulsifiers: A mini review. *Journal of Biomedical and Pharmaceutical Research*, 4: 26-35.

Murthy, S.N. & Shivakumar, H.N. (2010). Topical and transdermal drug delivery. *In: Handbook of Non-Invasive Drug Delivery Systems*. 1-36. William Andrew Publishing.

Naughton, P.J., Marchant, R., Naughton, V. & Banat, I.M. (2019). Microbial biosurfactants: Current trends and applications in agricultural and biomedical industries. *Journal of Applied Microbiology*, 127(1): 12-28.

Navon-Venezia, S., Zosim, Z., Gottlieb, A., Legmann, R., Carmeli, S., Ron, E.Z. et al. (1995). Alasan, a new bioemulsifier from *Acinetobacter radioresistens*. *Applied and Environmental Microbiology*, 61(9): 3240-3244.

Nawale, L., Dubey, P., Chaudhari, B., Sarkar, D. & Prabhune, A. (2017). Anti-proliferative effect of novel primary cetyl alcohol derived sophorolipids against human cervical cancer cells HeLa. *PloS One*, 12(4): 1-14.

Ndlovu, T., Rautenbach, M., Vosloo, J.A., Khan, S. & Khan, W. (2017). Characterisation and antimicrobial activity of biosurfactant extracts produced by *Bacillus amyloliquefaciens* and *Pseudomonas aeruginosa* isolated from a wastewater treatment plant. *AMB Express*, 7(1): 108.

Ohadi, M., Shahravan, A., Dehghannoudeh, N., Eslaminejad, T., Banat, I.M. & Dehghannoudeh, G. (2020). Potential use of microbial surfactant in microemulsion drug delivery system: A systematic review. *Drug Design, Development and Therapy*, 14: 541.

Ortiz, A., Teruel, J.A., Espuny, M.J., Marqués, A., Manresa, Á. & Aranda, F.J. (2006). Effects of dirhamnolipid on the structural properties of phosphatidylcholine membranes. *International Journal of Pharmaceutics*, 325(1-2): 99-107.

Owen, D.R. & Fan, L. (2013). U.S. Patent No. 8,431,523. Washington, DC: U.S. Patent and Trademark Office.

Patel, P., Bhatt, S., Patel, H., Marcelino, L.A. & Sayyed, R.Z. (2022). Biosurfactant – A biomolecules and its potential applications. *In:* Sayyed, R.Z. and Enshasy, H.E. (Eds), *Biosurfatnats: Production and Applications in Food and Agriculture.* Vol II. 133-149. CRC Press, Taylor & Francis Group, USA.

Park, S.Y., Kim, J.H., Lee, Y.J., Lee, S.J. & Kim, Y. (2013). Surfactin suppresses TPA-induced breast cancer cell invasion through the inhibition of MMP-9 expression. *International Journal of Oncology*, 42(1): 287-296.

Percival, S.L., Emanuel, C., Cutting, K.F. & Williams, D.W. (2012. Microbiology of the skin and the role of biofilms in infection. *International Wound Journal*, 9(1): 14-32.

Perfumo, A., Banat, I.M., Canganella, F. & Marchant, R. (2006). Rhamnolipid production by a novel thermophilic hydrocarbon-degrading *Pseudomonas aeruginosa* AP02-1. *Applied Microbiology and Biotechnology*, 72(1): 132.

Quinn, G.A., Maloy, A.P., Banat, M.M. & Banat, I.M. (2013). A comparison of effects of broad-spectrum antibiotics and biosurfactants on established bacterial biofilms. *Current Microbiology*, 67(5): 614-623.

Rahimi, K., Lotfabad, T.B., Jabeen, F. & Ganji, S.M. (2019). Cytotoxic effects of mono- and di-rhamnolipids from *Pseudomonas aeruginosa* MR01 on MCF-7 human breast cancer cells. *Colloids and Surfaces B: Biointerfaces*, 181: 943-952.

Ramalingam, V., Varunkumar, K., Ravikumar, V. & Rajaram, R. (2016). Development of glycolipid biosurfactants for inducing apoptosis in HeLa cells. *RSC Advances*, 6(68): 64087-64096.

Rebello, S., Asok, A.K., Mundayoor, S. & Jisha, M.S. (2014). Surfactants: Toxicity, remediation and green surfactants. *Environmental Chemistry Letters*, 12(2): 275-287.

Rivardo, F., Martinotti, M.G., Turner, R.J. & Ceri, H. (2011). Synergistic effect of lipopeptide biosurfactant with antibiotics against *Escherichia coli* CFT073 biofilm. *International Journal of Antimicrobial Agents*, 37(4): 324-331.

Rivardo, F., Turner, R.J., Allegrone, G., Ceri, H. & Martinotti, M.G. (2009). Anti-adhesion activity of two biosurfactants produced by *Bacillus* spp. prevents biofilm formation of human bacterial pathogens. *Applied Microbiology and Biotechnology*, 83(3): 541-553.

Rodrigues, L., Van der Mei, H.C., Teixeira, J. & Oliveira, R. (2004). Influence of biosurfactants from probiotic bacteria on formation of biofilms on voice prostheses. *Applied and Environmental Microbiology*, 70(7): 4408-4410.

Rodrigues, L., Banat, I.M., Teixeira, J. & Oliveira, R. (2006). Biosurfactants: Potential applications in medicine. *Journal of Antimicrobial Chemotherapy*, 57(4): 609-618.

Rodrigues, L., Van der Mei, H., Banat, I.M., Teixeira, J. & Oliveira, R. (2006). Inhibition of microbial adhesion to silicone rubber treated with biosurfactant from *Streptococcus thermophilus* A. *FEMS Immunology and Medical Microbiology*, 46(1): 107-112.

Roy, A., Mahata, D., Paul, D., Korpole, S., Franco, O.L. & Mandal, S.M. (2013). Purification, biochemical characterization and self-assembled structure of a fengycin-like antifungal peptide from *Bacillus thuringiensis* strain SM1. *Frontiers in Microbiology*, 4: 332.

Rufino, R.D., de Luna, J.M., Sarubbo, L.A., Rodrigues, L.R.M., Teixeira, J.A.C. & de Campos-Takaki, G.M. (2011). Antimicrobial and anti-adhesive potential of biosurfactants produced by *Candida* species. *Colloids and Surfaces B: Biointerfaces*, 84(1): 1-5.

Sabaté, D.C. & Audisio, M.C. (2013). Inhibitory activity of surfactin, produced by different *Bacillus subtilis* subsp. *subtilis strains*, against *Listeria monocytogenes* sensitive and bacteriocin-resistant strains. *Microbiological Research*, 168(3): 125-129.

Sajid, M., Khan, M.S.A., Cameotra, S.S. & Al-Thubiani, A.S. (2020). Biosurfactants: Potential applications as immunomodulator drugs. *Immunology Letters*, 223: 71-77.

Sak, K. (2012. Chemotherapy and dietary phytochemical agents. *Chemotherapy Research and Practice*, 1-11.

Sana, S., Datta, S., Biswas, D. & Sengupta, D. (2018). Assessment of synergistic antibacterial activity of combined biosurfactants revealed by bacterial cell envelop damage. Biochimica et Biophysica Acta (BBA)-Biomembranes, 1860(2): 579-585.

Santos, V.S.V., Silveira, E. & Pereira, B.B. (2018). Toxicity and applications of surfactin for health and environmental biotechnology. *Journal of Toxicology and Environmental Health*, Part B, 21(6-8): 382-399.

Saranraj, P., Sayyed, R.Z., Hamzah, K.J., Asokan, N., Sivasakthivelan, P. & Al-Tawaha, A.R.M.A. (2022b). *In:* Sayyed, R.Z. (Eds), *Biosurfactants: Production and Applications in Bioremediation/Reclamation*. 1-18. CRC Press, Taylor & Francis Group, USA.

Satpute, K.S., Banpurkar, G.A., I.M. Banat, Sangshetti, N.J., H.R. Patil & Gade, N.W. (2016). Multiple roles of biosurfactants in biofilms. *Current Pharmaceutical Design*, 22(11): 1429-1448.

Seydlová, G. & Svobodová, J. (2008). Review of surfactin chemical properties and the potential biomedical applications. *Open Medicine*, 3(2): 123-133.

Shah, I., Hamid, B., Zaman, M., Fatima, S., Farooq, S., Datta, R. et al. (2022). Microbial biosurfactants: An eco-friendly approach for bioremediation of contaminated environments. *In:* Sayyed, R.Z. (Eds), *Biosurfactants: Production and Applications in Bioremediation/ Reclaimation*. 197-207. CRC Press, Taylor & Francis Group, USA.

Shah, V., Doncel, G.F., Seyoum, T., Eaton, K.M., Zalenskaya, I., Hagver, R. & Gross, R. (2005). Sophorolipids, microbial glycolipids with anti-human immunodeficiency virus and sperm-immobilizing activities. *Antimicrobial Agents and Chemotherapy*, 49(10): 4093-4100.

Sharma, S., Sharma, N., Kumar, S. & Gupta, G.D. (2009). Liposomes: A review. *Journal of Pharmacy Research*, 2(7): 1163-1167.

Shim, G.Y., Kim, S.H., Han, S.E., Kim, Y.B. & Oh, Y.K. (2009). Cationic surfactin liposomes for enhanced cellular delivery of siRNA. *Asian J. Pharmaceut. Sci.*, 4: 207-214.

Shim, Y.S., Baek, J.W., Kang, M.J., Oh, Y.J., Yang, S. & Il, T.H. (2016). Reference values for the triglyceride to high-density lipoprotein cholesterol ratio and non-high-density lipoprotein cholesterol in Korean children and adolescents: The Korean National Health and Nutrition Examination Surveys 2007-2013. *Journal of Atherosclerosis and Thrombosis*, 23: 1334-1344.

Singh, P., Patil, Y. & Rale, V. (2019). Biosurfactant production: Emerging trends and promising strategies. *Journal of Applied Microbiology*, 126(1): 2-13.

Singh, P.K., Mukherji, R., Joshi-Navare, K., Banerjee, A., Gokhale, R., Nagane, S. et al. (2013). Fluorescent sophorolipid molecular assembly and its magnetic nanoparticle loading: A pulsed laser process. *Green Chem.*, 15: 943-956.

Sivapathasekaran, C., Das, P., Mukherjee, S., Saravanakumar, J., Mandal, M. & Sen, R. (2010). Marine bacterium derived lipopeptides: Characterization and cytotoxic activity against cancer cell lines. *International Journal of Peptide Research and Therapeutics*, 16(4): 215-222.

Smith, M.L., Gandolfi, S., Coshall, P.M. & Rahman, P.K. (2020). Biosurfactants: A Covid-19 perspective. *Frontiers in Microbiology*, 11.

Sotirova, A., Spasova, D., Vasileva-Tonkova, E. & Galabova, D. (2009). Effects of rhamnolipid-biosurfactant on cell surface of Pseudomonas aeruginosa. *Microbiological Research*, 164(3): 297-303.

Splendiani, A., Livingston, A.G. & Nicolella, C. (2006). Control of membrane-attached biofilms using surfactants. *Biotechnology and Bioengineering*, 94(1): 15-23.

Sriram, M.I., Gayathiri, S., Gnanaselvi, U., Jenifer, P.S., Raj, S.M. & Gurunathan, S. (2011).

Novel lipopeptide biosurfactant produced by hydrocarbon degrading and heavy metal tolerant bacterium *Escherichia fergusonii* KLU01 as a potential tool for bioremediation. *Bioresource Technology*, 102(19): 9291-9295.

Steller, S. & Vater, J. (2000). Purification of the fengycin synthetase multienzyme system from *Bacillus subtilis* b213. *Journal of Chromatography B: Biomedical Sciences and Applications*, 737(1-2): 267-275.

Suzuki, M., Kitagawa, M., Yamamoto, S., Sogabe, A., Kitamoto, D., Morita, T. & Imura, T. (2011). U.S. Patent No. 7,989,599. Washington, DC: U.S. Patent and Trademark Office.

Slifka, M.K. & Whitton, J.L. (2000). Clinical implications of dysregulated cytokine production. *J. Mol. Med.* https://doi.org/10.1007/s001090000086

Takahashi, M., Morita, T., Fukuoka, T., Imura, T. & Kitamoto, D. (2012). Glycolipid biosurfactants, mannosylerythritol lipids, show antioxidant and protective effects against H_2O_2-induced oxidative stress in cultured human skin fibroblasts. *Journal of Oleo Science*, 61(8): 457-464.

Thanomsub, B., Pumeechockchai, W., Limtrakul, A., Arunrattiyakorn, P., Petchleelaha, W., Nitoda, T. et al. (2006). Chemical structures and biological activities of rhamnolipids produced by *Pseudomonas aeruginosa* B189 isolated from milk factory waste. *Bioresource Technology*, 97(18): 2457-2461.

Thomas, D., Zachariah, S., Elamin, A.E.E. & Hashim, A.L.O. (2017). Limitations of serum creatinine as a marker of renal function. *Sch. Acad. J. Pharm.*, 6: 168-170.

Thornburg, C.C., Cowley, E.S., Sikorska, J., Shaala, L.A., Ishmael, J.E., Youssef, D.T. et al. (2013). Apratoxin H and apratoxin A sulfoxide from the Red Sea *cyanobacterium Moorea producens*. *Journal of Natural Products*, 76(9): 1781-1788.

Toren, A., Navon-Venezia, S., Ron, E.Z. & Rosenberg, E. (2001). Emulsifying activities of purified alasan proteins from *Acinetobacter radioresistens* KA53. *Applied and Environmental Microbiology*, 67(3): 1102-1106.

Ueno, Y., Hirashima, N., Inoh, Y., Furuno, T. & Nakanishi, M. (2007). Characterization of biosurfactant-containing liposomes and their efficiency for gene transfection. *Biological and Pharmaceutical Bulletin*, 30(1): 169-172.

Uzoigwe, C., Burgess, J.G., Ennis, C.J. & Rahman, P.K. (2015). Bioemulsifiers are not biosurfactants and require different screening approaches. *Frontiers in Microbiology*, 6: 245.

Varvaresou, A. & Iakovou, K. (2015). Biosurfactants in cosmetics and biopharmaceuticals. *Letters in Applied Microbiology*, 61(3): 214-223.

Vecino, X., Cruz, J.M., Moldes, A.B. & Rodrigues, L.R. (2017). Biosurfactants in cosmetic formulations: Trends and challenges. *Critical Reviews in Biotechnology*, 37(7): 911-923.

Waghmode, S., Swami, S., Sarkar, D., Suryavanshi, M., Roachlani, S., Choudhari, P. et al. (2020). Exploring the pharmacological potentials of biosurfactant derived from *Planococcus maritimus* SAMP MCC 3013. *Current Microbiology*, 77(3): 452-459.

Wang, C.L., Liu, C., Niu, L.L., Wang, L.R., Hou, L.H. & Cao, X.H. (2013). Surfactin-induced apoptosis through ROS–ERS–Ca 2+-ERK pathways in HepG2 cells. *Cell Biochemistry and Biophysics*, 67(3): 1433-1439.

Wu, Y.S., Ngai, S.C., Goh, B.H., Chan, K.G., Lee, L.H. & Chuah, L.H. (2017). Anticancer activities of surfactin and potential application of nanotechnology assisted surfactin delivery. *Frontiers in Pharmacology*, 8: 761.

Wyllie, A.H., Kerr, J.F.R. & Currie, A.R. (1980). Cell death: The significance of apoptosis. *In:* Bourne, G.H., Danielli, J.F., Jeon, K.W. (Eds). *International Review of Cytology*. 251–306. Academic, London.

Xiao, J., Li, Y. & Huang, Q. (2016). Recent advances on food-grade particles stabilized Pickering emulsions: Fabrication, characterization and research trends. *Trends in Food Science and Technology*, 55: 48-60.

Yan, X., Gu, S., Cui, X., Shi, Y., Wen, S., Chen, H. & Ge, J. (2019). Antimicrobial, anti-adhesive and anti-biofilm potential of biosurfactants isolated from *Pediococcus acidilactici* and *Lactobacillus plantarum* against *Staphylococcus aureus* CMCC26003. *Microbial Pathogenesis*, 127: 12-20.

Yang, Y., Fang, Z., Chen, X., Zhang, W., Xie, Y., Chen, Y. et al. (2017). An overview of Pickering emulsions: Solid-particle materials, classification, morphology, and applications. *Front Pharmaco.*, 8: 287.

Yuliani, H., Perdani, M.S., Savitri, I., Manurung, M., Sahlan, M., Wijanarko, A. et al. (2018). Antimicrobial activity of biosurfactant derived from *Bacillus subtilis* C19. *Energy Procedia.*, 153: 274-278.

Zaman, M., Hassan, S., Fatima, S., Hamid, B., Farooq, S., Qayoom, I. et al. (2022). Biosurfactants production and applications in food. *In:* Sayyed, R.Z. and Enshasy, H.E. (Eds), *Biosurfactants: Production and Applications in Food and Agriculture*. 225-241. Vol. II. CRC Press, Taylor & Francis Group. USA.

Zang, J., Feng, M., Zhao, J. & Wang, J. (2018). Micellar and bicontinuous microemulsion structures show different solute–solvent interactions: A case study using ultrafast nonlinear infrared spectroscopy. *Physical Chemistry Chemical Physics*, 20(30): 19938-19949.

CHAPTER

7

Bio-emulsions Used in Pharmaceutical and Cosmetic Industries

Lavina A. Pinto[1], Smita S. Bhuyan[2], Neelima J. Kulkarni[3], and Shilpa S. Mujumdar*[1]

[1] Department of Microbiology, Modern College of Arts, Science and Commerce (Autonomous), Shivajinagar, Pune - 411005, Maharashtra, India

[2] CRI - The Clinical Research Institute GmbH, Arnulfstrasse 19, 80335 Munich, Germany

[3] Department of Microbiology, P.E.S. Modern College of Arts, Science and Commerce (Autonomous), Ganeshkhind, Pune - 411016, Maharashtra, India

1. Introduction

Emulsions are blends of hydrophobic and hydrophilic liquids, required wildly to combine aqueous and oily ingredients (Khan et al., 2011). Emulsifiers form stable emulsions which is the prerequisite for a drug which has polar and non-polar components (Singh et al., 2013). Emulsions make thermodynamically unstable systems which can get stabilized by addition of emulsifiers (Khan et al., 2011). Emulsifiers stabilize the system by covering globes of the dispersed phase with a thin film. Pharmaceutical emulsions are generally made up of lotions (low viscosity) and creams (high viscosity). In these preparations the size of partials usually ranges from 0.1 to 100 μm (Leon and Herberet, 2009, Khan et al., 2011, Singh et al., 2019).

The bioemulsifiers (BE) and biosurfactants (BS) are used in pharmaceuticals mainly to enhance the solubility of water-insoluble drugs. Their solubilizing and wetting properties enable drug incorporation and delivery easy. They stabilize the dispersed phase and encapsulate the drug during the manufacturing process, hence have utmost significance (Fracchia et al., 2010, Alizadeh-Sani, 2018, Ali et al., 2022a, Ravinder et al., 2022).

Bioemulsifers and biosurfactants are surface-active molecules produced by microorganisms that display emulsifying properties. The terms biosurfactants and

* Corresponding author: hodmicro@moderncollegepune.edu.in

bioemulsifiers are often used interchangeably but they have marked differences in their physicochemical properties and physiological roles (Uzoigwe et al., 2015). Their amphiphilic nature helps in solubilizing both hydrophobic and hydrophilic curing agents. Synthetic surfactants show toxicity and other side effects hence are not preferred in manufacturing pharmaceutical formulations. However, low toxicity, effectiveness in low concentration, biodegradable and biocompatible nature of bioemulsifiers make them suitable in pharmaceutical and cosmetic formulations (Uzoigwe et al., 2015, Saranraj et al., 2022a). BE and BS show high stability at extreme salinity, pH and temperature (Ali et al., 2022b, Sadiq et al., 2022). In addition, antioxidant, antibacterial antifungal and nutritional properties of BE and BS eliminate the inclusion of preservatives and increase alimentary value (Singh et al., 2013, Bhattacharya et al., 2017).

2. Classification of Bioemulsifiers and Biosurfactants

BE and BS are commonly classified according to their chemical composition and molecular weight. These include low-weight molecular compounds such as glycolipids and lipopeptides and high molecular-weight compounds such as polysaccharides, proteins, and lipoproteins. An acid, peptide, cations, anions, mono-polysaccharide and di-polysaccharides commonly form the hydrophilic moiety whereas the unsaturated and saturated hydrocarbon chains or fatty acids form the hydrophobic moiety of BE and BS (Lourith et al., 2009, Uzoigwe et al., 2015). The most commonly classified BE and BS are as follows.

2.1 Glycolipids

The term glycolipid is used to describe any compound containing one or more saccharide residue bound by a glycosidic linkage to a hydrophobic lipid moiety (Thanomsub et al., 2007). Glycolipids are present in the membrane of bacteria, fungi, plants, and animals. Most known glycolipids are of bacterial origin. Glycoglycerolipids, glycosphingolipids, saccharolipids or acylamino sugars are different types of glycolipids (Lourith et al., 2008). The chemical structure of the glycolipids affects its application in pharmaceutical and cosmetic preparations. The nature of the sugar moiety of glycolipids determines the extent of hydration and in-turn controls the formation of bilayers. The length of the hydrophobic chains influences the phase transition temperature of surfactant molecules, e.g., the longer acyl chain of glycolipids contributes to its higher phase transition temperature leading to increased encapsulation efficiency and emulsion stability (Faivre and Rosilio, 2010). Glycolipids, mainly rhamnolipids and MELs are used in pharmaceuticals and cosmetic formulations because of their good detergency, emulsifying, foaming and dispersing properties. These can also promote oral drug absorption. Since the carbohydrate head groups of glycolipids interact with saccharide receptors with high specificity, they are involved in cell recognition and antibody responses. They also exhibit bioadhesion properties and thus can be used for targeted drug delivery.

Besides rhamnolipids and MEL, other glycolipids used in pharmaceutical and cosmetic preparations include sorbitan and sophorolipids. Sorbitan is widely used in topical cosmetic preparations of creams and emulsions and also as solubilizing and wetting agents (Borzeix et al., 1995). However, they can be toxic when ingested and can cause mild skin irritation. Sophorolipids possess antibacterial, antioxidant, antiviral, antifungal, anti-mycoplasma, and anticancer activities and additionally stimulate dermal fibroblasts. Also, sophorolipids are effective modulators of immune response. They are used as moisturizer softeners, wetting and foaming agents in cosmetics as a result of their excellent hygroscopic properties (Battacharya et al., 2017). They are used in preparations of lotions, body washes, hair products, lip color, eye shadow, acne treatment, deodorants, and skin smoothing and anti-wrinkle products. Sophorolipid BS produced by *Candida bombicola was* reported to have anti-HIV, spermicidal and cytotoxic activities (Shah et al., 2005). They also induced vaginal cell toxicity hence their applicability as long-term microbicidal contraception is under consideration. In addition, glycolipids provide extra benefits because of their antioxidant, antimicrobial and antiviral activities, macrophage-activating, cell-differentiation or fibrinolytic effects (Faivre and Rosilio, 2010).

2.2 Lipoproteins

Lipoproteins are amphiphilic cyclic peptides containing a lipid linked to a polypeptide or amino acid chain (Mnif and Ghribi, 2015). Lipopeptides such as fengycin and iturin are composed of ten amino acids each and surfactin is composed of seven amino acids. Most studied lipoproteins that are used effectively in pharmaceutical and cosmetic formulations are surfactin, iturin, fengycin and gramicidin. They are produced by members of genera *Pseudomonas, Streptomyces, Bacillus* and *Aspergillus* (Marti-Mestres and Nielloud, 2002, Ohadi et al., 2020). Lipopeptides have been used in anti-wrinkle, cleansing and whitening cosmetics. Lipopeptides have shown excellent washability and low skin irritation (Kanlayavattanakul and Lourith, 2010). Some lipopeptides have been used to deliver a melanocyte-stimulating hormone. Lipopeptides possess antibacterial and antifungal properties which can be used for the treatment and prevention of microbial infections and product preservation (Banat, 2019, Fernandes et al., 2007). Lipoproteins also possess antiviral and anti-mycoplasma properties. They also cause inhibition of fibrin clot formation and hemolysis responsible for the formation of ionic channels in lipid bilayer membranes (Rebello et al., 2014, Ohadi et al., 2020).

Surfactin, one of the extensively studied biosurfactants, is used in microemulsion formulations to form a stable drug delivery system. Shim et al. (2009) have shown that surfactin with liposomes can increase the given gene silencing process through a more effective delivery system. This results in the rise of the cellular uptake of small interfering RNA (siRNA), leading to enhancing the particular knockdown impact. According to their investigation cationic surfactin liposomes demonstrate enhanced siRNA delivery to the HeLa cells then surfactin-free liposomes. This is because of the high biocompatibility of surfactin (Shim et al., 2009). Surfactin and its analogues have also been reported to have antiviral activity against herpes viruses and retroviruses. The inhibitory action may be because of physico-chemical interactions between the virus envelope and the surfactant. Surfactin shows antitumor activities by suppressing cell survival signalling, inhibition of platelet aggregation properties and anti-inflammatory properties (Seydlová and Svobodová, 2008, Naughton et al., 2019).

Iturin was produced by *Bacillus subtilis* and found to be active even after autoclaving in the pH range of 5 to 11. It has a large antifungal spectrum but limited antibacterial activity against *Micrococcus* and *Sarcina* strains. Thus, iturin can be used as an alternative to common fungicide drugs. Moreover, Iturin exhibited good lytic activity on yeast spheroplasts and human erythrocytes (Santos et al., 2018).

Fengycin has high anti-fungal activity and possesses low haemolytic activity. It is assumed that fengycin has the ability to disintegrate the cell membrane by pore formation and modifying the structure of the lipid membrane (Marti-Mestres and Nielloud, 2012, Banat et al., 2019).

2.3 Glycoproteins

Glycoproteins consist of oligosaccharide moieties covalently attached to amino acid side chains of protein molecules. Bioemulsifers, such as emulsan, alasan, liposan and mannoproteins are among the widely used glycoprotein bioemulsifiers. Alasan produced by *Acinetobacter radioresistens* is a high-molecular-weight complex of polysaccharide and protein (Frelichowska et al., 2009). The emulsifying activity of the purified polysaccharide (apo-alasan) is very low while the protein portion of alasan has proven to be its active component (Banat et al., 2019). Protein-protein interactions may play an important role in producing the surface-active complex. The role of the polysaccharide is unclear but it is believed that it plays a role in the release of proteins into the medium and in protecting the protein complex against proteolytic activities.

Emulsan produced by *Acinetobacter calcoaceticus* RAG 1 is a highly efficient emulsifying agent for hydrocarbons in water and is effective even at lower concentrations in the range 0.001 to 0.01%. It is one of the strongest emulsion stabilizers reported till date. It is also a belief that it plays a role as a bacteriophage receptor on the cell surface of *Acinetobacter calcoaceticus* RAG-1 (Morita et al., 2013a; 2013b). Emulsan was used in the production of nanoparticles as vehicles for the delivery of antitumor agent, pheophorbide-a. Thus, it has a potential to be used in innovative drug delivery systems. Liposan is used to stabilize O/W emulsions. Mannoproteins are able to form stable emulsions with various hydrocarbons, organic solvents and waste oils and can potentially be used as cleaning agents (Banat et al., 2019).

3. Types of Emulsions

3.1 Oil-in-Water (O/W) Emulsions

Oil-in-water (O/W) emulsions consist of oil droplets dispersed throughout the aqueous phase. The particle size of the dispersed phase commonly ranges from 0.1 to 100 μm (Fig. 1). O/W emulsions are non-greasy and can be easily removed from the skin surface. External applications of O/W emulsions include a cooling effect and internal application includes masking the bitter taste of oil. Fats or oils for oral administration are always formulated as O/W emulsions. Fats or oils are either given as nutritional supplements or as vehicles for oil soluble drugs (Khan et al., 2011).

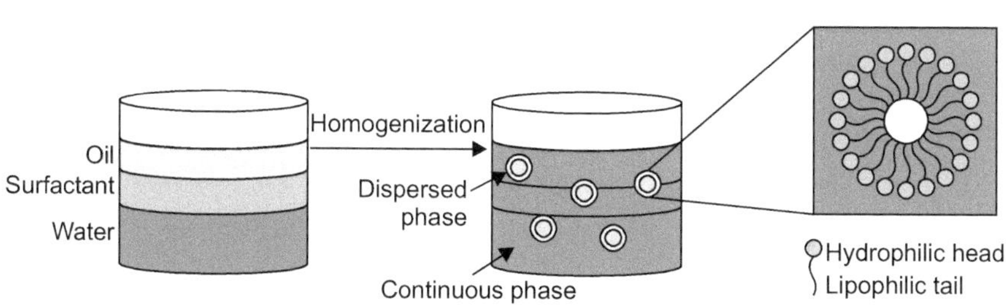

Fig. 1: Oil-in-water (O/W) emulsion.

3.2 Water-in-Oil (W/O) Emulsions

In this type of emulsion water is dispersed as globules in the oil continuous phase (Fig. 2). They are useful for cleansing the skin of oil soluble dirt, but its greasy texture hampers its application in cosmetics. The stability of the emulsion can be affected by HLB (Hydrophilic-lipophilic balance). The balance between the hydrophobic and hydrophilic moiety of the BE determines its properties and performance. HLB ranges from 0 to 20. Low HLB values (3 to 6) lead to the formation of W/O emulsions, while high HLB values (8 to 18) lead to the formation of O/W emulsions (Khan et al., 2011).

W/O emulsions are not water washable and are used externally to prevent evaporation of the moisture from the skin surface e.g., cold cream, hydrating creams and serums. They are preferred for formulations of products like cream meant for external use. W/O emulsions have an effect on the absorption of drugs; oil soluble drugs are released more quickly from W/O emulsion (Khan et al., 2011).

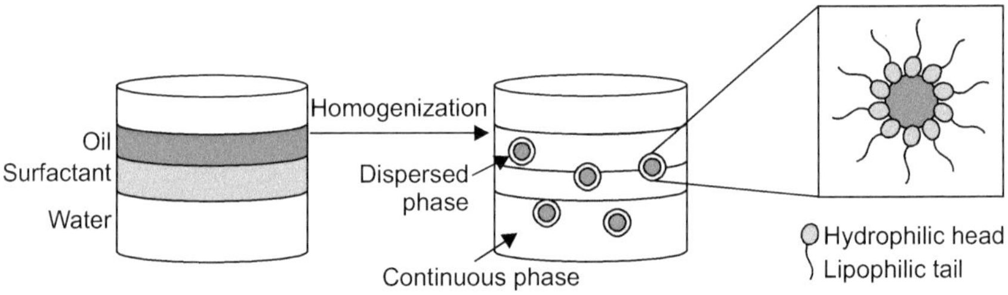

Fig. 2: Water-in-oil (W/O) emulsion.

3.3 Multiple Emulsions

In multiple emulsions, both oil in water and water in oil emulsions exist simultaneously. In these systems the oil-in-water or water-in-oil emulsions are dispersed in another liquid medium. Multiple emulsions are of two types which include oil-in-water-in-oil (O/W/O) emulsions and water-in-oil-in-water (W/O/W) emulsions (Figs. 3A and 3B). O/W/O emulsions consist of very small droplets of oil dispersed in the water globules of a W/O emulsion. Whereas a W/O/W emulsion consists of droplets of water dispersed in the oil phase of an O/W emulsion (Khan et al., 2011). Multiple emulsion systems offer protection to the entrapped substances as well as promote incorporation of several active ingredients in the different compartments of the emulsions. Prolonged release of the drug or active substance can be obtained by means of multiple emulsions. In cosmetics, skin moisturisers generally make use of multiple emulsions. In pharmaceutical preparations multiple emulsions can be used for taste masking, adjuvant vaccines, an immobilization of enzymes and sorbent reservoirs of overdose treatments, and sometimes for the augmentation of external skin or dermal absorption. Thermodynamic instability and complex structure pose a major threat to the use of multiple emulsions in cosmetic and pharmaceutical preparations (Khan et al., 2011).

3.4 Micro-emulsions

Micro-emulsions are emulsions of water, oil and surfactants in presence of a co-surfactant. These are also known as swollen emulsions, micellar emulsions or nano-emulsions (Brown et al., 2013). The dispersed particles of micro-emulsions are in

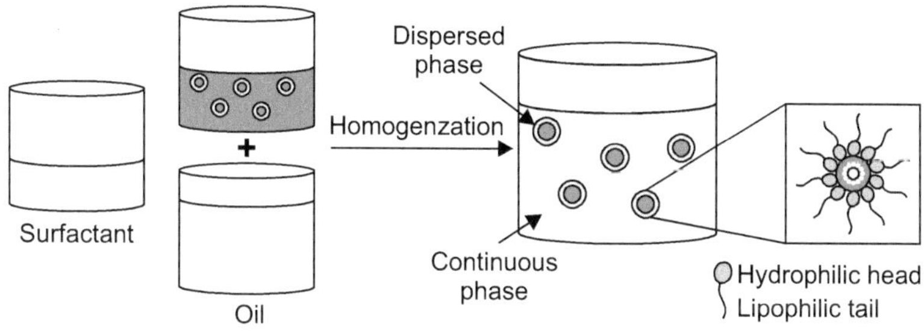

Fig. 3(A): Oil-in-water-in-oil (O/W/O) emulsion.

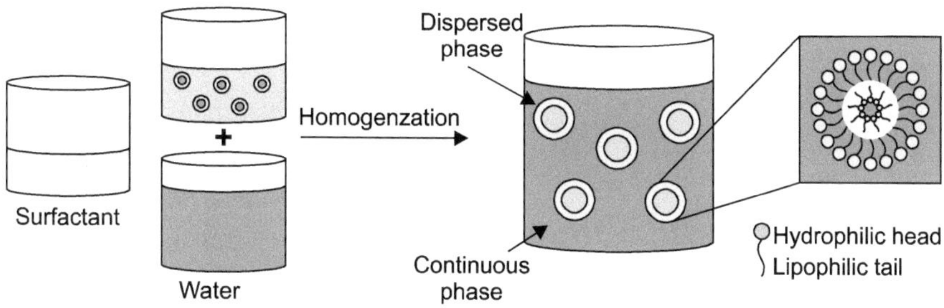

Fig. 3(B): Water-in-oil-in-water (W/O/W) emulsion.

nanometric sizes (5-50 nm), which gives micro-emulsions their unique property of transparency or translucency. The stability of microemulsions is contributed to by low interfacial tension which can be achieved by using a combination of surfactant and cosurfactant. Based on their structure, micro-emulsions are classified as O/W, W/O and bicontinuous microemulsions (Fig. 4). The bicontinuous microemulsions are a homogeneous phase with hydrophilic and hydrophobic balance. Monolayers of surfactant molecules form the junction of the aqueous and oil phases. The volume ratio of water and oil is used to tune the internal geometric structure and chemical composition of a microemulsion system yielding either a micellar or a bicontinuous structure. The bicontinuous structure can be used for emulsion polymerization to prepare size-controlled nanoparticles (Zang et al., 2018).

Micro-emulsions act as potential drug carrier systems for drug administration through oral, parenteral and topical routes. They exhibit improved drug solubilization and bioavailability and also easy drug release from the formulations. Micro-emulsions consisting of synthetic surfactants show higher risks of gastrointestinal irritation due to the presence of high amounts of surfactants. There is a possibility of micro-emulsions disrupting the stratum corneum leading to skin irritations. Thus, using biosurfactants prove advantageous over synthetic ones (Ohadi et al., 2020).

Mannosylerythritol lipid biosurfactants can be used to prepare stable W/O microemulsions without the addition of a co-surfactant. Rhamnolipids extracts from *Pseudomonas aeruginosa* PA1 can be used as co-surfactants in the microemulsion synthesis. Rhamnolipids from *P. aeruginosa* strain BS-161R were used to synthesize silver nanoparticles along with glycolipids as stabilizers. These stabilized silver nanoparticle microemulsions exhibited good antibiotic activity against both Gram-positive and Gram-negative pathogens and *Candida albicans*. Trehalose lipid was successfully used to create O/W microemulsion for the synthesis of novel nPMMA (polymethyl methacrylate nanoparticles) that are non-toxic and biocompatible. Surfactin is used in the preparation of microemulsions drug delivery systems of vitamin E and docosahexaenoic acid (Ohadi et al., 2020).

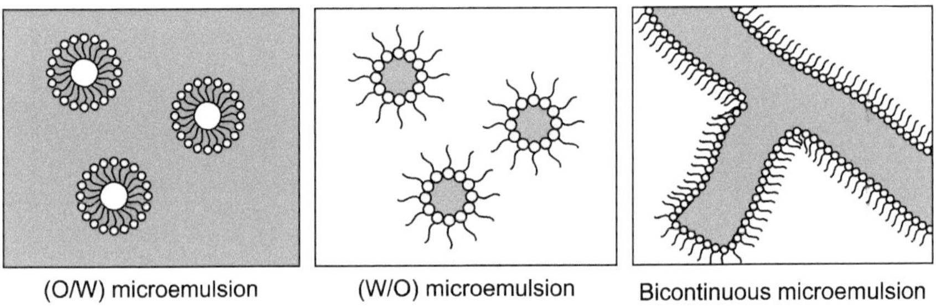

Fig. 4: Types of micro-emulsions.

3.5 Pickering Emulsions

Pickering emulsions make use of a solid particle which is adsorbed onto the interface of the two immiscible or partially miscible phases (Fig. 5). These solid particles prevent the dispersed phase from coalescing and impart high stability to the emulsions (Yang et al., 2017). The most commonly used synthetic solid particles in pickering emulsions include silica, clay, carbon black, hydroxyapatite and polystyrene. Biological solid particles include starch, cellulose, chitin, whey protein, soy protein and bacterial proteins. These particles offer more efficient stabilization than surfactant adsorption as they irreversibly attach to the oil–water interface. In pharmaceutical formulations, pickering emulsions prove to be a better drug delivery and drug releasing vehicle because of their dense packaging and interactions with the loaded drug (Yang et al., 2017, Frelichowska et al., 2009).

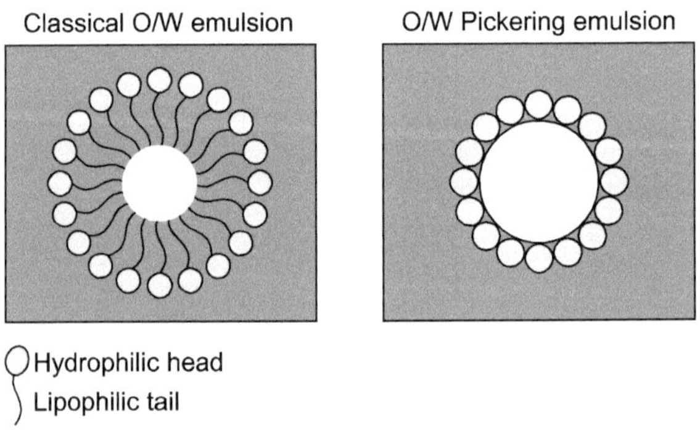

Fig. 5: Pickering emulsions.

4. Conclusion

Bioemulsifers and biosurfactants are amphiphilic molecules produced by microorganisms that exhibit surface activity and emulsifying properties. BE and BS are used in pharmaceuticals and cosmetic industries for various applications such as vehicles for oil soluble drugs, to promote oral drug absorption, for targeted drug delivery, as solubilizing and wetting agents, as well as modulators of immune response, to treat microbial infections. Different types of emulsions such as O/W emulsion, W/O emulsion, multiple emulsions, microemulsions and pickering emulsions are used for various applications in Pharmaceutical and Cosmetic industries depending on their properties. W/O emulsions are not water washable and are used externally to prevent evaporation of the moisture from the skin surface e.g., cold cream, hydrating creams, and serums. Fats or oils are either given as nutritional supplements or as vehicles for oil soluble drugs and are always formulated as O/W emulsions. Prolonged release of the drug or active substance can be obtained by means of multiple emulsions.

References

Ali, S.A.M., Sayyed, R.Z., Mir, M.I., Hameeda, B., Khan, Y., Alkhanani, M.F. et al. (2022a). Induction of systemic resistance in maize and antibiofilm activity of surfactin from Bacillus velezensis MS20. *Front. Microbiol.*, 13: 879739. https://doi.org/10.3389/fmicb.2022.879739

Alizadeh-Sani, M., Hamishehkar, H., Khezerlou, A., Azizi-Lalabadi, M., Azadi, Y., Nattagh-Eshtivani, E. et al. (2018). Bioemulsifiers derived from microorganisms: Applications in the drug and food industry. *Advanced Pharmaceutical Bulletin.*, 8(2): 1-191.

Banat, I.M. & Thavasi, R. (2019). Microbial Biosurfactants and their Environmental and Industrial Applications. CRC Press, Taylor and Francis Group. UK.

Bhattacharya, B., Ghosh, T.K. & Das, N. (2017). Application of biosurfactants in cosmetics and pharmaceutical industry. *Scholars Academic Journal of Pharmacy*, 6: 320-329.

Borzeix, F., Hillion, G., Marchal, C. & Stoltz, C. (1995). Use of sophorolipids in cosmetic and dermatological compositions. European Patent No. EP0820273B1.

Brown, P., Butts, C.P. & Eastoe, J. (2013). Stimuli-responsive surfactants. *Soft Matter.*, 9(8): 2365-2374.

Faivre, V. & Rosilio, V. (2010). Interest of glycolipids in drug delivery: From physicochemical properties to drug targeting. *Expert Opinion on Drug Delivery*, 7(9): 1031-1048.

Fernandes, A.P.V., Arruda, I.R., Fernando, A., Santos, B.A., Araújo, A.A.A., Maior, A.M.A. et al. (2007). Antimicrobial activity of surfactants produced by *Bacillus subtilis* R14 against multidrug-resistant bacteria. *Brazilian Journal of Microbiology*, 38: 704-709.

Fracchia, L., Cavallo, M., Allegrone, G. & Martinotti, M.G. (2010). A Lactobacillus-derived biosurfactant inhibits biofilm formation of human pathogenic *Candida albicans* biofilm producers. *Appl. Microbiol. Biotechnol.*, 2: 827-837.

Frelichowska, J., Bolzinger, M.A., Valour, J.P., Mouaziz, H., Pelletier, J. & Chevalier, Y. (2009). Pickering w/o emulsions: Drug release and topical delivery. *International Journal of Pharmaceutics*, 368(1-2): 7-15.

Kanlayavattanakul, M. & Lourith, N. (2010). Lipopeptides in cosmetics. *International Journal of Cosmetic Science*, 32(1): 1-8.

Khan, B.A., Akhtar, N., Khan, H.M.S., Waseem, K., Mahmood, T., Rasul, A. & Khan, H. (2011). Basics of pharmaceutical emulsions: A review. *African Journal of Pharmacy and Pharmacology*, 5(25): 2715-2725.

Leon, L. & Herberet, A.L. (2009). The Theory and Practice of Industrial Pharmacy. CBS Publishers, New Delhi, India. pp. 502-531.

Lourith, N. & Kanlayavattanakul, M. (2009). Natural surfactants used in cosmetics: Glycolipids. *International Journal of Cosmetic Science*, 31(4): 255-261.

Marti-Mestres, G. & Nielloud, F. (2002). Emulsions in health care applications—An overview. *Journal of Dispersion Science and Technology*, 23(1-3): 419-439.

Mnif, I. & Ghribi, D. (2015). Review lipopeptides biosurfactants: Mean classes and new insights for industrial, biomedical, and environmental applications. *Peptide Science*, 104(3): 129-147.

Morita, T., Fukuoka, T., Imura, T. & Kitamoto, D. (2013a). Production of mannosylerythritol lipids and their application in cosmetics. *Applied Microbiology and Biotechnology*, 97(11): 4691-4700.

Morita, T., Kawamura, D., Morita, N., Fukuoka, T., Imura, T., Sakai, H. & Kitamoto, D. (2013b). Characterization of mannosylerythritol lipids containing hexadecatetraenoic acid produced from cuttlefish oil by *Pseudozyma churashimaensis* OK96. *Journal of Oleo Science*, 62(5): 319-327.

Naughton, P.J., Marchant, R., Naughton, V. & Banat, I.M. (2019). Microbial biosurfactants: Current trends and applications in agricultural and biomedical industries. *Journal of Applied Microbiology*, 127(1): 12-28.

Ohadi, M., Shahravan, A., Dehghannoudeh, N., Eslaminejad, T., Banat, I.M. & Dehghannoudeh, G. (2020. Potential use of microbial surfactant in microemulsion drug delivery system: A systematic review. *Drug Design, Development and Therapy*, 14: 541.

Ravinder, R., Manasa, M., Roopa, D., Bukhari, N.A., Hatamleh, A.A., Khan, M.Y. et al. (2022). Biosurfactant producing multifarious *Streptomyces puniceus* RHPR9 of *Coscinium fenestratum* rhizosphere promotes plant growth in chilli. *Plos One*, 17(3): e0264975. https://doi.org/10.1371/ journal.pone.0264975

Rebello, S., Asok, A.K., Mundayoor, S. & Jisha, M.S. (2014). Surfactants: Toxicity, remediation and green surfactants. *Environmental Chemistry Letters*, 12(2): 275-287.

Sadiq, M., Amayri, M.A., Paramaiah, C., Mai, N.H., Ngo, T.Q. & Phan, T.T.H. (2022). How green finance and financial development promote green economic growth: Deployment of clean energy sources in South Asia. *Environmental Science and Pollution Research*, 29(43): 65521-65534.

Santos, V.S.V., Silveira, E. & Pereira, B.B. (2018). Toxicity and applications of surfactin for health and environmental biotechnology. *Journal of Toxicology and Environmental Health, Part B*, 21(6-8): 382-399.

Saranraj, P., Sayyed, R.Z., Sivasakthivelan, P., Hasan, M.S., Al-Tawaha, A.R.M.A. & Amala, K. (2022a). Microbial biosurfactants: Methods of investigation, characterization, current market value and applications. *In:* Sayyed, R.Z. (Ed.), *Biosurfactants: Production and Applications in Bioremediation/Reclamation*. 19-34. CRC Press, Taylor & Francis Group, USA.

Seydlová, G. & Svobodová, J. (2008). Review of surfactin chemical properties and the potential biomedical applications. *Open Medicine*, 3(2): 123-133.

Shah, V., Doncel, G.F., Seyoum, T., Eaton, K.M., Zalenskaya, I., Hagver, R. et al. (2005). Sophorolipids, microbial glycolipids with anti-human immunodeficiency virus and sperm-immobilizing activities. *Antimicrobial Agents and Chemotherapy*, 49(10): 4093-4100.

Shim, G.Y., Kim, S.H., Han, S.E., Kim, Y.B. & Oh, Y.K. (2009). Cationic surfactin liposomes for enhanced cellular delivery of siRNA. *Asian J. Pharmaceut. Sci.*, 4: 207-214.

Singh, P., Patil, Y. & Rale, V. (2019). Biosurfactant production: Emerging trends and promising strategies. *Journal of Applied Microbiology*, 126(1): 2-13.

Singh, P.K., Mukherji, R., Joshi-Navare, K., Banerjee, A., Gokhale, R., Nagane, S. et al. (2013). Fluorescent sophorolipid molecular assembly and its magnetic nanoparticle loading: A pulsed laser process. *Green Chem*, 15: 943-957.

Thanomsub, B., Pumeechockchai, W., Limtrakul, A., Arunrattiyakorn, P., Petchleelaha, W., Nitoda, T. et al. (2006). Chemical structures and biological activities of rhamnolipids produced by *Pseudomonas aeruginosa* B189 isolated from milk factory waste. *Bioresource Technology*, 97(18): 2457-2461.

Uzoigwe, C., Burgess, J.G., Ennis, C.J. & Rahman, P.K. (2015). Bioemulsifiers are not biosurfactants and require different screening approaches. Frontiers in *Microbiology*, 6: 245.

Yang, T., Zheng, J., Zheng, B.S., Liu, F., Wang, S., & Tang, C.H. (2018). High internal phase emulsions stabilized by starch nanocrystals. *Food Hydrocolloids*, 82: 230-238.

Zang, J., Feng, M., Zhao, J. & Wang, J. (2018). Micellar and bicontinuous microemulsion structures show different solute–solvent interactions: A case study using ultrafast nonlinear infrared spectroscopy. *Physical Chemistry Chemical Physics*, 20(30): 19938-19949.

CHAPTER

8

Microbial Biosurfactants and Bioemulsifiers: Properties, Types and Applications in Cosmetic Industry

Dhritiksha M. Baria, Nidhi Y. Patel, Shivani Yagnik Raval, Rakeshkumar R. Panchal, Kiransinh N. Rajput, and Vikram H. Raval*

Department of Microbiology and Biotechnology, University School of Sciences, Gujarat University, Ahmedabad, Gujarat, India

1. Introduction

Surface active substances are broadly used in many personalized and domestic products. Industrialization and globalization made their use more frequent as it gives a better product efficacy. Reduction of surface tension is required for major cosmetic products, food ingredients, detergents and washing products. It is also required in a few pharmaceutical base materials. The demand for surfactants and emulsifiers has been reported to be 13 million tons per annum approximately; such high demand makes it a crucial requirement and being a part of the personalized and hygiene product sector, their toxicity levels are crucial in a larger context (Yang et al., 2023). The synthetic chemical surfactants are getting replaced gradually by biological surface active compounds due to higher sustainability, better digestibility and eco friendly properties. A larger part of industrial demands has changed towards such biological products known as biosurfactants and bioemulsifiers. These biological compounds show better efficacy and their biodegradable nature and non-toxic property makes them popular; specifically, in personalized and direct products such as food, cosmetics and pharmaceutics. According to reports biosurfactants are used in ten to forty folds lesser concentrations (10-200 mg/L) than chemical ones specifically for micelle formations; this makes them more efficient and easier to use (Yao et al., 2021).

Biosurfactants are divided into two major categories – high molecular weight and low molecular weight compounds (Mujumdar et al., 2019). Glycolipids are well known and widely used low molecular weight compounds showing exemplary

* Corresponding author: vhraval@gujaratuniversity.ac.in; vikramhraval@gmail.com

results for emulsifications attributed to their amphoteric nature. Other compounds are sorpholipids, mannoselerythritol lipids (MELs) and surfactins used for cosmetic and surface cleaning products. The high molecular weight molecules like lipoproteins and lipopolysaccharides are used for higher emulsifying- stable products (Shakeri et al., 2020).

Majorly biosurfactants and bioemulsifiers are produced by bacteria and fungi. *Bacillus* spp. and *Bervibacteria* spp. are reported for the production of few lipopolysaccharides showing surface active properties. Lipoproteins are synthesized by *Nocardia, Pseudomonas, Rhodococcus* spp., *Thaiolacillus* spp. and *Arthrobacter* spp. A few species of yeast like *Pseudozyma* also produce lipoproteins with better efficacy (Devale et al., 2023).

Biosurfactants show emulsification, wetting, micelle formation, phase separation, surface activity and decrease of viscosity; these properties make it more versatile for use with bio availability and zero risk. Initially they were utilized in the detergent industry. Ongoing research on different compounds and their applicative properties extended their utilization in other industries too. The food, pharmaceutical and cosmetic industries are huge customers of biosurfactant and emulsifying compounds in the present scenario (Panjiar et al., 2015). The chapter contains information on their use in cosmetic industries specifically. The legal and regulatory issues are also reviewed. The production and purification aspect of biosurfactants for cosmetic products has been emphasized.

2. Sources of Biosurfactants and Bioemulsifiers

Different secondary metabolites such as steroids, alkaloids, peptides, terpenoids, biopolymers, and polyketides are produced by microorganisms (Khalifa et al., 2021, Baria et al., 2022). Biosurfactants and bioemulsifiers are also secondary metabolites mainly produced by various bacteria, fungi, algae and plants. The enhanced industrial demand of biosurfactants over chemical surfactants has created a supply-demand gap and opportunity for their large-scale production. There are known natural biosurfactants from plants also while the scale up studies and industrial applications are mainly piloted with microorganisms due to ease of management (Devale et al., 2023). Microorganisms show growth and production of biosurfactants and bioemulsifiers with raw sources like oil and by-products, readily available plant residues like unwanted crop portions and a cost-effective way for better production yields (Nogueira et al., 2020). Bacteria and Fungi are majorly studied and preferred as bio surfactant and bio emulsifier producers.

2.1 Bacterial Biosurfactants and Bioemulsifiers

Bacteria are capable of producing versatile products and some of the multifunctional products produced are biosurfactants and bioemulsifiers. Being biodegradable, bioavailable with comparable potency of surface-active molecules produced by bacteria, these products are replacing chemical based ones. The drawback of these bio-derived products are the purification steps and their cost (Shakeri et al., 2020). The product efficacies are found better but lower yield percentages lead to high costs. Furthermore, the purification steps are different for varying compounds. Yet various bacterial species like *Bacillus fusiformis, Lysinibacillus* spp., *Serratia rubidea, Rhodococcus erythropolis, Geobacillus* spp. *Pseudomonas* spp., *Thioxidans and Acenatobacter* spp. produce various biosurfactants and bioemulsifiers using raw waste (Shakeri et al., 2020, Khedher et al., 2017). The alteration of raw material usage shows promising results, like use of petroleum waste, used cooking oil, agricultural waste and dairy industry waste (Devale et al., 2023). Various biosurfactants and bioemulsifiers and bacteria are described in Table 1 with their composition and role in multiple industries. Yet there is a scope in this research area for more promising products with temperature and pH stability to maintain their properties.

2.2 Fungal Biosurfactants and Bioemulsifiers

Bacterial strains are broadly considered and explored for production of biosurfactants with regard to major aspects as compared to fungi; even lesser explored fungal species show cost-effective production of biosurfactants. Many fungal species including *Penicillium* sp., *Aspergillus flavus AF612, Geotrichum* sp. CLOA40, *Trichosporon mycotoxinivorans* CLA2, *Myroides odoratus* JCM7458, and *M. odoramitimus* JCM7460 are known for various trehalolipids and phospholipids (Reis et al., 2018, Mohy and Hossam, 2023). Fungal species form many emulsifiers and surfactants majorly with lipoprotein and fatty acid constituents (Table 1).

2.3 Yeast Biosurfactants and Bioemulsifiers

Various yeasts such as *Saccharomyces cerevisiae, Candida bombicola, Vibrio alginolyticus* 3B-2, *C. krusei, Candida tropicalis* BPU1 and *Aureobasidium pullulans* L3-GPY produce mannoprotein, heteropolymer type of biosurfactant (Vijaykumar and Saravanan, 2015, Santos et al., 2020, Cardoso et al., 2021). *Candida ishiwadae, Candida bombicola,* are known for producing various lipopolysaccharhide and sophorolipids biosurfactants. These products are patented in various countries (Luft, 2022).

3. Types of Biosurfactants and Bioemulsifiers

Biologically derived surface-active substances are mainly categorized into two types on the basis of their basic properties and molecular weights. There are two main functional properties; surface tension reduction and emulsification (micelle formation). Majorly high molecular weight compounds show only the emulsification property and are called bioemulsifiers while low molecular weight compounds may show only surface activity or both surface activity and emulsifying nature, the latter form a large group called biosurfactants. Biosurfactants and bioemulsifiers are further divided on the basis of their basic structural components (Tao et al., 2019).

3.1 Biosurfactants

Biosurfactants are mainly divided into four major categories Glycolipids, fatty acids and other lipids, lipopeptides and polypeptides. Their structural stability and functions at moderate pH, lower to moderate temperatures make them useful in many industrial applications (Thraeib et al., 2022).

3.1.1 Glycolipids

Carbohydrates bonded with long chain fatty acids through glyosidic bonds are known as Glycolipids. Major sub-categories are Rhamnolipids (mono or di Rhamnose

Table 1: Source of microbial surfactants, types, composition and role

Microorganisms	Biosurfactant/ Bioemulsifier	Composition	Role of BS/BE	References
Bacteria				
Bacillus cerus	Lipoproteins	Lipid and protein	Antimicrobial activity	(Yuliani et al., 2018)
Bacillus licheniformis	Lipopeptide	Lipopeptide: lichenysin (1006 to 1034 Da),	Powerful surface-active agent compared with other surfactants, good antibacterial activity	(Bhattacharya et al., 2017)
Pseudomonas aeruginosa	Rhamnolipids	Hydrophobic moiety containing 1 or 2 β-hydroxyl fatty acids	Surface activity and emulsifying activity	(Wittgens and Rosenau, 2018)
Corynebacterium kutscheri	Glycolipopeptide	Carbohydrate (40%), lipid (27%) and protein (29%)	Emulsification of different hydrocarbons	(Tripathi et al., 2018)
Acinetobacter calcoaceticus RAG-1	Emulsan	Polysaccharide and protein: backbone *O*-ester and *N*-acyl linkages	Effective emulsion stabilizers: at low concentration, powerful emulsifier	(Rashedi et al., 2018)
Acinetobacter calcoaceticus A2	Biodispersan	Glucosamine, 6-methylaminohexose, galactosamine uronic acid, unidentified amino sugar	Good substrate specificity, dispersion of limestone and titanium dioxide, better dispersion of water	(Edosa et al., 2018)
Acinetobacter radioresistens KA-53	Alasan	protein polysaccharide (1 MDa) Contains covalent bound alanine	Lowers surface tension (69–42 mN/m) Effective surfactant Good stabilizer oil-in-water emulsions	(Bhosale and Pinjari, 2021)

Contd...

Contd...

Microorganisms	Biosurfactant/ Bioemulsifier	Composition	Role of BS/BE	References
Azotobacter chroococcum	Lipopeptide	Lipid and protein (31.3:68.7)	Emulsification of waste motor lubricant oil, crude oil, diesel, kerosene, naphthalene, anthracene and xylene	(Citarasu et al., 2021)
Rhodococcus sp. H13-A	Glycolipid	Glycolipid Free amino groups Neutral lipids	Surface active properties	(Kuyukina and Ivshina, 2019)
Pseudomonas fluorescence	Trehaloselipid	o-dialkyl monoglycerides–protein	High yield with gasoline; emulsification against gasoline	(Saikia et al., 2021)
Fungi				
Aspergillus flavus AF612	Novel "Uzmaq"	methoxy phenyl oxime glycosides	Antifungal activity	(Ishaq et al., 2015)
Fusarium sp. *BS-8*	trehalolipids	Lipids + Peptides	Increase the pseudosolubility of hydrophobic compounds ability to decrease interfacial tension	(Reis et al., 2018)
Myroides odoramitimus JCM7460	Phospholipids and fatty acids	Lipid + Fatty acids	Good surface-active agent	(Mohy and Hossam, 2023)
Myroides odoratus JCM7458	Phospholipids and fatty acids	Lipid + Fatty acids	Good surface-active agent	
Yeast				
Saccharomyces cerevisiae	Mannoprotein	Carbohydrate/ polysaccharide + lipid	Good anti-adhesive property	(Rasheed and Haydar, 2023)
Candida lipolytica	Liposan	83% carbohydrate and 17% protein	Stabilization of hydrocarbon in water emulsions	(Rajasimman et al., 2021)

Contd...

Contd...

Microorganisms	Biosurfactant/ Bioemulsifier	Composition	Role of BS/BE	References
Candida antarctica SY16	Mannosylerythritol lipid	partially acylated derivatives of 4-O-β-D-mannopyranosyl-D-erythritol and fatty acids	Antimicrobial and biomedical property	(Mnif et al., 2018; Coelho et al., 2020)
Candida glabrata UCP 1002	Heteropolymer	45% protein, 20% lipid and 10% carbohydrate	Emulsification of kerosene and vegetable oil	(Cardoso et al., 2021)
Candida sphaerica UCP0995	Lunasan	Protein-carbohydrate-lipid complex	Remediation of motor oil	(Santos et al., 2020)
Candida bombicola	sophorolipids	sugar moiety linked to a fatty acid chain	Oil mobilization	(Celligoi et al., 2020)

moiety), Sophorolipids (disaccharide Sophorose moiety) and Trehalolipids (Shakeri et al., 2020). Structurally quite different such glycolipids are being used in the cosmetics and pharmaceutical industries and formulations. Rhamnolipids are used as the main component of anti-wrinkle agents. *Pseudomonas* spp., *S. bombicolla, Rhodococcus* spp., *Candida* spp. are few thoroughly studied species and show enhanced production of glycolipids as biosurfactants (Vijaykumar and Saravanan, 2015).

3.1.2 Fatty Acids and Other Lipids

Fatty acids, neutral fats and phospholipids are found as a basic component of some important biosurfactants. These are economically obtained by oxidation of hydrocarbons (alkanes) through microbial action (Tao et al., 2019). Lecithine and lysolecithine are examples of phospholipids as biosurfactants. These biosurfactants are mainly used as carriers for gene transfer and drug delivery systems. Many fatty acids are used in food industries. Corynimycolic acid, Spiculisporic acid and phosphatidylethanolamine are produced by *Corynebacterium* spp., *Penicillium* spp. and *Acenatobacter* spp. respectively and are being used as ingredients in various food and cosmetic formulas (Gurkok and Ozdal, 2021).

3.1.3 Lipopeptides

Surfactins, serravatin, iturin, lichenysins, polymyxin and gramicidin are the lipopeptidal biosurfactants used widely in medical and cosmetic products for showing antimicrobial activity. They preserve products from contamination, and in derma cosmetic products, lipopeptides act as components with additional antiseptic and anti-inflammatory properties (Yao et al., 2021). Few of them are reported for anti-tumour activity also. *B. serratia, B. polymixa, B. subtilis, Pseudomonas, B. lichenyformis* are main producers of these forms biosurfactants (Gurkok and Ozdal, 2021).

3.2 Bioemulsifiers

Bioemulsifiers are compounds produced by microbes with high stability in emulsification properties, characterized by high molecular weight and the ability to lower surface tension. The specifications of bioemulsifiers enable their use for texture smoothing (in creams and ointments), maintaining viscosity of materials by balancing water/liquid holding capacity, and serving as carriers for specific materials in the pharmaceutical, food, and cosmetics industries. Bioemulsifiers (Fig. 1) are mainly divided into two categories: polymeric and particulate emulsifiers. Well known polymeric compounds are emulsan, biodispersan, liposan, mannoprotein, xanthan gum, uronic acid bioemulsifier and particulate emulsifiers are vesicles and whole cells (dispersan) (Thraeib et al., 2022, Gurkok and Ozdal, 2021).

3.2.1 Emulsan and Alasan

Extracellular heteropolysaccharide emulsan is a combination of two biopolymers, including 20% high molecular weight exopolysaccharide and 80% lipopolysaccharide. Emulsan extracted in the late 1970s from an *Arthrobacter* sp. RAG-1 hydrocarbon-degrading organism (later identified as *Acinetobacter venetianus* RAG-1). Emulsan enhances the stability of alginate microspheres and it is also used for ease of biomolecule release systems in various concentrations. A few emulsan–alginate formulations are granted patents as drug delivery vehicles. Alasan consists of an

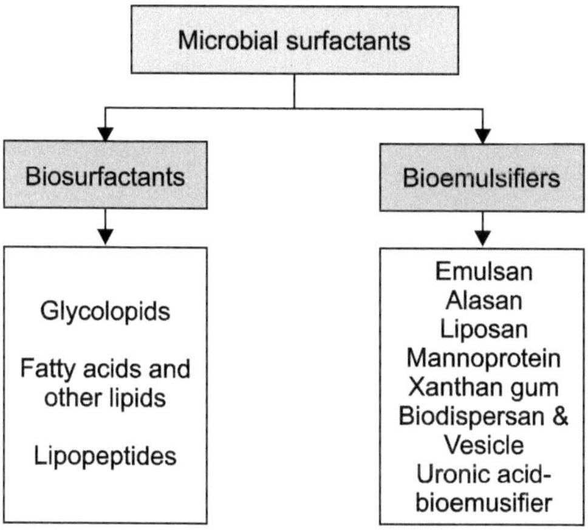

Fig. 1: Types of microbial surfactants.

alanine rich protein portion covalently bonded with polysaccharides (Thraeib et al., 2022).

3.2.2 Liposan

Candida lipolytica was reported first for the production of a water-soluble lipid-protein complex molecule known as liposan, composed of approximately 80% carbohydrates and 20% proteins. It is mainly used in the food and cosmetic industries. Its patented products are dietary supplements to reduce weight and maintain blood cholesterol. It is specifically used to enhance hydration in lip balms (Alizadeh et al., 2018).

3.2.3 Mannoprotein

Polymeric compounds composed of mannose and protein as main components, such as mannopyranosyl-protein complexes produced by organisms like yeast, exhibit specific enzymatic properties in living cells and are proven to be excellent emulsifiers (Gurkok and Ozdal, 2021).

3.2.4 Xanthan Gum

Xanthan gum is an exopolysaccharide isolated from organisms like *Xanthomonas citri* and *Xanthomonas campestris*. Xanthan gum (Fig. 1) shows emulsifying activity and is acceptable as a food additive as an emulsifier and thickener. It is also used for *in situ* oil displacement from nearly empty oil wells; the method is well known as microbial enhanced oil recovery (MEOR) (Panjiar et al., 2015).

3.2.5 Biodispersan and Vesicles

Biodispersan is a liquid formulation consisting of a combination of bacterial species showing emulsifying and dispersing effects. These formulations are primarily used for cleaning wastewater from paper industries. Vesicles produced by *Acenatobacter* spp. have been studied for their emulsifying activity with alkanes. These are made up of proteins, phospholipids and lipopolysaccharides and have a diameter of, 20-30 nm (Mujumdar et al., 2019, Yao et al., 2021).

3.2.6 Uronic Acid Bioemulsifier

Uronic acids containing emulsifiers are basically glycoproteins with high molecular weight which gives them stability upon emulsification. They are primarily produced from bacteria like *Antarctobacter* spp. and *Halomonas* spp. and used in food and cosmetic products (Yang et al., 2023).

4. Properties of Biosurfactants Relevant to Cosmetic Formulation

The formulation of the emulsion or nano emulsion has a significant impact on the distribution of active ingredients in the cosmetics industry. In addition to permitting a variety of aesthetic appearances and textures (transparent, opaque, light, or rich/dense), surfactants play a critical role in the formation of these stable systems (Sonneville-Aubrun et al., 2018). However, to minimize skin irritations, potential allergic responses, or inflammation brought on by a direct interaction of surfactants with skin keratinocytes, it is essential to carefully choose surfactants for cosmetic formulations (Vecino et al., 2017). The stratum corneum's intercellular lipids can also be depleted by some surfactants, which can cause skin dryness (Cleary, 2016)

Biosurfactants have been suggested as a replacement for chemically produced surfactants in the cosmetics sector because of their effects on viscosity and product consistency through emulsification, foaming, water binding ability, spreading, and wetting. These surfactants are used in a variety of products, including insect repellents, antacids, bath products, acne pads, anti-dandruff products, contact lens solutions, baby products, mascara, lipsticks, toothpaste, and dentine cleansers. They also serve as emulsifiers, foaming agents, solubilizers, wetting agents, cleaners, antimicrobial agents, mediators of enzyme action, and cleaners (Sarubbo et al., 2022).

The critical micelle concentration (CMC), the hydrophilic-lipophilic balance (HLB), as well as the ionic performance, among other parameters relate to the composition of biosurfactants as being significant in the cosmetic industry because these properties determine the type or use of biosurfactants in cosmetic formulations (Bezerraa et al., 2020).

4.1 Critical Micelle Concentration (CMC)

The CMC is the lowest biosurfactant concentration that results in the greatest decrease in water surface tension. It is frequently used to gauge the effectiveness of biosurfactants. The lowest CMC values are found in the literature for lipopeptides (surfactin) and glycolipids such as rhamnolipids. For instance, using paneer whey as the production medium a rhamnolipid from *P. aeruginosa* SR17 with a CMC of 110 mg/L was obtained (Patoway et al., 2016). Moreover, a rhamnolipid with a CMC of 27 mg/L was obtained by *P. aeruginosa* SWP-4 using used cooking oil as the carbon source (Lan et al., 2015).

4.2 Hydrophilic-lipophilic Balance (HLB)

HLB value is a significant factor for the proper incorporation of biosurfactants in cosmetic goods since it allows for the prediction of their biosurfactant emulsifying capacity (Nour et al., 2018). A biosurfactant can function as an emulsifier, wetting agent, antifoaming agent, among other things, depending on the HLB values. Greater lipophilic biosurfactants have lower HLB values than hydrophilic biosurfactants, which have higher HLB values. Greater lipophilic biosurfactants have lower HLB values than hydrophilic biosurfactants, which have higher HLB values.

Consequently, biosurfactants with greater solubility in water will stabilize oil in water emulsions (O/W) more effectively whereas biosurfactants with more solubility in oil would be better stabilizers of water in oil emulsions (W/O). More lipophilic emulsifiers with excellent stabilizer capabilities in W/O emulsions are associated with lower HLB values. The biosurfactants with greater HLB values, on the other hand, are more hydrophilic and sufficient to stabilize O/W emulsions. For dermatological applications, W/O emulsions with HLB values between 1 and 4 are preferred because the lipid coating on the skin favors oil-soluble active components. These cosmetic compositions have an occlusive and protecting function. However, O/W emulsions feel less greasy, with HLB values between 8 and 16 and preferred by consumers. These formulations typically comprise a liquid or semi-solid component (Esmaeili et al., 2021).

According to studies, sophorolipids produced by Torulopsis yeasts act as anionic biosurfactants to form stable emulsions when administered in their acid or monovalent metallic salt forms. These sophorolipids with HLB values ranging from 13 to 15, are ideal for the preparation of some skincare and personal care formulations (Liepins et al., 2021).

Alkylene oxides were used to react with the sophorolipids produced from *Torulopsis bombicolu* resulting in a group of long chain alkyl-sophorolipids. These altered chemical Compounds that enhance the natural moisturizing factor were reported. An oleylsophorolipid had an HLB rating of 7-8 and demonstrated particular skin compatibility (Bhatt et al., 2022).

4.3 Ionic Property

The ionic behavior of biosurfactants is also a chief factor if their application in cosmetic formulations is envisaged. Surfactants and biosurfactants can be categorized as anionic, cationic, non-ionic or amphoteric based on their polar head charge. Surfactants typically have good emulsifier capabilities. Comparing anionic surfactants to cationic, non-ionic; anionic shows the best wetting, foaming, and emulsifying capabilities. However, anionic surfactants are more irritating to the skin and eyes than non-ionic and amphoteric surfactants. However, cationic surfactants have demonstrated potent antibacterial and emulsifying effects. A recent study, evaluated the adsorption of a lipopeptide biosurfactant that was naturally derived from a stream of the maize wet milling industry (Rincón-Fontán et al., 2016). They found that the biosurfactant was amphoteric and was bound by both cationic and anionic resins. Additionally, when this biosurfactant was applied to hair at levels close to the CMC, the hair was able to adsorb the biosurfactant with a maximum capacity of 3679μg/g (Vecino et al., 2017).

4.4 Emulsifying Property

Biosurfactants have the capacity to operate as both an emulsifier and a demulsifier. Oil-in-water and water-in-oil emulsions are the two types of emulsions. Emulsions made with two different phase solutions are typically unstable. Biosurfactants facilitate the mixing of two immiscible liquids and enable the dispersion of one liquid into another, resulting in micellar solubilization of large molecules. Biosurfactants and bioemulsifiers are characterized by a broad emulsifying property through lowered surface tension of water and oil. The hydrocarbon structure of biosurfactants influences their capacity as emulsifiers. Moreover, the ability of biosurfactants to identify numerous hydrocarbons, including methyl-naphthalene, heavy and light crude oil, diesel, benzene, kerosene, and many other types of hydrocarbons (Wu et al. 2022). The emulsification activity of a biosurfactant produced by marine bacteria against a variety of hydrocarbons, including benzene, xylene, kerosene, diesel, and gas, was studied by (Peele et al., 2016). Based on their evaluation, the produced biosurfactant may be able to maintain the emulsions for more than a week. Additionally, studies show that adding gum Arabic to biosurfactants makes them more effective (Inès et al., 2023).

4.5 Foaming Capacity

Biosurfactants are concentrated at the gas-liquid interface, where they create foam by bubbling through the liquid. The bubble formation methods investigate the foaming characteristics (Fig. 2) of surface-active molecules, such as bovine serum albumin (BSA), sodium dodecyl sulfate (SDS), and surfactin. Foams are colloidal systems with a continuous phase of liquid or solid and a dispersed phase of gas. Polyhedral cells are one of two common structures in foams that can occur in spherical bubbles and foams. Additionally, because of gravity, its structure has a tendency to shift. The foams take on a spherical shape when they are liquid, but when they are dry, they take on a polyhedral shape (Bezerra et al., 2018).

4.6 Spreading and Wetting Property

The fundamental mechanism of wetting is the replacement of the liquid phase by another fluid from the surface of a solid. The angle at which a liquid drop interacts with a solid surface serves as a measure of a surfactant's capacity to wet surfaces. Instead of being linked to the surface tension, biosurfactants might operate as wetting agents by penetrating the pores. When rebuilding dry substances like powders, beads, or reagents in solid-phase devices, wetting agents are essential (Nour et al., 2018).

4.3 Other Properties

4.3.1 Biodegradability, Biocompatibility and Digestibility

The cosmetic, pharmaceutical, and food industries use biosurfactants due to their biocompatibility and digestibility. Biosurfactants are biological compounds that are likely to be produced as a secondary metabolite or in association with microbial development. They are easily degraded by bacteria and other microbes as compared to chemical surfactants. Chemical surfactants affect the environment; therefore, biodegradable biosurfactants have emerged as a potent alternative in cosmetic formulations (Tiso et al., 2017). They are utilized in the food industry because they are easily digestible by humans and in the cosmetics industry due to their compatibility with many compounds (Gudiña and Rodrigues, 2019).

4.3.2 Stability in Extreme Conditions

Microorganisms are found over a wide range due to their ability to survive under

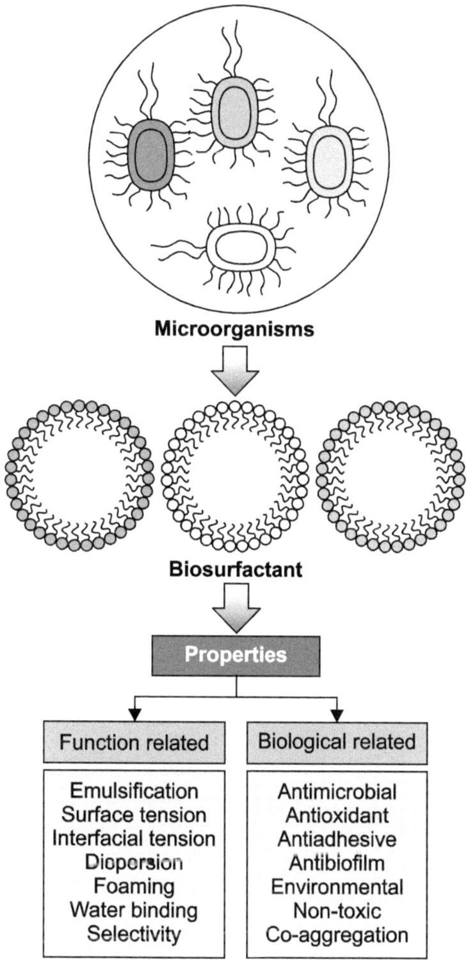

Fig. 2: Various functional and biological properties of biosurfactants and bioemulsifiers.

extreme conditions (Raval et al., 2018). Similarly, metabolites produced by such microbes also have stability under extreme conditions. In a pH range of 5 to 12, *Bacillus licheniformis* JF-2 lipopeptides maintain their stability at temperatures around 75°C for 140 hours. Moreover, while chemically generated surfactants can only be rendered inactive by 2% of salt, biosurfactants can withstand salt concentrations of up to 10% (Joshi et al., 2016).

4.3.3 Low Toxicity

A growing concern over the allergenicity of synthetic products has brought biosurfactants to light (Fenibo et al., 2019). They reduce the number of toxic substances in cosmetics, pharmaceuticals, and food, can be used in these industries (Akbari et al., 2018).

4.3.4 Surface and Interfacial Activities

In comparison to conventional surfactants, biosurfactants exhibit increased potency and productivity with a reduced mass and produce less surface tension. While surface and interfacial tension (oil/water) are roughly 1 and 30 ml/m in order, CMC of biosurfactants (competence measurement) ranges from 1 to 2000 mg/l (Joshi et al., 2016).

4.3.5 Specificity

Biosurfactants, which are Complex Organic Molecules (COMs) containing functional groups, have specialized functionalities that are of great interest due to the detoxification of particular pollutants and their use in cosmetics, pharmaceuticals, and food products (Vijayakumar and Saravanan, 2015).

4.3.6 Availability

Biosurfactants are used in various industries as they are produced by microorganisms using raw and industrial wastes which are easily available (Domínguez Rivera et al., 2019).

5. Biosurfactant/Bioemulsifier Extraction, Purification and Characterization for Cosmetic Formulations

The recovery of the economy and the fact that over 60% of the expenses of production are incurred in the downstream processes guarantee the commercial viability of a bioprocess. Biosurfactants are simpler to recover than other secondary metabolites because of their physicochemical characteristics, such as the ability to form micelles or surfaces. The majority of organic solvents used to extract biosurfactants are harmful, so researchers have developed less toxic, less expensive solvents to replace them in order to lower recovery costs. It is not possible to recover and purify the biosurfactant using just one downstream procedure. As a result, multi-step recovery procedures are adopted, allowing for superior quality of recovered products at various stages through a succession of purification and concentration phases (Pardhi et al., 2022).

The formulation of biosurfactants and bioemulsifiers in the cosmetic industry requires appropriate purification. The solvent extraction method can be used to recover biosurfactants from fermented culture broth and is frequently used as an intermediate step in the bioseparation of biotechnological products. It allows the culture broth to concentrate more easily and completes separation by floating proteins in the culture supernatant (Rangarajan and Clarke, 2016, Fenibo et al., 2019). Various solvents such as chloroform, methanol, ethanol, heptane, ethyl acetate and dichloromethane (Table 2) are used for extraction of biosurfactants. Crystallization (Table 3) is another method used to purify the biosurfactant after extraction or precipitation of the biosurfactant combination from the culture broth and subsequent dissolution in an organic phase. Rhamnolipid crystallizes when a solvent like hexane is added due to its reduced solubility in the solvent. Additionally, the ultrafiltration (Table 3) method is used for the concentration, recovery and purification of biomolecules of different molecular weight and size (Najmi et al., 2018).

Among various chromatography techniques, Thin-layer Chromatography (TLC), High-performance Liquid Chromatography (HPLC), size exclusion chromatography,

Table 2: Solvents used in purification of different microorganisms and analytical method

Type of biosurfactant	Microorganisms	Solvent used in purification	Analytical method	References
Sophorolipid	*Candida bombicola*	Ethyl acetate & pentane	Methanolysis	(Delbeke et al., 2016)
Glycolipid	*Ustilago maydis*	Chloroform:Methanol extraction	Biomass Free fatty acids Lipase activity	(Becker et al., 2022)
Glycolipoprotein	*Aspergillus ustus*	Dichloromethane, Ethyl acetate, and diethyl ether	Protein Carbohydrate- Phenol–sulphuric acid method Lipid - For free fatty acid using the method	(Adetunji, 2017)
Sophorolipid	*Trichosporon ashii*	Chloroform and methanol extraction	Carbohydrate- Anthrone reagent method Lipid Protein	(Ismail et al., 2019)
Protein-lipid polysaccharide complex	Candida lipolytica	Ethyl acetate extraction	Sugars- Phenol–sulfuric acid method Lipid- Gravimetric estimation	(Saini et al., 2021)

Table 3: Different strategies for extraction and purification of biosurfactant/bioemusifier with recovery and purity percentage

Microorganisms	Biosurfactant/ bioemulsifier	Extraction and purification method	Recovery (%)	Purity (%)	References
Bacillus megaterium	Surfactin	Macro porous resin column	-	91.6	(Dhanarajan et al., 2015)
Bacillus subtilis	Surfactin	Membrane-based separation	-	95	(Coutte et al., 2017)
Candida bombicola	Sophorolipid	Foam fractionation	86	90	(Dolman et al., 2019)
Pseudomonas aeruginosa	Rhamnolipid	Calcium precipitation	92	87	(Invally et al., 2019)
Pseudomonas aeruginosa	Rhamnolipid	Lyophilization	-	95	(Li et al., 2019)
Achromobacter sp.	Rhamnolipid	Acid precipitation	89.05	90	(Joy et al., 2020)
Pseudomonas putida	Rhamnolipid	Anti-solvent precipitation	94	99	(Biselli et al., 2020)
Bacillus lichenformis	Rhamnolipid	Ultrafiltration	95	99	(Rajasimman et al., 2021)
Bacillus subtilis ATCC 21,332	Surfactin	Hybrid adsorption and ultrafiltration	95	88	(Tran et al., 2022)
Starmerella bombicola	Sophorolipid	Crystallization	-	98.1	(Sivagiri et al., 2022)
Pseudozyma aphidis	Mannosylerythritol lipid (MEL)	Phase separation	93	-	(Venkataraman et al., 2022)
Pseudomonas libanensis	Viscosin	Solid-phase extraction	90	96	(Ciurko et al., 2023)

and ion-exchange chromatography are used for the purification of biosurfactants. Thin-layer chromatography is used to assess the accuracy and precision of the surfactin lipopeptides solubility in various solvents (Dlamini et al., 2020). The structural characterization of the biosurfactants aids in determining their potential uses in various industries. TLC was used to separate the phospholipids, rhamnolipids, and lipopeptides using a chloroform: methanol: water solvent system (Nwaguma et al.,, 2016).

Depending on the kind of biosurfactant, various chromatographic and spectrophotometric procedures are frequently used singly or in combination for biosurfactant characterization. The lipopeptide-type biosurfactants are typically separated from one another and identified using HPLC. The HPLC for glycolipids must be connected to either a mass spectrometer (MS) or an evaporative light scattering detector (ELSD). It was found that ultra-HPLC-MS and other HPLC-coupled devices operate more quickly than a qualitative HPLC. Surfactin isomers from *Bacillus subtilis* SZMC 6179J were recently studied using HPLC-ESI-MS (electrospray ion-mass spectrometry) (Bartal et al., 2018).

In size exclusion chromatography the Sephadex G-25 packed matrix (pH 7.5) was washed off using numerous column volumes of sodium phosphate buffer (20 mM). The DEAE active fraction with biosurfactant was applied to the surface of a Sephadex column (G-25), and fractions were collected based on molecular weight during the elution process using a 20 mM sodium phosphate buffer (pH 7.5) (Meena et al., 2021). Ion exchange chromatography is used to purify the biosurfactant. After dialysis of the biosurfactant; 5ml of a dialysed sample is filled into a diethyl aminoethyl (DEAE) cellulose-packed glass column. The activation is carried out with 0.5 M NaOH and equilibrated with sodium phosphate buffer (20 mM, pH 7.5) (Meena et al., 2020). Unbound protein was removed with the aid of a low ionic strength buffer, and the bound biosurfactant was removed using a stepwise gradient with varying molarity of NaCl in the sodium phosphate buffer (Meena et al., 2021). Compared to other downstream processing techniques, the chromatographic method of purification yields a highly purified output and allows for the purification of materials based on solubility, molecular weight, and ionic characteristics.

Lipopeptides produced by *Bacillus* spp. and *Virgibacillus salaries* were analyzed by Fourier Transform-Infrared Spectroscopy (FT-IR) using a traditional KBr disk (Elazzazy et al., 2015, López-Prieto et al., 2019). A new FT-IR technique, known as the attenuated total reflectance (ATR) crystal accessory, has just been developed. It produces quicker and more accurate readings. For better performance, mass spectrophotometry (MS) is frequently combined with other methods such as liquid chromatography-ESI-MS (LC-ESI-MS), electrospray ion-MS (ESI-MS), secondary ion-MS (SIMS), gas chromatography-MS (GC-MS), and matrix-assisted laser desorption/ionization-time of flight-MS (MALDI TOF-MS). Lichenysin-A and aneurinifactin, two recently found biosurfactants, are isolated and identified using MALDI TOF-MS (Joshi et al., 2016, Balan et al., 2017).

6. Applications of Biosurfactants and Bio Emulsifiers in Cosmetics

Biosurfactants are basically surface-active substances that originated in the biological system as secondary metabolites. It is amphiphilic in nature, due to which it can interact with hydrophilic as well as hydrophobic interfaces. However, the hydrophilic moiety of a biosurfactant consists of carbohydrates, amino acids, phosphates, cyclic peptides, carboxylic acids, etc. Owing to their physicochemical properties, biosurfactants find suitable applications in the cosmetic sector due to their multifunctional properties (Nour et al., 2018), viz. foaming, water-binding capacity, emulsification, wetting, de-emulsification, etc., which make effective use of these properties in the formulation of various cosmetic products (Adetunji & Olaniran, 2021). It maintains skin health due to the presence of useful fatty acids, which are useful in the prevention of skin dryness, & free radical scavenging ability, which confers protection from UV damage. The diverse biosurfactant active constituents enable its use in hair regeneration and in various skin repair mechanisms. The *in-vitro* studies show the compatibility with human skin cells due to important aspects of various types of biosurfactants rhamnolipids, sophorolipids and mannosylerythritollipids (MELs) (Roy, 2017). The chemical constituents of biosurfactants (Ambaye et al., 2021) enable their use in various forms, such as powders, creams, shampoos, lotions, and other cosmetic products, due to emulsifying, foaming, wetting, and solubilizing properties.

6.1 Glycolipids—Skin Moisturizing Capability

Amongst the types of biosurfactants, Glycolipids are the most useful type of low molecular weight biosurfactants. Their structural variation may be present due to a fatty acid chain along with a carbohydrate portion corresponding to a distinct component which varies according to its nature. Further glycolipids (Bezerra et al., 2018) can be sub categorized into various lipids types such as rhamnose, trehalose, sophorose, cellobiose, mannosylerythritol, lipomannosyl-mannitols, lipomannans & lipoarabionmannanes etc. Because of properties such as unique physico-chemical properties including surface tension reduction, and various biological abilities such as low toxicity many glycolipids are important candidates for cosmetic industries. They can be used in the form of moisturizers, which are complex mixtures of substances used for nourishment of the external layers of the skin by hydrating it and reducing evaporation.

Many studies explore the health benefits of glycolipid biosurfactants on skin cells types and layers due to their potential in skincare applications (Kumar et al., 2023). A recent study shows the importance of glycolipid extracts in bio medicinal and cosmetics applications (Karnwal et al., 2023). The crude glycolipid exhibits high antioxidant radical scavenging abilities in case of *Pseudomonas aeruginosa* which enable its use in the cosmetics sector. Additionally, the anticancer activity of glycolipid types of biosurfactants against various cancer cell lines shows the functional role of biosurfactants in therapeutics as well. The antimicrobial activities of glycolipid biosurfactants are found due to its efficacy in destabilizing membranes by generating ionic channels and various pores in them. The glycolipid biosurfactant may regulate, activate & or inhibit many enzymes' activities having biotechnological potential (Ahmadi-Ashtiani et al., 2020). They are used particularly in cosmetics (Ghazi Faisal et al., 2023) due to their moisturizing capability. This anti-adhesive attribute the adhesion of bacteria also can be prevented.

6.2 Rhamnolipids in Reduction of Skin Inflammation

Rhamnolipids are one of the most explored microbial biosurfactants which are amphipathic in nature and also referred to as "oily glycolipids". They consist of two moieties i.e. Rhamnose moiety which is also known as glycon and lipid moiety which is referred as a glycon. Rhamnose is hydrophilic and consists of a rhamnose counterpart linked by a-1,2 glycosidic linkage & on the contrary the lipid moiety is hydrophobic in nature and consists of saturated and unsaturated fatty acids with ester linkages (Palanisamy et al., 2023). Due to glycosidic bonds both the moieties are interlinked. The gram-negative bacteria particularly *Pseudomonas aeruginosa* is the dominant sp. Which produces four varied types of rhamnolipids (Adu et al., 2020) which have potential applications in many sectors.

Due to rhamnolipids' potential to reduce surface tension between various types of phases they can be used as a potential source of secondary metabolites in cosmetic formulations, detergent industry etc. The rhamnolipids have the ability to cease the outer membrane protein viz. OmpA; apparently it confers other properties such as antimicrobial, anticancer, immunomodulatory and antioxidant (Kumar at al., 2023). Due to their lower range of surface tension ability rhamnolipids have been recently explored. They have also been explored for critical micellar concentrations & high emulsification indexes of about 70% in a short duration. Rhamnolipids are used particularly as an active ingredient for the preparation of various formulations which are useful for dermatological diseases & for re-epithelization of mucous membrane layers, and for gum disease treatment. Recently a partially purified surfactant sourced from *Bacillus subtilis* was explored for the formation of a nano emulsion in which xanthan gum is used (Karnwal et al., 2023) along with emulsan to form pseudoplastic behavior which leads to its application in the cosmetic industry. Many commercially available products include RelipidiumTM a moisturizer which can be used for body and face nourishment, manufactured in Monheim, Germany (Kumar et al., 2023).

Additionally, the researchers compare the effect of naturally occurring/derived glycolipid biosurfactants, particularly rhamnolipids, with synthetic surfactants on human keratinolytic cells, which are very useful for cosmetic formulations (Adu et al., 2023a). Another study shows the viability of cells and synthesis of cytokines. These are not affected much by mono rhamnolipids. However, their effect may vary depending upon their chemical constituents. Further di-rhamnolipids at a non-inhibitory level drastically reduce the inflammation and also enhance cytokines production which are anti-inflammatory substances possessing potential for use in skin care product formulationsas compared to synthetic surfactants (Adu et al., 2023b). The rhamnolipids are also found to be useful in the treatment of psoriasis which is a tropical skin infection. The antioxidant radical scavenging ability of rhamnolipids by 1-diphenyl-2-picrylhydrazyl radical scavenging assays (DPPH) from *Pseudomonas aeruginosa* was reported in a dose or concentration dependent manner.

In a recent study conducted by (Thapliyal and Negi, 2023) mentioned the use of the rhamnolipid bio-surfactant from *Pseudomonas aeruginosa* for the formulation of a shampoo comprising of about 2% of rhamnolipid in water.

6.3 Soporolipids in Repairing Skin Cells and Acne Treatment

Soporolipids (SLPs) are the type of biosurfactants used as a potent substance in the treatment of various infections such as acne t, dandruff, and body odors. They may be used as desquamating as well as depigmenting substances which particularly eliminate the surface protective layer of the epidermis which plays an important role in wound healing (Gaur et al., 2022). They are one of the suitable candidates for cellulite treatment as well, where they reduce the subcutaneous fat layer by increasing

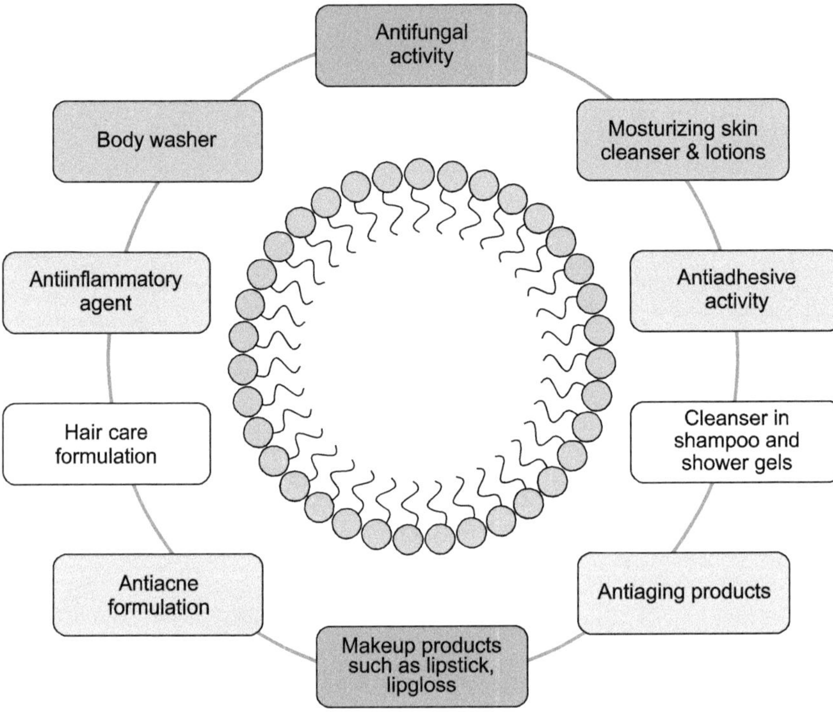

Fig. 3: Applications of biosurfactants and bioemulsifiers in cosmetic formulations.

leptin synthesis in the adipocyte's cells. They are also useful for repairing, rejuvenating and protecting the skin due to their ability to stimulate skin fibroblast synthesis and regulation. The products incorporating sophorolipids are available as lip creams, eye shadows, compressed powders, & other cosmetic solution formulations. SLPs are used in formulating several beauty products such as bath products, acne pads, antidandruff substances, pencil-shaped lip linears, lip sticks, lip creams and toothpastes (Fig. 3) (Drakontis and Amin, 2020). Many cosmetics products are commercially available manufactured under the name Ecover Belgium by Group Soliance, France. Another product used in cosmetics is manufactured by Kaneka Co. Japan Cosmetics. MG Intobio Co. Ltd., South Korea also manufactures soap with improved properties (Moldes et al., 2020). SLPs have been used as an active ingredient in various cosmetic formulations for body and skin applications. They are explored due to lower toxicity towards human keratinolytic cells. They also promote collagen synthesis in the dermis layer which leads to improvement in rejuvenation and toning of skin cells. Due to leptin synthesis by adipocyte cells, they reduce fat deposition which makes them important for cellulitis treatment. Aging of skin and appearance of wrinkles occurs due to increased elastase activity particularly from the dermal fibroblasts. SLPs reduce free radicals and inhibit elastase activity. They also act as a macrophage initiator. They are also useful as fibrinolytic agents for wound treatment, for desquamating & as depigmenting agents for treatment of brown patches on the skin. Also due to bactericides and bacteriostatic capability they are used for dandruff control, acne treatment and as an active agent in deodorant formulations. Product named Sopholiance TM is used in the formulation of deodorants, shower gels and marketed by Givaudan Active Beauty Paris France & Kanebo skin care products used in the formulation of moisturizers, cleansers etc. marketed by Kanebo Cosmetics Japan. The antiwrinkle, wound healing and immunomodulatory effects of sophorolipids were discovered (Maeng et al., 2018) by using fermented hydrolyzed horse oil on human epidermal cells; apparently the authors also demonstrated the dose dependent level of gene expression of the collagen gene which was found to inhibit elastase enzyme related to skin aging. It also increases m-RNA expression levels of various lipopolysaccharides responsible for protein synthesis. Due to an immensely significant role in wound healing treatment & skin grafting, sophorolipids are explored for formulations. Maeng et al., 2018, reported the wound healing ability of sophorolipids. This was a key step ahead for the identification of a potent active ingredient in formulations to enhance chronic wound healing properties and for skin grafting reduction. Kao Co. Ltd Japan is the major manufacturer of sophorolipids and other compounds such as humectants which can be used as skin & hair moisturizers as well as in lipstick formulations which helps in hair restoration and protects the skin against damage (Adu et al., 2020). The best example is the development of new sunscreen formulations which provide protection against UV damage and also nourishe the skin cells. The researcher Rincón-Fontán formulated the greener sunscreen formulation which is based on mining silicate type of material (mica). Additionally similar authors in different works reported the mica along with the biosurfactant for the stabilization of Pickering emulsions consisting of vitamin E in a triangular design. It was also found that the addition of biosurfactant extract may improve emulsion volume up to about 70% (Rincón-Fontán et al., 2019).

The researchers prepared a toothpaste formulation by using a biosurfactant obtained from *Nocardiopsis* VITSISB, which could replace the use of sodium lauryl sulphate and reduce the harmful effects of chemicals (Vecino et al., 2017). The mentioned work shows that the role of the biosurfactant in the formulation is most efficient and less toxic as compared to chemically synthesized surfactants. Farias et al., (2019) also formulated a mouth wash containing biosurfactant as an active ingredient with lower toxicity.

6.4 MELs with Multiple Benefits for Skin

Mannosylerythritol lipids possess a unique liquid crystal forming and moisturizing ability and gain much significance as an active component in cosmetic formulations (Adu et al., 2020). The MELs from yeast belong to the genus *Pseudozyme.* They are explored for their unique properties (De Andrade et al., 2017). The study was conducted using Sodium lauryl sulfate which damages skin cells. It was found that using mannosylerythritol lipids enhances the recovery of cell viability with a higher recovery rate of about 80 % as compared to that of natural ceramide. The MELs were reported to increase stratum corneum content in the skin cells and further perspiration on the skin surface was suppressed by them (Coelho et al., 2020). The study reported (El-Khordagui et al., 2021) the use of MELs from yeast cells for the recovery of damaged hair and lastly they provide flexibility, tensile strength and a smooth appearance to hair. The MELs from *Ustilago scitaminea* show ceramide-like skin care ability in a human skin model. The MELs also show a potent moisturizing ability of cultured human skin cells having SDS. The patent for the incorporation of MELs in skin care formulations as an anti wrinkle agent which is very useful for the treatment of skin related problems. MELs are considered as the most proficient biosurfactants which can be used effectively in cosmetics' formulations. Varieties of cosmetic products such as eye shadows, lipsticks, soaps, sprinkle powders, body sprays, massage oil, nail polish, lip liner etc. formulations involve the use of MELs (Palanisamy et al., 2023). Particularly MEL-A is a potent hair growth stimulator which activates papilla cells. Further an increase in interest on the stress associated with the hydroxyl radical and superoxide anion effects in skin care product formulations is gaining attention due to their moisturizing properties (Fig. 3) (Coelho et al., 2020). Due to the significant role of squalene as a scavenging molecule gains attention on exploration of MELs as antioxidant molecules. MELs had shown DPPH radical scavenging ability in a dose dependent manner and are considered as the strongest scavenging substances. In the formulation of cosmetic foundations which are skin colored materials which can be applied to the skin to obtain a uniform complexion and to cover dark spots, MELs play an important role. Various functional components can be added to skin care products to enhance skin health and to ensure its proper functioning. To further enhance the functional properties of cosmetic products many forms of metal oxide such as iron and titanium oxides etc. can be used as the main raw ingredients. Further for improvement in dispersibility of the active components these metal oxides are coated with lipophilic substances (Bezerra et al., 2018). MELs with a special capacity for self-assembly at different surface interfaces can replace the conventional usage of different synthetic compounds as coating materials, which diminishes the moisturizing property of the compound toward skin. Alternatively, MELs coated with metal oxide particles can be used which provides benefits in terms of nourishment in the foundation which leads to improvement & a novel formulation with better efficacy. The manufacturer from, Japan Daito Kasei Kogyo Co. Ltd. is currently marketing (Fracchia et al., 2019) useful cosmetic products from biosurfactants in powder form.

Bio emulsifiers are substances that have a higher molecular weight than biosurfactants. They are complex mixtures of various heteropolysaccharides, lipopolysaccharides, lipoproteins etc (Mujumdar et al., 2019). Many other types of surfactants originating from biological systems have been reported (Floris et al., 2018). The use of particular rhamnolipids and sophorolipids as emulsifying agents in lip gloss formulations was manufactured by Evonik industries, and as a result both show an excellent stabilizing ability when mixed with silicone oil (Adu et al., 2023a). Apart from its other personalized health care products, an antidandruff shampoo was prepared (Karnwal et al., 2023) by combining various chemical additives, MELs and SLSs with antidandruff agents, along with salicylic acid and benzoic acid.

7. Regulation for Cosmetic Safety for Use of Biosurfactants and Bioemulsifiers

Currently, the only European Regulation available on cosmetic products is the Regulation No 1223/2009. The cosmetic trade intends to stream safe and effective products, which can meet the consumers' needs, by following this regulation or any other precise regulation before it is placed on the market. The Regulation No 1223/2009 covers surfactants as a constituent in cosmetic products; though, there is no discussion in this article about biosurfactants. However, if their usage as ingredients in cosmetics is intended, both surfactants (synthesized chemically) and biosurfactants (produced biologically) should adhere to the same standards. Moreover, the International Nomenclature Cosmetic Ingredient (INCI) records is included in Article 33 of Regulation No. 1223/2009 and uses a common nomenclature for cosmetic components. The INCI mandates the use of consistent, orderly, and global nomenclature to identify compounds in cosmetic products. Ingredients are given INCI names in this way depending on their chemical constituents and structure (Vecino et al., 2017).

Additionally, in accordance with Article 13 of Regulation No. 1223/2009, distributors of cosmetic products are required to inform regulatory agencies via the Cosmetic Products Notification Portal (CPNP) of the cosmetic products they have placed on the European market. Additionally, CosIng is a database of cosmetic ingredients that provides details on cosmetic components and substances based on Regulation No. 1223/2009, including their legal requirements and limitations. BS produced by *L. paracasei*, which functions as a key emulsifier agent in O/W emulsion systems when combined with natural antioxidant extract and essential oils, is a natural component in cosmetic formulations. Over 97% of cells proliferated in the presence of BS or O/W emulsions containing the natural antioxidant extract and BS, is better than the results attained with SDS. These findings provide new prospects for the use of biosurfactants in creams and other cosmetic applications, provided that they maintain consumer safety and adhere to Regulation (EC) No. 1223/2009 on cosmetic products (Ferreira et al., 2017).

Currently, it is apparent the cosmetics sector has a high level of interest in natural goods, largely as a result of increased consumer concerns about environmental sustainability and health. As a result, the cosmetics business is looking for healthier and more sustainable products. However, the severe safety evaluations necessary for "nonnatural compounds" in the cosmetic regulations should also apply to these natural chemicals. In view of this, several organizations, like ECOCERT, promote and assist businesses towards certification. These organizations assist companies in obtaining a label designating their products as natural or organic cosmetics with safety requirements. To qualify for an organic or natural cosmetic label, the 10% or 5% of all ingredients and 95% or 50% of the vegetable ingredients, respectively, must come from ecological farming (Vecino et al., 2017).

8. Conclusion

Biosurfactants and bioemulsifiers are "multifunctional materials" since they have a wide range of uses in different industries such as textiles, detergents, paint, food and cosmetics. They are substitutions of chemical surfactants due to their outstanding functional properties such as low toxicity, surface/interfacial activity, skin-compatibility and specificity. The skin-friendly attributes of biosurfactants allow their use in the cosmetic sector. The requirement of a purification strategy depends on its application. For instance, in cosmetic and pharmaceutical applications, biosurfactants must follow different standard regulations. However, for environmental cleanup, the final yield should be microbial load free. It is still difficult to increase biosurfactant production for industrial use. This chapter covers promising characteristics of microbial surfactants in relation to their potential to improve the efficacy of cosmetic and personal care products. The requirement of different biosurfactants and bioemulsifiers is steadily increasing in the cosmetic industry and the demand can be satisfied by developing better production, purification and detection methods.

Acknowledgements

Authors DMB and NYP acknowledge the "SHODH - ScHeme of Developing High-Quality Research" by the Department of Higher Education, Government of Gujarat with (Reference No. 201901380189) & (Reference No. 202001380174), respectively. SYR acknowledge the Research Associateship from Gujarat State Biotechnology Mission, GoG. All authors acknowledge the infrastructural support and facilities provided by the Department of Microbiology and Biotechnology, Gujarat University, Ahmedabad.

References

Adetunji, A.I. & Olaniran, A.O. (2021). Production and potential biotechnological applications of microbial surfactants: An overview. *Saudi J. Biol. Sci.*, 28(1): 669-679. https://doi.org/10.1016/j.sjbs.2020.10.058

Adetunji, A.I. (2017). Treatment of lipid-rich wastewater using free and immobilized bioemulsifier and hydrolytic enzymes from indigenous bacterial isolates. A.I. Thesis, University of KwaZulu-Natal (Westville campus), Durban, South Africa.

Adu, S.A., Naughton, P.J., Marchant, R. & Banat, I.M. (2020). Microbial biosurfactants in cosmetic and personal skincare pharmaceutical formulations. *Pharmaceutics*, 12(11): 1099. https://doi.org/10.3390/pharmaceutics12111099

Adu, S.A., Twigg, M.S., Naughton, P.J., Marchant, R. & Banat, I.M. (2023a). Glycolipid biosurfactants in skincare applications: Challenges and recommendations for future exploitation. *Molecules*, 28(11): 4463. https://doi.org/10.3390/molecules28114463

Adu, S.A., Twigg, M.S., Naughton, P.J., Marchant, R. & Banat, I.M. (2023b). Characterisation of cytotoxicity and immunomodulatory effects of glycolipid biosurfactants on human keratinocytes. *Appl. Microbiol. Biotechnol.*, 107(1): 137-152. https://doi.org/10.1007/s00253-022-12302-5

Ahmadi-Ashtiani, H.R., Baldisserotto, A., Cesa, E., Manfredini, S., Sedghi Zadeh, H., Ghafori Gorab, M. et al. (2020). Microbial biosurfactants as key multifunctional ingredients for sustainable cosmetics. *Cosmetics*, 7(2): 46. https://doi.org/10.3390/cosmetics7020046

Akbari, S., Abdurahman, N.H., Yunus, R.M., Fayaz, F. & Alara, O.R. (2018). Biosurfactants—A new frontier for social and environmental safety: A mini review. *Biotechnol. Res. Innov.*, 2(1): 81-90. https://doi.org/10.1016/j.biori.2018.09.001

Alizadeh-Sani, M., Hamishehkar, H., Khezerlou, A., Azizi-Lalabadi, M., Azadi, Y., Nattagh-Eshtivani, E. et al. (2018). Bioemulsifiers derived from microorganisms: Applications in the drug and food industry. *Adv. Pharm. Bull.*, 8(2): 191. https://doi.org/10.15171/apb.2018.023

Ambaye, T.G., Vaccari, M., Prasad, S. & Rtimi, S. (2021). Preparation, characterization and application of biosurfactant in various industries: A critical review on progress, challenges and perspectives. *Environ. Technol. Innov.*, 24: 102090. https://doi.org/10.1016/j.eti.2021.102090

Balan, S.S., Kumar, C.G. & Jayalakshmi, S. (2017). Aneurinifactin, a new lipopeptide biosurfactant produced by a marine *Aneurinibacillus aneurinilyticus* SBP-11 isolated from Gulf of Mannar: Purification, characterization and its biological evaluation. *Microbiol. Res.*, 194: 1-9. https://doi.org/10.1016/j.micres.2016.10.005

Baria, D.M., Patel, N.Y., Yagnik, S.M., Panchal, R.R., Rajput, K.N. & Raval, V.H. (2022). Exopolysaccharides from marine microbes with prowess for environment cleanup. Environ. *Sci. Pollut. Res.*, 29(51): 76611-76625. https://doi.org/10.1007/s11356-022-23198-z

Bartal, A., Vigneshwari, A., Bóka, B., Vörös, M., Takács, I. Kredics, L. et al. (2018). Effects of different cultivation parameters on the production of surfactin variants by a *Bacillus subtilis* strain. *Molecules*, 23(10): 2675. https://doi.org/10.3390/molecules23102675

Becker, F., Linne, U., Xie, X., Hemer, A.L., Bölker, M., Freitag, J. et al. (2022). Import and export of mannosylerythritol lipids by *Ustilago maydis*. *Mbio.*, 13(5): e02123-22. https://doi.org/10.1128/mbio.02123-22

Bezerra, K.G.O., Rufino, R.D., Luna, J.M. & Sarubbo, L.A. (2018). Saponins and microbial biosurfactants: Potential raw materials for the formulation of cosmetics. *Biotechnol. Prog.*, 34(6): 1482-1493. https://doi.org/10.1002/btpr.2682

Bezerraa, K.G., Durvala, I.J., Silvab, I.A. & Fabiola, C.G. (2020). Emulsifying capacity of biosurfactants from *Chenopodium quinoa* and *Pseudomonas aeruginosa* UCP 0992 with focus of application in the cosmetic industry. *Chem. Eng.*, 79: 211-216. https://doi.org/10.3303/CET2079036

Bhatt, T., Bhimani, A., Detroja, A., Gevariya, D. & Sanghvi, G. (2022). Cosmetic application of surfactants from marine microbes. *In:* Kim, S.K. and Shin, K.H. (Eds), *Marine Surfactants*. 271-293. CRC Press.

Bhattacharya, B., Ghosh, T.K. & Das, N. (2017). Application of bio-surfactants in cosmetics and pharmaceutical industry. *Sch. Acad. J. Pharm.*, 6(7): 320-329. https://doi.org/10.21276/sajp

Bhosale, H. & Pinjari, R. (2021). Review on microbial surfactants as promising bioactive molecules. *Research and Development in Pharmaceutical Science*, 2: 11.

Biselli, A., Willenbrink, A.L., Leipnitz, M. & Jupke, A. (2020). Development, evaluation, and optimisation of downstream process concepts for rhamnolipids and 3-(3-hydroxyalkanoyloxy) alkanoic acids. *Sep. Purif. Technol.*, 250: 117031. https://doi.org/10.1016/j.seppur.2020.117031

Cardoso, S.L., Dantas, R., Costa, C.S., Campos, E.S. & Tambourgi, E.B. (2021). The antarctic yeast *Candida sake*: Investigation of fermentation parameters for biosurfactant production. *Chem. Eng. Trans.*, 87: 589-594. https://doi.org/10.3303/CET2187099

Celligoi, M.A.P., Silveira, V.A., Hipolito, A., Caretta, T.O. and Baldo, C. (2020). Sophorolipids: A review on production and perspectives of application in agriculture. *Span. J. Agric. Res.*, 18(3): 28. https://doi.org/10.5424/sjar/2020183-15225

Citarasu, T., Thirumalaikumar, E., Abinaya, P., Babu, M.M. and Uma, G. (2021). Biosurfactants from halophilic origin and their potential applications. *In:* Adetunji, C.O. and Asiri, A.M. (Eds), *Green Sustainable Process for Chemical and Environmental Engineering and Science*. 481-521. Elsevier.

Ciurko, D., Chebbi, A., Kruszelnicki, M., Czapor-Irzabek, H., Urbanek, A.K., Polowczyk, I. et al. (2023). Production and characterization of lipopeptide biosurfactant from a new strain of *Pseudomonas antarctica* 28E using crude glycerol as a carbon source. *RSC Advances*, 13(34): 24129-24139. https://doi.org/10.1039/D3RA03408A

Cleary, J. (2016). Stratum corneum lipid composition alters the heterogeneous growth of *Staphylococcus aureus*. State University of New York, Binghamton.

Coelho, A.L.S., Feuser, P.E., Carciofi, B.A.M., de Andrade, C.J. & de Oliveira, D. (2020). Mannosylerythritol lipids: Antimicrobial and biomedical properties. *Appl. Microbiol. Biotechnol.*, 104(6): 2297-2318. https://doi.org/10.1007/s00253-020-10354-z

Coutte, F., Lecouturier, D., Dimitrov, K., Guez, J.S., Delvigne, F., Dhulster, P. & Jacques, P. (2017). Microbial lipopeptide production and purification bioprocesses, current progress and future challenges. *Biotechnol. J.*, 12(7): 1600566. https://doi.org/10.1002/biot.201600566

De Andrade, C.J., De Andrade, L.M., Rocco, S.A., Sforça, M.L., Pastore, G.M. & Jauregi, P. (2017). A novel approach for the production and purification of mannosylerythritol lipids (MEL) by *Pseudozyma tsukubaensis* using cassava wastewater as substrate. *Sep. Purif. Technol.*, 180: 157-167. https://doi.org/10.1016/j.seppur.2017.02.045

Delbeke, E.I., Roelants, S.L., Matthijs, N., Everaert, B., Soetaert, W., Coenye, T. et al. (2016).

Sophorolipid amine oxide production by a combination of fermentation scale-up and chemical modification. *Ind. Eng. Chem. Res.*, 55(27): 7273-7281. https://doi.org/10.1021/acs.iecr.6b00629

Devale, A., Sawant, R., Pardesi, K., Perveen, K., Khanam, M.N., Shouche, Y. and Mujumdar, S. (2023). Production and characterization of bioemulsifier by *Parapedobacter indicus*. *Front. Microbiol.*, 14: 1111135. https://doi.org/10.3389/fmicb.2023.1111135

Dhanarajan, G., Rangarajan, V. & Sen, R. (2015). Dual gradient macroporous resin column chromatography for concurrent separation and purification of three families of marine bacterial lipopeptides from cell free broth. *Sep. Purif. Technol.*, 143: 72-79. https://doi.org/10.1016/j.seppur.2015.01.025

Dlamini, B., Rangarajan, V. & Clarke, K.G. (2020). A simple thin layer chromatography-based method for the quantitative analysis of biosurfactant surfactin vis-a-vis the presence of lipid and protein impurities in the processing liquid. *Biocatal. Agric. Biotechnol.*, 25: 101587. https://doi.org/10.1016/j.bcab.2020.101587

Dolman, B.M., Wang, F. and Winterburn, J.B. (2019). Integrated production and separation of biosurfactants. *Process Biochem.*, 83: 1-8. https://doi.org/10.1016/j.procbio.2019.05.002

Domínguez Rivera, Á., Martínez Urbina, M.Á. and Lópezy López, V.E. (2019). Advances on research in the use of agro-industrial waste in biosurfactant production. *World J. Microbiol. Biotechnol.*, 35(10): 155. https://doi.org/10.1007/s11274-019-2729-3

Drakontis, C.E. & Amin, S. (2020). Biosurfactants: Formulations, properties, and applications. *Curr. Opin. Colloid Interface Sci.*, 48: 77-90. https://doi.org/10.1016/j.cocis.2020.03.013

Edosa, T.T., Jo, Y.H., Keshavarz, M. and Han, Y.S. (2018). Biosurfactants: Production and potential application in insect pest management. *Trends Entomol.*, 14(79): 79-87.

Elazzazy, A.M., Abdelmoneim, T.S. and Almaghrabi, O.A. (2015). Isolation and characterization of biosurfactant production under extreme environmental conditions by alkali-halo-thermophilic bacteria from Saudi Arabia. *Saudi J. Biol. Sci.*, 22(4): 466-475. https://doi.org/10.1016/j.sjbs.2014.11.018

El-Khordagui, L., Badawey, S.E. & Heikal, L.A. (2021). Application of biosurfactants in the production of personal care products, and household detergents and industrial and institutional cleaners. *In: Green Sustainable Process for Chemical and Environmental Engineering and Science*. pp. 49-96. Elsevier. https://doi.org/10.1016/B978-0-12-823380-1.00005-8

Esmaeili, H., Mousavi, S.M., Hashemi, S.A., Lai, C.W., Chiang, W.H. & Bahrani, S. (2021). Application of biosurfactants in the removal of oil from emulsion. *In:* Adetunji, C.O. and Asiri, A.M. (Eds), *Green Sustainable Process for Chemical and Environmental Engineering and Science*. 107-127. Elsevier.

Farias, J.M., Stamford, T.C.M., Resende, A.H.M., Aguiar, J.S., Rufino, R.D., Luna, J.M. et al. (2019). Mouthwash containing a biosurfactant and chitosan: An eco-sustainable option for the control of cariogenic microorganisms. *Int. J. biol. Macromol.*, 129: 853-860. https://doi.org/10.1016/j.ijbiomac.2019.02.090

Fenibo, E.O., Douglas, S.I. & Stanley, H.O. (2019). A review on microbial surfactants: Production, classifications, properties and characterization. *J. Adv. Microbiol.*, 18(3): 1-22. https://doi.org/10.9734/JAMB/2019/v18i330170

Ferreira, A., Vecino, X., Ferreira, D., Cruz, J.M., Moldes, A.B. & Rodrigues, L.R. (2017). Novel cosmetic formulations containing a biosurfactant from Lactobacillus paracasei. *Colloids Surf. B: Biointerfaces*, 155: 522-529. https://doi.org/10.1016/j.colsurfb.2017.04.026

Floris, R., Rizzo, C. & Giudice, A.L. (2018). Biosurfactants from marine microorganisms. *In: Metabolomics – New Insights into Biology and Medicine*. 1-16.

Fracchia, L., Ceresa, C. & Banat, I.M. (2019). Biosurfactants in cosmetic, biomedical and pharmaceutical industry. *In: Microbial Biosurfactants and Their Environmental and Industrial Applications*. 258-288.

Gaur, V.K., Sirohi, R., Pandey, A.K., Varjani, S. & Pandey, A. (2022). Sustainable technologies for the production of sophorolipids from renewable wastes. *In: Biomass, Biofuels, Biochemicals*. pp. 275-294. Elsevier. https://doi.org/10.1016/B978-0-323-89855-3.00014-5

Ghazi Faisal, Z., Sallal Mahdi, M. & Alobaidi, K.H. (2023). Optimization and chemical characterization of biosurfactant produced from a novel *Pseudomonas guguanensis* strain Iraqi ZG. *KM. Int. J. Microbiol.* https://doi.org/10.1155/2023/1571991

Gudiña, E.J. and Rodrigues, L.R. (2019). Research and production of biosurfactants for the food industry. *In:* Molina, G., Gupta, V.K., Singh, B.N. and Gathergood, N. (Eds), *Bioprocessing for Biomolecules Production*. 125-143. Wiley.

Gürkök, S. & Özdal, M. (2021). Microbial biosurfactants: Properties, types, and production. *Anatol. J. Biol.*, 2(2): 7-12.

Inès, M., Mouna, B., Marwa, E. & Dhouha, G. (2023). Biosurfactants as emerging substitutes of their synthetic counterpart in detergent formula: Efficiency and environmental friendly. *J. Polym. Environ.*, 31: 2779-2791. https://doi.org/10.1007/s10924-023-02778-1

Invally, K., Sancheti, A. & Ju, L.K. (2019). A new approach for downstream purification of rhamnolipid biosurfactants. *Food Bioprod. Process.*, 114: 122-131. https://doi.org/10.1016/j.fbp.2018.12.003

Ishaq, U., Akram, M.S., Iqbal, Z., Rafiq, M., Akrem, A., Nadeem, M. et al. (2015). Production and characterization of novel self-assembling biosurfactants from *Aspergillus flavus*. *J. Appl. Microbiol.*, 119(4): 1035-1045. https://doi.org/10.1111/jam.12929

Ismail, D.A., Youssif, M.A. & Negm, N.A. (2019). Biosurfactants: Properties and applications. *In:* Biresaw, G. and Mittal, K.L. (Eds), *Surfactants in Tribology*. 25-56. CRC Press.

Joshi, S.J., Al-Wahaibi, Y.M., Al-Bahry, S.N., Elshafie, A.E., Al-Bemani, A.S., Al-Bahri, A. et al. (2016). Production, characterization, and application of *Bacillus licheniformis* W16 biosurfactant in enhancing oil recovery. *Front. Microbiol.*, 7: 1853. https://doi.org/10.3389/fmicb.2016.01853

Joy, S., Khare, S.K. & Sharma, S. (2020). Synergistic extraction using sweep-floc coagulation and acidification of rhamnolipid produced from industrial lignocellulosic hydrolysate in a bioreactor using sequential (fill-and-draw) approach. *Process Biochem.*, 90: 233-240. https://doi.org/10.1016/j.procbio.2019.11.014

Karnwal, A., Shrivastava, S., Al-Tawaha, A.R.M.S., Kumar, G., Singh, R., Kumar, A. et al. (2023). Microbial biosurfactant as an alternate to chemical surfactants for application in cosmetics industries in personal and skin care products: A critical review. *BioMed Res. Int.*, https://doi.org/10.1155/2023/2375223

Khalifa, S.A., Shedid, F.S., Saied, E.M., Jassbi, A.R., Jamebozorgi, F.H. Rateb, M.E. et al. (2021). Cyanobacteria—From the oceans to the potential biotechnological and biomedical applications. *Mar. Drugs*, 19(5): 241. https://doi.org/10.3390/md19050241

Khedher, S.B., Boukedi, H., Dammak, M., Kilani-Feki, O., Sellami-Boudawara, T., Abdelkefi-Mesrati, L. et al. (2017). Combinatorial effect of *Bacillus amyloliquefaciens* AG1 biosurfactant and *Bacillus thuringiensis* Vip3Aa16 toxin on *Spodoptera littoralis* larvae. *J. Invertebr. Pathol.*, 144: 11-17. https://doi.org/10.1016/j.jip.2017.01.006

Kumar, R., Barbhuiya, R.I., Bohra, V., Wong, J.W., Singh, A., and Kaur, G. (2023). Sustainable rhamnolipids production in the next decade–advancing with *Burkholderia thailandensis* as a potent biocatalytic strain. *Microbiol. Res.*, 127386. https://doi.org/10.1016/j.micres.2023.127386

Kumari, R., Singha, L.P. & Shukla, P. (2023). Biotechnological potential of microbial bio-surfactants, their significance and diverse applications. *FEMS Microbes*, xtad015. https://doi.org/10.1093/femsmc/xtad015

Kuyukina, M.S. and Ivshina, I.B. (2019). Production of trehalolipid biosurfactants by *Rhodococcus*. *In:* Alvarez, H. (Eds), *Biology of Rhodococcus. Microbiology of Monographs*. 271-298. Springer, Cham.

Lan, G., Fan, Q., Liu, Y., Chen, C., Li, G., Liu, Y. and Yin, X. (2015). Rhamnolipid production from waste cooking oil using *Pseudomonas* SWP-4. Biochem. *Eng. J.*, 101: 44-54. https://doi.org/10.1016/j.bej.2015.05.001

Li, Z., Zhang, Y., Lin, J., Wang, W. and Li, S. (2019). High-yield di-rhamnolipid production by Pseudomonas aeruginosa YM4 and its potential application in MEOR. *Molecules*. 24(7): 1433. https://doi.org/10.3390/molecules24071433

Liepins, J., Balina, K., Soloha, R., Berzina, I., Lukasa, L.K. & Dace, E. (2021). Glycolipid biosurfactant production from waste cooking oils by yeast: Review of substrates, producers and products. *Fermentation*. 7(3): 136. https://doi.org/10.3390/fermentation7030136

López-Prieto, A., Martínez-Padrón, H., Rodríguez-López, L., Moldes, A.B. and Cruz, J.M. (2019). Isolation and characterization of a microorganism that produces biosurfactants in corn steep water. *CyTA-J. Food*, 17(1): 509-516. https://doi.org/10.1080/19476337.2019.1607909

Luft, L. (2022). Fungal polysaccharides as biosurfactants and bioemulsifiers. *In:* Deshmukh, S.K., Deshpande, M.V., Sridhar, K.R. (Eds),*Fungal Biopolymers and Biocomposites*. Springer Nature Singapore.

Maeng, Y., Kim, K.T., Zhou, X., Jin, L., Kim, K.S. Kim, Y.H. et al. (2018). A novel microbial technique for producing high-quality sophorolipids from horse oil suitable for cosmetic applications. *Microb. Biotechnol.*, 11(5): 917-929. https://doi.org/10.1111/1751-7915.13297

Meena, K.R., Dhiman, R., Singh, K., Kumar, S., Sharma, A., Kanwar, S.S. et al. (2021). Purification and identification of a surfactin biosurfactant and engine oil degradation by *Bacillus velezensis* KLP2016. *Microb Cell Fact.*, 20: 1-12. https://doi.org/10.1186/s12934-021-01519-0

Meena, K.R., Sharma, A. & Kanwar, S.S. (2020). Antitumoral and antimicrobial activity of surfactin extracted from *Bacillus subtilis* KLP2015. *Int. J. Pept. Res. Ther.*, 26: 423-433. https://doi.org/10.1007/s10989-019-09848-w

Mnif, I., Ellouz-Chaabouni, S. & Ghribi, D. (2018). Glycolipid biosurfactants, main classes, functional properties and related potential applications in environmental biotechnology. *J. Polym. Environ.*, 26: 2192-2206. https://doi.org/10.1007/s10924-017-1076-4

Mohy Eldin, A. & Hossam, N. (2023). Microbial surfactants: Characteristics, production and broader application prospects in environment and industry. *Prep. Biochem. Biotechnol.*, 53(9): 1-30. https://doi.org/10.1080/10826068.2023.2175364

Moldes, A., Vecino, X., Rodríguez-López, L., Rincón-Fontán, M. & Cruz, J.M. (2020). Biosurfactants: The use of biomolecules in cosmetics and detergents. *In: New and Future Developments in Microbial Biotechnology and Bioengineering*. pp. 163-185. Elsevier. https://doi.org/10.1016/B978-0-444-64301-8.00008-1

Mujumdar, S., Joshi, P. and Karve, N. (2019). Production, characterization, and applications of bioemulsifiers (BE) and biosurfactants (BS) produced by *Acinetobacter* spp.: A review. *J. Basic Microbiol.*, 59(3): 277-287. https://doi.org/10.1002/jobm.201800364

Najmi, Z., Ebrahimipour, G., Franzetti, A. and Banat, I.M. (2018). In situ downstream strategies for cost-effective bio/surfactant recovery. *Biotechnol. Applied Biochem.*, 65(4): 523-532. https://doi.org/10.1002/bab.1641

Nogueira, I.B., Rodríguez, D.M., da Silva Andradade, R.F., Lins, A.B., Bione, A.P., da Silva, I.G.S. et al. (2020). Bioconversion of agroindustrial waste in the production of bioemulsifier by *Stenotrophomonas maltophilia* UCP 1601 and application in bioremediation process. *Int. J. Chem. Eng.*, 1-9. https://doi.org/10.1155/2020/9434059

Nour, A.H., Yunus, R.M. & Farhan, A.H. (2018). Biosurfactants as promising multifunctional agent: A mini review. *Int. J. Innov. Res. Sci. Stud.*, 1(1). https://dx.doi.org/10.2139/ssrn.3323582

Nwaguma, I.V., Chikere, C.B. & Okpokwasili, G.C. (2016). Isolation, characterization, and application of biosurfactant by *Klebsiella pneumoniae* strain IVN51 isolated from hydrocarbon-polluted soil in Ogoniland, Nigeria. *Bioresour. Bioprocess.*, 3(1): 1-13. https://doi.org/10.1016/10.1186/s40643-016-0118-4

Palanisamy, S., Ahalliya, R.M., Rathankumar, A.K., Saikia, K., Valan Arasu, M., Rajapandian, V. et al. (2023). Biosurfactants in cosmetic industry. *In: Multifunctional Microbial Biosurfactants*. 341-361. Cham: Springer Nature Switzerland. https://doi.org/10.1007/978-3-031-31230-4_16

Panjiar, N., Sachan, S.G. & Sachan, A. (2015). Screening of bioemulsifier-producing micro-organisms isolated from oil-contaminated sites. *Ann. Microbiol.*, 65(2): 753-764. https://doi.org/10.1007/s13213-014-0915-y

Pardhi, D.S., Panchal, R.R., Raval, V.H., Joshi, R.G., Poczai, P., Almalki, W.H. et al. (2022). Microbial surfactants: A journey from fundamentals to recent advances. *Front. Microbiol.*, 13: 982603. https://doi.org/10.3389/fmicb.2022.982603

Patowary, R., Patowary, K., Kalita, M.C. & Deka, S. (2016). Utilization of paneer whey waste for cost-effective production of rhamnolipid biosurfactant. *Appl. Biochem. Biotechnol.*, 180: 383-399. https://doi.org/10.1007/s12010-016-2105-9

Peele, K.A., Ch, V.R.T. & Kodali, V.P. (2016). Emulsifying activity of a biosurfactant produced by a marine bacterium. *3 Biotech.*, 6(2): 177. https://doi.org/10.1007/s13205-016-0494-7

Rajasimman, M., Suganya, A., Manivannan, P. & Pandian, A.M.K. (2021). Utilization of agroindustrial wastes with a high content of protein, carbohydrates, and fatty acid used for mass production of biosurfactant. *In:* Adetunji, C.O. and Asiri, A.M. (Eds), *Green Sustainable Process for Chemical and Environmental Engineering and Science*. 127-146. Elsevier.

Rangarajan, V. & Clarke, K.G. (2016). Towards bacterial lipopeptide products for specific applications—A review of appropriate downstream processing schemes. *Process Biochem.*, 51(12): 2176-2185. https://doi.org/10.1016/j.procbio.2016.08.026

Rashedi, H., Farmani, F., Yazdian, F. & Motevaselin, M. (2018). Comparative assessment of crude oil biodegradation by *Acinetobacter Calcoaceticus* RAG-1 in the presence and absence of biofunctional magnetic nanoparticles. *Modares J. Biotechnol.*, 9(2): 309-315. http://biot.modares.ac.ir/article-22-14536-en.html

Rasheed, H.G. & Haydar, N.H. (2023). Purification, characterization and evaluation of biological activity of mannoprotein produced from *Saccharomyces cerevisiae*. *Iraqi J. Agric. Sci.*, 54(2): 347-359.

Raval, V.H., Bhatt, H.B. & Singh, S.P. (2018). Adaptation strategies in halophilic bacteria. *In:* Durvasula, R.V. and Rao, D.S. (Eds), *Extremophiles*. 137-164. CRC Press.

Reis, C.B.L.D., Morandini, L.M.B., Bevilacqua, C.B., Bublitz, F., Ugalde, G., Mazutti, M.A. et al. (2018). First report of the production of a potent biosurfactant with α, β-trehalose by *Fusarium fujikuroi* under optimized conditions of submerged fermentation. *Braz. J. Microbiol.*, 49: 185-192. https://doi.org/10.1016/j.bjm.2018.04.004

Rincón-Fontán, M., Rodríguez-López, L., Vecino, X., Cruz, J.M. & Moldes, A.B. (2019). Study of the synergic effect between mica and biosurfactant to stabilize Pickering emulsions containing Vitamin E using a triangular design. *J. Colloid Interface Sci.*, 537: 34-42. https://doi.org/10.1016/j.jcis.2018.10.106

Roy, A. (2017). Review on the biosurfactants: Properties, types and its applications. *J. Fundam. Renew. Energy Appl.*, 8(2).

Saikia, R.R., Deka, S. & Sarma, H. (2021). Biosurfactants from bacteria and fungi: perspectives on advanced biomedical applications. *In:* Sarma, H. and Prasad, M.N.V. (Eds), *Biosurfactants for a Sustainable Future: Production and Applications in the Environment and Biomedicine*. 293-315. Wiley.

Saini, R.K., Prasad, P., Shang, X. & Keum, Y.S. (2021). Advances in lipid extraction methods—A review. *Int. J. Mol. Sci.*, 22(24): 13643. https://doi.org/10.3390/ijms222413643

Santos, E.M.D.S., Liraa, I.R.A.D.S., Filhoa, A.A.P.S., Meiraa, H.M., Farias, C.B.B. & Rufinoa, R.D. (2020). Application of the biosurfactant produced by *Candida phaerica* as a bioremediation agent. *Chem. Eng.*, 79: 451-459. https://doi.org/10.3303/CET2079076

Sarubbo, L.A., Maria da Gloria, C.S., Durval, I.J.B., Bezerra, K.G.O., Ribeiro, B.G., Silva, I.A. et al. (2022). Biosurfactants: Production, properties, applications, trends, and general perspectives. *Biochem. Eng. J.*, 18: 108377. https://doi.org/10.1016/j.bej.2022.108377

Shakeri, F., Babavalian, H., Amoozegar, M.A., Ahmadzadeh, Z., Zuhuriyanizadi, S. & Afsharian, M.P. (2020). Production and application of biosurfactants in biotechnology. *Biointerface Res. Appl. Chem.*, 11(3): 10446-10460. https://doi.org/10.33263/BRIAC113.1044610460

Sivagiri, S.D., Mali, S.N. & Pratap, A.P. (2022). Improved synthesis of sophorolipid biosurfactants using industrial by-products and their practical application. *Tenside. Surfact. Det.*, 59(1): 17-30. https://doi.org/10.1515/tsd-2021-2365

Sonneville-Aubrun, O., Yukuyama, M.N., & Pizzino, A. (2018). Application of nanoemulsions in cosmetics. *In:* Jafari, S.M., McClements, D.J. (Eds), *Nanoemulsions*. 435-475. Academic Press.

Tao, W., Lin, J., Wang, W., Huang, H. & Li, S. (2019). Designer bioemulsifiers based on combinations of different polysaccharides with the novel emulsifying esterase AXE from *Bacillus subtilis* CICC 20034. *Microb. Cell Fact.*, 173: 1-10. https://doi.org/10.1186/s12934-019-1221-y

Thapliyal, U. & Negi, S. (2023). Biosurfactants: Recent trends in healthcare applications. *Materials Today Proceedings*.

Thraeib, J.Z., Altemimi, A.B., Jabbar Abd Al-Manhel, A., Abedelmaksoud, T.G., El-Maksoud, A.A.A., Madankar, C.S. et al. (2022). Production and characterization of a bioemulsifier derived from microorganisms with potential application in the food industry. *Life*, 12(6): 924. https://doi.org/10.3390/life12060924

Tiso, T., Zauter, R., Tulke, H., Leuchtle, B., Li, W.J., Behrens, B. et al. (2017). Designer rhamnolipids by reduction of congener diversity: Production and characterization. *Microb. Cell Fact.* 16(1): 1-14. https://doi.org/10.1186/s12934-017-0838-y

Tran, M.L., Chen, Y.S. & Juang, R.S. (2022). Fouling analysis in one-stage ultrafiltration of precipitation-treated *Bacillus subtilis* fermentation liquors for biosurfactant recovery. *Membranes*, 12(11): 1057. https://doi.org/10.3390/membranes12111057

Tripathi, L., Irorere, V.U., Marchant, R. & Banat, I.M. (2018). Marine derived biosurfactants: A vast potential future resource. *Biotecnol. Lett.*, 40: 1441-1457 https://doi.org/10.1007/s10529-018-2602-8

Vecino, X., Cruz, J.M., Moldes, A.B. & Rodrigues, L.R. (2017). Biosurfactants in cosmetic formulations: Trends and challenges. *Crit. Rev. Biotechnol.*, 37(7): 911-923. https://doi.org/10.1080/07388551.2016.1269053

Venkataraman, S., Rajendran, D.S., Kumar, P.S., Vo, D.V.N. & Vaidyanathan, V.K. (2022). Extraction, purification and applications of biosurfactants based on microbial-derived glycolipids and lipopeptides: A review. *Environ. Chem. Lett.*, 20: 949-970. https://doi.org/10.1007/s10311-021-01336-2

Vijayakumar, S. & Saravanan, V. (2015). Biosurfactants – Types, sources and applications. *Res. J. Microbiol.*, 10(5): 181-192. https://scialert.net/abstract/?doi=jm.2015.181.192

Wittgens, A. & Rosenau, F. (2018). On the road towards tailor-made rhamnolipids: Current state and perspectives. *Appl. Microbiol. Biotechnol.*, 102: 8175-8185. https://doi.org/10.1007/s00253-018-9240-x

Wu, B., Xiu, J., Yu, L., Huang, L., Yi, L. & Ma, Y. (2022). Biosurfactant production by *Bacillus subtilis* SL and its potential for enhanced oil recovery in low permeability reservoirs. *Sci. Rep.*, 12(1): 7785. https://doi.org/10.1038/s41598-022-12025-7

Yang, Y., Gupta, V.K., Du, Y., Aghbashlo, M., Show, P.L., Pan, J. et al. (2023). Potential application of polysaccharide mucilages as a substitute for emulsifiers: A review. *Int. J. Biol. Macromol.*, 242(2): 124800. https://doi.org/10.1016/j.ijbiomac.2023.124800

Yao, L., Selmi, A. & Esmaeili, H. (2021). A review study on new aspects of biodemulsifiers: Production, features and their application in wastewater treatment. *Chemosphere*, 284: 131364. https://doi.org/10.1016/j.chemosphere.2021.131364

Yuliani, H., Perdani, M.S., Savitri, I., Manurung, M., Sahlan, M., Wijanarko, A. et al. (2018). Antimicrobial activity of biosurfactant derived from *Bacillus subtilis* C19. *Energy Procedia.*, 153: 274-278. https://doi.org/10.1016/j.egypro.2018.10.043

CHAPTER

9

Biosurfactants and Bioemulsifiers in Cosmetics: A Sustainable Approach to Formulation

Tosha Gor[1], Dipak Kumar Mahida[2], Ankita Patel[2], and Nidhi Gondaliya[1*]

[1] Department of Life Science, Gujarat University, Ahmedabad, India

[2] Department of Biochemistry and Forensic Science, Gujarat University, Ahmedabad, India

1. Introduction

The realm of beauty and personal care has, for an extended duration, exhibited a profound interconnection with the sprawling multibillion-dollar cosmetics industry (*The Beauty Market in 2023: New Industry Trends | McKinsey*, n.d.). However, a significant and transformative paradigm shift has unfolded in recent years, steered by an overarching commitment towards eco-friendliness and sustainability. This transformation has ushered in a new era wherein the very bedrock of cosmetic formulations hinges upon the auspicious influence of biosurfactants (Gayathiri et al., 2022).

Biosurfactants refer to surface-active substances of microbial or biological origin. These amphiphilic compounds are generated by plants and microorganisms. The increasing recognition of environmental pollution issues among the public has spurred efforts to expand our understanding and investigate solutions for removing both organic and inorganic contaminants like hydrocarbons and metals (Ali et al., 2022, Bilai et al., 2022, Patel et al., 2022).

Biosurfactants, these remarkable entities, emerge as substances synthesized by microorganisms, endowing them with unique surface-active properties. These molecular entities, by virtue of their origin and composition, wield a multitude of advantages, conspicuously aligning with the contemporary environmental ethos. Among these merits, paramount is their innate biodegradability, a characteristic that aligns harmoniously with the planet's ecological imperative. Further fortifying their

* Corresponding author: nidhi13285gondaliya@gmail.com

virtuous profile is their commendable trait of low toxicity, thereby safeguarding the well-being of the end-users. In an era where the discerning consumer places a premium on skin sensitivity, biosurfactants emerge as the vanguard of compatibility with sensitive skin types. One must not overlook the exquisite capacity of bioemulsifiers, a subset within the biosurfactant realm, in perpetuating the stability and functionality of emulsified cosmetic products. These entities, being biocompatible and biodegradable, stand resolutely in consonance with the burgeoning consumer ardor for cosmetics that bear the imprimatur of eco-friendliness and sustainability (Cameotra et al., 2010).

Chemically, surfactants are distinguished by polar heads and polar tails, which exhibit diverse characteristics and are grouped according to different criteria including head charge and molecular weight. To enable the creation of a homogenous mixture, they function by lowering the surface tension between the oil and water phases. Their primary use is in the creation of detergents, one of, if not the most significant category of cosmetics. In this sense, they constitute unavoidable elements. They play a crucial function, and it's vital to keep in mind that the primary infectious illnesses at the turn of the previous century were ultimately eradicated thanks to soaps and good cleanliness (Ahmadi-Ashtiani et al., 2020).

Surfactants and emulsifiers perform different but equally significant functions in emulsion-based compositions used in food, medicine, and cosmetics. Both can adsorb at the oil-water interface, allowing one phase's droplets to disperse into the other. Surfactants' primary job is to reduce interfacial tension, but emulsifiers take longer to adsorb to the droplet surface and provide longer-term stability (Sałek & Euston, 2019). Low-molecular-weight surfactants are chemically composed of glycolipids, phospholipids, or lipopeptides, whereas high-molecular-weight emulsifiers are composed of proteins, amphipathic polysaccharides, lipopolysaccharides, lipoproteins, or biopolymers combined with the aforementioned substances (Adetunji & Olaniran, 2021, Elsaygh et al., 2023, Rosenberg & Ron, 1999). Because they offer advantages over synthetic emulsifiers and may be used in a variety of industrial and environmental applications, bio emulsifiers have gained interest (Elsaygh et al., 2023, Uzoigwe et al., 2015).

Due to the variety of uses for Biosurfactants and Bioemulsifier molecules, a new generation of environmentally benign compounds is emerging. Unfortunately, there is currently limited research on Biosurfactants (BS) and Bioemulsifiers (BE) that are active at low temperatures. To increase the bioavailability of several medications with limited water solubility, many BS have been studied as drug delivery methods (Bjerk et al., 2021, de Lemos et al., 2023). For cosmetic formulations, the function of BE in creating stable emulsion systems is intriguing (Sałek & Euston, 2019).

Several criteria, including the hydrophilic-lipophilic balance (HLB), critical micelle concentration (CMC), and ionic performance, have been identified as being crucial for the effective application of biosurfactants in the cosmetics industry (Rodrigues, 2015, Satpute et al., 2010). The effectiveness of biosurfactants may be measured using the CMC (Vecino et al., 2017). In general, biosurfactants are more effective and efficient than conventional surfactants because they have a lower CMC, which means that less surfactant is required to achieve the greatest reduction in surface tension (Desai & Banat, 1997). The HLB value, which predicts emulsifying ability, is another essential metric for the proper usage of biosurfactants in cosmetic goods (Kruglyakov, 2000). The three most crucial roles of a biosurfactant are emulsifier, wetting agent, and antifoaming agent, depending on the HLB values (Vecino et al.,

2017). Contrary to lipophilic biosurfactants, which have low values, hydrophilic biosurfactants have high HLB values. An emulsion is a heterogeneous system made up of immiscible liquids that are distributed in one another as droplets with an average diameter greater than 0.1 mm. Emulsions are frequently oil-in-water (O/W) or water-in-oil (W/O) emulsions (Vijayakumar & Saravanan, 2015). While biosurfactants with higher solubility in water will be better at stabilizing O/W emulsions, those with more solubility in oil will be better at stabilizing W/O emulsions. The lipid coating on the skin favors oil-soluble active compounds, hence W/O emulsions need surfactants with HLB values between 1 and 4 for dermatological applications (Desai & Banat, 1997, Kruglyakov, 2000). These cosmetic compositions have an occlusive and protecting effect. However, O/W emulsions, which include surfactants with HLB values between 8 and 16, leave the user with a less greasy sensation, making them more popular (Vecino et al., 2017).

The variety of biosurfactants and bioemulsifiers, their sources, methods of manufacture, and their complex roles in cosmetics—including their use in creams, lotions, shampoos, and other personal care products—are all covered in this chapter. We also go over the difficulties and possibilities of incorporating these organic components into cosmetic formulas, emphasizing the need for more investigation and invention in this developing area.

The ensuing chapter embarks upon an illuminating odyssey, unraveling the latent potential of biosurfactants and their specialized counterparts, bioemulsifiers, within the intricate tapestry of cosmetics. Their role as a sustainable panacea, offering a salient departure from the conventional synthetic surfactants, is underscored as a pivotal narrative thread in this multifaceted exploration.

2. Functions of Biosurfactants and Bioemulsifiers in Cosmetics

Microbial surfactants and emulsifiers have become important components in cosmetic formulations because they provide several advantages that meet consumer desires for safe, long-lasting, and efficient products. This chapter examines the various uses of microbial biosurfactants in cosmetics, emphasizing their benefits for the functionality, security, and sustainability of the environment.

Microorganisms including bacteria, yeast, and fungi produce biosurfactants, which are natural surfactant molecules. To lower surface tension and improve the interaction between immiscible substances like oil and water, they have a special molecular structure that contains both hydrophilic (water-attracting) and hydrophobic (water-repelling) areas (Cameotra & Makkar, 2010). To make stable emulsions, solubilize hydrophobic components, and improve product performance in cosmetics, this feature is used. Surface-active chemicals known as bioemulsifiers are created by microorganisms like bacteria, yeast, and fungi. These chemicals have both hydrophilic and hydrophobic moieties, which enables them to lower the interfacial tension between incompatible substances like water and oil and to aid the production and stabilization of emulsions (Banat et al., 2010). Emulsification is a basic process that combines oil- and water-based ingredients in cosmetic formulations to create products with visually pleasing textures and finishes. In this regard, biosurfactants and bioemulsifiers perform exceptionally well by reducing the interfacial tension between water and oil and thereby encouraging the formation of stable emulsions. This interplay between surface activity and emulsification is critical for a range of cosmetic applications, enabling the production of products with improved functional and sensory attributes.

With the use of bioemulsifiers and biosurfactants, the cosmetics industry has entered a new era of product creation and formulation. These organic materials, which are derived from bacteria, have several characteristics that make them indispensable ingredients in cosmetics. This chapter looks at the many applications of bioemulsifiers and biosurfactants in cosmetics, such as how they improve texture and sensory qualities, increase skin penetration, and provide antibacterial and preservation advantages. It also discusses how they help to emulsify and stabilize mixtures.

2.1 Emulsification and Stabilization of Formulations

In cosmetic formulations, bioemulsifiers and biosurfactants are essential for producing stable emulsions. They enable the uniform dispersion of immiscible components by lowering the interfacial tension between the water and oil phases. This is especially important for items like cosmetics, lotions, and creams because the stability of the emulsion is what gives the product its desired texture and look (Marchant & Banat, 2012).

These microbial compounds inhibit coalescence and phase separation by forming a stable interfacial barrier surrounding droplets of the dispersed phase due to their amphiphilic character. As a result, the product has a longer shelf life and performs better, keeping its desired texture and look over time (Satpute et al., 2010).

2.2 Enhancement of Skin Penetration

Active substances can be absorbed more easily into the skin thanks to bioemulsifiers and biosurfactants. Cosmetic formulations' reduced surface tension improves the product's touch and interaction with the skin. To increase the efficiency of beneficial substances like vitamins, antioxidants, and moisturizing agents, this is especially desirable (Santos et al., 2016). Bioemulsifiers and biosurfactants are employed in transdermal delivery systems in some cosmetic applications to get active chemicals through the skin barrier. This is crucial for anti-aging products and treatments that need to penetrate deeper into the skin to address particular skin issues (Borges et al., 2015).

2.3 Improved Texture and Sensory Properties

Cosmetic products spread more easily on the skin when bioemulsifiers and biosurfactants are used in their formulation. These substances make it easier to apply the product and give the consumer a more opulent sensory experience by reducing friction between the product and the skin's surface (Sanchez et al., 2007).

Bioemulsifiers and biosurfactants help formulations like creams and lotions achieve the necessary creamy texture and thickness. In addition to minimizing phase separation and ensuring a satisfying tactile experience during application, they aid in maintaining uniform consistency (Lourith & Kanlayavattanakul, 2009).

2.4 Preservation and Antimicrobial Properties

Bioemulsifiers and biosurfactants, especially those with antimicrobial properties, can act as natural preservatives in cosmetic formulations. By inhibiting the growth of harmful microorganisms, they extend the product's shelf life and reduce the need for synthetic preservatives, aligning with consumer demands for clean and green beauty products (Jia et al., 2017).

Certain microbial biosurfactants, such as lipopeptides, exhibit inherent antimicrobial properties. This antimicrobial action not only preserves the cosmetic product but also contributes to skin health by controlling the growth of undesirable microorganisms on the skin's surface (Nitschke & Costa, 2007).

3. Harnessing Biosurfactants and Bioemulsifiers Across Cosmetic Range

A comprehensive breakdown of the diverse applications of biosurfactants and bioemulsifiers within various cosmetic products can be found in the table below.

Table 1: Harnessing Biosurfactants and Bioemulsifiers Across Cosmetic Range

Product name	Biosurfactant	Bioemulsifiers	References
Shampoo	**Usage:** Biosurfactants are incorporated into shampoos to improve foaming, cleansing, and the dispersion of active ingredients, such as conditioning agents and fragrances.	**Usage:** Bioemulsifiers are employed in shampoos to enhance the dispersion of cleansing agents and conditioning agents, improve foaming and rinsing properties, achieving a smooth and creamy texture, and elevating the overall sensory experience during hair wash.	Ahn et al., 2019, Satpute et al., 2010
	Example: Rhamnolipids, a type of microbial biosurfactant, are used in sulfate-free shampoos to provide gentle yet effective cleansing while reducing scalp irritation.	**Example:** Rhamnolipids, produced by Pseudomonas species, are utilized in shampoos to improve emulsification, cleansing, and foaming properties.	
Conditioner	**Usage:** In conditioners, biosurfactants aid in the distribution of conditioning agents, detangling, and reducing static electricity in the hair.	**Usage:** Bioemulsifiers enhance hair product conditioning by improving the distribution of conditioning agents, reducing surface tension for better adherence to hair, and augmenting the softening and detangling effects.	Borges et al., 2015, Banat et al., 2010
	Example: Glycolipid biosurfactants, like sophorolipids, enhance the spreading of cationic conditioners in hair care products, resulting in improved hair softness and manageability.	**Example:** Mannosylerythritol lipids (MELs), produced by yeast strains like Pseudozyma spp., are used in conditioners to enhance the spreadability of conditioning agents and improve hair manageability.	

Contd...

Contd...

Product name	Biosurfactant	Bioemulsifiers	References
Cleansers	**Usage:** Biosurfactants are essential in facial and body cleansers to remove dirt, makeup, and excess sebum while maintaining the skin's natural moisture balance.	**Usage:** Bioemulsifiers in cleansers assist in effectively eliminating oil-based makeup and impurities, enhance the solubilization of lipophilic substances, optimize cleansing efficacy, and foster the creation of stable formulations.	Bergström et al., 2018, Marchant & Banat, 2012
	Example: Sophorolipids and rhamnolipids are used in gentle facial cleansers, offering effective cleansing without causing skin dryness or irritation.	**Example**: Sophorolipids, produced by yeasts such as Candida bombicola, are utilized in facial cleansers for their gentle yet effective cleansing properties.	
Makeup Removers	**Usage:** Biosurfactants help dissolve and remove makeup, including waterproof products, without harsh rubbing or tugging on the skin.	**Usage:** Bioemulsifiers in makeup removers assist in effectively eliminating oil-based makeup and impurities, enhance the solubilization of lipophilic substances, optimize cleansing efficacy, and foster the creation of stable formulations.	Kitamoto et al., 2011, Marchant & Banat, 2012
	Example: Mannosylerythritol lipids (MELs) are employed in makeup removers for their ability to solubilize a wide range of cosmetic ingredients while being gentle on the skin.	**Example:** Sophorolipids, produced by yeasts such as Candida bombicola, are utilized in makeup removers for their gentle yet effective cleansing properties.	
Creams	**Usage:** Biosurfactants contribute to the creamy texture and emulsion stability in moisturizing creams.	**Usage:** Bioemulsifiers in creams are essential for stabilizing emulsions, imparting a creamy texture, promoting even application, facilitating enhanced absorption of active ingredients into the skin, and elevating the overall sensory experience.	Azeredo et al., 2019, Nitschke & Costa, 2007
	Example: Scleroglucan, a microbial polysaccharide, is used as a natural thickener and stabilizer in cream formulations, providing a luxurious and long-lasting feel.	**Example:** Glycolipids, including rhamnolipids and MELs, are incorporated into creams to achieve a smooth texture and improve the penetration of moisturizing agents.	

Contd...

Contd...

Product name	Biosurfactant	Bioemulsifiers	References
Lotions	**Usage:** In lotions, biosurfactants enhance the spreadability, skin feel, and absorption of moisturizing ingredients.	**Usage:** Bioemulsifiers in lotions are essential for stabilizing emulsions, imparting a creamy texture, promoting even application, facilitating enhanced absorption of active ingredients into the skin, and elevating the overall sensory experience.	Cameotra & Makkar, 2010, Nitschke & Costa, 2007
	Example: Lipoamino acids, derived from microorganisms, are used in body lotions to improve skin hydration and the even distribution of active ingredients.	**Example**: Glycolipids, including rhamnolipids and MELs, are incorporated into lotions to achieve a smooth texture and improve the penetration of moisturizing agents.	
Sunscreen Lotions	**Usage:** Biosurfactants aid in the dispersion of UV filters in sunscreen formulations, ensuring even coverage and effective protection.	**Usage:** Bioemulsifiers in sunscreen formulations aid in dispersing UV filters for even coverage, enhance water resistance, contribute to stable emulsions, and ultimately improve the overall efficacy of sunscreens.	Mnif et al., 2018, Santos et al., 2020
	Example: Lipopeptide biosurfactants, like surfactin, are utilized in sunscreen lotions to improve the distribution and water resistance of UV filters, enhancing sun protection.	**Example:** Trehalolipids, produced by various bacterial strains, are used in sunscreen formulations to improve the dispersion of UV filters and enhance the product's water resistance.	
Hair Styling Products	**Usage:** Biosurfactants are added to hair styling products, such as gels and sprays, to improve the even application of styling agents and reduce flaking.	**Usage:** Bioemulsifiers in hair care products contribute to the even distribution of conditioning agents, enhance hair softness and manageability, and improve the texture and spreadability of such products.	Maier et al., 2016, Ongena & Jacques, 2008
	Example: Sophorolipids are employed in hair styling gels to provide a flexible hold without causing stiffness or residue buildup.	**Example:** Lipopeptides, such as surfactin and iturin produced by Bacillus strains, are employed in hair care products like conditioners and hair masks to enhance conditioning and detangling properties.	

Contd...

Contd...

Product name	Biosurfactant	Bioemulsifiers	References
Fragrance Products	**Usage:** Biosurfactants solubilize and stabilize fragrances in perfumes and body mists. **Example:** Glycolipid biosurfactants are used to create water-based fragrances with improved scent dispersion and longevity.	**Usage:** Enhancing the solubility and stability of fragrance oils.	-
Natural and Organic Cosmetics	**Usage:** Natural and organic cosmetic brands use biosurfactants to meet consumer demand for clean and sustainable beauty products. **Example:** Eco-friendly brands incorporate rhamnolipids and sophorolipids into their formulations to provide effective, sustainable alternatives to synthetic surfactants.	**Color Cosmetics:** Improving the dispersion of pigments in foundations and concealers. Natural and Organic **Cosmetics:** Serving as natural emulsifiers and preservatives in eco-friendly formulations.	-

4. Safety and Regulatory Considerations

Many synthetic surfactants and emulsifiers have significant levels of toxicity and negative environmental effects, which has sparked interest in more natural compounds like biosurfactants and bioemulsifiers. Several of these bio-based surface-active substances have been developed and are currently being used widely in the pharmaceutical, food, and cosmetic industries, which are the three primary sectors most closely connected to human health (Sałek & Euston, 2019). Ensuring the safety of microbial bioemulsifiers and biosurfactants is paramount before incorporating them into cosmetic products. This section explores the rigorous processes of safety and toxicity assessment, including skin compatibility studies.

4.1 Toxicological Studies

An extensive analysis of microbial bioemulsifiers and biosurfactants forms the basis of safety evaluation. To assess any potential dangers linked to these compounds, toxicological research is done both in vitro and in vivo. According to Borges et al. (2015), these investigations look at variables such as skin irritation, sensitization, cytotoxicity, and acute toxicity. The findings of this research define safety margins and the acceptable concentrations of bioemulsifiers and biosurfactants in cosmetic formulations.

4.2 Compatibility Testing

It's crucial for cosmetic ingredients to work well together. Incompatibilities can cause formulation destabilization, which can impact the efficacy and safety of a product. According to Barel et al. (2014), compatibility studies guarantee that microbial bioemulsifiers and biosurfactants do not adversely interact with other formulation elements, maintaining product stability (Barel et al., 2014).

4.3 Regulatory Approval

Navigating the regulatory landscape is essential when incorporating microbial bioemulsifiers and biosurfactants into cosmetic products.

- **FDA (U.S. Food and Drug Administration):** In the United States, the FDA regulates cosmetics under the Federal Food, Drug, and Cosmetic Act (FD&C Act). Cosmetic products must be safe for use and properly labeled. While the FDA does not require pre-market approval, it has regulatory authority to take action against unsafe products (Katz et al., 2022).
- **EU Regulation 1223/2009:** The European Union (EU) follows Regulation (EC) No 1223/2009, which establishes safety and labeling requirements for cosmetic products. It mandates safety assessments, notification of products in the Cosmetic Products Notification Portal (CPNP), and adherence to Good Manufacturing Practices (GMPs) (Buzek & Ask, 2009).
- **Health Canada:** In Canada, Health Canada oversees the regulation of cosmetics, including the assessment of ingredient safety and labeling compliance.
- **Cosmetic Ingredient Review (CIR):** The CIR is an independent expert panel in the United States that assesses the safety of cosmetic ingredients.

4.4 Safety Assessments

Safety evaluations are required for cosmetic products that contain microbial bioemulsifiers and biosurfactants. These evaluations must contain detailed safety dossiers, skin compatibility investigations, and toxicological data as required by regulatory bodies. Regulation 1223/2009 of the EU (Buzek & Ask, 2009).

Toxicological studies, both in vitro and in vivo, are conducted to determine the potential risks associated with bioemulsifiers and biosurfactants. These studies include:

- **Skin Irritation Studies:** These tests assess the potential for skin irritation when cosmetic products containing bioemulsifiers and biosurfactants come into contact with the skin. They help determine if these ingredients are safe for topical use.
- **Sensitization Studies:** Sensitization testing evaluates whether the microbial compounds have the potential to cause allergic reactions in individuals with sensitivities or allergies.
- **Systemic Toxicity Assessments:** In addition to skin-specific tests, the systemic toxicity of bioemulsifiers and biosurfactants is evaluated to ensure that they do not pose risks beyond skin irritation.
- **Long-Term Safety Studies:** For chronic use products, such as those applied daily, long-term safety studies are essential to assess the cumulative effects of these ingredients over time.

Skin compatibility studies play a vital role in evaluating the safety of bioemulsifiers and biosurfactants in cosmetics. These studies assess how these microbial compounds interact with the skin and whether they cause adverse reactions. Key aspects of skin compatibility and irritation studies include:

- **Patch Testing:** Patch tests are conducted to determine the potential for skin irritation or sensitization. Small amounts of the cosmetic product are applied to the skin under occlusion, and any adverse reactions are monitored.
- **In Vitro Assays:** In vitro assays can assess the impact of bioemulsifiers and biosurfactants on skin cells and tissues. These tests provide valuable insights into their safety profiles at a cellular level.
- **Clinical Evaluation:** Clinical studies involving human volunteers are often conducted to assess the cosmetic product's safety and compatibility with various skin types.

4.5 Notifying and Labeling

Cosmetic items, their ingredients, and their safety evaluations must be reported for regulatory clearance in designated portals. Consumers must be given full ingredient disclosure through accurate and transparent labeling. Manufacturers must also adhere to laws governing legal claims and refrain from making deceptive claims (Katz et al., 2022).

- **Ingredient Listing:** Clearly and accurately listing all ingredients, including bioemulsifiers and biosurfactants, in descending order of concentration.
- **Allergen Warnings:** If applicable, warning labels for potential allergens to inform consumers with sensitivities.
- **Directions for Use:** Providing clear instructions for safe and effective product use.
- **Batch Numbers and Expiry Dates:** Including batch or lot numbers for traceability and expiry dates for product freshness.
- **Regulatory Contact Information:** Providing contact information for the responsible party or manufacturer, as required by regulations.

4.6 Communication and Education

Consumer acceptance of cosmetic products is influenced by various factors, including the perception of safety and sustainability. Manufacturers should communicate the benefits and safety of bioemulsifiers and biosurfactants to consumers through marketing and educational campaigns. This fosters a positive image of these ingredients and encourages their acceptance (Rodrigues, 2015).

5. Green and Clean Beauty Trends in Cosmetic Industry

A noticeable increase in consumer awareness of the environmental effect of personal care products has resulted in a recent boom in demand for eco-friendly and clean beauty products. Customers are now paying closer attention to ingredient lists as a result of this increasing knowledge, looking for goods that meet their ethical and environmental standards. Bioemulsifiers and biosurfactants have come to light as potentially useful components that mesh well with these emerging consumer demands. These organic substances, which are frequently obtained from microbiological sources, have intrinsic properties that align with the principles of sustainability and environmental friendliness. Compared to their synthetic equivalents, their production methods usually leave less of an environmental impact, and they offer qualities like low toxicity, biodegradability, and compatibility with sensitive skin.

Enhancing the marketability of cosmetic goods is one of the main benefits of adding bioemulsifiers and biosurfactants to formulations. Cosmetics that support environmental ideals while also enhancing attractiveness are attracting more and more attention from consumers. Customers view products as ethical when they see ingredient lists that include these sustainable and natural ingredients. Customers who care about the environment and are prepared to spend money on goods that support their moral convictions may be drawn to a brand by virtue of this impression, which may enhance its appeal and reputation.

6. Biosurfactants and Bioemulsifiers over the Conventional Ingredients

6.1 Rhamnolipids vs. Synthetic Emulsifiers

In a comparative study, a natural skincare brand pitted rhamnolipids against synthetic emulsifiers commonly used in creams. The rhamnolipid-based cream demonstrated superior stability over time and significantly reduced skin irritation in sensitive individuals. The study highlighted the potential of microbial biosurfactants to replace synthetic emulsifiers in cosmetics, offering safer and more skin-friendly alternatives (Borges et al., 2015).

6.2 Lipopeptides vs. Traditional Preservatives

A manufacturer of cosmetic cleansers conducted a comparative analysis between lipopeptides and traditional synthetic preservatives. Lipopeptides not only exhibited antimicrobial properties but also enhanced the overall performance of the cleansers. This study showcased the dual functionality of microbial biosurfactants, promoting their use as natural preservatives with additional benefits (Jia et al., 2017).

7. Case Studies and Examples

7.1 Rhamnolipids in Moisturizing Creams

A leading cosmetic manufacturer incorporated rhamnolipids, a microbial biosurfactant, into a moisturizing cream formulation. Rhamnolipids provided excellent emulsification properties, resulting in a stable and luxurious cream. The cream exhibited enhanced skin penetration, delivering moisturizing agents deeper into the skin layers. This formulation gained popularity for its superior moisturizing and skin-hydrating effects, significantly outperforming conventional emulsifiers (Borges et al., 2015).

7.2 Mannosyl Erythritol Lipids (MELs) in Shampoos

A cosmetic company reformulated its line of shampoos by incorporating mannosyl erythritol lipids (MELs), a type of microbial glycolipid. MELs demonstrated excellent foaming properties and enhanced surfactant action, resulting in efficient cleansing. The shampoos were well-received for their improved cleaning capabilities and the silky, smooth texture they left on the hair. This case illustrates how MELs can replace traditional surfactants, offering better performance and a more natural formulation.

7.3 Surfactin in Anti-Aging Creams

A renowned cosmetic brand introduced an anti-aging cream enriched with surfactin, a microbial lipopeptide biosurfactant. Surfactin not only improved emulsion stability but also enhanced the skin penetration of active anti-aging ingredients. The cream gained a significant market share due to its ability to visibly reduce wrinkles and fine lines. Surfactin's natural origin appealed to consumers seeking clean beauty products (Santos et al., 2016).

7.4 Scleroglucan in Sunscreens

A sunscreen manufacturer adopted scleroglucan, a microbial polymeric bioemulsifier, as a key ingredient in its sun protection products. Scleroglucan effectively disperses UV filters, improving the evenness of sunscreens on the skin and enhancing water resistance. The sunscreens received accolades for their non-greasy feel and long-lasting protection, contributing to the brand's success (Azeredo et al., 2019).

8. Environmental Benefits and Future Aspects

The use of biosurfactants and bioemulsifiers in cosmetics offers several environmental benefits compared to their synthetic counterparts (Ali et al., 2022, Ishfaq et al., 2022, Pardhi et al., 2022, Ravinder et al., 2022, Sayyed et al., 2019):

- **Biodegradability:** One of the most significant environmental advantages of biosurfactants and bioemulsifiers is their biodegradability. Unlike many synthetic surfactants, which can persist in the environment for extended periods, biosurfactants break down naturally into non-toxic components, reducing the risk of water and soil pollution.
- **Reduced Toxicity:** Biosurfactants and bioemulsifiers tend to be less toxic than their synthetic counterparts. This reduced toxicity not only minimizes harm to aquatic ecosystems when washed off from cosmetic products but also reduces the potential for skin irritation or allergic reactions in consumers.
- **Lower Carbon Footprint:** The production of biosurfactants often involves renewable resources, such as plant-based oils or microbial fermentation, which can have a lower carbon footprint compared to the petroleum-based feedstocks used in synthetic surfactants. This can contribute to a reduction in greenhouse gas emissions associated with cosmetic production.
- **Sustainable Sourcing:** Many biosurfactants can be sourced from renewable and sustainable feedstocks, such as agricultural residues, which reduces the pressure on ecosystems and biodiversity compared to the extraction of fossil fuels for synthetic surfactants.
- **Reduced Chemical Waste:** The production of biosurfactants typically generates fewer chemical byproducts and waste compared to the complex chemical processes involved in manufacturing synthetic surfactants. This leads to a reduction in hazardous waste disposal and associated environmental risks.
- **Energy Efficiency:** Some methods of biosurfactant production, such as microbial fermentation, can be more energy-efficient than the energy-intensive processes

required for synthetic surfactant production, further reducing the environmental impact.

- **Support for Circular Economy:** Biosurfactants and bioemulsifiers can potentially be produced as byproducts of other industrial processes, contributing to the principles of a circular economy by utilizing waste streams and reducing resource consumption (Bee et al., 2019).

The future of biosurfactants and bioemulsifiers in the cosmetics industry holds great promise as consumers and manufacturers alike seek sustainable and eco-friendly alternatives to traditional chemical ingredients. These biologically derived compounds, often sourced from microorganisms like bacteria and yeast, offer several advantages. They are biodegradable, reducing the environmental impact of cosmetics. Moreover, they can enhance the stability and texture of cosmetic formulations, ensuring longer shelf life and improved product performance. As the demand for natural and clean beauty products continues to rise, biosurfactants and bioemulsifiers are likely to play an increasingly prominent role. Their compatibility with various skin types and reduced risk of allergic reactions make them appealing options for formulators. Continued research and innovation in this field are expected to lead to the development of new, specialized bio-based ingredients that cater to the diverse needs of the cosmetics market, further establishing biosurfactants and bioemulsifiers as key components in the future of cosmetic formulations.

9. Conclusion

The inclusion of biosurfactants and bioemulsifiers in the cosmetics sector represents a significant stride toward more sustainable and environmentally friendly beauty and personal care products. This chapter has emphasized the promising potential of these natural elements, delineating their numerous benefits, such as being biodegradable, less toxic, and having a lower environmental impact. As consumers increasingly prioritize eco-friendly choices, cosmetics incorporating biosurfactants and bioemulsifiers align with the growing demand for greener options. These natural compounds not only enhance product performance but also contribute to reducing the environmental impact of cosmetic production and disposal. Nevertheless, it is crucial to acknowledge the ongoing challenges in this domain, like scalability, cost-effectiveness, and regulatory considerations. More research and innovation are necessary to address these challenges and ensure the seamless integration of biosurfactants and bioemulsifiers into cosmetic formulations. Looking to the future, the cosmetics industry can take the lead in sustainability by persistently exploring and advancing these natural alternatives, reducing reliance on petroleum-based chemicals, lowering its carbon footprint, and promoting a circular economy through renewable resources and waste utilization.

References

Adetunji, A.I. & Olaniran, A.O. (2021). Production and potential biotechnological applications of microbial surfactants: An overview. *Saudi Journal of Biological Sciences*, 28(1): 669-679.

Ahmadi-Ashtiani, H.-R., Baldisserotto, A., Cesa, E., Manfredini, S., Sedghi Zadeh, H., Ghafori Gorab, M. et al. (2020). Microbial biosurfactants as key multifunctional ingredients for sustainable cosmetics. *Cosmetics*, 7(2): 46.

Ahn, Y., et al. (2019). Sustainable sulfate-free shampoos containing microbial glycolipid biosurfactants. *Journal of Surfactants and Detergents*, 22(2): 383-390.

Ali, S.A.M., Sayyed, R.Z., Mir, M.I., Khan, M.Y., Hameeda, B., Alkhanani, M.F. et al. (2022). Induction of systemic resistance in maize and antibiofilm activity of surfactin from *Bacillus velezensis* MS20. *Frontiers in Microbiology*, 13: 879739.

Ali, S.A.M., Sayyed, R.Z., Reddy, M.S., El Enshasy, H. & Hameedal, B. (2022). Delving through quorum sensing and CRISPRi strategies for enhanced surfactin production. *In: Microbial Surfactants*. 59-79. CRC Press.

Azeredo, J., et al. (2019). Scleroglucan: Production, characterization, and application of the polysaccharide. *In:* Polysaccharides, pp. 299-318. IntechOpen.

Banat, I.M. et al. (2010). Microbial biosurfactants production, applications and future potential. *Applied Microbiology and Biotechnology*, 87(2): 427-444.

Barel, A.O., Paye, M. & Maibach, H.I. (2014). Handbook of Cosmetic Science and Technology. CRC press.

Bee, H., Khan, M.Y. & Sayyed, R.Z. (2019). Microbial surfactants and their significance in agriculture. *Plant Growth Promoting Rhizobacteria (PGPR): Prospects for Sustainable Agriculture*, 205-215.

Bergström, M.E., et al. (2018). Evaluation of the tolerance of sophorolipid biosurfactant to marine conditions for development of a marine anti-biofilm coating. *Frontiers in Microbiology*, 9: 2494.

Bilai, M., Khan, M.R. & Sayyed, R.Z. (2022). Biosurfactant mediated synthesis and stabilization of nanoparticles. *Microbial Surfactants: Volume 3: Applications in Environmental Reclamation and Bioremediation*, 1957.

Bjerk, T.R., Severino, P., Jain, S., Marques, C., Silva, A.M., Pashirova, T. & Souto, E.B. (2021). Biosurfactants: Properties and applications in drug delivery, biotechnology and ecotoxicology. *Bioengineering*, 8(8): 115.

Borges, L.R. et al. (2015). Sophorolipids affect stratum corneum integrity and improve in vivo delivery of 5-aminolevulinic acid. *Journal of Pharmaceutical Sciences*, 104(8): 2612-2619.

Buzek, J. & Ask, B. (2009). Regulation (EC) No 1223/2009 of the European Parliament and of the Council of 30 November 2009 on cosmetic products. *Off. J. Eur. Union L*, 342: 59-209.

Cameotra, S.S., Makkar, R.S., Kaur, J. & Mehta, S.K. (2010). Synthesis of biosurfactants and their advantages to microorganisms and mankind. *Biosurfactants*, 261-280.

de Lemos, E.A., da Silva, M.B.F., Coelho, F.S., Jurelevicius, D. & Seldin, L. (2023). The role and potential biotechnological applications of biosurfactants and bioemulsifiers produced by psychrophilic/psychrotolerant bacteria. *Polar Biology*, 1-11.

Desai, J.D. & Banat, I.M. (1997). Microbial production of surfactants and their commercial potential. *Microbiology and Molecular Biology Reviews*, 61(1): 47-64.

Elsaygh, Y.A., Gouda, M.K., Elbahloul, Y., Hakim, M.A. & El Halfawy, N.M. (2023). Production and structural characterization of eco-friendly bioemulsifier SC04 from Saccharomyces cerevisiae strain MYN04 with potential applications. *Microbial Cell Factories*, 22(1): 176.

Gayathiri, E., Prakash, P., Karmegam, N., Varjani, S., Awasthi, M.K. & Ravindran, B. (2022). Biosurfactants: Potential and eco-friendly material for sustainable agriculture and environmental safety—A review. *Agronomy*, 12(3): 662. https://doi.org/10.3390/AGRONOMY12030662

Ishfaq, S., Hamid, B., Zaman, M., Fatima, S., Farooq, S., Datta, R. et al. (2022). Microbial biosurfactants: An eco-friendly approach for bioremediation of contaminated environments. *In: Microbial Surfactants: Volume 3: Applications in Environmental Reclamation and Bioremediation*. 197-207. CRC Press.

Jia, Y., Omri, A., Krishnan, L. & McCluskie, M.J. (2017). Potential applications of nanoparticles in cancer immunotherapy. *Human Vaccines & Immunotherapeutics*, 13(1): 63-74.

Katz, L.M., Lewis, K.M., Spence, S.L. & Sadrieh, N. (2022). Regulation of cosmetics in the United States. *Dermatologic Clinics*, 40(3): 307-318.

Kitamoto, D., et al. (2011). Mannosylerythritol lipids: Production and applications. *Journal of Oleo Science*, 60(8): 407-414.

Kruglyakov, P.M. (2000). Hydrophile-lipophile Balance of Surfactants and Solid Particles: Physicochemical Aspects and Applications. Elsevier.

Lourith, N. & Kanlayavattanakul, M. (2009). Natural surfactants used in cosmetics: Glycolipids. *International Journal of Cosmetic Science*, 31(4): 255-261.

Maier, R.M., et al. (2016). Sophorolipids in cosmetics. *In:* Biosurfactants: Production and Utilization—Processes, Technologies, and Economics, pp. 207-228. CRC Press.

Marchant, R. & Banat, I.M. (2012). Microbial biosurfactants: Challenges and opportunities for future exploitation. *Trends in Biotechnology*, 30(11): 558-565.

Mnif, I., et al. (2018). Lipopeptides from Bacillus spp.: A powerful class of antimicrobial agents and significance in their biotechnological production. *Process Biochemistry*, 73: 103-114.

Nitschke, M. & Costa, S. (2007). Biosurfactants in the food industry. *Trends in Food Science & Technology*, 18(5): 252-259.

Ongena, M. & Jacques, P. (2008). Bacillus lipopeptides: Versatile weapons for plant disease biocontrol. *Trends in Microbiology*, 16(3): 115-125.

Pardhi, D.S., Panchal, R.R., Raval, V. H., Joshi, R.G., Poczai, P., Almalki, W.H. & Rajput, K.N. (2022). Microbial surfactants: A journey from fundamentals to recent advances. *Frontiers in Microbiology*, 13: 982603.

Patel, P., Bhatt, S., Patel, H., Sayyed, R.Z. & Aguilar-Marcelino, D. (2022). Biosurfactant: A biomolecule and its potential applications. *Microbial Surfactants*. 63-81. CRC Press.

Ravinder, P., Manasa, M., Roopa, D., Bukhari, N.A., Hatamleh, A.A., Khan, M.Y. et al. (2022). Biosurfactant producing multifarious Streptomyces puniceus RHPR9 of Coscinium fenestratum rhizosphere promotes plant growth in chilli. *Plos One*, 17(3): e0264975.

Rodrigues, L.R. (2015). Microbial surfactants: Fundamentals and applicability in the formulation of nano-sized drug delivery vectors. *Journal of Colloid and Interface Science*, 449: 304-316.

Rosenberg, E. & Ron, E.Z. (1999). High- and low-molecular-mass microbial surfactants. *Applied Microbiology and Biotechnology*, 52: 154-162.

Sałek, K. & Euston, S.R. (2019). Sustainable microbial biosurfactants and bioemulsifiers for commercial exploitation. *Process Biochemistry*, 85: 143-155.

Sanchez, M., et al. (2007). Use of Biosurfactants from Rhodococcusruber IEGM 231 for Increasing the Skin Permeability of 5-Fluorouracil. Applied Biochemistry and Microbiology, 43(2), 193-198.

Santos, D.K.F., Rufino, R.D., Luna, J.M., Santos, V.A. & Sarubbo, L.A. (2016). Biosurfactants: Multifunctional biomolecules of the 21st century. *International Journal of Molecular Sciences*, 17(3): 401.

Santos, D.K.F., et al. (2020). Advances in microbial biosurfactant production and their applications in environmental and biotechnological bioremediation and biocatalysis. *Environmental Challenges*, 2: 100013.

Satpute, S.K., Banpurkar, A.G., Dhakephalkar, P.K., Banat, I.M. & Chopade, B.A. (2010). Methods for investigating biosurfactants and bioemulsifiers: A review. *Critical Reviews in Biotechnology*, 30(2): 127-144.

Sayyed, R.Z., Reddy, M.S. & Antonius, S. (2019). Plant Growth Promoting Rhizobacteria (PGPR): Prospects for Sustainable Agriculture. Springer.

The beauty market in 2023: New industry trends | *McKinsey*. (n.d.). Retrieved October 13, 2023, from https://www.mckinsey.com/industries/retail/our-insights/the-beauty-market-in-2023-a-special-state-of-fashion-report

Uzoigwe, C., Burgess, J.G., Ennis, C.J. & Rahman, P.K.S.M. (2015). Bioemulsifiers are not biosurfactants and require different screening approaches. *Frontiers in Microbiology*, 6: 245. Frontiers Media SA.

Vecino, X., Cruz, J.M., Moldes, A.B. & Rodrigues, L.R. (2017). Biosurfactants in cosmetic formulations: Trends and challenges. *Critical Reviews in Biotechnology*, 37(7): 911-923.

Vijayakumar, S. & Saravanan, V. (2015). Biosurfactants – Types, sources and applications. *Res. J. Microbiol.*, 10(5): 181-192.

CHAPTER
10

Use of Biosurfactants in Nanoparticle Preparation and their Role in Drug Delivery

Sadia Zafar[1*], Maria Bilal[1], Inam Mehdi Khan[1], Areeba Ijaz[1], Ambreen Aslam[2], Anis Ali Shah[1], Muhammad Arslan Ashraf[3], and Rizwan Rasheed[3]

[1] Department of Botany, Division of Science and Technology, University of Education, Lahore, Pakistan

[2] Department of Physics, Division of Science and Technology, University of Education, Lahore, Pakistan

[3] Department of Botany, Government College University Faisalabad, Faisalabad, Pakistan

1. Introduction

The challenge in pharmaceutical sciences lies in developing drug delivery systems to enhance the oral bioavailability of medications with low water solubility (Rodrigues LR, 2015). Strategies like MDDS, incorporating lipids, surfactants, cosurfactants, and/or cosolvents, aim to overcome this obstacle, particularly for hydrophobic drugs (Wu et al., 2017). MDDS, designed for various delivery methods, exhibit small size and a globular shape, facilitating the solubilization of hydrophobic drugs (Karasulu HY, 2008). In contemporary industries, surfactants play a crucial role, with a projected global market of USD 52.4 billion by 2025, growing at a yearly rate of 3.5% due to factors like population growth and increased product awareness (Zakharova et al., 2019, Pashirova et al., 2020). The demand for surfactants has surged during the COVID-19 pandemic, especially in products like hand sanitizers, attributed to their anti-allergic as well as emulsification quality (Subramaniam et al., 2020).

Surfactants, molecules that exhibit both hydrophilic and hydrophobic components, can be of microbial or synthetic origin. The surrounding impact of artificial surfactants in various industries, including oil, pharmaceuticals, medical, and agriculture, has raised concerns regarding their usage (Johnson et al., 2021,

* Corresponding author: sadia.zafar@ue.edu.pk

Liu, 2020). This has prompted the search for environmentally friendly alternatives, including biobased polymeric surfactants (Bhadani et al., 2020, Foley et al., 2012). To address the demand for sustainability and biodegradability, technologies utilizing microbial sources have emerged. Biosurfactants, derived from microorganisms, offer benefits such as low toxicity, biodegradability, and the use of renewable resources. They excel in dissolution, segregation, emulsion formation, and can cut down on interfacial and surface tension, aiding in the bioactives' adsorption via biological membranes (Nitschke and Silva, 2018).

Biosurfactant-forming fungi, filamentous yeast, and other microbes are widely studied for their ability to produce various compounds such as glycerol, phospholipids, lipopeptides, oleic acids, saponins, and aliphatic polyglycosides as byproducts. Lipophilic groups consist of proteins or peptides with hydrophobic regions, while hydrophilic groups include monosaccharides, disaccharides, polysaccharides, and amino acids. Regardless of circumstances and systems, biosurfactants consistently demonstrate the capacity to lower the interfacial and surface tension of water and oil components (Singh et al., 2019).

Bacterial biosurfactants, due to their antiviral, antibacterial, and antifungal characteristics, show promise in medicine and biomedicine (Inès and Dhouha, 2015). However, the use of pathogenic bacterial species like Pseudomonas and Bacillus raises toxicological risks and biosafety concerns (Singh et al., 2019). Recent research explores non-pathogenic alternatives such as sophorolipids from Candida bombicola and yeast-like fungi like Starmerella bombicola to mitigate these hazards (De Graeve et al., 2018).

Lower molecular weight biosurfactants, including phospholipids, glycerols, and low-molecular-weight lipopeptides (LPs), demonstrate industrialization potential by lowering interfacial and surface tension. Glycolipids like mannosylerythritol lipids, rhamnolipids, trehalolipids, and cellobiose lipids are of particular interest (Jahan et al., 2020). While microemulsion drug delivery systems (MDDS) are widely used, challenges like formulation flaws, rigidity, and wrapping persist. The use of biosurfactants in microemulsion formulation aims to address issues such as inadequate solubility, reduced drug loading capacity, and increased risk of gastrointestinal irritation associated with high concentrations of surfactants (Gibaud and Attivi, 2012).

Biosurfactants in microemulsion formulations, as opposed to traditional surfactants, can enhance protection, reduce toxicity, and lower irritation in the gastric environment (Rodrigues LR, 2015, Rodrigues et al., 2006). This shift toward biosurfactants aligns with their antimicrobial, antitumor, and reparative properties (Ohadi et al., 2017, Gupta et al., 2017). Biosurfactants are categorized into two primary groups based on their chemical structures: lipopeptides and glycolipids (Ohadi et al., 2018). Biosurfactant-based microemulsions are thermodynamically stable in nature, and the development of their isotropic systems holds promise for drug delivery systems (DDS). Creating nanomaterials with distinct anatomy and high homogeneity is a challenging hurdle in drug delivery systems (DDS) (Kiran et al., 2011, Palanisamy and Raichur, 2009). Sanitary, harmless, size-controlled, and ecologically friendly methods are essential for the production of nanoparticles that are both stable and uniform (Sadiq et al., 2022, Kiran et al., 2011).

The use of oil-water biosurfactant mixtures in microemulsion techniques can facilitate the production of nanoparticles that are both uniform and stable (Hazra et al., 2015). For instance, the reverse microemulsion technique utilized glycolipid biosurfactants as a bioemulsifier to synthesize silver nanoparticles (Kumar et al., 2010). The primary objective of the current study aims to assess the safety and effectiveness of utilizing biosurfactants in drug delivery systems.

2. Production of Biosurfactants Derived from Microbial Sources

Biosurfactants can be produced extracellularly using biocatalyst enzymes, through fermentation processes, enzyme-substrate reactions, and the activity of microorganisms like Bacillus and Pseudomonas in Fig. 1. Two separate pathways exist for synthesizing the hydrophobic and hydrophilic portions of biosurfactants, involving de novo synthesis or induction by the substrate, with both potentially being substrate-dependent (Ali et al., 2022b, Arima et al., 1968).

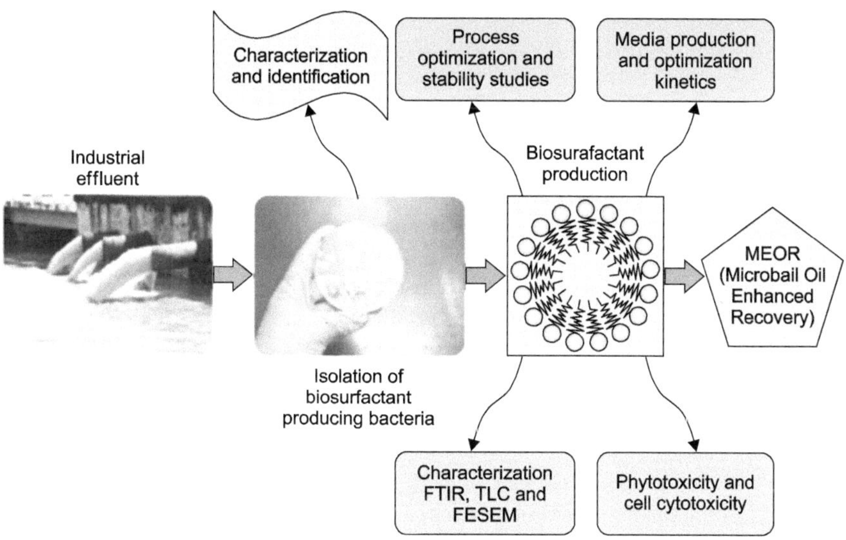

Fig. 1: Production of biosurfactants by microbial sources from industrial effluent.

2.1 Glycolipids

They represent a prevalent type of biosurfactant, with examples such as mannosylerythritol lipids (MELs), trehalolipids, sophorolipids, and rhamnolipids. These glycolipid biosurfactants comprise monosaccharides and disaccharides, along with hydroxy- or aliphatic acids with extended chain lengths (Drakontis and Amin, 2020). Rhamnolipid is a significant glycolipid with superior physicochemical characteristics, primarily produced by species of Pseudomonas and Burkholderia. Its synthesis involves the hydrophobic acid portion derived from fatty acids and the sugar component comprising rhamnose from D-glucose (Abdel-Mawgoud et al., 2014). P. aeruginosa and Burkholderia species are among the few bacteria with the enzymes essential for the biosynthesis of rhamnolipids, involving enzymes like rhamnolipidA, rhamnolipidB, rhamnolipidC, rhamnolipidG, and rhamnolipidI (Kiss et al., 2017).

2.2 Lipopeptide Biosurfactants

In lipopeptide biosurfactants surfactin synthetase catalyzes a non-ribosomal pathway that produces surfactin, along with other lipopeptide biosurfactants like arthrofactin, lichenysin, and iturin (Ndlovu et al., 2017, Das et al., 2008). Bacillus subtilis is the primary bacterium used in surfactin synthesis, and genetic engineering has been employed to enhance its yield (Wu et al., 2019). Solid-state fermentation (SSF) is an alternative method for surfactin manufacturing, involving an organism's growth on or inside solid supports without access to free water. Pseudomonas aeruginosa has also been reported to produce biosurfactants, specifically lipopeptides, produced using sustainable resources such as peanuts and lubricating oil (Thavasi et al., 2011).

2.3 Biosurfactants/Bioemulsifiers Exhibiting a Substantial Molecular Weight

Biosurfactants/bioemulsifiers with elevated molecular mass, known as biogenic emulsion stabilizers, are produced by yeast, fungi, and bacteria. These compounds consist of composite mixtures of lipoproteins, heteropolysaccharides, and proteins (Uzoigwe et al., 2015). *Acinetobacter* spp., one of the earliest known producers of BEs, yields commercially available bio-emulsifiers like Emulsan and Alasan (Mujumdar et al., 2019). Emulsan, a polymer composed of lipoheteropolysaccharide, contains D-galactose-amine and is created during the stability phase using fermentation techniques like batch processing, continuous culture (chemo-stat), immobilized cell system, and self-cycling fermentation configuration. Saccharomyces species and Kluyveromyces marxianus are reported to produce bio-emulsifiers within their cell walls, released through pressurized heat treatments (Alizadeh-Sani et al., 2018).

3. Properties and Phase Behavior of Biosurfactants in Microemulsion-based Drug Delivery System

The characteristics and phase behavior of biosurfactants are used in drug delivery systems like microemulsions, which offer advantages for integrating both hydrophilic lipophilic active pharmaceutical ingredients (APIs). However, for APIs that don't fit into these categories, such as minerals, creating a colloidal DDS product based on microemulsions may be challenging, forming suspensions (Gudiña et al., 2013). Synthetic surfactants, though used in pharmaceutical formulations, have shown some harmful effects. Recently, there has been a discussion about substituting microbial surfactants for synthetic ones in pharmaceutically approved microemulsion systems to enhance safety (Shah et al., 2022, Bee et al., 2019, Rodrigues LR, 2015).

Microbial surfactants, derived from various microorganisms, exhibit diverse amphipathic molecules and distinct chemical designs. Microbial surfactants are grouped according to their chemical make-up and molecular weight, in contrast to chemically synthesized surfactants, which are categorized by head groups. This classification comprises large-molecule surfactants like polysaccharides, proteins, and lipoproteins, as well as small- molecule surfactants like glycolipids and lipid peptides. Usually, amphiphilic and polyphilic polymers prove more beneficial for emulsion stabilization, while microbial surfactants with smaller molecular weights have more straightforward designs and ideal surface-active properties (Fracchia et al., 2019).

The characteristics and robustness of microemulsion-based Drug Delivery Systems (DDS) are challenging to predict. However, it is widely acknowledged that having prior knowledge of the system's features and a successful microemulsion formulation by a highly simplified process of rational element selection through a comprehensive evaluation of its components, including surface characteristics and the hydrophilic–lipophilic balance (HLB) helps in predicting them (Rodrigues LR, 2015).

4. Glycolipid Biosurfactants as Ingredients for Microemulsion Systems

4.1 Mannosylerythritol Lipid

It is demonstrated that MEL-A and MEL-D can produce stable Water-in-Oil (W/O) microemulsions without the need for salt or co-surfactant addition, as demonstrated in studies by Fukuoka et al., 2012, and Worakitkanchanaku et al., 2008. Furthermore, research has explored the aqueous-phase behavior, MEL-B's vesicle-forming activity and self-assembling ability support the idea that mannosylerythritol lipids (MELs) have unique emulsifying activity and can form stable W/O microemulsions even in the absence of salts or co-surfactants (Worakitkanchanaku et al., 2008).

4.2 Rhamnolipids and Sophorolipids

Rhamnolipids serve as co-surfactants in microemulsion synthesis. Additionally, the conformational changes of rhamnolipid molecules at the Oil/Water (O/W) interface affect the phase behavior and microstructure of these microemulsions (Xie et al., 2007). The creation of stable methyl methacrylate emulsions using rhamnolipid extracts from Pseudomonas aeruginosa PA1 is recommended (Mendes and Collaborators, 2015). They have been employed as stabilizers in Water/Oil (W/O) microemulsions for enzymatic reactions (Moya-Ramírez et al., 2017).

Glycolipid biosurfactants, such as rhamnolipids, have been utilized in the process of creating uniform, stable nanoparticles such as silver nanoparticles, showing antibiotic activity against various pathogens (Ali et al., 2022a, Kiran et al., 2010). Silver nanoparticles were synthesized using rhamnolipids derived from *P. aeruginosa* strain BS-161R with antimicrobial properties (Kumar et al., 2010). In addition to silver nanoparticles, rhamnolipids have been involved in the synthesis of nickel nanoparticles by means of the micro-emulsion method (Palanisamy and Raichur et al., 2009). Furthermore, metal-bound nanoparticle synthesis and stabilization have been effectively accomplished with the use of rhamnolipids and sophorolipids. contributing to the development of stable, alcohol-free microemulsions (Hazra et al., 2014).

4.3 Trehalose Lipid

Trehalose lipid has been effectively employed in an atomized microemulsion process modified for oil/water (O/W) synthesizing novel polymethyl methacrylate nanoparticles (nPMMA). The biosurfactant quantity needed for this formation is significantly lower than that utilized in conventional microemulsion polymerization systems. This suggests that the resulting nanoparticles are non-toxic and biocompatible (Hazra et al., 2014).

5. Using Lipopeptide Biosurfactants in Microemulsion Systems

5.1 Surfactin

Surfactin has proven successful in the formation of microemulsions. It has been demonstrated that using microemulsion formulations rather than synthetic surfactants can decrease toxicity and increase the physicochemical stability of the microemulsion formulation. To improve their pharmaceutical performance, surfactin, for example, was used to prepare a Microemulsion Drug Delivery System (MDDS) that included vitamin E and docosahexaenoic acid (DHA) (He et al., 2017).

5.2 Emulsifying Action

Emulsions, as kinetically stabilized systems, rely on the balance of various factors such as the makeup of fluid phases (water and oil), emulsifier properties, as well as preparation conditions. Their composition, stability as well as appearance are influenced by these elements and external factors like temperature and pressure during preparation. Emulsions can undergo breakdown through mechanisms like skimming, flocculation, coagulation, Ostwald maturation, and coalescence. Creaming, driven by density differences in oil and aqua phases, may lead to phase separation as emulsion droplets migrate due to gravity. Natural emulsifiers, like biosurfactants, play a crucial role in rapidly adsorbing on oil droplet surfaces, reducing interfacial tension, and facilitating droplet breakdown. Quillaja saponin extract and biosurfactant pseudofactin II from *P. fluorescens* BD5 are examples of effective emulsifiers, with pseudofactin II outperforming synthetic surfactants in emulsifying hydrocarbons and vegetable oils (Kaizu and Alexandridis, 2017, Arima et al., 1968, McClements and Gumus, 2016, Janek et al., 2010).

5.3 Physicochemical Characterization

Microorganisms have the capacity to generate diverse biosurfactants, each with distinct bioactivity. Beyond their source, the processes used in their manufacture and purification influence the physicochemical characteristics of these biosurfactants. It is essential to comprehend these qualities for accurately determining their industrial applications (Arima et al., 1968, Saranraj et al., 2022a). This section explores the key properties of biosurfactants that are essential for their effectiveness as emulsifiers.

5.4 Surface and Interfacial Tension

A vital characteristic of a bio-emulsifier to reduce interfacial and surface tension is a fundamental function of amphiphilic molecules in forming kinctically stabilized emulsions. At interfaces, these molecules get adsorbed, taking the place of water or oil molecules and lowering intermolecular forces. thereby minimizing surface or interfacial tension (Drakontis and Amin, 2020).

In comparison to chemical surfactants, biosurfactants have demonstrated that they are more efficient in lowering interfacial tension (Pereira et al., 2013). Surfactin, a prominent lipopeptide biosurfactant, exhibits substantial surface activity, lowering surface tension from 72 mN/m to 27 ± 2 mN/m (Alvarez et al., 2015). Under harsh conditions, it achieves interfacial tension values of 3.79 ± 0.27 mN/m and 0.32 ±

0.02 mN/m (Al-Wahaibi et al., 2014). The lipopeptide biosurfactin, arthrofactin from *Candida lipolytica* UCP 0988's biosurfactant and *Arthrobacter* sp. Strain MIS38 together display comparatively less activity on the surface due to their complex chemical structures, lacking a clear polarity distribution and containing branched or ring structures (Rufino et al., 2014).

The ability of lipopeptide surfactin to form spherical structures facilitates minimized aggregation number structures and dense packing at interfaces. Saponins, another type of biosurfactant, exhibit unique surface properties attributed to strong hydrogen bonds between saccharide groups in the interfacial layer and dense molecular packing at the interface, resulting in denser surface layers than common amphiphiles (Morikawa et al., 1993).

These distinct properties of biosurfactants play a crucial role in strategies for attaching themselves to cell membranes and biomolecules. A closer look at how their attributes compare, including surface activity and the values of the critical micellization concentrations (CMC)s, is available in the review (Otzen, 2017). In certain cases, contaminants in surfactant compositions may be connected to low CMC values and high surface activity.

5.5 Self-Assembly

In aqueous solvents, surfactants spontaneously form micelles when their concentration is high, resulting in nanostructures through the thermodynamically stable process of micellization which surpass the Critical Micelle Concentration (CMC) (Kumar and Ngueagni, 2021). Because of their weak Van der Waals interactions and hydrophobic effect, biosurfactants have a tendency to self-assemble spontaneously. Surface tension reduction is the primary determinant of a biosurfactant's effectiveness, with the goal of decreasing the surface tension of water from 72 mN/m to about 30 mN/m (Hu et al., 2020).

The size and shape of micelles are influenced by factors like biosurfactant concentration, temperature, pH, pressure, and salt content. Repulsive forces between micelles and head groups limit the quantity of associated biosurfactants (Arima et al., 1968). Beyond the CMC value, increasing surfactant concentration encourages the production of a large number of micelles. Both rhamnolipids and surfactin exhibit the creation of micelles with a small number of aggregations, and they can further organize into bubble structures (Aveyard et al., 2003).

5.6 Solubilization

Hydrophobic compounds can be dissolved by amphiphiles that self-assemble in aqueous solutions, such as oil, by residing in the hydrophobic domains of the nanostructure. Factors such as concentration, pH, and the presence of salts and additives (electrolytes) can change the micelle size and affect the solubility of hydrophobic organic compounds in the presence of biosurfactants (Penfold et al., 2018).

Rhamnolipids enhance the process of making amphiphilic molecules more hydrophobic in order to solubilize hydrophobic compounds. When the pH rises, biosurfactants often form vesicles and micelles, which hinders the dissolution of other molecules. The functionalities of biosurfactants vary depending on the substrate, impacting the rate at which various hydrocarbons are soluble or emulsified (Shao et al., 2014).

In general, biosurfactants work better as solubilizing agents than artificial surfactants. In contrast to artificial surfactants such as sodium dodecyl sulphate (SDS), polysorbate (Tween 80), Triton's a X-100, and rhamnolipid, a biosurfactant derived from Rhodococcus erythropolis HX-2 demonstrated greater solubilization for petroleum-based substances and polycyclic aromatic hydrocarbons (Percebom and Towesend, 2018).

6. Applications of Biosurfactants

Applications for biosurfactants are numerous, ranging from food and beverages to bioremediation, pharmaceuticals, biomedicine, and nanotechnology, and their biodegradability and minimal toxicity make them beneficial for the environment (Table 1) (Zaman et al., 2022). They are exemplary in degrading hydrocarbons in polluted water and soil and have promising applications in drug delivery systems (Jimoh and Lin, 2019). Marchant et al. (2014) and Chakraborty et al. (2014), have reported that microorganisms use a variety of carbon and energy sources for growth. A mixture of carbon sources and insoluble substrates allows the production of different substances, including biosurfactants. Various molecular arrangements and surface activities of biosurfactants can be produced by bacteria, yeasts, and some filamentous fungi (Santos et al., 2016). Residues such as whey, dairy products, glycerol, molasses, animal fats and fermented alcohol, olive oil production byproducts, cassava processing wastewater, and potatoes are examples of materials used for biosurfactant production (Silva et al., 2018). These amphipathic molecules, mainly secondary metabolites, contribute to the survival of microorganisms through the addition of nutrients and interactions with hosts and acting as biocides (Ravinder et al., 2022, Gudiña et al., 2016).

Table 1: Biosurfactants and their applications

Type of biosurfactant	Microorganism	Application	Reference
Glycolipids	*Rhodococcus* spp.	Bioremediation	(Tripathi et al., 2014)
	Tsukamurella sp.,	Antimicrobial agent	(Mnif et al., 2018)
	Arthrobacter sp.	Antimicrobial agent	(Abdel-Mawgoud et al., 2018)
Glycolipoprotein	*Aspergillus niger*	Antimicrobial activity	(Ansari et al., 2018)
Lichenisina	*Bacillus licheniformis*	Antimicrobial activity Hemolytic and chelating agent	(Gomaa et al., 2013)
Lipopeptides	*Bacillus subtilis*	Bacterial growth inhibition Biomedical application	(do Valle Gomes et al., 2012)
Surfactin	*Kurtzmanomyces* sp.	Biomedical application	(Chen et al., 2015)
Manosileritritol lipids	*Candida antartica*	Anti-inflammatory secretion inhibitor and RBL-2H3 cell mediators	(Silva et al., 2018)
Sophorolipids	*Candida bombicola,*	Emulsifier	(Solaiman et al., 2016)

Contd...

Contd...

Type of biosurfactant	Microorganism	Application	Reference
	Candida apicola	Emulsifier	(Solaiman et al., 2016)
Rhamnolipids	*Renibacterium salmoninarum*	Bioremediation	(Mahjoubi et al., 2018)
	Pseudomonas aeruginosa	Bioremediation	(Jadhav et al., 2018)
	Pseudomonas chlororaphis	Biocontrol Agent	(Jadhav et al., 2011)
	Pseudomonas putida	Bioremediation	(Jadhav et al., 2018)
Sophorolipids	*Candida bombicola*	anticancer agents, precursor of bioplastic	(Cavalero & Cooper, 2003)
Polyketide tderivative	*Penicillium chrysogenum*	Bioremediation	(Gao et al., 2011)
Mannosylerythritol lipids	*Pseudozyma rugulosa*	Antimicrobial, Bioremediation	(Morita et al., 2006)
Carbohydrate-lipid complex	*Pseudomonas fluorescens*	Industrial application, emulsifier	(Nerurkar et al., 2009)
Subtilisin	*Bacillus subtilis*	Antimicrobial	(Sutyak et al., 2008)
Peptide lipids	*Bacillus licheniformis*	Antimicrobial	(Begley et al., 2009)
Phospholipids	*Acinetobacter* sp.	Bioremediation	(Kosaric, 2001)

Due to their microbial origin and favorable characteristics, biosurfactants are researched for potential therapeutic applications (Marchant and Banat, 2012, Patel et al., 2019, Patel et al., 2022). With properties such as low level of toxicity, ability to emulsify and demulsify, resistance to temperature changes and variations in ionic strength, biological degradation, and antibacterial properties. biosurfactants are gaining interest for applications related to food, beauty products, advanced bioremediation, medication delivery (Saranraj et al., 2022b, Chen et al., 2015). They are divided into two groups: low molecular mass biosurfactants (like glycolipids, and lipopeptides) and high molecular mass biosurfactants (like proteins, polysaccharides, and lipoproteins). Because of their simpler structures, low molecular mass variants frequently show notable activity (Ikegami et al., 2020, Chong and Li, 2017).

6.1 Applications of Biosurfactants in Field of Biotechnology

The manufacturing of biosurfactants for the market faces challenges such as high costs for raw materials and production , coupled with relatively low output. Research is actively addressing this by focusing on increasing yield while minimizing raw material expenses.Optimization studies have explored how variables such as temperature, pH, oxygen level, and the type and amount of carbon and nitrogen sources can influence the production of biosurfactants by microorganisms. Additionally, recent efforts have emphasized using renewable substrates for biosurfactant production (Domínguez-Rivera et al., 2019).

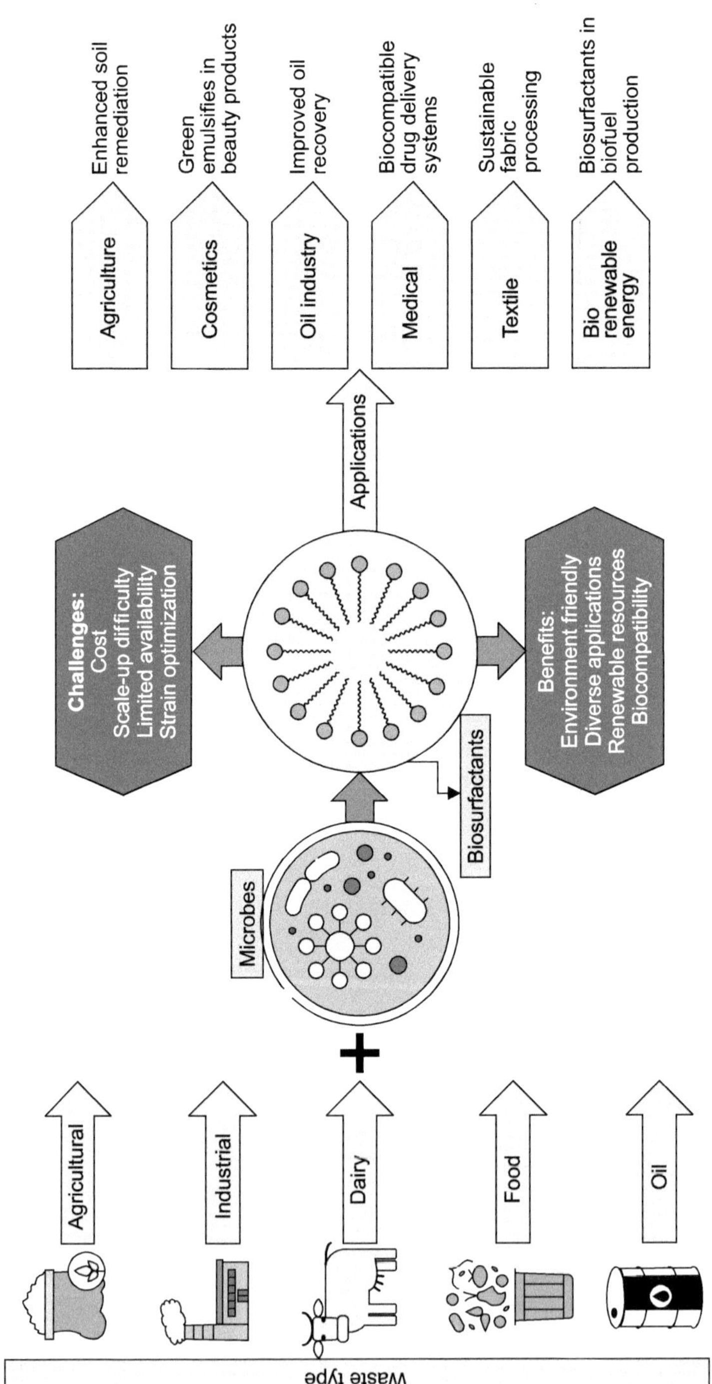

Fig. 2: Applications of biosurfactants.

Industrial starch production generates starch-rich wastewater, which can serve as a possible substrate for the synthesis of biosurfactants. Grains such as maize, cassava, wheat, and potatoes are commonly used for this purpose. Fox and Bala (2000) studied the utilization of potato substrates for surfactant production from *B. subtilis*, reporting a surface tension of 28.3 mN/m and a CMC of 100 mg/L. Fat in chicken, oil of sunflower, molasses of sugarcane, juice of sugarcane have also been employed for sophorolipid production by *C. bombicola*, with efficient biosurfactant production achieved using specific industrial waste, such as fish waste and bagasse from sugarcane as carbon sources.

6.2 Pharmaceutical Applications of Biosurfactants

Biosurfactants exhibit diverse roles in the pharmaceutical sector such as those related to immune modulating, haemorrhagic, spermicidal, antiviral, antiadhesive, cancer fighting, and spermicidal properties. Surfactin, for instance, demonstrates various actions that include death related genes, coagulating, hypercholesterolemia, anti-adhesive, antibacterial, cancer – fighting, and antimicrobial properties. Lipopeptide biosurfactants from *Acinetobacter junii* (AjL) have shown retardation in opposition to *Candida utilis*, presenting potential as a new drug. Additionally, biosurfactants have promising applications in cosmetics, offering reduced toxicity, skin compatibility, and nourishing properties. For instance, a biosurfactant from Lactobacillus paracasei, developed for cosmetics, showed similar results to traditional surfactants with no observed toxicity in fibroblasts. Biosurfactants are also employed in drug delivery systems, addressing the challenge of improving the oral bioavailability of drugs with low aqueous solubility. Hydrogels containing antimicrobial biosurfactants are proposed for wound healing against drug-resistant skin infections, and some commercially available products already utilize these innovative molecules in skincare applications (Adu et al., 2020).

7. Conclusion

Several scholarly works have documented the unique characteristics of biosurfactants, suas their antibacterial, emulsifying, and anti-adhesive capabilities. Biosurfactants can be created chemically or by the action of microbes, and their potential uses in a variety of industries have been suggested. The production of novel biosurfactants by microorganisms such as fungi, yeast, and bacteria enhances antimicrobial aspects, particularly in food and medicine safety. These biosurfactants including glycolipids, surfactants, and high molecular mass biosurfactants, contribute to animal and plant defense of plants and animals, and disease control and treatment. These biosurfactants may aid in the detection of distinct molecules according to their physical,chemical and structural characteristics Their ability to self-assemble and amphiphilic nature provide new opportunities for the creation of antibacterial formulations that offer pharmaceutically interesting bioactives. These systems for drug delivery can be developed as semisolid hydrogels for topical administration, which can be used to heal wounds and combat drug-resistant skin infections. There have been numerous more documented instances of using biosurfactants to lower the price of biotechnological products, one such instance is the improvement of growth media by the use of *B. subtilis* surfactants.

Regarding the ecotoxicological effect, a number of biosurfactants derived from microorganisms such as lipopeptides, glycolipids, mannosylerythritol lipid, rhamnolipids, sophorolipids and phospholipids have been found to be useful in preventing the growth of phytopathogens because of their antimicrobial and antibiofilm qualities. Thus, it seems sensible that biosurfactants can be thought of as a viable substitute with a variety of industrial uses.

References

Abdel-Mawgoud, A.M. & Stephanopoulos, G. (2018). Simple glycolipids of microbes: Chemistry, biological activity and metabolic engineering. *Synth. Syst. Biotechnol.*, 3(1): 3-19.

Abdel-Mawgoud, A.M., Lépine, F. & Déziel, E. (2014). A stereospecific pathway diverts β-oxidation intermediates to the biosynthesis of rhamnolipid biosurfactants. *Chem. Biol.*, 21(1): 156-164.

Adu, S.A., Naughton, P.J., Marchant, R. & Banat, I.M. (2020). Microbial biosurfactants in cosmetic and personal skincare pharmaceutical formulations. *Pharmaceutics*, 12(11): 1099.

Ali, S.A.M., Sayyed, R.Z., Mir, M.I., Hameeda, B., Khan, Y., Alkhanani, M.F. et al. (2022a). Induction of systemic resistance in maize and antibiofilm activity of surfactin from *Bacillus velezensis* MS20. *Front. Microbiol.*, 13: 879739. https://doi.org/10.3389/fmicb.2022.879739

Ali, S.A.M., Sayyed, R.Z., Reddy, M.S., Enshasy, H.E. & Hameeda, B. (2022b). Delving through quorum sensing and CRISPRi strategies for enhanced surfactin production. *In:* Sayyed, R.Z. (Eds), *Biosurfatnats: Production and Applications in Bioremediation/Reclamation*. 59-79. CRC Press, Taylor & Francis Group, USA.

Alizadeh-Sani, M., Hamishehkar, H., Khezerlou, A., Azizi-Lalabadi, M., Azadi, Y. Nattagh-Eshtivani, E. et al. (2018). Bioemulsifiers derived from microorganisms: Applications in the drug and food industry. *Adv. Pharm. Bull.*, 8(2): 191.

Alvarez, V.M., Jurelevicius, D., Marques, J.M., de Souza, P.M., de Araújo, L., Barros, T.G. et al. (2015). Bacillus amyloliquefaciens TSBSO 3.8, a biosurfactant-producing strain with biotechnological potential for microbial enhanced oil recovery. *Colloids Surf. B Biointerfaces*, 136: 14-21.

Al-Wahaibi, Y., Joshi, S., Al-Bahry, S., Elshafie, A., Al-Bemani, A. & Shibulal, B. (2014). Biosurfactant production by Bacillus subtilis B30 and its application in enhancing oil recovery. *Colloids Surf. B Biointerfaces*, 114: 324-333.

Ansari, A., Pervez, S., Javed, U., Abro, M.I., Nawaz, M.A., Qader, S.A.U. et al. (2018). Characterization and interplay of bacteriocin and exopolysaccharide-mediated silver nanoparticles as an antibacterial agent. *Inter. J. biol. Macromol.*, 115: 643-650.

Arima, K., Kakinuma, A. & Tamura, G. (1968). Surfactin, a crystalline peptide lipid surfactant produced by Bacillus subtilis: Isolation, characterization and its inhibition of fibrin clot formation. *Biochem. Biophys. Res. Commun.*, 31(3): 488-494.

Aveyard, R., Binks, B.P. & Clint, J.H. (2003). Emulsions stabilized solely by colloidal particles. *Adv. Colloid Interface Sci.*, 100: 503-546.

Bee, H., Khan, M.Y. & Sayyed, R.Z. (2019). Microbial surfactants and their significance in agriculture. *In:* Sayyed Reddy Antonious (Ed.), *PGPR: Prospects for Sustainable Agriculture*. 205-216. Springer-Nature, Singapore.

Begley, M., Cotter, P.D., Hill, C. & Ross, R.P. (2009). Identification of a novel two-peptide lantibiotic, lichenicidin, following rational genome mining for Lan M proteins. *Appl. Environ. Microbiol.*, 75: 5451-5460.

Bhadani, A., Kafle, A., Ogura, T., Akamatsu, M., Sakai, K., Sakai, H. et al. (2020). Current perspective of sustainable surfactants based on renewable building blocks. *Curr. Opin. Colloid Interface Sci.*, 45: 124-135.

Cavalero, D.A. & Cooper, D.G. (2003). The effect of medium composition on the structure and physical state of sophorolipids produced by *Candida bombicola* ATCC 22214. *J. Biotechnol.*, 103(1): 31-41.

Chakraborty, J. & Das, S. (2014). Biosurfactant-based bioremediation of toxic metals. *In: Microbial Biodegradation and Bioremediation*. 167-201. Elsevier.

Chen, W.C., Juang, R.S. & Wei, Y.H. (2015). Applications of a lipopeptide biosurfactant, surfactin, produced by microorganisms. *Biochem. Eng. J.*, 103: 158-169.

Chong, H. & Li, Q. (2017). Microbial production of rhamnolipids: Opportunities, challenges and strategies. *Microb. Cell Factories*, 16: 1-12.

Das, P., Mukherjee, S. & Sen, R. (2008). Genetic regulations of the biosynthesis of microbial surfactants: An overview. *Biotechnol. Genet. Eng. Rev.*, 25(1): 165-186.

De Graeve, M., De Maeseneire, S.L., Roelants, S.L. & Soetaert, W. (2018). Starmerella bombicola, an industrially relevant, yet fundamentally underexplored yeast. *FEMS Yeast Res.*, 18(7): foy072.

do Valle Gomes, M.Z. & Nitschke, M. (2012). Evaluation of rhamnolipid and surfactin to reduce the adhesion and remove biofilms of individual and mixed cultures of food pathogenic bacteria. *Food Control*, 25(2): 441-447.

Domínguez Rivera, Á., Martínez Urbina, M.Á. & Lópezy López, V.E. (2019). Advances on research in the use of agro-industrial waste in biosurfactant production. *World J. Microbiol. Biotechnol.*, 35(10): 155.

Drakontis, C.E. & Amin, S. (2020). Biosurfactants: Formulations, properties, and applications. *Curr. Opin. Colloid Interface Sci.*, 48: 77-90.

Foley, P., Beach, E.S. & Zimmerman, J.B. (2012). Derivation and synthesis of renewable surfactants. *Chem. Soc. Rev.*, 41(4): 1499-1518.

Fox, S.L. & Bala, G.A. (2000). Production of surfactant from Bacillus subtilis ATCC 21332 using potato substrates. *Bioresour. Technol.*, 75(3): 235-240.

Fracchia, L., Ceresa, C. & Banat, I.M. (2019). Biosurfactants in cosmetic, biomedical and pharmaceutical industry. *In:* Banat, I.M., Thavasi, R. (Eds), *Microbial Biosurfactants and Their Environmental and Industrial Applications*. 258-288.

Fukuoka, T., Yanagihara, T., Ito, S., Imura, T., Morita, T., Sakai, H., Kitamoto, D. et al. (2012). Reverse vesicle formation from the yeast glycolipid biosurfactant mannosylerythritol lipid-D. *J. Oleo Sci.*, 61(5): 285-289.

Gao, Y.Z., Li, Q.L., Ling, W.T. & Zhu, X.Z. (2011). Arbuscular mycorrhizal phytoremediation of soils contaminated with phenanthrene and pyrene. *J. Hazard. Mat.*, 185: 703-709.

Gibaud, S. & Attivi, D. (2012). Microemulsions for oral administration and their therapeutic applications. *Expert Opin. Drug Deliv.*, 9(8): 937-951.

Gomaa, E.Z. (2013). Antimicrobial activity of a biosurfactant produced by Bacillus licheniformis strain M104 grown on whey. *Braz. Arch. Biol. Technol.*, 56: 259-268.

Gudiña, E.J., Rangarajan, V., Sen, R. & Rodrigues, L.R. (2013). Potential therapeutic applications of biosurfactants. *Trends Pharmacol. Sci.*, 34(12): 667-675.

Gudiña, E.J., Teixeira, J.A. & Rodrigues, L.R. (2016). Biosurfactants produced by marine microorganisms with therapeutic applications. *Mar. Drugs*, 14(2): 38.

Gupta, S., Raghuwanshi, N., Varshney, R., Banat, I.M., Srivastava, A.K., Pruthi, P.A. & Pruthi, V. (2017). Accelerated in vivo wound healing evaluation of microbial glycolipid containing ointment as a transdermal substitute. *Biomed. Pharmacother.*, 94: 1186-1196.

Hazra, C., Kundu, D., Chatterjee, A., Chaudhari, A. & Mishra, S. (2014). Poly(methyl methacrylate)(core)–biosurfactant (shell) nanoparticles: Size controlled sub-100 nm synthesis, characterization, antibacterial activity, cytotoxicity and sustained drug release behavior. *Colloids and Surfaces A: Physicochemical and Engineering Aspects*, 449: 96-113.

Hazra, C., Kundu, D. & Chaudhari, A. (2015). Lipopeptide biosurfactant from *Bacillus clausii*

BS02 using sunflower oil soapstock: Evaluation of high throughput screening methods, production, purification, characterization and its insecticidal activity. *RSC Adv.*, 5(4): 2974-2982.

He, Z., Zeng, W., Zhu, X., Zhao, H., Lu, Y., & Lu, Z. (2017). Influence of surfactin on physical and oxidative stability of microemulsions with docosahexaenoic acid. *Colloids Surf B Biointerfaces*, 151: 232-239.

Hu, X., Qiao, Y., Chen, L.Q., Du, J.F., Fu, Y.Y., Wu, S. et al. (2020). Enhancement of solubilization and biodegradation of petroleum by biosurfactant from Rhodococcus erythropolis HX-2. *Geomicrobiol. J.*, 37(2): 159-169.

Ikegami, K., Hirose, Y., Sakashita, H., Maruyama, R. & Sugiyama, T. (2020). Role of polyphenol in sugarcane molasses as a nutrient for hexavalent chromium bioremediation using bacteria. *Chemosphere*, 250: 126267.

Inès, M. & Dhouha, G. (2015). Glycolipid biosurfactants: Potential related biomedical and biotechnological applications. *Carbohydr. Res.*, 416: 59-69.

Jadhav, J., Dutta, S., Kale, S. & Pratap, A. (2018). Fermentative production of rhamnolipid and purification by adsorption chromatography. *Prep. Biochem. Biotechnol.*, 48(3): 234-241.

Jadhav, M., Kalme, S., Tamboli, D. & Govindwar, S. (2011). Rhamnolipid from Pseudomonas desmolyticum NCIM-2112 and its role in the degradation of Brown 3REL. *J. Basic Microbiol.*, 51(4): 385-396.

Jahan, R., Bodratti, A.M., Tsianou, M. & Alexandridis, P. (2020). Biosurfactants, natural alternatives to synthetic surfactants: Physicochemical properties and applications. *Adv. Colloid Interface Sci.*, 275: 102061.

Janek, T., Łukaszewicz, M., Rezanka, T. & Krasowska, A. (2010). Isolation and characterization of two new lipopeptide biosurfactants produced by Pseudomonas fluorescens BD5 isolated from water from the Arctic Archipelago of Svalbard. *Bioresour. Technol.*, 101(15): 6118-6123.

Jimoh, A.A. & Lin, J. (2019). Biosurfactant: A new frontier for greener technology and environmental sustainability. *Ecotoxicol. Environ. Saf.*, 184: 109607.

Johnson, P., Trybala, A., Starov, V. & Pinfield, V.J. (2021). Effect of synthetic surfactants on the environment and the potential for substitution by biosurfactants. *Adv. Colloid Interface Sci.*, 288: 102340.

Kaizu, K. & Alexandridis, P. (2016). Effect of surfactant phase behavior on emulsification. *J. Colloid Interface Sci.*, 466: 138-149.

Karasulu, H.Y. (2008). Microemulsions as novel drug carriers: The formation, stability, applications and toxicity. *Expert Opin. Drug Deliv.*, 5(1): 119-135.

Kiran, G.S., Sabu, A. & Selvin, J. (2010). Synthesis of silver nanoparticles by glycolipid biosurfactant produced from marine Brevibacterium casei MSA19. *J. Biotechnol.*, 148(4): 221-225.

Kiran, G.S., Selvin, J., Manilal, A. & Sujith, S. (2011). Biosurfactants as green stabilizers for the biological synthesis of nanoparticles. *Crit. Rev. Biotechnol.*, 31(4): 354-364.

Kiss, K., Ng, W.T. & Li, Q. (2017). Production of rhamnolipids-producing enzymes of Pseudomonas in *E. coli* and structural characterization. *Front. Chem. Sci. Eng.*, 11: 133-138.

Kosaric, N. (2001). Biosurfactants and their application for soil bioremediation. *Food Technol. Biotechnol.*, 39: 295-304.

Kumar, C.G., Mamidyala, S.K., Das, B., Sridhar, B., Devi, G.S. & Karuna, M.S. (2010). Synthesis of biosurfactant-based silver nanoparticles with purified rhamnolipids isolated from *Pseudomonas aeruginosa* BS-161R. *J. Microbiol. Biotechnol.*, 20(7): 1061-1068.

Kumar, P.S. & Ngueagni, P.T. (2021). A review on new aspects of lipopeptide biosurfactant: Types, production, properties and its application in the bioremediation process. *J. Hazard. Mater.*, 407: 124827.

Kurozuka, A., Onishi, S., Nagano, T., Yamaguchi, K., Suzuki, T., & Minami, H. (2017). Emulsion polymerization with a biosurfactant. *Langmuir*, 33(23): 5814-5818.

Liu, L. (2020). Penetration of surfactants into skin. *J. Cos. Sci.*, 71(2): 91-109.

Mahjoubi, M., Cappello, S., Souissi, Y., Jaouani, A. & Cherif, A. (2018). Microbial bioremediation of petroleum hydrocarbon–contaminated marine environments. *Recent Insights Petroleum Sci. Eng.*, 325: 325-350.

Marchant, R. & Banat, I.M. (2012). Microbial biosurfactants: Challenges and opportunities for future exploitation. *Trends Biotechnol.*, 30(11): 558-565.

Marchant, R., Funston, S., Uzoigwe, C., Rahman, P.K.S.M. & Banat, I.M. (2014). Production of biosurfactants from nonpathogenic bacteria. *Biosurfactants: Prod. Util. Process. Technol. Econ.*, 73-82.

McClements, D.J. & Gumus, C.E. (2016). Natural emulsifiers—Biosurfactants, phospholipids, biopolymers, and colloidal particles: Molecular and physicochemical basis of functional performance. *Advances in Colloid and Interface Science*, 234: 3-26.

Mendes, A.N., Filgueiras, L.A., Pinto, J.C. & Nele, M. (2015). Physicochemical properties of rhamnolipid biosurfactant from *Pseudomonas aeruginosa* PA1 to applications in microemulsions. *J. Biomaterials Nanobiotechnol.*, 6(01): 64.

Mnif, I., Ellouz-Chaabouni, S. & Ghribi, D. (2018). Glycolipid biosurfactants, main classes, functional properties and related potential applications in environmental biotechnology. *J. Polym. Environ.*, 26: 2192-2206.

Morikawa, M., Daido, H., Takao, T., Murata, S., Shimonishi, Y., & Imanaka, T. (1993). A new lipopeptide biosurfactant produced by *Arthrobacter* sp. strain MIS38. *J. Bacteriol.*, 175(20): 6459-6466.

Morita, T., Konishi, M., Fukuoka, T., Imura, T. & Kitamoto, D. (2006). Discovery of *Pseudozyma rugulosa* NBRC 10877 as a novel producer of the glycolipid biosurfactants, mannosylerythritol lipids, based on rDNA sequence. *Appl. Microbiol. Biotechnol.*, 73: 305-313.

Moya-Ramírez, I., García-Román, M. & Fernández-Arteaga, A. (2017). Rhamnolipids: Highly compatible surfactants for the enzymatic hydrolysis of waste frying oils in microemulsion systems. *ACS Sustain. Chem. Eng.*, 5(8): 6768-6775.

Mujumdar, S., Joshi, P. & Karve, N. (2019). Production, characterization, and applications of bioemulsifiers (BE) and biosurfactants (BS) produced by *Acinetobacter* spp.: A review. *J. Basic Microbiol.*, 59(3): 277-287.

Nagarajan, R. & Ruckenstein, E. (1991). Theory of surfactant self-assembly: A predictive molecular thermodynamic approach. *Langmuir*, 7(12): 2934-2969.

Ndlovu, T., Rautenbach, M., Khan, S. & Khan, W. (2017). Variants of lipopeptides and glycolipids produced by *Bacillus amyloliquefaciens* and *Pseudomonas aeruginosa* cultured in different carbon substrates. *AMB Express*, 7(1): 1-13.

Nerurkar, A.S., Hingurao, K.S. & Suthar, H.G. (2009). Bioemulsfiers from marine microorganisms. *J. Sci. Indus. Res.*, 68: 273-277.

Nitschke, M. & Silva, S.S.E. (2018). Recent food applications of microbial surfactants. *Crit. Rev. Food Sci. Nutr.*, 58(4): 631-638.

Ohadi, M., Forootanfar, H., Rahimi, H.R., Jafari, E., Shakibaie, M., Eslaminejad, T. et al. (2017). Antioxidant potential and wound healing activity of biosurfactant produced by *Acinetobacter junii* B6. *Curr. Pharm. Biotechnol.*, 18(11): 900-908.

Otzen, D.E. (2017). Biosurfactants and surfactants interacting with membranes and proteins: Same but different? *Biochim. Biophys. Acta Biomembr.*, 1859(4): 639-649.

Palanisamy, P. & Raichur, A.M. (2009). Synthesis of spherical NiO nanoparticles through a novel biosurfactant mediated emulsion technique. *Mat. Sci. Eng. C.*, 29(1): 199-204.

Pashirova, T.N., Sapunova, A.S., Lukashenko, S.S., Burilova, E.A., Lubina, A.P. et al. (2020). Synthesis, structure-activity relationship and biological evaluation of tetracationic gemini Dabco-surfactants for transdermal liposomal formulations. *Inter. J. Pharm.*, 575: 118953.

Patel, P., Bhatt, S., Patel, H., Marcelino, L.A. & Sayyed, R.Z. (2022). Biosurfactant – A biomolecules and its potential applications. *In:* Sayyed, R.Z. and Enshasy, H.E. (Eds), *Biosurfatnats: Production and Applications in Food and Agriculture*. Vol. II, 133-149. CRC Press, Taylor & Francis Group, USA.

Patel, S., Homaei, A., Patil, S. & Daverey, A. (2019). Microbial biosurfactants for oil spill remediation: Pitfalls and potentials. *Appl. Microbiol. Biotechnol.*, 103(1): 27-37.

Penfold, J., Thomas, R.K., Tucker, I., Petkov, J.T., Stoyanov, S.D., Denkov, N. et al. (2018). Saponin adsorption at the air-water interface—Neutron reflectivity and surface tension study. *Langmuir*, 34(32): 9540-9547.

Percebom, A.M., Towesend, V.J., de Andrade, M.D.P.S. & Gramatges, A.P. (2018). Sustainable self-assembly strategies for emerging nanomaterials. *Curr. Opin. Green Sustain. Chem.*, 12: 8-14.

Pereira, J.F., Gudiña, E.J., Costa, R., Vitorino, R., Teixeira, J.A., Coutinho, J.A. et al. (2013). Optimization and characterization of biosurfactant production by *Bacillus subtilis* isolates towards microbial enhanced oil recovery applications. *Fuel*, 111: 259-268.

Ravinder, R., Manasa, M., Roopa, D., Bukhari, N.A., Hatamleh, A.A., Khan, M.Y. et al. (2022). Biosurfactant producing multifarious *Streptomyces puniceus* RHPR9 of *Coscinium fenestratum* rhizosphere promotes plant growth in chilli. *Plos One*, 17(3): e0264975. https://doi.org/10.1371/ journal.pone.0264975

Rodrigues, L.R. (2015). Microbial surfactants: Fundamentals and applicability in the formulation of nano-sized drug delivery vectors. *J. Colloid Interface Sci.*, 449: 304-316.

Rodrigues, L., Banat, I.M., Teixeira, J. & Oliveira, R. (2006). Biosurfactants: Potential applications in medicine. *J. Antimicrobe. Chemother.*, 57(4): 609-618.

Rufino, R.D., de Luna, J.M., de Campos Takaki, G.M. & Sarubbo, L.A. (2014). Characterization and properties of the biosurfactant produced by *Candida lipolytica* UCP 0988. *Electron. J. Biotechnol.*, 17(1): 34-38.

Sadiq, M.B., Khan, M.R. & Sayyed, R.Z. (2022). Biosurfactant mediated synthesis and stabilization of nanoparticles. *In:* Sayyed, R.Z. (Eds), *Biosurfatnats: Production and Applications in Bioremediation/Reclamation*. 158-168. CRC Press, Taylor & Francis Group, USA.

Santos, D.K.F., Rufino, R.D., Luna, J.M., Santos, V.A. & Sarubbo, L.A. (2016). Biosurfactants: Multifunctional biomolecules of the 21st century. *Inter. J. Mol. Sci.*, 17(3): 401.

Saranraj, P., Sayyed, R.Z., Sivasakthivelan, P., Hasan, M.S., Al-Tawaha, A.R.M.A., & Amala, K. (2022a). Microbial biosurfactants: Methods of investigation, characterization, current market value and applications. *In:* Sayyed, R.Z. (Eds), *Biosurfactants: Production and Applications in Bioremediation/Reclamation*. 19-34. CRC Press, Taylor & Francis Group, USA.

Saranraj, P., Sayyed, R.Z., Kokila, M., Sudha, A., Sivasakthivelan, P., Devi, M.D. et al. (2022b). Plant growth-promoting and biocontrol metabolites produced by Endophytic *Pseudomonas fluorescence*. *In:* Secondary Metabolites and Volatiles of PGPR in Plant-growth Promotion (pp. 349-381). Cham: Springer International Publishing.

Shah, I., Hamid, B., Zaman, M., Fatima, S., Farooq, S., Datta, R. et al. (2022). Microbial biosurfactants: An eco-friendly approach for bioremediation of contaminated environments. *In:* Sayyed, R.Z. (Eds), *Biosurfactants: Production and Applications in Bioremediation/ Reclamation*. 197-207. CRC Press, Taylor & Francis Group, USA.

Shao, Y., Wang, Y., Xu, X., Wu, X., Jiang, Z., He, S. & Qian, K. (2014). Occurrence and source apportionment of PAHs in highly vulnerable karst system. *Science of the Total Environment*, 490: 153-160.

Silva, A.B., Bastos, A.S., Justino, C.I., da Costa, J.P., Duarte, A.C. & Rocha-Santos, T.A. (2018). Microplastics in the environment: Challenges in analytical chemistry – A review. *Analytica Chimica Acta*, 1017: 1-19.

Singh, P., Patil, Y. & Rale, V. (2019). Biosurfactant production: Emerging trends and promising strategies. *J. Appl. Microbiol.*, 126(1): 2-13.

Solaiman, D., Ashby, R., Birbir, M.E.R.A.L. & Caglayan, P. (2016). Antibacterial activity of sophorolipids produced by *Candida bombicola* gram-positive and gram-negative bacteria isolated from salted hides. *J. Am. Leather Chem. Assoc.*, 111(10): 358-364.

Subramaniam, M.D., Venkatesan, D., Iyer, M., Subbarayan, S., Govindasami, V., Roy, A. et al. (2020). Biosurfactants and anti-inflammatory activity: A potential new approach towards COVID-19. *Curr. Opin. Environ. Sci. Health*, 17: 72-81.

Sutyak, K.E., Wirawan, R.E., Aroutcheva, A.A. & Chiindas, M.L. (2008). Isolation of the *Bacillus subtilis* antimicrobial peptide subtilosin from the dairy product derived *Bacillus amyloliquefaciens*. *J. Appl. Microbiol.*, 104: 1067-1074.

Thavasi, R., Subramanyam Nambaru, V.R.M., Jayalakshmi, S., Balasubramanian, T. & Banat, I.M. (2011). Biosurfactant production by *Pseudomonas aeruginosa* from renewable resources. *Indian J. Microbiol.*, 51: 30-36.

Tripathi, M. & Garg, S.K. (2014). Dechlorination of chloro-organics, decolorization, and simultaneous bioremediation of Cr 6+ from real tannery effluent employing indigenous *Bacillus cereus* isolate. *Environ. Sci. Pollut. Res.*, 21: 5227-5241.

Uzoigwe, C., Burgess, J.G., Ennis, C.J. & Rahman, P.K. (2015). Bioemulsifiers are not biosurfactants and require different screening approaches. *Front. Microbiol.*, 6: 245.

Worakitkanchanakul, W., Imura, T., Fukuoka, T., Morita, T., Sakai, H., Abe, M. et al. (2008). Aqueous-phase behavior and vesicle formation of natural glycolipid biosurfactant, mannosylerythritol lipid-B. *Colloids Surf. B. Biointerfaces*, 65(1): 106-112.

Worakitkanchanakul, W., Imura, T., Morita, T., Fukuoka, T., Sakai, H., Abe, M. et al. (2008). Formation of W/O microemulsion based on natural glycolipid biosurfactant, mannosylerythritol Lipid-A. *J. Oleo Sci.*, 57(1): 55-59.

Wu, Q., Zhi, Y. & Xu, Y. (2019). Systematically engineering the biosynthesis of a green biosurfactant surfactin by *Bacillus subtilis* 168. *Metab. Eng.*, 52: 87-97.

Wu, Y.S., Ngai, S.C., Goh, B.H., Chan, K.G., Lee, L.H., & Chuah, L.H. (2017). Anticancer activities of surfactin and potential application of nanotechnology assisted surfactin delivery. *Front. Pharmacol.*, 8: 761.

Xie, Y., Ye, R. & Liu, H. (2007). Microstructure studies on biosurfactant-rhamnolipid/n-butanol/water/n-heptane microemulsion system. *Colloids Surf. A. Physicochem. Eng. Asp.*, 292(2-3): 189-195.

Zakharova, L.Y., Pashirova, T.N., Doktorovova, S., Fernandes, A.R., Sanchez-Lopez, E., Silva, A.M. et al. (2019). Cationic surfactants: Self-assembly, structure-activity correlation and their biological applications. *Inter. J. Mol. Sci.*, 20(22): 5534.

Zaman, M., Hassan, S., Fatima, S., Hamid, B., Farooq, S., Qayoom, I. et al. (2022). Biosurfactants production and applications in food. *In:* Sayyed, R.Z. and Enshasy, H.E. (Eds), *Biosurfactants: Production and Applications in Food and Agriculture.* Vol. II, 225-241. CRC Press, Taylor & Francis Group, USA.

CHAPTER

11

Biosurfactant Nanoparticles and Drug Delivery

Poonam Kumari[1,3*], Shilpa Kumari[2], Madan Lal[3], and Sushma Sharma[3]

[1] Department of Physics, ACBS, Eternal University, Baru Sahib, Sirmour, India

[2] IEC University Baddi, Pinjore-Nalagarh Highway, Solan, India

[3] Department of Plant Pathology, Dr. Khem Singh Gill Akal College of Agriculture, Eternal University, Baru Sahib, India

1. Introduction

Modern medicine places a high priority on drug distribution since it affects how efficiently and effectively therapeutic interventions work. Due to their potential to solve a number of medication delivery problems, nanoparticles have attracted a lot of attention as prospective drug carriers. On the other hand, biosurfactants are organic, biocompatible compounds having surface-active characteristics (Galabova et al., 2014, Fracchia et al., 2015). A class of naturally occurring surface-active substances known as biosurfactants is created by various microbial, plant, and animal species (Sobrinho et al., 2013). Based on chemical structure, they were divided into two categories, such as Minimal Molecular Weight Biosurfactants: Lipids, glycolipids, or lipopeptides make up biosurfactants with relatively low molecular weights. Rhamnolipids, sophorolipids, and trehalolipids are a few examples; and Excellent Molecular Weight Larger molecules known as biosurfactants are occasionally made up of proteins, lipoproteins, or polysaccharides. Some lipoproteins and glycoproteins, for instance, can have surfactant characteristics (Benhur et al., 2020, Bagheri et al., 2022, Adu et al., 2023, Kaur et al., 2023). Biosurfactants are classified according to a number of criteria, such as their chemical make-up, place of origin, kind, purpose, effect on the environment, and manufacturing hosts. These various natural substances offer environmentally benign alternatives to synthetic surfactants and have a wide range of industrial applications (Ali et al., 2022). This chapter examines the integration of biosurfactants and nanoparticles to improve therapeutic results and address medication delivery issues. Biosurfactant nanoparticles have special benefits for drug

* Corresponding author: punamnisha8789@gmail.com; p19phy@gmail.com

delivery, such as better drug loading abilities, improved biocompatibility, and lower toxicity (Vecino et al., 2017, Ali et al., 2022). The advantages of using biosurfactant nanoparticles for drug delivery applications are discussed in this chapter.

An overview of biosurfactants, their characteristics, and prospective uses in medication delivery is given in this chapter. It covers drug loading and release processes, diverse drug delivery applications, safety issues, and future prospects in this developing subject. It also includes the production and characterization of biosurfactant nanoparticles.

2. Biosurfactants: Nature and Properties

2.1 Nature of Biosurfactants

Biosurfactants are naturally occurring chemicals that are frequently created by microorganisms like bacteria, yeast, and fungi (Roy, 2017, Karnwal et al., 2023). However, they are also present in certain higher organisms, such as humans and some plants and animals. This section identifies and categorizes biosurfactants according to their chemical makeup and place of origin. The topic of biosurfactants generated from various natural sources, such as microbial, plant-based, and marine sources, is covered (Patel et al., 2022, Ravinder et al., 2022). Based on their chemical makeup, biosurfactants can be divided into a number of classes and have a varied range of structural characteristics (Sani et al., 2023). The most prevalent classes of biosurfactants include fatty acids, glycolipids, lipopeptides, and phospholipids (Shekhar et al., 2015, Vijayakumar et al., 2015, Kumari et al., 2023). The high biodegradability of biosurfactants is one of their main benefits. Biosurfactants can be degraded by natural processes, making them more environmentally benign than synthetic surfactants, which can linger in the environment and pollute it.

2.2 Properties of Biosurfactants

Because of their surface-active characteristics, biosurfactants can help liquids spread out across solid surfaces by lowering their surface tension. In processes like emulsification, foam generation, and wetting, this feature is crucial. Emulsification is the process of creating and maintaining emulsions, which are mixes of immiscible liquids like oil and water (Drakontis et al., 2020). Biosurfactants are efficient in this process (Luft et al., 2020, Adetunji et al., 2021, da Silva et al., 2021). The food, cosmetics, and pharmaceutical industries are just a few that value this feature. Stable foams made by Products like detergents, firefighting foam, and cosmetics can benefit from the usage of biosurfactants (Bilai et al., 2022, Ravinder et al., 2022). Biosurfactants are suitable for usage in pharmaceutical, cosmetic, and food applications since they are typically regarded as safe and biocompatible. Some biosurfactants have antimicrobial qualities that can be used to their advantage as antimicrobial agents, especially in pharmaceutical and medical applications. Biosurfactants are less damaging to the environment than synthetic surfactants because of their biodegradability. They can aid in reducing how damaging industrial operations are to the environment (Ishfaq et al., 2022, Saranraj et al., 2022). Biosurfactants are used in a variety of industries, including food and beverages, cosmetics, medicines, agriculture, bioremediation, petroleum, and more (Nikolova et al., 2021). They can improve a variety of processes' productivity and long-term viability. Biosurfactants are surface-active, naturally occurring compounds with a variety of characteristics and possible uses. They are appealing for a variety of industrial and environmental reasons due to their biodegradability, biocompatibility, and other functional qualities, and continuing research is working to maximize their potential and enhance their manufacturing procedures.

Listed below are a few typical sources of biosurfactants:

Bacteria: It is known that a wide variety of bacteria make biosurfactants. *Pseudomonas, Bacillus, Clostridium*, and *Lactobacillus* are a few common bacterial genera. The human body, water, soil, and other surroundings all contain these bacteria.

Yeasts: Some yeast species, like Candida, are able to produce biosurfactants (Gayathiri et al., 2022). Due to their ease of growing, yeasts are frequently utilized in biotechnological processes to produce biosurfactants.

Fungi: It is also known that filamentous fungi like Aspergillus and Trichoderma create biosurfactants (Sanches et al., 2021). These fungi are frequently isolated from plant matter and soil.

Plants: Natural surfactants known as saponins are produced by some plants. For instance, the fruit of the soapberry tree (Sapindus), which generates saponins, can be utilized to make natural surfactants (Yekeen et al., 2020).

Animals: As a component of their defense mechanisms, several animals, especially insects, create biosurfactants. For instance, ants and some bug species manufacture biosurfactants to fend off predators.

Debris Streams: The waste products and leftovers of different industrial operations can also be used to make biosurfactants. For instance, when they are put through the proper microbial fermentation processes, agricultural waste, waste from food processing, and waste from oil refineries can all serve as sources of biosurfactants.

Depending on the originating organism and the conditions under which it is grown, the precise type of biosurfactant and its qualities can differ significantly. To make biosurfactants more economically and environmentally sustainable, researchers are always looking into new sources and improving production techniques.

Although biosurfactants have many benefits, it's crucial to remember that they can be produced less cheaply and effectively than chemical surfactants on a big scale. However, it is anticipated that current research and technological advancements will address these issues and broaden the applicability of biosurfactants across numerous industries.

3. Nanoparticles in Drug Delivery

The use of nanoparticles in drug delivery has become a potential strategy to increase the potency and precision of pharmaceuticals. These minuscule particles, which are typically between 1 and 100 nanometers in size, can be designed to encapsulate, target, and release medications in a precise and regulated manner.

Key benefits and features of employing nanoparticles in medicine administration: Immunotherapeutic drugs, such as vaccinations and immunomodulators, can be delivered using nanoparticles to improve the body's immune response against diseases like cancer (Liu et al., 2019).

For example, polymers, lipids, and metals can be used to create nanoparticles with specific features including controlled release kinetics, stability, and biocompatibility (Kumar et al., 2019).

Drug delivery using nanoparticles: production challenges, potential toxicity issues, and legal issues. In order to improve the efficacy and safety of therapies for a variety of medical illnesses, this chapter will address these concerns and broaden the applications of nanotechnology in drug delivery.

3.1 Types of Nanoparticles Used in Drug Delivery

Drug delivery nanoparticles are created to enclose, safeguard, and transport therapeutic substances to certain target areas within the body. Different materials can be used to create these nanoparticles, and each one has distinct advantages and features. Following are some typical kinds of nanoparticles utilized in medicine delivery as shown in Fig. 1.

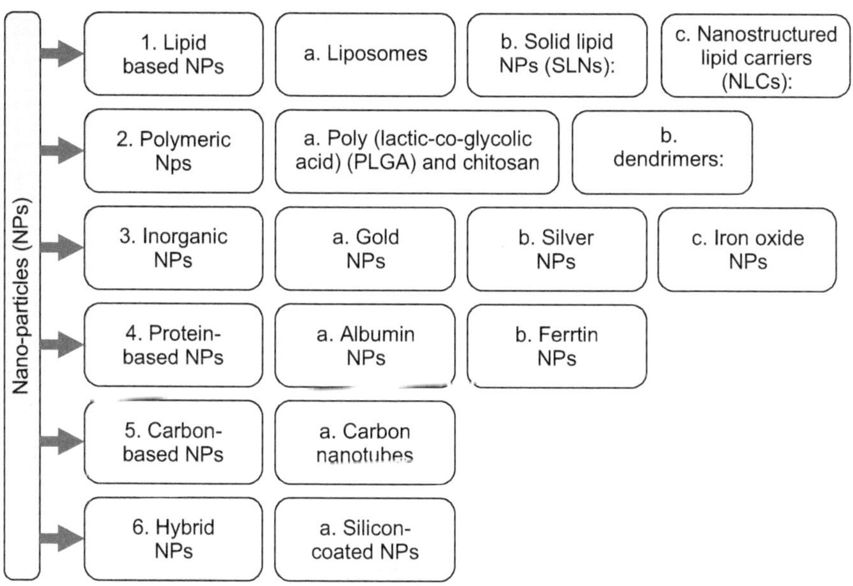

Fig. 1: Some typical kinds of nanoparticles utilized in drug delivery.

Lipid-based nanoparticles: Lipid bilayers make up the spherical vesicles known as liposomes. They are flexible drug carriers because they can encapsulate both hydrophobic and hydrophilic medications. Liposomes have been used in numerous drug delivery applications because they are biocompatible (García-Pinel et al., 2019).

3.1.1 Solid Lipid Nanoparticles (SLNs)

Lipid nanoparticles called SLNs have a strong lipid core. Compared to liposomes, they provide regulated medication release and enhanced stability (Duan et al., 2020).

3.1.2 Nanostructured Lipid Carriers (NLCs)

An adaptation of SLNs called NLCs has a mixture of solid and liquid lipids. Greater drug loading and better drug release control are made possible by this combination (Salvi et al., 2019).

3.1.3 Iron Oxide Nanoparticles

Iron oxide nanoparticles are magnetic and can be guided to specific locations within the body using external magnetic fields. They are used in drug delivery and Magnetic Resonance Imaging (MRI) (Dadfar et al., 2019).

3.1.4 Dendrimers

Dendrimers are highly branched, three-dimensional polymers that can encapsulate drugs within their structure. They have a well-defined structure and can be customized for drug delivery applications (Nikzamir et al., 2021).

3.1.5 Gold Nanoparticles

Gold nanoparticles have unique optical properties and can be functionalized for drug delivery and imaging. They are often used in cancer therapy and diagnostics (Bai et al., 2020).

3.1.6 Hybrid Nanoparticles

Silica-coated nanoparticles: Silica-coated nanoparticles are hybrid nanoparticles often referred to as "core-shell" nanoparticles (Hanske et al., 2020).

3.1.7 Carbon-Based Nanoparticles

Carbon nanotubes: Carbon nanotubes are cylindrical carbon structures that can be used to transport drugs and imaging agents. They have unique mechanical and electrical properties (Khanna, 2019).

3.1.8 Silver Nanoparticles

Silver nanoparticles exhibit antimicrobial properties and can be used for localized drug delivery in wound healing and infection control. Polymeric Nanoparticles: These nanoparticles are typically made from biodegradable polymers such as poly (lactic-co-glycolic acid) (PLGA) and chitosan. They can be engineered for sustained drug release and are suitable for various drug types (Renu et al., 2020).

These are only a few examples of the various kinds of nanoparticles that are employed in the delivery of drugs. The exploration of biosurfactant nanoparticles as novel materials and design approaches to produce nanoparticles suited to certain drug delivery difficulties and therapeutic objectives continue in this chapter. The medicine being given, the target tissue or organ, and the intended release kinetics are only a few examples of the variables that influence the choice of nanoparticle type.

One of the most important aspects of healthcare and pharmaceuticals is medication distribution, which presents a number of difficulties for researchers and medical experts. These difficulties are: Specific delivery, stability of a drug bioavailability, medicine solubility, resistance to drugs regulated release, biocompatibility and patient adherence. Through the development of novel drug delivery technologies including nanoparticles, liposomes, microneedles, and tailored delivery systems, researchers and pharmaceutical companies are constantly attempting to overcome these problems.

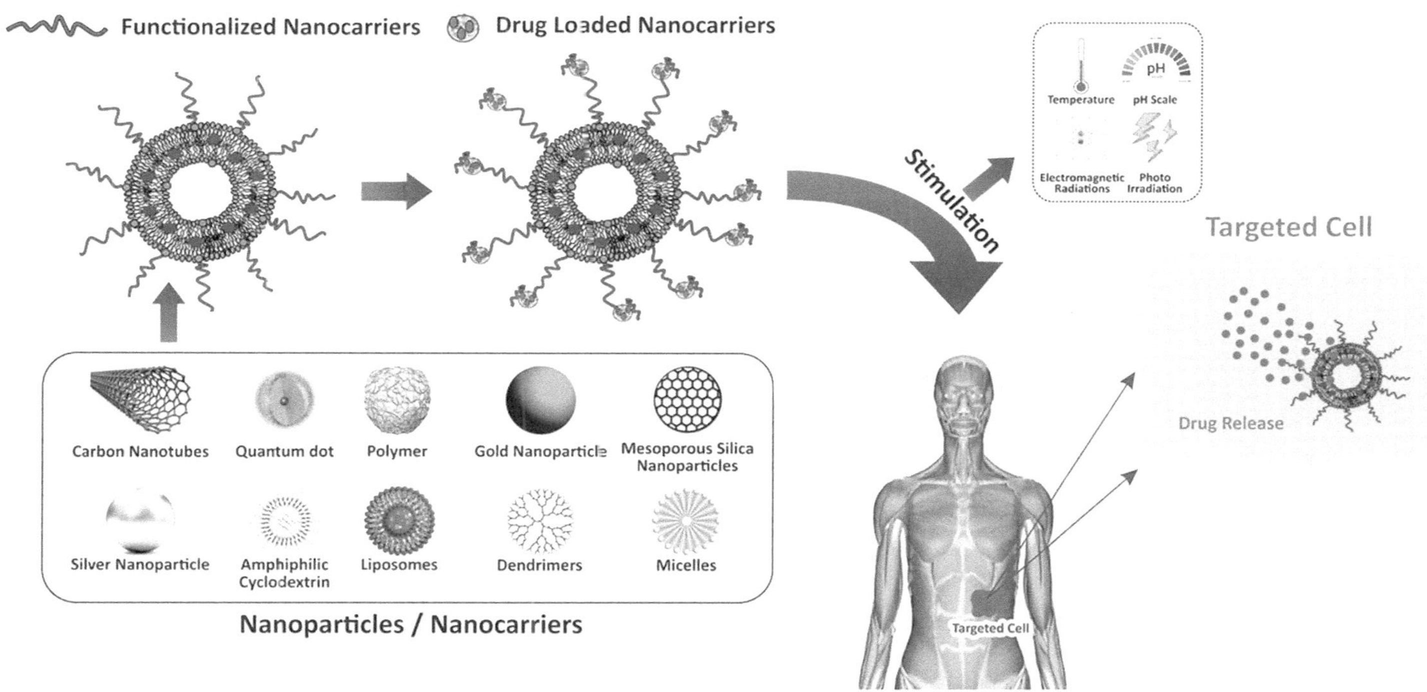

Fig. 2: Nanoparticles used in drug delivery are designed to encapsulate, protect, and deliver therapeutic agents to specific target sites within the body (Yetisgin et al., 2020).

For drug efficacy and safety to improve, as well as for patient care in general, these barriers must be removed. Nanoparticles for targeted drug delivery represented by Fig. 2.

4. Biosurfactant Nanoparticles

Nanoparticles that have been coated or stabilized with biosurfactants are referred to as biosurfactant nanoparticles. Microorganisms, such as bacteria, fungus, and yeast, create chemicals called biosurfactants that are surface-active. They are helpful in a variety of applications, including the stabilization of nanoparticles, because they can lower the surface tension of liquids. An overview of biosurfactant nanoparticles is provided Table 1 given below.

Table 1: Types of biosurfactant nanoparticles and their applications in drug delivery

Lipid-Based Nanoparticles	Nanoparticles' high surface energy can cause aggregation and settling, but biosurfactants can reduce this tendency and enhance their stability in suspension (Christopher et al., 2019).
Biosurfactant Sources	Biosurfactants, derived from natural sources like bacteria, yeast, and fungi, are considered more environmentally friendly than synthetic alternatives (Jahan et al., 2020).
Biocompatibility	Biosurfactants, being biocompatible and less toxic than synthetic surfactants, are ideal for biomedical and pharmaceutical applications, including drug delivery systems, imaging agents, and diagnostics (Çelik et al., 2020).
Environmental Applications	Biosurfactant nanoparticles are utilized in environmental remediation processes for efficient removal of pollutants from water and soil due to their ability to solubilize hydrophobic compounds (Rasheed et al., 2020).
Enhanced Oil Recovery	Biosurfactants are utilized in the oil and gas industry for enhanced oil recovery (EOR), by reducing oil-water interfacial tension and facilitating oil displacement (Wu et al., 2022).
Food and Cosmetics:	Biosurfactants and nanoparticles are utilized in the food and cosmetic industries as emulsifying agents, foaming agents, and stabilizers in various products (Alara et al., 2023).
Biodegradability	Biosurfactants, due to their biodegradability, have a lower environmental impact compared to synthetic surfactants (Johnson et al., 2021).

4.1 Methods of Biosurfactant Nanoparticle Fabrication

In order to develop stable and useful nanostructures, biosurfactants are combined with nanoparticles during the synthesis of biosurfactant nanoparticles. Depending on the particular biosurfactant and nanoparticle of interest, various techniques can be used to accomplish this. Here are a few typical procedures for creating biosurfactant nanoparticles as shown in Fig. 3.

Biosurfactants and nanoparticle precursors are combined in a solution for the co-precipitation process. In order to assist the precipitation of nanoparticles in the

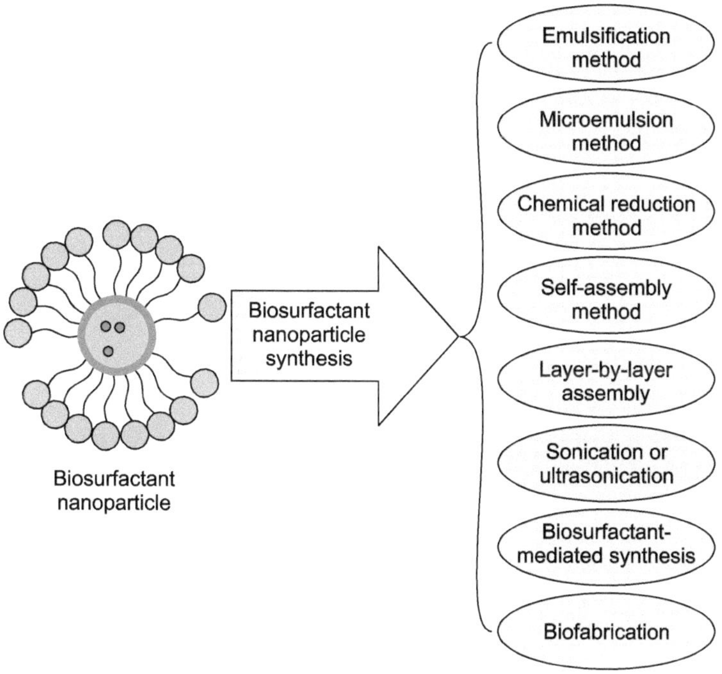

Fig. 3: Typical synthesis methods for biosurfactant nanoparticles development.

presence of biosurfactants, appropriate parameters (e.g., pH, temperature) are altered. During the creation of the nanoparticles, the biosurfactants adsorb onto their surfaces, stabilizing them (Johnson et al., 2021).

4.1.1 Emulsification Method

Biosurfactants are combined with an aqueous phase containing nanoparticle precursors and a nonpolar organic phase (such as oil). In order to produce a stable emulsion, vigorous mechanical agitation or sonication is used. The biosurfactant molecules bind the nanoparticles as they form within the emulsion, preventing aggregation (Rangel-Muñoz et al., 2020).

4.1.2 Microemulsion Method

This method involves the use of a microemulsion system, typically consisting of water, oil, surfactant (biosurfactant), and co-surfactant (Ohadi et al., 2020). Nanoparticle precursors are introduced into the microemulsion, and under suitable conditions (e.g., temperature), nanoparticles nucleate and grow in the confined spaces of the microemulsion droplets.

4.1.3 Chemical Reduction Method

Metal nanoparticles are created using biosurfactants as reducing agents. When biosurfactants are combined with metal ion precursors, the metal ions are reduced to create nanoparticles while being stabilized (Christopher et al., 2019).

4.1.4 Self-Assembly Method

It is possible for some biosurfactants to self-assemble into micelles or other structures. A solution of biosurfactant micelles can be supplemented with nanoparticles, which causes the nanoparticles to become encapsulated within the micelles (Vellu et al., 2022).

4.1.5 Layer-by-Layer Assembly

In this technique, layers of biosurfactant and polymers or nanoparticles with opposing charges are alternately placed on a substrate. The nanoparticle layers can be functionalized and stabilized with assistance from the biosurfactant layers (Santos et al., 2019).

4.1.6 Sonication or Ultrasonication

Nanoparticle suspensions may be combined with biosurfactants, and the nanoparticles may then be ultrasonically dispersed and coated with biosurfactant molecules (Liao et al., 2021).

4.1.7 Biosurfactant-Mediated Synthesis

Certain biosurfactants are useful for the creation of specific kinds of nanoparticles because they possess reducing and stabilizing capabilities. For instance, due to their capacity for reduction, rhamnolipids have been employed to create silver nanoparticles (Elakkiya et al., 2020).

4.1.8 Biofabrication

In some circumstances, nanoparticle production by microorganisms that naturally create biosurfactants is possible. The culture medium can then be used to collect these nanoparticles (Iravani et al., 2020).

The particular biosurfactant, nanoparticle material, and desired features of the biosurfactant nanoparticles all influence the approach selection. Each technique has benefits and drawbacks, and scientists choose the best one based on the application and properties they want their finished nanoparticles to have.

5. Characterization Techniques of Biosurfactant Nanoparticles

It is critical to maintain the stability, size, shape, surface characteristics, and drug-loading capability of biosurfactant nanoparticles utilized in drug delivery. Here are a few typical techniques for characterization as shown in Fig 4.

5.1 Dynamic Light Scattering (DLS)

DLS is used to measure the size distribution and zeta potential of biosurfactant nanoparticles. It provides information about the average particle size and their dispersion in a solution (Tosi et al., 2020).

5.2 Transmission Electron Microscopy (TEM)

TEM allows for high-resolution imaging of nanoparticles, providing information

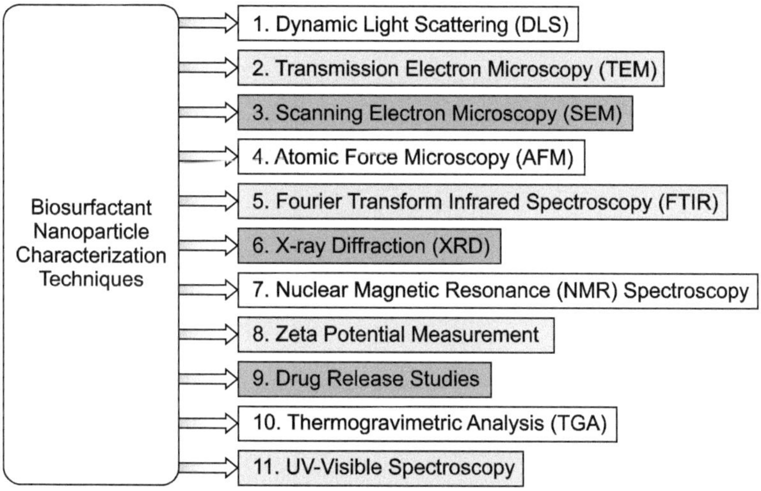

Fig. 4: Characterization techniques of biosurfactant nanoparticles utilized in drug delivery.

about their size, shape, and morphology at the nanoscale. It helps in visualizing the internal structure of the nanoparticles (Lin et al., 2021).

5.3 Scanning Electron Microscopy (SEM)

SEM is another imaging technique that provides surface morphology information. It can complement TEM by giving a different perspective on the external characteristics of the nanoparticles (Liu et al., 2020).

5.4 Atomic Force Microscopy (AFM)

AFM is used to study the surface topography and mechanical properties of biosurfactant nanoparticles. It can provide information on surface roughness and particle height (Magazzù et al., 2023).

5.5 Fourier Transform Infrared Spectroscopy (FTIR)

FTIR is used to identify the functional groups present in biosurfactant molecules and their interactions with drug molecules. It can help confirm the encapsulation of drugs within the nanoparticles (Zarnowiec et al., 2015).

5.6 X-ray Diffraction (XRD)

XRD is used to determine the crystallinity of biosurfactant nanoparticles. It can indicate whether the particles are amorphous or crystalline in nature (Falsafi et al., 2020).

5.7 Nuclear Magnetic Resonance (NMR) Spectroscopy

NMR can be employed to analyze the chemical structure of biosurfactants and their interactions with drug molecules. It provides insights into the molecular composition and conformation (Bundle et al., 1974).

5.8 UV-Visible Spectroscopy

UV-Visible spectroscopy can be used to study the optical properties of biosurfactant nanoparticles. It can help in quantifying the drug-loading capacity and assessing drug release kinetics (Macheroux, 1999).

5.9 Thermogravimetric Analysis (TGA)

TGA is used to determine the thermal stability and decomposition temperature of biosurfactant nanoparticles. It can also provide information about the percentage of drug encapsulated (Macheroux, 1999).

5.10 Zeta Potential Measurement

Zeta potential measurements can provide information about the surface charge of biosurfactant nanoparticles. It is crucial for understanding their stability and potential for aggregation (Salopek et al., 1992).

5.11 Drug Release Studies

These studies involve assessing the release profile of the drug from the biosurfactant nanoparticles under various conditions, such as pH, temperature, and time. This helps in understanding the drug delivery behavior (Nimesh et al., 2006).

5.12 In vitro and In vivo Biological Assays

To evaluate the efficacy of drug delivery, various biological assays can be conducted, including cell viability assays, cell uptake studies, and animal experiments (Kumar et al., 2017).

5.13 Stability Studies

Long-term stability studies are crucial to assess the storage stability of biosurfactant nanoparticles. These studies monitor changes in particle size, drug release, and other properties over time (Erhayem et al., 2014).

Characterization techniques play a critical role in designing and developing biosurfactant nanoparticles for drug delivery applications, as they provide valuable information on the physical, chemical, and biological properties of these nanoparticles. These insights are essential for optimizing their performance and safety as drug delivery carriers.

6. Physicochemical Properties of Biosurfactant Nanoparticles

Biosurfactant nanoparticles are a type of nanomaterial that combines the unique properties of biosurfactants with the advantages of nanoparticles. Biosurfactants are surface-active molecules produced by microorganisms, plants, and animals. They have gained significant attention due to their potential applications in various fields, including biotechnology, pharmaceuticals, environmental remediation, and nanotechnology. When biosurfactants are formed into nanoparticles, they exhibit specific physicochemical properties that make them useful for various applications. Some of these properties are as follows.

6.1 Surface Activity

Biosurfactants, in their nanoparticle form, retain their surface-active properties. This means they can lower the surface tension of liquids and increase their spreading and wetting capabilities. This property is valuable in emulsification, foaming, and dispersing applications (Płaza et al., 2014).

6.2 Biodegradability

Biosurfactants, being natural products produced by microorganisms, are typically biodegradable. This makes biosurfactant nanoparticles environmentally friendly compared to synthetic surfactants, which can be persistent and harmful (Parthipan et al., 2022).

6.3 Low Toxicity

Biosurfactants are generally considered non-toxic or low in toxicity, which is advantageous for applications in the pharmaceutical and food industries. Biosurfactant nanoparticles can be designed for drug delivery systems with reduced side effects (Kiran et al., 2011).

6.4 Stability and Self-Assembly

Biosurfactant nanoparticles can exhibit excellent stability, which is crucial for their use in various applications. Biosurfactants can self-assemble into nanoparticles due to their amphiphilic nature. This self-assembly behavior is essential for the formation of stable emulsions and can be exploited in drug delivery systems where controlled release of bioactive compounds is required (Baccile et al., 2021).

6.5 Compatibility

Biosurfactant nanoparticles are often compatible with biological systems, making them suitable for applications in medicine and biotechnology. They can be used for drug delivery, targeting specific cells or tissues (Sangeetha et al., 2013).

6.6 Size and Morphology Control

The size and morphology of biosurfactant nanoparticles can be controlled during the synthesis process. This control allows for tailoring the nanoparticles to specific applications, such as drug delivery or nanocarriers (Sangeetha et al., 2013).

6.7 Interfacial Properties

Biosurfactant nanoparticles have unique interfacial properties, which can be harnessed for stabilizing emulsions, improving solubility, and enhancing the bioavailability of poorly water-soluble drugs (Cui et al., 2024).

6.8 Biocompatibility

Biosurfactants are often biocompatible, making them suitable for use in biological and medical applications. They can be used to encapsulate drugs or other bioactive compounds for targeted delivery to cells or tissues (Sangeetha et al., 2013).

6.9 Antimicrobial Properties

Some biosurfactants exhibit antimicrobial properties, which can be useful in applications related to microbial control or infection prevention.

It's important to note that the specific physicochemical properties of biosurfactant nanoparticles can vary depending on the type of biosurfactant used, the method of synthesis, and the intended application. Researchers continue to explore and optimize biosurfactant nanoparticles for a wide range of applications, and their versatility makes them a promising area of study in nanotechnology and biotechnology.

7. Applications of Biosurfactant Nanoparticles in Drug Delivery

Biosurfactant nanoparticles have garnered significant interest in the field of drug delivery due to their unique properties and advantages. These nanoparticles, derived from natural biosurfactants, offer several promising applications in drug delivery systems as shown in Fig. 5.

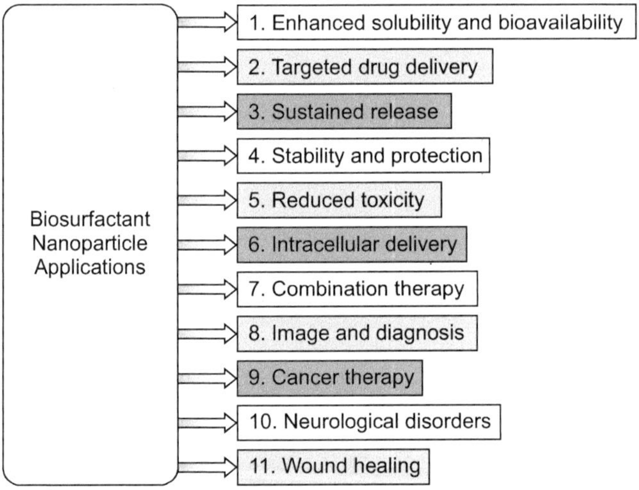

Fig. 5: Applications of biosurfactant nanoparticles in drug delivery.

7.1 Enhanced Solubility and Bioavailability

Biosurfactant nanoparticles can encapsulate hydrophobic drugs, improving their solubility in aqueous environments. This enhanced solubility leads to better bioavailability, ensuring that a larger fraction of the administered drug is absorbed by the body.

7.2 Targeted Drug Delivery

Biosurfactant nanoparticles can be functionalized to carry specific ligands or antibodies on their surfaces. This allows for targeted drug delivery to particular cells or tissues, minimizing off-target effects and reducing the required drug dosage.

Antimicrobial Drug Delivery: Some biosurfactants possess inherent antimicrobial properties, making them suitable for delivering antimicrobial drugs or antimicrobial peptides to treat infections (Bjerk et al., 2021).

7.3 Vaccine Delivery

Biosurfactant nanoparticles have been explored for the delivery of vaccines and immunotherapeutic agents. They can improve the stability and immunogenicity of vaccines, leading to enhanced immune responses (Hussein et al., 2022).

Ocular Drug Delivery: Biosurfactant nanoparticles can be formulated into eye drops or ophthalmic gels to improve drug penetration into the eye and prolong drug retention on the ocular surface(Sangeetha et al., 2013, Nunes et al., 2022).

7.3.1 Sustained Release

By controlling the release kinetics of drugs from biosurfactant nanoparticles, it is possible to achieve sustained drug release over an extended period. This can improve patient compliance and reduce the frequency of drug administration (Katiyar et al., 2019).

7.3.2 Stability and Protection

Biosurfactant nanoparticles can protect sensitive drugs from degradation, such as enzymatic breakdown or chemical degradation. This protection ensures the stability of the drug during storage and transportation (Sultana et al., 2020).

7.3.3 Reduced Toxicity

Biosurfactants are generally biocompatible and less toxic compared to synthetic surfactants. When used as drug carriers, biosurfactant nanoparticles can reduce the potential toxicity associated with certain drug formulations (Bjerk et al., 2021).

7.3.4 Intracellular Delivery

Biosurfactant nanoparticles can facilitate the intracellular delivery of drugs, particularly macromolecules like proteins and nucleic acids. This is crucial for gene therapy and the delivery of biologics (Hussein et al., 2022).

7.4 Combination Therapy

Biosurfactant nanoparticles can carry multiple drugs or therapeutic agents simultaneously. This enables combination therapy, where different drugs with complementary mechanisms of action are delivered together to enhance therapeutic efficacy (Bekeredjian et al., 2005).

7.5 Imaging and Diagnosis

Biosurfactant nanoparticles can be labeled with contrast agents or imaging probes, making them suitable for diagnostic purposes. They can serve as carriers for imaging agents, allowing for simultaneous drug delivery and disease monitoring (Yi et al., 2019).

7.5.1 Cancer Therapy

Biosurfactant nanoparticles can be used for targeted drug delivery in cancer therapy. They can accumulate in tumor tissues through the enhanced permeability and retention (EPR) effect, allowing for localized drug delivery (Hussein et al., 2022).

7.5.2 Neurological Disorders

Biosurfactant nanoparticles can potentially cross the blood-brain barrier (BBB) and deliver drugs to the central nervous system, making them relevant for the treatment of neurological disorders (Nunes et al., 2022).

7.5.3 Wound Healing

Biosurfactant nanoparticles can be employed in wound dressings or ointments to deliver growth factors, antimicrobial agents, or other wound-healing compounds to enhance the healing process (Arif et al., 2021).

Research into biosurfactant nanoparticles for drug delivery continues to evolve, and their versatility and biocompatibility make them a promising platform for developing innovative drug delivery systems with improved therapeutic outcomes and reduced side effects.

8. Biocompatibility and Toxicity Considerations

Biosurfactant nanoparticles have gained significant attention in drug delivery due to their potential to enhance drug solubility, stability, and targeting. However, when considering their application, it's crucial to assess their biocompatibility and toxicity to ensure the safety of the drug delivery system.

Due to its special qualities, such as their biocompatibility, low toxicity, and capacity to self-assemble into nanoscale structures, biosurfactant nanoparticles have drawn a lot of interest in drug delivery applications. However, there may be toxicity issues with biosurfactant nanoparticles in medication delivery systems, just like with any other nanomaterial. To ensure the safe and successful use of these nanoparticles in medical applications, it is imperative to assess these worries. Here are some important things to think about:

8.1 Biocompatibility

Biosurfactants are generally considered more biocompatible than synthetic surfactants, which reduces the risk of acute toxicity reactions. However, the biocompatibility of biosurfactant nanoparticles may still vary depending on the specific type of biosurfactant used and its concentration (Arif et al., 2021, Nitschke et al., 2022).

Nanoparticle Toxicity: The size, shape, and surface properties of nanoparticles, including biosurfactant nanoparticles, can influence their toxicity. Small nanoparticles may have a higher likelihood of crossing biological barriers and interacting with cellular components. Surface modifications can be applied to reduce toxicity and enhance biocompatibility (Katiyar et al., 2020).

8.2 Toxicity of Biosurfactants

Some biosurfactants are produced by microorganisms and can vary in composition and structure. It's important to assess the toxicity of the biosurfactant itself before encapsulating drugs in nanoparticles. This can be done through in vitro and in vivo studies to evaluate their impact on cells and organisms.

Drug Toxicity: In drug delivery applications, it's essential to consider the toxicity of the drug being delivered, as it can have a more significant impact on overall safety than the nanoparticles themselves. Encapsulation within nanoparticles can alter the drug's pharmacokinetics and toxicity profile.

Route of Administration: The route of administration for drug-loaded biosurfactant nanoparticles can influence their toxicity. For example, nanoparticles delivered intravenously may have different toxicity profiles than those administered orally or topically.

Biodegradability: The biodegradability of biosurfactant nanoparticles is an important factor in their safety profile. Nanoparticles that can be broken down and cleared from the body are generally considered safer than those that persist for extended periods.

Long-term Effects: Long-term exposure to biosurfactant nanoparticles may have cumulative effects that need to be studied over extended periods to assess chronic toxicity (Kiran et al., 2011, Płaza et al., 2014, Thakur et al., 2020, Rane et al., 2021).

The physicochemical parameters such as particle size, shape, surface charge and chemistry, composition, and subsequent stability play a crucial role in determining the biocompatibility and toxicity of these nanoparticles as shown in Fig 6. Biosurfactants can be used to synthesize nanoparticles with low toxicity and increased biodegradability.

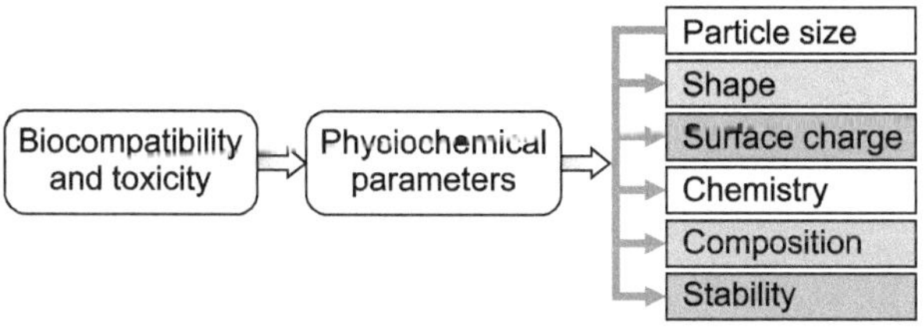

Fig. 6: The biocompatibility and toxicity of nanoparticles are significantly influenced by physicochemical parameters like particle size, shape, surface charge, chemistry, composition, and stability.

Although biosurfactant nanoparticles have several benefits for medication administration, their potential toxicity needs to be thoroughly assessed through rigorous preclinical research. The toxicity of the medicine being delivered, the route of administration, and the long-term consequences should all be considered by researchers in addition to the toxicity of the nanoparticles themselves. This strategy minimizes potential patient harm while ensuring the safe and efficient use of biosurfactant nanoparticles in medication administration applications.

9. Challenges and Future Perspectives

Biosurfactant nanoparticles have gained significant attention in the field of drug delivery due to their unique properties and potential advantages. However, they also come with certain challenges and offer various future perspectives, which are essential to consider for their successful application in drug delivery systems. Here, we will discuss both the challenges and future perspectives of biosurfactant nanoparticles in drug delivery.

9.1 Challenges

While biosurfactants are typically thought of as biocompatible, their safety profiles for usage in drug delivery must be carefully examined. Thorough evaluations of toxicity and potential negative effects on cells and tissues are required.

Manufacturing Cost: Scaling up production for pharmaceutical uses may be difficult and expensive when producing biosurfactants. It is necessary to find methods to optimize the production of biosurfactants while lowering costs.

Biomedical Imaging: Biosurfactant nanoparticles can be engineered to incorporate imaging agents, allowing for simultaneous drug delivery and real-time monitoring of treatment responses

Stability: Biosurfactant nanoparticles may suffer from issues related to stability, such as aggregation or degradation over time. Ensuring their stability during storage and administration is critical.

Controlled Drug Release: Achieving precise control over drug release rates from biosurfactant nanoparticles can be challenging. Developing methods for fine-tuning drug release kinetics is crucial for tailored therapeutic outcomes.

Specific Targeting: Achieving targeted drug delivery to specific cells or tissues is a key goal in drug delivery. Developing targeting strategies that can be incorporated into biosurfactant nanoparticles is a significant challenge.

Limited Drug Loading: Biosurfactant nanoparticles may have limitations in terms of their drug-loading capacity. Strategies to enhance drug loading efficiency without compromising stability and functionality are needed.

Regulatory Approval: Biosurfactants used in drug delivery systems will need to undergo rigorous regulatory evaluation to ensure their safety and efficacy, which can be a time-consuming and costly process.

9.2 Future Perspectives

Improved Formulations: Ongoing research and development activities may result in the development of innovative biosurfactant-based formulations with improved stability, controlled-release features, and drug-loading capabilities.

Advanced Analytical Techniques: By developing advanced analytical techniques, it will be possible to characterize biosurfactant nanoparticles more thoroughly and better comprehend their structural and functional characteristics.

Combination Therapies: Biosurfactant nanoparticles can be explored for combination therapies, where multiple drugs or therapeutic agents are delivered simultaneously for synergistic effects.

Targeted Drug Delivery: Advancements in targeting ligands and surface modifications can enhance the specificity of biosurfactant nanoparticles, allowing for targeted drug delivery to specific cells or tissues.

Personalized Medicine: The development of biosurfactant-based drug delivery systems that can be tailored to individual patient needs based on their genetics or disease profiles holds significant promise for personalized medicine.

Environmental Sustainability: Research into environmentally friendly production methods for biosurfactants can contribute to their sustainable use in drug delivery and reduce their ecological impact.

Biosurfactant nanoparticles offer exciting prospects for drug delivery and they also face challenges that need to be addressed through ongoing research and development. Overcoming these challenges and capitalizing on the future perspectives will enable the effective utilization of biosurfactant nanoparticles in improving drug delivery systems, ultimately benefitting patient outcomes.

10. Conclusion

With many benefits over conventional drug delivery methods, biosurfactant nanoparticles hold enormous potential for the future of drug delivery. Here is a summary of their potential advantages.

Improved Drug Solubility: Biosurfactant nanoparticles can make medications that aren't very soluble in water more soluble, boosting their bioavailability and therapeutic effectiveness. This is crucial because it enables lower doses and fewer negative effects for medications with poor wter solubility.

Targeted Medication Delivery: By functionalizing these nanoparticles to target particular cells or tissues, it is possible to reduce side effects and increase medication delivery to the site of action. This makes personalized and precise medical techniques possible.

Biocompatibility and Reduced Toxicity: Biosurfactants are generally biocompatible and biodegradable, reducing the risk of toxicity and adverse reactions compared to synthetic surfactants and delivery systems.

Sustained Release: Biosurfactant nanoparticles can provide controlled and sustained drug release, maintaining therapeutic levels over an extended period. This can enhance patient compliance and reduce the frequency of drug administration.

Reduced Immunogenicity: Biosurfactants derived from natural sources are less likely to trigger immune responses in the body, making them suitable for long-term drug delivery applications.

Eco-Friendly and Sustainable: The production of biosurfactants often involves renewable resources and is environmentally friendly, aligning with the growing demand for sustainable healthcare technologies.

Possibility of Combination Therapies: Biosurfactant nanoparticles can encapsulate a variety of medications or therapeutic agents, opening the door to combination therapies that simultaneously target several facets of a disease. Despite the enormous potential of biosurfactant nanoparticles, there are still obstacles to be overcome, including scalability, stability, and the creation of standardized production techniques. They must also undergo extensive testing and receive regulatory approval in order to guarantee their safety and effectiveness in clinical settings.

In conclusion, biosurfactant nanoparticles are a promising drug delivery method that have the potential to revolutionize the industry by making drugs more soluble, allowing them to be delivered to precise areas, and lowering their toxicity. For these technologies to reach their full potential and provide patients with cutting-edge drug delivery options, more research and development is required in this field.

References

Adetunji, A.I. & Olaniran, A.O. (2021). Production and potential biotechnological applications of microbial surfactants: An overview. *Saudi Journal of Biological Sciences*, 28(1): 669-679. https://doi.org/10.1016/j.sjbs.2020.10.058

Adu, S.A., Twigg, M.S., Naughton, P.J., Marchant, R. & Banat, I.M. (2023). Glycolipid biosurfactants in skincare applications: challenges and recommendations for future exploitation. *Molecules*, 28(11): 4463. https://doi.org/10.3390/molecules28114463

Alara, O.R., Abdurahman, N.H., Alara, J.A., Ukaegbu, C.I., Tade, M.O. & Ali, H.A. (2023). Biosurfactants as Emulsifying Agents in Food Formulation. Advancements in Biosurfactants Research. 157-170. Springer. https://doi.org/10.1007/978-3-031-21682-4_8

Ali, S.A.M., Sayyed, R., Mir, M.I., Khan, M., Hameeda, B. et al. (2022). Induction of systemic resistance in maize and antibiofilm activity of surfactin from Bacillus velezensis MS20. *Frontiers in Microbiology*, 13: 879739. https://doi.org/10.3389/fmicb.2022.879739

Ali, S.A.M., Sayyed, R., Reddy, M., El Enshasy, H. & Hameeda, B. (2022). Delving through quorum sensing and CRISPRi strategies for enhanced surfactin production. *Microbial Surfactants*, 59-79. CRC Press. 10.1201/9781003260165-4

Arif, M., Sharaf, M., Samreen, Khan, S., Chi, Z. & Liu, C.-G. (2021). Chitosan-based nanoparticles as delivery-carriers for promising antimicrobial glycolipid biosurfactants to improve the eradication rate of Helicobacter pylori biofilm. *Journal of Biomaterials Science, Polymer Edition*, 32(6): 813-832. https://doi.org/10.1080/09205063.2020.1870323

Baccile, N., Seyrig, C., Poirier, A., Alonso-de Castro, S., Roelants, S.L. & Abel, S. (2021). Self-assembly, interfacial properties, interactions with macromolecules and molecular modelling and simulation of microbial bio-based amphiphiles (biosurfactants). A tutorial review. *Green Chemistry*, 23(11): 3842-3944. https://doi.org/10.1039/D1GC00097G

Bagheri, H., Mohebbi, A., Amani, F.S. & Naderi, M. (2022). Application of low molecular weight and high molecular weight biosurfactant in medicine/biomedical/pharmaceutical industries. *Green Sustainable Process for Chemical and Environmental Engineering and Science*, 1-60. Elsevier. https://doi.org/10.1016/B978-0-323-85146-6.00027-9

Bai, X., Wang, Y., Song, Z., Feng, Y., Chen, Y., Zhang, D. & Feng, L. (2020). The basic properties of gold nanoparticles and their applications in tumor diagnosis and treatment. *International Journal of Molecular Sciences*, 21(7): 2480. https://doi.org/10.3390/ijms21072480

Bekeredjian, R., Grayburn, P.A. & Shohet, R.V. (2005). Use of ultrasound contrast agents for gene or drug delivery in cardiovascular medicine. *Journal of the American College of Cardiology*, 45(3): 329-335. 10.1016/j.jacc.2004.08.067

Benhur, A.M., Pingali, S. and Amin, S. (2020). Application of biosurfactants and biopolymers in sustainable cosmetic formulation design. *Journal of Cosmetic Science*, 71(6).

Bilai, M., Khan, M.R. and Sayyed, R. (2022). Biosurfactant mediated synthesis and stabilization of nanoparticles. *Microbial Surfactants: Volume 3: Applications in Environmental Reclamation and Bioremediation.* 1957.

Bjerk, T.R., Severino, P., Jain, S., Marques, C., Silva, A.M., Pashirova, T. et al. (2021). Biosurfactants: Properties and applications in drug delivery, biotechnology and ecotoxicology. *Bioengineering*, 8(8): 115. https://doi.org/10.3390/bioengineering8080115

Bundle, D.R., Smith, I.C. and Jennings, H.J. (1974). Determination of the structure and conformation of bacterial polysaccharides by carbon 13 nuclear magnetic resonance: Studies on the group-specific antigens of neisseria meningitidis serogroups a and x. *Journal of Biological Chemistry*, 249(7): 2275-2281. https://doi.org/10.1016/S0021-9258(19)42828-7

Çelik, P.A., Manga, E.B., Çabuk, A. and Banat, I.M. (2020). Biosurfactants' potential role in combating COVID-19 and similar future microbial threats. *Applied Sciences*, 11(1): 334. https://doi.org/10.3390/app11010334

Christopher, F.C., Ponnusamy, S.K., Ganesan, J.J. & Ramamurthy, R. (2019). Investigating the prospects of bacterial biosurfactants for metal nanoparticle synthesis – A comprehensive review. *IET Nanobiotechnology*, 13(3): 243-249. https://doi.org/10.1049/iet-nbt.2018.5184

Cui, S., McClements, D.J., He, X., Xu, X., Tan, F., Yang, D. et al. (2024). Interfacial properties and structure of Pickering emulsions co-stabilized by different charge emulsifiers and zein nanoparticles. *Food Hydrocolloids*, 146: 109285. https://doi.org/10.1016/j.foodhyd.2023.109285

da Silva, A.F., Banat, I.M., Giachini, A.J. and Robl, D. (2021). Fungal biosurfactants, from nature to biotechnological product: Bioprospection, production and potential applications. *Bioprocess and Biosystems Engineering*, 44(10): 2003-2034. https://doi.org/10.1007/s00449-021-02597-5

Dadfar, S.M., Roemhild, K., Drude, N.I., von Stillfried, S., Knüchel, R., Kiessling, F. et al. (2019). Iron oxide nanoparticles: Diagnostic, therapeutic and theranostic applications. *Advanced Drug Delivery Reviews*, 138: 302-325. https://doi.org/10.1016/j.addr.2019.01.005

Drakontis, C.E. and Amin, S. (2020). Biosurfactants: Formulations, properties, and applications. *Current Opinion in Colloid & Interface Science*, 48: 77-90. https://doi.org/10.1016/j.cocis.2020.03.013

Duan, Y., Dhar, A., Patel, C., Khimani, M., Neogi, S., Sharma, P. et al. (2020). A brief review on solid lipid nanoparticles: Part and parcel of contemporary drug delivery systems. *RSC Advances*, 10(45): 26777-26791. https://doi.org/10.1039/D0RA03491F

Elakkiya, V.T., Suresh Kumar, P., Alharbi, N.S., Kadaikunnan, S., Khaled, J.M. & Govindarajan, M. (2020). Swift production of rhamnolipid biosurfactant, biopolymer and synthesis of biosurfactant-wrapped silver nanoparticles and its enhanced oil recovery. *Saudi Journal of Biological Sciences*, 27(7): 1892-1899. https://doi.org/10.1016/j.sjbs.2020.04.001

Erhayem, M. & Sohn, M. (2014). Stability studies for titanium dioxide nanoparticles upon adsorption of Suwannee River humic and fulvic acids and natural organic matter. *Science of the Total Environment*, 468: 249-257. https://doi.org/10.1016/j.scitotenv.2013.08.038

Falsafi, S.R., Rostamabadi, H. and Jafari, S.M. (2020). X-ray diffraction (XRD) of nanoencapsulated food ingredients. *Characterization of Nanoencapsulated Food Ingredients*, 271-293. Elsevier. https://doi.org/10.1016/B978-0-12-815667-4.00009-2

Fracchia, L., Banat, J.J., Cavallo, M. & Banat, I.M. (2015). Potential therapeutic applications of microbial surface-active compounds. *AIMS Bioengineering*, 2(3): 144-162. https://doi.org/10.3934/bioeng.2015.3.144

Galabova, D., Sotirova, A., Karpenko, E. and Karpenko, O. (2014). Role of microbial surface-active compounds in environmental protection. *The Role of Colloidal Systems in Environmental Protection*, 41-83. Elsevier. https://doi.org/10.1016/B978-0-444-63283-8.00003-X

García-Pinel, B., Porras-Alcalá, C., Ortega-Rodríguez, A., Sarabia, F., Prados, J., Melguizo, C. et al. (2019). Lipid-based nanoparticles: Application and recent advances in cancer treatment. *Nanomaterials*, 9(4): 638. https://doi.org/10.3390/nano9040638

Gayathiri, E., Prakash, P., Karmegam, N., Varjani, S., Awasthi, M.K. & Ravindran, B. (2022). Biosurfactants: Potential and eco-friendly material for sustainable agriculture and environmental safety—A review. *Agronomy*, 12(3): 662. https://doi.org/10.3390/agronomy12030662

Hanske, C., Sanz-Ortiz, M.N. & Liz-Marzán, L.M. (2020). Silica-coated plasmonic metal nanoparticles in action. *Colloidal Synthesis of Plasmonic Nanometals*, 755-820. eBook ISBN9780429295188

Hussein, H.A. & Abdullah, M.A. (2022). Biosurfactant as a vehicle for targeted antitumor and anticancer drug delivery. *Green Sustainable Process for Chemical and Environmental Engineering and Science*, 299-317. Elsevier. https://doi.org/10.1016/B978-0-323-85146-6.00019-X

Iravani, S. & Varma, R.S. (2020). Bacteria in heavy metal remediation and nanoparticle biosynthesis. *ACS Sustainable Chemistry & Engineering*, 8(14): 5395-5409. https://doi.org/10.1021/acssuschemeng.0c00292

Ishfaq, S., Hamid, B., Zaman, M., Fatima, S., Farooq, S., Datta, R. et al. (2022). 11 Microbial biosurfactants an eco-friendly approach for bioremediation of contaminated environments. *Microbial Surfactants: Volume 3: Applications in Environmental Reclamation and Bioremediation*, 197-207. CRC Press.

Jahan, R., Bodratti, A.M., Tsianou, M. & Alexandridis, P. (2020). Biosurfactants, natural alternatives to synthetic surfactants: Physicochemical properties and applications. *Advances in Colloid and Interface Science*. 275: 102061. https://doi.org/10.1016/j.cis.2019.102061

Johnson, P., Trybala, A., Starov, V. & Pinfield, V.J. (2021). Effect of synthetic surfactants on the environment and the potential for substitution by biosurfactants. *Advances in Colloid and Interface Science*, 288: 102340. https://doi.org/10.1016/j.cis.2020.102340

Karnwal, A., Shrivastava, S., Al-Tawaha, A.R.M.S., Kumar, G., Singh, R., Kumar, A. et al. (2023). Microbial biosurfactant as an alternate to chemical surfactants for application in cosmetics industries in personal and skin care products: A critical review. *BioMed. Research International*, 2023. https://doi.org/10.1155/2023/2375223

Katiyar, S.S., Ghadi, R., Kushwah, V., Dora, C.P. & Jain, S. (2020). Lipid and biosurfactant based core shell-type nanocapsules having high drug loading of paclitaxel for improved breast cancer therapy. *ACS Biomaterials Science & Engineering*, 6(12): 6760-6769. https://doi.org/10.1021/acsbiomaterials.0c01290

Katiyar, S.S., Kushwah, V., Dora, C.P. & Jain, S. (2019). Novel biosurfactant and lipid core-shell type nanocapsular sustained release system for intravenous application of methotrexate. *International Journal of Pharmaceutics*, 557: 86-96. https://doi.org/10.1016/j.ijpharm.2018.12.043

Kaur, H., Kumar, P., Cheema, A., Kaur, S., Singh, S. & Dubey, R.C. (2023). Biosurfactants as promising surface-active agents: Current understanding and applications. *Multifunctional Microbial Biosurfactants*, 271-306. Springer. https://doi.org/10.1007/978-3-031-31230-4_13

Khanna, S. (2019). Carbon nanotubes: Properties and applications. *Carbon Nanotubes and Nanoparticles*, 195-216.

Kiran, G.S., Selvin, J., Manilal, A. and Sujith, S. (2011). Biosurfactants as green stabilizers for the biological synthesis of nanoparticles. *Critical Reviews in Biotechnology*, 31(4): 354-364. https://doi.org/10.3109/07388551.2010.539971

Kumar, S., Nehra, M., Dilbaghi, N., Marrazza, G., Hassan, A. A. & Kim, K.-H. (2019). Nano-based smart pesticide formulations: Emerging opportunities for agriculture. *Journal of Controlled Release*. 294: 131-153. https://doi.org/10.1016/j.jconrel.2018.12.012

Kumar, V., Sharma, N. and Maitra, S. (2017). In vitro and in vivo toxicity assessment of nanoparticles. *International Nano Letters*, 7(4): 243-256. 10.1007/s40089-017-0221-3

Kumari, R., Singha, L.P. and Shukla, P. (2023). Biotechnological potential of microbial bio-surfactants, their significance and diverse applications. *FEMS Microbes*. xtad015. https://doi.org/10.1093/femsmc/xtad015

Liao, S., Ghosh, A., Becker, M.D., Abriola, L.M., Cápiro, N.L. et al. (2021). Effect of rhamnolipid biosurfactant on transport and retention of iron oxide nanoparticles in water-saturated quartz sand. *Environmental Science: Nano*, 8(1): 311-327. https://doi.org/10.1039/D0EN01033B

Lin, Y., Zhou, M., Tai, X., Li, H., Han, X. & Yu, J. (2021). Analytical transmission electron microscopy for emerging advanced materials. *Matter*, 4(7): 2309-2339. https://doi.org/10.1016/j.matt.2021.05.005

Liu, J., Zhang, R. and Xu, Z.P. (2019). Nanoparticle-based nanomedicines to promote cancer immunotherapy: Recent advances and future directions. *Small*, 15(32): 1900262. https://doi.org/10.1002/smll.201900262

Liu, Y., Harlow, J. & Dahn, J. (2020). Microstructural observations of "single crystal" positive electrode materials before and after long term cycling by cross-section scanning electron microscopy. *Journal of the Electrochemical Society*, 167(2): 020512. 10.1149/1945-7111/ab6288

Luft, L., Confortin, T.C., Todero, I., Zabot, G.L. and Mazutti, M.A. (2020). An overview of fungal biopolymers: Bioemulsifiers and biosurfactants compounds production. *Critical Reviews in Biotechnology*, 40(8): 1059-1080. https://doi.org/10.1080/07388551.2020.1805405

Macheroux, P. (1999). UV-visible spectroscopy as a tool to study flavoproteins. *Flavoprotein Protocols*. 1-7. https://link.springer.com/protocol/10.1385/1-59259-266-X:1

Magazzù, A. & Marcuello, C. (2023). Investigation of soft matter nanomechanics by atomic force microscopy and optical tweezers: A comprehensive review. *Nanomaterials*, 13(6): 963. https://doi.org/10.3390/nano13060963

Nikolova, C. & Gutierrez, T. (2021). Biosurfactants and their applications in the oil and gas industry: Current state of knowledge and future perspectives. *Frontiers in Bioengineering and Biotechnology*, 9: 626639. https://doi.org/10.3389/fbioe.2021.626639

Nikzamir, M., Hanifehpour, Y., Akbarzadeh, A. & Panahi, Y. (2021). Applications of dendrimers in nanomedicine and drug delivery: A review. *Journal of Inorganic and Organometallic Polymers and Materials*, 31: 2246-2261. 10.1007/s10904-021-01925-2

Nimesh, S., Manchanda, R., Kumar, R., Saxena, A., Chaudhary, P., Yadav, V. et al. (2006). Preparation, characterization and in vitro drug release studies of novel polymeric nanoparticles. *International Journal of Pharmaceutics*, 323(1-2): 146-152. https://doi.org/10.1016/j.ijpharm.2006.05.065

Nitschke, M. and Marangon, C.A. (2022). Microbial surfactants in nanotechnology: Recent trends and applications. *Critical Reviews in Biotechnology*, 42(2): 294-310. https://doi.org/10.1080/07388551.2021.1933890

Nunes, J.C., Magalhães, F.F., Araújo, M.T., Almeida, M.R., Freire, M.G. and Tavares, A.P. (2022). Expansion of targeted drug-delivery systems using microbially sourced biosurfactants. *Green Sustainable Process for Chemical and Environmental Engineering and Science*, 105-120. Elsevier. https://doi.org/10.1016/B978-0-323-85146-6.00034-6

Ohadi, M., Shahravan, A., Dehghannoudeh, N., Eslaminejad, T., Banat, I.M & Dehghannoudeh, G. (2020). Potential use of microbial surfactant in microemulsion drug delivery system: A systematic review. *Drug Design, Development and Therapy*. 541-550. https://doi.org/10.2147/DDDT.S232325

Parthipan, P., Cheng, L., Dhandapani, P., Elumalai, P., Huang, M. & Rajasekar, A. (2022). Impact of biosurfactant and iron nanoparticles on biodegradation of polyaromatic hydrocarbons (PAHs). *Environmental Pollution*, 306: 119384. https://doi.org/10.1016/j.envpol.2022.119384

Patel, P., Bhatt, S., Patel, H., Sayyed, R. & Aguilar-Marcelino, D. (2022). Biosurfactant a biomolecule and its potential applications. *In:* Sayyed, R.Z. (Ed.), *Microbial Surfactants*. 63-81. CRC Press.

Płaza, G.A., Chojniak, J. & Banat, I.M. (2014). Biosurfactant mediated biosynthesis of selected metallic nanoparticles. *International Journal of Molecular Sciences*, 15(8): 13720-13737. https://doi.org/10.3390/ijms150813720

Rane, A.N., Geetha, S. & Joshi, S.J. (2021). Biosurfactants: Production and role in synthesis of nanoparticles for environmental applications. *Biosurfactants for a Sustainable Future: Production and Applications in the Environment and Biomedicine*. 183-206. https://doi.org/10.1002/9781119671022.ch9

Rangel-Muñoz, N., González-Barrios, A.F., Pradilla, D., Osma, J.F. & Cruz, J.C. (2020). Novel bionanocompounds: Outer membrane protein a and laccase co-immobilized on magnetite nanoparticles for produced water treatment. *Nanomaterials*, 10(11): 2278. https://doi.org/10.3390/nano10112278

Rasheed, T., Shafi, S., Bilal, M., Hussain, T., Sher, F. & Rizwan, K. (2020). Surfactants-based remediation as an effective approach for removal of environmental pollutants—A review. *Journal of Molecular Liquids*, 318: 113960. https://doi.org/10.1016/j.molliq.2020.113960

Ravinder, P., Manasa, M., Roopa, D., Bukhari, N.A., Hatamleh, A.A., Khan, M.Y. et al. (2022). Biosurfactant producing multifarious Streptomyces puniceus RHPR9 of *Coscinium fenestratum* rhizosphere promotes plant growth in chilli. *Plos One*, 17(3): e0264975.

Renu, S., Shivashangari, K.S. & Ravikumar, V. (2020). Incorporated plant extract fabricated silver/poly-D, l-lactide-co-glycolide nanocomposites for antimicrobial based wound healing. *Spectrochimica Acta Part A: Molecular and Biomolecular Spectroscopy*, 228: 117673. https://doi.org/10.1016/j.saa.2019.117673

Roy, A. (2017). Review on the biosurfactants: Properties, types and its applications. *J. Fundam. Renew. Energy Appl.*, 8(2). 10.4172/2090-4541.1000248

Salopek, B., Krasic, D. & Filipovic, S. (1992). Measurement and application of zeta-potential. *Rudarsko-geolosko-naftni zbornik*, 4(1): 147. https://hrcak.srce.hr/file/39012?ref=Guzels.TV

Salvi, V.R. & Pawar, P. (2019). Nanostructured lipid carriers (NLC) system: A novel drug targeting carrier. *Journal of Drug Delivery Science and Technology*, 51: 255-267. https://doi.org/10.1016/j.jddst.2019.02.017

Sanches, M.A., Luzeiro, I.G., Alves Cortez, A.C., Simplício de Souza, É., Albuquerque, P.M., Chopra, H.K. et al. (2021). Production of biosurfactants by Ascomycetes. *International Journal of Microbiology*, 2021. https://doi.org/10.1155/2021/6669263

Sangeetha, J., Thomas, S., Arutchelvi, J., Doble, M. & Philip, J. (2013). Functionalization of iron oxide nanoparticles with biosurfactants and biocompatibility studies. *Journal of Biomedical Nanotechnology*, 9(5): 751-764. https://doi.org/10.1166/jbn.2013.1590

Sani, A., Qin, W.-Q., Li, J.-Y., Liu, Y.-F., Zhou, L., Yang, S.-Z. & Mu, B.-Z. (2023). Structural diversity and applications of lipopeptide biosurfactants as biocontrol agents against phytopathogens: A review. *Microbiological Research*, 127518.

Santos, A., Pereira, I., Ferreira, C., Veiga, F. & Fakhrullin, R. (2019). Layer-by-layer assembly for Nanoarchitectonics. *Advanced Supramolecular Nanoarchitectonics*, 89-121. https://doi.org/10.1016/B978-0-12-813341-5.00005-X

Saranraj, P., Sayyed, R., Sivasakthivelan, P., Hasan, M.S., Rahman, A., Al-Tawaha, M. et al. (2022). 2 Microbial biosurfactants methods of investigation, characterization, current market value and applications. *Microbial Surfactants: Volume 3: Applications in Environmental Reclamation and Bioremediation*, 19-34. CRC Press.

Shekhar, S., Sundaramanickam, A. & Balasubramanian, T. (2015). Biosurfactant producing microbes and their potential applications: A review. *Critical Reviews in Environmental Science and Technology*, 45(14): 1522-1554. https://doi.org/10.1080/10643389.2014.955631

Sobrinho, H.B., Luna, J.M., Rufino, R.D., Porto, A. & Sarubbo, L.A. (2013). Biosurfactants: Classification, properties and environmental applications. *Recent Developments in Biotechnology*, 11(14): 1-29.

Sultana, S., Alzahrani, N., Alzahrani, R., Alshamrani, W., Aloufi, W., Ali, A. et al. (2020). Stability issues and approaches to stabilised nanoparticles based drug delivery system. *Journal of Drug Targeting*, 28(5): 468-486. https://doi.org/10.1080/1061186X.2020.1722137

Thakur, S., Singh, A., Sharma, R., Aurora, R. & Jain, S.K. (2020). Biosurfactants as a novel additive in pharmaceutical formulations: Current trends and future implications. *Current Drug Metabolism*, 21(11): 885-901. https://doi.org/10.2174/1389200221666201008143238

Tosi, M.M., Ramos, A.P., Esposto, B.S. & Jafari, S.M. (2020). Dynamic light scattering (DLS) of nanoencapsulated food ingredients. *Characterization of Nanoencapsulated Food Ingredients*, 191-211. Elsevier. https://doi.org/10.1016/B978-0-12-815667-4.00006-7

Vecino, X., Cruz, J., Moldes, A. & Rodrigues, L. (2017). Biosurfactants in cosmetic formulations: Trends and challenges. *Critical Reviews in Biotechnology*, 37(7): 911-923. https://doi.org/10.1080/07388551.2016.1269053

Vellu, J.P., Nasir, H.M., Setapar, S.H.M., Ahmad, A., Chuo, S.C. & ALShammari, M.B. (2022). Insights into the application of reverse micelles in cosmetic formulations: A review. *Journal of Cosmetic Science*, 73(5).

Vijayakumar, S. & Saravanan, V. (2015). Biosurfactants – Types, sources and applications. *Res. J. Microbiol.*, 10(5): 181-192. https://scialert.net/abstract/?doi=jm.2015.181.192

Wu, B., Xiu, J., Yu, L., Huang, L., Yi, L. & Ma, Y. (2022). Biosurfactant production by Bacillus subtilis SL and its potential for enhanced oil recovery in low permeability reservoirs. *Scientific Reports*, 12(1): 7785. https://doi.org/10.1038/s41598-022-12025-7

Yekeen, N., Malik, A.A., Idris, A.K., Reepei, N.I. & Ganie, K. (2020). Foaming properties, wettability alteration and interfacial tension reduction by saponin extracted from soapnut (Sapindus Mukorossi) at room and reservoir conditions. *Journal of Petroleum Science and Engineering*, 195: 107591. https://doi.org/10.1016/j.petrol.2020.107591

Yetisgin, A.A., Cetinel, S., Zuvin, M., Kosar, A. & Kutlu, O. (2020). Therapeutic nanoparticles and their targeted delivery applications. *Molecules*, 25(9): 2193. https://doi.org/10.3390/molecules25092193

Yi, G., Son, J., Yoo, J., Park, C. & Koo, H. (2019). Rhamnolipid nanoparticles for in vivo drug delivery and photodynamic therapy. *Nanomedicine: Nanotechnology, Biology and Medicine*, 19: 12-21. https://doi.org/10.1016/j.nano.2019.03.015

Zarnowiec, P., Lechowicz, L., Czerwonka, G. & Kaca, W. (2015). Fourier transform infrared spectroscopy (FTIR) as a tool for the identification and differentiation of pathogenic bacteria. *Current Medicinal Chemistry*, 22(14): 1710-1718. https://www.ingentaconnect.com/content/ben/cmc/2015/00000022/00000014/art00008#expand/collapse

CHAPTER

12

Biosurfactants as Potential Drug Delivery Systems with Special Reference to Nanovesicle Liposomes

Sharav A. Desai*, Vipul P. Patel, Sandip Nagare, and Kunal Bhosle

Department of Pharmaceutics, Sanjivani College of Pharmaceutical Education & Research, Savitribai Phule Pune University, Kopargaon, India

1. Introduction

Biosurfactants are a diverse group of surface-active compounds produced mainly by microorganisms such as bacteria, yeast, fungi, and specific plants. These compounds have a distinct molecular structure that includes both hydrophilic and hydrophobic regions. The hydrophilic portion typically consists of polar or charged molecules, whereas the hydrophobic portion consists of non-polar or hydrophobic chains. This amphiphilic arrangement enables biosurfactants to reduce surface tension at the interface of various phases, thereby facilitating processes such as emulsification and dispersion of immiscible substances. Their properties include lowering surface tension, improving absorption and spreading, stabilizing emulsions, generating stable foams, increasing the solubility of hydrophobic compounds, being biodegradable, and exhibiting low toxicity. Due to these properties, biosurfactants are utilised in numerous industries, such as pharmaceuticals for drug delivery and formulations, the food industry for emulsification and foaming, environmental remediation for oil spill treatment and soil bioremediation, agriculture for soil improvement and plant growth promotion, cosmetics and personal care for mildness and emulsification, and the oil and gas industry for enhanced oil recovery and emulsification in processing. Current research continues to investigate new applications, with an emphasis on the sustainable and environmentally friendly character of biosurfactants in comparison to synthetic surfactants.

Modern pharmaceuticals depend heavily on drug delivery systems, which considerably influence the efficacy, safety, and convenience of drug treatments.

* Corresponding author: sharavdesaibpharm@sanjivani.org.in

Here are a few aspects that highlight the significance of drug delivery systems in the pharmaceutical industry.

1.1 Enhanced Efficacy and Targeted Delivery

Drug delivery systems permit the precise and targeted administration of medications to particular organs, tissues, or cells in the body. This targeted approach improves the treatment's efficacy by ensuring that the substance reaches the intended site of action at the optimal concentration.

By precisely regulating the release rate and dosage of the drug, drug delivery systems contribute to the minimization of potential side effects. This is especially crucial when the substance may be toxic or harmful at higher concentrations or when it is widely distributed throughout the body (Li et al., 2019).

It is possible to simplify drug delivery systems, making it easier for patients to adhere to prescribed treatments. For example, extended-release formulations have reduced dosing frequency, thereby increasing patient compliance and improving overall treatment outcomes.

1.2 Optimised Pharmacokinetics

Drug delivery systems can modulate a drug's pharmacokinetics and can influence factors such as absorption, distribution, metabolism, and excretion. Through this optimisation it is possible to assure a consistent and predictable drug concentration within the body, thereby improving therapeutic outcomes (Adepu & Ramakrishna, 2021a).

1.3 Protection and Stability of Drugs

Certain drug delivery systems are known to protect drugs from degradation and thereby ensure their stability and effectiveness during storage and transport. This is especially important for medications that are sensitive to light, heat, humidity, or chemical reactions (Patra et al., 2018).

1.4 Customization and Personalization of Treatment

Drug delivery systems today can be designed to fulfil the unique requirements of each patient. By doing variables such as age, weight, medical condition, and response to treatment will be considered. Individualised drug administration can optimise therapy, thereby maximising its benefits (Baryakova et al., 2023).

1.5 Innovative Therapeutic Approaches

Drug delivery systems are now allowing for the creation of innovative therapeutic approaches, like gene therapy, nanomedicine, and immunotherapy. These novel approaches have the capacity to completely transform disease management and treatment outcomes (Yao et al., 2020).

For example a drug delivery system, with a sustained-release formulation, prolongs the duration of drug activity within the body reduces the need for frequent administration and prolonging the therapeutic effect.

1.6 Facilitation of Challenging Drug Molecules

Drug delivery systems today are capable of effective delivery of medications that are complex to administer, including large molecules like proteins, peptides, and nucleic acids. These may have limited oral bioavailability or require specific delivery methods.

Overall, drug delivery systems considerably improve the efficacy, safety, and patient experience of pharmaceutical treatments. This makes them a fundamental component for modern medical care (Adepu & Ramakrishna, 2021a).

The association between biosurfactants and drug delivery is a growing area with huge potential for enhancing drug delivery systems. Biosurfactants, with surface-active properties produced by microorganisms, have unique properties that are known to promote enhanced drug delivery efficiency. Biosurfactants today play a significant role in optimising drug formulations through mechanisms such as increased solubility, encapsulation, and bioavailability.

Cooper et al. (1981) found that biosurfactants are capable of considerably enhancing the solubility and bioavailability of hydrophobic drugs and thereby mitigate a common challenge in drug formulation. Biosurfactants can encapsulate hydrophobic drugs by forming micelles or other structures, thereby increasing their solubility and bioavailability and ultimately increasing their therapeutic efficacy (Mulligan, 2005).

In addition, biosurfactants also play a crucial role in the formation of nanovesicles, including liposomes. These nanovesicles are capable of encapsulating pharmaceuticals and enabling targeted drug delivery, controlled drug release, and drug protection from degradation (Collitt et al., 2023). Particularly liposomal drug delivery systems have become known for enhancing drug delivery efficacy (Torchilin, 2005).

In addition to these physical characteristics, biosurfactants commonly exhibit biocompatibility and low toxicity. This makes them suitable for use in drug delivery systems (Marchant & Banat, 2012). This biocompatibility is required for pharmaceutical applications, where patient safety and minimal adverse effects are of utmost importance.

Incorporating biosurfactants into drug delivery systems not only enhances drug efficacy, but is also consistent with sustainable practises (Banat et al., 2010). Biosurfactants are typically derived from renewable resources and can be manufactured using eco-friendly processes. This encourages the development of environmentally friendly pharmaceutical technologies, thereby nurturing sustainability in the pharmaceutical industry.

In conclusion, the incorporation of biosurfactants into drug delivery systems represents a promising strategy for enhancing drug solubility, stability, targeted delivery, and overall effectiveness. Biosurfactants contribute to the advancement of drug delivery technologies by virtue of their unique properties and capabilities, providing sustainable and efficient solutions for drug administration.

2. Biosurfactants: Nature and Properties

Biosurfactants are a group of amphiphilic compounds generated by living organisms, primarily microorganisms such as bacteria, yeast, fungi, and certain plants. These molecules have a distinct molecular structure with hydrophilic and hydrophobic regions. This enables them to reduce the surface tension at the interface of immiscible phases, such as oil and water, thereby facilitating the emulsification and dispersion of substances that would otherwise remain separate (Satpute et al., 2010).

2.1 Different Types of Biosurfactants

2.1.1 Glycolipids

Glycolipid biosurfactants consist of a carbohydrate moiety (such as glucose or galactose) linked to a lipid moiety. The lipids, rhamnolipids, trehalolipids, and sophorolipids are notable examples (Mukherjee et al., 2006). *Pseudomonas aeruginosa* is a well-known producer of rhamnolipids.

2.1.2 Lipopeptides

Lipopeptides are characterized by a peptide chain linked to a lipid group. Examples include surfactin, iturin, and fengycin, which Bacillus species frequently produce (Satpute et al., 2010)

2.1.3 Phospholipids

Phospholipid biosurfactants typically consist of a phosphate group, glycerol and two fatty acids. The phospholipid biosurfactant phosphatidylcholine is a part of cell membranes (Desai & Banat, 1997).

2.1.4 Polymeric Biosurfactants

These are compounds with a high molecular weight that consist of repeating units. Examples include emulsan and alasan. These are produced by *Acinetobacter* species (Singh et al., 2007).

2.1.5 Proteins and Peptides

Microorganisms produce proteins and peptides with biosurfactant properties. Examples include bio emulsan and alasan (Rodrigues et al., 2006a).

2.1.6 Saponins

Saponins are biosurfactants composed of a hydrophilic sugar portion linked to a hydrophobic triterpene or steroid moiety. Plants are major sources of saponins. Saponin biosurfactants are typically obtained from Yucca plant extracts (Harborne, 1997).

2.1.7 Free Fatty Acids and Neutral Lipids

Some free fatty acids and neutral lipids are biosurfactants. These compounds are produced by microorganisms or secreted during the microbial degradation of complex organic compounds.

Understanding the various forms of biosurfactants is crucial for identifying their unique properties in a variety of applications, such as environmental remediation, pharmaceuticals, food industry, agriculture, and cosmetics. The selection of a particular biosurfactant is frequently influenced by the intended application and the desired surfactant properties, such as stability, emulsification capacity, and biodegradability.

2.2 Characteristics of Biosurfactants that Make them Suitable for Drug Delivery

Biosurfactants, which are amphiphilic compounds derived from microorganisms or

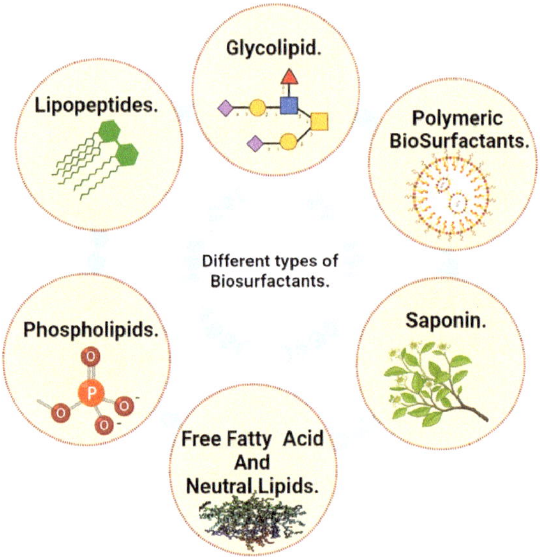

Fig. 1: Different types of biosurfactants.

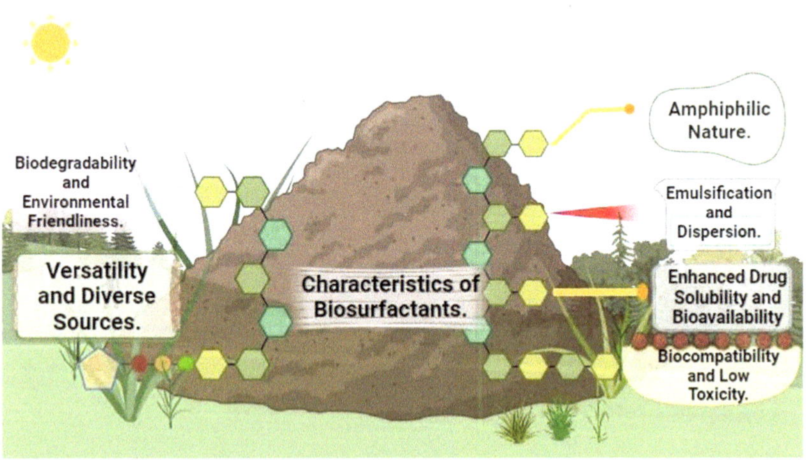

Fig. 2: Characteristics of biosurfactants.

natural sources, possess properties that make them ideal for drug delivery applications. These characteristics substantially affect the effectiveness and efficiency of drug delivery systems.

2.2.1 Amphiphilic Nature

Amphiphilic nature, surface activity, emulsification capabilities, encapsulation potential, drug protection, enhanced drug solubility and bioavailability, biocompatibility, and biodegradability are the key properties that make biosurfactants excellent candidates for drug delivery (Satpute et al., 2010, Rodrigues et al., 2006a). Biosurfactants have an amphiphilic molecular structure containing both hydrophilic and hydrophobic regions. This property enables biosurfactants to interact effectively with both aqueous and lipid phases, making them suitable for solubilizing and stabilizing pharmaceuticals of varying hydrophobicity (Kosaric & Vardar-Sukan, 2015).

Surface Activity and Surface Tension Reduction: Biosurfactants exhibit remarkable surface activity, thereby decreasing the surface tension of liquids. This property improves drug solubility and bioavailability, particularly in oral and topical drug delivery (P. Singh & Cameotra, 2004).

2.2.2 *Emulsification and Dispersion*

It is possible to use biosurfactants for emulsification and disperse immiscible phases to stabilize emulsions. This is required for drug delivery because it allows the formulation of drug-loaded emulsions that improve drug dispersion and absorption, especially for hydrophobic drugs (Singh et al., 2007).

2.2.3 *Encapsulation and Drug Protection*

Biosurfactants can form micellar or vesicular structures that encapsulate medications. This encapsulation can protect the drug against degradation and assures its stability during storage and transport. In addition, it permits controlled drug release. It can improve the therapeutic efficacy of the administered drug (Santos et al., 2014).

2.2.4 *Enhanced Drug Solubility and Bioavailability*

Enhanced Drug Solubility and Bioavailability of Biosurfactants improve the solubility and bioavailability of hydrophobic pharmaceuticals by facilitating their dispersion and absorption. Biosurfactants increase drug solubility by forming micelles or other structures, thereby enhancing the bioavailability of poorly soluble pharmaceuticals (Bolmal et al., 2020).

2.2.5 *Biocompatibility and Low Toxicity*

Compared to synthetic surfactants, biosurfactants are typically more biocompatible and less toxic. This property assures their safety for use in drug delivery systems, minimizing adverse effects on biological systems (Marchant & Banat, 2012).

2.2.6 *Biodegradability and Environmental Friendliness*

Biosurfactants are biodegradable, aligning with environmentally benign and sustainable pharmaceutical technologies. Their biodegradability renders them an environmentally favorable drug delivery option, highlighting their potential for sustainable healthcare practices (Van Hamme et al., 2006).

2.2.7 *Versatility and Diverse Sources*

Biosurfactants can be derived from numerous natural sources, such as microorganisms and vegetation. This adaptability enables a wide variety of biosurfactants with diverse properties, making them suitable for a variety of drug formulations and delivery systems (Nitschke & Costa, 2007).

Table 1: Examples of biosurfactants along with their properties and applications

Biosurfactant	Source	Properties and applications	References
Rhamnolipids	*Pseudomonas aeruginosa, Pseudomonas chlororaphis, Burkholderia thailandensis*, etc.	Glycolipid biosurfactants with excellent surface activity. Applications in bioremediation, enhanced oil recovery, and pharmaceuticals due to their emulsifying properties and low toxicity.	(Chen et al., 2017)
Surfactin	*Bacillus subtilis, Bacillus licheniformis*	Lipopeptide biosurfactant with antimicrobial properties. Applications in pharmaceuticals, cosmetics, and various biotechnological processes.	(Banat et al., 2000)
Sophorolipids	*Candida bombicola, Candida apicola*	Glycolipid biosurfactants with excellent emulsifying properties. Applications in enhanced oil recovery, cosmetics, food, and pharmaceuticals due to their biodegradability and low toxicity.	(Nguyen & Sabatini, 2011)
Trehalose Lipids	*Mycobacterium* species, *Rhodococcus* species	Glycolipid biosurfactants with potential applications in drug delivery, emulsification, and bioremediation due to their ability to stabilize emulsions and their biocompatibility.	(Janek et al., 2018)
Biosurfactant Mannosylerythritol Lipids (MELs)	*Candida* species	Glycolipid biosurfactants with excellent emulsifying properties. Applications in pharmaceuticals, food, cosmetics, and oil recovery due to their surface-active behavior and low cytotoxicity.	(Nakahara et al., 1990)
Saponins	Plants like Yucca, Quillaja, Ginseng	Plant-derived biosurfactants with diverse biological activities. Applications in pharmaceuticals, agriculture, food, and cosmetics due to their emulsifying, foaming, and antioxidant properties.	(Rai et al., 2021)
Emulsan	*Acinetobacter calcoaceticus*	Polymeric biosurfactant with emulsifying properties. Applications in enhanced oil recovery and environmental remediation.	(Araújo et al., 2019)

Utilizing these properties, biosurfactants have enormous potential for enhancing drug delivery, enabling the development of effective and efficient drug formulations that will ultimately enhance patient therapeutic outcomes.

3. Drug Delivery Systems in Pharmaceuticals

Drug delivery systems are developed with the purpose of improving the effectiveness of therapy as well as the safety of medications. This is accomplished by effectively transporting chemicals to specific locations within the body. These systems incorporate a diverse range of technologies and methodologies to guarantee the accurate, continuous, and regulated delivery of pharmaceuticals. The following table represents the common drug delivery systems in use.

These drug delivery systems provide a variety of alternatives for optimizing drug administration, improving therapeutic outcomes, and reducing side effects. As a result, treatments can be customized to meet the specific requirements of individual patients. Ongoing research and technological developments are being made with the goals of further improving already existing systems and creating cutting-edge medication delivery technologies that will transform the healthcare industry.

3.1 Challenges and Limitations of Conventional Drug Delivery Systems

Conventional drug delivery systems, despite their widespread application and demonstrated efficacy, have a number of drawbacks and obstacles that, if not addressed, could compromise their effectiveness and the results of therapeutic interventions. The optimal distribution of pharmaceuticals becomes difficult as a result of these issues, resulting in the need for the investigation of more modern technologies for drug delivery. The following are some of the key drawbacks of conventional medication delivery systems:

Addressing these challenges is crucial to improving drug delivery systems and enhancing their therapeutic potential. Innovative drug delivery technologies aim to overcome these limitations and provide targeted, controlled, and sustained drug release, ultimately optimizing therapeutic outcomes.

4. Biosurfactants as Drug Delivery Systems

Biosurfactants, owing to their unique features and numerous applications, have attracted great attention as possible drug delivery methods. These naturally occurring chemicals have the potential to act as carriers for medicinal medicines, hence enhancing the soluble, stable, and bioavailable properties of the drug. Biosurfactants, due to their unique molecular structure and properties, can enhance drug delivery through various mechanisms. These mechanisms influence drug solubility, stability, bioavailability, and the overall efficacy of drug delivery systems. Here are the key mechanisms by which biosurfactants enhance drug delivery:

4.1 Improved Drug Solubility and Dispersion

Biosurfactants have amphiphilic properties, with both hydrophobic and hydrophilic regions, resulting in enhanced drug solubility and dispersion. This characteristic

Table 2: Table of existing drug delivery systems

Drug delivery system	Route of administration	Examples	Applications
Oral	Tablets, capsules, oral disintegrating tablets, buccal and sublingual delivery	Immediate-release, sustained-release, enteric-coated formulations	Treatment of a wide range of systemic and local conditions
Injectable	Intravenous, intramuscular, subcutaneous, intrathecal, intraventricular	Solutions, suspensions, emulsions	Rapid delivery of drugs into the bloodstream or target tissues
Transdermal	Patches, creams, gels	Nicotine patches, pain creams, antifungal gels	Treatment of localized conditions and systemic conditions with poor oral bioavailability
Pulmonary	Inhalers, nebulizers	Asthma inhalers, COPD nebulizers, surfactant therapy	Treatment of respiratory conditions
Nasal	Nasal sprays	Decongestants, allergy medications, pain relievers	Treatment of localized conditions and systemic conditions with rapid onset
Ophthalmic	Eye drops, ointments	Antibiotics, anti-inflammatory drugs, glaucoma medications	Treatment of eye conditions
Intradermal	Microneedle patches	Vaccines, biologics	Non-invasive delivery of drugs into the skin
Implantable	Drug-eluting implants	Contraceptive implants, pain pumps, insulin pumps	Local delivery of drugs over an extended period
Nanoparticle	Liposomes, polymeric nanoparticles, nanocapsules	Doxorubicin liposomes, paclitaxel nanoparticles, siRNA nanocapsules	Targeted delivery of drugs, improved solubility and bioavailability

Table 3: Challenges and limitations of conventional drug delivery systems

Challenge/Limitation	Description	References
Lack of targeted delivery	Conventional drug delivery methods often lack specificity, resulting in non-specific distribution of drugs to healthy tissues. This lack of targeting can diminish overall therapeutic efficacy and increase the risk of side effects.	(Wen et al., 2015)
Drug degradation and metabolism	Drugs may undergo degradation and metabolism in the gastrointestinal tract or liver before reaching the target site, reducing their bioavailability and efficacy. Oral administration is particularly susceptible to these issues.	(Walters, 2012)
Limited drug solubility	Poor drug solubility poses a significant challenge, especially for orally administered drugs, leading to incomplete absorption and reduced bioavailability. This limitation impacts the effective delivery of the drug to the target site.	(Leuner & Dressman, 2000)
Variable absorption	Drug absorption variability among individuals can result from factors such as gastrointestinal pH, motility, and food presence. This variation affects the predictability and consistency of drug levels in the body.	(Dressman & Reppas, 2000)
Patient compliance	Complex dosing regimens or administration procedures can decrease patient compliance, affecting the effectiveness of the treatment. Patient forgetfulness or discomfort with the administration method can be barriers to adherence.	(Brown & Bussell, 2011)
Short half-life of drugs	Drugs with a short half-life often necessitate frequent dosing to maintain therapeutic levels, which can inconvenience patients and impact their daily routines.	(Jones et al., 2011)
Invasive administration	Invasive methods of administration, such as injections, can cause discomfort and adverse reactions at the injection site, affecting patient acceptance and compliance.	(Mehrotra et al., 2011)
Limited duration of action	Some conventional drug formulations provide a rapid drug release, resulting in a short duration of action. This necessitates frequent dosing, which can be inconvenient for patients and may impact compliance.	(Sultana et al., 2022)
Potential for toxicity	Conventional drug delivery systems can lead to systemic drug exposure, increasing the risk of toxicity, especially for drugs with a narrow therapeutic index.	(Adepu & Ramakrishna, 2021b)
Development of resistance	The prolonged use of conventional drug delivery systems, especially in antibiotics, can contribute to the development of drug resistance in pathogens, reducing the efficacy of treatments over time.	(Nainu et al., 2021)

enables them to solubilize poorly water-soluble drugs by producing micelles or other colloidal structures, thereby enhancing drug solubility and dispersion in aqueous solutions (Karlapudi et al., 2018).

4.2 Micellar Solubilization and Encapsulation

Biosurfactants can form micelles in aqueous solutions when present in concentrations above their critical micelle concentration (CMC). These micelles can encapsulate hydrophobic pharmaceuticals within their hydrophobic centers, protecting the drug from degradation and increasing its stability (Makkar & Cameotra, 1999).

4.3 Enhanced Permeability and Absorption

Biosurfactants can make biological barriers like cell membranes and nasal linings more permeable making it easier for drugs to get into the body. This is very important for medicines that don't work well in the body (Soberón-Chávez et al., 2021).

4.4 Facilitated Drug Transport Across Biological Membranes

Biosurfactants have the capability of interacting with biological membranes and alter their characteristics to enhance medication transportation through these membranes. This interaction has the potential to improve the absorption and transportation of drugs to specific target locations (Thakur et al., 2021).

4.5 Reduction of Surface Tension and Enhanced Wetting

Biosurfactants can lower a liquid's surface tension, which improves its spreading and wetting properties. This can make it easier for the drug to stick to the target area, which can lead to better drug absorption (Wan et al., 2011).

4.6 Controlled Drug Release

Biosurfactant-derived drug delivery systems have demonstrated the ability to achieve controlled and sustained release of medications. The encapsulation and controlled release of pharmaceuticals is made possible by the amphiphilic properties exhibited by biosurfactants. This unique characteristic allows for the optimization of therapeutic outcomes by facilitating the gradual release of the encapsulated drugs (Odeniyi et al., 2019).

4.7 Targeted Drug Delivery and Site-Specific Release

In order to achieve targeted medication delivery and site-specific drug release, biosurfactants can be modified or changed in a number of different ways. It is possible for biosurfactants to preferentially deliver drugs to particular tissues or cells by attaching ligands to them or changing their structure. This makes it possible for site-specific drug release (Mallakpour et al., 2021).

4.8 Synergistic Effects with Other Drug Delivery Systems

Biosurfactants have the potential to improve the effectiveness of several drug delivery systems, including nanoparticles and liposomes. The drug transport capabilities of carriers can be enhanced through the augmentation of drug solubility and stability using biosurfactants (Čerpnjak et al., 2013).

Utilizing these mechanisms, biosurfactants have demonstrated considerable potential for enhancing drug delivery across multiple routes of administration, including oral, topical, and parenteral, thereby enhancing therapeutic efficacy and patient outcomes.

5. Nanovesicle Liposomes: An Emerging Drug Delivery System

Nanovesicle liposomes are nanoscale, spherical vesicles composed of a lipid bilayer structure, resembling cell membranes. Due to their ability to encapsulate a wide variety of hydrophilic and hydrophobic drugs, enhancing drug solubility, bioavailability, and targeted delivery, they are utilised in drug delivery and therapeutic applications (Bangham et al., 1965).

5.2 Nanovesicle Liposome Structure

The structure of nanovesicle liposomes is characterised by a phospholipid-dominated lipid bilayer. The main structural components include:

5.2.1 Phospholipids

Phospholipids are amphiphilic molecules with hydrophilic heads and hydrophobic tails that constitute the structure of lipid bilayers (Cevc & Vierl, 2010).

5.2.2 Aqueous Core

An inner space within the lipid bilayer, provides an environment for encapsulating hydrophilic pharmaceuticals or biomolecules (Sercombe et al., 2015).

5.2.3 Hydrophilic Shell

The outer layer of a vesicle is composed of hydrophilic phospholipid cores, which are required to facilitate interactions with the surrounding aqueous environment (Torchilin, 2005).

5.2.4 Hydrophobic Core

It is the interior space within the bilayer that is suitable for encapsulating hydrophobic drugs (Torchilin, 2005).

There is a frequent requirement to modify nanovesicle liposomes for specific drug delivery purposes. Such modifications may be for targeted drug release or enhanced stability or any other specific requirement. It is made possible by lipid composition or surface functionalization modification (Noble et al., 2014).

5.3 Properties of Nanovesicle Liposomes as Drug Carriers

5.3.1 Size Modification and Homogeneity

It is possible to customize nanovesicle liposomes to specific diameters. The diameters can range from tens to hundreds of nanometres on average. It will make it possible for drug encapsulation and enhanced targeting (Khodaverdi et al., 2022).

5.3.2 Drug Encapsulation Efficiency

Nanovesicle liposomes have a high drug encapsulation efficiency because of the presence of bilayer structures. They allow for encapsulation of both hydrophilic and hydrophobic compounds (Xu et al., 2012).

5.3.3 Biocompatibility and Biodegradability

Lipids used in nanovesicle liposomes are biocompatible and biodegradable that helps in minimising potential toxicity and allowing for safe drug delivery (Bulbake et al., 2017).

5.3.4 Surface Modification and Functionalization

Nanovesicle liposomes can be easily modified through ligand conjugation or charge modification. Modification can be useful in providing specific characteristics like targeted drug delivery or prolonged release (Pattni et al., 2015).

5.3.5 Stability and Shelf-Life

Nanovesicle liposomes are known to increase the stability and shelf-life of encapsulated pharmaceuticals. They are also involved in preventing their degradation and preserving their therapeutic efficacy (Wang & Grainger, 2019).

5.3.6 Enhanced Permeability and Retention (EPR) Effect

Nanovesicle liposomes can passively target drugs to tumor sites with permeable vasculature, exploiting the EPR effect. This enhances drug delivery to diseased tissues while decreasing systemic exposure. The EPR effect is a phenomenon in which tumors have leaky blood vessels and impaired lymphatic drainage. This allows nanovesicle liposomes to accumulate in tumors to a greater extent than in healthy tissues. Once in the tumor, the liposomes can release their drug cargo, which can then target and kill cancer cells. This reduces the amount of drug that is exposed to healthy tissues, which can help to reduce side effects (Maeda et al., 2000).

5.3.7 Facilitated Drug Release

Nanovesicle liposomes can be engineered to accomplish controlled and sustained drug release, thereby maximising the drug's therapeutic efficacy and minimising the risk of adverse effects (Barenholz, 2012).

5.4 Advantages of Liposome Nanovesicles as Drug Carriers

Following are the advantages associated with using nanovesicles as drug carriers.

5.4.1 Versatile Drug Delivery

Nanovesicle liposomes can encapsulate a variety of drugs, including small molecules, peptides, proteins, and nucleic acids, facilitating their efficient delivery to target sites (Abdelkader et al., 2011).

5.4.2 Enhanced Bioavailability

The encapsulation of hydrophobic drugs in the liposomal core and hydrophilic drugs in the aqueous core can substantially enhance drug solubility and bioavailability (Allen & Cullis, 2013).

5.4.3 Targeted Drug Delivery

Nanovesicle liposomes enable targeted drug delivery via surface modifications, ensuring drug delivery to specific tissues or cells, minimising off-target effects, and enhancing therapeutic efficacy (V. P. Torchilin, 2005).

5.4.4 Reduced Toxicity

By targeting drug delivery to specific tissues, nanovesicle liposomes can reduce drug-related toxicity by minimising systemic exposure (Liu et al., 2022).

5.4.5 Combination Therapy

Nanovesicle liposomes can increase treatment efficiency by simultaneously delivering combined drugs. They enable targeting multiple pathways or components implicated in a disease (Nekkanti & Kalepu, 2015).

5.4.6 Customised Release Profiles

The structure of nanovesicle liposomes permits customised release profiles, such as immediate release, sustained release, or triggered release, allowing for versatility in drug delivery (Pattni et al., 2015).

5.4.7 Clinical Translation

Nanovesicle liposomes have been successfully transformed into clinical applications and have already demonstrated their potential applications in a variety of therapeutic areas, such as oncology, infectious diseases, and cardiovascular disorders (Barenholz, 2012).

Nanovesicle liposomes offer a versatile and effective platform for improving drug delivery and enhancing therapeutic outcomes by leveraging their properties and advantages.

6. Integration of Biosurfactants with Nanovesicle Liposomes

Integrating the unique properties of biosurfactants and nanovesicle liposomes to improve drug delivery systems is a promising method involving the incorporation of biosurfactants and nanovesicle liposomes. This integration offers numerous benefits and prospective applications within the pharmaceutical industry.

6.1 Synergies and Benefits of Combining Biosurfactants with Nanovesicle Liposomes Enhanced Drug Solubility and Loading

Biosurfactants can enhance the solubility and loading of drugs within nanovesicle liposomes. Their amphiphilic nature aids in the solubilization of hydrophobic drugs in the lipid bilayer while facilitating the encapsulation of hydrophilic drugs in the aqueous core, thereby increasing drug loading capacity (Karlapudi et al., 2018).

6.2 Improved Stability and Shelf Life

Biosurfactants contribute to stabilising the lipid bilayer of nanovesicle liposomes, thereby enhancing their stability and extending their shelf life. The presence of biosurfactants reduces the surface tension of liposomes and inhibits their aggregation, thereby preserving their structural integrity (Arafat et al., 2017).

6.3 Facilitated Drug Release and Controlled Release Profiles

Biosurfactants can modify the release kinetics of medications from nanovesicle liposomes. Biosurfactants may impact the permeability of the liposomal membrane, resulting in controlled and sustained drug release, which is essential for optimising therapeutic efficacy (Hazra et al., 2014).

6.4 Targeted Drug Delivery and Enhanced Cellular Uptake

By conjugating specific ligands and antibodies to biosurfactant-modified nanovesicle liposomes, targeted drug delivery can be engineered. The increased affinity for target cells or tissues enhances drug localization and cellular uptake, thereby optimising drug delivery to the desired sites (Kang et al., 2020).

6.5 Biocompatibility and Reduced Toxicity

The incorporation of biocompatible biosurfactants into nanovesicle liposomes guarantees decreased toxicity and improves biocompatibility. This is crucial for assuring the safety of drug delivery systems, especially for in vivo applications (Panchariya et al., 2021)

6.6 Versatile Formulations for Diverse Drugs

Synergistic combinations enable the development of formulations adaptable to a wide variety of drugs with varying properties. This adaptability is essential for the encapsulation of various therapeutics, such as hydrophilic, hydrophobic, and amphiphilic medications (Jahangirian et al., 2017).

6.7 Potential for Natural and Sustainable Formulations

Biosurfactants, which are frequently derived from natural sources, are compatible with green and sustainable practices. Nangare & Patil (2020) report that integrating biosurfactants with nanovesicle liposomes enables the development of environmentally benign drug delivery systems, thereby contributing to eco-conscious and sustainable pharmaceutical practices.

6.8 Enhanced Bioavailability and Therapeutic Efficacy

The combination increases the bioavailability of the drug, allowing for more efficient absorption and delivery of the drug to the target sites. Increased drug availability correlates to increased therapeutic efficacy and enhanced patient outcomes (Karlapudi et al., 2018).

6.9 Possibility for Dual Drug Delivery Systems

The integration can facilitate the development of dual drug delivery systems by encapsulating various drugs within nanovesicle liposomes. This dual-delivery strategy enables synergistic therapeutic effects by administering multiple medications to treat complex diseases or conditions (Jahangirian et al., 2017).

In conclusion, the incorporation of biosurfactants with nanovesicle liposomes is a potent strategy for improving drug delivery systems, as it offers enhanced drug solubility, stability, controlled release, targeted delivery, and decreased toxicity. This comprehensive approach has the potential to significantly advance drug delivery technologies and improve therapeutic outcomes in a variety of medical applications.

6.10 Studies and Research Demonstrating Enhanced Drug Delivery Using this Combination

The research paper "*Synthesis, Characterization of Liposomes Modified with Biosurfactant MEL-A Loading Betulinic Acid and Its Anticancer Effect in HepG2 Cell*" investigated liposomes modified with biosurfactant MEL-A loading betulinic acid and its anticancer effect in HepG2 cells. In the published paper the author has discussed the synthesis and characterization of liposomes. He also has discussed its effect on cell proliferation and apoptosis. The results of the pater indicated that liposomes can inhibit the proliferation of HepG2 cells and they also can promote apoptosis in the cell. The liposomes modified with MEL-A are found to be more effective than those without MEL-A. The authors concluded that biosurfactant modified liposomes have great potential for the treatment of liver cancer (Shu et al., 2019).

The article titled "*Surface Functionalization of Piperine-Loaded Liposomes with Sophorolipids Improves Drug Loading and Stability*" has presented the use of amphiphilic biosurfactants, more specifically sophorolipids, in drug delivery systems. The study showed that modified piperine-loaded liposomes along with sophorolipids are capable of enhancing their stability and dispersion. The methods used in the work involved preparing sophorolipid-modified liposomes and measuring particle sizes and entrapment efficiencies (EEs) of piperine. Various conditions are used to evaluate the stability of liposomes, including serum, storage at 4°C, different NaCl concentrations and pH levels. Moreover, authors have investigated the in vitro release profile of piperine. The results of the study showed that the incorporation of sophorolipids is capable of improving the stability and dispersion of the liposomes, as evidenced by reduced particle sizes, lower polydispersity indices (PDIs), and increased serum stability. Sophorolipids also increased the entrapment efficiencies of piperine in the liposomes and stabilized them during storage at 4°C. It is also been found that under high NaCl concentrations and various pH conditions, sophorolipid-modified liposomes have exhibited favorable in vitro stability in relation to particle sizes and entrapment efficiencies. Furthermore, these modified liposomes demonstrate sustained in vitro drug release profiles (Chen & Zhang, 2022).

The article titled "*Rhamnolipids form drug-loaded nanoparticles for dermal drug delivery*" explores the use of bacterial biosurfactants, specifically rhamnolipids, as a natural approach to solubilize and transport hydrophobic molecules for dermal drug delivery. Eight distinct rhamnolipids were investigated for their potential as drug carrier systems. These rhamnolipids were found to self-assemble into spherical nanoparticles with precise average sizes ranging from 5 to 100 nm, exhibiting low polydispersity and stability at various concentrations. The study focused on loading hydrophobic drug molecules like Nile red, dexamethasone, and tacrolimus into these nanoparticles, achieving drug loadings of up to 30% (w/w). Nile red-loaded nanoparticles were evaluated for dermal drug delivery using human skin, demonstrating efficient penetration into the stratum corneum and lower epidermis. Importantly, rhamnolipid nanocarriers were deemed non-toxic to human fibroblasts, suggesting their suitability as candidates for dermal drug delivery applications (Müller et al., 2017).

The article titled "*Fungal chitosan from Agaricus bisporus (Lange) Sing. Chaidam increased the stability and antioxidant activity of liposomes modified with biosurfactants and loading betulinic acid*" explores the use of fungal chitosan derived from Agaricus bisporus in enhancing the stability and antioxidant properties of liposomes loaded with betulinic acid and modified with biosurfactants. The study found that fungal chitosan, which was structurally similar to commercial chitosan from aquatic sources, could effectively coat the betulinic acid-loaded liposomes. This coating resulted in improvements in liposome characteristics such as size, zeta potential, and encapsulation efficiency. In the present article, the study has highlighted the importance of biosurfactants, more specifically mannosylerythritol lipid A (MEL-A), in increasing the antioxidant property of these liposomes. The result in the article indicated that the liposomes coated with fungal chitosan down sized with higher zeta potential, and superior antioxidant effects compared to those coated with commercial chitosan. These results indicated the potential of biosurfactant-modified liposomes coated with fungal chitosan as a promising delivery system for bioactive compounds and has also emphasized the importance of biosurfactants in increasing the performance of such delivery systems (Wu et al., 2019).

The article titled "*Characterization of biosurfactant-containing liposomes and their efficiency for gene transfection*" has discussed the importance of biosurfactants, specifically Mannosylerythritol Lipid A (MEL-A), in non-viral gene transfection using cationic liposomes. Prior research had presented the improved gene transfection efficiency with MEL-A even though the molecular mechanisms behind this enhancement remained unclear. The study investigated the effect of various MELs (MEL-A, MEL-B, and MEL-C) on significant transfection processes, including DNA encapsulation and membrane fusion with anionic liposomes. The findings revealed that MEL-A-containing liposomes enhanced both processes, whereas MEL-B and MEL-C-containing liposomes primarily improved one of them. This suggests that specific physicochemical properties in MEL-A-containing liposomes contribute to enhancing the efficiency of liposome-mediated gene transfection, shedding light on the underlying mechanisms (Ueno et al., 2007).

The article titled "*Rhamnolipid-Based Liposomes as Promising Nano-Carriers for Enhancing the Antibacterial Activity of Peptides Derived from Bacterial Toxin-Antitoxin Systems*" addresses the pressing issue of antimicrobial resistance and the need for innovative antimicrobial agents, including peptides derived from bacterial toxin-antitoxin (TA) systems. The study focuses on ParELC3, a synthetic peptide derived from the ParE toxin known for inhibiting bacterial topoisomerases. However, ParELC3 faces challenges in penetrating bacterial membranes, rendering it inactive against bacteria. To overcome these limitations, the researchers used rhamnolipid-based liposomes to deliver ParELC3 across bacterial membranes allowing it to reach its intracellular target, the topoisomerases. Their findings demonstrated that ParELC3-loaded liposomes are found to be more effective in inhibiting the growth of standard *E. coli* and *S. aureus* strains compared to free peptide, without causing cytotoxicity to human cells. These findings suggested the potential of rhamnolipid-based liposomes as nanocarrier systems to enhance the bioactivity of peptides, highlighting their role in developing novel antimicrobial solutions (Sanches et al., 2021).

7. Challenges and Future Directions

7.1 Current Challenges in Implementing Biosurfactant-based Nanovesicle Liposomes

Implementing biosurfactant-based nanovesicle liposomes in drug delivery systems presented a number of challenges that delayed their widespread adoption and clinical translation. These challenges may have evolved, so it is essential to consult the latest research for current information. The primary challenges are identified as follows.

7.2 Standardization of Production Methods

It is difficult to develop consistent and repeatable methods for the production of biosurfactant-based nanovesicle liposomes. Standardization is essential for efficient drug delivery because production variation can impact the liposomes' quality, stability, and performance.

7.3 Scalability of Production

It is a formidable obstacle to increase the production of biosurfactant-based nanovesicle liposomes for commercial use. Transitioning from pilot -scale to industrial -scale production while preserving product quality and cost-effectiveness requires significant research and investment.

7.4 Biocompatibility and Safety Concerns

Concerns Regarding Biocompatibility and Safety Ensuring the biocompatibility and safety of biosurfactants is a significant obstacle. To evaluate the potential toxicological effects and long-term safety of biosurfactant-based nanovesicle liposomes in biological systems, extensive research is required.

7.5 Storage and Shelf Stability

Storage and Shelf Stability: It is crucial to maintain stability during storage. During storage, biosurfactant-based nanovesicle liposomes may encounter difficulties associated with aggregation, fusion, or degradation, thereby diminishing their shelf life and, consequently, their efficacy.

7.6 Controlled Drug Release

It is difficult to achieve controlled and sustained medication release from nanovesicle liposomes containing the biosurfactant. The task of fine-tuning drug release profiles to match therapeutic requirements while avoiding abrupt releases remains complex.

7.7 Targeted Delivery and Specificity

Targeted Delivery and Specificity: Using biosurfactant-based nanovesicle liposomes to improve targeted drug delivery to specific tissues or cells is a challenge. To optimize therapeutic outcomes, it is essential to achieve precise targeting while minimizing off-target effects.

7.8 In Vivo Efficacy and Pharmacokinetics

In Vivo Efficacy and Pharmacokinetics: It is crucial for the clinical translation of biosurfactant-based nanovesicle liposomes to demonstrate their in vivo efficacy and comprehend their pharmacokinetics. Understanding how these liposomes operate in complex biological environments should be the focus of research.

7.9 Cost-Effectiveness and Accessibility

The overall cost of production and accessibility of biosurfactant-based nanovesicle liposomes must be addressed. For widespread adoption in clinical contexts, it is vital to strike a balance between effectiveness and affordability.

7.10 Regulatory Approval and Compliance

Regulatory Approval and Compliance: It is a significant challenge to meet regulatory requirements and ensure compliance with established standards. To gain regulatory approval for the use of these formulations, comprehensive preclinical and clinical studies are required.

Addressing these challenges is fundamental to realizing the full potential of biosurfactant-based nanovesicle liposomes in drug delivery and advancing their application in clinical settings.

8. Potential Future Advancements and Applications in Drug Delivery Using Biosurfactants and Nanovesicle Liposomes

In the domain of drug delivery, biosurfactants and nanovesicle liposomes offer an abundance of opportunities for innovation and innovative applications. Here are some prospective future developments and applications that could have a substantial impact on drug delivery:

8.1 Combination Therapies and Dual Drug Delivery

Future research may concentrate on using biosurfactant-coated nanovesicle liposomes to simultaneously deliver multiple medications. This strategy could be particularly advantageous in the treatment of complex diseases where combination therapies offer synergistic effects, enhanced efficacy, and reduced adverse effects.

8.2 Personalized Medicine and Targeted Delivery

Advances in the comprehension of patient-specific variations could lead to the individualization of drug formulations. Individualizing biosurfactant-based nanovesicle liposomes for each patient, considering genetic, metabolic, or disease-specific factors, may improve treatment efficacy.

8.3 Gene Therapy Delivery

Biosurfactant-based nanovesicle liposomes could be engineered to transport genetic materials for gene therapy, such as RNA or DNA. The potential of the combination to remain secure and deliver genetic payloads to specific cells or tissues may revolutionize gene-based therapies.

8.4 Vaccination and Immunotherapy

Increasing the use of biosurfactant-based nanovesicle liposomes to deliver vaccines or immunotherapeutic agents could improve their stability, antigen presentation, and targeted delivery, thereby enhancing vaccine efficacy and immune response.

8.5 Intracellular Delivery for Rare Diseases

Biosurfactant-coated nanovesicle liposomes could be investigated for effective intracellular drug delivery, especially for the treatment of rare genetic diseases that require precise delivery of therapeutics to cellular compartments.

8.6 Bioavailability Enhancement for Poorly Soluble Drugs

Future advancements may concentrate on the use of biosurfactant-based nanovesicle liposomes to enhance the solubility and bioavailability of poorly soluble drugs, thereby expanding the range of drugs that can be effectively delivered using this combination.

8.7 Theranostic Applications

Integration of diagnostic agents within biosurfactant-based nanovesicle liposomes could pave the way for theranostic applications, allowing simultaneous diagnosis and targeted drug delivery for improved treatment monitoring and efficacy.

8.8 Intranasal Drug Administration for CNS Disorders

In order to target the central nervous system (CNS), biosurfactant-based nanovesicle liposomes can be modified for intranasal delivery. By facilitating drug delivery across the blood-brain barrier, this strategy could be useful in the treatment of CNS diseases.

8.9 Environmental Remediation

Beyond pharmaceuticals, biosurfactant-coated nanovesicle liposomes may have applications in environmental remediation, particularly in the delivery of agents for cleansing soil and water sources of pollutants and contaminants.

8.10 Printing (3D) and Customized Formulations

Utilizing 3D printing technology, biosurfactant-based nanovesicle liposomes could be incorporated into personalized drug formulations, providing precise dosing, controlled release, and patient-specific treatment protocols.

These prospective advancements and applications demonstrate the wide-ranging opportunities for biosurfactant-based nanovesicle liposomes in revolutionizing drug delivery, making treatments more effective, targeted, and personalized. For the most up-to-date and comprehensive information on potential advancements and applications, it is recommended to consult reputable scientific journals and databases containing recent research articles, reviews, and cutting-edge studies.

9. Conclusion

Within the field of medication delivery, biosurfactants, which are naturally generated surface-active chemicals, hold great promise in addition to a high degree of diversity. Their value rests in the singular qualities they possess as well as the numerous ways in which they could be applied to improve the administration of medications. Biosurfactants have the potential to improve the solubility, stability, and bioavailability of pharmaceuticals, particularly those with a low degree of solubility. They are able to help in the encapsulation of medications, which in turn promotes the medications' dispersion and absorption within biological systems. In addition, biosurfactants have the ability to increase the permeability of medications across biological barriers, such as the blood-brain barrier, which makes it easier to deliver drugs precisely where they are needed in the body. In addition, these biocompatible and biodegradable molecules provide a sustainable and environmentally favorable alternative to synthetic surfactants, which is in line with the growing need for green and environmentally friendly solutions in the field of medication delivery. Exploiting the potential of biosurfactants has the potential to revolutionize drug delivery, hence paving the way for therapeutic treatments that are more effective, targeted, and efficient across a wide range of medical applications.

References

Abdelkader, H., Ismail, S., Kamal, A. & Alany, R.G. (2011). Design and evaluation of controlled-release niosomes and discomes for naltrexone hydrochloride ocular delivery. *Journal of Pharmaceutical Sciences*, 100(5). https://doi.org/10.1002/jps.22422

Adepu, S. & Ramakrishna, S. (2021a). Controlled drug delivery systems: Current status and future directions. *Molecules*, 26(9). https://doi.org/10.3390/molecules26195905

Allen, T.M. & Cullis, P.R. (2013). Liposomal drug delivery systems: From concept to clinical applications. *Advanced Drug Delivery Reviews*, 65(1). https://doi.org/10.1016/j.addr.2012.09.037

Arafat, M., Kirchhoefer, C., Mikov, M., Sarfraz, M. & Löbenberg, R. (2017). Nanosized liposomes containing bile salt: A vesicular nanocarrier for enhancing oral bioavailability of BCS class III drug. *Journal of Pharmacy and Pharmaceutical Sciences*, 20. https://doi.org/10.18433/J3CK88

Araújo, H.W.C., Andrade, R.F.S., Montero-Rodríguez, D., Rubio-Ribeaux, D., Alves Da Silva, C.A. & Campos-Takaki, G.M. (2019). Sustainable biosurfactant produced by Serratia marcescens UCP 1549 and its suitability for agricultural and marine bioremediation applications. *Microbial Cell Factories*, 18(1). https://doi.org/10.1186/s12934-018-1046-0

Banat, I.M., Franzetti, A., Gandolfi, I., Bestetti, G., Martinotti, M.G., Fracchia, L. et al. (2010). Microbial biosurfactants production, applications and future potential. *Applied Microbiology and Biotechnology*, 87(2). https://doi.org/10.1007/s00253-010-2589-0

Banat, I.M., Makkar, R.S. & Cameotra, S.S. (2000). Potential commercial applications of microbial surfactants. *Applied Microbiology and Biotechnology*, 53(5). https://doi.org/10.1007/s002530051648

Bangham, A.D., Standish, M.M. & Watkins, J.C. (1965). Diffusion of univalent ions across the lamellae of swollen phospholipids. *Journal of Molecular Biology*, 13(1). https://doi.org/10.1016/S0022-2836(65)80093-6

Barenholz, Y. (2012). Doxil® – The first FDA-approved nano-drug: Lessons learned. *Journal of Controlled Release*, 160(2). https://doi.org/10.1016/j.jconrel.2012.03.020

Baryakova, T.H., Pogostin, B.H., Langer, R. & McHugh, K.J. (2023). Overcoming barriers to patient adherence: The case for developing innovative drug delivery systems. *Nature Reviews Drug Discovery*, 22(5). https://doi.org/10.1038/s41573-023-00670-0

Bolmal, U.B., Subhod, R.P., Gadad, A.P. & Patill, A.S. (2020). Formulation and evaluation of carbamazepine tablets using biosurfactant in ternary solid dispersion system. *Indian Journal of Pharmaceutical Education and Research*, 54(2). https://doi.org/10.5530/ijper.54.2.35

Brown, M.T. & Bussell, J.K. (2011). Medication adherence: WHO cares? *Mayo Clinic Proceedings*, 86(4). https://doi.org/10.4065/mcp.2010.0575

Bulbake, U., Doppalapudi, S., Kommineni, N. & Khan, W. (2017). Liposomal formulations in clinical use: An updated review. *Pharmaceutics*, 9(2). https://doi.org/10.3390/pharmaceutics9020012

Čerpnjak, K., Zvonar, A., Gašperlin, M. & Vrečer, F. (2013). Lipid-based systems as a promising approach for enhancing the bioavailability of poorly water-soluble drugs. *Acta Pharmaceutica*, 63(4). https://doi.org/10.2478/acph-2013-0040

Cevc, G. & Vierl, U. (2010). Nanotechnology and the transdermal route. A state of the art review and critical appraisal. *Journal of Controlled Release*, 141(3). https://doi.org/10.1016/j.jconrel.2009.10.016

Chen, H. & Zhang, Q. (2022). Surface functionalization of piperine-loaded liposomes with sophorolipids improves drug loading and stability. *Journal of Pharmaceutical Innovation*. https://doi.org/10.1007/s12247-022-09687-1

Chen, J., Wu, Q., Hua, Y., Chen, J., Zhang, H. & Wang, H. (2017). Potential applications of biosurfactant rhamnolipids in agriculture and biomedicine. *Applied Microbiology and Biotechnology*, 101(23–24). https://doi.org/10.1007/s00253-017-8554-4

Collitt, S.D., Sekhon, K.K., Gomari, S.R. & Rahman, P.K.S.M. (2023). Biosurfactants and their significance in altering reservoir wettability for enhanced oil recovery. *Challenges and Recent Advances in Sustainable Oil and Gas Recovery and Transportation*. https://doi.org/10.1016/B978-0-323-99304-3.00003-0

Cooper, D.G., Macdonald, C.R., Duff, S.J.B. & Kosaric, N. (1981). Enhanced production of surfactin from Bacillus subtilis by continuous product removal and metal cation additions. *Applied and Environmental Microbiology*, 42(3). https://doi.org/10.1128/aem.42.3.408-412.1981

Desai, J.D. & Banat, I.M. (1997). Microbial production of surfactants and their commercial potential. *Microbiology and Molecular Biology Reviews*, 61(1). https://doi.org/10.1128/mmbr.61.1.47-64.1997

Dressman, J.B. & Reppas, C. (2000). In vitro-in vivo correlations for lipophilic, poorly water-soluble drugs. *European Journal of Pharmaceutical Sciences*, 11(Suppl. 2). https://doi.org/10.1016/S0928-0987(00)00181-0

Elhissi, A. (2020). Liposomes for pulmonary drug delivery: The role of formulation and inhalation device design. *Current Pharmaceutical Design*, 23(3). https://doi.org/10.2174/1381612823666161116114732

Erel-Akbaba, G. & Akbaba, H. (2021). A comparative study of cationic liposomes for gene delivery. *Journal of Research in Pharmacy*, 25(4). https://doi.org/10.29228/jrp.30

Ferreira, M., Ogren, M., Dias, J.N.R., Silva, M., Gil, S. & Tavares, L. (2021). Liposomes as antibiotic delivery systems: A promising nanotechnological strategy against antimicrobial resistance. *Molecules*, 26(7). https://doi.org/10.3390/molecules26072047

Harborne, J.B. (1997). Saponins used in traditional and modern medicine and saponins used in food and agriculture. *Phytochemistry*, 46(7). https://doi.org/10.1016/s0031-9422(97)80034-9

Hazra, C., Kundu, D., Chatterjee, A., Chaudhari, A. & Mishra, S. (2014). Poly(methyl methacrylate) (core)-biosurfactant (shell) nanoparticles: Size controlled sub-100 nm synthesis, characterization, antibacterial activity, cytotoxicity and sustained drug release behavior. *Colloids and Surfaces A: Physicochemical and Engineering Aspects*, 449(1). https://doi.org/10.1016/j.colsurfa.2014.02.051

Jahangirian, H., Lemraski, E.G., Webster, T.J., Rafiee-Moghaddam, R. & Abdollahi, Y. (2017). A review of drug delivery systems based on nanotechnology and green chemistry: Green nanomedicine. *International Journal of Nanomedicine*, 12. https://doi.org/10.2147/IJN.S127683

Janek, T., Krasowska, A., Czyznikowska, Z. & Łukaszewicz, M. (2018). Trehalose lipid biosurfactant reduces adhesion of microbial pathogens to polystyrene and silicone surfaces: An experimental and computational approach. *Frontiers in Microbiology*, 9(Oct). https://doi.org/10.3389/fmicb.2018.02441

Jones, R. Do, Jones, H.M., Rowland, M., Gibson, C.R., Yates, J.W.T., Chien, J.Y. et al. (2011). PhRMA CPCDC initiative on predictive models of human pharmacokinetics, part 2: Comparative assessment of prediction methods of human volume of distribution. *Journal of Pharmaceutical Sciences*, 100(10). https://doi.org/10.1002/jps.22553

Kang, S., Duan, W., Zhang, S., Chen, D., Feng, J. & Qi, N. (2020). Muscone/RI7217 co-modified upward messenger DTX liposomes enhanced permeability of blood–brain barrier and targeting glioma. *Theranostics*, 10(10). https://doi.org/10.7150/thno.41322

Karlapudi, A.P., Venkateswarulu, T.C., Tammineedi, J., Kanumuri, L., Ravuru, B.K. Kodali, V.P. et al. (2018). Role of biosurfactants in bioremediation of oil pollution – A review. *Petroleum*, 4(3). https://doi.org/10.1016/j.petlm.2018.03.007

Khodaverdi, H., Zeini, M.S., Moghaddam, M.M., Vazifedust, S., Akbariqomi, M. & Tebyaniyan, H. (2022). Lipid-based nanoparticles for the targeted delivery of anticancer drugs: A review. *Current Drug Delivery*, 19(10). https://doi.org/10.2174/1567201819666220117102658

Kiaie, S.H., Mojarad-Jabali, S., Khaleseh, F., Allahyari, S., Taheri, E., Zakeri-Milani, P. et al. (2020). Axial pharmaceutical properties of liposome in cancer therapy: Recent advances and perspectives. *International Journal of Pharmaceutics*, 581. https://doi.org/10.1016/j.ijpharm.2020.119269

Kosaric, N. & Vardar-Sukan, F. (2015). Biosurfactants: Production and utilization processes, technologies, and economics.

Lestini, B.J., Sagnella, S.M., Xu, Z., Shive, M.S., Richter, N.J., Jayaseharan, J. et al. (2002). Surface modification of liposomes for selective cell targeting in cardiovascular drug delivery. *Journal of Controlled Release*, 78(1–3). https://doi.org/10.1016/S0168-3659(01)00505-3

Leuner, C. & Dressman, J. (2000). Improving drug solubility for oral delivery using solid dispersions. *European Journal of Pharmaceutics and Biopharmaceutics*, 50(1). https://doi.org/10.1016/S0939-6411(00)00076-X

Li, C., Wang, J., Wang, Y., Gao, H., Wei, G. Huang, Y. et al. & (2019). Recent progress in drug delivery. *Acta Pharmaceutica Sinica B*, 9(6). https://doi.org/10.1016/j.apsb.2019.08.003

Liu, P., Chen, G. & Zhang, J. (2022). A review of liposomes as a drug delivery system: Current status of approved products, regulatory environments, and future perspectives. *Molecules*, 27(4). https://doi.org/10.3390/molecules27041372

Maeda, H., Wu, J., Sawa, T., Matsumura, Y. & Hori, K. (2000). Tumor vascular permeability and the EPR effect in macromolecular therapeutics: A review. *Journal of Controlled Release*, 65(1–2). https://doi.org/10.1016/S0168-3659(99)00248-5

Makkar, R.S. & Cameotra, S.S. (1999). Biosurfactant production by microorganisms on unconventional carbon sources. *Journal of Surfactants and Detergents*, 2(2). https://doi.org/10.1007/s11743-999-0078-3

Mallakpour, S., Azadi, E. & Hussain, C.M. (2021). Recent advancements in synthesis and drug delivery utilization of polysaccharides-based nanocomposites: The important role of nanoparticles and layered double hydroxides. *International Journal of Biological Macromolecules*, 193. https://doi.org/10.1016/j.ijbiomac.2021.10.123

Marchant, R. & Banat, I.M. (2012). Microbial biosurfactants: Challenges and opportunities for future exploitation. *Trends in Biotechnology*, 30(11). https://doi.org/10.1016/j.tibtech.2012.07.003

Mehrotra, R., Mishra, S., Singh, M. & Singh, M. (2011). The efficacy of oral brush biopsy with computer-assisted analysis in identifying precancerous and cancerous lesions. *Head and Neck Oncology*, 3(1). https://doi.org/10.1186/1758-3284-3-39

Microbial Biosurfactants and their Environmental and Industrial Applications. (2019). *In: Microbial Biosurfactants and their Environmental and Industrial Applications*. https://doi.org/10.1201/b21950

Mishra, G.P., Bagui, M., Tamboli, V. & Mitra, A.K. (2011). Recent applications of liposomes in ophthalmic drug delivery. *Journal of Drug Delivery*, 2011. https://doi.org/10.1155/2011/863734

Mukherjee, S., Das, P. & Sen, R. (2006). Towards commercial production of microbial surfactants. *Trends in Biotechnology*, 24(11). https://doi.org/10.1016/j.tibtech.2006.09.005

Müller, F., Hönzke, S., Luthardt, W.O., Wong, E.L., Unbehauen, M. Bauer, J. et al. (2017). Rhamnolipids form drug-loaded nanoparticles for dermal drug delivery. *European Journal of Pharmaceutics and Biopharmaceutics*, 116. https://doi.org/10.1016/j.ejpb.2016.12.013

Mulligan, C.N. (2005). Environmental applications for biosurfactants. *Environmental Pollution*, 133(2). https://doi.org/10.1016/j.envpol.2004.06.009

Nainu, F., Permana, A.D., Djide, N.J.N., Anjani, Q.K., Utami, R.N. Rumata, N.R. et al. (2021). Pharmaceutical approaches to antimicrobial resistance: Prospects and challenges. *Antibiotics*, 10(8). https://doi.org/10.3390/antibiotics10080981

Nakahara, T., Tabuchi, T., Kitamoto, D. & Haneishi, K. (1990). Production of mannosylerythritol lipids by Candida antarctica from vegetable oils. *Agricultural and Biological Chemistry*, 54(1). https://doi.org/10.1271/bbb1961.54.37

Nangare, S.N. & Patil, P.O. (2020). Green synthesis of silver nanoparticles: An eco-friendly approach. *Nano Biomedicine and Engineering*, 12(4). https://doi.org/10.5101/nbe.v12i4.p281-296

Nekkanti, V. & Kalepu, S. (2015). Recent advances in liposomal drug delivery: A review. *Pharmaceutical Nanotechnology*, 3(1). https://doi.org/10.2174/22117385036661507091739O5

Nguyen, T.T. & Sabatini, D.A. (2011). Characterization and emulsification properties of rhamnolipid and sophorolipid biosurfactants and their applications. *International Journal of Molecular Sciences*, 12(2). https://doi.org/10.3390/ijms12021232

Nitschke, M. & Costa, S.G.V.A.O. (2007). Biosurfactants in food industry. *Trends in Food Science and Technology*, 18(5). https://doi.org/10.1016/j.tifs.2007.01.002

Noble, G.T., Stefanick, J.F., Ashley, J.D., Kiziltepe, T. & Bilgicer, B. (2014). Ligand-targeted liposome design: Challenges and fundamental considerations. *Trends in Biotechnology*, 32(1). https://doi.org/10.1016/j.tibtech.2013.09.007

Odeniyi, M., Omoteso, O., Adepoju, A. & Jaiyeoba, K. (2019). Starch nanoparticles in drug delivery: A review. *Polymers in Medicine*, 48(1). https://doi.org/10.17219/pim/99993

Olivier, J.C. (2005). Drug transport to brain with targeted nanoparticles. *NeuroRx*, 2(1). https://doi.org/10.1602/neurorx.2.1.108

Panchariya, V., Bhati, V., Madhyastha, H., Madhyastha, R., Prasad, J. Sharma, P. et al. (2021). Chromatic intervention and biocompatibility assay for biosurfactant derived from Balanites aegyptiaca (L.) Del. *Scientific Reports*, 11(1). https://doi.org/10.1038/s41598-021-83573-7

Patra, J.K., Das, G., Fraceto, L.F., Campos, E.V.R., Rodriguez-Torres, M.D.P., Acosta-Torres, L.S. (2018). Nano based drug delivery systems: Recent developments and future prospects. *Journal of Nanobiotechnology*, 16(1). https://doi.org/10.1186/s12951-018-0392-8

Pattni, B.S., Chupin, V.V. & Torchilin, V.P. (2015). New developments in liposomal drug delivery. *Chemical Reviews*, 115(19). https://doi.org/10.1021/acs.chemrev.5b00046

Pujol-Autonell, I., Mansilla, M.J., Rodriguez-Fernandez, S., Cano-Sarabia, M., Navarro-Barriuso, J., Ampudia, R.M. et al. (2017). Liposome-based immunotherapy against autoimmune diseases: Therapeutic effect on multiple sclerosis. *Nanomedicine*, 12(11). https://doi.org/10.2217/nnm-2016-0410

Rai, S., Acharya-Siwakoti, E., Kafle, A., Devkota, H.P. & Bhattarai, A. (2021). Plant-derived saponins: A review of their surfactant properties and applications. *Sci.*, 3(4). https://doi.org/10.3390/sci3040044

Rodrigues, L.R., Teixeira, J.A. & Oliveira, R. (2006a). Low-cost fermentative medium for biosurfactant production by probiotic bacteria. *Biochemical Engineering Journal*, 32(3). https://doi.org/10.1016/j.bej.2006.09.012

Sanches, B.C.P., Rocha, C.A., Bedoya, J.G.M., Da Silva, V.L., Da Silva, P.B., Fusco-Almeida, A.M. et al. (2021). Rhamnolipid-based liposomes as promising nano-carriers for enhancing the antibacterial activity of peptides derived from bacterial toxin-antitoxin systems. *International Journal of Nanomedicine*, 16. https://doi.org/10.2147/IJN.S283400

Santos, D.K.F., Brandão, Y.B., Rufino, R.D., Luna, J.M., Salgueiro, A.A., Santos, V.A. et al. (2014). Optimization of cultural conditions for biosurfactant production from Candida lipolytica. *Biocatalysis and Agricultural Biotechnology*, 3(3). https://doi.org/10.1016/j.bcab.2014.02.004

Satpute, S.K., Banat, I.M., Dhakephalkar, P.K., Banpurkar, A.G. & Chopade, B.A. (2010). Biosurfactants, bioemulsifiers and exopolysaccharides from marine microorganisms. *Biotechnology Advances*, 28(4). https://doi.org/10.1016/j.biotechadv.2010.02.006

Sercombe, L., Veerati, T., Moheimani, F., Wu, S.Y., Sood, A.K. & Hua, S. (2015). Advances and challenges of liposome assisted drug delivery. *Frontiers in Pharmacology*, 6(Dec). https://doi.org/10.3389/fphar.2015.00286

Shu, Q., Wu, J. & Chen, Q. (2019). Synthesis, characterization of liposomes modified with biosurfactant MEL-A loading betulinic acid and its anticancer effect in HepG2 cell. *Molecules*, 24(21). https://doi.org/10.3390/molecules24213939

Singh, A., Van Hamme, J.D. & Ward, O.P. (2007). Surfactants in microbiology and biotechnology: Part 2. Application aspects. *Biotechnology Advances*, 25(1). https://doi.org/10.1016/j.biotechadv.2006.10.004

Singh, P. & Cameotra, S.S. (2004). Potential applications of microbial surfactants in biomedical sciences. *Trends in Biotechnology*, 22(3). https://doi.org/10.1016/j.tibtech.2004.01.010

Soberón-Chávez, G., González-Valdez, A., Soto-Aceves, M.P. & Cocotl-Yañez, M. (2021). Rhamnolipids produced by Pseudomonas: From molecular genetics to the market. *Microbial Biotechnology*, 14(1). https://doi.org/10.1111/1751-7915.13700

Sudhakar, K., Fuloria, S., Subramaniyan, V., Sathasivam, K.V., Azad, A.K., Swain, S.S. et al. (2021). Ultraflexible liposome nanocargo as a dermal and transdermal drug delivery system. *Nanomaterials*, 11(10). https://doi.org/10.3390/nano11102557

Sultana, A., Zare, M., Thomas, V., Kumar, T.S.S. & Ramakrishna, S. (2022). Nano-based drug delivery systems: Conventional drug delivery routes, recent developments and future prospects. *Medicine in Drug Discovery*, 15. https://doi.org/10.1016/j.medidd.2022.100134

Thakur, P., Saini, N.K., Thakur, V.K., Gupta, V.K., Saini, R.V. & Saini, A.K. (2021). Rhamnolipid the glycolipid biosurfactant: Emerging trends and promising strategies in the field of biotechnology and biomedicine. *Microbial Cell Factories*, 20(1). https://doi.org/10.1186/s12934-020-01497-9

Torchilin, V. (2005). Lipid-core micelles for targeted drug delivery. *Current Drug Delivery*, 2(4). https://doi.org/10.2174/1567201054370221

Torchilin, V.P. (2005). Recent advances with liposomes as pharmaceutical carriers. *Nature Reviews Drug Discovery*, 4(2). https://doi.org/10.1038/nrd1632

Ueno, Y., Hirashima, N., Inoh, Y., Furuno, T. & Nakanishi, M. (2007). Characterization of biosurfactant-containing liposomes and their efficiency for gene transfection. *Biological and Pharmaceutical Bulletin*, 30(1). https://doi.org/10.1248/bpb.30.169

Van Hamme, J.D., Singh, A. & Ward, O.P. (2006). Physiological aspects. Part 1: Surfactants in microbiology and biotechnology. *Biotechnology Advances*, 24(6). https://doi.org/10.1016/j.biotechadv.2006.08.001

Walters, W.P. (2012). Going further than Lipinski's rule in drug design. *Expert Opinion on Drug Discovery*, 7(2). https://doi.org/10.1517/17460441.2012.648612

Wan, J., Chai, L., Lu, X., Lin, Y. & Zhang, S. (2011). Remediation of hexachlorobenzene contaminated soils by rhamnolipid enhanced soil washing coupled with activated carbon selective adsorption. *Journal of Hazardous Materials*, 189(1–2). https://doi.org/10.1016/j.jhazmat.2011.02.055

Wang, Y. & Grainger, D.W. (2019). Lyophilized liposome-based parenteral drug development: Reviewing complex product design strategies and current regulatory environments. *Advanced Drug Delivery Reviews*, 151–152. https://doi.org/10.1016/j.addr.2019.03.003

Wen, H., Jung, H. & Li, X. (2015). Drug delivery approaches in addressing clinical pharmacology-related issues: Opportunities and challenges. *AAPS Journal*, 17(6). https://doi.org/10.1208/s12248-015-9814-9

Wu, J., Niu, Y., Jiao, Y. & Chen, Q. (2019). Fungal chitosan from Agaricus bisporus (Lange) Sing. Chaidam increased the stability and antioxidant activity of liposomes modified with biosurfactants and loading betulinic acid. *International Journal of Biological Macromolecules*, 123. https://doi.org/10.1016/j.ijbiomac.2018.11.062

Xu, X., Costa, A. & Burgess, D.J. (2012). Protein encapsulation in unilamellar liposomes: High encapsulation efficiency and a novel technique to assess lipid-protein interaction. *Pharmaceutical Research*, 29(7). https://doi.org/10.1007/s11095-012-0720-x

Yao, Y., Zhou, Y., Liu, L., Xu, Y., Chen, Q., Wang, Y. et al. (2020). Nanoparticle-based drug delivery in cancer therapy and its role in overcoming drug resistance. *Frontiers in Molecular Biosciences*, 7. https://doi.org/10.3389/fmolb.2020.00193

Zhu, M., Wang, X., Zhou, Y., Li, M., Wang, Y., Wang, Y. et al. (2023). Breast tumor-targeted drug delivery via polymer nanocarriers: Endogenous and exogenous strategies. *Journal of Applied Polymer Science*. https://doi.org/10.1002/app.5422

CHAPTER

13

Biosurfactants as Antimicrobial Agents

Sonal Sanjay Kelkar, Karishma Ramchand Bhojwani, and Ekta Eknath Kamble*

PES Modern College of Arts, Science, and Commerce (Autonomous), Shivajinagar, Pune, India

1. Introduction

After the discovery of antibiotics, the 1940s saw their introduction as a treatment for severe infections. Combatants' bacterial illnesses during World War II were successfully managed with penicillin. However, penicillin resistance quickly developed into a significant clinical issue, and by the 1950s, many of the advancements made in the previous decade were in jeopardy. New beta-lactam antibiotics were identified, produced, and implemented as a response, regarding public trust (Ventola, 2015). The key players in virulence circuit control and host pathogen interaction coordination are quorum sensing signals and global regulators. In order to combat microbial pathogenicity, addressing these regulators offers a potential trend. Bacterial cells can flourish in biofilms, which are matrices made of polysaccharides, proteins, and DNA. Antibiotics are not effective against the cell populations found inside biofilm matrices (El-Mowafy et al., 2019).

In biomedical research, pathogenic bacteria and their persistent pathogenicity pose significant concerns due to inhabitant multidrug-resistant microorganisms. Antibiotic resistance has increased the mortality rate to an insurmountable level, much like the force of globalisation. As technology advances and the world gets bigger, novel microbial species appear as a result of antigenic migration brought on by catastrophic random mutation. The surfaces of medical devices are prone to microbial colonization and biofilm formation.

The pipeline for new antibiotics is drying up as antimicrobial resistance is on the rise and discovery of novel antimicrobial agents is at an all-time low. According to a news release by the WHO in 2019, 7,00,000 deaths occur globally each year due to antimicrobial resistance. If no steps are taken to change this statistic another 10

* Corresponding author: ekta.kamble90@gmail.com

million are estimated to succumb to it by 2050 (WHO, 2019). The issue of resistant hospital and community acquired infections has become so severe that WHO has published a list of priority pathogens against which antimicrobial agents for which antimicrobial agents are urgently needed (Tacconelli et al., 2018). Since antibiotics are failing in treating infections, exploring alternative methods of antimicrobial therapy is the need of the hour. Mulani et al. (2019) have summarized the alternative methods of treatment that are actively being studied for their potential use in treating infections. These include antibiotic combination therapy, bacteriophage therapy, photodynamic light therapy, use of nanoparticles, and antimicrobial peptides. Other methods explore use of molecules targeting antimicrobial resistance enzymes, antimicrobial-resistant bacteria, drug delivery systems, physiochemical methods, and unconventional strategies (Murugaiyan et al., 2022).

"Surfactant," or surface-active agent, is the root of the word "biosurfactant". The useful molecules referred to as surfactants, which are produced chemically, tend to emulsify substances, and lessen surface tension (Jamal, 2019). The group of biomolecules known as biosurfactants (BSs) is of microbial origin, and structurally diverse but exhibits considerable emulsifying and surface-active properties. These molecules are amphiphilic in that they have a hydrophilic polar head and hydrocarbon tail which is hydrophobic (Santos et al., 2016). An important characteristic of surfactants is the critical micelle concentration (CMC) which is the lowest surfactant concentration necessary to achieve the lowest surface tension. Amphipathic molecules combine after they approach the CMC, with the hydrophilic parts of the molecule facing outward and the hydrophobic parts facing inward. Since these surfactants with low CMC values are deemed to be more favourable for applied usage than those with higher CMC values, further development of surfactant after attaining the CMC values do not result in a further drop in surface tension (Sarubbo et al., 2022). Different bacteria, yeast, and filamentous fungi, such as *Bacillus, Pseudomonas, Burkholderia, Mycobacterium, Rhodococcus, Lactobacillus, Arthrobacter, Gordonia, Nocardia,* and *Acinetobacter*, can either release them into the extracellular space or place them on the surface of microbial cells. BS receive little attention among the secondary metabolites generated by microbes connected to human health. They have a more sophisticated role in the complex human environment than just acting as biosurfactants; they can also regulate the microbiota through the quorum sensing system and antibacterial activity. In addition to other methods of battling illnesses like bacterial and fungal infections, biosurfactants such as glycolipids can be an effective tool for preventing or destroying biofilms. Sodium dodecyl sulphate, an anionic surfactant, has been shown to destroy biofilms by promoting the creation of a central hollow inside them. The disintegration of *Pseudomonas fluorescens* biofilms was accelerated by the application of strong shear stress and cetyltrimethylammonium bromide (a cationic surfactant). Biofilm detachment was shown to be triggered by non-ionic surfactants, polyoxy ethlylene sorbitan monolaurate (Tween-20) and ethoxylated p-tert-octyl phenol (Triton X-100). It has been suggested that some biosurfactants, which are surface-active chemicals produced by microorganisms, have been found to have antibiofilm properties (El-Mowafy et al., 2019).

As an outcome biosurfactants are now being recognised as potentially bioactive compounds because of their distinctive structural makeup, biological roles, low toxicity, increased biodegradability, and adaptability. The exhaustion of oil supplies necessary for the manufacture of synthetic surfactants is another factor driving the hunt for biological surfactants. As a consequence, microbial surfactants become even more intriguing (Vijayakumar & Saravanan, 2015). However, the way that biosurfactants commonly work in treating infections is concentrated in their capacity to lower surface tension. In addition, although having low toxicity, biosurfactants can kill living things at specific doses. Recent advancements in biosurfactant production have enhanced their usefulness and research interest. This chapter describes the various types of biosurfactants studied for their antibacterial, antifungal and antiviral activities.

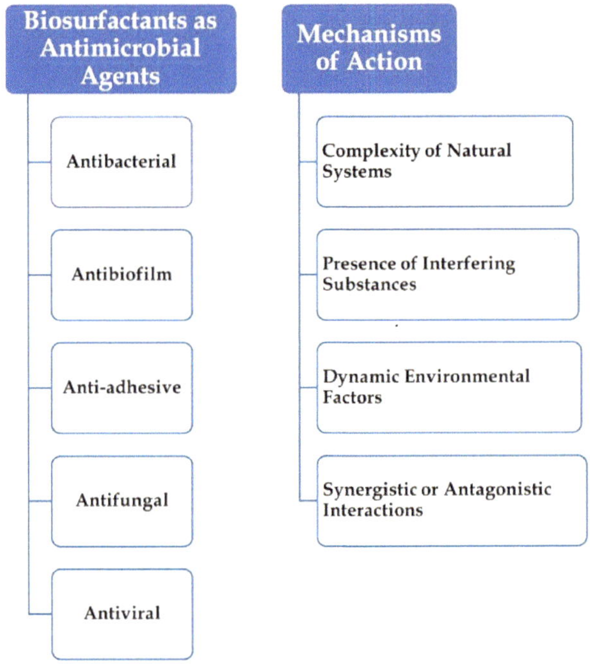

Fig. 1: Biosurfactants as antimicrobial agents and their mode of actions

2. Role of Biosurfactants in Physiology

Many different bacteria can create biosurfactants through excretion or attachment to cells, particularly when grown on substrates that are not soluble in water. It has been hypothesised that biosurfactants have a role in the emulsification of insoluble substrates, even if their role in microbial cells remains to be fully understood. Both interactions with emulsified droplets and proximity among the cell and hydrocarbon droplets have been described. The primary physiological function of biosurfactants lies in the fact that they lower the surface tension within phases. Biosurfactants make substrates more palatable to microbes for ingestion and metabolism, enabling them to grow on materials that are insoluble in water (Sobrinho et al., 2014).

3. Properties of Biosurfactants

In assessing the effectiveness and selecting microorganisms with the potential to produce these agents, the physicochemical characteristics of biosurfactants—such as reduced surface tension, solubility, emulsifying capacity, stabilising capacity, foaming capacity, low CMC, and detergency—are crucial. Almost all biosurfactants have several traits, many of which have advantages over traditional surfactants, despite the variability in their chemical makeup and properties.

4. Classification of Biosurfactants

Biosurfactants are categorized according to their chemical make-up and molecular weight.

4.1 Classification Based on Molecular Weight

4.1.1 Low Molecular-Weight Biosurfactants

At the air/water interfaces, these substances reduce surface and interfacial tension. Glycolipids or lipopeptides are usually the low molecular weight biosurfactants. Rhamnolipids, trehalolipids, and sophorolipids, which are disaccharides that are acylated with long-chain fatty acids or hydroxyl fatty acids, are the glycolipids that have received the most research attention to date (Shah et al., 2016).

4.1.2 High-Molecular Weight Biosurfactants

These are primarily known as bioemulsan. They are more proficient at sustaining oil in water emulsions. They function at low concentrations and are extremely effective emulsifiers. They carry significant substrate specificity as well. Several bacterial species from multiple genera create an exo-cellular polymeric surfactant comprised proteins, lipopolysaccharides, polysaccharides, and/or complicated combinations of these biopolymers (Shah et al., 2016).

4.2 Classification Based on Chemical Structure

4.2.1 Glycolipids

Glycolipids are microbial surface-active substances developed by a variety of bacteria. They are composed of a carbohydrate moiety coupled to fatty acids. Due to their biodegradability, variety, and bioactivity, these naturally occurring amphiphiles are currently receiving a lot of attention. Due to their self-assembly proficiency, anticancer and anti-inflammatory capabilities, and antimicrobial properties, including antibacterial, antifungal, and antiviral activities, glycolipids have proven to be effective therapeutic agents in treating a variety of disorders. Additionally, they have been shown to act as scavengers in preventing the development of biofilm and rupturing mature biofilm, demonstrating their suitability as anti-adhesive coating elements for medical insertional substances, and resulting in a decrease in widespread hospital infections (Shu et al., 2021).

4.2.2 Lipopeptides and Lipoprotein

The second largest type of biosurfactants are made up of lipopeptides and lipoproteins. They are mostly made of bacteria, such as those in the genera *Bacillus* and *Pseudomonas*. A fatty acid (between the range of C12 and C18) is joined to a peptide chain (within the range of 4 to 12 amino acids) to create lipopeptides. These can have a linear hydrophilic head or a lactone ring when they are cyclic lipopeptides.

The most well-known cyclic lipopeptides are surfactins, viscosins, lichenysins, amphisins, iturins, fengycins and putisolvins. Surfactin is unique among them for its superior surface qualities, which include a 40-unit reduction in the surface tension of water (measured in mN/m), as well as its biological activity against a few pathogenic microbes. Apparently, the antibacterial, antifungal, antiviral, and even anti-inflammatory properties of surfactin are real (Vecino et al., 2021).

The *Serratia* genus produces biosurfactants, which are low-molecular-weight, non-ionic lipopeptides having an antibacterial effect. These substances include serrawettin W1, W2, and W3, as well as stephensiolides A through K (cyclic lipopeptides distinguished by a macrolactone ring). The *Bacillus* genus has been studied in order to find new antimicrobial BS using the same methodology. For particulars, *Bacillus subtilis* develops a lipopeptide BS referred to as surfactin, which is marked as a cyclic heptapeptide associated to a β-hydroxy fatty acid (Clements et al., 2019).

4.2.3 Fatty Acids, Phospholipids, and Neutral Lipids

When growing on n-alkane substrate, several bacteria and yeast create significant amounts of fatty acids and phospholipids type of surfactants. Alkanes undergo microbial oxidations that result in the production of fatty acids, which have been used as surfactants. In addition to these straight chain acids, microorganisms also create complex fatty acids with OH groups and alkyl branches. Corynomuolic acids are an example of such complex acids (Shah et al., 2016).

4.2.4 Polymeric Microbial Surfactants

These surfactants have repeating units of heterosaccharide that contain proteins. The polymeric biosurfactants emulsan, liposan, mannoprotein, and polysaccharide protein complexes are the best studied within this category (Desai & Banat, 1997, Shah et al., 2016).

4.2.5 Particulate Biosurfactant

An effect on alkane absorption in microbial cells is produced by the partitioning of external membrane vesicles by particulate biosurfactants into a microemulsion. Vesicles produced by the *Acinetobacter* spp. have a buoyant density of 1.158 cubic gcm, a diameter of 20 to 50 nm, and are made up of proteins, phospholipids, and lipopolysaccharides (Karanth et al., 1999).

4.3 Classification Based on Ionic Nature

4.3.1 Cationic Biosurfactants

Antimicrobial surfactants are cationic, which include quaternary ammonium compounds (quats), making them the most well-known class of antimicrobial surfactants. The use of specific quats, such as alkyl dimethyl benzyl ammonium chloride (ADBAC), alkyl dimethyl ethylbenzyl ammonium chloride (ADEBAC), and didecylammonium chloride (DDAC) are widespread for disinfecting hard surfaces. Cationic surfactants have a strong affinity for the interfaces of all three microbial classes i.e., bacteria, viruses, and fungi, particularly when applied under alkaline to neutral pH conditions, due to the negative charge of microbial surfaces (Falk, 2019).

Researchers discovered that exposing *Candida albicans* to dioctadecyldimethyl ammonium bromide and hexadecyltrimethyl ammonium bromide resulted in a charge reversal from negative to positive, which was associated with decreased viability (Vieira & Carmona-Ribeiro, 2006).

4.3.2 Anionic Biosurfactants

Anionic surfactants have been used as antimicrobials since the 1930s, with acid/anionic formulae being particularly effective at a low pH of 2-3. The effectiveness of anionic surfactants is enhanced by their alkyl chain length, which should be C12-C16 to match the lipid structure of the cell wall and the presence of polyvalent cations in the formula (Block, 1983). The alkyl chains of anionic surfactants interact with the lipid structure of the cell wall. Microorganisms, such as bacteria and fungi, have cell walls composed of phospholipids. Anionic surfactants with alkyl chains that match the lipid structure of the cell wall (C12-C16) can more easily integrate into the lipid bilayer. This integration disrupts the cell wall's structure, compromising its integrity and causing leakage of intracellular components. The length of the alkyl chain is crucial because it determines the compatibility and effectiveness of the surfactant in interacting with the lipid bilayer. Alkyl chains of C12-C16 are considered optimal for this purpose. Anionic surfactants work well in acidic environments because they can switch the cell charge from negative to positive, making them more effective. Surfactants from strong acids like sulphates and sulfonates work especially well in this range, but those without acids can still work too (Falk, 2019). Previous studies have shown that linear alkylbenzene sulfonate (LAS) and sodium lauryl sulphate (SLS) have been found to be effective in deactivating viruses. SLS is known for its ability to denature proteins and is commonly used in cell lysing solutions for assays (Inácio et al., 2011; Tsujimura et al., 2015). In addition, it has been shown to disinfect *S. aureus*. LAS is approved for use as a food contact sanitizer by the EPA and has been shown to have antimicrobial efficacy in laundry applications.

5. Mechanism of Action

Biosurfactants are polymers produced by microorganisms that are amphiphilic, i.e., have both hydrophilic and hydrophobic properties enabling them to interact with the interface, which is the boundary between two phases in a heterogeneous system of two immiscible liquids. When organic molecules are present in the aqueous phase, they tend to accumulate at the solid contact in any interfacial situation and form a conditioning layer that changes the original surface's wettability and surface energy. Biosurfactants can also interact with interfaces and influence bacterial adherence and detachment of the biofilms, like organic conditioning films (Naughton et al., 2019, Pardhi et al., 2022).

The adsorption of charged biosurfactants on interfaces involves various interactions due to their amphiphilic nature. Most natural interfaces are negatively charged, although positive charges may be present in rare cases. Therefore, when examining the interactions of ionic biosurfactants with interfaces, it is crucial to consider the ionic conditions and pH. Studies have shown that positively charged biomaterial surfaces can inhibit the growth of Gram-negative bacteria but not Gram-positive bacteria. In addition, the molecular structure of a surfactant influences its behaviour at surfaces (Rodrigues et al., 2006). However, it is essential to note that the situation in natural systems differs from laboratory conditions (Qazi et al., 2020).

5.1 Complexity of Natural Systems

Natural systems contain a wide range of components, including various organic and inorganic substances, microorganisms, biofilms, and other biological entities. The interactions among these components can significantly influence the behaviour and fate of surfactants or any other chemical substance. Laboratory conditions, on the other hand, often simplify the system to study specific aspects, which may not fully represent the intricacies of natural environments.

5.2 Presence of Interfering Substances

Natural systems can contain a multitude of interfering substances such as organic matter, salts, minerals, and pollutants, present in varying concentrations, which can interact with surfactants and modify their adsorption, aggregation, or stability at interfaces. In laboratory experiments, controlling and replicating the exact composition of natural systems becomes challenging, leading to potential discrepancies between laboratory results and real-world observations.

5.3 Dynamic Environmental Factors

Natural systems are subject to dynamic environmental conditions, such as changes in temperature, pH, pressure, flow rates, and microbial activity. These factors can impact the functioning of surfactants, including their adsorption, desorption, and biodegradation rates. Laboratory experiments typically involve controlled and static conditions, which may not capture the full range of variations and complexities found in natural environments.

5.4 Time Scales and Long-Term Effects

Natural systems operate on longer time scales compared to laboratory experiments. The performance of surfactants and their effects on interfaces or microorganisms may evolve over extended periods, which may not be adequately captured in short-term laboratory studies. Additionally, long-term exposure and accumulation of surfactants in natural systems can lead to different outcomes and consequences compared to shorter-term laboratory experiments.

5.5 Synergistic or Antagonistic Interactions

Natural systems often involve the simultaneous presence of multiple interacting substances and microorganisms. Synergistic or antagonistic interactions among these components can affect the conduct and effectiveness of surfactants. Laboratory studies, focusing on isolated interactions or simplified systems, may not consider the complexity of such interactions, leading to discrepancies when compared to real-world scenarios.

The mechanism of action for the antimicrobial activity of quaternary ammonium compounds (quats) and anionic biosurfactants, alike, extends beyond mere adsorption to the interface. The process involves several steps (Falk, 2019, Gerba, 2015):

1. Anionic biosurfactants adsorb onto the surface of microbial cells.

2. They penetrate the lipid bilayer of the cell membrane, leading to increased permeability. This disruption can result in leakage of cellular contents of the microbial cell.
3. Leakage of lower molecular weight intracellular materials, including ions, metabolites, and other cellular components leads to a disruption of the intracellular environment and inhibit essential cellular processes.
4. They disrupt the structure and function of proteins by denaturing or unfolding them, thereby inhibiting their normal enzymatic activities. Additionally, they can interact with nucleic acids, leading to damage or degradation of DNA or RNA molecules, which hampers the microbial cell's ability to replicate and function properly.
5. By destabilizing the microbial cell, compromising its integrity and functionality to an extent that it cannot survive or replicate that eventually results in cell lysis and death.

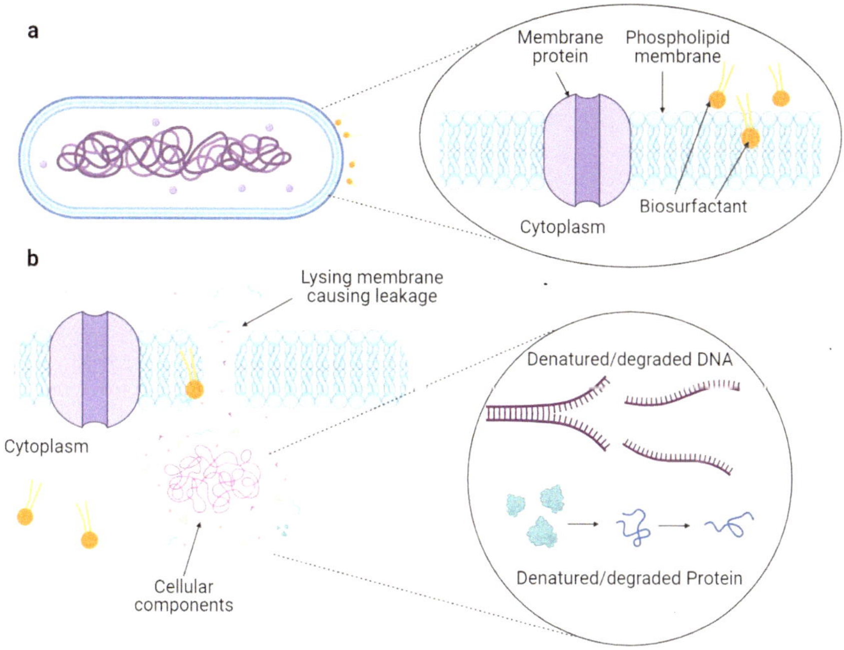

Fig. 2: Mechanism of action of antimicrobial activity of biosurfactants. (a) Biosurfactant molecules penetrate the cell membrane lysing it. (b) The resultant pores cause leakage of cellular components that are degraded due to biosurfactant action.

The effectiveness of quats is highest in their monomer form, as it allows for easier interaction with cell interfaces. Therefore, higher critical micelle concentration (CMC) values lead to higher concentrations of monomers (Falk, 2019).

5.6 Amphoteric Biosurfactants

Amine oxides and betaines are a type of surfactant that can have a positive charge at acidic pH, making them effective against cell membranes. They have shown efficacy against *S. aureus* and *E. coli*, particularly at myristic or palmitic chain lengths (Birnie et al., 2000). A combination of cocoamidopropyl betaine and alkyl dimethyl amine oxide has been found to have a synergy in antimicrobial performance, but their inhibitory concentration is higher than benzalkonium chloride (Falk, 2019). Heterocyclic amine oxides have also demonstrated antimicrobial activity against Bacillus, yeast such as the *Candida* species, and molds such as the *Penicillium* spp, *Aspergillus fumigatus*, with longer alkyl chain lengths showing stronger interaction with cell membranes (Subik et al., 1977). Glycine-based surfactants, such as dodicin, have shown efficacy against bacteria and some viruses (Falk, 2019).

5.7 Non-Ionic Biosurfactants

Glycolipids like sophorolipids and rhamnolipids have antimicrobial activity and their efficacy varies depending on the type of microbe being targeted (Falk, 2019). The sophorolipid compound is effective in damaging the cell membrane of Gram-positive bacteria *B. subtilis*, as evidenced by the release of intracellular enzyme; malate dehydrogenase. It is also able to inhibit the growth of mold. However, the compound is not effective against *E. coli*, which is a Gram-negative bacterium (Kim et al., 2002).

Surfactin, a cyclic polypeptide descendant of the *Bacillus* species, has surface-active properties (Heerklotz & Seelig, 2001) and has shown antibacterial (Sabaté & Audisio, 2013) and antiviral activity (Jung et al., 2012). It can lyse many types of cell membranes, which makes it a capable candidate for cancer treatment as well. Surfactin can potentially be used to kill cancer cells, but it needs to be delivered in a way that specifically targets cancer cells and does not harm healthy cells (Chen et al., 2022).

Free fatty acids have been recognized for their antibacterial properties and are used as food preservatives to prevent the growth of microorganisms. Studies have shown that fatty acids can inhibit the growth of various bacteria, including both Gram-positive and Gram-negative bacteria. In vitro studies have also shown that the antibacterial activity of fatty acids is comparable to that of antimicrobial polypeptides, such as bacteriocins. This is due to the ability of fatty acids to target and disrupt bacterial cell membranes, leading to cell death. Free fatty acids disrupt electron transport and oxidative phosphorylation, leading to cell membrane lysis and interference in other cellular metabolic pathways. In this way, fatty acids play an important role in the human body's defense mechanisms, particularly in the skin and mucosal surfaces. Lauric acid, myristic acid, palmitic acid, sapienic acid, and cis-8-octadecenoic acid are among the most abundant fatty acids found in these areas. They have been shown to have antibacterial properties and can help prevent the growth of harmful bacteria on the skin. Deficiencies in these fatty acids may make the skin more susceptible to colonization by bacteria such as *Staphylococcus aureus*. They possess bactericidal properties and can also hinder bacterial activity. Saturated fatty acids with a carbon chain length greater than five can halt the swarming behavior of the urinary tract pathogen, *Proteus mirabilis*. Additionally, certain fatty acids can reduce the activity of toxins, enzymes, and hemolysins that can counteract the efficacy of drugs. Lipase synthesised by *Candida rugosa* can produce fatty acids from coconut oil that can be used against both Gram-negative and Gram-positive bacteria. Particularly lauric acid has been found to be effective against bacteria like *S. aureus*, *E. coli*, *Helicobacter pylori* and yeast such as *Candida*, *Saccharomyces* species, with monoenoic and dienoic (cis-configuration) acids showing improved efficacy over saturated acids. Micro-efficacy assessments involve evaluating the impact of a treatment of antimicrobial agents like biosurfactants in this case on cellular processes, molecular interactions, or microorganism viability and function. Upon esterification, micro-efficacy reduces due to changes in chemical properties, reduced bioavailability and altered molecular interactions. When a compound is derivatized to aldehyde, alcohol, or amine form, it can enhance its micro-efficacy by increasing its reactivity, solubility, or ability to interact with target molecules or cellular components. These studies provide valuable insights into the effectiveness of a treatment at the microscopic level, which can help with further research, development, and optimization of interventions for specific microorganisms or diseases (Syldatk & Hausmann, 2010).

6. Biosurfactant as Antibacterial Agent

The use of biosurfactants has proven to be effective in impeding or slowing down the growth of typical microorganisms found in food products. The mode of action is determined by the physicochemical properties of the bioactive substance. Various mechanisms of action of biosurfactants as an antibacterial have been proposed, such as changes in permeability, destabilization, and rupture of the cell membrane, solubilization of membrane lipid component or the disruption of protein conformation, leading to the impairment of critical functions such as energy generation and transport (Syldatk & Hausmann, 2010).

Lipopeptides produced by *B. subtilis* and *B. licheniformis*, Mannosylerythritol lipids (MELs) developed by *Candida antarctica*, rhamnolipids made by *P. aeruginosa* have been successfully found to exhibit antibacterial properties. Biosurfactants resulting from *B. licheniformis* were characterised for their surface activity and structural makeup, and their continuous synthesis was detailed. A biosurfactant developed by *B. licheniformis* that compares favourably to conventional surfactants, lichenysin A, was also shown to have antibacterial properties.

Previous studies demonstrated that rhamnolipids derived from *P. aeruginosa* could reduce the proliferation of certain food-related pathogens, such as *Bacillus cereus*, *Staphylococcus aureus*, and *Micrococcus luteus* when added to the culture (Rodrigues et al., 2006). Biosurfactant produced by *Bacillus* sp. was successful in inhibiting growth of *E. coli*. In a study conducted by Yuliani in 2018, against five pathogens, namely, *Staphylococcus aureus, Escherichia coli*, *Pseudomonas aeruginosa*, *Salmonella enterica typhi*, and *Listeria monocytogenes* they were found to be susceptible to biosurfactant produced by *Bacillus subtilis* (Yuliani et al., 2018).

It was revealed currently that lichenysin has chelating capabilities, which could help clarify why lipopeptides disturb membranes. An explanation for the surfactin-induced porosity that underlies the antibiotic and hemolytic effect of these lipopeptides was found in another study that described the molecular process of membrane perforation by surfactin. Additionally, it was concluded from this work that, even at concentrations well below the threshold for solubilization, the characteristics of the membrane barrier are likely to be compromised where surfactin oligomers interact with phospholipids. This lipopeptide's antibacterial effect may very likely be primarily mediated by structural variations brought on by these features. The next batch of antibiotics might consist of surfactin-type peptides, which can quickly operate on membrane integrity instead of other crucial cellular functions (Shu et al., 2021).

Table 1: Role of biosurfactants as antimicrobial agent

Microorganisms	Biosurfactant	Activity	Reference
P. aeruginosa	Rhamnolipid	Antimicrobial activity against *M. tuberculosis*. Anti-adhesive activity against several bacteria and yeast strains.	(Rodrigues et al., 2006)
B. subtilis	Surfactin	Anti-bacterial, Anti-adhesive and anti-biofilm activities against uropathogenic strains. Anti-fungal activities. Antiviral activity against HSV virus, simian immunodeficiency virus, encephalomyocarditis virus, etc.	(Rodrigues et al., 2006)
B. pumilus	Pumilacidin (surfactin analogue)	Antiviral activity against human immunodeficiency virus 1 and herpes simplex virus.	(Vollenbroich et al., 1997)
Rhodococcus erythropolis	Trehalose lipid	Antiviral activity against HSV and influenza virus	(Syldatk & Hausmann, 2010)
Streptococcus thermophilus	Glycolipid	Anti-microbial, immunological, neurological properties. Anti-adhesive against *Rothia dentocariosa, S. epidermis* and *S. salivarius*.	(Busscher et al., 1994)
B. subtilis	Iturin	Anti-bacterial, Anti-adhesive and anti-biofilm activities against uropathogenic strains. Anti-fungal activities against profound mycosis. Effect on morphology and membrane structure of yeast cells. Antifungal activity against mold; *Aspergillus flavus*.	(Rodrigues et al., 2006)
Candida bombicola	Sophorolipids	Anti-bacterial and Anti-biofilm activity *B. subtilis* strains. Antifungal activities against *Fusarium* spp.	(Díaz De Rienzo et al., 2015)

7. Biosurfactant as Anti-Biofilm Agents

A biofilm in nature is an intricate three-dimensional structure made of an extracellular polymeric substance (EPS) matrix within which are embedded single, dual, or multiple microbial species that synthesized it (Otto, 2012). It is a major virulence factor that aids bacteria to survive hostile environmental conditions including antibiotics to which they are, therefore, resistant. Around 80% of persistence and recurrent microbial illnesses in humans are brought about by bacterial biofilms (Khatoon et al., 2018). In the food industry, bacterial biofilms can lead to contamination, disease transmission, and the spoilage of food products. Therefore, preventing the formation of biofilms on food surfaces is critical to ensure high-quality products for consumers (Idrees et al., 2021). To address this, physical, chemical, or biological methods can be used, including the development of innovative packaging materials. Biosurfactants, owing to their significant surface activity, have proven effective in preventing the formation of biofilms.

Biosurfactants have been shown to decrease hydrophobic interactions, leading to a reduction in surface hydrophobicity and microbial adhesion. This suggests that biosurfactants can play a significant role in disrupting biofilm formation by altering the chemical and physical conditions of the environment and controlling microbial interactions with interfaces. As a result, biosurfactants offer a promising approach to preventing and combating the formation of biofilms, particularly in the food industry where biofilms pose a significant threat to food safety and quality. Research shows that a biosurfactant from *Lactobacillus paracasei* demonstrated antibiofilm activity against *Candida albicans, Staphylococcus aureus, Staphylococcus epidermidis, and Streptococcus agalactiae*. Another biosurfactant derived from *Bacillus licheniformis* reduced *E. coli* biofilm formation by 54%. Additionally, a biosurfactant produced by *Nocardiopsis* sp. MSA13 significantly interrupted *Vibrio alginolyticus* biofilm formation, and a glycolipid from *Brevibacterium casei* inhibited biofilm production by *Vibrio spp., E. coli,* and *Pseudomonas* spp. (Syldatk & Hausmann, 2010).

Lipopeptide biosurfactants are the most notable classes of biosurfactants that have exhibited promising results in reducing adhesion to various materials and inhibiting the growth of different microorganisms (Chen et al., 2022). In addition to exhibiting antibacterial activity, lipopeptide biosurfactants have also been found to be effective against fungi, viruses, and other microorganisms (Rodrigues et al., 2006). In a study conducted by Morikawa et al. (1993) a biosurfactant named arthrofactin, produced by *Arthrobacter* species, exhibited a significantly higher efficacy compared to surfactin, with a seven-fold increase in effectiveness.

Rhamnolipids, mainly produced by *Pseudomonas aeruginosa*, have gained significant attention due to their diverse applications, environmentally-friendly nature, and effectiveness under extreme conditions. They have been predominantly used in environmental protection, but their potential in biomedical sciences has also been explored (Rodrigues et al., 2006). Despite being biologically safe, only a few studies have been conducted on their application as therapeutic agents. However, rhamnolipids have been identified as a suitable alternative to synthetic medicines and antimicrobial agents, which highlights their potential as safe and effective therapeutic agents. Glycolipids have shown to be effective in reducing microbial adhesion to surfaces, even under varying pH and the presence of electrolytes. They can interfere with the initial attachment of microorganisms to surfaces, preventing the formation of biofilms (Syldatk & Hausmann, 2010).

Sophorolipids also show potential in inhibiting biofilm formation, with studies reporting that they can reduce biofilm formation by up to 90%. They achieve this by interfering with cell signaling pathways that are essential for biofilm formation. Sophorolipids are considered safe and have been approved for use in various industries, including food and pharmaceuticals, by regulatory agencies such as the US FDA. They are classified into acidic and lactonic forms based on their structural composition, and the lactonic forms are primarily responsible for their bioactivity, which includes antimicrobial, anticarcinogenic, and antifungal properties. Additionally, they have low cytotoxicity, making them an attractive alternative to synthetic surfactants in various applications (Díaz De Rienzo et al., 2015).

8. Biosurfactant as Anti-Adhesive Agents

When surfaces are treated with surfactants prior to exposure to microorganisms, the surfactants can be adsorbed onto the surfaces. This can affect microbial adhesion to surfaces for colonization in two ways (Syldatk & Hausmann, 2010):

1. By curbing the capacity of microorganisms to form biofilms, as surfactants can intrude with cellular metabolism and either boost or hinder the adhesive forces that conserve the mechanical stability of the biofilm.
2. By leading to less cohesive characteristics, which may cause detachment of the biomass from the surface of adhesion.

The mechanism of action of biosurfactants on biofilm disruption is still an area of ongoing research, and the exact details are not fully understood. However, one hypothesis states that biosurfactants can alter the charge-charge properties within biofilms, which can potentially decrease the chances of bacteria acquiring antibiotic resistance through spontaneous mutations. The charge-charge interactions between the biosurfactants and the biofilm components may disrupt the cohesive forces within the biofilm, leading to the detachment of bacteria from the biofilm structure. This disruption can make the bacteria more susceptible to antimicrobial agents, including antibiotics. Furthermore, by altering the charge properties of the biofilm, biosurfactants may interfere with the communication and signalling processes among bacteria within the biofilm. This disruption of quorum sensing, a cell-to-cell communication system used by bacteria to coordinate collective behaviours, can hinder the development of antibiotic resistance mechanisms, such as the production of efflux pumps or biofilm-specific resistance genes (Díaz De Rienzo et al., 2015).

9. Biosurfactant as Antifungal Agent

Magnaporthe grisea, commonly known as rice blast fungus, is an important model organism in the study of fungal Phyto-pathogenicity and host-parasite interactions. This fungal pathogen causes a devastating disease that affects rice, leading to various names such as rice rotten neck, rice seedling blight, blast of rice, oval leaf spot of graminea, pitting disease, ryegrass blast, Johnson spot, neck blast, wheat blast, and Imochi. The antifungal properties of surfactin have been studied in previous research work which have demonstrated that treatment with surfactin CS30-1 results in a 40% reduction in *M. grisea* growth, with a gradual decline in growth observed with an increasing concentration of surfactin CS30-1. Conversely, when subjected to purified surfactin CS30-2. *M. grisea* growth exhibited a similar declining trend as the concentration of surfactin CS30-2 increased. Both surfactin CS30-1 and surfactin CS30-2 were found to elicit significant ultrastructural and morphological modifications in the hyphae of *M. grisea*. However, antifungal activity of surfactin CS30-1 was found to be superior to that of surfactin CS30-2. These findings provide insights into the mechanisms underlying the antifungal activity of surfactins and their potential for controlling *M. grisea* infections (Wu et al., 2019).

Most of the research regarding the antimicrobial activity of sophorolipids has focused on their efficacy against bacteria, with limited information available on their antifungal activity. Consequently, there is a need for further investigation to explore new sophorolipids that exhibit antifungal activity. A study was conducted to evaluate the antifungal activity of sophorolipids against plant and human pathogens

Colletotrichum gloeosporioides (ITCC 6434), *Fusarium verticilliodes* (MTCC 10556), *Fusarium oxysporum* f. sp. pisi (ITCC 4814), *Corynespora cassiicola* (ITCC 6748), and *Trichophyton rubrum* (MTCC 8477). The tested sophorolipids demonstrated encouraging antifungal activities against *Colletotrichum gloeosporioides*, *Fusarium oxysporum* f. sp. pisi, and *Fusarium verticillioides*, as evidenced by their low minimum inhibitory concentration (MIC) values. These results indicated that the sophorolipids were effective in inhibiting the growth of these fungal pathogens at relatively low concentrations. However, it is important to note that no inhibitory activity was observed against *Corynespora cassiicola* at the concentrations evaluated in this study. This suggested that the tested sophorolipids may not possess significant antifungal effects against this specific pathogen under the conditions and concentrations utilized in the study (Sen et al., 2017).

10. Biosurfactant as Antiviral Agent

Research has shown that biosurfactants possess anti-bacterial as well as antifungal activities. Protection of hosts from viral infections needs new and highly efficacious treatment strategies. Biosurfactants could be potent antiviral agents, as suggested by earlier research, owing to their excellent medical applications. Biosurfactants being amphipathic can interact with the hydrophobic domain of the virus (lipid membrane) and thereby can destabilize the membrane structure of the virus leading to the death of the virus cell. In addition, biosurfactants' micelle (aggregate) forming property enhances their utility in directly targeting the virus and it is quite useful in drug delivery as a liposome (Subramaniam et al., 2020). These features prove to be expedient in the discovery of antiviral drugs. Biosurfactants such as rhamnolipids and sophorolipids have been shown to have good anti-viral properties (Borsanyiova et al., 2016).

Cyclosporin A is a biopeptide produced by fungi *Tolypocladium inflatum* that inhibits the propagation of the influenza virus by interfering with the viral cycle at later stages. Lipopeptide as an adjuvant is employed for the stimulation of the immune system to produce antibodies. Synthetic lipopeptide vaccines can induce virus specific cytotoxic T-lymphocytes against the influenza nucleoprotein epitope (Smith et al., 2020).

Considering these facts, the use of biosurfactants in medicine could be an appropriate therapeutic strategy in combating coronaviruses and other such enveloped viruses. Coronaviruses belong to the family Coronaviridae, which already includes two well-known viruses: SARS-CoV (Severe Acute Respiratory Syndrome Coronavirus) and MERS-CoV (Middle East Respiratory Syndrome Coronavirus). The most prominent coronavirus at present is SARS-CoV-2, the virus responsible for the COVID-19 pandemic. Coronaviruses are enveloped viruses, meaning they possess a lipid membrane that encloses their essential proteins and positive-sense mRNA genome. This lipid membrane plays a crucial role in the virus's ability to infect host cells. The spike proteins present on the surface of coronaviruses are particularly important for the virus's entry into host cells. These spike proteins specifically bind to receptors, known as ACE2 receptors, present on the surface of human cells. This binding event facilitates the virus's entry into the host cell. Once inside the host cell, the coronavirus exploits the host cellular machinery to replicate its own genetic material and produce more viral proteins. This process allows the virus to spread rapidly within the infected cells, leading to significant damage (Çelik et al., 2020).

Acute Respiratory Distress Syndrome (ARDS) caused by Covid-19 manifests as the accumulation of fluid in the alveoli, impairing the efficient transfer of oxygen across the alveolar membranes into the bloodstream and leading to hypoxia. The primary cause of this syndrome is the dysfunction of surfactant due to the attack of SARS-CoV. The surfactant layer, located within the alveoli helps them maintain proper shape, and undergoes disruption when the Coronavirus infects the lungs. Consequently, the alveoli become filled with fluids, contributing to the development of hypoxia. In this context, biosurfactants hold promise as potential therapeutic agents to address this condition by aiding in the solubilization of the alveolar substrate. Moreover, biosurfactants could also find application in the development of vaccines for Covid-19 and other viral infections (Khodavirdipour et al., 2022).

11. Conclusion

In conclusion, biosurfactants are a promising alternative to chemical surfactants in various fields, including antimicrobial applications. They are produced from microorganisms and have a wide range of applications due to their unique properties, such as biodegradability, non-toxicity, and efficient antimicrobial activity. Different strains of microorganisms can produce different types of biosurfactants with varying structures and properties, which makes them highly versatile.

Research in the field of biosurfactants is ongoing, and it is expected that more novel and efficient biosurfactants will be discovered in the near future. The use of biosurfactants as an antimicrobial agent can contribute to the development of environmentally friendly and sustainable practices in various industries, such as food, cosmetics, pharmaceuticals, and oil. Therefore, it is important to continue exploring the potential of biosurfactants as an alternative to chemical surfactants to ensure the safety of the environment and human health.

References

Birnie, C.R., Malamud, D. & Schnaare, R.L. (2000). Antimicrobial evaluation of N-alkyl betaines and N-alkyl-N, N-dimethylamine oxides with variations in chain length. *Antimicrobial Agents and Chemotherapy*, 44(9): 2514–2517. https://doi.org/10.1128/AAC.44.9.2514-2517.2000

Block, S.S. (1983). Disinfection, Sterilization and Preservation (3rd ed.). Lea & Febiger.

Borsanyiova, M., Patil, A., Mukherji, R., Prabhune, A. & Bopegamage, S. (2016). Biological activity of sophorolipids and their possible use as antiviral agents. *Folia Microbiologica*, 61(1): 85–89. https://doi.org/10.1007/s12223-015-0413-z

Busscher, H.J., Neu, T.R. & van der Mei, H.C. (1994). Biosurfactant production by thermophilic dairy streptococci. *Applied Microbiology and Biotechnology*, 41(1): 4–7. https://doi.org/10.1007/BF00166073

Çelik, A., Manga, E.B., Çabuk, A. & Banat, I.M. (2020). Biosurfactants' potential role in combating COVID-19 and similar future microbial threats. *Applied Sciences 2021*, 11: 334, 11(1): 334. https://doi.org/10.3390/APP11010334

Chen, X., Lu, Y., Shan, M., Zhao, H., Lu, Z. & Lu, Y. (2022). A mini-review: Mechanism of antimicrobial action and application of surfactin. *World Journal of Microbiology and Biotechnology*, 38(8): 143. https://doi.org/10.1007/s11274-022-03323-3

Clements, T., Ndlovu, T., Khan, S. & Khan, W. (2019). Biosurfactants produced by Serratia species: Classification, biosynthesis, production and application. *Applied Microbiology and Biotechnology*, 103(2): 589–602. https://doi.org/10.1007/s00253-018-9520-5

Desai, J.D. & Banat, I.M. (1997). Microbial production of surfactants and their commercial potential. *Microbiology and Molecular Biology Reviews*, 61(1): 47–64. https://doi.org/10.1128/mmbr.61.1.47-64.1997

Díaz De Rienzo, M.A., Banat, I.M., Dolman, B., Winterburn, J. & Martin, P.J. (2015). Sophorolipid biosurfactants: Possible uses as antibacterial and antibiofilm agent. *New Biotechnology*, 32(6): 720–726. https://doi.org/10.1016/j.nbt.2015.02.009

El-Mowafy, M., Elgaml, A. & Shaaban, M. (2019). New approaches for competing microbial resistance and virulence. *In:* M. Blumenberg, M. Shaaban & A. Elgaml (Eds), *Microorganisms* (Ch. 10). IntechOpen. https://doi.org/10.5772/intechopen.90388

Falk, N.A. (2019). Surfactants as antimicrobials: A brief overview of microbial interfacial chemistry and surfactant antimicrobial activity. *Journal of Surfactants and Detergents*, 22(5): 1119–1127. Wiley-Blackwell. https://doi.org/10.1002/jsde.12293

Gerba, C.P. (2015). Quaternary ammonium biocides: Efficacy in application. *Applied and Environmental Microbiology*, 81(2): 464–469. https://doi.org/10.1128/AEM.02633-14

Heerklotz, H. & Seelig, J. (2001). Detergent-like action of the antibiotic peptide surfactin on lipid membranes. *Biophysical Journal*, 81(3): 1547–1554. https://doi.org/10.1016/S0006-3495(01)75808-0

Idrees, M., Sawant, S., Karodia, N. & Rahman, A. (2021). Staphylococcus aureus Biofilm: Morphology, genetics, pathogenesis and treatment strategies. *International Journal of Environmental Research and Public Health*, 18(14). https://doi.org/10.3390/IJERPH18147602

Inácio, Â.S., Mesquita, K.A., Baptista, M., Ramalho-Santos, J., Vaz, W.L.C. & Vieira, O.V. (2011). In vitro surfactant structure-toxicity relationships: Implications for surfactant use in sexually transmitted infection prophylaxis and contraception. *PloS One*, 6(5). https://doi.org/10.1371/JOURNAL.PONE.0019850

Jamal, P. (2019). Biosurfactant as the next antimicrobial agents in pharmaceutical applications. *Biomedical Journal of Scientific & Technical Research*, 13(3): 9950–9952. https://doi.org/10.26717/bjstr.2019.13.002394

Jung, M., Lee, S., Kim, H. & Kim, H. (2012). Recent studies on natural products as anti-HIV agents. *Current Medicinal Chemistry*, 7(6): 649–661. https://doi.org/10.2174/0929867003374822

Karanth, N.G.K., Deo, P.G. & Veenanadig, N.K. (1999). Microbial production of biosurfactants and their importance. *Current Science*, 77(1): 116–126.

Khatoon, Z., McTiernan, C.D., Suuronen, E.J., Mah, T.F. & Alarcon, E.I. (2018). Bacterial biofilm formation on implantable devices and approaches to its treatment and prevention. *Heliyon*, 4(12): e01067. https://doi.org/10.1016/J.HELIYON.2018.E01067

Khodavirdipour, A., Chamanrokh, P., Alikhani, M.Y. & Alikhani, M.S. (2022). Potential of *Bacillus subtilis* against SARS-CoV-2 – A sustainable drug development perspective. *Frontiers in Microbiology*, 13. https://doi.org/10.3389/fmicb.2022.718786

Kim, K., Yoo, D., Kim, Y., Lee, B., Shin, D. & Kim, E.K. (2002). Characteristics of sophorolipid as an antimicrobial agent. *Journal of Microbiology and Biotechnology*, 12(2): 235–241.

Morikawa, M., Daido, H., Takao, T., Murata, S., Shimonishi, Y. & Imanaka, T. (1993). A new lipopeptide biosurfactant produced by *Arthrobacter* sp. strain MIS38. *Journal of Bacteriology*, 175(20): 6459–6466. https://doi.org/10.1128/jb.175.20.6459-6466.1993

Mulani, M.S., Kamble, E.E., Kumkar, S.N., Tawre, M.S. & Pardesi, K.R. (2019). Emerging strategies to combat ESKAPE pathogens in the era of antimicrobial resistance: A review. *Frontiers in Microbiology*, 10(Apr): 539. https://doi.org/10.3389/fmicb.2019.00539

Murugaiyan, J., Anand Kumar, P., Rao, G.S., Iskandar, K., Hawser, S., Hays, J.P. et al. (2022). Progress in alternative strategies to combat antimicrobial resistance: Focus on antibiotics. *Antibiotics*, 11(2). Multidisciplinary Digital Publishing Institute (MDPI). https://doi.org/10.3390/antibiotics11020200

Naughton, P.J., Marchant, R., Naughton, V. & Banat, I.M. (2019). Microbial biosurfactants: Current trends and applications in agricultural and biomedical industries. *Journal of Applied Microbiology*, 127(1): 12–28. https://doi.org/10.1111/JAM.14243

Otto, M. (2012). *Staphylococcal Infections: Mechanisms of Biofilm Maturation and Detachment as Critical Determinants of Pathogenicity*. https://doi.org/10.1146/annurev-med-042711-140023

Pardhi, D.S., Panchal, R.R., Raval, V.H., Joshi, R.G., Poczai, P., Almalki, W.H. et al. (2022). Microbial surfactants: A journey from fundamentals to recent advances. *Frontiers in Microbiology*, 13: 982603. Frontiers Media S.A. https://doi.org/10.3389/fmicb.2022.982603

Qazi, M.J., Schlegel, S.J., Backus, E.H.G., Bonn, M., Bonn, D. & Shahidzadeh, N. (2020). Dynamic surface tension of surfactants in the presence of high salt concentrations. *Langmuir*, 36(27): 7956–7964. https://doi.org/10.1021/acs.langmuir.0c01211

Rodrigues, L.R., Banat, I.M., Van Der Mei, H.C., Teixeira, J.A. & Oliveira, R. (2006). Interference in adhesion of bacteria and yeasts isolated from explanted voice prostheses to silicone rubber by rhamnolipid biosurfactants. *Journal of Applied Microbiology*, 100(3): 470–480. https://doi.org/10.1111/j.1365-2672.2005.02826.x

Sabaté, D.C. & Audisio, M.C. (2013). Inhibitory activity of surfactin, produced by different *Bacillus subtilis* subsp. subtilis strains, against Listeria monocytogenes sensitive and bacteriocin-resistant strains. *Microbiological Research*, 168(3): 125–129. https://doi.org/10.1016/J.MICRES.2012.11.004

Santos, D.K.F., Rufino, R.D., Luna, J.M., Santos, V.A. & Sarubbo, L.A. (2016). Biosurfactants: Multifunctional biomolecules of the 21st century. *International Journal of Molecular Sciences*, 17(3): 1–31. https://doi.org/10.3390/ijms17030401

Sarubbo, L.A., Silva, M. da G.C., Durval, I.J.B., Bezerra, K.G.O., Ribeiro, B.G., Silva, I.A. et al. (2022). Biosurfactants: Production, properties, applications, trends, and general perspectives. *Biochemical Engineering Journal*, 181: 108377. https://doi.org/10.1016/J.BEJ.2022.108377

Sen, S., Borah, S.N., Bora, A. & Deka, S. (2017). Production, characterization, and antifungal activity of a biosurfactant produced by Rhodotorula babjevae YS3. *Microbial Cell Factories*, 16(1): 95. https://doi.org/10.1186/s12934-017-0711-z

Shah, N., Nikam, R., Gaikwad, S., Sapre, V. & Kaur, J. (2016). Biosurfactant: Types, detection methods, importance and applications. *Indian Journal of Microbiology Research*, 3(1): 5. https://doi.org/10.5958/2394-5478.2016.00002.9

Shu, Q., Lou, H., Wei, T., Liu, X. & Chen, Q. (2021). Contributions of glycolipid biosurfactants and glycolipid-modified materials to antimicrobial strategy: A review. *Pharmaceutics*, 13(2): 1–22. Multidisciplinary Digital Publishing Institute (MDPI). https://doi.org/10.3390/pharmaceutics13020227

Smith, M.L., Gandolfi, S., Coshall, P.M. & Rahman, P.K.S.M. (2020). Biosurfactants: A Covid-19 Perspective. *Frontiers in Microbiology*, 11, 553872. https://doi.org/10.3389/FMICB.2020.01341/BIBTEX

Sobrinho, Humberto B.S., Luna, Juliana M., Rufino, Raquel D., Porto, Ana Lucia F. & Sarubbo, L.A. (2014). *Biosurfactants : Classification, Properties and Environmental Applications*. (Issue January).

Subik, J., Takacsova, G., Psenak, M. & Devinsky, F. (1977). Antimicrobial activity of amine oxides: Mode of action and structure-activity correlation. *Antimicrobial Agents and Chemotherapy*, 12(2): 139–146. https://doi.org/10.1128/AAC.12.2.139

Subramaniam, M.D., Venkatesan, D., Iyer, M., Subbarayan, S., Govindasami, V., Roy, A. et al. (2020). Biosurfactants and anti-inflammatory activity: A potential new approach towards COVID-19. *Current Opinion in Environmental Science & Health*, 17: 72–81. https://doi.org/10.1016/J.COESH.2020.09.002

Syldatk, C. & Hausmann, R. (2010). Microbial biosurfactants. *In:* Inamuddin, M.I. Ahamed & R. Prasad (Eds), *European Journal of Lipid Science and Technology*, 112(6): 615–616. Springer Singapore. https://doi.org/10.1002/ejlt.201000294

Tacconelli, E., Carrara, E., Savoldi, A., Harbarth, S., Mendelson, M., Monnet, D.L. et al. (2018). Discovery, research, and development of new antibiotics: The WHO priority list of antibiotic-resistant bacteria and tuberculosis. *The Lancet Infectious Diseases*, 18(3): 318–327. https://doi.org/10.1016/S1473-3099(17)30753-3

Tsujimura, K., Murase, H., Bannai, H., Nemoto, M., Yamanaka, T. & Kondo, T. (2015). Efficacy of five commercial disinfectants and one anionic surfactant against equine herpesvirus type 1. *The Journal of Veterinary Medical Science*, 77(11): 1545. https://doi.org/10.1292/JVMS.15-0030

Vecino, X., Rodríguez-López, L., Rincón-Fontán, M., Cruz, J.M. & Moldes, A.B. (2021). Nanomaterials synthesized by biosurfactants. *Comprehensive Analytical Chemistry*, 94: 267–301. Elsevier. https://doi.org/10.1016/bs.coac.2020.12.008

Ventola, C.L. (2015). The antibiotic resistance crisis: Causes and threats. *P & T Journal*, 40(4): 277–283. https://doi.org/Article

Vieira, D.B. & Carmona-Ribeiro, A.M. (2006). Cationic lipids and surfactants as antifungal agents: Mode of action. *Journal of Antimicrobial Chemotherapy*, 58(4): 760–767. https://doi.org/10.1093/JAC/DKL312

Vijayakumar, S. & Saravanan, V. (2015). Biosurfactants – types, sources and applications. *Research Journal of Microbiology*, 10(5): 181–192. https://doi.org/10.3923/jm.2015.181.192

Vollenbroich, D., Pauli, G., Özel, M. & Vater, J. (1997). Antimycoplasma properties and application in cell culture of surfactin, a lipopeptide antibiotic from Bacillus subtilis. *Applied and Environmental Microbiology*, 63(1): 44–49. https://doi.org/10.1128/aem.63.1.44-49.1997

WHO. (2019). New report calls for urgent action to avert antimicrobial resistance crisis. *Joint News Release*, Vol. 29. https://www.who.int/news/item/29-04-2019-new-report-calls-for-urgent-action-to-avert-antimicrobial-resistance-crisis

Wu, S., Liu, G., Zhou, S., Sha, Z. & Sun, C. (2019). Characterization of antifungal lipopeptide biosurfactants produced by marine Bacterium Bacillus sp. CS30. *Marine Drugs*, 17(4): 199. https://doi.org/10.3390/MD17040199

Yuliani, H., Perdani, M.S., Savitri, I., Manurung, M., Sahlan, M., Wijanarko, A. & Hermansyah, H. (2018). Antimicrobial activity of biosurfactant derived from *Bacillus subtilis* C19. *Energy Procedia*, 153: 274–278. https://doi.org/10.1016/J.EGYPRO.2018.10.043

CHAPTER

14

Exploring the Potential of Glycolipids from *Pseudomonas* as Microbial Surfactants and their Applications in Health and Medicine

Loknath Samanta[1*], Laccy Phurailatpam[2], Amrutha Lakshmi M.[3], and Ritu Mawar[4]

[1] Department of Biotechnology, Mizoram University, Aizawl, Mizoram, India
[2] Department of Botany, Dayalbagh Educational Institute, Agra, India
[3] ICAR-Indian Institute of Oil Palm Research, Pedavegi, India
[4] ICAR-Central Arid Zone Research Institute, Jodhpur, India

1. Introduction

Microbial surfactants are structurally varied amphipathic surface-active compounds which include both hydrophobic as well as hydrophilic components. Such distinct physical characteristics of microbial biological surfactants (BSs) form a barrier between several liquid substances with distinct intensity of polarity and hydrogen bonds (Santos et al., 2016). The hydrophobic part of BS consists of hydrocarbon chains which include long chains of both saturated and unsaturated fatty acids. However, the hydrophilic part is diverse in nature as it can either be composed of polysaccharides, amphoteric, ionic, non-ionic or amino acids (Silva et al., 2014, Mao et al., 2015). In general, microbial biosurfactants are divided into two broad categories namely low molecular weight and high molecular weight categories (Bagheri et al., 2022). The low molecular weight BSs are well known for their properties of excellent surface activity thereby reducing interfacial and surface tension in the middle of distinct phases. They also possess low critical micelle concentration (CMC) and balance emulsions (Uzoigwe et al., 2015). They include several low molecular weight lipids (500 to 1500 Da) such as neutral lipids, phospholipids, glycolipids, or fatty acids. (Adetunji & Olaniran, 2021). Microbial surfactants possess a multiple range of well-defined diverse structural characteristics such as corrosion, dispersion, surface activity, emulsification, wetting, foaming, cleansing etc. (Sarubbo et al., 2022). Such properties

* Corresponding author: amrutha.m@icar.gov.in

make them capable of being an appropriate tool for numerous biotechnological applications like microbe-imparted oil recovery, bioremediation, emulsion-balancing factor, pharmaceutics, cosmetics and food industries (Araújo et al., 2020, Adetunjiet & Olaniran 2021, Ali et al., 2022).

Among the various groups of BSs investigated so far, the most common BSs are the glycolipids from *Pseudomonas aeruginosa* (Ceresa et al., 2021a). Rhamnolipids type of glycolipid biosurfactant isolated from *P. aeruginosa* can be significantly utilized for bioremediation. Additionally, they can successfully enhance the accumulation of hydrocarbons (Liu et al., 2018, Shah et al., 2022, Saranraj et al., 2022a,b). Certain recent investigations have also reported the isolation of bacteria producing *Pseudomonas* sp. from the roots of reed plants with hydrocarbon degrading properties (Saeed et al., 2021, Sadiq et al., 2022). Therefore, based on the literature survey, we can observe that the *Pseudomonas* genus is one of the most potential groups of microbes that can be employed for isolation of biosurfactants. Rhamnolipids are one of the most deeply investigated amphipathic microbial surfactants which had been declared as "oily glycolipids" by Bergstrom et al. in 1946 (Thakur et al., 2021). It is composed of two distinct components namely, rhamnose which is the glycon part and lipid which is the aglycon part (Chebbi et al., 2021). The rhamnose component consists of mono or di (L) rhamnose connected by a hydrophillic α-1,2-glycosidic linkage. The lipid component consists of saturated/unsaturated chains of beta-hydroxy fatty acids, typically ranging from C8–C24 in length. These chains are connected through ester bonds and exhibit hydrophobic properties. The rhamnose and lipid components are finally linked with glycosidic bonds (Jiang et al., 2021).

Microbial biosurfactants are multifaceted molecules manifesting several prospective applications in the fields of pharmaceutics and agriculture (Bee et al., 2019, Patel et al., 2022, Zaman et al., 2022). They are frequently utilized in commercial sectors as substitutes for chemical surfactants. These include various types of anionic surfactants like carboxylate fluoro, cationic surfactants like quaternary pyridinium and ammonium, alkyl ether phosphates, sodium lauryl sulfate, zwitterionic surfactants and alkyl benzene sulfonate. Such substitutions are specifically required for their economical production, particularly within pharmaceutics (Karnwal, 2023). Biosurfactants function as agents that disrupt the stability of the plasma membrane by interfering with the hydrophobic structural makeup of lipid membranes. Due to their detergent-like characteristics, BSs have the ability to merge with the lipid bilayer and result in the development of pores on the basis of their concentration. Due to their distinctive chemical compositions, BSs display diverse physical, chemical as well as biological characteristics among which are notably impactful immunological properties (Sajid et al., 2020). BSs are generally viewed as superior in effectiveness, selectivity, environmental friendliness, and stability compared to numerous other synthetic compounds. Hence, they have garnered significant attention in both fundamental scientific research and industrial investigations. This chapter delves into the multifaceted realm of glycolipids sourced from *Pseudomonas*, aiming to elucidate their potential as microbial surfactants and their diverse applications within the domains of health and medicine.

2. Glycolipid Biosurfactants

Glycolipids are extensively studied BSs with low molecular weight, and they are produced from various sources such as frying wastes, hydrocarbons, olive oil wastes and industrial wastes (Liepins et al., 2021). The structural component of glycolipid BSs includes a hydrophilic part composed of carbohydrates like rhamnose, glucose, mannose, galactose, trehalose, and sophorose. Additionally, there is a hydrophobic moiety with a long fatty acid chain (Srivastava et al., 2022). Efficiency of BSs against bacteria, viruses, mycoplasma, and fungi is indeed attributed to their ability to destabilize biological membranes through the formation of ion channels and pores (Liepins et al., 2021). Glycolipids' disruptive action on membranes can compromise the structural integrity of microorganisms, leading to their inhibition or destruction. Indeed, glycolipid biosurfactants encompass a diverse range of subclasses, each with their unique structure and properties (Balleza et al., 2019). Some of the notable subdivisions include rhamnolipids, galactosyl-diglyceride, sophorolipids, monoacylglycerol, trehalose lipids, cellobiose lipids, mannosylerythritol lipids, diglycosyl diglycerides, lipomannosyl-mannitols, lipoarabinomannanes, and lipomannans (Ahmadi-Ashtiani et al., 2020). Certainly, rhamnolipids have been the focus of extensive study in the recent past due to their favorable characteristics (Zeng et al., 2018). They exhibit a lower range in both surface tension (28 to 30 mN/m) and critical micellar concentration (10 to 200 mg/L). Additionally, they possess high emulsifying indexes ranging from 60 to 70%, and their production is notably high, occurring within a very short time duration. These attributes make rhamnolipids particularly attractive for various industrial and environmental applications (Dobler et al., 2020).

3. Rhamnolipid (RL), the Chief Glycolipid Biosurfactant

Rhamnolipid is a type of glycolipid biosurfactant produced by the bacteria *Pseudomonas aeruginosa* which is well-known for its pathogenic effects on plants and animals including humans. They are sugar lipids, chemical compounds which are among the well-known microbial surfactants (Mnif and Ghribi, 2016). RL was first reported as a novel biomolecule in 1946, when *P. aeruginosa* exhibited antibiotic properties by producing pyolipic acid against *Mycobacterium tuberculosis* (Eslami et al., 2020). The physiological properties of RL may vary considerably on the basis of the producing microbes, their carbon source and culture set-up, including the media used. Generally, RLs can be characterized as mono- (one rhamnose ring) as well as di-rhamnolipids (two rhamnose rings linked through alpha-1,2-glycosidic linkage) (El-Housseiny et al., 2020). Its lipid component may have one, two or three saturated or unsaturated beta-hydroxy fatty acid chains with variable length of approximately C8-C16. The most well-studied microbe in the context of rhamnolipid production is *Pseudomonas aeruginosa*. Rhamnolipid biosynthesis by this bacterium has gone through considerable processes, genetic optimization and rescaling to achieve the current scenario where RL derived products are being commercialised (Sarubbo et al., 2022). In spite of its promising functional properties, the application of RL is fairly limited in several industries such as beauty products including food and skin-care. As a result a concerted search for harmless rhamnolipid sources continues. A number of non-disease causing *Pseudomonas* strains were found to synthesize RLs (Devale et al., 2022). Likewise, *P. fluorescens*, *P. putida*, *P. alcaligenes*, *P. stutzeri*, *P. chlororaphis* were reported to produce different types of RLs (Salek et al., 2022). Other than *Pseudomonas* strains, other bacterial strains like *Burkholderia thailandensis*, *Marinobacter* sp., *Acinetobacter calcoaceticus*, *Burkholderia plantarii*, *Enterobacter asburiae* were also reported to secrete RLs (Salek et al., 2022).

4. Applications of Gycolipids Biosurfactant

Indeed, glycolipids have a wide range of applications due to their diverse properties. The ability of glycolipid to reduce surface tension makes them valuable in various industries (Mnif and Ghribi, 2016). Their role in reducing surface tension between solid/liquid, liquid/liquid, and liquid/gas interfaces is particularly advantageous in applications such as detergents, cosmetics and cleaning agents (Gürkök and Özdal, 2021). Rhamnolipids which is a type of glycolipid unveiled a diverse range of properties, expanding their potential applications (Mnif et al., 2018). Beyond their surface-active characteristics, rhamnolipids have been observed to influence the folding of outer membrane protein OmpA. Moreover, RLs exhibit certain beneficial characteristics such as immunomodulation, anticancer, antimicrobial and capability to produce nanoparticles (Yela et al., 2016) (Fig. 1).

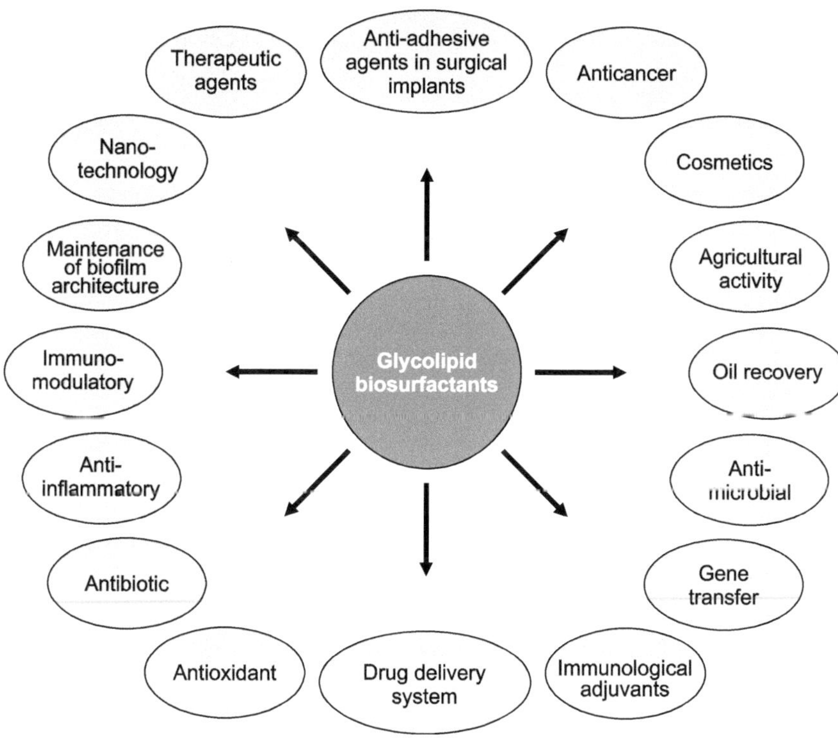

Fig. 1: Applications of different types of glycolipids derived from several strains of *Pseudomonas* in therapeutics, biomedicine and pharmaceutical industries.

4.1 Effects of Glycolipid Biosurfactants on Immune System

Glycolipid biosurfactants have been identified as compounds that influence both the humoral as well as cellular immune systems (Sajid et al., 2020). The immunoregulatory function of numerous biosurfactants is mainly exploited by microbes to begin host infection. For example, RLs impart immune evasion to *P. aeruginosa* by preventing the synthesis of antimicrobial peptide (e.g., human beta defensin-2), hampering phagocytic function and by promoting lysis of cells such as macrophages and neutrophils (Jurado-Martín et al., 2021). Nevertheless, the immunosuppressive actions of selective biosurfactants can be employed for the medication of several immune-related diseases. An earlier study reported that treatment with sophorolipids significantly saved rats from the harmful consequence of infectious shock by reducing the secretion of pro-inflammatory cytokines and nitric oxide (Subramaniam et al., 2020). As a reaction to the effect of sophorolipids, IgE secreting myeloma cells retard the activities of STAT3, PAX5 and TLR-2, thereby reducing the expression of IL-6 gene and IgE secretion. This type of consequence showed that sophorolipids could alleviate the deleterious results of IgE- regulated immune responses (Pal et al., 2023).

Rhamnolipid, primarily produced by *P. aeruginosa,* is an endotoxin (Wood et al., 2023). Many of the earlier studies reported several immunostimulatory functions of RL (Table 1).

Table 1: Reports on various applications of glycolipids in medical and pharmaceutics from various strains of *Pseudomonas* species

Types of biosurfactants	Producing microbes	Applications	References
Glycolipids	*Pseudomonas otitidis*	Pharmaceutical industries, food and cosmetics	Singh and Tiwari, 2016
Rhamnolipid	*Pseudomonas* sp. PS-17	Antibacterial	Sotirova et al., 2008
Rhamnolipid	*P. aeruginosa* PTCC 13401	Nanoparticle synthesis in medical sector	Bayee et al., 2020
Mono & di-RLs	*P. aeruginosa*	Anti-microbial activity in medical sector	Ndlovu et al., 2017
Rhamnolipid	*P. aeruginosa* OBP1	Antibacterial	Ceresa et al., 2019
Rhamnolipid	*P. aeruginosa*	Nano-emulsions for drug delivery mechanism against SCC7 tumour cells	Yi et al., 2019
Rhamnolipid	*P. aeruginosa* DS10-129	Anti-biofilm effects by Disruption of initial adhesion	Rodrigues et al., 2006
Mono & di-RLs	*P. aeruginosa* MR01	Cytotoxic actions on breast cancer cells of humans	Rahimi et al., 2019
Rhamnolipid	*P. aeruginosa*	Antibacterial activity against food pathogens: *B. cereus* and *S. aureus*	de Freitas Ferreira et al., 2018
Rhamnolipid	*P. aeruginosa* PA1	Anti-biofilm	de Araujo et al., 2016
Rhamnolipid	*P. aeruginosa* strain B5	Restriction of spore germination and hyphal growth	Goswami et al., 2015
Rhamnolipid	*P. aeruginosa* DS9	Restriction of spore germination and mycelial growth	Kim et al., 2000

Contd...

Contd...

Types of biosurfactants	Producing microbes	Applications	References
Rhamnolipid	*P. aeruginosa* SS14	Inhibiting spore germination and mycelial growth	Sen et al., 2019
Rhamnolipid	*P. aeruginosa* MN1	Anti-biofilm effects	Abdollahi et al., 2020
Rhamnolipid	*P. aeruginosa* DSVP20	Anti-biofilm effects	Singh et al., 2013

However, recent studies have observed that RLs behave as a principal factor for persistence of *P. aeruginosa* within the host organism (Moradali et al., 2017). RLs are heat tolerant types of glycolipids with several properties. They are cytotoxicity, antimicrobial nature, antibiofilm formation, antiamoebic, antiviral, zoo sporicidal, algicidal, haemolytic and anti-adhesive (Jiang et al., 2014). Many of the previous investigations had reported several immune regulatory activities of RLs like chemotactic recruition of neutrophil cells and enhanced oxidative activities, liberation of histamine from mast cells, elevated secretion of serotonin and 12-hydroxyeicosatetraenoic acid which is a potential chemical attractant for neutrophil from platelets of human beings (Sajid et al., 2020). Moreover, RLs also halt phagocytic activities of macrophages by hampering the procedure of internalization thereby, preventing the fusion of lysosomes with particle enveloped phagosomes within a macrophage (McSorley, 2016). Secretion of RL by fully grown biofilm cells of *P. aeruginosa* is enhanced when they come across polymorphonuclear leukocytes. Nevertheless, at the time of biofilm development, *P. aeruginosa* produces lesser RL (Alhede et al., 2014). RLs cannot be identified by TLR-2 and TLR-4 (Toll like receptor), however, they can pass through the lipid bilayer and accumulate within the cell cytoplasm.

According to earlier studies, RLs have the property to lyse macrophage and necrosis of polymorphonuclear neutrophils (PMNs) and suppress exhibition of antimicrobial peptides (Sajid et al., 2020). Another study also reported that RL secreted by *P. aeruginosa* at the time of pathogenesis have the ability to lyse immune cells like PMNs and monocyte- derived macrophages (Watters et al., 2016). On the other hand, an investigation by Gennip et al., 2009 had shown that deactivation of rhlA gene of *P. aeruginosa* inhibited the synthesis of RL thereby, resulting in disabling the defence towards PMNs. Furthermore, a recent investigation had revealed the significant role of RL in the existence of *P. aeruginosa* within the host organism. From all the above reports, we can conclude that RL is responsible for necrosis of PMNs and defend the cells of *P. aeruginosa* from the effect of PMNs (Wood et al., 2023).

4.2 Role of Glycolipid Surfactant in Biomedical and Pharmaceutical Applications

Biosurfactants generally regulate several biological functions which may include locomotion, metabolism and survival of microbial organisms. BSs also enhance the availability of surface area and biological availability of hydrophobic water-insoluble substrates and they also play a significant role in elimination of heavy metals from the atmosphere. Such molecules also stimulate the affinity or detachment of microbes towards substrates, microbial locomotion, conditioning of cell surfaces and accumulation at interfaces or substrates where the interaction occurs (Ceresa et al., 2021a). Moreover, tolerance against harmful molecules, substrate accession and cellular demarcation also constitute the general characteristics of microbial biosurfactants (Fracchia et al., 2012). As we have already mentioned before, RLs play several roles in the survival of microbial organisms. RLs are one of the critical components which are exploited for preserving the architecture of biofilm. RLs are also regarded as a significant virulent element present in *Pseudomonas* strains. As a consequence of natural mechanisms, RLs have evolved to enhance assimilation of hydrophobic substrates by bacteria (Van Hamme et al., 2006).

BSs not only provide a selective advantage to their microbial producers but they also impart antimicrobial activities towards other microbes which do not synthesize such molecules. They can function as virulent elements as well as quorum-sensing molecules which modulate the exhibition of several other virulence factors like promotion of biofilm architecture, preservation and finally the dispersion of biofilms. Moreover, BSs are also critical for maintenance of gas-channels, exchange of nutrients across biofilms, and diffusion structure and biofilm surfaces (Raaijmakers et al., 2010, Ceresa et al., 2021a). Recently, many investigations reported that BSs possess multiple promising biological characteristics which can be exploited by medical and pharmaceutical industries. The mechanism of activities of biosurfactants on the cells of microbes may include interaction or detachment to or from substrates, bringing changes in wettability and surface energy, decrease hydrophobicity and upregulate permeability via liberation of LPS and the development of transmembrane pores. As a result, BSs destroy the overall integrity of the membrane resulting in lysis of cells, impaired membrane actions like energy production and transportation process and metabolite seepage (de Jesús Cortés-Sánchez et al., 2013, Mandal et al., 2013).

4.3 Glycolipid Biosurfactants as Antitumor Agents

Rhamnolipids are known to have promising effects on both human as well as animal cancerous cells (Thakur et al., 2021). Another form of RL, Dirhamnolipids have been reported to delay the division process and growth of insect cell line C6/36 as well as breast cancer cell line MCF-7. RLs produced by B1898 have also been proved to recover wounds, identify the cytoskeleton of phagocytic as well as nonphagocytic cells and can modify their structure (Christova et al., 2013). Such types of beneficial features of RLs, were minutely investigated in many cell types isolated from tumors of various origins (Thakur et al., 2021). One of the earlier studies investigated on RL-1 which is a mono-rhamnolipid and RL-2, a di-rhamnolipid from BN10 strains regarding their antitumor properties against JMSU-1, HL-60, BV-173, SKW-3, human cancer cell lines (Christova et al., 2013). The cytotoxicity experiment proved that mono-rhamnolipid, RL-1 is far more potent in comparison to mono-rhamnolipid, RL-2. Moreover, RL-1 had more lethal toxic consequences against BV-173 and SKW-3 cancer cell lines, but reduced toxic effects against HL-60 and JMSU-1 cancer cell lines. Furthermore, RL-2 was exploited in higher concentrations in order to reveal their cytotoxic consequence in comparison to RL-1. The report further mentioned that RL-1 imparts modifications in the structure of blood cancer cells by giving their plasma membrane, a blebbing like shape, condensation of chromosomes, appearance of apoptotic cells and nuclear disintegration resulting in programmed cell death of BV-173 cells (Christova et al., 2013). The most interesting part of the study is that researchers found higher concentrations of RL-1 cause an excessive expression of *Bcl-2 and c-myc* genes of BV-173, which ultimately cause frequent division and anti-apoptosis of BV-173 cells (Christova et al., 2013). Such a unique quality makes RL-1 more potent to be exploited as an anticancer agent in pharmaceutics and biomedical sectors. Likewise, another investigation reported the isolation of RL1 and RL2 from MR01 strain of *P. aeruginosa.* Through the cytotoxicity test against MCF-7 cells, it was observed that proliferation of these cells was observed to be halted by both RLs. However, RL1 was found to have more potential in destroying tumour cells (Rahimi et al., 2019). Their significant effect on cell feasibility was a consequence of their communication with cell membranes. Moreover, phase-contrast microscopy had proved that modification in cell morphology of MCF-7 cells was observed after treating them with mono- and di-rhamnolipids at concentrations of 25, 50, 100 µg/mL for 48 hours (Reddy et al., 2013, Rahimi et al., 2019). Such treated MCF-7 cells are also found to have a shrunken spherical morphology and the control cells have cuboidal or polygonal structures (Reddy et al., 2013, Rahimi et al., 2019). It had also been reported that RLs treating MCF-7 cells were found to lose their interaction with neighbouring cells resulting in the loss of their substrate which ultimately leads to flotation (Rahimi et al., 2019). All such morphological changes are symptoms of apoptotic cells from which we can conclude that RLs are promising inhibitors of MCF-7 growth (Ramya et al., 2017). Another study revealed that rhamnolipid biosurfactants produced by *P. aeruginosa* inhibited the proliferation of growth of HeLa cancer cells at a concentration of 5 µg/mL due to their cytotoxic actions (Lotfabad et al., 2010). On the basis of all the above reports, it is very clear that RLs can be beneficially exploited as prospective therapeutic antitumor agents.

4.4 Glycolipid Biosurfactants in Oxidative Stress Responses

RLs are reported to decrease oxidative stress in *Solanum lycopersicum*, which occur due to infection by *Alternaria alternata* by imparting tolerance as well as defence mechanisms in the host plant (Yan et al., 2016). Higher levels of oxidative stress in plants often causes deterioration of the fruit. However, they provide lower-level help in the induction of defence mechanisms towards the pathogen within the fruit (Thakur et al., 2021). One of the studies had reported that when cherry tomatoes were treated with both RLs and pathogen, within 12 hours, the intracellular level of hydrogen peroxide (H_2O_2) increases, however after 12 hours, extreme reduction in the level of H_2O_2 was observed (Yan et al., 2016)). Therefore, there should be a balance between reactive oxygen species (ROS) production and degradation in the early phase. RLs also upregulate the concentration of both superoxide dismutase and catalase which quench the high quantity of ROS. Glutathione (GSH) was also found to have the property of eliminating intracellular ROS by stimulating the activation of glutathione reductase which is responsible for conversion of oxidized glutathione to glutathione. During the exposure of tomato fruit to RLs and pathogens, an elevated quantity of GSH within a 12–36 hours interval after treatment was observed. However, reduced concentration of H_2O_2 as well as GSH was seen when the tomato was treated with *A. alternata* along with RLs after an interval of 12 hours (Yan et al., 2016). Therefore, RLs can impart tolerance in cherry tomatoes against the pathogenic effect of *A. alternata* via their response mechanism to oxidative stress by inducing the generation of antioxidant enzymes to remove the extreme level of ROS.

4.5 Glycolipids in Drug Delivery

Currently, a key area of focus in nano-research involves the biosynthesis of nanoparticles using microbes. This environmentally friendly approach, known as green chemistry, establishes a connection between nanotechnology and microbial biotechnology. To create an environmentally friendly method for producing bioactive nanoparticles, BSs have surfaced as a promising alternative (Singh et al., 2019). These substances not only aid in the synthesis of nanoparticles but also contribute to stabilizing them s. Several biosurfactants have been explored for their potential use in nanoparticle synthesis, and among them, rhamnolipids have been employed by various research groups as a stabilizing agent (Worakitsiri et al., 2011, Hazra et al., 2013, Yi et al., 2019). Recognizing the potential of rhamnolipids, researchers have explored their role as nanoparticles for drug delivery. In this regard, a pioneering study conducted by Müller et al. investigated the use of various rhamnolipids, synthesized chemically or commercially, as nano-carriers for drug delivery to the skin in an ex vivo system (Müller et al., 2017). The researchers employed rhamnolipid nanoparticles loaded with various hydrophobic drugs such as Nile red, dexamethasone, or tacrolimus for their skin delivery investigations. The authors explicitly showed that loading these rhamnolipids with hydrophobic drugs is feasible, achieving up to 30% of drug loading. Moreover, through ex-vivo study, it was illustrated that fluorophore Nile reds laden on RL nanoparticles, when applied to excised human skin, could effectively transfer Nile red into the skin without inducing toxic effects even at concentrations greater than critical micelle concentrations (CMCs). This study introduces a novel perspective on employing rhamnolipids as a substitutive drug delivery system. In 2019, a sophisticated investigation utilized rhamnolipid nanoparticles laden with a hydrophobic photosensitizer, "pheophorbide a," and inoculated it intravenously to mice bearing SCC7 tumours. Remarkably, the results revealed not only a substantial aggregation of "pheophorbide a"-loaded nanoparticles (Pba-RLNP) in the tumour tissue but also effective tumour suppression through photodynamic therapy (Müller et al., 2017). Regarding the scope of the prospective exploitation of RLs, discussed here, it can be suggested that RLs nanoparticles could augment cancer chemotherapy by leveraging their anticancer activities and enhancing permeability and retention effects thus, resulting in the accumulation of drugs within cancer cells.

4.6 Role of Glycolipids Biosurfactants as Antimicrobial Agents

The most frequently investigated classes of biosurfactants with antimicrobial activity are lipopeptides and glycolipids (Cochrane et al., 2016). Specifically, Polymyxin A and Polymyxin B derived from *Bacillus polymyxa*, as well as fengycin, surfactin, bacillomycins iturin, and mycosubtilins, secreted by *Bacillus subtilis*, along with pumilacidin from *Bacillus pumilus*, viscosin from *Pseudomonas fluorescens* and lichenysin from *Bacillus licheniformis* are recognized as common antimicrobial lipopeptides (Ceresa et al., 2021a). The most extensively studied examples of gycolipids include rhamnolipids from *P. aeruginosa* and sophorolipids from *Candida bombicola* (Benincasa et al., 2004). One recent study reported that fengycin from *Bacillus amyloliquefaciens* can significantly trigger considerable changes in the surface topography of the potential human pathogen *P. aeruginosa* resulting in reduced cell height as well as loss of intracellular components (Medeot et al., 2020). The combination of rhamnolipids and surfactin generated from *P. aeruginosa* and *B.*

amyloliquefaciens respectively, exhibited significant antimicrobial functions against a wide range of potential and pathogenic microbes. They may consist of antibiotic-tolerant bacterial species like *Staphylococcus aureus* and *Escherichia coli*, and yeast such as *Candida albicans* (Ndlovu et al., 2017). Another recent study demonstrated the antifungal activity of a RL from *P. aeruginosa* against *Trichophyton rubrum*. This research also revealed that a purified microbial surfactant with a concentration of 0.5 mg/mL efficiently caused loss of cell membrane integrity, reduced hyphal growth, induced alterations in hyphal morphology in vitro and inhibited the germination of spores thus successfully curing experimentally induced cutaneous dermatophytosis within 21 days when treated on the surface of infected mice (Sen et al., 2019).

In numerous studies, the antimicrobial mechanism of activities of biosurfactants has been attributed to their well-recognized disruptive actions on cell membranes, owing to their amphiphilic nature. Nevertheless, there is increasing evidence of the role of biosurfactants in quorum sensing signalling (Khan et al., 2019, Yan et al., 2019). Analytical studies on the biosynthesis of RLs by a strain of *P. aeruginosa* isolated from manure proved that its co-inoculation with a selective bacterial consortium significantly enhanced the synthesis of RLs (Ceresa et al., 2021a). This improvement was observed in terms of maximum yield in comparison to the axenic culture. Such an outcome seemed to be linked to interspecies communication through quorum sensing, specifically involving AI-2 signalling molecules. This highlights the importance of interspecies communication in biosurfactant generation (Woźniak-Karczewskaet al., 2017). All such hopeful results underscore the necessity to investigate the role of biosurfactants in microbial competitive interactions.

4.7 Role of Biosurfactants as Antiadhesive and Antibiofilm Agents

Recently, research has been conducted on silicone elastomer discs coated with rhamnolipid R89 made up of mono- (75%) and di- (25%) RL families. This particular rhamnolipid is produced by the clinical strain of *P. aeruginosa*. According to their findings, Ceresa et al. (2019) observed that silicone discs coated with rhamnolipid R89 led to a reduction in both biofilm biomass and metabolic activity. This reduction was significant, approximately 71% for *S. aureus* and 65% for *Stayphylococcus epidermidis*, over a 72-hour period. Remarkably, the coating did not impact cell viability, thereby preserving the required biocompatibility and leaching of products. Furthermore, it was demonstrated that the presence of R89 solutions effectively disseminated pre-formed biofilms of *S. aureus* and *S. epidermidis* up to 93%. Such dissemination was attributed to the antimicrobial property of the RL mixture. Antibiofilm function of three sophorolipid mixtures, namely, SLA which is an acidic congenator, SL18, which is a lactonic congenator and SLV, which is composed of the two was assessed against biofilm formation and pre-formed biofilms of *S. aureus*, *C. albicans* and *P. aeruginosa* (Ceresa et al., 2021b). In co-inoculations, biosurfactants halted the construction of microbial biofilms by 90–95%. The absorption of various concentrations of biosurfactants on silicone significantly restricted *S. aureus* and *C. albicans* biofilm formation, reaching up to 72%, in a concentration-reliant fashion. However, this experimental technique was unsuccessful against biofilm formation of *P. aeruginosa*. Moreover, when these sophorolipids mixtures were treated with 24-hour-old biofilms, all three congenator mixtures exhibited biofilm disintegrating actions, resulting in a reduction of 70%, 75%, and 80% for *S. aureus*, *P. aeruginosa*, and *C. albicans*, respectively (Ceresa et al., 2021b).

4.8 Antiviral Activity of Biosurfactants

Over the past three decades, biosurfactants have been recognized for their antiviral characteristics towards several types of enveloped viruses. Such repressive actions of BSs are attributed to the establishment of ionic channels in lipid envelopes as well as viral capsids, disruption of proteins involved in viral adsorption and diffusion procedures, and suppression of the fusion of viral membrane (Yuan et al., 2018, Ceresa et al., 2021a). Rhamnolipid PS-17, both in its free form and in combination with alginate, has been identified as a promising anti-herpes simplex virus agent. It demonstrates blockage of the viral cytopathic function and the repression of viral replication in a quantity reliant manner, with concentrations lower than the CMC of herpes virus (Ceresa et al., 2021a). A recent investigation highlighted the capability of rhamnolipids 222B to deactivate two different types of enveloped viruses, namely bovine coronavirus and Herpes Simplex Virus 1 (HSV-1). Their findings indicated that rhamnolipid 222B at concentrations of 0.009% and 0.0045% could deactivate 6 and 4 log PFU/ml of HSV-1 in 5–10 minutes, respectively, while maintaining non-cytotoxicity at or below 0.005% concentration (Jin et al., 2021). Furthermore, the researchers investigated the potential application of RLs as coatings on both fabric as well as plastic surfaces for antiviral safeguards and masks. On the basis of their investigation, applying 50 μL of rhamnolipid 222B at 0.005% on 1 cm^2 of plastic surfaces or mask fabrics could deactivate approximately ~10^3 Plaque-Forming Units (PFU) of HSV-1 within three-five minutes. This suggests the potential use of rhamnolipid coatings on masks to avoid or decrease the spreading of several enveloped viral strains.

The ongoing pandemic caused by an extremely dreadful respiratory syndrome–coronavirus-2 or SARS-CoV-2 has opened up possibilities for innovative pharmaceutical and biomedical exploitations of biosurfactants. The virulence of SARS-CoV-2 relies on the strength of its lipidic envelope, which encapsulates essential proteins and RNA (Smith et al., 2020). The amphiphilic property of BSs enables them to directly come across the lipid membrane of the virus, disrupting the viral structure and thereby, reducing its infectivity (Sandeep & Rajasree, 2017). Moreover, the ability of BSs to form micellar structures at their CMC might be critical for their exploitation as liposomes in drug delivery to the site of infection. This property helps to preserve the functionality of the drug from the harsh circumstances in the body (Ceresa et al., 2021a). The applications of BSs can also be considered as an immediate therapy for Acute Respiratory Distress Syndrome (ARDS) through solubilization of the alveolar substrate and enhancement of the clearance of liquid from this particular area (Smith et al., 2020). A recent assessment of existing evidence regarding the role of biosurfactants in the creation of Microemulsion Drug Delivery Systems (MDDS) to enhance the biological availability of hydrophobic drugs was conducted by Ohadi et al. (2020). They concluded that BSs serve as a potential biosource for MDDS because of their outstanding self-assembling and emulsifying functions.

5. Conclusion

Biosurfactants are emerging as surface-active molecules with significant potential for various applications in the biomedical and pharmaceutical industries. Their unique properties make them promising candidates for diverse uses including drug delivery, antimicrobial treatments, and other innovative biomedical applications. They are highly charismatic substances owing to their notable antimicrobial properties, effective against viruses, bacteria and fungi. Additionally, they exhibit anti-adhesive and biofilm destructive functions, making them versatile candidates for various applications in combating microbial threats. Their utilization both independently or in combination with other antimicrobial or chemotherapeutic drugs, holds promise for future strategies aimed at preventing and countering microbial infections, as well as inhibiting biofilm formation and proliferation. This approach may open new avenues for effective interventions against a range of microbial challenges. Moreover, biosurfactants have lately attracted attention within the scientific community as a promising new generation of pharmaceuticals. They are being considered for inclusion in anticancer, immunomodulatory, wound healing, cosmetic, and drug delivery agents, highlighting their potential versatility in various therapeutic and cosmetic applications. It's crucial to emphasize that many of these properties of biosurfactants can interact or affect each other, potentially leading to side effects in different applications. Therefore, thorough investigation and understanding of these interactions are essential for ensuring the safe and effective use of biosurfactants in various applications. The adoption of biosurfactants at a commercial scale is both appropriate and mandatory to mitigate the adverse environmental impact of conservative synthetic surfactants. However, challenges related to the economic value of their prospective implementation and their accessibility still need to be addressed and resolved for widespread use and impact.

References

Abdollahi, S., Tofighi, Z., Babaee, T., Shamsi, M., Rahimzadeh, G., Rezvanifar, H. et al. (2020). Evaluation of anti-oxidant and anti-biofilm activities of biogenic surfactants derived from *Bacillus amyloliquefaciens* and *Pseudomonas aeruginosa*. *Iranian Journal of Pharmaceutical Research*, 19(2): 115.

Adetunji, A.I. & Olaniran, A.O. (2021). Production and potential biotechnological applications of microbial surfactants: An overview. *Saudi Journal of Biological Sciences*, 28(1): 669-679.

Ahmadi-Ashtiani, H.R., Baldisserotto, A., Cesa, E., Manfredini, S., Sedghi Zadeh, H., Ghafori Gorab, M. et al. (2020). Microbial biosurfactants as key multifunctional ingredients for sustainable cosmetics. *Cosmetics*, 7(2): 46.

Alhede, M., Bjarnsholt, T., Givskov, M. & Alhede, M. (2014). *Pseudomonas aeruginosa* biofilms: Mechanisms of immune evasion. *Advances in Applied Microbiology*, 86: 1-40.

Ali, S.A.M., Sayyed, R.Z., Reddy, M.S., Enshasy, H.E. & Hameeda, B. (2022). Delving through quorum sensing and CRISPRi strategies for enhanced surfactin production. *In:* Sayyed, R.Z. (Ed.), *Biosurfatnats: Production and Applications in Bioremediation/Reclamation*. 59-79. CRC Press, Taylor & Francis Group, USA.

Araújo, W.J., Oliveira, J.S., Araújo, S.C.S., Minnicelli, C.F., Silva-Portela, R.C.B., Da Fonseca, M.M. et al. (2020). Microbial culture in minimal medium with oil favors enrichment of biosurfactant producing genes. *Frontiers in Bioengineering and Biotechnology*, 962.

Bagheri, H., Mohebbi, A., Amani, F.S. & Naderi, M. (2022). Application of low molecular weight and high molecular weight biosurfactant in medicine/biomedical/pharmaceutical industries. *Green Sustainable Process for Chemical and Environmental Engineering and Science*, 1-60. Academic Press.

Balleza, D., Alessandrini, A. & Beltrán García, M.J. (2019). Role of lipid composition, physicochemical interactions, and membrane mechanics in the molecular actions of microbial cyclic lipopeptides. *The Journal of Membrane Biology*, 252(2-3): 131-157.

Bayee, P., Amani, H., Najafpour, G.D. & Kariminezhad, H. (2020). Experimental investigations on behaviour of rhamnolipid biosurfactant as a green stabilizer for the biological synthesis of gold nanoparticles. *International Journal of Engineering*, 33(6): 1054-1060.

Bee, H., Khan, M.Y. & Sayyed, R.Z. (2019). Microbial surfactants and their significance in agriculture. *In:* Sayyed Reddy Antonious (Ed.), *PGPR: Prospects for Sustainable Agriculture.* 205-216. Springer-Nature, Singapore.

Benincasa, M., Abalos, A., Oliveira, I. & Manresa, A. (2004). Chemical structure, surface properties and biological activities of the biosurfactant produced by *Pseudomonas aeruginosa* LBI from soapstock. *Antonie Van Leeuwenhoek*, 85: 1-8.

Ceresa, C., Fracchia, L., Fedeli, E., Porta, C. & Banat, I.M. (2021). Recent advances in biomedical, therapeutic and pharmaceutical applications of microbial surfactants. *Pharmaceutics*, 13(4): 466.

Ceresa, C., Rinaldi, M., Tessarolo, F., Maniglio, D., Fedeli, E., Tambone, E. (2021). Inhibitory effects of lipopeptides and glycolipids on *C. albicans – Staphylococcus* spp. dual-species biofilms. *Frontiers in Microbiology*, 11: 545654.

Ceresa, C., Tessarolo, F., Maniglio, D., Tambone, E., Carmagnola, I., Fedeli, E. et al. (2019). Medical-grade silicone coated with rhamnolipid R89 is effective against *Staphylococcus* spp. biofilms. *Molecules*, 24(21): 3843.

Chebbi, A., Franzetti, A., Duarte Castro, F., Gomez Tovar, F.H., Tazzari, M., Sbaffoni, S. & Vaccari, M. (2021). Potentials of winery and olive oil residues for the production of rhamnolipids and other biosurfactants: A step towards achieving a circular economy model. *Waste and Biomass Valorization*, 12: 4733-4743.

Christova, N., Tuleva, B., Kril, A., Georgieva, M., Konstantinov, S., Terziyski, I. et al. (2013). Chemical structure and in vitro antitumor activity of rhamnolipids from *Pseudomonas aeruginosa* BN10. *Applied Biochemistry and Biotechnology*, 170: 676-689.

Cochrane, S.A. & Vederas, J.C. (2016). Lipopeptides from *Bacillus* and *Paenibacillus* spp.: A gold mine of antibiotic candidates. *Medicinal Research Reviews*, 36(1): 4-31.

de Araujo, L.V., Guimarães, C.R., da Silva Marquita, R.L., Santiago, V.M., de Souza, M.P., Nitschke, M. et al. (2016). Rhamnolipid and surfactin: Anti-adhesion/antibiofilm and antimicrobial effects. *Food Control*, 63: 171-178.

de Freitas Ferreira, J., Vieira, E.A. & Nitschke, M. (2019). The antibacterial activity of rhamnolipid biosurfactant is pH dependent. *Food Research International*, 116: 737-744.

de Jesús Cortés-Sánchez, A., Hernández-Sánchez, H. & Jaramillo-Flores, M.E. (2013). Biological activity of glycolipids produced by microorganisms: New trends and possible therapeutic alternatives. *Microbiological Research*, 168(1): 22-32.

Devale, A., Sawant, R. & Mujumdar, S. (2022). 12 Rhamnolipids (RLs) green solution for global hydrocarbon pollution. *Microbial Surfactants: Volume 3: Applications in Environmental Reclamation and Bioremediation*, 2025.

Dobler, L., Ferraz, H.C., de Castilho, L.V.A., Sangenito, L.S., Pasqualino, I.P., Dos Santos, A.L.S. et al. (2020). Environmentally friendly rhamnolipid production for petroleum remediation. *Chemosphere*, 252: 126349.

El-Housseiny, G.S., Aboshanab, K.M., Aboulwafa, M.M. & Hassouna, N.A. (2020). Structural and Physicochemical Characterization of Rhamnolipids produced by *Pseudomonas aeruginosa* P6. AMB Express, 10: 1-12.

Eslami, P., Hajfarajollah, H. & Bazsefidpar, S. (2020). Recent advancements in the production of rhamnolipid biosurfactants by *Pseudomonas aeruginosa*. RSC Advances, 10(56): 34014-34032.

Fracchia, L., Cavallo, M., Martinotti, M.G. & Banat, I.M. (2012). Biosurfactants and bioemulsifiers biomedical and related applications – Present status and future potentials. *Biomedical Science, Engineering and Technology*, 14(1): 1-49.

Goswami, D., Borah, S.N., Lahkar, J., Handique, P.J. & Deka, S. (2015). Antifungal properties of rhamnolipid produced by *Pseudomonas aeruginosa* DS9 against *Colletotrichum falcatum*. *Journal of Basic Microbiology*, 55(11): 1265-1274.

Gürkök, S. & Özdal, M. (2021). Microbial biosurfactants: Properties, types, and production. *Anatolian Journal of Biology*, 2(2): 7-12.

Hazra, C., Kundu, D., Chaudhari, A. & Jana, T. (2013). Biogenic synthesis, characterization, toxicity and photocatalysis of zinc sulfide nanoparticles using rhamnolipids from *Pseudomonas aeruginosa* BS01 as capping and stabilizing agent. *Journal of Chemical Technology & Biotechnology*, 88(6): 1039-1048.

Jiang, L., Shen, C., Long, X., Zhang, G. & Meng, Q. (2014). Rhamnolipids elicit the same cytotoxic sensitivity between cancer cell and normal cell by reducing surface tension of culture medium. *Applied Microbiology and Biotechnology*, 98: 10187-10196.

Jiang, N., Dillon, F.M., Silva, A., Gomez-Cano, L. & Grotewold, E. (2021). Rhamnose in plants – From biosynthesis to diverse functions. *Plant Science*, 302: 110687.

Jin, L., Black, W. & Sawyer, T. (2021). Application of environment-friendly rhamnolipids against transmission of enveloped viruses like SARS-CoV2. *Viruses*, 13(2): 322.

Jurado-Martín, I., Sainz-Mejías, M. & McClean, S. (2021). *Pseudomonas aeruginosa*: An audacious pathogen with an adaptable arsenal of virulence factors. *International Journal of Molecular Sciences*, 22(6): 3128.

Karnwal, A. (2023). Prospects of Microbial bio-surfactants to endorse prolonged conservation in the pharmaceutical and agriculture industries. *Chemistry Select*, 8(26): e202300401.

Khan, F., Oloketuyi, S.F. & Kim, Y.M. (2019). Diversity of bacteria and bacterial products as antibiofilm and antiquorum sensing drugs against pathogenic bacteria. *Current Drug Targets*, 20(11): 1156

Kim, B.S., Lee, J.Y. & Hwang, B.K. (2000). In vivo control and in vitro antifungal activity of rhamnolipid B, a glycolipid antibiotic, against *Phytophthora capsici* and *Colletotrichum orbiculare*. *Pest Management Science: Formerly Pesticide Science*, 56(12): 1029-1035.

Landman, D., Georgescu, C., Martin, D.A. & Quale, J. (2008). Polymyxins revisited. *Clinical Microbiology Reviews*, 21(3): 449-465.

Liepins, J., Balina, K.., Soloha, R., Berzina, I., Lukasa, L.K. & Dace, E. (2021). Glycolipid biosurfactant production from waste cooking oils by yeast: Review of substrates, producers and products. *Fermentation*, 7(3): 136.

Liu, G., Zhong, H., Yang, X., Liu, Y., Shao, B. & Liu, Z. (2018). Advances in applications of rhamnolipids biosurfactant in environmental remediation: A review. *Biotechnology and Bioengineering*, 115(4): 796-814.

Lotfabad, T.B., Abassi, H., Ahmadkhaniha, R., Roostaazad, R., Masoomi, F., Zahiri, H.S. et al (2010). Structural characterization of a rhamnolipid-type biosurfactant produced by *Pseudomonas aeruginosa* MR01: Enhancement of di-rhamnolipid proportion using gamma irradiation. *Colloids and Surfaces B: Biointerfaces*, 81(2): 397-405.

Mandal, S.M., Barbosa, A.E. & Franco, O.L. (2013). Lipopeptides in microbial infection control: Scope and reality for industry. *Biotechnology Advances*, 31(2): 338-345.

Mao, X., Jiang, R., Xiao, W. & Yu, J. (2015). Use of surfactants for the remediation of contaminated soils: A review. *Journal of Hazardous Materials*, 285: 419-435.

McSorley, J.C. (2016). Pre-clinical studies on novel lipid therapeutics with anti-virulence and antibiotic synergising potential against *Pseudomonas aeruginosa* isolates from bronchiectatic airways (Thesis).

Medeot, D.B., Fernandez, M., Morales, G.M. & Jofré, E. (2020). Fengycins from *Bacillus amyloliquefaciens* MEP218 exhibit antibacterial activity by producing alterations on the cell surface of the pathogens *Xanthomonas axonopodis* pv. vesicatoria and *Pseudomonas aeruginosa* PA01. *Frontiers in Microbiology*, 10: 3107.

Mnif, I. & Ghribi, D. (2016). Glycolipid biosurfactants: Main properties and potential applications in agriculture and food industry. *Journal of the Science of Food and Agriculture*, 96(13): 4310-4320.

Mnif, I., Ellouz-Chaabouni, S. & Ghribi, D. (2018). Glycolipid biosurfactants, main classes, functional properties and related potential applications in environmental biotechnology. *Journal of Polymers and the Environment*, 26: 2192-2206.

Moradali, M.F., Ghods, S. & Rehm, B.H. (2017). *Pseudomonas aeruginosa* lifestyle: A paradigm for adaptation, survival, and persistence. *Frontiers in Cellular and Infection Microbiology*, 7: 39.

Müller, F., Hönzke, S., Luthardt, W.O., Wong, E.L., Unbehauen, M., Bauer, J. et al. (2017). Rhamnolipids form drug-loaded nanoparticles for dermal drug delivery. *European Journal of Pharmaceutics and Biopharmaceutics*, 116: 31-37.

Ndlovu, T., Rautenbach, M., Vosloo, J.A., Khan, S. & Khan, W. (2017). Characterisation and antimicrobial activity of biosurfactant extracts produced by *Bacillus amyloliquefaciens* and *Pseudomonas aeruginosa* isolated from a wastewater treatment plant. *AMB Express*, 7(1): 1-19.

Ohadi, M., Shahravan, A., Dehghannoudeh, N., Eslaminejad, T., Banat, I.M. & Dehghannoudeh, G. (2020). Potential use of microbial surfactant in microemulsion drug delivery system: A systematic review. Drug Design, Development and Therapy, 541-550.

Pal, S., Chatterjee, N., Das, A.K., McClements, D.J. & Dhar, P. (2023). Sophorolipids: A comprehensive review on properties and applications. *Advances in Colloid and Interface Science*, 102856.

Patel, P., Bhatt, S., Patel, H., Marcelino, L.A., and Sayyed, R.Z. (2022). Biosurfactant – A biomolecules and its potential applications. *In:* Sayyed, R.Z. and Enshasy, H.E. (Eds), *Biosurfactants: Production and Applications in Food and Agriculture.* Vol. II, 133-149. CRC Press, Taylor & Francis Group, USA.

Raaijmakers, J.M., De Bruijn, I., Nybroe, O. & Ongena, M. (2010). Natural functions of lipopeptides from *Bacillus* and *Pseudomonas*: More than surfactants and antibiotics. *FEMS Microbiology Reviews*, 34(6): 1037-1062.

Rahimi, K., Lotfabad, T.B., Jabeen, F. & Ganji, S.M. (2019). Cytotoxic effects of mono- and di-rhamnolipids from *Pseudomonas aeruginosa* MR01 on MCF-7 human breast cancer cells. *Colloids and Surfaces B: Biointerfaces*, 181: 943-952.

Ramya, N., Priyadharshini, X.X., Prakash, R. & Dhivya, R. (2017). Anti-cancer activity of *Trachyspermum ammi* against MCF7 cell lines mediates by p53 and Bcl-2 mRNA levels. *J Phytopharmacol*, 6(2): 78-83.

Reddy, A.S., Malek, S.U.A., Ibrahim, H. & Sim, K.S. (2013). Cytotoxic effect of Alpinia scabra (Blume) Náves extracts on human breast and ovarian cancer cells. *BMC Complementary and Alternative Medicine*, 13: 1-14.

Rodrigues, L.R., Banat, I.M., Van der Mei, H.C., Teixeira, J.A. & Oliveira, R. (2006). Interference in adhesion of bacteria and yeasts isolated from explanted voice prostheses to silicone rubber by rhamnolipid biosurfactants. *Journal of Applied Microbiology*, 100(3): 470-480.

Sadiq, M.B., Khan, M.R. & Sayyed, R.Z. (2022). Biosurfactant mediated synthesis and stabilization of nanoparticles. *In:* Sayyed, R.Z. (Eds), *Biosurfactants: Production and Applications in Bioremediation/Reclamation.* 158-168. CRC Press, Taylor & Francis Group, USA.

Saeed, Q., Xiukang, W., Haider, F.U., Kučerik, J., Mumtaz, M.Z., Holatko, J. et al. (2021). Rhizosphere bacteria in plant growth promotion, biocontrol, and bioremediation of contaminated sites: A comprehensive review of effects and mechanisms. *International Journal of Molecular Sciences*, 22(19): 10529.

Sajid, M., Khan, M.S.A., Cameotra, S.S. & Al-Thubiani, A.S. (2020). Biosurfactants: Potential applications as immunomodulator drugs. *Immunology Letters*, 223: 71-77.

Sałek, K., Euston, S.R. & Janek, T. (2022). Phase behaviour, functionality, and physicochemical characteristics of glycolipid surfactants of microbial origin. *Frontiers in Bioengineering and Biotechnology*, 10: 816613.

Sandeep, L. & Rajasree, S. (2017). Biosurfactant: Pharmaceutical perspective. *J. Anal. Pharm. Res*, 4(00105): 10-15406.

Santos, D.K.F., Rufino, R.D., Luna, J.M., Santos, V.A. & Sarubbo, L.A. (2016). Biosurfactants: Multifunctional biomolecules of the 21st century. *International Journal of Molecular Sciences*, 17(3): 401.

Saranraj, P., Sayyed, R.Z., Sivasakthivelan, P., Hasan, M.S., Al-Tawaha, A.R.M.A. & Amala, K. (2022a). Microbial biosurfactants: Methods of investigation, characterization, current market value and applications. *In:* Sayyed, R.Z. (Eds), *Biosurfactants: Production and Applications in Bioremediation/Reclamation.* 19-34. CRC Press, Taylor & Francis Group, USA.

Saranraj, P., Sayyed, R.Z., Hamzah, K.J., Asokan, N., Sivasakthivelan, P., & Al-Tawaha, A.R.M.A. (2022b). *In:* Sayyed, R.Z. (Eds), *Biosurfactants: Production and Applications in Bioremediation/Reclamation.* 1-18. CRC Press, Taylor & Francis Group, USA.

Sarubbo, L.A., Mariada Gloria, C.S., Durval, I.J.B., Bezerra, K.G.O., Ribeiro, B.G., Silva, I.A. et al. (2022). Biosurfactants: Production, properties, applications, trends, and general perspectives. *Biochemical Engineering Journal*, 181: 108377.

Sen, S., Borah, S.N., Kandimalla, R., Bora, A. & Deka, S. (2019). Efficacy of a rhamnolipid biosurfactant to inhibit *Trichophyton rubrum* in vitro and in a mice model of dermatophytosis. *Experimental Dermatology*, 28(5): 601-608.

Shah, I., Hamid, B., Zaman, M., Fatima, S., Farooq, S., Datta, R. et al. (2022). Microbial biosurfactants: An eco-friendly approach for bioremediation of contaminated environments. *In:* Sayyed, R.Z. (Eds), *Biosurfactants: Production and Applications in Bioremediation/ Reclamation.* 197-207. CRC Press, Taylor & Francis Group, USA.

Silva, E.J., e Silva, N.M.P.R., Rufino, R.D., Luna, J.M., Silva, R.O. & Sarubbo, L.A. (2014). Characterization of a biosurfactant produced by *Pseudomonas cepacia* CCT6659 in the presence of industrial wastes and its application in the biodegradation of hydrophobic compounds in soil. *Colloids and Surfaces B: Biointerfaces*, 117: 36-41.

Singh, N., Pemmaraju, S.C., Pruthi, P.A., Cameotra, S.S. & Pruthi, V. (2013). Candida biofilm disrupting ability of di-rhamnolipid (RL-2) produced from *Pseudomonas aeruginosa* DSVP20. *Applied Biochemistry and Biotechnology*, 169: 2374-2391.

Singh, P. & Tiwary, B.N. (2016). Isolation and characterization of glycolipid biosurfactant produced by a *Pseudomonas otitidis* strain isolated from Chirimiri coal mines, India. *Bioresources and Bioprocessing*, 3: 1-16.

Singh, P., Patil, Y. & Rale, V. (2019). Biosurfactant production: Emerging trends and promising strategies. *Journal of Applied Microbiology*, 126(1): 2-13.

Smith, M.L., Gandolfi, S., Coshall, P.M. & Rahman, P.K. (2020). Biosurfactants: A Covid-19 perspective. *Frontiers in Microbiology*, 11: 1341.

Sotirova, A.V., Spasova, D.I., Galabova, D.N., Karpenko, E. & Shulga, A. (2008). Rhamnolipid–biosurfactant permeabilizing effects on gram-positive and gram negative bacterial strains. *Current Microbiology*, 56: 639-644.

Srivastava, R.K., Bothra, N., Singh, R., Sai, M.C., Nedungadi, S.V. & Sarangi, P.K. (2022). Microbial originated surfactants with multiple applications: A comprehensive review. *Archives of Microbiology*, 204(8): 452.

Subramaniam, M.D., Venkatesan, D., Iyer, M., Subbarayan, S., Govindasami, V., Roy, A. et al. (2020). Biosurfactants and anti-inflammatory activity: A potential new approach towards COVID-19. *Current Opinion in Environmental Science & Health*, 17: 72-81.

Thakur, P., Saini, N.K., Thakur, V.K., Gupta, V.K., Saini, R.V. & Saini, A.K. (2021). Rhamnolipid the Glycolipid Biosurfactant: Emerging trends and promising strategies in the field of biotechnology and biomedicine. *Microbial Cell Factories*, 20: 1-15.

Uzoigwe, C., Burgess, J.G., Ennis, C.J. & Rahman, P.K. (2015). Bioemulsifiers are not biosurfactants and require different screening approaches. *Frontiers in Microbiology*, 6: 245.

Van Gennip, M., Christensen, L.D., Alhede, M., Phipps, R., Jensen, P.Ø., Christophersen, L. et al. (2009). Inactivation of the rhlA gene in *Pseudomonas aeruginosa* prevents rhamnolipid production, disabling the protection against polymorphonuclear leukocytes. *Apmis*, 117(7): 537-546.

Van Hamme, J.D., Singh, A. & Ward, O.P. (2006). Physiological aspects: Part 1 in a series of papers devoted to surfactants in microbiology and biotechnology. *Biotechnology Advances*, 24(6): 604-620.

Watters, C., Fleming, D., Bishop, D. & Rumbaugh, K.P. (2016). Host responses to biofilm. *Progress in Molecular Biology and Translational Science*, 142: 193-239.

Wood, S.J., Goldufsky, J.W., Seu, M.Y., Dorafshar, A.H. & Shafikhani, S.H. (2023). *Pseudomonas aeruginosa* cytotoxins: Mechanisms of cytotoxicity and impact on inflammatory responses. *Cells*, 12(1): 195.

Worakitsiri, P., Pornsunthorntawee, O., Thanpitcha, T., Chavadej, S., Weder, C. & Rujiravanit, R. (2011). Synthesis of polyaniline nanofibers and nanotubes via rhamnolipid biosurfactant templating. *Synthetic Metals*, 161(3-4): 298-306.

Woźniak-Karczewska, M., Myszka, K., Sznajdrowska, A., Szulc, A., Zgoła-Grześkowiak, A., Ławniczak, Ł. et al. (2017). Isolation of rhamnolipids-producing cultures from faeces: Influence of interspecies communication on the yield of rhamnolipid congeners. *New Biotechnology*, 36: 17-25.

Yan, F., Hu, H., Lu, L. & Zheng, X. (2016). Rhamnolipids induce oxidative stress responses in cherry tomato fruit to *Alternaria alternata*. *Pest Management Science*, 72(8): 1500-1507.

Yan, X., Gu, S., Cui, X., Shi, Y., Wen, S., Chen, H. & Ge, J. (2019). Antimicrobial, anti-adhesive and anti-biofilm potential of biosurfactants isolated from *Pediococcus acidilactici* and *Lactobacillus plantarum* against *Staphylococcus aureus* CMCC26003. *Microbial Pathogenesis*, 127: 12-20.

Ycla, A.C.A., Martínez, M.A.T., Piñeros, G.A.R., López, V.C., Villamizar, S.H., Vélez, V.L.N. et al. (2016). A comparison between conventional *Pseudomonas aeruginosa* rhamnolipids and *Escherichia coli* transmembrane proteins for oil recovery enhancing. *International Biodeterioration & Biodegradation*, 112: 59-65.

Yi, G., Son, J., Yoo, J., Park, C. & Koo, H. (2019). Rhamnolipid nanoparticles for in vivo drug delivery and photodynamic therapy. *Nanomedicine: Nanotechnology, Biology and Medicine*, 19: 12-21.

Yuan, L., Zhang, S., Wang, Y., Li, Y., Wang, X. & Yang, Q. (2018). Surfactin inhibits membrane fusion during invasion of epithelial cells by enveloped viruses. *Journal of Virology*, 92(21): 10-1128.

Zaman, M., Hassan, S., Fatima, S., Hamid, B., Farooq, S., Qayoom, I. et al. (2022). Biosurfactants production and applications in food. *In:* Sayyed, R.Z. and Enshasy, H.E. (Eds), *Biosurfactants: Production and Applications in Food and Agriculture*. Vol. II, 225-241. CRC Press, Taylor & Francis Group, USA.

Zeng, Z., Liu, Y., Zhong, H., Xiao, R., Zeng, G., Liu, Z. et al. (2018). Mechanisms for rhamnolipids-mediated biodegradation of hydrophobic organic compounds. *Science of the Total Environment*, 634: 1-11.

CHAPTER
15

Glycolipids from *Pseudomonas* and their Applications in Health and Medicine

Suneeta Panicker*

Dr. D.Y. Patil Arts, Commerce and Science College, Behind Yeshwantrao Chavan Memorial Hospital, Sant Tukaram Nagar, Pimpri, Pune, India

1. Introduction

Glycolipids (GLs) are a broad category of structurally heterogeneous biological compounds that are made by a wide range of organisms, including microbes, plants, animals, and people (Holst, 2008). They contain lipid and glycosyl moieties, as their name would imply. The IUPAC refers to compounds with one or more monosaccharide residues linked to a hydrophobic moiety by a glycosidic linkage as Glycolipids (Chester, 1997). The term "GLs" can be used to refer to both glycoside and non glycoside GLs, depending on whether the sugar and lipid residues are joined by glycosidic (such as O- or N-glycosidic linkages) or non-glycosidic (such as ester or amide links) linkages. Mono-, di-, oligo-, or polysaccharides like: such as glucose, cellobiose, or glycan, respectively, alcohol sugars/polyols like: mannitol, erythritol, or arabinol, etc., amino sugars like; desosamine, etc., or sugar acids like: glucuronic acids, can make up the glycosyl residue. The lipid residue of GLs includes a variety of sterols, hopanoids, carotenoids, polyketides, fatty acids, fatty alcohols, fatty amino alcohols, and fatty amino alcohols with various substitutions, chain lengths, saturation levels, branching, and di-/oligo-polymerizations (Abdel-Mawgoud and Stephanopoulos, 2018).

A broad classification of GLs includes simple and complex GLs (Leray, 2012). Simple GLs (SGLs), also known as saccharolipids (Merrill and Vu, 2016), are two-component GLs with glycosyl and lipid moieties that are physically connected to one another. However, complex glycolipids (CGLs) are structurally more heterogeneous because they also contain residues such as glycerol (glycoglycerolipids), peptide (glycopeptidolipids), acylated-sphingosine (glycosphingolipids), or other residues in

* Corresponding author: suneetapanicker@gmail.com

addition to the glycosyl and lipid moieties. Even though they only comprise glycosyl and lipid moieties, polysaccharide-containing GLs are categorised as complex glycolipids due to the complexity of their polysaccharide residues. Simple glycolipids (SGLs), which contain both hydrophilic glycosyl and lipophilic lipid residues, are amphiphilic compounds (Abdel-Mawgoud and Stephanopoulos, 2018).

Glycolipids include a wide variety of compounds: glycosphingolipids (cerebrosides, globosides, gangliosides, sulfatides, and others), glycoglycerolipids, glycophosphospholipids (e.g., phosphatidylinositols), glycosylated prenols (e.g., dolichol-phospho-glycans), glycosylated sterols, glycosylated polyketides, and saccharolipids (Zeidan and Hannun, 2007). According to the structure of the lipid moiety, glycolipids—a general name for complex carbohydrates made up of a glycan moiety and a lipid moiety—are typically separated into two categories: glycosphingolipids (GSLs) and glycoglycerolipids. They have biological specificity due to their particular carbohydrate structures, such as the presence or lack of sialic acid. They support the transfer and exchange of materials by serving as the foundation of biological structures like cell membranes in both eukaryotic and prokaryotic cells. These substances, such as the well-known phosphatidylcholine, sphingosine, and glycolic acids, can be found in superior cells (Lang, 2002).

Glycolipids can perform bulk functions by influencing the biophysical characteristics of membranes interior and exterior. They achieve this by providing ion impermeability and sidedness, balancing the proportions of lipids that are prone to forming bilayers and those that are not, and influencing the density of surface charge and hydration. In substances like ceramides, glycosylceramides, sphingolipids, glycosphingolipids, and sphingosines, glycolipids have also shown their biological activity. These substances mediate anti-proliferative responses like inhibition of cell growth, proliferation, and differentiation, interruption of the cell cycle, signal transduction, senescence transformation, inflammation, and apoptosis. These structural, amphipathic compounds have been shown to have anti-microbial activity among other biological functions (Cortés-Sánchez et al., 2013).

Glycolipids make up the most common group of biosurfactants and are researched more due to their high production yield and broad range of possible applications. These compounds are often created by combining various kinds of mono, di, tri, and tetra-saccharide carbohydrates with hydrophobic portions of one or more long-chain aliphatic fatty acids. Rhamnolipids, sophorolipids, trehalolipids, and mannosylerythritol lipids are the primary glycolipids (Islas et al., 2010).

Glycolipids are microbial surface-active substances produced by various bacteria. They consist of a carbohydrate moiety linked to fatty acids. Exhibiting considerable structural diversity, glycolipids can reduce surface tension and interfacial tension at the surface and interface, respectively. The most well-known glycolipids include rhamnolipids, trehalolipids, mannosylerythritol lipids, and cellobiose lipids. They have attracted a lot of attention in the real world as biopesticides for managing plant diseases and safeguarding stored goods. Glycolipid biosurfactants enable the preservation of plants and plant crops from pest invasion due to their antifungal activity toward phytopathogenic fungi and larvicidal and mosquitocidal capacities. The emulsifying and antibacterial properties of glycolipids make them excellent candidates for use as food additives and preservatives. Furthermore, because it enables the bioconversion of wastes into useful chemicals and lowers production costs, the valorization of food byproducts through the creation of glycolipid biosurfactants has drawn a lot of attention. In general, the retention of glycolipids from fermentation media is necessary for their usage in numerous domains (Mnif and Ghribi, 2016).

2. Glycolipids from Bacteria

Bacterial glycolipids are intricate amphipathic compounds with a hydrophobic moiety, such as an acylglycerol, a sphingoid backbone, a ceramide, or a prenyl phosphate, connected to one or more monosaccharide residues. Together with lipopolysaccharides (LPSs) and capsular polysaccharides, they form a wider group of compounds known as glycoconjugates that are tethered to the cell membrane by a lipid moiety (Chester, 1997).

Bacterial glycolipids are important for the structure and function of bacterial membranes. They are also crucial for activating the host's immune system. Glycolipids are crucial parts of biological membranes that serve a wide range of purposes. They might function as receptors, cell aggregators and dissociators. They also maybe be in charge of particular cellular interactions, and be involved in signal transduction. Bacterial species also produce a wide range of glycolipids, and it may be assumed that just a portion of this variety has been studied thus far.

Diacyldiglucosylglycerols are the most prevalent bacterial glycolipids, mostly found in Gram-positive bacteria. Phosphoglycolipids is another important glycolipid found in Gram-positive bacteria.

In the five different categories of biosurfactants, glycolipids are most prevalent. In addition to the surface activity that is inherent in their chemical structure, glycolipids also exhibit a wide variety of biological activities. The potential applications of these compounds are as antimicrobial, anti-adhesive, anti-biofilm, anti-oxidant, anti-inflammatory, immunomodulatory and cellular differentiation agents (Banat et al., 2010). Glycolipids from various bacteria like Pseudomonas, *Caminus Sympodiomycopsis Pseudozyma* have been studied for these properties (Cortés-Sánchez et al., 2013).

Simple glycolipids (SGLs) have highly exciting biological effects on a variety of other organisms extending from viruses to human cells. Even though the exact mechanism underlying these bioactivities is unknown, it has been postulated that the majority of its bioactivities are caused by their surface activities (Lukic et al., 2016).

Biosurfactants are more widely used than synthetic surfactants due to their greater biodegradability, decreased toxicity, thermostability, and tolerance for harsh environments (Varvaresou and Iakovou, 2015).

The variety of glycolipids is a result of the variation in the carbohydrate moiety and fatty acid chains, and the subclass often includes rhamnolipids, sophorolipids, mannosyl- and cellobiose-derived lipids, trehalolipids, xylolipids, and others (Mnif and Ghribi, 2016). The majority of glycolipid-producing bacteria, with the exception of those that share essential traits with biosurfactants, are found in samples that have been contaminated by oil, suggesting that they may have applications in bioremediation as well as sustainable production. Glycolipids have a wide range of possible applications due to their numerous biological characteristics.

Surprisingly, a wealth of research shows that all types of glycolipids have excellent antimicrobial activity against bacteria, fungi, and viruses, and the physiological changes induced by glycolipids have been thoroughly studied with an eye towards both the planktonic and biofilm states of microorganisms (Chen et al., 2017).

2.1 Rhamnolipids (RLs)

The most often researched compound because of its potent surface activity and high production yields is the rhamnolipid (RL), a glycolipid-type biosurfactant, which is mostly produced by *Pseudomonas aeruginosa* (Eslami et al., 2020). However, it has been documented that numerous other non-Pseudomonas isolates are also capable of producing RLs, which accounts for the structural diversity of RLs.

The basic structure of RL consists of a rhamnose moiety and a lipid moiety. The diversity of RL structure is influenced by factors such as the amount of rhamnose,as well as the length and number of carbon chains. This diversity is further compounded by variations in RL producing organisms, resulting in differences in the chemical structures of RL. Numerous thorough and in-depth studies on the microorganism-inhibiting effects of rhamnolipids involving planktonic and biofilm cells have been carried out. Rhamnolipids were reported to have an impact on the significant pathogens like Staphylococcus aureus, Staphylococcus epidermidis, and Listeria monocytogenes. Cell lysis and the subsequent release of cellular components are responsible for inducing bacterial mortality. This was earlier supported by visual evidence like SEM and TEM images. Because of the distinct amphiphilic property that enables the interaction between rhamnolipids and phospholipids, the cell membrane gets altered and damaged with an increase in cell permeability and reduction of cell surface hydrophobicity (de Freitas Ferreira et al., 2019).

Numerous studies have demonstrated the anti-biofilm characteristics of rhamnolipids. Biofilm is thought to be the sessile population of planktonic microorganisms stuck at the solid surface. Biofilms are a major worry for the food processing industries because they can contaminate pipes, corrode equipment, and ultimately destroy food. Furthermore, oral bacteria have always posed a hazard to oral health and hygiene, and rhamnolipids have been investigated for their ability to destroy bacterial biofilms that cause oral disease. Rhamnolipids decrease the adhesion of Streptococcus mutans to polystyrene surfaces, and they also destroy the pre-formed biofilm on these surfaces, which is one of the main etiological agents in dental caries (Abdollahi et al., 2020). Rhamnolipids from non-pathogenic Burkholderia thailandensis E264 showed strong ability to destroy mature biofilm of some oral pathogens (Streptococcus oralis, Actinomyces naeslundii, Neisseria mucosa, and Streptococcus sanguinis) (Elshikh et al., 2017). This suggested their potential use in related oral applications against oral-bacteria biofilms. In conclusion, prevention of the first adhesion stage is a result of rhamnolipids' biofilm-inhibiting behaviour. The physicochemical interactions between rhamnolipids and Pseudomonas aeruginosa biofilm layers were examined in order to determine the precise role that rhamnolipids play in the biofilm development process (Kim et al., 2015). It was shown that numerous crucial elements in rhamnolipid-mediated biofilm reduction included a reduction in free surface energy on the membrane, interactions with some EPS proteins, and a loss in EPS quantity.

Rhamnolipids are also very effective at inhibiting the growth of fungus, including both hyphal growth and spore germination. Rhamnolipids have demonstrated potent antifungal efficacy against Phytophthora capsici, a plant-harming fungus. This effectiveness is attributed to their ability to lyse zoospores and inhibit both zoospore germination and hypha formation (Kim et al., 2000). Rhamnolipid's ability to inhibit *Colletotrichum falcatum* in vitro and in vivo was further validated by Goswami et al., 2015. The breakdown of the fungal membrane gave rise to the antifungal mode, which enabled its use as a substitute fungicide to manage sugarcane red rot disease. Rhamnolipids, on the other hand, can fight off hazardous fungi that are derived from humans or animals and successfully stop the fungi's hyphal growth and spore germination. This is a potentially effective treatment for dermatophytic infections, the most common superficial mycoses worldwide (Sen et al., 2019). Rhamnolipids are highly effective in reducing fungal biofilms, which are notoriously resistant and have low nutritional requirements, making them challenging to eliminate. In a dose-dependent manner rhamnolipids reduced the amount of pre-formed Candida albicans biofilm by 90% on polystyrene surfaces (Singh et al., 2013). Rhamnolipids, for instance, have demonstrated a substantial dose-dependent anticancer effect on human breast cancer cells MCF-7 (Rahimi et al., 2019). Additionally, as potential immunomodulators, glycolipids have anti-inflammatory effects on human immune-related disorders.

2.2 Sophorolipids (SLs)

Sophorolipids were first discovered in 1961. It is another characteristic glycolipid which has undergone extensive research and has been commercialized by several businesses due to its uniform product yield in high quantities (Jezierska et al., 2018). The production of sophorolipids by numerous yeast species, including *Starmerella bombicola, Candida riodocensis, Candida stellate,* and *Wickerhamiella domercqiae*, growing on carbohydrates and lipophilic substrates, has been demonstrated over the past decades (Kurtzman et al., 2010). They have a hydrophobic fatty acid tail with 16 or 18 carbon atoms and a hydrophilic carbohydrate head called sophorose.

Sophorolipids can be classified into two primary forms: acidic form and lactonic form due to their hydrophobic fatty acid tail of 16 or 18 carbon atoms and hydrophilic carbohydrate head sophorose. The sophorose moiety has a lengthy chain of hydroxyl fatty acids that are glycosidically connected to it, and the carboxylic tail of fatty acid can either be free (for an acidic form) or esterified at the 6′- or 6″-position (for a lactonic form). The carbon number, unsaturation, and hydroxylation of the fatty acid chain in sophorolipids also represent the diversity in the structure, depending on the different types of carbon sources in microbial fermentation (Shin et al., 2010). Therefore, the degree of lactonization is the crucial variable and the structural difference has significantly influenced the biological and physicochemical activities. In general, lactonized sophorolipid exhibits stronger application potential, superior surface activity, and antibacterial properties, while the acidic form shows superior forming capacity and solubility (Van Bogaert et al., 2011).

Sophorolipids exhibit a wide range of characteristics, including emulsifying, lubricating micelle formation, detergency, dispersibility, and foaming.

They also exhibit strong antibacterial activity, which has been extensively researched and used in a wide range of fields. It has been demonstrated that Gram-positive bacteria, such as *Enterococcus faecium, Staphylococcus aureus,* and *Streptococcus mutans*, appear to be more sensitive to sophorolipids than Gram-negative bacteria, such as *Proteus mirabilis*, *Escherichia coli*, and *Salmonella enterica* subsp. *enterica* (Fontoura et al., 2020). Nonetheless, certain studies showed both types of strains to have a powerful killing effect and elucidated the underlying mechanism. The subsequent research verified that sophorolipid-induced production of reactive oxygen species (ROS) led to cell death (Gaur et al., 2019).

Streptococcus mutans, Lactobacillus fermentum, Streptococcus salivarius, and *Streptococcus sobrinus*, five representative species of oral bacteria known to cause caries, were successfully inhibited, demonstrating the great potential of sophorolipids in oral health and hygiene (Solaiman et al., 2017). Additionally, sophorolipids have helped to diminish *Clostridium perfringens* and *Campylobacter jejuni* microbiological contamination in the chicken industry, helping to reduce significant economic losses (Silveira et al., 2019).

Recent studies have shown that sophorolipids have been successful in removing the biofilm of clinical strains (*Staphylococcus aureus, Pseudomonas aeruginosa*, and *Candida albicans*) from medical-grade silicone discs (Ceresa et al., 2020). Biofilm is a problem that needs to be resolved right away since it frequently adheres to the surface of materials like medical devices, causing persistent microbial contamination and treatment resistance. Thus, Sophorolipids have enormous potential to be exploited as antimicrobial agents in a range of industries, including poultry, food preservation, pharmaceuticals, and medical equipment and instruments.

Sophorolipids also target pathogenic fungi and have a considerable impact on spore germination, mycelial proliferation, and biofilm development. They display a broad antifungal spectrum that includes *Corynespora cassiicola, Fusarium verticilliodes, Fusarium oxysporum f.* sp. *pisi, Colletotrichum gloeosporioides*, and *Trichophyton rubrum. Candida albicans* hyphal growth and biofilm formation were inhibited by sophorolipids, and hypha-specific genes were thought to be down-regulated to achieve this effect (Haque et al., 2016).

In one investigation, it was discovered that sophorolipids dramatically boosted the generation of ROS and the expression of genes associated to oxidative stress in *Candida albicans*, which finally caused membrane perforation and necrosis, which resulted in cell death (Haque et al., 2019). Sophorolipids were exposed to biological control of plant disease, and the results revealed that their pH solubility had influenced their efficacy (Haque et al., 2016). Sophorolipids have a restraining effect on spore germination and hyphal tip growth of several plant pathogens. Sophorolipids are likely to be used to treat cutaneous mycoses and have antifungal and anti-biofilm effects on Trichophyton mentagrophytes, a zoonotic dermatophyte (Chen et al., 2020).

In animal models, sophorolipids may modify the immune response to lower sepsis-related mortality; the underlying process involves lowering nitric oxide and controlling inflammatory cytokines (Hardin et al., 2007).

2.3 Mannosylerythritol Lipids (MELs)

A class of glycolipid biosurfactants known as mannosylerythritol lipids (MELs) are mostly generated by the Pseudozyma and Ustilago species (Arutchelvi et al., 2008). Pseudozyma antarctica, Pseudozyma tsukubaensis, and Pseudozyma hubeiensis were just a few of the strains that were isolated as MEL-producers based on glucose and oil substrates after the 1980s (Saika et al., 2016). The mannose molecule was connected to an erythritol residue at C-10 and two fatty acids at C-20 and C-30 to form the basic chemical structure (Saika et al., 2018). These compounds can be grouped into four groups based on the varying positions and numbers of acetyl groups: MEL-A (diacetylated at C-40 and C-60), MEL-B (monoacetylated at C-60), MEL-C (monoacetylated at C-40), and MEL-D (deacetylated) (Kitamoto et al., 2001). The literature indicates that the structure of MEL is primarily influenced by the carbon source, and structural diversity confers a variety of interfacial and biological properties, such as emulsibility, non-toxicity, anti-cancer and antioxidant activity, biodegradability, and environmental compatibility, showing their great application prospects in the medical, environmental, cosmetic, pharmaceutical, and food industries (Morita et al., 2015). In its practical applications, MELs' bactericidal properties are crucial. Significant antibacterial efficacy was demonstrated by MELs, particularly against Gram-positive bacteria (Kitamoto, 1993). The solubilizing activity of MEL-A and MEL-B on bilayer biomembranes was suggested as the cause of their strong inhibitory effects on Micrococcus luteus (Fukuoka et al., 2007). MEL prevented most gram-positive bacteria from proliferating but did not prevent gram-negative bacteria. MELs were also found to show moderate antifungal effects on Candida magnolia (Recke et al., 2013).

MELs are expected to be a unique, secure replacement for food preservation during food storage since they have so far been reported to have outstanding sterilizing action on pathogenic bacteria. Investigations into the antibacterial activity of MELs against *Bacillus cereus* and *Staphylococcus aureus* concluded that the mode of action required rupturing the cell membrane, leaking cellular contents, collapsing the entire cytoskeleton, and inducing cell apoptosis (Shu et al., 2020). Additional transcriptome analysis revealed that the ABC transporter system was enriched in the differentially expressed genes, confirming that the dysfunction of transmembrane protein played a significant role in MEL-mediated cell death (Liu et al., 2020). It is important to note that MELs can lessen the lingering pollution brought on by fungus and microbial biofilms.

Researchers looked into the suppressive effects of MELs on the early infection behaviors of a number of phytopathogenic fungal conidia, such as *Blumeria graminis* f. sp. tritici (wheat powdery mildew fungi), *Colletotrichum dematium* (mulberry anthracnose fungi), *Glomerella cingulata* (strawberry anthracnose fungi), and *Magnaporthe grisea* (rice blast fungi) likely because they prevent conidial germination. Thus it was predicted that MELs would be used in the future as cutting-edge agricultural chemical insecticides (Yoshida et al., 2015).

In addition, it is clear that MELs had strong anti-biofilm action against *Staphylococcus aureus* through attachment inhibition and biofilm dispersal, which was attributed to the role of biosurfactants in microbial adhesion and desorption (Shu et al., 2020).

2.4 Glycolipids from *Pseudomonas*

Pseudomonas aeruginosa, a Gram negative bacterium produces four well-known rhamnolipids viz. 3-[3-(2-O-α-L-Rhamnopyranosyl-α-L-rhamnopyranosyloxy) decanoyloxy] decanoic acid (Rha2-C10-C10), 3-[(6-Deoxy-α-L-mannopyranosyl)oxy] decanoic acid (Rha-C10), 3-[3-(α-L-Rhamnopyranosyloxy) decanoyloxy] decanoic acid (Rha-C10-C10) and 3-[(2-O-α-LRhamnopyranosyl- α-L-rhamnopyranosyl)oxy] decanoic acid (Rha2-C10) (Gudiña et al., 2015).

Diverse carbon sources may impact the availability of the fundamental building blocks necessary for the manufacture of rhamnolipids, which is why various PA strains produce diverse forms of rhamnolipid (Dobler et al., 2016).

The producing microorganism, ingredients of the medium, and the cultivation conditions, all have an impact on the final proportion of rhamnolipid congeners. The qualities of the biosurfactant are determined by the homologue ratio, and it has been

Table 1: Different *P. aeruginosa* strains showing rhamnolipid production yield, critical micelle concentration (CMC) and surface tension (ST)

Microorganism	Fermentation mode	Min. STa (mN m^{-1})	CMC (mgL^{-1})	Max Yield (g L^{-1})	Reference
P. aeruginosa S6	Batch	33.9	50	0.18	(Yin et al., 2009)
P. aeruginosa (P20)	Batch	-	-	7.5	(Shah et al., 2016)
P. aeruginosa DR1	Batch	30	80	2.8	(Reddy et al., 2016)
P. aeruginosa MA01	Batch	32.5	10.1	12	(Abbasi et al., 2012)
P. aeruginosa LBI	Batch	24	120	15.8	(Benincasa et al., 2004)
P. aeruginosa S2	Fed-batch	30	-	9.4	(Chen et al., 2007)
P. aeruginosa ATCC 9027	Batch	30	40	0.3	(Zhang and Miller, 1992)
P. aeruginosa #112	Batch	30	13	5.1	(Gudi˜na et al., 2016)

observed that even little variations in the composition of the mixture can significantly affect its physicochemical properties (Guo et al., 2009).

One of the glycolipid-type biosurfactants, rhamnolipids (RLs) are mostly produced by *Pseudomonas aeruginosa* and are among the most often researched compounds because of their potent surface activity and high production yields (Morikawa et al., 2000). Two hydroxydecanoic acids and two rhamnose moieties were found in RL isolated from *P. aeruginosa* and were joined by a glycosidic linkage (Pinzon and Ju, 2009). *Pseudomonas* species are thought to be the predominant RL producers, and *P. aeruginosa* has so far been fully examined as a primary source of RLs with titers > 100 g/L (D´eziel et al., 1999).

Glycolipids are typically found as storage lipids in intracellular inclusions, but they can also be found in Gram-positive bacteria, *P. diminuta*, and *P. vesicularis* membranes (Wilkinson, 1988). Multi Omic approaches have shown that when the supply of phospholipids is restricted, *P. aeruginosa* makes substitute glycolipids to take their place. This phospholipid to glycolipid transition has an important impact on antibiotic susceptibility because glycolipids offer protection when exposed to the antimicrobial peptide polymyxin B. As a result, the previously reported modification of LPS reflects a very old resistance mechanism, but glycolipid-mediated resistance to polymyxin B represents a new resistance mechanism (Chung et al., 2017).

However, the function of the membrane under Pi stress may be significantly influenced by the presence of glycolipids in the membrane. *Pseudomonas aeruginosa* developed novel glycolipids in response to phosphorus stress. The stringent conservation of this lipid remodelling pathway across all P. aeruginosa isolates to date and its high expression in the metatranscriptome of cystic fibrosis (CF) patients point to a critical function for lipid remodelling in the ecophysiology of this bacterium. These glycolipids may shield the bacteria from injury by cationic antimicrobial peptides, suggesting a new resistance mechanism to polymyxin B. It is interesting to note that lipid remodelling as a response to survive phosphate stress comes with trade-offs in terms of antibiotic resistance (Jones et al., 2021).

When toluene is used as the only source of carbon and energy during growth, *P. putida* creates glycolipids. Using thin layer chromatography on silica gel GF254, three glycolipids have been recovered from the spent media from batch and chemostat cultures. The mono- and di-rhamnolipids, which serve as biosurfactants, are thought to be the source of the glycolipids. Toluene absorption is thought to be aided by the release of glycolipid into the media, which emulsifies toluene. The population study demonstrates competition and strain variation in continuous culture for *P. putida*. When grown on toluene rather than succinate as the carbon source, *P. putida* exhibits a shift in the composition of outer membrane proteins. Production of glycolipids depends on culture techniques, nutrition, environmental factors, and perhaps species strain. The toluene batch culture contained two glycolipids, labeled Gl and G2, which were found to be present. This is in line with findings showing *P. putida* produces more glycolipids while growing on toluene or other xenobiotic substrates (Choe et al., 1992). The glycolipid was thought to be used in toluene encapsulation by glycolipid micelles, offering a toluene absorption mechanism that would guard the cell membrane from degradation by toluene's direct action. Toluene was thought to induce the formation of glycolipids.

It was discovered that the newly discovered indigenous strain BN10 of *Pseudomonas aeruginosa* produced glycolipid (i.e., rhamnolipid-type) biosurfactants.

Nuclear magnetic resonance and mass spectroscopy were combined to determine the chemical structure of two representative rhamnolipidic fractions, RL-1 and RL-2, which were separated on silica gel columns. Their cytotoxic impact on the cancer cell lines HL-60, BV-173, SKW-3, and JMSU-1 was further examined. In terms of potency, RL-1 was more effective than RL-2, inhibiting cellular viability by 50% at lower concentrations. Additionally, the outcomes of the fluorescence staining investigation showed that RL-1 restricted the growth of BV-173 pre-B human leukemia cells by inducing apoptotic cell death. These results imply that RL-1 may have potential as a novel and promising therapeutic drug in biomedicine (Christova et al., 2013).

2.4.1 Application of Rhamnolipids from Pseudomonas

Rhamnolipids are rhamnose-containing glycolipid biosurfactants that are produced and secreted by Pseudomonas aeruginosa. The production of rhamnolipids relies on enzymes involved in the synthesis of the exopolysaccharide alginate, as well as on key metabolic pathways such fatty acid synthesis and dTDP-activated sugars.

A very intricate genetic regulatory system that also regulates several virulence-related features in *P. aeruginosa* governs the synthesis of these surfactants. Rhamnolipids have a variety of possible uses in industry and the environment, including the creation of fine chemicals, the evaluation of surfaces and surface coatings, as additives in environmental remediation, and as biological control agents. The commercial-scale manufacturing of rhamnolipids must be cost-effective to realize the vast range of applications.

The application of a biosurfactant depends on a number of physicochemical characteristics. The critical micelle concentration (CMC), is that concentration of surfactants above which micelles develop and surfactant monomers and micelles exist in dynamic equilibrium. This is a crucial characteristic of a surfactant. Low CMC values (11–20 mg/l) were noted in mixtures primarily comprising mono-rhamnolipid with C10 fatty acids. A high CMC (230 mg/l) was recorded for rhamnolipid mixtures with a larger proportion of congeners with unsaturated fatty acids (Guo et al., 2009).

A study by Pornsunthorntawee et al. (2008) showed that Rhamnolipids produced by *P. aeruginosa* SP4 were found to have excellent emulsification properties for vegetable oils (palm oil, soybean oil, coconut oil, and olive oil). However, this does not apply to short chain hydrocarbons (pentane, hexane, heptane, toluene, and 1-chlorobutane).

In the absence of biosurfactant, phenanthrene solubilization was 0.69 mg/l, according to Zhang et al. (1997), but it increased to 35 mg/l in the presence of isolated rhamnolipids from *P. aeruginosa* ATCC 9027 (0.30-0.35 mmol/l). Rhamnolipid concentrations below their CMC did not exhibit solubilization activity for PAHs, according to a study by Noordman et al. (2000).

Rhamnolipids exhibit outstanding antifungal activity against *Aspergillus niger* and *Gliocadium virens* (16 g/mL), as well as against *C. globosum, P. crysogenum,* and *A. pullulans* (32 g/mL), while the growth of the phytopathogenic fungi *B. Cinerea* and *R. solani* was inhibited at 18 g/mL (Abalos et al., 2001).

2.4.1.1 Antitumor Property of Rhamnolipids

Cancer is a terrible disease, and research on RLs has revealed that they have a powerful impact on both human and animal cancer cells (Christova et al., 2013).

Insect cell line C6/36 and breast cancer cell line MCF-7 growth was inhibited by rhamnolipids generated from *P. aeruginosa* B189. Additionally, RLs isolated from B1898 were able to modify their shape, recognize the cytoskeleton of phagocytic and nonphagocytic cells, and heal wounds (Thanomsub et al., 2006). Further tests of these RL characteristics were conducted on several cell types obtained from malignancies of various origins (Christova et al., 2013). According to a study, rhamnolipid-type biosurfactant (MR01 biosurfactant) made from *P. aeruginosa* MR01 suppressed the proliferation of HeLa cancer cells at a concentration of 5 g/mL due to its cytotoxic properties (Lotfabad et al., 2010). All these investigations help in concluding that rhamnolipids have the potential to be effective therapeutic anticancer agents.

2.4.1.2 Immunomodulatory Property of Rhamnolipids

Rhamnolipids have a variety of effects, which has led to new research on their potential as immunomodulators. Rhamnolipids, which are biosurfactants, can activate immune cells and cause the release of pro-inflammatory cytokines (McClure and Schiller, 1992). For instance, rhamnolipids from Burkholderia plantarii could stimulate human mononuclear cells to secrete tumor necrosis factor (TNF). When rhamnolipids are preincubated with monocytes, the opsonization of zymosan and the oxidative responses of monocytes can both be multiplied (Kharami et al., 1989). This biosurfactant has been linked to the stimulation of platelet production of serotonin and 12-hydroxyeicosatetraenoic acid (König et al., 1992) as well as the stimulation of mast cell histamine release (Bergmann et al., 1989). A notable decrease in the phagocytosis of the bacteria was seen during in vitro infection of mice peritoneal macrophages that had been preincubated with rhamnolipids (McClure and Schiller, 1996). When administered intratracheally into the rat's lungs at the physiological concentration, authors also looked at how rhamnolipids affected the internalization of zymosan particles. Their findings suggested that RLs have an inhibitory effect on alveolar macrophages in terms of zymosan particle internalization (McClure and Schiller, 1996). Numerous studies have demonstrated that rhamnolipids can lyse and necrose polymorphonuclear leukocytes and macrophages, respectively (Van Gennip et al., 2009). Rhamnolipids also prevent diacylglycerol (DAG) from interacting with protein kinase C, which reduces the production of human beta defensin-2, according to a study by Dossel et al. (2012).

2.4.1.3 Antifungal Property of Rhamnolipids

Fruit and vegetable post-harvest deterioration is brought on by fungi, which reduces global food supply by 5–10% (Rodrigues et al., 2017). To treat postharvest decay, chemical fungicides are utilized, but these have drawbacks such as toxicity, groundwater contamination, risks to human health, prolonged degradation intervals, and the emergence of fungicide-resistant strains (Pekmezovic et al., 2015). Rhamnolipids are now preferred over chemical fungicides because they have antifungal activity against fungi like *Alternaria alternata*, *Mucor circinelloides*, and *Verticillium dahlia*, inhibit fungicide-resistance against chemical pesticides, and stimulate plant immunity against plant pathogens (Sha and Meng, 2016).

2.4.1.4 Oxidative Stress Reduction Property of Rhamnolipids

Through the induction of resistance and defensive mechanisms in cherry tomatoes, rhamnolipids have also been known to lessen the oxidative stress brought on by

Alternaria alternata (Yan et al., 2016). High levels of oxidative stress cause fruit to rot, while low levels cause defense mechanisms in fruit to be activated against the fungus infection (Asselbergh et al., 2007). RLs can trigger resistance in the plant against pathogens through the development of antioxidant enzymes to remove the excess ROS in response to oxidative stress.

2.4.1.5 Rhamnolipids Against Pathogenic Microorganisms in Food

There is a need for natural preservatives since bacterial pathogens are becoming resistant to artificial preservatives (de Freitas Ferreira et al., 2019). These preservatives could prevent the growth of pathogens without compromising the quality of the food. Due to their antibacterial properties against a variety of microbes, rhamnolipids can be utilized as an alternative to prevent food contamination (Srey et al., 2013). RLs can be utilized in the food sector to manage the *L. monocytogenes*-borne illnesses due to their antibacterial characteristic and their synergism with nisin. By increasing their antimicrobial action in an acidic environment, rhamnolipids can inhibit the growth of Gram-positive bacteria in acidic food, which allows the food industry to employ them for food spoilage control without compromising food quality.

2.4.1.6 Anti-biofilm Property Against *Staphylococcus aureus*

S. aureus has the capacity to form a biofilm in foods such as dairy, fish, poultry, ready-to-eat food, and meat, which could cause damage, contamination, and food degradation (Srey et al., 2013). Rhamnolipids function as anti-biofilm agents to inhibit the development of biofilm on food (do Valle Gomes and Nitschke, 2012). Research suggests that rhamnolipids prevent the growth of biofilms on two grounds: (1) the makeup of the biofilm matrix, and (2) nutritional media, indicating their possible use in the food industry, particularly in the dairy sector.

2.4.1.7 Outer Membrane Protein (OMP) Modulation by RLs

Gram negative bacteria produce OMPs in the cytoplasm, also known as integral membrane proteins. OMPs play a variety of roles. For example, they function as siderophore receptors, enzymes (proteases, palmitoyl transferases, and lipases), nutrient absorption channels, and protein translocators (Rollauer et al., 2015). OMPs include porins (OmpC, OmpD, OmpE, and OmpF), Braun's lipoprotein, and OmpA (a heat-modifiable protein) (Isibasi et al., 1988). Research has demonstrated that the (transmembrane OmpA) TM-OmpA is folded by RL micelles. Rhamnolipids additionally boosted TM-OmpA's thermostability in comparison to SDS while decreasing it in comparison to non-ionic surfactant DDM (Andersen and Otzen, 2014).

2.4.1.8 Potential Use of Rhamnolipid Nanoparticles in the Delivery of Drugs

Biosurfactants have emerged as a possible option, not only for the synthesis but also for the stability of nanoparticles, in the quest to create environmentally acceptable methods for the production of bioactive nanoparticles. Rhamnolipids, a type of biosurfactant, has been used by a number of groups (Hazra et al., 2013) as a stabilizing agent during the manufacture of nanoparticles. The researchers have eloquently shown that it is possible to load these rhamnolipids with hydrophobic medicines (up to 30% of drug loading). A novel avenue for using these rhamnolipids as an alternate drug delivery mechanism has been made possible by numerous investigations.

Rhamnolipid nanoparticles are expected to improve cancer treatment through their anti-cancer properties and improved permeability and retention effects, which cause medications to clump together inside cancer cells.

Through their incorporation into the lipid bilayer, rhamnolipids operate as antibiofilm and antiadhesive agents to disrupt the formation of biofilm by upsetting the cohesiveness of the cell membranes. In order to stop food from going bad, rhamnolipids can also be used as a biofilm-controlling technique. The post-harvest deterioration of fruits and vegetables can be avoided because of their antifungal qualities. Various applications of RLs are listed in the Table 2.

Table 2: Applications for rhamnolipids in cited literature

Application	Example	Reference
Bioremediation	Desorption of contaminants from soil	(Juwarkar et al., 2008)
	Impacts on microbial adhesion/microbial mobility	(Singh et al., 2007)
	Bioremediation of petroleum	(Campos et al., 2013)
	Bioremediation of pesticides in agriculture	(Ławniczak et al., 2013)
	Remediation of oil-contaminated water	(Mulligan, 2005)
Pest control	Enhancing the pesticide and agrochemical solubility	(Paulino et al., 2016)
	Control plant diseases	(Mnif and Ghribi, 2016)
Oil recovery	Microbial enhanced oil recovery (MEOR)	(Zhao et al., 2016)
	Increase amount of recoverable oil aided by rhamnolipid	(Paulino et al., 2016)
	Microbial de-emulsification of oil emulsions	(Long et al., 2013)
	Oil-processing operations	(Singh et al., 2007)
Medical use	Low toxicity, biocompatibility and digestibility	(Thanomsub et al., 2006)
	Prevent biofilm formation	(De Rienzo et al., 2016)
	Anticancer agents	(De Rienzo et al., 2016)
Food processing	Improvement in the stability of dough, volume, texture and conservation	(Long et al., 2014)
	As antimicrobial agent preventing food spoilage	(Campos et al., 2013)
Mining processing	Enhanced metal extraction from the mining	(Campos et al., 2013)
Nanoparticles	Nanoparticle synthesis using microemulsion method	(Kiran et al., 2016)
	Drug delivery	(Yi et al., 2019)
UF membranes cleaning	Great potential in industrial application as membrane cleaner	(Long et al., 2014)
Microbial fuel cells	Promoting power density output of microbial fuel cells	(Zheng et al., 2015)
Cosmetic and pharmacy	High emulsifying activity	(Haba et al., 2003)

Although RLs have demonstrated many benefits in various applications, their poor yield and challenging manufacturing process has prevented mainstream manufacture of RLs until 2016. However, Evonik Industry was the first business to produce RLs on a large scale. They produced rhamnolipids using butane and recombinant Pseudomonas putida. A pioneering and the first business to develop a cost-effective method to generate RL on an industrial scale (5000 tons annually) is the German biotech company Biotensidon GmbH. Other businesses making RLs in industrial facilities include TeeGene Biotech (UK), AGAE Technologies LLC, Jeneil Biosurfactant, Paradigm Biomedical Inc., and numerous Chinese businesses like Shaanxi Pioneer Biotech Company. Different strains of P. aeruginosa are shown to produce RLs at a commercial level, few of which are depicted in Table 1.

According to a study by Dusane et al. (2011), *P. aeruginosa* strain-produced rhamnolipid is capable of invading both Gram-positive and Gram-negative bacterial strains, including *E. coli, B. subtilis, S. aureus,* and *S. epidermidis*. Another study found that *B. bronchiseptica* biofilms on polystyrene could be quickly destroyed by rhamnolipids generated by *P. aeruginosa* PAO1. 70 (Ivshina et al., 1998). *P. aeruginosa* 47T2 NCIB 40 was able to create a rhamnolipid in a different investigation by Haba et al. (2003) that performed admirably as an emulsifier. According to Wood et al. (2018), *P. aeruginosa* supernatants significantly inhibited the growth of sulfate-reducing bacteria (SRB) biofilms, which are the primary cause of metal corrosion in drilling equipment and oil wells. Kim et al. (2011) published the initial study on the insecticidal properties of rhamnolipids. They noted that *Pseudomonas* sp. EP-3's rhamnolipid had anti-Green *Peach aphid* (*Myzus persicae*) properties. Rhamnolipids' antifungal activity against seven plant pathogens has been examined in another investigation. The outcomes demonstrated the potent antifungal activity of rhamnolipid from *Pseudomonas aeruginosa* ZJU211 against two Oomycetes, three Ascomycota, and two Mucor species (Sha et al., 2012). Similar to this, it has been shown that *P. aeruginosa* DS9-produced rhamnolipids have antifungal properties against the *F. sacchari* that causes pokkah boeng illness (Goswami et al., 2014).

P. aeruginosa-derived RLs play a significant role in the cosmetic and pharmaceutical industries because of their emulsifying ability, solubilizing biodegradability, low toxicity, and detergency properties, which can ensure the safety of cosmetics and drug delivery systems. Since the chemical synthesis of surfactants reveals adverse effects on people's health the texture consistency of health care and cosmetic goods including antacids, acne pads, contact lens solutions, deodorants, and others is based on high emulsifying activity (Lourith and Kanlayavattanakul, 2009). Additionally, Haba et al., 2003 investigated the rhamnolipids produced by *Pseudomonas aeruginosa* 47T2 NCBIM 40044 for their capacity to emulsify. It was discovered that only employing linseed oil coupled with RL47T2 resulted in the development of a strong and stable emulsion after other types of oils had been tested (Haba et al., 2003).

Juwarkar et al. (2008) demonstrated the potential of *P. aeruginosa* BS2's rhamnolipid biosurfactant for bioremediation of soil contaminated with several metals (Cr, Pb, Cd, Ni, and Cu). The practicality of employing rhamnolipid was demonstrated by the use of columns; however it varied for various metals. Rhamnolipid biosurfactant (extracted from *P. aeruginosa* PA1) was found to have the capacity to remove oil contamination from sandy soils, according to Santa et al.'s findings in another investigation (Santa et al., 2007).

According to Lan et al. (2015), *Pseudomonas* sp. SWP-4's increased production of rhamnolipids successfully reduced the viscosity of crude oil and effectively improved oil recovery. According to Li et al. (2002), *P. aeruginosa* (P-1) rhamnolipid can reduce crude oil viscosity by 38.5% while increasing oil recovery by 11.2%.

Rhamnolipid (derived from *P. aeruginosa* B189) extracted from milk factory waste was tested by Thanomsub et al. (2006) for its efficacy in insect cell line and breast cancer therapy. At their minimal inhibitory concentrations (MIC) (6.25 and 50 mg mL^{-1}), the two rhamnolipids Rha-Rha-C10-C10 and Rha-Rha-C10-C12 generated by the strain described displayed inhibition activity in the spread of breast cancer.

Farias et al. (2014) used the microemulsion method to study the creation of silver nanoparticles using rhamnolipid generated by a strain of *P. aeruginosa* UCP0992 in a cheap medium. The produced nanoparticle had a size of 1.13 nm and could remain stable for at least three months without the addition of a passivator.

The only study on the use of *P. aeruginosa*-produced rhamnolipid in the intravenous administration of rhamnolipid nanoparticles for photodynamic treatment was published by Yi et al. (2019). In this study, pheophorbide was used to create rhamnolipid nanoparticles with a diameter of roughly 136.1 nm and good water solubility. The findings demonstrated that tumor growth was stopped after the loaded nanoparticle was injected into a mouse model of the SCC7 tumor.

3. Conclusion

Rhamnolipid biosurfactants are the most thoroughly investigated glycolipids since they exhibit lesser toxicity, higher surface activity, and stability over a wide range of temperatures, pHs, and salinities. Thus they are ecological alternatives to synthetic surfactants. The fact that they can be used in environmentally friendly technologies since they are biodegradable is the most essential benefit.

Due to a better environmental friendliness when compared to chemical surfactants, glycolipids, a type of typical biosurfactants, are renowned for their diverse bioactivities and have the potential to be used in a wide range of practical applications Deep understanding of the primary antimicrobial behaviors of glycolipids reveals a number of underlying mechanisms when targeting microorganisms in their planktonic or biofilm state, which provides a theoretical foundation for utilizing novel, safe antimicrobial agents used in the food, cosmetics, and medical industries.

The following causes may be responsible for a single cell's caused death. Cell lysis and apoptosis are thought to be preceded by membrane permeabilization. Particularly, glycolipids are able to integrate into the lipid bilayer relying on their distinctively comparable structure, causing the membrane skeleton to physically deform and collapse along with a functional conformational change of associated proteins. Glycolipids have been shown to have rendered the disorder of movement across the membrane and material supplement by regulating the different expressions of related genes, which might be an internal reason accounting for the final cell apoptosis. ATP-binding cassette (ABC) transporters, the primary member of membrane proteins, are responsible for transmembrane migration of material and energy transformation via ATP hydrolysis. Correlational studies also found that bacteria and fungi produce ROS in addition to membrane-mediated mechanisms. In general, ROS generation is boosted by redox-cycling agents, membrane disruptors, and antibiotics while ROS is produced by normal cells during oxygen respiration and metabolism. Therefore, oxidative stress develops when a cell is unable to detoxify the excessively generated ROS, resulting in necrosis or apoptosis, which causes cell death. On the other hand, removing persistent and resistant biofilm-induced refractory microbial contamination has grown into a significant challenge that needs to be resolved immediately. The primary goal of a glycolipid-involved method to combat microbial biofilm is to prevent the sticky stage of biofilm formation and encourage the dispersal stage. The intrinsic antibacterial and physiochemical properties of glycolipids are responsible for the anti-adhesive activity, which results in a decrease in cell density and alterations in the hydrophilic/hydrophobic properties of treated surfaces. The extracellular matrix (EPS) detaches from the biofilm matrix and the entire biofilm community disintegrates as a result of the glycolipids' penetration of the biofilm matrix. Despite their exceptional antibacterial effects, glycolipids interact negatively with the human immune system. Antibiotics are becoming increasingly ineffective as a result of antimicrobial resistance, which currently poses a severe threat to public health and hygiene. Nevertheless, creating new antibiotics is expensive, time-consuming, and complicated, and there is a chance that new drug resistance will develop. As a result, the high potential of glycolipids to reduce microbial resistance will be a hot area and present a novel method for inactivating drug-resistant germs through the interaction of antibiotics and glycolipids, which will make a breakthrough in the antimicrobial sector.

References

Abalos, A., Pinazo, A., Mrosa, I., Magaly, C.H., García, F. & Manresa, A. (2001). Physicochemical and antimicrobial properties of new rhamnolipids produced by *Pseudomonas aeruginosa* AT10 from soybean oil refinery wastes. *Langmuir*, 17: 1367-1371. 10.1021/la0011735.

Abbasi, H., Hamedi, M.M.T.B., Lotfabad, H.S., Zahiri, H., Sharafi, F., Masoomi, A.A. et al. (2012). Biosurfactant-producing bacterium, *Pseudomonas aeruginosa* MA01 isolated from spoiled apples: Physicochemical and structural characteristics of isolated biosurfactant. *Journal of Bioscience and Bioengineering*, 113(2): 211–219. https://doi.org/10.1016/j.jbiosc.2011.10.002

Abdel-Mawgoud, A.M. & Stephanopoulos, G. (2018). Simple glycolipids of microbes: Chemistry, biological activity and metabolic engineering. *Synthetic and Systems Biotechnology*, 3(1): 3–19. https://doi.org/10.1016/j.synbio.2017.12.001

Abdollahi, S., Tofighi, Z., Babaee, T., Shamsi, M., Rahimzadeh, G., Rezvanifar, H. et al. (2020). Evaluation of anti-oxidant and anti-biofilm activities of biogenic surfactants derived from *Bacillus amyloliquefaciens* and *Pseudomonas aeruginosa*. *Iranian Journal of Pharmaceutical Research*, 19(2): 115–126. https://doi.org/10.22037/IJPR.2020.1101033

Andersen, K.K. & Otzen, D.E. (2014). Folding of outer membrane protein A in the anionic biosurfactant rhamnolipid. *FEBS Letters*, 588(10): 1955–1960. https://doi.org/10.1016/j.febslet.2014.04.004

Arutchelvi, J.I., Bhaduri, S., Uppara, P.V. & Doble, M. (2008). Mannosylerythritol lipids: A review. *Journal of Industrial Microbiology and Biotechnology*, 35(12): 1559–1570. https://doi.org/10.1007/s10295-008-0460-4

Asselbergh, B., Curvers, K., França, S.C., Audenaert, K., Vuylsteke, M. et al. (2007). Resistance to Botrytis cinerea in sitiens, an abscisic acid-deficient tomato mutant, involves timely production of hydrogen peroxide and cell wall modifications in the epidermis. *Plant Physiology*, 144(4): 1863–1877. https://doi.org/10.1104/pp.107.099226

Banat, I.M., Franzetti, A., Gandolfi, I., Bestetti, G., Martinotti, M.G. et al. (2010). Microbial biosurfactants production, applications and future potential. *Applied Microbiology and Biotechnology*, 87(2): 427–444. https://doi.org/10.1007/s00253-010-2589-0

Benincasa, M., Abalos, A., Oliveira, I. & Manresa A. (2004). Chemical structure, surface properties and biological activities of the biosurfactant produced by Pseudomonas aeruginosa LBI from soapstock. *Antonie van Leeuwenhoek*, 85(1): 1–8. https://doi.org/10.1023/B:ANTO.0000020148.45523.41

Bergmann, U., Scheffer, J., Köller, M., Schönfeld, W., Erbs, G. et al. (1989). Induction of inflammatory mediators (histamine and leukotrienes) from rat peritoneal mast cells and human granulocytes by Pseudomonas aeruginosa strains from burn patients. *Infection and Immunity*, 57(7): 2187–2195. https://doi.org/10.1128/iai.57.7.2187-2195.1989

Campos, J.M., Stamford, T.L., Sarubbo, L.A., de Luna, J.M., Rufino, R.D. & Banat, I.M. (2013). Microbial biosurfactants as additives for food industries. *Biotechnology Progress*. 29(5): 1097–1108. https://doi.org/10.1002/btpr.1796

Ceresa, C., Fracchia, L., Williams, M., Banat, I.M. & Díaz De Rienzo, M.A. (2020). The effect of sophorolipids against microbial biofilms on medical-grade silicone. *Journal of Biotechnology*, 309: 34–43. https://doi.org/10.1016/j.jbiotec.2019.12.019.

Chen, J., Wu, Q., Hua, Y., Chen, J., Zhang, H. & Wang, H. (2017). Potential applications of biosurfactant rhamnolipids in agriculture and biomedicine. *Applied Microbiology and Biotechnology*, 101(23–24): 8309–8319. https://doi.org/10.1007/s00253-017-8554-4

Chen, J., Liu, X., Fu, S., An, Z., Feng, Y., Wang, R. & Ji, P. (2020). Effects of sophorolipids on fungal and oomycete pathogens in relation to pH solubility. *Journal of Applied Microbiology*, 128(6): 1754–1763. https://doi.org/10.1111/jam.14594

Chen, S.Y., Lu, W.B., Wei, Y.H., Chen, W.M. & Chang, J.S. (2007). Improved production of biosurfactant with newly isolated Pseudomonas aeruginosa S2. *Biotechnology Progress*, 23(3): 661–666. https://doi.org/10.1021/bp0700152

Chester, M.A. (1997). Nomenclature of glycolipids (IUPAC recommendations 1997). *Pure and Applied Chemistry*, 69(12): 2475–2488. https://doi.org/10.1351/pac199769122475

Choe, B.Y., Krishna, N.R. and Pritchard, D.G. (1992). Proton NMR study on rhamnolipids produced by Pseudomonas aeruginosa. *Magnetic Resonance in Chemistry*, 30(10): 1025–1026. https://doi.org/10.1002/mrc.1260301019

Christova, N., Tuleva, B., Kril, A., Georgieva, M., Konstantinov, S., Terziyski, I. et al., (2013). Chemical structure and in vitro antitumor activity of rhamnolipids from Pseudomonas aeruginosa BN10. *Applied Biochemistry and Biotechnology*, 170(3): 676–689. https://doi.org/10.1007/s12010-013-0225-z

Chung, E.S., Lee, J.Y., Rhee, J.Y. & Ko, K.S. (2017). Colistin resistance in Pseudomonas aeruginosa that is not linked to arnB. *Journal of Medical Microbiology*, 66(6): 833–841. https://doi.org/10.1099/jmm.0.000456

Cortés-Sánchez, A.D.J., Hernández-Sánchez, H. & Jaramillo-Flores, M.E. (2013). Biological activity of glycolipids produced by microorganisms: New trends and possible therapeutic alternatives. *Microbiological Research*, 168: 22–32. https://doi.org/10.1016/j.micres.2012.07.002

de Freitas Ferreira, J., Vieira, E.A. & Nitschke, M. (2019). The antibacterial activity of rhamnolipid biosurfactant is pH dependent. *Food Research International*, 116: 737–744. https://doi.org/10.1016/j.foodres.2018.09.005

de Rienzo Diaz, M.A., Stevenson, P.S., Marchant, R. & Banat, I.M. (2016). Effect of biosurfactants on Pseudomonas aeruginosa and Staphylococcus aureus biofilms in a BioFlux channel. *Applied Microbiology and Biotechnology*, 100(13): 5773–5779. https://doi.org/10.1007/s00253-016-7310-5

Déziel, E.F., Lépine, F., Dennie, D., Boismenu, D., Mamer, O.A. & Villemur, R. (1999). Liquid chromatography/mass spectrometry analysis of mixtures of rhamnolipid produced by Pseudomonas aeruginosa strain 57RP grown on mannitol or naphthalene. *Biochimica et Biophysica Acta*, 1440(2–3): 244–252. https://doi.org/10.1016/s1388-1981(99)00129-8

do Valle Gomes Zezzi, M.Z. & Nitschke, M. (2012). Evaluation of rhamnolipid and surfactin to reduce the adhesion and remove biofilms of individual and mixed cultures of food pathogenic bacteria. *Food Control*, 25(2): 441–447. https://doi.org/10.1016/j.foodcont.2011.11.025

Dobler, L., Vilela, L.F., Almeida, R.V. and Neves, B.C. (2016). Rhamnolipids in perspective: Gene regulatory pathways, metabolic engineering, production and technological forecasting. *New Biotechnology*, 33(1): 123–135. https://doi.org/10.1016/j.nbt.2015.09.005

Dössel, J., Meyer-Hoffert, U., Schröder, J.M. & Gerstel, U. (2012). Pseudomonas aeruginosa derived V rhamnolipids subvert the host innate immune response through manipulation of the human beta-defensin-2 expression. *Cellular Microbiology*, 14(9): 1364–1375. https://doi.org/10.1111/j.1462-5822.2012.01801.x

Dusane, D.H., Pawar, V.S., Nancharaiah, Y.V., Venugopalan, V.P., Kumar, A.R. & Zinjarde, S.S. (2011). Anti-biofilm potential of a glycolipid surfactant produced by a tropical marine strain of Serratia marcescens. *Biofouling*, 27(6): 645–654. https://doi.org/10.1080/08927014.2011.594883

Elshikh, M., Funston, S., Chebbi, A., Ahmed, S., Marchant, R. & Banat, I.M. (2017). Rhamnolipids from non-pathogenic Burkholderia thailandensis E264: Physicochemical characterization, antimicrobial and antibiofilm efficacy against oral hygiene related pathogens. *New Biotechnology*, 36: 26–36. https://doi.org/10.1016/j.nbt.2016.12.009

Eslami, P., Hajfarajollah, H. & Bazsefidpar, S. (2020). Recent advancements in the production of rhamnolipid biosurfactants by Pseudomonas aeruginosa. *RSC Advances*, 10(56): 34014–34032. https://doi.org/10.1039/d0ra04953k

Farias, C.B.B., Silva, A.F., Rufino, R.D., Luna, J.M., Souza, J.E.G. & Sarubbo, L.A. (2014). Synthesis of silver nanoparticles using a biosurfactant produced in low-cost medium as stabilizing agent. *Electronic Journal of Biotechnology*, 17(3): 122–125. https://doi.org/10.1016/j.ejbt.2014.04.003

Fontoura, I.C.C.D., Saikawa, G.I.A., Silveira, V.A.I., Pan, N.C., Amador, I.R., Baldo, C. et al. (2020). Antibacterial activity of sophorolipids from Candida bombicola against human pathogens. *Brazilian Archives of Biology and Technology*, 63: e20180568. https://doi.org/10.1590/1678-4324-2020180568

Fukuoka, T., Morita, T., Konishi, M., Imura, T., Sakai, H. & Kitamoto, D. (2007). Structural characterization and surface-active properties of a new glycolipid biosurfactant, mono-acylated mannosylerythritol lipid, produced from glucose by Pseudozyma antarctica. *Applied Microbiology and Biotechnology*, 76(4): 801–810. https://doi.org/10.1007/s00253-007-1051-4, PubMed: 17607573

Gaur, V.K., Regar, R.K., Dhiman, N., Gautam, K., Srivastava, J.K., Patnaik, S. et al. (2019). Biosynthesis and characterization of sophorolipid biosurfactant by Candida spp.: Application as food emulsifier and antibacterial agent. *Bioresource Technology*, 285: 121314. https://doi.org/10.1016/j.biortech.2019.121314

Goswami, D., Handique, P.J. and Deka, S. (2014). Rhamnolipid biosurfactant against Fusarium sacchari—The causal organism of pokkah boeng disease of sugarcane. *Journal of Basic Microbiology*, 54(6): 548–557. https://doi.org/10.1002/jobm.201200801

Goswami, D., Borah, S.N., Lahkar, J., Handique, P.J. & Deka, S. (2015). Antifungal properties of rhamnolipid produced by Pseudomonas aeruginosa DS9 against Colletotrichum falcatum. *Journal of Basic Microbiology*, 55(11): 1265–1274. https://doi.org/10.1002/jobm.201500220

Gudiña, E.J., Rodrigues, A.I., Alves, E., Domingues, M.R., Teixeira, J.A. & Rodrigues, L.R. (2015). Bioconversion of agro-industrial by-products in rhamnolipids toward applications in enhanced oil recovery and bioremediation. *Bioresource Technology*, 177: 87–93. https://doi.org/10.1016/j.biortech.2014.11.069

Gudiña, E.J., Rodrigues, A.I., de Freitas, V., Azevedo, Z., Teixeira, J.A. & Rodrigues, L.R. (2016). Valorization of agro-industrial wastes towards the production of rhamnolipids. *Bioresource Technology*, 212: 144–150. https://doi.org/10.1016/j.biortech.2016.04.027

Guo, Y.P., Hu, Y.Y., Gu, R.R. and Lin, H. (2009). Characterization and micellization of rhamnolipidic fractions and crude extracts produced by Pseudomonas aeruginosa mutant

MIG-N146. *Journal of Colloid and Interface Science*, 331(2): 356–363. https://doi.org/10.1016/j.jcis.2008.11.039

Haba, E., Pinazo, A., Jauregui, O., Espuny, M.J., Infante, M.R. & Manresa, A. (2003). Physicochemical characterization and antimicrobial properties of rhamnolipids produced by Pseudomonas aeruginosa 47T2 NCBIM 40044. *Biotechnology and Bioengineering*. 81(3): 316–322. https://doi.org/10.1002/bit.10474

Haque, F., Alfatah, M., Ganesan, K. & Bhattacharyya, M.S. (2016). Inhibitory effect of sophorolipid on Candida albicans biofilm formation and hyphal growth. *Scientific Reports*, 6: 23575. https://doi.org/10.1038/srep23575

Haque, F., Verma, N.K., Alfatah, M., Bijlani, S. & Bhattacharyya, M.S. (2019). Sophorolipid exhibits antifungal activity by ROS mediated endoplasmic reticulum stress and mitochondrial dysfunction pathways in Candida albicans. *RSC Advances*, 9(71): 41639–41648. https://doi.org/10.1039/c9ra07599b

Hardin, R., Pierre, J., Schulze, R., Mueller, C.M., Fu, S.L., Wallner, S.R. et al. (2007). Sophorolipids improve sepsis survival: Effects of dosing and derivatives. *Journal of Surgical Research*, 142(2): 314–319. https://doi.org/10.1016/j.jss.2007.04.025

Hazra, C., Kundu, D., Chaudhari, A. & Jana, T. (2013). Biogenic synthesis, characterization, toxicity and photocatalysis of zinc sulphide nanoparticles using rhamnolipids from Pseudomonas aeruginosa BS01 as capping and stabilizing agent. *Journal of Chemical Technology and Biotechnology*, 88(6): 1039–1048. https://doi.org/10.1002/jctb.3934

Holst, O. (2008). Glycolipids: Occurrence, significance, and properties. *In:* B.O. Fraser-Reid, K. Tatsuta and J. Thiem (Eds), *Glycoscience: Chemistry and Chemical Biology*, 1603e27. Springer, Berlin Heidelberg.

Isibasi, A., Ortiz, V., Moreno, J., Paniagua, J., Vargas, M., González, C. et al. (1988). The role of outer membrane proteins from gram-negative bacteria as vaccines with special emphasis in typhoid fever: Monoclonal antibodies against *S. typhi porins. In:* Cañedo, L.E., Todd, L.E., Packer, L., Jaz, J. (Eds), *Cell Function and Disease*. 281–292. Springer, Boston, MA.

Islas, D.J., Medina, M.S.A. & Gracida, R.J.N. (2010). Properties, applications and production of biosurfactants: A review. *Rev. Int. Contam. Ambient*., 26(1): 65-84.

Ivshina, I.B., Kuyukina, M.S., Philp, J.C. & Christofi, N. (1998). Oil desorption from mineral and organic materials using biosurfactant complexes produced by Rhodococcus species. *World Journal of Microbiology and Biotechnology*, 14(5): 711–717. https://doi.org/10.1023/A:1008885309221

Jezierska, S., Claus, S. & Van Bogaert, I. (2018). Yeast glycolipid biosurfactants. *FEBS Letters*, 592(8): 1312–1329. https://doi.org/10.1002/1873-3468.12888

Jones, R.A., Shropshire, H., Caimeng, Z., Murphy, A., Lidbury, I., Wei, T. et al. (2021). Phosphorus stress induces the synthesis of novel glycolipids in Pseudomonas aeruginosa that confer protection against a last-resort antibiotic. *ISME Journal*, 15(11): 3303–3314. https://doi.org/10.1038/s41396-021-01008-7

Juwarkar, A.A., Dubey, K.V., Nair, A. and Singh, S.K. (2008). Bioremediation of multi-metal contaminated soil using biosurfactant – A novel approach. *Indian Journal of Microbiology*, 48(1): 142–146. https://doi.org/10.1007/s12088-008-0014-5

Kharami, A., Bibi, Z., Nielsen, H., Høiby, N. and Döring, G. et al. (1989). Effect of Pseudomonas aeruginosa rhamnolipid on human neutrophil and monocyte function. *APMIS*, 97(12): 1068–1072. https://doi.org/10.1111/j.1699-0463.1989.tb00519.x

Kim, B.S., Lee, J.Y. and Hwang, B.K. (2000). In vivo control and in vitro antifungal activity of rhamnolipid B, a glycolipid antibiotic, against Phytophthora capsici and Colletotrichum orbiculare. *Pest Management Science*, 56(12): 1029–1035. https://doi.org/10.1002/1526-4998(200012)56:12<1029::AID-PS238>3.0.CO;2-Q

Kim, L.H., Jung, Y., Yu, H.W., Chae, K.J. & Kim, I.S. (2015). Physicochemical interactions between rhamnolipids and Pseudomonas aeruginosa biofilm layers. *Environmental Science and Technology*, 49(6): 3718–3726. https://doi.org/10.1021/es505803c

Kim, S.K., Kim, Y.C., Lee, S., Kim, J.C., Yun, M.Y. & Kim, I.S. (2011). Insecticidal activity of rhamnolipid isolated from Pseudomonas sp. EP-3 against green peach aphid (*Myzus* persicae). *Journal of Agricultural and Food Chemistry*, 59(3): 934–938. https://doi.org/10.1021/jf104027x

Kiran, G.S., Ninawe, A.S., Lipton, A.N., Pandian, V. & Selvin, J. (2016). Rhamnolipid biosurfactants: Evolutionary implications, applications and future prospects from untapped marine resource. *Critical Reviews in Biotechnology*, 36(3): 399–415. https://doi.org/10.3109/07388551.2014.979758

Kitamoto, D. (1993). Surface active properties and antimicrobial activities of mannosylerythritol lipids as biosurfactants produced by Candida antarctica. *Scientific World Journal*, 29: 91–96.

Kitamoto, D., Yanagishita, H., Endo, A., Nakaiwa, M., Nakane, T. & Akiya, T. (2001). Remarkable antiagglomeration effect of a yeast biosurfactant, diacylmannosylerythritol, on ice-water slurry for cold thermal storage. *Biotechnology Progress*, 17(2): 362–365. https://doi.org/10.1021/bp000159f

König, B., Bergmann, U. and König, W. (1992). Induction of inflammatory mediator release (serotonin and 12-hydroxyeicosatetraenoic acid) from human platelets by Pseudomonas aeruginosa glycolipid. *Infection and Immunity*, 60(8): 3150–3155. https://doi.org/10.1128/iai.60.8.3150-3155.1992

Kurtzman, C.P., Price, N.P.J., Rayand, K.J. & Kuo, T.M. (2010). Production of sophorolipid biosurfactants by multiple species of the Starmerella (Candida) bombicola yeast clade. *FEMS Microbiology Letters*, 311(2): 140–146. https://doi.org/10.1111/j.1574-6968.2010.02082.x

Lan, G., Fan, Q., Liu, Y., Liu, Y., Yin, X. & Luo, M. (2015). Effects of the addition of waste cooking oil on heavy crude oil biodegradation and microbial enhanced oil recovery using Pseudomonas sp. SWP-4. *Biochemical Engineering Journal*, 103: 219–226. https://doi.org/10.1016/j.bej.2015.08.004

Lang, S. (2002). Biological amphiphiles (microbial biosurfactants). *Current Opinion in Colloid and Interface Science*, 7(1–2): 12–20. https://doi.org/10.1016/S1359-0294(02)00007-9

Ławniczak, L., Marecik, R. & Chrzanowski, L. (2013). Contributions of biosurfactants to natural or induced bioremediation. *Applied Microbiology and Biotechnology*, 97(6): 2327–2339. https://doi.org/10.1007/s00253-013-4740-1

Leray, C. (2012). Simple lipids with two different components. *In:* Leray, C. (Ed.), *Introduction to Lipidomics: From Bacteria to Man* (1st ed.). 69.e210. CRC Press, Boca Raton.

Li, Q., Kang, C., Wang, H., Liu, C. & Zhang, C. (2002). Application of microbial enhanced oil recovery technique to Daqing Oilfield. *Biochemical Engineering Journal*, 11(2–3): 197–199. https://doi.org/10.1016/S1369-703X(02)00025-6

Liu, X., Shu, Q., Chen, Q., Pang, X., Wu, Y., Zhou, W. et al. (2020). Antibacterial efficacy and mechanism of mannosylerythritol lipids-A on Listeria monocytogenes. *Molecules*. 25(20): 4857. https://doi.org/10.3390/molecules25204857

Long, X., Zhang, G., Shen, C., Sun, G., Wang, R., Yin, L. & Meng, Q. (2013). Application of rhamnolipid as a novel bio demulsifier for destabilizing waste crude oil. *Bioresource Technology*, 131: 1–5. https://doi.org/10.1016/j.biortech.2012.12.128

Long, X., Meng, Q. & Zhang, G. (2014). Application of biosurfactant rhamnolipid for cleaning of UF membranes. *Journal of Membrane Science*, 457: 113–119. https://doi.org/10.1016/j.memsci.2014.01.044

Lotfabad, T.B., Abassi, H., Ahmadkhaniha, R., Roostaazad, R., Masoomi, F., Zahiri, H.S. et al. (2010). Structural characterization of a rhamnolipid-type biosurfactant produced by Pseudomonas aeruginosa MR01: Enhancement of di-rhamnolipid proportion using gamma irradiation. *Colloids and Surfaces. B, Biointerfaces*, 81(2): 397–405. https://doi.org/10.1016/j.colsurfb.2010.06.026

Lourith, N. & Kanlayavattanakul, M. (2009). Natural surfactants used in cosmetics: Glycolipids. *International Journal of Cosmetic Science*, 31(4): 255–261. https://doi.org/10.1111/j.1468-2494.2009.00493.x

Lukic, M., Pantelic, I. & Savic, S. (2016). An overview of novel surfactants for formulation of cosmetics with certain emphasis on acidic active substances. *Tenside Surfactants Detergents*, 53(1): 7–19. https://doi.org/10.3139/113.110405

McClure, C.D. & Schiller, N.L. (1992). Effects of Pseudomonas aeruginosa rhamnolipids on human monocyte-derived macrophages. *Journal of Leukocyte Biology*, 51(2): 97–102. https://doi.org/10.1002/jlb.51.2.97

McClure, C.D. & Schiller, N.L. (1996). Inhibition of macrophage phagocytosis by Pseudomonas aeruginosa rhamnolipids in vitro and in vivo. *Current Microbiology*, 33(2): 109–117. https://doi.org/10.1007/s002849900084

Merrill, Jr. A.H. & Vu, M.N. (2016). Glycolipids. *In:* Ralph A. Bradshaw, Philip D. Stahl (Eds), *Encyclopedia of Cell Biology*. 180.e93. Academic Press, Waltham.

Mnif, I. & Ghribi, D. (2016). Glycolipid biosurfactants: Main properties and potential applications in agriculture and food industry. *Journal of the Science of Food and Agriculture*, 96(13): 4310–4320. https://doi.org/10.1002/jsfa.7759

Morikawa, M., Hirata, Y. & Imanaka, T. (2000). A study on the structure-function relationship of lipopeptide biosurfactants. *Biochimica et Biophysica Acta*, 1488(3): 211–218. https://doi.org/10.1016/s1388-1981(00)00124-4

Morita, T., Fukuoka, T., Imura, T. & Kitamoto, D. (2015). Mannosylerythritol lipids: Production and applications. *Journal of Oleo Science*, 64(2): 133–141. https://doi.org/10.5650/jos.ess14185

Mulligan, C.N. (2005). Environmental applications for biosurfactants. *Environmental Pollution*, 133(2): 183–198. https://doi.org/10.1016/j.envpol.2004.06.009

Noordman, W.H., Bruining, J.W., Wietzes, P., Janssen, D.B. & Bruining, J.W. (2000). Facilitated transport of a PAH mixture by a rhamnolipid biosurfactant in porous silica matrices. *Journal of Contaminant Hydrology*, 44(2): 119–140. https://doi.org/10.1016/S0169-7722(00)00097-8

Paulino, B.N., Pessôa, M.G., Mano, M.C., Molina, G., Neri-Numa, I.A. & Pastore, G.M. (2016). Current status in biotechnological production and applications of glycolipid biosurfactants. *Applied Microbiology and Biotechnology*, 100(24): 10265–10293. https://doi.org/10.1007/s00253-016-7980-z

Pekmezovic, M., Rajkovic, K., Barac, A., Senerović, L. & Arsic Arsenijevic, V. (2015). Development of kinetic model for testing antifungal effect of Thymus vulgaris L. and Cinnamomum cassia L. essential oils on Aspergillus flavus spores and application for optimization of synergistic effect. *Biochemical Engineering Journal*, 99: 131–137. https://doi.org/10.1016/j.bej.2015.03.024

Pinzon, N.M. & Ju, L.K. (2009). Improved detection of rhamnolipid production using agar plates containing methylene blue and cetyl trimethylammonium bromide. *Biotechnology Letters*, 31(10): 1583–1588. https://doi.org/10.1007/s10529-009-0049-7

Pornsunthorntawee, O., Wongpanit, P., Chavadej, S., Abe, M. & Rujiravanit, R. (2008). Structural and physicochemical characterization of crude biosurfactant produced by Pseudomonas aeruginosa SP4 isolated from petroleum-contaminated soil. *Bioresource Technology*, 99(6): 1589–1595. https://doi.org/10.1016/j.biortech.2007.04.020

Rahimi, K., Lotfabad, T.B., Jabeen, F. & Mohammad Ganji, S. (2019). Cytotoxic effects of mono- and di-rhamnolipids from Pseudomonas aeruginosa MR01 on MCF-7 human breast cancer cells. *Colloids and Surfaces. B, Biointerfaces*, 181: 943–952. https://doi.org/10.1016/j.colsurfb.2019.06.058

Recke, V.K., Beyrle, C., Gerlitzki, M., Hausmann, R., Syldatk, C., Wray, V. et al. (2013). Lipase-catalyzed acylation of microbial mannosylerythritol lipids (biosurfactants) and their characterization. *Carbohydrate Research*, 373: 82–88. https://doi.org/10.1016/j.carres.2013.03.013

Reddy, S.K., Khan, M.Y., Archana, K., Reddy, M.G. & Hameeda, B. (2016). Utilization of mango kernel oil for the rhamnolipid production by Pseudomonas aeruginosa DR1 towards its application as biocontrol agent. *Bioresource Technology*, 221: 291–299. https://doi.org/10.1016/j.biortech.2016.09.041

Rodrigues, A.I., Gudiña, E.J., Teixeira, J.A. & Rodrigues, L.R. (2017). Sodium chloride effect on the aggregation behaviour of rhamnolipids and their antifungal activity. *Scientific Reports*, 7(1): 12907. https://doi.org/10.1038/s41598-017-13424-x

Rollauer, S.E., Sooreshjani, M.A., Noinaj, N. & Buchanan, S.K. (2015). Outer membrane protein biogenesis in Gram-negative bacteria. *Philosophical Transactions of the Royal Society of London. Series B, Biological Sciences*, 370(1679): 20150023. https://doi.org/10.1098/rstb.2015.0023

Saika, A., Koike, H., Fukuoka, T. & Morita, T. (2018). Tailor-made mannosylerythritol lipids: Current state and perspectives. *Applied Microbiology and Biotechnology*, 102(16): 6877–6884. https://doi.org/10.1007/s00253-018-9160-9

Saika, A., Koike, H., Fukuoka, T., Yamamoto, S., Kishimoto, T. et al. (2016). A gene cluster for biosynthesis of mannosylerythritol lipids consisted of 4-O-beta-D-mannopyranosyl-(2R,3S)-erythritol as the sugar moiety in a Basidiomycetous yeast Pseudozyma tsukubaensis. *Plos One*, 11(6): e0157858. https://doi.org/10.1371/journal.pone.0157858

Santa Anna, L.M., Soriano, A.U., Gomes, A.C., Menezes, E.P., Gutarra, M.L., Freire, D.M. et al. (2007). Use of biosurfactant in the removal of oil from contaminated sandy soil. *Journal of Chemical Technology and Biotechnology*, 82(7): 687–691. https://doi.org/10.1002/jctb.1741

Sen, S., Borah, S.N., Kandimalla, R., Bora, A. & Deka, S. (2019). Efficacy of a rhamnolipid biosurfactant to inhibit Trichophyton rubrum in vitro and in a mice model of dermatophytosis. *Experimental Dermatology*, 28(5): 601–608. https://doi.org/10.1111/exd.13921

Sha, R. & Meng, Q. (2016). Antifungal activity of rhamnolipids against dimorphic fungi. *Journal of General and Applied Microbiology*, 62(5): 233–239. https://doi.org/10.2323/jgam.2016.04.004

Sha, R., Jiang, L., Meng, Q., Zhang, G. & Song, Z. (2012). Producing cell-free culture broth of rhamnolipids as a cost-effective fungicide against plant pathogens. *Journal of Basic Microbiology*, 52(4): 458–466. https://doi.org/10.1002/jobm.201100295

Shah, M.U.H., Sivapragasam, M., Moniruzzaman, M. & Yusup, S.B. (2016). A comparison of recovery methods of rhamnolipids produced by Pseudomonas aeruginosa. *Procedia Engineering*, 148: 494–500. https://doi.org/10.1016/j.proeng.(2016.06.538

Shin, J.D., Lee, J., Kim, Y.B., Han, I.S. & Kim, E.K. (2010). Production and characterization of methyl ester sophorolipids with 22-carbonfatty acids. *Bioresource Technology*, 101(9): 3170–3174. https://doi.org/10.1016/j.biortech.2009.12.019

Shu, Q., Wei, T., Lu, H., Niu, Y. & Chen, Q. (2020). Mannosylerythritol lipids: Dual inhibitory modes against Staphylococcus aureus through membrane-mediated apoptosis and biofilm disruption. *Applied Microbiology and Biotechnology*, 104(11): 5053–5064. https://doi.org/10.1007/s00253-020-10561-8

Silveira, V.A.I., Nishio, E.K., Freitas, C.A.U.Q., Amador, I.R., Kobayashi, R.K.T., Caretta, T. et al. (2019). Production and antimicrobial activity of sophorolipid against Clostridium perfringens and Campylobacter jejuni and their additive interaction with lactic acid. *Biocatalysis and Agricultural Biotechnology*, 21: 101287. https://doi.org/10.1016/j.bcab.2019.101287

Singh, A., Van Hamme, J.D. & Ward, O.P. (2007). Surfactants in microbiology and biotechnology: Part 2. Application aspects. *Biotechnology Advances*, 25(1): 99–121. https://doi.org/10.1016/j.biotechadv.2006.10.004

Singh, N., Pemmaraju, S.C., Pruthi, P.A., Cameotra, S.S. & Pruthi, V. (2013). Candida biofilm disrupting ability of di-rhamnolipid (RL-2) produced from Pseudomonas aeruginosa DSVP20. *Applied Biochemistry and Biotechnology*, 169(8): 2374–2391. https://doi.org/10.1007/s12010-013-0149-7

Solaiman, D.K.Y., Ashby, R.D. & Uknalis, J. (2017). Characterization of growth inhibition of oral bacteria by sophorolipid using a microplate-format assay. *Journal of Microbiological Methods*, 136: 21–29. https://doi.org/10.1016/j.mimet.2017.02.012

Srey, S., Jahid, I.K. & Ha, S.D. (2013). Biofilm formation in food industries: A food safety concern. *Food Control*, 31(2): 572–585. https://doi.org/10.1016/j.foodcont.2012.12.001

Thanomsub, B., Pumeechockchai, W., Limtrakul, A., Arunrattiyakorn, P., Petchleelaha, W. et al. (2006). Chemical structures and biological activities of rhamnolipids produced by Pseudomonas aeruginosa B189 isolated from milk factory waste. *Bioresource Technology*, 97(18): 2457–2461. https://doi.org/10.1016/j.biortech.2005.10.029

Van Bogaert, I.N.A., Zhang, J. & Soetaert, W. (2011). Microbial synthesis of sophorolipids. *Process Biochemistry*, 46(4): 821–833. https://doi.org/10.1016/j.procbio.2011.01.010

Van Gennip, M., Christensen, L.D., Alhede, M., Phipps, R., Jensen, P.Ø., Christophersen, L. et al. (2009). Inactivation of the rhlA gene in Pseudomonas aeruginosa prevents rhamnolipid production, disabling the protection against polymorphonuclear leukocytes. *APMIS*, 117(7): 537–546. https://doi.org/10.1111/j.1600-0463.2009.02466.x

Varvaresou, A. & Iakovou, K. (2015). Biosurfactants in cosmetics and biopharmaceuticals. *Letters in Applied Microbiology*, 61(3): 214–223. https://doi.org/10.1111/lam.12440

Wilkinson, S.G. (1988). Gram-negative bacteria. *In:* C. Ratledge and S.G. Wilkinson (Eds), *Microbial Lipids*. Vol.1, 333–348. Academic Press, San Diego.

Wood, T.L., Gong, T., Zhu, L., Miller, J., Miller, D.S., Yin, B. & Wood, T.K. (2018). Rhamnolipids from Pseudomonas aeruginosa disperse the biofilms of sulfate-reducing bacteria. *Biofilms and Microbiomes*, 4: 22. https://doi.org/10.1038/s41522-018-0066-1

Yan, F., Hu, H., Lu, L. & Zheng, X. (2016). Rhamnolipids induce oxidative stress responses in cherry tomato fruit to Alternaria alternata. *Pest Management Science*, 72(8): 1500–1507. https://doi.org/10.1002/ps.4177

Yi, G., Son, J., Yoo, J., Park, C. & Koo, H. (2019). Rhamnolipid nanoparticles for in vivo drug delivery and photodynamic therapy. *Nanomedicine: Nanotechnology, Biology and Medicine*. 19: 12–21. https://doi.org/10.1016/j.nano.2019.03.015

Yin, H., Qiang, J., Jia, Y., Ye, J., Peng, H., Qin, H. et al. (2009). Characteristics of biosurfactant produced by Pseudomonas aeruginosa S6 isolated from oil-containing wastewater. *Process Biochemistry*, 44(3): 302–308. https://doi.org/10.1016/j.procbio.2008.11.003

Yoshida, S., Koitabashi, M., Nakamura, J., Fukuoka, T., Sakai, H., Abe, M. et al. (2015). Effects of biosurfactants, mannosylerythritol lipids, on the hydrophobicity of solid surfaces and infection behaviours of plant pathogenic fungi. *Journal of Applied Microbiology*, 119(1): 215–224. https://doi.org/10.1111/jam.12832

Zeidan, Y.H. & Hannun, Y.A. (2007). Translational aspects of sphingolipid metabolism. *Trends in Molecular Medicine*, 13(8): 327–336. https://doi.org/10.1016/j.molmed.2007.06.002

Zhang, Y. & Miller, R.M. (1992). Enhanced octadecane dispersion and biodegradation by a Pseudomonas rhamnolipid surfactant (biosurfactant). *Applied and Environmental Microbiology*, 58(10): 3276–3282. https://doi.org/10.1128/aem.58.10.3276-3282.1992

Zhang, Y., Maier, W.J. & Miller, R.M. (1997). Effect of rhamnolipids on the dissolution, bioavailability and biodegradation of phenanthrene. *Environmental Science and Technology*, 31(8): 2211–2217. https://doi.org/10.1021/es960687g

Zhao, F., Zhou, J., Han, S., Ma, F., Zhang, Y. & Zhang, J. (2016). Medium factors on anaerobic production of rhamnolipids by Pseudomonas aeruginosa SG and a simplifying medium for in situ microbial enhanced oil recovery applications. *World Journal of Microbiology and Biotechnology*, 32(4): 54. https://doi.org/10.1007/s11274-016-2020-9

Zheng, T., Xu, Y.S., Yong, X.Y., Li, B., Yin, D., Cheng, Q.W. et al. (2015). Endogenously enhanced biosurfactant production promotes electricity generation from microbial fuel cells. *Bioresource Technology*, 197: 416–421. https://doi.org/10.1016/j.biortech.2015.08.136

CHAPTER

16

Lipopeptides in Medicine

Ashish Jain*, Shrikrishna Bhagat, and Udaykumar Reddy

Department of Microbiology, Smt. C.H.M. College, Ulhasnagar, University of Mumbai, Maharashtra, India

1. Introduction

Humans have relied on natural sources, predominantly plants, for fundamental needs, notably medicinal remedies to combat various ailments. Natural products exhibit remarkable diversity in origin and bioactivity, motivating extensive exploration as potential therapeutic agents. Their intricate structures often challenge straightforward chemical synthesis, necessitating semi-synthetic approaches to retain their original efficacy while enhancing drug-like properties (Cragg & Newman, 2013). The discovery of penicillin marked a pivotal shift towards harnessing microorganisms, catalyzing revolutionary advancements in modern medicine (Pham et al., 2019). In 2017, WHO declared the global health crisis over the relentless evolution of antimicrobial resistance (AMR), and highlighted the critical need for innovative frameworks free from resistance mechanisms, cross-resistance, and bacterial resistance tendencies for the multi-drug resistant (MDR) pathogens (Kwon & Powderly, 2021, O'Neill, 2016). Antimicrobial peptides (AMPs), lipopeptides, and their synthetic counterparts have gained attention for their potent bactericidal properties and low propensity for resistance development (Mahlapuu et al., 2016, Chen & Lu, 2020). These naturally occurring compounds, found in all multicellular organisms, exhibit a cell membrane-targeting mode of action. Some AMPs also possess immunomodulatory properties to combat bacterial virulence factors (Lei et al., 2019).

Lipopeptides (LPs) and lipoprotein biosurfactants are non-ribosomally synthesized antimicrobial peptides (AMPs) incorporating lipid moieties. Lipopeptides (LPs) possess an amphiphilic structure comprising a fatty acid component and a cyclic oligopeptide segment. This particular classification distinguishes itself through its distinct isoforms, marked by differences in the composition of the peptide portion, fatty acid chain stretch, and the linkage between these constituents (Bender et al., 1999). These bioactive secondary metabolites, derived from diverse microorganisms, exhibit profound therapeutic and biotechnological potential. *Bacillus* and

* Corresponding author: ajcdri@gmail.com

Pseudomonas species have historically been recognized as prominent producers of lipopeptides (LPs). Nonetheless, contemporary scientific investigations have directed their attention toward members of the Streptomyces group, among other microbial groups, as emerging and noteworthy LP producers (Zambry et al., 2021). Surfactin, fengycin, and iturin are the major lipopeptides produced by the *Bacillus* and show a wide range of applications (Ongena & Jacques, 2008). Because of their ability to disrupt cell membranes, researchers are increasingly intrigued by the potential of lipopeptides (LPs) for novel applications in areas such as antimicrobial and antiviral agents, anti-cancer treatments, cosmetics, and immunotherapies (Chen et al., 2015, Sharma et al., 2014). Microbial-origin recombinant lipopeptides, exemplified by Lipo-N-ter, serve as a groundbreaking adjuvant, showcasing their innovative potential with the capacity to elicit heightened anti-tumor responses surpassing those achieved by synthetic lipopeptides, frequently employed for immune enhancement in vaccines. LPs find utility in eco-friendly crop protection as fungicides, mitigating environmental concerns associated with conventional chemical pesticides. These multifaceted compounds represent a burgeoning field of research, holding immense promise across various domains, including human medicine and sustainable agriculture. This chapter provides an in-depth exploration of lipopeptides, elucidating their sources, inherent properties, medicinal applications, and the promising prospects they hold as a foundation for drug development in the field of medicine.

2. Lipopeptides

Lipopeptides are biologically derived agents with surface activity primarily synthesized by *Bacillus* species. They are distinguished by their addition of an identified fatty acid that is linked to an amino-acid moiety and has a molecular mass averaging 1-1.2 kDa (Meena & Kanwar, 2015). The significant structural and functional diversity observed in lipopeptides stems from a wide range of changes in amino acid structure, their quantities, and their arrangements, coupled with variations in the constituents of fatty acids. These variations have led to the identification of several major classes of lipopeptides. Antimicrobial lipopeptides are primarily synthesized by bacteria and fungi through the non-ribosomal pathway and incorporate one or more lipid chains into their molecular structure. Many lipopeptides adopt a cyclic conformation and contain non-native amino acid residues, which enhance their resistance to enzymatic degradation. Antibacterial lipopeptides can be positively charged cationic (polymyxin B, colistin, octapeptin, and trichogin. or anionic (daptomycin, tsushimycin, glumamycin, aspartocin, laspartomycin, surfactin, crystallomycin,) and they demonstrate a wide-ranging effectiveness against a wide spectrum of microbes (Jerala, 2007). Lipopeptides are an extraordinary class of self-assembling molecules capable of forming peptide-functionalized nano-sized structures. The self-assembly of the lipid component leads to its segregation from the surrounding aqueous environment. This naturally occurring pattern plays a pivotal role in various biological processes and has substantial importance in pharmaceuticals (Xu et al., 2015).

2.1 Classes of Lipopeptides

2.1.1 Non-Ribosomal Lipopeptides

Cyclic lipopeptides are non-ribosomal lipid peptides (NRLPs) produced by non-ribosomal peptide synthetases (NRPSs), which are responsible for the production of micro-heterogeneous structures with varying biological activity. Lipopolypeptides (LPs) are generated through the consecutive arrangement of amino acid residues in Non-Ribosomal Peptides (NRPs), utilizing both repetitive and non-repetitive pathways (Mahlstedt & Walsh, 2010, Raaijmakers et al., 2010). A single species of *Bacillus* may produce a single type or all groups of lipopeptides depending on the genetic constitution and expression (Strieker et al., 2010). *Bacillus subtilis* produces NRPs as secondary products. These NRPs hold significant industrial appeal due to their applications in antibiotics, anti-tumor agents, antiviral agents, and immunomodulation. Complexes of considerable size involving enzymes, ranging from 100 to 1600 kDa, play a pivotal role in NRP production, yielding diverse linear, cyclic, or branched structures comprising OH groups and amino acids. Post-synthetic modifications like N-methylation, glycosylation, acylation, and heterocyclic ring formation further diversify these NRPs. These non-ribosomal peptides (NRPs) can be categorized into two major classes: a) thiotemplate NRPs (such as Lipopeptides) and b) non-thiotemplate NRPs. This classification is based on the synthesis pathway they follow. (Wang et al., 2015, Sumi et al., 2015)

Non-ribosomal lipopolypeptide (LP) families include iturin, surfactin, fengycin, and kurstakin, and ongoing research continues to uncover novel non-ribosomal LPs, broadening their potential applications (Ali et al., 2022). Figure 1 illustrates typical representatives of common lipopeptides (LPs) families.

2.1.2 Surfactin

Surfactin has been identified as one of the most potent biosurfactants ever discovered. The structure of it was elucidated after it was purified from *Bacillus subtilis* in 1968, and derivatives have since been extracted from other sources. Surfactin has anti-microbial and cancer-fighting abilities, along with being used as a biosurfactant. The surfactin family is comprised of heptapeptide forms of surfactin lichenysin, esperin, surfactin, pumilacidin, lichenycin, and pumilacidin and it is considered to be the first family of CLP. Almost all the species of Bacilli are known to produce surfactin. Both linear and cyclic forms of surfactin are known to occur (Liu et al., 2015) A complex of enzymes, collectively known as surfactin synthetase, consisting of four enzymatic subunits (Srf A, Srf B, Srf C, and Srf D), is involved in the non-ribosomal pathway used for the synthesis of surfactin. The Srf operon, responsible for synthesis, consists of seven modules from Srf A, Srf B, and Srf C, while Srf D participates in the initial reaction (Wu et al., 2017). Surfactin synthesis in Bacillus involves a regulatory network with key components: ComX, ComA/ComP, and RapC. ComA is activated by ComX via ComP, binding to the srfA promoter. RapC dephosphorylates ComA, influenced by intracellular pentapeptide PhrC levels. Low PhrC leads to increased RapC activity, promoting surfactin synthesis, while high PhrC inhibits it (Cosby et al., 1998, Cosby & Zuber, 1997). Surfactin holds diverse activities like anti-viral, anti-microbial, anti-adherent, and cytotoxic due to which it has high applicability in cosmetics, food processing, oil recovery, and pharmaceuticals, in laundry, and as a bio-pesticide (Meena & Kanwar, 2015).

2.1.3 Iturin

In the 1970s to 1990s, scientists made significant strides in uncovering the structures of diverse lipopeptides (LPs) found in bacterial species, particularly within the Bacilli class, like *Paenibacillus koreensis*. Among these, one remarkable example is iturin

A, cyclic lipopeptides. It is made up of an acyl chain with a variable size ranging from C14 to C17, which is connected to a heptapeptide. This heptapeptide features a distinctive chiral pattern with the sequence 'LDDLLDL,' a characteristic that was first identified in 1994 (Pilz et al., 2023).

The Iturin family has five principal categories, such as bacillomycin D, F, and L, mycosubtilin, and iturin. For Iturin biosynthesis and Bacillomycin D production, researchers have pinpointed the operon's starting point for Bacillomycin (bmy) production. During early stationary growth, DegU interacts with the sA-regulated promoter, activating gene expression. DegU, a two-component response regulator in Bacillus subtilis, oversees various cellular functions. The regulatory protein DegQ, with pleiotropic effects, positively impacts bmy expression, albeit through an unidentified interaction with DegU. Additionally, ComA is crucial for full Bacillomycin gene expression and regulates degQ gene expression. These interactions and regulatory mechanisms are integral to the Iturin biosynthesis pathway (Koumoutsi et al., 2007).

2.1.4 Fengycin

Fengycin is synthesized by the operation of a group of five biosynthetic genes, namely ppsA, ppsB, ppsC, ppsD, and ppsE, along with 5 multifaceted peptide synthetases (Fen 1-5). These decapeptides, including Fengycin A and B (plipastatins and members of the third family), are characterized by a linear, iso, or anti-iso molecular structure of b-hydroxyl fatty acid chain (C14-C18), which can be either unsaturated or saturated. The gathering of the peptide part of these LPs is facilitated by Fen 1–5 enzymes; however, the synthesis of fengycin comes to a halt when Fen 4 is disrupted. (Batool et al., 2011).

These lipopeptides have a unique structure that includes the third amino acid (Tyrosine) and the last amino acid (Isoleucine) form the lactone ring. Despite this, there are only subtle structural dissimilarities between the two groups. In Fengycin, the third position features D-Tyrosine, the eighth position has glutamine, and the ninth position has L-Tyrosine, whereas, in Plipastatin, the third position has L-Tyrosine, the eighth position has glutamine, and the ninth position has D-Tyrosine. The biosynthesis of Fengycin is encoded by the open reading frame (ORF) of Fengycin synthetase, which consists of five modules arranged in the following order: FenC, FenD, FenE, FenA, and FenB. FenC, D, and E each contribute two amino acids, whereas FenA adds three amino acids, and FenB contributes only one amino acid to the growing oligopeptide chain (Wu et al., 2007).

2.1.5 Kurstakin

In the early 2000s, MALDI technology enabled *Bacillus* strain differentiation, leading to the discovery of Kurstakin LPs named after *Bacillus thuringiensis* subsp. Kurstaki. Its operon was found in 2009/2010. Kurstakin has antifungal properties and is used in targeted agricultural biocontrol against lepidopteran larvae. Its composition comprises a heptapeptide sequence (TGASHQQ) connected to fatty acid chains through an amide bond, accompanied by a distinctive lactone ring positioned near the Carboxy terminus and the fourth amino acid residue (Ser) (Abderrahmani et al., 2011).

Naturally, filamentous fungi synthesize Echinocandins (echinocandin B and Pneumocandin B), which exhibit both fungistatic and fungicidal properties. The primary structure of Echinocandins is a cyclical hexapeptide with N-acylation (Szymanski et al., 2022). Cyclomarins, (type A, B, C, and D,) are cyclic heptapeptides composed of both proteogenic and non-proteinogenic amino acids, with the addition of a hydrophobic group (prenylation). These variants distinguish themselves at specific positions. Their remarkable antibacterial attributes, which were subsequently identified, offer significant promise for a range of uses, including the management of tuberculosis (Taylor et al., 2022). In a recent discovery, linear versions of pelgipeptin B′ and C′, were extracted from *P. elgii*. These analogues exhibit peptide sequence similarities with paenipeptin, another novel non-ribosomal lipopeptide family (Huang et al., 2017).

Another family of lipopeptides (LPs), tridecaptins, exhibit robust anti-microbial capabilities, notably against methicillin-resistant *Staphylococcus aureus* (MRSA) strains (da Costa et al., 2022). These findings contribute to the ongoing search for therapeutics with potent bioactivity and reduced systemic toxicity.

3. Ribosomal Lipopeptides (RLPs)

Recently, a discovery has been made regarding Lipolanthine, a new class of lipopeptides (LPs) originating from ribosomes. These LPs feature a lanthipeptide component linked to a bis-methylated guanidino fatty acid (MGFA) on its N-terminus. However, fully comprehending these compounds remains a complex task. The synthesis of RLPs typically involves the initial ribosomally-derived RiPP production, followed by subsequent post-translational changes and modifications, including processes like macrocyclization, prenylation, and acylation (Wiebach et al., 2018). In the production of different ribosome-originated LPs, distinctive enzyme families have been discovered; one example is ComX, which is known for its distinct isoprenyl-modified tryptophan part. Additionally, new prenyltransferase groups, including aromatic prenyltransferases, have been identified in cyanobactins. Many of these LPs undergo prenylation using dimethylallyl pyrophosphate, which attaches to the oxygen molecules of threonine, serine, and tyrosine (McIntosh et al., 2011). Goadvionins (Kozakai et al., 2020), Albopeptins (Oikawa et al., 2022), Selidamides (Hubrich et al., 2022), Cynobactins (Martins & Vasconcelos, 2015) are the major ribosomally synthesized lipopeptides are found.

4. Ultra-short Synthetic Lipopeptides

Researchers published a series of short, lysine-rich lipopeptides with peptide lengths ranging from a single amino acid to four residues and identified a few promising candidates (Makovitzki et al., 2006). Significantly, any alteration involving just one amino acid responsible for the whole antibacterial effectiveness loss, underscores the importance of the peptide sequence in antibacterial activity (Makovitzki et al., 2006). Numerous lipopeptides enriched with lysine, including the oligo acyl lysyl series and double-headed lipopeptides, are documented in the scientific literature (Meir et al., 2017, Małuch et al., 2020). Most short lipopeptides identified to date are enriched in lysine; however, some groups have recently investigated arginine-rich short lipopeptides targeting Gram-positive bacteria (Sikorska et al., 2018). In recent research work, lipopeptides containing arginine and tryptophan residues were also discovered to be potent adjuvants for enhancing the effectiveness of conventional antibiotics against MDR Gram-positive infections (Zhong et al., 2021, Małuch et al., 2020). The wide array of both naturally occurring and artificially created AMPs has been a valuable model for developing short peptides or their lipid-modified counterparts with antibacterial properties (Fjell et al., 2012). According to scientific literature, most lipopeptides exhibit efficacy towards gram-positive pathogenic microbes, while only a few display effectiveness against gram-negative ones (Sikorska et al., 2018, Zhong et al., 2021). Certain investigations have pointed out that short di-lipid conjugated lipopeptides rich in lysine can act on the membrane and make it permeable, which improves the effectiveness of antibiotics selective for gram-positive bacteria, such as novobiocin for gram-negative (Ramirez et al., 2012). Deciphering the potential of short and ultra-short lipopeptides to act as antibiotics/antibiotic adjuvants is still nascent and holds tremendous prospects for further exploration.

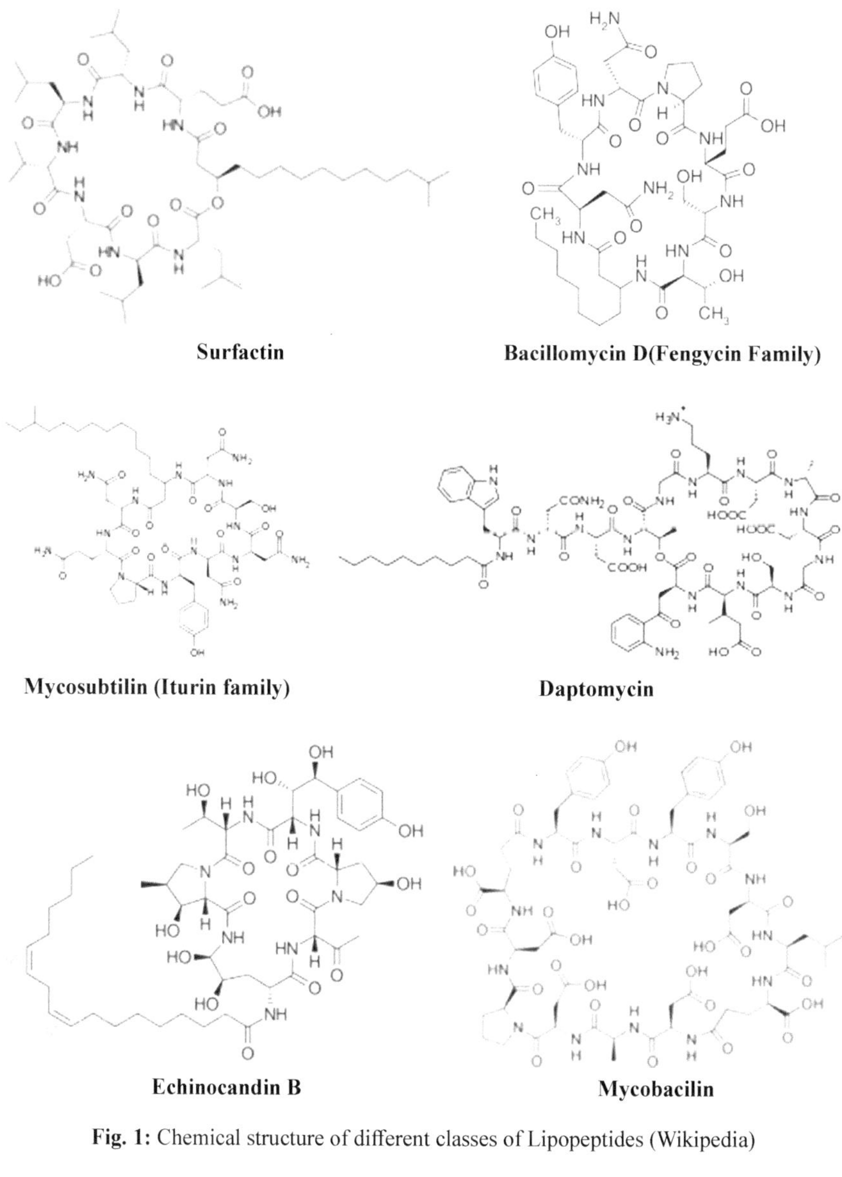

Fig. 1: Chemical structure of different classes of Lipopeptides (Wikipedia)

5. Application of Lipopeptides

Lipopeptides (LPs) possess diverse and significant biological characteristics, making them valuable in numerous sectors, including surfactants, the food and beverage industry for preservation, personal care, cosmetics, and the pharmaceutical industry, with attention on their physicochemical attributes and bioactive capacities (Zaman et. al., 2022). Figure 2 illustrates the wide-ranging applications of LPs in different fields.

Lipopeptides (LPs) are commonly incorporated into cosmetics and personal care items due to their flexible chemical and physical characteristics. The demand for them as cosmetic ingredients stems primarily from their low level of toxicity, biological compatibility, and skin-repairing traits (Adu et al., 2020, Patel et al., 2022). An innovative moisturizing foundation formulation utilizes a gel-like material comprising lipopeptides featuring a repetitive AA structure combined with a lipid component. This formulation serves as a gelator and moisturizer in contact lens applications (Draelos, 2018). The antioxidant properties and the ability of lipopeptides to scavenge reactive oxygen species (ROS) make them suitable for prolonging the shelf life of various products, such as skin care items and food foodstuffs. Researchers have shown that at low LP dosage mixture of different lipopeptides reduces lipid peroxidation and carotene–bleaching, so it can be used as a food additive or preservative, which can protect the nutritional value of the food containing lipids (Jemil et al., 2017). Besides this LPs have numerous therapeutic applications we are focusing on in detail.

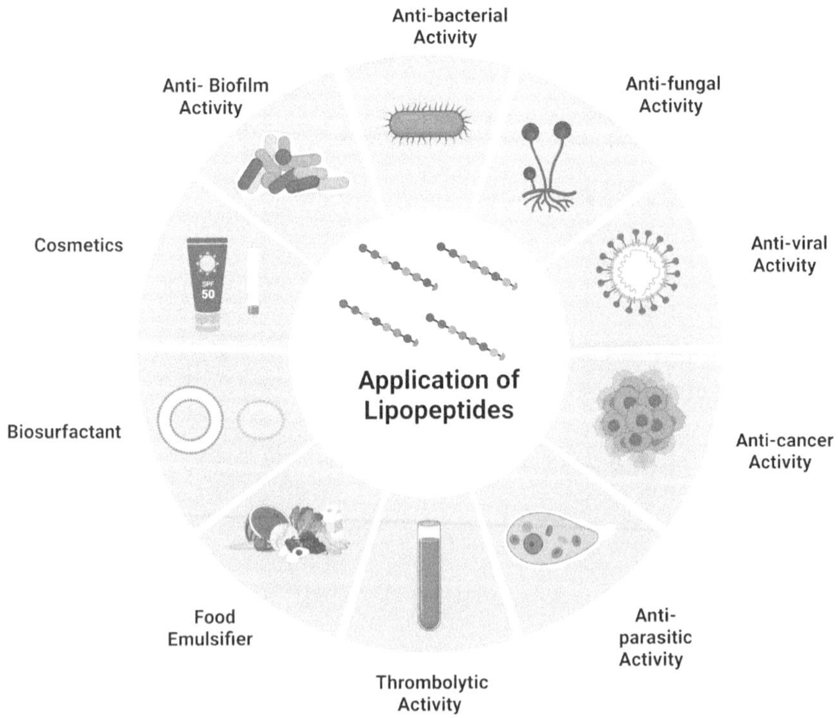

Fig. 2: Applications of the lipopeptides

5.1 Application of Lipopeptides in Therapeutics

The identification of the LPs' anti-mycoplasma and anti-viral characteristics in the late 1990s prompted the proposal to utilize them for safeguarding the integrity of biotechnology and pharmaceutical industry products. Here, we are discussing some major applications of LPs in therapeutics and medicine and recent research in the same field. Table 1 discussed in detail the different sources of different LPs and their applications in various therapeutic areas.

5.1.1 Antibacterial Lipopeptides

The challenge of antibiotic resistance has been present since the inception of antibiotics in the 1940s when penicillin was initially employed as a medicinal treatment (Von Döhren, 2009). Daptomycin and polymyxin B are a pair of lipopeptide antibiotics that have recently received approval from the FDA for addressing multidrug-resistant infectious diseases. Meanwhile, various other lipopeptide antibiotics are currently undergoing distinct phases of clinical and preclinical evaluation (Pirri et al., 2009). Daptomycin was the first cyclic lipopeptide antibiotic used for treating severe blood infections and skin infections (Nakhate et al., 2013). The bacterial LPs are less harmful to humans and animals, have a high rate of biodegradation, cause less irritation, and work well on human skin (Kanwar et al., 2017). Polymyxins are cyclic, positively charged lipopeptides that target the bacterial membranes of both Gram-positive and Gram-negative bacteria. The polymyxin synthetase is encoded by pmxA, B, and E of the five genes in the gene cluster; however, pmxC and D have roles in trafficking (Aleti et al., 2015, Choi et al., 2009). Polymyxin's antibacterial action is achieved by binding to lipopolysaccharides (LPs) through an electrostatic interface with its N-terminal fatty acyl extremity (Deris et al., 2014).

A marine variant *B. laterosporus* generates a linear lipopeptide known as tauramamide, which exhibits potent antibacterial properties against *Enterococcus* specie (Desjardins et al., 2007), whereas cyclic lipopeptides laterocidine and relacidines, show activity against pathogenic strains of *E. coli*, and *P. aeruginosa* (Li et al., 2020, Li et al., 2020). Researchers isolated WAP-8294A lipopeptides in 1997 from *Lysobacter* sp. Since then, the WAP-8294 and several lipopeptide analogues have been identified and subjected to phase I/II clinical trials due to their high antibacterial efficacy against MRSA (Kato et al., 1997, Expósito et al., 2015). Several non-ribosomal lipopeptides produced by *Paenibacillus* spp. and *Bacillus subtilis* exert antibacterial action against phytopathogenic flora, and other foodborne bacteria, including *Campylobacter* spp. (Caulier et al., 2019). Polymyxins B and polymyxins E are found to be more effective against the WHO-declared critically major pathogens, which include *A. baumannii, P. aeruginosa,* and *S. maltophilia* (Velkov et al., 2019)

Cationic peptides form robust interactions with the bacterial membrane, which is negatively charged in nature. Altering the fatty acid components of these lipopeptides results in the creation of an α-helix peptide arrangement. The process of protein lipidation, involving the attachment of a lipid group to a peptide series, escalates the probability of achieving the correct conformation for effective interaction with the membrane bilayer. Ultrashort lipopeptides in this group operate in a manner akin to antimicrobial peptides (AMPs) (Mangoni & Shai, 2011).

Table 1: Sources of lipopeptides and their therapeutic applications

Source of lipopeptides	**Lipopeptides produced**	**Properties**	**Application**	Reference
B. amyloliquefaciens strain fiply 3A	Bacillomycin D	Anti-cancer	Induces concentration-dependent apoptosis in cancer cell lines A498 (renal carcinoma), A549 (alveolar adenocarcinoma), and HCT-15 (colon adenocarcinoma) without any effect on the normal cell line of L-132 (pulmonary epithelial cells)	(Hajare et al., 2013)
B. subtilis AM1	Surfactin	Antibacterial	Prevents biofilm formation of Legionella sp.	(Loiseau et al., 2015)
B. safensis F4	Surfactin	Biosurfactant	Antibacterial, anti-planktonic, anti-adhesive against biofilm-forming Staphylococcus epidermidis and Anti-tumor activity against breast cancer cell lines and mouse melanoma cells	(Abdelli et al., 2019)
B. atrophaeus AKLSR1	Surfactin	Biosurfactant	Anti-proliferative property brings apoptosis in lung cancer cell line A549	(Routhu et al., 2019)
B. subtilis I'1a	Iturin	Biosurfactants	Inhibits the growth of uropathogenic bacteria isolated from the biofilm present on urinary catheters	(Bernat et al., 2016)
B. atrophaeus OSY-7LA	Surfactin Subtilosin Plipastatin	Antimicrobial activity against gram positive bacteria	Inhibit the growth of *Listeria monocytogenes*, methicillin resistant *Staphylococcus aureus, Bacillus cereus* and also helps in food preservation	(Guo et al., 2016)
B. mojavensis A21	Lipopeptide mixture	Biosurfactant	Anticoagulant activity potential for therapeutic and biomedical application	(Ben et al., 2015)
Bacillus sp. DT001	Pumilacidin	Anti-proliferative	Inhibits the growth of parasite *Plasmodium falciparum* by affecting mitochondrial function and reduces calcium level in cytoplasm	(Torres-Mendoza et al., 2018)
B. amyloliquefaciens WH1	Surfactin (WH1 fungin)	Immunomodulator	Oral administration of WH1 fungin for the prevention of Type 1 Diabetes mellitus	(Gao et al., 2014)
Bacillus sp.	Fengycin	Antimicrobial and Antiviral	Inhibits the replication of Equine arteritis virus (EAV) and feline herpes virus type1 (FHV-1) and exhibits virucidal activity against EAV	(Scopel et al., 2014)

Contd...

Contd...

Source of lipopeptides	**Lipopeptides produced**	**Properties**	**Application**	Reference
B. amyloliquefaciens 41B-1	Iturin	Antifungal	Leads to ROS production, Hog1 mitogen-activated protein kinase stimulation, and weaknesses in cell wall integrity and microsclerotial germnation in Verticillium dahliae causing verticillium wilt in cotton	(Han et al., 2015)
B. subtilis strain EA-CB0015	Fengycin	Antifungal	Membrane perturbation	(González-Jaramillo et al., 2017)
Synthetic	Myr-WD	Mycobacterium protein coronin 1	Restrict the entry of H1N1 & murine coronavirus into cell	(Sardar et al., 2021)
Synthetic	Myr-HBVpreS/2–78	HB virus' envelope L-protein sequence	Used in developing HBV vaccine (inhibitor)	(Schieck et al., 2010)
Symploca hydnoides	Trikoramide A	RiLP, unusual C-prenylated cyclotryptophan	Cytotoxic activity	(Phyo et al., 2019)
Synthetic	Fibroblast Stimulating lipopeptide-1 (FSL1)	Diacetylated LP; N-terminus of a 44-kDa lipoprotein of Mycoplasma salivarium	Used as adjuvant for vaccine development of Enterovirus 71 (EV71)	(Shibata et al., 2000)
B. amyloliquefaciens DSM 23117	Iturin A, fengycin	Anti-cancer	Inhibit the human alveolar adenocarcinoma cancer cell line viz. A549.	(Pretorius et al., 2015)
B. subtilis	Iturin A	Anticancer	Preventing chronic myelogenous leukemia in vitro via paraptosis, apoptosis, and by targeting of autophagy pathways	(Zhao et al., 2018)
Bacillus velezensis NST6	Bacillomycin D	Bactericidal	Antibacterial effect against *Staphylococcus epidermidis, Staphylococcus aureus* and methicillin-resistant *S. aureus.*	(Nam et al., 2021)
Bacillus amyloliquefaciens	Bacillomycin D	Antifungal	Interaction with cell membrane & pore formation and ROS accumulation	(Gu et al., 2017)

5.1.2 Antifungal Lipopeptides

With the increasing prevalence of resistance to traditional antifungal therapies, there is a growing demand for alternative antifungal agents. Bacillomycin F efficiently targets *A. niger*, and *Candida* spp., while these peptides have activity against dermatomycoses, it also produces considerable hemolysis (Latoud et al., 1986). Echinocandins are a fascinating example of a lipopeptide with obvious medicinal benefits. Large lipopeptides, which are naturally found and partially synthesized, were initially recognized in 1974 as potent antifungal composites with a pronounced lethality against *Candida* species and, to a somewhat lesser degree, *Aspergillus* spp. (Pirri et al., 2009). Echinocandins work by non-competitively inhibiting 1-3-D-glucan synthase, inhibiting the production of 1-3-D-glucan, which leads to impairment of the cell wall integrity leading to osmotic disruption triggering fungicidal action (Diekema et al., 2005). The study conducted by researchers to understand the effectiveness of pseudofactin and surfactin evaluated the adherence of *C. albicans* strains to polystyrene microplates. When microplates were conditioned before adding cells, the biosurfactants reduced the tested strain's adherence by 35–90% (Biniarz et al., 2015). After seven days of incubation, researchers discovered that increasing the concentration of surfactins inhibits the total mycelia growth of *A. flavus* (Mnif & Ghribi, 2015). Bacillus subtilis produced new lipopeptide homologs with Bacillomycin showing antifungal activity against a variety of fungal pathogens, and it completely inhibited these pathogens, making it one of the most promising antifungal lipopeptides (Ramachandran et al., 2018). WH1fungin, a surfactin family lipopeptide produced by *Bacillus amyloliquefaciens*, exhibits fungicidal properties via a unique mechanism. It binds to the mitochondrial membrane ATPase, thereby blocking its function. Through caspase-related pathways, this inhibition eventually leads to cytochrome C-mediated apoptosis (Qi et al., 2010)

5.1.3 Anti-viral Lipopeptides

In vivo, peptide-based treatments often exhibit a limited duration of effectiveness, but researchers have demonstrated that attaching lipids to these peptides can enhance their anti-viral properties and fine-tune their pharmacokinetic profiles. To delve deeper into this, scientists generated a range of lipopeptides (S. Xia et al., 2020). LP-based vaccines potentially work by using the lipid component to bind to antigen-presenting cell (APC) membranes, enabling entry into the dendritic cell cytoplasm. This allows the presentation of CTL-activating epitopes to CD8+ T cells via dendritic cells for direct priming, bypassing macrophages.

LP-11, a lipopeptide bonded to a fatty acid, demonstrated effective inhibitory action towards a range of primary HIV-1 strains and clinically resistant mutant forms. Particularly noteworthy was its notably heightened ex-vivo antiviral effectiveness and extended half-life. Furthermore, LP-11 displayed increased alpha-helicity and thermal resilience, retaining its structural integrity even when exposed to elevated temperatures and humidity (Chong et al., 2016). Recent discoveries have revealed that lipopeptides based on the fusion inhibitor T-20 (enfuvirtide) exhibit significantly amplified anti-HIV capabilities. In a recent investigation, a series of novel lipopeptides were created, with variations in the length of their fatty acid chains. Within this collection, one lipopeptide named LP-80, modified with stearic acid, stood out as displaying the most remarkable activity against HIV (Chong et al., 2019). Lipopeptides derived from enfuvirtide

(T-20) and conjugated with fatty acids have exhibited substantially enhanced anti-HIV action. A series of fusion inhibitors modified with cholesterol were examined, leading to noteworthy discoveries. Among these, novel cholesterylated inhibitors, particularly LP-83 and LP-86, demonstrated exceptional potency in inhibiting various strains of HIV-1, HIV-2, and SIV (Zhu et al., 2019). Furthermore, a recently developed T20-based lipopeptide, known as LP-40, has exhibited significantly enhanced activity against HIV (Zhu et al., 2018). Cholesterol-conjugated lipopeptide EK1C2A displays remarkably potent inhibitory activity against HIV-1. This effectiveness is attributed to a shared action with pan-coronavirus (CoV) fusion inhibitors and HIV-1 fusion inhibitors (Lan et al., 2021). A study by researchers demonstrated that seven specific lipopeptides bind to nsp12, a key protein in SARS-CoV-2. This binding site is the same as the binding site for remdesivir with stronger affinities. Ferrocin A, an iron-chelating compound, exhibited the highest binding affinity. These outcomes have shown that these lipopeptides can be a promising tool to combat Coronavirus infection (B. Xia et al., 2021). Likewise, scientists have developed IPB02, derived from the HR2 sequence shows promising results in preventing coronavirus infection (Zhu et al., 2020).

5.1.4 Lipopeptides in Cancer Therapy

Maintaining the equilibrium between cellular growth and cell demise is critical for normal cell development. Excessive cell division and proliferation can lead to tumorigenesis and the onset of cancer. Various LPs possess anti-tumor properties that help regulate the cell cycle progression and restore its balance.

Derived from *Bacillus megaterium*, Iturin A has been shown to boost the expression of BAX while concurrently repressing Bcl-2, Mcl-1, and Bcl-xl proteins. Moreover, Iturin A restricts the binding of phosphate to AKT and subsequently downregulates the expression of FoxO3a and GSK3B proteins. This dual mechanism ultimately arrests the cell cycle in the G0/G1 phase and induces apoptosis in breast cancer cells (Dey et al., 2015). Iturin demonstrates significant anti-cancer activity against breast cancer leukemia. The effect of Iturin was evaluated on different types of cancer cells by using various concentrations during 48 hour time durations to assess its anti-cancer effects (Hwang et al., 2009). Surfactin is a potent lipopeptide recognized for its versatility as a bioactive molecule, exhibiting significant anti-tumor properties (Sachdev & Cameotra, 2013). In vitro, studies have shown that Surfactin also has an impact on the growth and progression of human colon cancer (Sivapathasekaran et al., 2010). By exerting influence over signaling pathways that govern cell survival, Surfactin acts as an anti-cancer agent, resulting in the interruption of the cell cycle and its subsequent arrest (Kim et al., 2007). Surfactin effectively hinders cancer cell invasion, metastasis, and colony formation by regulating vital regulatory proteins. They curtail the expression of MMP2 and MMP9 genes, influencing the regulation of AP-1, PI3K/Akt, NF-κB, and ERK1/2. Surfactin triggers apoptosis through two distinct pathways: the intrinsic mitochondrial pathway and the caspase pathway. These routes involve various factors such as ERK1/2, PI3K/Akt, calcium ion production, up-regulation of Bax/Bad, down-regulation of Bcl-2, and the release of Cytochrome-c, ultimately leading to the activation of the caspase cascade and the promotion of apoptosis. This process contributes to maintaining immune homeostasis (Wu et al., 2017). Studies have revealed that the lipopeptide surfactin initiates apoptosis in breast cancer cells. This lipopeptide has been shown to induce apoptosis in MCF7 cells through a process that engages the Ros/JNK (Jun N-terminal kinase) pathway, resulting in the activation of caspase and mitochondria (Cao et al., 2010, Wang et al., 2013). The PFII lipopeptide biosurfactant has emerged as a promising contender for melanoma treatment. The research found that apoptotic death in melanoma cells exposed to PFII revealed that PFII micelles caused plasma membrane permeabilization, distinguishing its pro-apoptotic mechanism from other lipopeptides, these findings strongly suggest its ability as a promising anti-melanoma candidate (Janek et al., 2013). Fellutamides A and B, also recognized as 210–211, which are the pioneering discovery of anti-cancer lipopeptides originating from *Pseudomonas fellutanum* (Liu et al., 2017). Compounds 210 and 211 demonstrated substantial cytotoxic activity against murine leukemia P388 cells. Recently, a marine fungus, *P. purpurogenum* G59, yielded various analouges to this lipopeptide (Zhang et al., 2016). Marine cyanobacterium *Hormothamnion enteromorphoides* produces laxaphycin A along with B4 shows anti-tumor activity against human colon cancer and synergistically inhibits colorectal cancer cells (Cai et al., 2018). Bacillomycin D induces apoptosis and suppresses the growth rate of alveolar cancerous cells which reveals that this lipopeptide exhibits highly potent anti-tumor activity against cancer cells, distinguishing it markedly from healthy cells (Hajare et al., 2013).

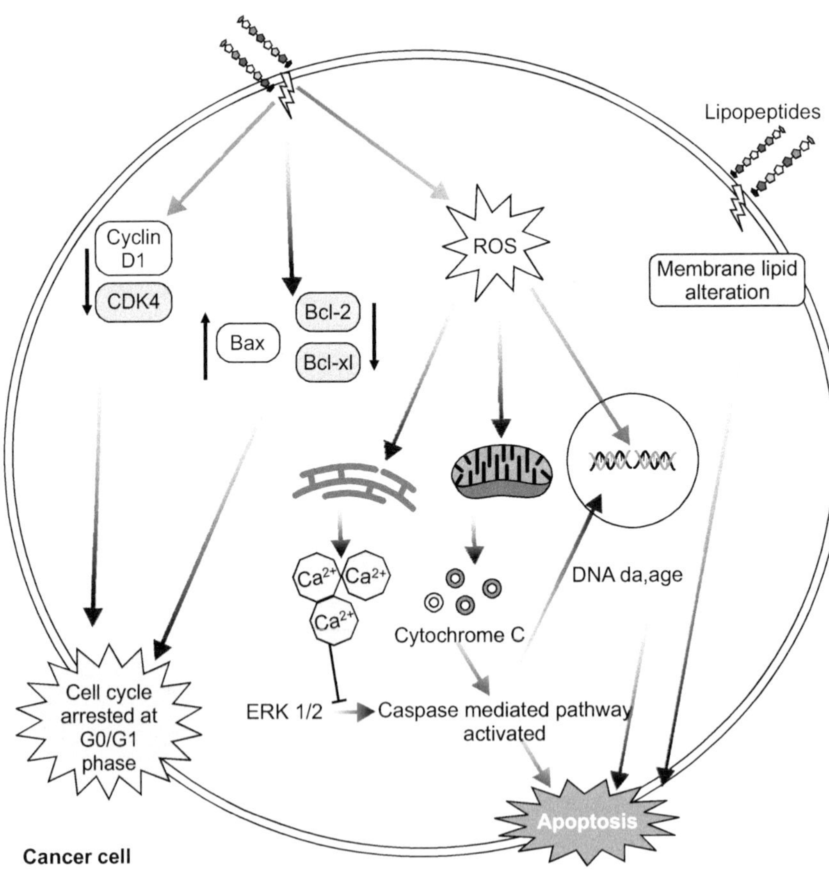

Fig. 3: Effect of Lipopeptides on cancer cell

Iturin and Fengycin, while effective against cancer cells, can be toxic to normal cells at higher doses, with surfactin showing greater anti-cancer potential. However, these lipopeptides, including surfactin, have limitations preventing direct use in cancer patients. The inhibition of cancer cell growth depends on dosage and duration of incubation. Further research aims to enhance their cytotoxic effects on cancer cells, minimize harm to normal cells, and pave the way for future anti-cancer drug development.

5.1.5 Toll-Like Receptor (TLR) Activator LPs

In the immune system, specifically, macrophages and dendritic cells are responsible for the detection of self and foreign pathogens causing infections and tissue damage. These immune cells detect foreign invaders through pattern recognition receptors (PRRs). These PRRs encompass various families, including Toll-like receptors (TLRs), C-type lectin receptors (CLRs), retinoic acid-inducible gene (RIG)-I-like receptors (RLRs), and NOD-like receptors (NLRs). From all these the Toll-like receptor family is the most extensively studied and takes part in innate immunity. The TLR family comprises 13 members, with 11 present in mammals (TLR1-11), while mice possess all, except TLR12 (TLR1-9, TLR11-13) (Takeuchi & Akira, 2010.)

Human Toll-like receptor family member TLR2 is essential and responsible for detecting various pathogen-associated molecular patterns (PAMPs). It achieves this by forming heterodimers with TLR1/TLR6, permitting TLR2 to identify a broad spectrum of ligands. For instance, in partnership with TLR2/1, it recognizes substances like lipoarabinomannan and triacyl lipopeptides (Pam3CSK4), and in conjunction with TLR2/6, it detects lipoteichoic acid and diacyl lipopeptides (macrophage activating lipopeptides) (Oliveira-Nascimento et al., 2012). Among natural ligands, bacterial lipoproteins are recognized as the most potent activators of TLR2. Pam3Cys, a lipopeptide originating in the membranes of gram-negative bacteria, triggers a Th2 immune response and leads to the activation of cytokines like TNF-α, IL-1, and IL-6 by stimulating TLR2 (Wenink et al., 2009). Pam2Cys analogues have been produced as possible adjuvants for cancer vaccines, and the effect of their structural makeup on TLR2 agonist activity has been studied (Lu et al., 2020). The potential of TLR2/6 agonists to lower viral loads has led to recent research on employing Pam2Cys analogs for the SARS-CoV-2 virus (COVID-19) therapies. The studies have demonstrated that this analogue effectively reduces viral transmission (Proud et al., 2021). In Combating HSV infection, lipopeptides have been engineered into nanoparticles, serving as potent self-adjuvant vaccines. These micelles have demonstrated a significant upsurge in various cytokines and TNF, both in vitro, not only due to the micelles but also the peptides used in the study (Accardo et al., 2014).

5.1.6 Anti-biofilm Lipopeptides

Microorganisms rely on surface attachment and the development of biofilms as critical survival strategies in a variety of terrestrial environments. Surfactins, in addition to their established antibacterial and anti-viral characteristics, have proven effective in hindering microbial adhesion and biofilm development. An illustrative example is the use of Surfactin derived from *Bacillus subtilis* as a pre-coating agent on vinyl urethral catheters, which led to a major decrease in biofilm development (Mireles et al., 2001). Likewise, a lipopeptide that has biosurfactant activity acquired from a *Bacillus circulans* strain demonstrated an antiadhesive effect against various bacterial species (Das et al., 2009).

Surfactin displays a notably specific antiadhesive influence, resulting in a significant hindrance of biofilm development by the two most pathogenic strains, *S. aureus,* and *E. coli*, on polystyrene planes (Rivardo et al., 2009). The antiadhesive properties of surfactin are attributed to its anionic nature, which creates electrostatic repulsion between bacteria and the surfactin molecules attached to the polystyrene surface. Consequently, Surfactin appears to hold significant promise as an antiadhesive compound with the potential to safeguard surfaces against microbial contamination (Zeraik et al., 2010). Newly discovered lipopeptides (LPs) possessing antifungal properties encompass pelgipeptins produced by *Paenibacillus elgii*. These LPs have been examined for their effectiveness in treating *Candida albicans* biofilms (Fulgêncio et al., 2021).

5.1.7 Anti-thrombolytic Lipopeptides

The plasminogen-plasmin system plays a crucial role in dissolving blood clots in various physiological and pathological processes that involve proteolysis. Plasminogen, which is the inactive form of plasminogen known as a zymogen, gets activated through proteolytic processes by enzymes like urokinase-type and tissue-type plasminogen activators. This stimulation of plasminogen and prourokinase is a pivotal mechanism for initiating and sustaining fibrinolytic properties. Surfactin, on the other hand, triggers the activation of prourokinase and induces a change in the conformation of plasminogen, which enhances the fibrinolysis in laboratory settings as well as within living organisms (Al-Ajlani et al., 2007). Surfactin C, such as in a rat model of pulmonary embolism, has been observed to enhance the dissolution of blood clots in combination with prourokinase. Furthermore, surfactin demonstrates the ability to inhibit platelet aggregation, prevent the formation of new fibrin clots, and enhances fibrinolysis, facilitating the dispersion of fibrinolytic agents. It's important to note that surfactin's antiplatelet effect is recognized for its influence on downstream signaling pathways rather than its detergent properties. Notably, surfactin presents certain advantages over alternative thrombolytic agents, with a notably lower occurrence of side effects. Consequently, surfactin holds promise as a long-term clot-dissolving agent. Its impact on the plasminogen-plasmin system and its potential as a therapeutic agent for clot dissolution render it a captivating subject of research in the field of thrombolytic treatments (Meena & Kanwar, 2015).

5.1.8 Anti-parasitic Lipopeptides

Serratamolide was initially identified in a *Serratia* species, and subsequently, a strain of *S. marcescens* known as serrawettin W1 displayed notable effectiveness (Clements-Decker et al., 2022). Several Serratia species produce cyclic lipopeptides known as Stephensiolides (Ganley et al., 2018). Comprising five amino acids connected to a C8 fatty acid sequence, Stephensiolide A exhibits antiparasitic effects on *Plasmodium falciparum* (Ganley et al., 2018). In an additional study, a sequence of peptides made from four antigens of the malaria parasite *P. falciparum* was palmitoylated at the lysine residue to enhance their immunogenicity (BenMohamed et al., 2004). Recently Iwasaki and co-researchers isolated anti-parasitic marine lipopeptide hoshinoamide C from Marine cyanobacterium *Caldora penicillata* which exhibited anti-malaria activity (Iwasaki et al., 2020). A novel cyclic lipopeptide with surfactin-like properties was extracted from the recently identified *Bacillus* sp. DE2B strain in Ghana. It demonstrates notable efficacy against *Trypanosoma brucei* and *Leishmania Donovan*, showcasing potent antiparasitic activity (Nartey et al., 2022).

6. Future Challenges

With the rise of antibiotic-resistant bacteria, lipopeptides are being explored as potent antimicrobial agents. LPs possess powerful moieties to combat widespread bacterial and fungal infections, offering new avenues for treating diseases that are otherwise difficult to manage. Recent progress in high-throughput isolation and screening methods has significantly enhanced the exploration for novel families of lipopeptides (LP) that possess potent bioactivity while minimizing systemic toxicity and side effects (Schlusselhuber et al., 2018). Between 2010 and 2022, researchers identified two distinct lipopeptide families: pelgipeptins and tridecaptins. These compounds are isolated from *Paenibacillus* spp. which show potent antifungal activity. Pelgipeptins have shown promise as potential biocontrol agents in agriculture (Kim et al., 2020, Bee et al., 2019). Genetic engineering represents a viable means to enhance both the quantity and variety of lipopeptides (LPs) generated during fermentation procedures. The primary strategies employed to boost LP production involve the management of transcriptional processes, ensuring an adequate supply of precursor molecules, suppressing pathways responsible for LP degradation, and engineering non-ribosomal peptide synthetases (NRPS). While the genetic modification of NRPS biosynthesis pathways has been explored in various organisms, *B. subtilis* remains the most prevalent in this regard. However, the realm of ribosomally synthesized lipopeptides (RiLP) engineering remains largely uncharted (Théatre et al., 2022). Furthermore, more research into the properties of specific lipopeptides congeners is required, as well as the development of appropriate purification process techniques. Chronic infections often involve biofilms, which are resistant to conventional treatments. Lipopeptides can modulate the immune system, making them candidates for immunotherapy. They may be used to boost the body's natural defence, making them applicable for treating autoimmune diseases, allergies, and certain cancers. Scientists are currently exploring the potential of lipopeptides in preventing viruses from entering host cells. This research may pave the way for the creation of innovative anti-viral treatments that can target a broad spectrum of viruses, including HIV, influenza, and newly emerging viral challenges. Lipopeptides are utilized in designing peptide-based vaccines, stimulating targeted immune responses against specific pathogens. Customized vaccines hold promise in preventing infectious diseases and certain types of cancer. Lipopeptides are being explored for their potential to induce apoptosis in cancer cells. By triggering this natural process, lipopeptides may become part of innovative cancer therapies, offering alternatives to traditional chemotherapy. The future of lipopeptides in medicine is bright, with ongoing research focusing on harnessing their unique properties for a wide array of therapeutic applications. Continued innovation and understanding of these molecules will undoubtedly lead to groundbreaking treatments, improving healthcare quality globally.

References

Abdelli, F., Jardak, M., Elloumi, J., Stien, D., Cherif, S., Mnif, S. et al. (2019). Antibacterial, anti-adherent and cytotoxic activities of surfactin(s) from a lipolytic strain Bacillus safensis F4. *Biodegradation*, 30(4): 287–300. https://doi.org/10.1007/s10532-018-09865-4

Abderrahmani, A., Tapi, A., Nateche, F., Chollet, M., Leclère, V., Wathelet, B. et al. (2011). Bioinformatics and molecular approaches to detect NRPS genes involved in the biosynthesis of kurstakin from Bacillus thuringiensis. *Applied Microbiology and Biotechnology*, 92(3): 571–581. https://doi.org/10.1007/s00253-011-3453-6

Accardo, A., Vitiello, M., Tesauro, D., Galdiero, M., Finamore, E., Martora, F. et al. (2014). Self-assembled or mixed peptide amphiphile micelles from herpes simplex virus glycoproteins as potential immunomodulatory treatment. *International Journal of Nanomedicine*, 9(1): 2137–2148. https://doi.org/10.2147/IJN.S57656

Adu, S.A., Naughton, P.J., Marchant, R. & Banat, I.M. (2020). Microbial biosurfactants in cosmetic and personal skincare pharmaceutical formulations. *Pharmaceutics*, 12(11): 1099. https://doi.org/10.3390/pharmaceutics12111099

Al-Ajlani, M.M., Sheikh, M.A., Ahmad, Z. & Hasnain, S. (2007). Production of surfactin from Bacillus subtilis MZ-7 grown on pharmamedia commercial medium. *Microbial Cell Factories*, 6: 17. https://doi.org/10.1186/1475-2859-6-17

Aleti, G., Sessitsch, A. & Brader, G. (2015). Genome mining: Prediction of lipopeptides and polyketides from Bacillus and related Firmicutes. *Computational and Structural Biotechnology Journal*, 13: 192–203. Elsevier B.V. https://doi.org/10.1016/j.csbj.2015.03.003

Ali, S.A.M., Sayyed, R.Z., Mir, M.I., Hameeda, B., Khan, Y., Alkhanani, M.F. et al. (2022). Induction of systemic resistance in maize and antibiofilm activity of surfactin from *Bacillus velezensis* MS20. *Front. Microbiol.*, 13: 879739. https://doi.org/10.3389/fmicb.2022.879739

Batool, M., Khalid, M.H., Hassan, M.N. & Fauzia Yusuf, H. (2011). Homology modeling of an antifungal metabolite plipastatin synthase from the Bacillus subtilis 168. *Bioinformation*, 7(8): 384–387. https://doi.org/10.6026/97320630007384

Bee, H., Khan, M.Y. & Sayyed, R.Z. (2019). Microbial surfactants and their significance in agriculture. *In:* Plant Growth Promoting Rhizobacteria (PGPR): Prospects for Sustainable Agriculture, 205-215. https://doi.org/10.1007/978-981-13-6790-8_18

Ben Ayed, H., Nasri, R., Jemil, N., Ben Amor, I., Gargouri, J., Hmidet, N. et al. (2015). Acute and sub-chronic oral toxicity profiles of lipopeptides from Bacillus mojavensis A21 and evaluation of their in vitro anticoagulant activity. *Chemico-biological Interactions*, 236: 1–6. https://doi.org/10.1016/j.cbi.2015.04.018

Bender, C.L., Alarcón-Chaidez, F. & Gross, D.C. (1999). Pseudomonas syringae phytotoxins: Mode of action, regulation, and biosynthesis by peptide and polyketide synthetases. *Microbiology and Molecular Biology Reviews: MMBR*, 63(2): 266–292. https://doi.org/10.1128/MMBR.63.2.266-292.1999

BenMohamed, L., Thomas, A. & Druilhe, P. (2004). Long-term multiepitopic cytotoxic-T-lymphocyte responses induced in chimpanzees by combinations of Plasmodium falciparum liver-stage peptides and lipopeptides. *Infection and Immunity*, 72(8): 4376–4384. https://doi.org/10.1128/IAI.72.8.4376-4384.2004

Bernat, P., Paraszkiewicz, K., Siewiera, P., Moryl, M., Płaza, G. & Chojniak, J. (2016). Lipid composition in a strain of Bacillus subtilis, a producer of iturin A lipopeptides that are active against uropathogenic bacteria. *World Journal of Microbiology & Biotechnology*, 32(10): 157. https://doi.org/10.1007/s11274-016-2126-0

Biniarz, P., Baranowska, G., Feder-Kubis, J. & Krasowska, A. (2015). The lipopeptides pseudofactin II and surfactin effectively decrease Candida albicans adhesion and hydrophobicity. *Antonie van Leeuwenhoek, International Journal of General and Molecular Microbiology*, 108(2): 343–353. https://doi.org/10.1007/s10482-015-0486-3

Cai, W., Matthew, S., Chen, Q.Y., Paul, V.J. & Luesch, H. (2018). Discovery of new A- and B-type laxaphycins with synergistic anti-cancer activity. *Bioorganic and Medicinal Chemistry*, 26(9): 2310–2319. https://doi.org/10.1016/j.bmc.2018.03.022

Cao, X. Hong, Wang, A. Hua, Wang, C. Ling, Mao, D. Zhi, Lu, M. Fang, Cui, Y. Qian et al. (2010). Surfactin induces apoptosis in human breast cancer MCF-7 cells through a ROS/

JNK-mediated mitochondrial/caspase pathway. *Chemico-Biological Interactions*, 183(3): 357–362. https://doi.org/10.1016/j.cbi.2009.11.027

Caulier, S., Nannan, C., Gillis, A., Licciardi, F., Bragard, C. & Mahillon, J. (2019). Overview of the anti-microbial compounds produced by members of the Bacillus subtilis group. *Frontiers in Microbiology*, 10(Issue FEB). Frontiers Media S.A. https://doi.org/10.3389/fmicb.2019.00302

Chen W.-C., Juang R.-S. & Wei Y.-H. (2015). Applications of a lipopeptide biosurfactant, surfactin, produced by microorganisms. *Biochem. Eng. J.*, 103: 158–169. https://doi.org/10.1016/j.bej.2015.07.009

Chen, C.H. & Lu, T.K. (2020). Development and challenges of anti-microbial peptides for therapeutic applications. *Antibiotics* (Basel, Switzerland), 9(1): 24. https://doi.org/10.3390/antibiotics9010024

Choi, S.K., Park, S.Y., Kim, R., Kim, S., Lee, C.H., Kim, J.F. et al. (2009). Identification of a polymyxin synthetase gene cluster of Paenibacillus polymyxa and heterologous expression of the gene in Bacillus subtilis. *Journal of Bacteriology*, 191(10): 3350–3358. https://doi.org/10.1128/JB.01728-08

Chong, H., Wu, X., Su, Y. & He, Y. (2016). Development of potent and long-acting HIV-1 fusion inhibitors. *AIDS*, 30(8): 1187–1196. https://doi.org/10.1097/QAD.0000000000001073

Chong, H., Xue, J., Zhu, Y., Cong, Z., Chen, T., Wei, Q. et al. (2019). Monotherapy with a low-dose lipopeptide HIV fusion inhibitor maintains long-term viral suppression in rhesus macaques. *PLoS Pathogens*, 15(2). https://doi.org/10.1371/journal.ppat.1007552

Clements-Decker, T., Kode, M., Khan, S. & Khan, W. (2022). Underexplored bacteria as reservoirs of novel antimicrobial lipopeptides. *Frontiers in Chemistry*, 10. Frontiers Media S.A. https://doi.org/10.3389/fchem.2022.1025979

Cosby, W.M. & Zuber, P. (1997). Regulation of Bacillus subtilis sigmaH (spo0H) and AbrB in response to changes in external pH. *Journal of Bacteriology*, 179(21): 6778–6787. https://doi.org/10.1128/jb.179.21.6778-6787.1997

Cosby, W.M., Vollenbroich, D., Lee, O.H. & Zuber, P. (1998). Altered Srf expression in Bacillus subtilis resulting from changes in culture pH is dependent on the Spo0K oligopeptide permease and the ComQX system of extracellular control. *Journal of Bacteriology*, 180(6): 1438–1445. https://doi.org/10.1128/JB.180.6.1438-1445.1998

Cragg, G.M. & Newman, D.J. (2013). Natural products: A continuing source of novel drug leads. *Biochimica et Biophysica Acta – General Subjects*, 1830(Issue 6): 3670–3695. https://doi.org/10.1016/j.bbagen.2013.02.008

da Costa, R.A., Andrade, I.E.P.C., Pinto, O.H.B., de Souza, B.D.P., Fulgêncio, D.L.A., Mendonça, M.L. et al. (2022). A novel family of non-secreted tridecaptin lipopeptide produced by Paenibacillus elgii. *Amino Acids*, 54(11): 1477–1489. https://doi.org/10.1007/s00726-022-03187-9

Das, P., Mukherjee, S. & Sen, R. (2009). Antiadhesive action of a marine microbial surfactant. *Colloids and Surfaces, B. Biointerfaces*, 71(2): 183–186. https://doi.org/10.1016/j.colsurfb.2009.02.004

Deris, Z.Z., Swarbrick, J.D., Roberts, K.D., Azad, M.A.K., Akter, J., Horne, A.S. et al. (2014). Probing the penetration of antimicrobial polymyxin lipopeptides into gram-negative bacteria. *Bioconjugate Chemistry*, 25(4): 750–760. https://doi.org/10.1021/bc500094d

Desjardine, K., Pereira, A., Wright, H., Matainaho, T., Kelly, M. & Andersen, R.J. (2007). Tauramamide, a lipopeptide antibiotic produced in culture by Brevibacillus laterosporus isolated from a marine habitat: Structure elucidation and synthesis. *Journal of Natural Products*, 70(12): 1850–1853. https://doi.org/10.1021/np070209r

Dey, G., Bharti, R., Dhanarajan, G., Das, S., Dey, K.K., Kumar, B.N.P. et al. (2015). Marine lipopeptide Iturin A inhibits Akt mediated GSK3β and FoxO3a signaling and triggers apoptosis in breast cancer. *Scientific Reports*, 5. https://doi.org/10.1038/srep10316

Diekema, D.J., Petroelje, B., Messer, S.A., Hollis, R.J. & Pfaller, M.A. (2005). Activities of available and investigational antifungal agents against Rhodotorula species. *Journal of Clinical Microbiology*, 43(1): 476–478. https://doi.org/10.1128/JCM.43.1.476-478.2005

Draelos, Z.D. (2018). The science behind skin care: Moisturizers. *Journal of Cosmetic Dermatology*, 17(2): 138–144. https://doi.org/10.1111/jocd.12490

Expósito, R.G., Postma, J., Raaijmakers, J.M. & De Bruijn, I. (2015). Diversity and activity of Lysobacter species from disease suppressive soils. *Frontiers in Microbiology*, 6(Nov). https://doi.org/10.3389/fmicb.2015.01243

Fjell, C.D., Hiss, J.A., Hancock, R.E. & Schneider, G. (2011). Designing anti-microbial peptides: Form follows function. *Nature Reviews. Drug Discovery*, 11(1): 37–51. https://doi.org/10.1038/nrd3591

Fulgêncio, D.L.A., da Costa, R.A., Guilhelmelli, F., Silva, C.M.S., Ortega, D.B., de Araujo, T.F. et al. (2021). *In vitro* antifungal activity of pelgipeptins against human pathogenic fungi and *Candida albicans* biofilms. *AIMS Microbiology*, 7(1): 28–39. https://doi.org/10.3934/microbiol.2021003

Ganley, J.G., Carr, G., Ioerger, T.R., Sacchettini, J.C., Clardy, J. & Derbyshire, E.R. (2018). Discovery of antimicrobial lipo depsipeptides produced by a Serratia sp. within mosquito microbiomes. *ChemBioChem*, 19(15): 1590–1594. https://doi.org/10.1002/cbic.201800124

Gao, Z., Zhao, X., Yang, T., Shang, J., Shang, L., Mai, H. et al. (2014). Immunomodulation therapy of diabetes by oral administration of a surfactin lipopeptide in NOD mice. *Vaccine*, 32(50): 6812–6819. https://doi.org/10.1016/j.vaccine.2014.08.082

González-Jaramillo, L.M., Aranda, F.J., Teruel, J.A., Villegas-Escobar, V. & Ortiz, A. (2017). Antimycotic activity of fengycin C biosurfactant and its interaction with phosphatidylcholine model membranes. *Colloids and Surfaces. B, Biointerfaces*, 156, 114–122. https://doi.org/10.1016/j.colsurfb.2017.05.021

Gu, Q., Yang, Y., Yuan, Q., Shi, G., Wu, L., Lou, Z. et al. (2017). Bacillomycin D produced by Bacillus amyloliquefaciens is involved in the antagonistic interaction with the plant pathogenic fungus Fusarium graminearum. *Applied and Environmental Microbiology*, 83(19): 1075–1092. https://doi.org/10.1128/AEM.01075-17

Guo, Y., Huang, E., Yang, X., Zhang, L., Yousef, A.E. & Zhong, J. (2016). Isolation and characterization of a Bacillus atrophaeus strain and its potential use in food preservation. *Food Control*, 60: 511–518. http://dx.doi.org/10.1016/j.foodcont.2015.08.029

Hajare, S.N., Subramanian, M., Gautam, S. & Sharma, A. (2013). Induction of apoptosis in human cancer cells by a Bacillus lipopeptide bacillomycin D. *Biochimie*, 95(9): 1722–1731. https://doi.org/10.1016/j.biochi.2013.05.015

Han, Q., Wu, F., Wang, X., Qi, H., Shi, L., Ren, A. et al. (2015). The bacterial lipopeptide iturins induce Verticillium dahliae cell death by affecting fungal signalling pathways and mediate plant defence responses involved in pathogen-associated molecular pattern-triggered immunity. *Environmental Microbiology*, 17(4): 1166–1188. https://doi.org/10.1111/1462-2920.12538

Huang, E., Yang, X., Zhang, L., Moon, S.H. & Yousef, A.E. (2017). New Paenibacillus strain produces a family of linear and cyclic antimicrobial lipopeptides: Cyclization is not essential for their antimicrobial activity. *FEMS Microbiology Letters*, 364(8): 10.1093/femsle/fnx049. https://doi.org/10.1093/femsle/fnx049

Hubrich, F., Bösch, N.M., Chepkirui, C., Morinaka, B.I., Rust, M., Gugger, M. et al. (2022). Ribosomally derived lipopeptides containing distinct fatty acyl moieties. *Proceedings of the National Academy of Sciences of the United States of America*, 119(3): e2113120119. https://doi.org/10.1073/pnas.2113120119

Hwang, Y.-H., Kim, M.-S., Song, I.-B., Park, B.-K., Lim, J.-H., Park, S.-C. et al. (2009). Subacute (28 day) toxicity of Surfactin C, a lipopeptide produced by Bacillus subtilis, in rats. *Journal of Health Science*, 55(3).

Iwasaki, A., Ohtomo, K., Kurisawa, N., Shiota, I., Rahmawati, Y., Jeelani, G. et al. (2020). Isolation, structure determination, and total synthesis of Hoshinoamide C, an antiparasitic lipopeptide from the marine cyanobacterium Caldora penicillata. *Journal of Natural Products*, 84(1): 126–135. https://doi.org/10.1021/acs.jnatprod.0c01209

Janek, T., Krasowska, A., Radwańska, A. & Łukaszewicz, M. (2013). Lipopeptide biosurfactant pseudofactin II induced apoptosis of Melanoma A 375 cells by specific interaction with the plasma membrane. *PLoS One*, 8(3). https://doi.org/10.1371/journal.pone.0057991

Jemil, N., Manresa, A., Rabanal, F., Ben Ayed, H., Hmidet, N. & Nasri, M. (2017). Structural characterization and identification of cyclic lipopeptides produced by Bacillus methylotrophicus DCS1 strain. *Journal of Chromatography. B, Analytical Technologies in the Biomedical and Life Sciences*, 1060: 374–386. https://doi.org/10.1016/j.jchromb.2017.06.013

Jerala, R. (2007). Synthetic lipopeptides: A novel class of anti-infectives. *Expert Opinion on Investigational Drugs*, 16(8): 1159–1169. https://doi.org/10.1517/13543784.16.8.1159

Kanwar, S.S., Raj Meena, K. & Sharma, A. (2017). Issue 11 | Article 1111 Citation: Meena K.R., Sharma, A., Kanwar, S.S. Microbial Lipopeptides and their Medical Applications. *Annals of Pharmacology and Pharmaceutics*, 2(11).

Kato, A., Nakaya, S., Ohashi, Y., Hirata, H., Fujii, K. & Harada, K.-I. (1997). WAP-8294A2, A novel anti-MRSA antibiotic produced by Lysobacter sp. *Journal of the American Chemical Society*, 119(28): 6680–6681. https://doi.org/10.1021/ja970895o

Kim, J., Le, K.D., Yu, N.H., Kim, J.I., Kim, J.C. & Lee, C.W. (2020). Structure and antifungal activity of pelgipeptins from Paenibacillus elgii against phytopathogenic fungi. *Pesticide Biochemistry and Physiology*, 163: 154–163. https://doi.org/10.1016/j.pestbp.2019.11.009

Kim, S. Young, Kim, J.Y., Kim, S.H., Bae, H.J., Yi, H., Yoon, S.H. et al. (2007). Surfactin from Bacillus subtilis displays anti-proliferative effect via apoptosis induction, cell cycle arrest and survival signaling suppression. *FEBS Letters*, 581(5): 865–871. https://doi.org/10.1016/j.febslet.2007.01.059

Koumoutsi, A., Chen, X.H., Vater, J. & Borriss, R. (2007). DegU and YczE positively regulate the synthesis of bacillomycin D by Bacillus amyloliquefaciens strain FZB42. *Applied and Environmental Microbiology*, 73(21): 6953–6964. https://doi.org/10.1128/AEM.00565-07

Kozakai, R., Ono, T., Hoshino, S., Takahashi, H., Katsuyama, Y., Sugai, Y. et al. (2020). Acyltransferase that catalyzes the condensation of polyketide and peptide moieties of goadvionin hybrid lipopeptides. *Nature Chemistry*, 12(9): 869–877. https://doi.org/10.1038/s41557-020-0508-2

Kwon, J.H. & Powderly, W.G. (2021). The post-antibiotic era is here. *Science* (New York, N.Y.), 373(6554): 471. https://doi.org/10.1126/science.abl5997

Lan, Q., Pu, J., Cai, Y., Zhou, J., Wang, L., Jiao, F. et al. (2021). Lipopeptide-based pan-CoV fusion inhibitors potently inhibit HIV-1 infection. *Microbes and Infection*, 23(8). Elsevier Masson s.r.l. https://doi.org/10.1016/j.micinf.2021.104840

Latoud, C., Peypoux, F., Michel, G., Genet, R. & Morgat, J.L. (1986). Interactions of antibiotics of the iturin group with human erythrocytes. *Biochimica et Biophysica Acta*, 856.

Lei, J., Sun, L., Huang, S., Zhu, C., Li, P., He, J. et al. (2019). The anti-microbial peptides and their potential clinical applications. *American Journal of Translational Research*, 11(7): 3919–3931.

Li, Z., Chakraborty, P., de Vries, R.H., Song, C., Zhao, X., Roelfes, G. et al. (2020). Characterization of two relacidines belonging to a novel class of circular lipopeptides that act against Gram-negative bacterial pathogens. *Environmental Microbiology*, 22(12): 5125–5136. https://doi.org/10.1111/1462-2920.15145

Li, Z., de Vries, R.H., Chakraborty, P., Song, C., Zhao, X., Scheffers, D.-J. et al. (2020). Novel modifications of non-ribosomal peptides from Brevibacillus laterosporus MG64 and investigation of their mode of action. *Applied and Environmental Microbiology*, 86(24). https://doi.org/10.1128/aem.01981-20

Liu, J.F., Mbadinga, S.M., Yang, S.Z., Gu, J.D. & Mu, B.Z. (2015). Chemical structure, property and potential applications of biosurfactants produced by Bacillus subtilis in petroleum recovery and spill mitigation. *International Journal of Molecular Sciences*, 16(3): 4814–4837. https://doi.org/10.3390/ijms16034814

Liu, S., Su, M., Song, S.J. & Jung, J.H. (2017). Marine-derived penicillium species as producers of cytotoxic metabolites. *Marine Drugs*, 15(10). MDPI AG. https://doi.org/10.3390/md15100329

Loiseau, C., Schlusselhuber, M., Bigot, R., Bertaux, J., Berjeaud, J.M. & Verdon, J. (2015). Surfactin from Bacillus subtilis displays an unexpected anti-Legionella activity. *Applied*

Microbiology and Biotechnology, 99(12): 5083–5093. https://doi.org/10.1007/s00253-014-6317-z

Lu, B.L., Williams, G.M., Verdon, D.J., Dunbar, P.R. & Brimble, M.A. (2020). Synthesis and evaluation of novel TLR2 agonists as potential adjuvants for cancer vaccines. *Journal of Medicinal Chemistry*, 63(5): 2282–2291. https://doi.org/10.1021/acs.jmedchem.9b01044

Mahlapuu, M., Håkansson, J., Ringstad, L. & Björn, C. (2016). Anti-microbial peptides: An emerging category of therapeutic agents. *Front. Cell. Infect. Microbiol.*, 6: 194. doi:10.3389/fcimb.2016.00194

Mahlstedt, S.A. & Walsh, C.T. (2010). Investigation of anticapsin biosynthesis reveals a four-enzyme pathway to tetrahydrotyrosine in Bacillus subtilis. *Biochemistry*, 49(5): 912–923. https://doi.org/10.1021/bi9021186

Makovitzki, A., Avrahami, D. & Shai, Y. (2006). Ultrashort antibacterial and antifungal lipopeptides. *Proceedings of the National Academy of Sciences of the United States of America*, 103(43): 15997–16002. https://doi.org/10.1073/pnas.0606129103

Małuch, I., Stachurski, O., Kosikowska-Adamus, P., Makowska, M., Bauer, M. et al. (2020). Double-headed cationic lipopeptides: An emerging class of antimicrobials. *International Journal of Molecular Sciences*, 21(23): 8944. https://doi.org/10.3390/ijms21238944

Mangoni, M.L. & Shai, Y. (2011). Short native anti-microbial peptides and engineered ultrashort lipopeptides: Similarities and differences in cell specificities and modes of action. *Cellular and Molecular Life Sciences: CMLS*, 68(13): 2267–2280. https://doi.org/10.1007/s00018-011-0718-2

Martins, J. & Vasconcelos, V. (2015). Cyanobactins from cyanobacteria: Current genetic and chemical state of knowledge. *Marine Drugs*, 13(11): 6910–6946. https://doi.org/10.3390/md13116910

McIntosh, J.A., Donia, M.S., Nair, S.K. & Schmidt, E.W. (2011). Enzymatic basis of ribosomal peptide prenylation in cyanobacteria. *Journal of the American Chemical Society*, 133(34): 13698–13705. https://doi.org/10.1021/ja205458h

Meena, K.R. & Kanwar, S.S. (2015). Lipopeptides as the antifungal and antibacterial agents: Applications in food safety and therapeutics. *BioMed Research International*, 2015, 473050. https://doi.org/10.1155/2015/473050

Meir, O., Zaknoon, F., Cogan, U. & Mor, A. (2017). A broad-spectrum bactericidal lipopeptide with anti-biofilm properties. *Scientific Reports*, 7(1): 2198. https://doi.org/10.1038/s41598-017-02373-0

Mireles, J.R. 2nd, Toguchi, A. & Harshey, R.M. (2001). Salmonella enterica serovar typhimurium swarming mutants with altered biofilm-forming abilities: Surfactin inhibits biofilm formation. *Journal of Bacteriology*, 183(20): 5848–5854. https://doi.org/10.1128/JB.183.20.5848-5854.2001

Mnif, I. & Ghribi, D. (2015). Lipopeptides biosurfactants: Mean classes and new insights for industrial, biomedical, and environmental applications. *Biopolymers*, 104(3): 129–147. John Wiley and Sons Inc. https://doi.org/10.1002/bip.22630

Nakhate, P.H., Yadav, V.K. & Pathak, A.N. (2013). Correspondence: A review on Daptomycin: The first US-FDA approved lipopeptide antibiotics. *Journal of Scientific and Innovative Research*, 2(5): 970–980. www.jsirjournal.com

Nam, J., Alam, S.T., Kang, K., Choi, J. & Seo, M.H. (2021). Anti-staphylococcal activity of a cyclic lipopeptide, C_{15}-bacillomycin D, produced by Bacillus velezensis NST6. *Journal of Applied Microbiology*, 131(1): 93–104. https://doi.org/10.1111/jam.14936

Nartey, A.P., Dofuor, A.K., Owusu, K.B., Camas, A.S., Deng, H., Jaspars, M. et al. (2022). Digyalipopeptide A, an antiparasitic cyclic peptide from the Ghanaian *Bacillus* sp. strain DE2B. *Beilstein Journal of Organic Chemistry*, 18: 1763–1771. https://doi.org/10.3762/bjoc.18.185

Oikawa, H., Mizunoue, Y., Nakamura, T., Fukushi, E., Yulu, J., Ozaki, T. et al. (2022). Structure and biosynthesis of the ribosomal lipopeptide antibiotic albopeptins. *Bioscience, Biotechnology, and Biochemistry*, 86(6): 717–723. https://doi.org/10.1093/bbb/zbac039

Oliveira-Nascimento, L., Massari, P. & Wetzler, L.M. (2012). The role of TLR2 in infection and immunity. *Frontiers in Immunology*, 3: 79. https://doi.org/10.3389/fimmu.2012.00079

O'Neill, J. (2016). Tackling drug-resistant infections globally: Final report and recommendations. incom

Ongena, M. & Jacques, P. (2008). Bacillus lipopeptides: Versatile weapons for plant disease biocontrol. *Trends in Microbiology*, 16(3): 115–125. https://doi.org/10.1016/j.tim.2007.12.009

Patel, P., Bhatt, S., Patel, H., Marcelino, L.A. & Sayyed, R.Z. (2022). Biosurfactant – A biomolecule and its potential applications. *In:* Sayyed, R.Z. and Enshasy, H.E. (Eds), *Biosurfactants: Production and Applications in Food and Agriculture*. Vol II, 133-149. CRC Press, Taylor & Francis Group, USA.

Pham, J.V., Yilma, M.A., Feliz, A., Majid, M.T., Maffetone, N., Walker, J.R. et al. (2019). A review of the microbial production of bioactive natural products and biologics. *Frontiers in Microbiology*, 10(Jun). https://doi.org/10.3389/fmicb.2019.01404

Phyo, M.Y., Ding, C.Y.G., Goh, H.C., Goh, J.X., Ong, J.F.M., Chan, S.H. et al. (2019). Trikoramide A, a prenylated cyanobactin from the marine cyanobacterium *Symploca hydnoides*. *Journal of Natural Products*, 82(12): 3482–3488. https://doi.org/10.1021/acs.jnatprod.9b00675

Pilz, M., Cavelius, P., Qoura, F., Awad, D. & Brück, T. (2023). Lipopeptides development in cosmetics and pharmaceutical applications: A comprehensive review. *Biotechnology Advances*, 67: 108210. https://doi.org/10.1016/j.biotechadv.2023.108210

Pirri, G., Giuliani, A., Nicoletto, S.F., Pizzuto, L. & Rinaldi, A.C. (2009). Lipopeptides as anti-infectives: A practical perspective. *Central European Journal of Biology*, 4(3): 258–273. https://doi.org/10.2478/s11535-009-0031-3

Pretorius, D., Van Rooyen, J. & Clarke, K.G. (2015). Enhanced production of antifungal lipopeptides by Bacillus amyloliquefaciens for biocontrol of postharvest disease. *New Biotechnology*, 32(2): 243–252. https://doi.org/10.1016/j.nbt.2014.12.003

Proud, P.C., Tsitoura, D., Watson, R.J., Chua, B.Y., Aram, M.J. Bewley, K.R. et al. (2021). Prophylactic intranasal administration of a TLR2/6 agonist reduces upper respiratory tract viral shedding in a SARS-CoV-2 challenge ferret model. *EBioMedicine*, 63. https://doi.org/10.1016/j.ebiom.2020.103153

Qi, G., Zhu, F., Du, P., Yang, X., Qiu, D., Yu, Z. et al. (2010). Lipopeptide induces apoptosis in fungal cells by a mitochondria-dependent pathway. *Peptides*, 31(11): 1978–1986. https://doi.org/10.1016/j.peptides.2010.08.003

Raaijmakers, J.M., De Bruijn, I., Nybroe, O. & Ongena, M. (2010). Natural functions of lipopeptides from Bacillus and Pseudomonas: More than surfactants and antibiotics. *FEMS Microbiology Reviews*, 34(6): 1037–1062. https://doi.org/10.1111/j.1574-6976.2010.00221.x

Ramachandran, R., Shrivastava, M., Narayanan, N.N., Thakur, R.L., Chakrabarti, A. & Roy, U. (2018). Evaluation of antifungal efficacy of three new cyclic lipopeptides of the class bacillomycin from Bacillus subtilis RLID 12.1. *Anti-microbial Agents and Chemotherapy*, 62(1). https://doi.org/10.1128/AAC.01457-17

Ramirez, D., Berry, L., Domalaon, R., Li, Y., Arthur, G., Kumar, A. et al. (2021). Dioctanoyl ultrashort tetrabasic β-peptides sensitize multidrug-resistant gram-negative bacteria to Novobiocin and Rifampicin. *Frontiers in Microbiology*, 12: 803309. https://doi.org/10.3389/fmicb.2021.803309

Rivardo, F., Turner, R.J., Allegrone, G., Ceri, H. & Martinotti, M.G. (2009). Anti-adhesion activity of two biosurfactants produced by Bacillus spp. prevents biofilm formation of human bacterial pathogens. *Applied Microbiology and Biotechnology*, 83(3): 541–553. https://doi.org/10.1007/s00253-009-1987-7

Routhu, S.R., Nagarjuna Chary, R., Shaik, A.B., Prabhakar, S., Ganesh Kumar, C. & Kamal, A. (2019). Induction of apoptosis in lung carcinoma cells by antiproliferative cyclic lipopeptides from marine algicolous isolate Bacillus atrophaeus strain AKLSR1. *Process Biochemistry*, 79: 142–154. https://doi.org/10.1016/j.procbio.2018.12.010

Sachdev, D.P. & Cameotra, S.S. (2013). Biosurfactants in agriculture. *Applied Microbiology and Biotechnology*, 97(3): 1005–1016. https://doi.org/10.1007/s00253-012-4641-8

Sardar, A., Lahiri, A., Kamble, M., Mallick, A.I. & Tarafdar, P.K. (2021). Translation of mycobacterium survival strategy to develop a lipo-peptide based fusion inhibitor*. *Angewandte Chemie* (International ed. in English), 60(11): 6101–6106. https://doi.org/10.1002/anie.202013848

Schieck, A., Müller, T., Schulze, A., Haberkorn, U., Urban, S. & Mier, W. (2010). Solid-phase synthesis of the lipopeptide Myr-HBVpreS/2-78, a hepatitis B virus entry inhibitor. *Molecules* (Basel, Switzerland), 15(7): 4773–4783. https://doi.org/10.3390/molecules15074773

Schlusselhuber, M., Godard, J., Sebban, M., Bernay, B., Garon, D., Seguin, V. et al. (2018). Characterization of Milkisin, a novel lipopeptide with anti-microbial properties produced by *Pseudomonas* sp. UCMA 17988 isolated from bovine raw milk. *Frontiers in Microbiology*, 9: 1030. https://doi.org/10.3389/fmicb.2018.01030

Scopel e Silva, D., de Castro, C.C., da Silva e Silva, F., Sant'anna, V., Vargas, G.D., de Lima, M. et al. (2014). Anti-viral activity of a Bacillus sp. P34 peptide against pathogenic viruses of domestic animals. *Brazilian Journal of Microbiology:* (Publication of the Brazilian Society for Microbiology), 45(3): 1089–1094. https://doi.org/10.1590/s1517-83822014000300043

Sharma, D., Mandal, S.M. & Manhas, R.K. (2014). Purification and characterization of a novel lipopeptide from Streptomyces amritsarensis sp. nov. active against methicillin-resistant Staphylococcus aureus. *AMB Express*, 4: 50. https://doi.org/10.1186/s13568-014-0050-y

Shibata, K., Hasebe, A., Into, T., Yamada, M. & Watanabe, T. (2000). The N-terminal lipopeptide of a 44-kDa membrane-bound lipoprotein of Mycoplasma salivarium is responsible for the expression of intercellular adhesion molecule-1 on the cell surface of normal human gingival fibroblasts. *Journal of Immunology* (Baltimore, Md.: 1950), 165(11): 6538–6544. https://doi.org/10.4049/jimmunol.165.11.6538

Sikorska, E., Stachurski, O., Neubauer, D., Małuch, I., Wyrzykowski, D., Bauer, M. et al. Short arginine-rich lipopeptides: From self-assembly to anti-microbial activity. *Biochimica et biophysica acta. Biomembranes*, 1860(11): 2242–2251. https://doi.org/10.1016/j.bbamem.2018.09.004

Sivapathasekaran, C., Das, P., Mukherjee, S., Saravanakumar, J., Mandal, M. & Sen, R. (2010). Marine bacterium derived lipopeptides: Characterization and cytotoxic activity against cancer cell lines. *International Journal of Peptide Research and Therapeutics*, 16(4): 215–222. https://doi.org/10.1007/s10989-010-9212-1

Strieker, M., Tanović, A. & Marahiel, M.A. (2010). Non-ribosomal peptide synthetases: Structures and dynamics. *Current Opinion in Structural Biology*, 20(2): 234–240. https://doi.org/10.1016/j.sbi.2010.01.009

Sumi, C.D., Yang, B.W., Yeo, I.C. & Hahm, Y.T. (2015). Anti-microbial peptides of the genus Bacillus: A new era for antibiotics. *Canadian Journal of Microbiology*, 61(2): 93–103. https://doi.org/10.1139/cjm-2014-0613

Szymański, M., Chmielewska, S., Czyżewska, U., Malinowska, M. & Tylicki, A. (2022). Echinocandins – Structure, mechanism of action and use in antifungal therapy. *Journal of Enzyme Inhibition and Medicinal Chemistry*, 37(1): 876–894. https://doi.org/10.1080/14756366.2022.2050224

Takeuchi, O. & Akira, S. (2010). Pattern recognition receptors and inflammation. *Cell*, 140(6): 805–820. https://doi.org/10.1016/j.cell.2010.01.022

Taylor, G., Frommherz, Y., Katikaridis, P., Layer, D., Sinning, I., Carroni, M. et al. (2022). Antibacterial peptide Cyclomarin A creates toxicity by deregulating the Mycobacterium tuberculosis ClpC1-ClpP1P2 protease. *The Journal of Biological Chemistry*, 298(8): 102202. https://doi.org/10.1016/j.jbc.2022.102202

Théatre, A., Hoste, A.C.R., Rigolet, A., Benneceur, I., Bechet, M., Ongena, M. et al. (2022). Bacillus sp.: A remarkable source of bioactive lipopeptides. *Advances in Biochemical Engineering/Biotechnology*, 181: 123–179. https://doi.org/10.1007/10_2021_182

Torres-Mendoza, D., Coronado, L.M., Pineda, L.M., Guzmán, H.M., Dorrestein, P.C., Spadafora, C. et al. (2018). Pumilacidins from the Octocoral-associated *Bacillus* sp. DT001 display anti-proliferative effects in *Plasmodium falciparum*. *Molecules* (Basel, Switzerland), 23(9): 2179. https://doi.org/10.3390/molecules23092179

Velkov, T., Thompson, P.E., Azad, M.A.K., Roberts, K.D. & Bergen, P.J. (2019). History, chemistry and antibacterial spectrum. *Advances in Experimental Medicine and Biology*, 1145: 15–36. Springer New York LLC. https://doi.org/10.1007/978-3-030-16373-0_3

Von Döhren, H. (2009). Antibiotics: Actions, origins, resistance, by C. Walsh. 2003. Washington, DC: ASM Press. 345 pp. (hardcover). *Protein Science*, 13(11): 3059–3060. https://doi.org/10.1110/ps.041032204

Wang, C.L., Liu, C., Niu, L.L., Wang, L.R., Hou, L.H. & Cao, X.H. (2013). Surfactin-induced apoptosis through ROS-ERS-Ca2+-ERK pathways in HepG2 cells. *Cell Biochemistry and Biophysics*, 67(3): 1433–1439. https://doi.org/10.1007/s12013-013-9676-7

Wang, T., Liang, Y., Wu, M., Chen, Z., Lin, J. & Yang, L. (2015). Natural products from Bacillus subtilis with anti-microbial properties. *Chinese Journal of Chemical Engineering*, 23(4): 744–754, https://doi.org/10.1016/j.cjche.2014.05.020

Wenink, M.H., Santegoets, K.C., Broen, J.C., van Bon, L., Abdollahi-Roodsaz, S., Popa, C. et al. (2009). TLR2 promotes Th2/Th17 responses via TLR4 and TLR7/8 by abrogating the type I IFN amplification loop. *Journal of Immunology* (Baltimore, Md.: 1950), 183(11): 6960–6970. https://doi.org/10.4049/jimmunol.0900713

Wiebach, V., Mainz, A., Siegert, M.J., Jungmann, N.A., Lesquame, G., Tirat, S. et al. (2018). The anti-staphylococcal lipolanthines are ribosomally synthesized lipopeptides. *Nature Chemical Biology*, 14(7): 652–654. https://doi.org/10.1038/s41589-018-0068-6

Wu, C.Y., Chen, C.L., Lee, Y.H., Cheng, Y.C., Wu, Y.C., Shu, H.Y. et al. (2007). Non-ribosomal synthesis of fengycin on an enzyme complex formed by fengycin synthetases. *The Journal of Biological Chemistry*, 282(8): 5608–5616. https://doi.org/10.1074/jbc.M609726200

Wu, Y.S., Ngai, S.C., Goh, B.H., Chan, K.G., Lee, L.H. & Chuah, L.H. (2017). Anti-cancer activities of surfactin and potential application of nanotechnology assisted surfactin delivery. *Frontiers in Pharmacology*, 8: 761. https://doi.org/10.3389/fphar.2017.00761

Xia, B., Luo, M., Pang, L., Liu, X. & Yi, Y. (2021). Lipopeptides against COVID-19 RNA-dependent RNA polymerase using molecular docking. *Biomedical Journal*, 44(6): S15–S24. https://doi.org/10.1016/j.bj.2021.11.010

Xia, S., Liu, M., Wang, C., Xu, W., Lan, Q., Feng, S. et al. (2020). Inhibition of SARS-CoV-2 (previously 2019 nCoV) infection by a highly potent pan-coronavirus fusion inhibitor targeting its spike protein that harbors a high capacity to mediate membrane fusion. *Cell Research*, 30(4): 343–355. https://doi.org/10.1038/s41422-020-0305-x

Xu, X., Jiang, Q., Zhang, X., Nie, Y., Zhang, Z., Li, Y. et al. (2015). Virus-inspired mimics: Self-assembly of dendritic lipopeptides into arginine-rich nanovectors for improving gene delivery. *Journal of Materials Chemistry, B*, 3(35): 7006–7010. https://doi.org/10.1039/c5tb01070e

Zaman, M., Hassan, S., Fatima, S., Hamid, B., Farooq, S., Qayoom, I. et al. (2022). Biosurfactants production and applications in food. *In:* Sayyed, R.Z. and Enshasy, H.E. (Eds), *Biosurfactants: Production and Applications in Food and Agriculture.* Vol II, 225-241. CRC Press, Taylor & Francis Group, USA.

Zambry, N.S., Rusly, N.S., Awang, M.S., Md Noh, N.A. & Yahya, A.R.M. (2021). Production of lipopeptide biosurfactant in batch and fed-batch Streptomyces sp. PBD-410L cultures growing on palm oil. *Bioprocess and Biosystems Engineering*, 44(7): 1577–1592. https://doi.org/10.1007/s00449-021-02543-5

Zeraik, A.E. & Nitschke, M. (2010). Biosurfactants as agents to reduce adhesion of pathogenic bacteria to polystyrene surfaces: Effect of temperature and hydrophobicity. *Current Microbiology*, 61(6): 554–559. https://doi.org/10.1007/s00284-010-9652-z

Zhang, Z., Guo, W., He, X., Che, Q., Zhu, T., Gu, Q. & Li, D. (2016). Peniphenylanes A-G from the deep-sea-derived fungus Penicillium fellutanum HDN14-323. *Planta Medica*, 82(9–10): 872–876. https://doi.org/10.1055/s-0042-102885

Zhao, H., Yan, L., Xu, X., Jiang, C., Shi, J., Zhang, Y. et al. (2018). Potential of Bacillus subtilis lipopeptides in anti-cancer I: Induction of apoptosis and paraptosis and inhibition of autophagy in K562 cells. *AMB Express*, 8(1): 78. https://doi.org/10.1186/s13568-018-0606-3

Zhong, C., Zhang, F., Zhu, N., Zhu, Y., Yao, J., Gou, S. et al. (2021). Ultra-short lipopeptides against gram-positive bacteria while alleviating anti-microbial resistance. *European Journal of Medicinal Chemistry*, 212: 113138. https://doi.org/10.1016/j.ejmech.2020.113138

Zhu, Y., Zhang, X., Ding, X., Chong, H., Cui, S., He, J. et al. (2018). Exceptional potency and structural basis of a T1249-derived lipopeptide fusion inhibitor against HIV-1, HIV-2, and simian immunodeficiency virus. *Journal of Biological Chemistry*, 293(14): 5223–5334. httpss://doi.org/10.1074/jbc.RA118.001729

Zhu, Y., Chong, H., Yu, D., Guo, Y., Zhou, Y. & He, Y. (2019). Design and characterization of cholesterylated peptide HIV-1/2 fusion inhibitors with extremely potent and long-lasting anti-viral activity. *Journal of Virology*, 93(11). https://doi.org/10.1128/jvi.02312-18

Zhu, Y., Yu, D., Yan, H., Chong, H. & He, Y. (2020). Design of potent membrane fusion inhibitors against SARS-CoV-2, an emerging coronavirus with high fusogenic activity. https://doi.org/10.1128/JVI

CHAPTER

17

Microbial Surfactant: New Edge in Pharmaceuticals

Burhan Hamid[1*], Neesa Majeed[1], Ali Mohd Yatoo[1,2], Parvaze Ahmad Wani[1], Zaffar Bashir[1], Sreedevi Sarsan[3], and Gulrez Nizami[4]

[1] Centre of Research for Development, University of Kashmir, Srinagar, India
[2] Department of Environmental Science, University of Kashmir, Srinagar, India
[3] Department of Microbiology, St. Pious X Degree and PG College for Women, Hyderabad, India
[4] Department of Chemistry, Mohammad Ali Jauhar University, Rampur, India

1. Introduction

It is now widely acknowledged that the environment is badly impacted by the extensive usage of synthetic surfactants (Liu et al., 2022). Synthetic surfactants, utilized across various industries, including pharmaceuticals, medical devices, agriculture, feed and food production, environmental cleanup, and petro-chemicals, are a subject of particular concern (Naughton et al., 2019, Johnson et al., 2021). There is increasing social and legal demand for these compounds to be biodegradable and produced responsibly using sustainable substrates due to environmental concerns worldwide (Abdel and Mansour, 2018). In response to these requirements, research has intensified, and more recently, new methods utilizing biosurfactants-biogenic surface-active chemicals of microbial origin have been developed (Inamuddin and Suvardhan, 2019).

The term "biosurfactants" refers to a class of structurally diverse biomolecules that exhibit strong emulsifying and surface-active properties (Anestopoulos et al., 2020). These are compounds possessing hydrophilic and hydrophobic components generated primarily by the surfaces of microbes or that may be expelled extracellularly (Araujo et al., 2018). With water loving and water repelling groups, biosurfactants have the capacity to aggregate different phases of liquids and are utilized to lower interfacial tension (Mustafa et al., 2022). Chemical surfactants, primarily those generated from petroleum, make up the majority of available commercial surfactants

* Corresponding author: burhan.cordct@uok.edu.in; peerzada19@gmail.com

(Kashif et al., 2022). The need to seriously explore biological surfactants as potential replacements for current products has increased (Johnson et al., 2021), nevertheless, by the rapid advancements in biotechnology and consumers' growing environmental consciousness paired with anticipated new legislation (Mallik and Banerjee, 2022). Bacterially produced surface compounds (active) have consequently gained substantial importance recently as a result of their lower toxic effect, higher rate of biodegradation, diversity and effectiveness under varying pH and temperature conditions (Ribeiro et al., 2020, Gayathiri et al., 2022).

Biosurfactants enhance the wetting, foaming, emulsification, solubilization, dispersion, and detergent functions (El-Khordagui et al., 2021). The wide variety of characteristics and uses of bio-surfactants justifies the possibility of their use in a number of industrial setups, including cosmetics (Banat et al., 2021), pharmaceutics, food, petrochemicals, mining, metallurgy (Matos, 2019), agrochemicals, fertilizers, drinks, and others (Das et al., 2022). Bio-surfactants can be used in the pharmaceutical sector as a green substitute (Bhadani et al., 2020, Mallik and Banerjee, 2022). They can increase a drug's solubility, especially if it is poorly water soluble (Corazza et al., 2022). This comprises an increasing amount of more recent bioactive compounds, such as oligonucleotides, peptides, proteins, vitamins, and tiny molecular therapies (Gomes et al., 2021). The ability to dissolve is a requirement for in-vivo medication absorption. Additionally, bio-surfactants increase thermodynamic activity and rate of diffusion (Banjare et al., 2018) as well as the stability of medications that are encapsulated (Jalali et al., 2022). They are especially crucial for allowing medications to penetrate through cell walls and membranes, skin, as well as other biological surfaces (Luengo et al., 2021). Suppositories and other semisolid dosage forms benefit from the fluidity and in-vivo dissolving provided by bio-surfactants, which act as plasticizers (Bhattacharya et al., 2017). They can be used as agents of dispersal and adsorbents for powders, granules, and nanoparticles (Dominguez et al., 2019). The creation of stable and homogeneous nanoparticles can be accomplished utilizing microemulsion techniques and an oil-water biosurfactant mixture (Onaizi et al., 2021). For example, the reverse microemulsion process was employed to create silver nanoparticles, and glycolipid biosurfactants served as bioemulsifiers (Kiran et al., 2011).

2. Biosurfactant Classification

The structure, composition, and microbiological source of biosurfactants are used to categorize them (Bjerk et al., 2021). Bio-surfactants can be divided into two main categories on the basis of their molecular weight. Low molecular weight surface-active compounds are effective for decreasing surface tension (Jahan et al., 2020). This part contains glycolipids, lipopeptides, and lipoproteins, as well as phospholipids, fatty acids and neutral acids (Bhattacharya et al., 2017). Polymers with higher molecular weight, often known as bio-emulsifiers, are more active stabilizing emulsion agents. Polymeric and particle biosurfactants are included in this category (Table 1) (Karlapudi et al., 2018, Zaman et al., 2022).

2.1 Glycolipids

Glycolipids are one of the most important biosurfactants which are composed of carbohydrates, hydroxy aliphatic acids or long-chain aliphatic acids (Chowdhury et al., 2019). Here, a glycosidic bond joins lipids and carbohydrate molecules

Table 1: Classification of biosurfactant

Biosurfactants type	Subclass	Type of microorganism	Reference
Glycolipids	Rhamnolipid	*Pseudomonas aeruginosa*	Hazra et al., 2014, Bhattacharya et al., 2017, Blount et al., 2020, Mir et al., 2022
	Treahalolipid	*Micrococcus luteus Rhodococcus Erythropolis*	
	Cellobiolipids	*Ustilago maydis*	
	Glycolipid	*Burkholderia cenocepacia*	
	Glucose, Fructose Sucrose lipids	*Corynebacterium, Nocardia* and *Brevibacterium*	
	Xylo Lipid	*Lactococcus Lactis*	
Lipopeptides and lipoproteins	Surfactin	*Bacillus subtilis*	Kural and Gursoy, 2011, He et al., 2017, Bhattacharya et al., 2017, Yang and Yousef, 2018, Borah et al., 2021
	Arthrofactin	*Arthrobacter* sp.	
	Gramicidin	*Brevibacillus brevis*	
	Polymixin	*Bacillus polymyxa*	
	Iturin	*Bacillus subtilis*	
Fatty acids, neutral lipids and phospholipids	Fatty acids	*Corynebacterium lupus*	Shekhar et al., 2015, Bhattacharya et al., 2017, Pirog et al., 2019, de Souza et al., 2022
	Oleic acid	*Issatchenkia orientalis*	
	Neutral lipids	*Nocardia erythropolis*	
	Phospholipids	*Acidithiobacillus thiooxidans*	
Polymeric biosurfactants	Emulsan	*Acinetobacter calcoaceticus*	Mujumdar et al., 2019, Abdolshahi et al., 2019, Saborimanesh and Mulligan, 2019
	Mannoprotein	*Saccharomyces cerevisiae*	
	Carbohydrate-protein-lipid	*Pseudomonas fluorescens*	
	Protein PA	*Pseudomonas aeruginosa*	
	Bioemulsan	*Gordonia* sp.	
Particulate biosurfactant	Fimbriae and Vesicles	*Acinetobacter calcoaceticus*	Santos et al., 2016, Walvekar et al., 2022
	Whole cell	Variety of bacteria	

together. Some of the examples of glycolipids are rhamnolipids, trehalose lipids and sophorolipids (Claus et al., 2021).

2.2 Lipoproteins and Lipopeptides

A class of macromolecules called lipopeptides and lipoproteins possess efficient bio-surfactant character. Surface active substances are cyclic lipopeptides with an amino acid or polypeptide chain attached to a lipid (Ismail et al., 2019). Cyclic lipopeptides with notable surface-active characteristics include gramicidins (a decapeptide antibiotic) and polymyxins (a lipopeptide antibiotic) (Saranraj et al., 2021). Surfactin, an important cyclic lipopeptide comprises seven amino acid rings connected to hydroxy-methyl tetradecanoic acid (Vigneshwaran et al., 2021). Another drug of this kind, lichenysin, has great temperature, pH, and salt stability and acts synergistically (Thakur et al., 2020).

2.3 Fatty Acids, Neutral Lipids and Phospholipids

Numerous fungi, yeasts, and bacteria release a significant amount of phospholipids, neutral lipids or fatty acids when they thrive on hydrophobic substrates like alkanes (Srivastava et al., 2022). Length of hydrocarbon directly correlates to the water loving and water repelling compounds. The balance between hydrophilicity and lipophilicity is directly correlated with the hydrocarbon length of the biosurfactant structure. For several biological applications, these kinds of bio-surfactants are necessary (Fenibo et al., 2019).

2.4 Polymeric Biosurfactants

These are high molecular weight polymeric biosurfactants, sometimes referred to as bio-emulsifiers, commonly found in polysaccharide-protein complexes like emulsan, liposan, alasan, and lipomanan, among others (Pnjanapongchai et al., 2022). Acinetobacter calcoaceticus produces the bio-emulsifier emulsan, an extracellular polyanionic lipopolysaccharide (Handore et al., 2022). Water soluble liposan being a bio-emulsifier is produced by yeast *Candida lipolytica.* 17% proteins and 83% carbohydrates make up its composition (Patel et al., 2022).

2.5 Particulate Biosurfactants

Particulate biosurfactants are external membrane vesicles that divide the hydrocarbons in microemulsions and are crucial for microbial cells to absorb alkanes (Bhadra et al., 2022). Vesicles of the *Acinetobacter* sp. strain HO1-N are composed of proteins, lipo-polysaccharides and phospholipids, having 1.158 g/cm^3 (buoyant density) and 20 to 50 nm (diameter) (Rocha et al., 2019).

3. Properties of Microbial Surfactant and Its Phase Behavior

The complexity of a logical selection of the constituents can result in a successful composition. It would be considerably lowered by previous information of the characteristics of a system as well as its components through genuine evaluation of different variables eg. characteristics of surface (i.e., interfacial and surface tensions, CMC), lipophilic-hydrophilic balance (HLB), Ninham-Israelachvili packing parameter, P (also known as critical packing parameter, CPP), and Winsor-R ratio.

3.1 Surface Properties

A given solution's surface tension will decrease when a surfactant is added because of the adsorption of surfactants on the interface. The interface adsorbs more molecules due to the low surface tension of diluted solutions (Qazi et al., 2020). The aggregation of unimers in bulk solution occurs because more molecules are adsorbed at the surface due to Gibbs energy. This results in a connection between non-polar chains to water, which is higher than the energy required for the repulsion of the interaction of head groups. Additionally,chain-packing constraints in the aggregate core, and establishment of the interfacial region, are related factors contributing to aggregate formation (Rodrigues, 2015). As a result, aggregation will be thermodynamically encouraged and will take place naturally. A micelle is a simple and widespread type of aggregate which is produced in solution during adsorption of the surfactant and self-aggregation. The CMC stands for the minimum concentration required for micelle formation. The surfactant concentration does not result in any change in the surface tension (Reeve et al., 2021).

3.2 Hydrophilic–Lypophilic Balance (HLB)

It is recognized that the HLB parameter has an impact on an emulsion's stability. It also observes the role of water loving and water repelling groups of surfactants in the emulsification of different molecules. The creation of W/O microemulsions (O = oil, W = water) is often favored by lower values of HLB (3-6), whereas the development of O/W microemulsions is favored by high HLB values (8-18). A co-surfactant is frequently needed to lower the effective HLB value of surfactants with very high HLB values (HLBs > 20). Only non-ionic compounds are eligible for the HLB value and are determined on a relative basis (Rodrigues, 2015).

3.3 Critical Packing Parameter (CPP)

Contact angle of water in association with hydrocarbons is reduced due to the formation of aggregates which are enclosed by chains whose orientation is towards the interior of the aggregate in solution. As was already indicated, a variety of different aggregate forms can also form in addition to micelles. Area of the head group of the surfactant and hydrophobic carbon chain influences aggregate type and its tendency to interact with water (Evans and Wennerstrom, 1999). To exemplify such a characteristic, the critical packing parameter (CPP) for surfactants is employed. The various surfactant moieties and the ambient factors have a big impact on the packing parameter. As a result, it is anticipated that surfactants with various CPPs may pack in various ways (Israelachvili, 2011).

3.4 Spontaneous Curvature (H_o)

As an alternative, aggregation can be explained in terms of the surfactant film's so-called spontaneous curvature, H_o, which is the desired mean curvature the film adopts in the absence of mechanical restrictions. The best curvature depends on percentage of non-polar and polar volume, making the H_o notion qualitatively similar to the CPP. The idea underlying it, however, depends on films' mechanical characteristics. In general, H_o is associated to a continuum with global physical qualities, whereas CPP is related to specific molecules (Rodrigues, 2015).

3.5 Winsor-R ratio

Ratio of energy of interaction between O and W phases of surfactant is called winsor-R ratio. There are three possible outcomes: $R < 1$, meaning the tendency of water to interact with surfactant is stronger compared to oil-surfactant interaction, resulting in Winsor Type I microemulsions; $R > 1$, meaning the tendency of water to interact with surfactant is less than the interaction between oil and surfactant resulting in the formation of Winsor Type II microemulsions; and when $R = 1$, there is balance in the interaction between water to surfactant and oil to surfactant forming Winsor Type III microemulsions (Rodrigues, 2015).

4. Pharmaceutical Application of Microbial Surfactant

A growing quantum of research in recent years has revealed that biosurfactants possess a number of biological qualities that the pharmaceutical and biomedical industries can use (Fig. 1, Table 2) (Kashif et al., 2022). The method of action of biosurfactants on the surfaces of microorganisms includes binding to and attaching to membranes, altering wettability and surface energy, and increasing permeability by releasing LPS and creating transmembrane holes (Bhadra et al., 2022). As a result, they compromise membrane integrity, causing cell lysis and metabolite leakage; they also impair protein structures and cause the loss of membrane activities like transport and energy production (Ceresa et al., 2021).

Fig. 1: Scope of microbial surfactants in pharmaceuticals.

Table 2: Application of microbial surfactant in pharmaceuticals

Microbial surfactant	Pharmaceutical application	References
Surfactin	Antifungal and antimicrobial properties. Preventing the development of clots. Ion channel development and hemolysis in lipid membranes. Tumor-fighting capability against carcinoma cells An effective method of drug delivery	Bucci et al., 2018, Roque et al., 2022, Yede et al., 2022
Rhamnolipid	Anti-adhesive properties against several yeast and bacterial strains. *Mycobacterium tuberculosis*-specific antimicrobial activity	
Iturin	The prevention of mycosis by antimicrobial and antifungal activities Improve the electrical conductivity of lipid biomolecule membranes. Eliminate yeast cells membrane structure	Dhuldhaj et al., 2021, dos Santos et al., 2022
Pumilacidin	Antiviral action *Herpes Simplex Virus* (HSV-1).	Saimmai et al., 2020
Mannosylerythritol lipids	Neurological, immunological, and antimicrobial qualities.	Coelho et al., 2020
Glycolipid	Anti-adhesive action against a number of yeast and bacterial strains.	Silva et al., 2021
Lichenysin	The membrane-disturbing effects of lipopeptides may be due to their chelating characteristics.	Ines and Dhouha, 2015

4.1 Anti-Cancer Property

Several classes of biosurfactants, such as glycolipids, have been examined, especially, mannosylerythritol Lipids-B, rhamnolipid, sophorose lipid, polyol lipid, succinoyl trehalose lipid-1, mannosylerythritol lipids-A and succinoyl trehalose lipid-3 (Maniglia et al., 2019). All of these categories of biosurfactants, with the exception of rhamnolipid, are reported to accelerate differentiation of a cell compared to cell proliferation in the human pro-myelocytic leukemia cell line (Govindarajan, 2018). A category of surfactant glycolipids called mannosyl erythritol lipid, produced by the yeast *Candida antarctica*, promotes granulocyte development. Differentiation of mouse malignant melanoma and apoptosis of cells results in its induction by this glycolipid (Magalhaes et al., 2021).

4.2 Antimicrobial Activity of Biosurfactants

The structural composition of biosurfactants shows toxicity to the permeability of plasma membranes in a manner akin to detergents (Goncalves et al., 2018). Two bio-surfactants derived from the microorganisms *Streptococcus thermophilus* and *Lactococcus lactis* have been tested for their antibacterial properties against a diverse range of microbial strains derived from explanted voice prostheses (Patel et al., 2021). It was discovered that, even at low concentrations, both bio-surfactants have strong antibacterial action (Salek and Euston, 2019). The anti-adhesive ability of bio-surfactants is well established. They act as antimicrobials by preventing bacterial adherence and biofilm formation on catheter materials and voice prostheses (Patel et al., 2021). It has been demonstrated that the lipopeptides released by *Bacillus licheniformis* and *Bacillus subtilis* as well as glycolipid biosurfactants derived from *Candida antartica*, *Pseudomonas aeruginosa*, and *Candida*, have strong antimicrobial properties (Saimmai et al., 2020). *Bacillus subtilis* R14 bio-surfactant of the lipopeptide type exhibited antibacterial efficacy against 29 bacterial strains. Similar to this, a marine *Bacillus circulans* produced another biosurfactant that had potent antibacterial action against gram-negative and gram-positive bacteria (Malakar and Deka, 2021).

4.3 Antiviral Activity

There have been reports about the antiviral properties of surfactin and its analogues (Kumar and Jain, 2020). Interaction between the surfactant and virus envelopes results in inhibition of enveloped viruses such as retroviruses and herpes viruses compared to non-enclosed viruses. An antibacterial lipopeptide generated by *B. subtilis* inactivated some viruses such as *pseudorabies* virus, *porcine parvovirus*, newcastle disease virus and bursal disease virus (Theatre et al., 2022). There was however no impact on the *porcine parvovirus* or *pseudorabies* virus (Theatre et al., 2022). Growth of herpes simplex virus types 1 and 2 were inhibited when Rhamnolipid produced by Pseudomonas sp. was mixed with alginate and their inhibition was dose dependent and manifested at doses below the dose of critical micelle (Kiousi, 2022).

4.4 Anti-human Immunodeficiency Virus and Sperm-immobilizing Activity

A female-controlled, efficient, and safe vaginal microbicide is required due to the prevalence of HIV/AIDS among women aged 15 to 49 (Bradley et al., 2019). Authors reported spermicidal, Anti-HIV and cytotoxic activity when cells interacted with sophorolipidbio-surfactant and its structural analogues produced by *Candida bombicola* (Bhattacharya et al., 2017). Diacetate ethyl ester derivative, a sophorolipid, was reported to be an active spermicidal and virucidal compound (Borsanyiova et al., 2016). Its anti-HIV virucidal and sperm-immobilizing properties are comparable to those of nonoxynol-9. Moreover, compounds also showed enhanced toxicity to vaginal cells to raise issues about its suitability for long-term microbicidal contraception (Zairi et al., 2009).

4.5 Anti-adhesive Activity of Biosurfactants

Bio-surfactants prevent infectious organisms from sticking to surgical tools or from growing into biofilms (Mujumdar et al., 2017). Surfactin pre-coating of vinyl urethral catheters before inoculation prevented growth of bacteria such as *Salmonella enterica*, *Escherichia coli*, *Salmonella typhimurium* and *Proteus mirabilis* from forming biofilms (Shekhar et al., 2015). Both the microbial population on prosthesis and the air flow tolerance that develops on vocal prostheses following biofilm development were dramatically reduced by microbial surfactants (Rathinam et al., 2022).

4.6 Pharmaceutical Emulsions

Emulsions are made either from the diversion of water-in-oil (W/O) or oil-in-water (O/W), which are biphasic liquid dosage forms typically designed for both exterior and internal use (Kiran et al., 2022). An emulsifying agent is necessary to stabilize this heterogeneous system since emulsions are thermodynamically unstable. Emulsion stabilization is achieved by utilizing bio-surfactants such as rhamnolipids and mannosylerythritol lipids released by *Pseudomonas aeruginosa* and *Candida antarctica* respectively (Fenibo et al., 2019).

4.7 Immunological Adjuvants

Vaccines when combined with immunological adjuvants enhance the immune response by increasing the production of antibodies for longer-lasting immunity, thereby reducing the amount of foreign material administered (Calina et al., 2020). These adjuvants do not themselves provide immunity; rather, they detoxify the foreign substance with the help of a strong immune build up by the adjuvants. As immunological adjuvants, some strong non-pyrogenic, non-toxic bacterial lipopeptides are being studied. Antigens possessing lower molecular weight such as herbicolin A, iturin AL and microcystin (MLR) coupled to poly-l-lysine (MLRPLL) enhanced humoral immune response significantly when combined with a conventional antigen in chickens and rabbits (Ceresa et al., 2021). A number of lipopeptides such as (N-palmitoyl-S- [2, 3-bis (palmitoyloxy)-(2R, S)-propyl-]-(R)-cysteinyl-serine) enhanced particular immune reaction against certain substances possessing a lower molecular weight. Anti pathogenic bacterial colonization and restoration of essential bacterial culture were reported for the members of the *Lactobacilli* and such bacteria can be used as an effective probiotic (Singh and Cameotra, 2004).

4.8 Dermatological Application of Microbial Surfactant

Numerous bioactive metabolites, such as biosurfactants, are thought to have promise for dermatological uses, such as wound healing. Zouari et al. 2016(b) examined the lipopeptide of *Bacillus subtilis* SPB1 for healing rat wounds and its antioxidant capability under *in vitro* conditions. While comparing the effect of control and treatment, the authors found significantly higher healing and closure of wounds in the experimental rats. The epidermal regeneration in the biopsies taken after SPB1 LP treatment was flawless. It has been hypothesized that the Lipopeptides' capacity to scavenge free radicals helps to reduce swellings and enhances the growth of tissue, wound resurfacing with a new epithelium, and differentiation of epidermis. Gupta et al. (2017) used an ointment containing a glycolipid released by *B. licheniformis* SV1 to examine accelerated healing of wounds in rat tissue *in vivo*. They discovered that the early stages of wound healing involved re-epithelization and fibroblast cell proliferation, while the later stages saw more rapid collagen deposition. According to some theories, the wound-healing abilities displayed by the LPs under investigation may be due to the capacity of surfactants to prevent the biofilm which lowers oxidative stress (Oryan et al., 2018).

Interaction of keratin, elastin and collagen with preservatives found in certain personal care products causes irritation of skin, allergies, damages skin lipids and adversely affects skin cells (Naughton et al., 2019). Biosurfactants are lipid and protein components that are compatible with the skin cell membrane (Adu et al., 2020). Formation of vesicles or micelles results due to flagella shedding and production of rhamnolipids by *P. aeruginosa*. Antimicrobial protein psoriasin production is promoted by rhamnolipid in the presence of flagellin in the skin of healthy humans as reported by Meyer-Hoffert et al. (2011). In order to cause the production of the antimicrobial protein psoriasin, which is capable of killing *P. aeruginosa*, flagellin activates keratinocytes. Hence, healthy skin can stop pathogen colonization before the pathogens may devise ways to interfere with the immune defense response.

4.9 Biosurfactants in Drug Delivery

The use of biosurfactants as drug delivery vehicles enhances applications like passive vaccination, especially in situations when there are few other medication therapy choices (Naugton et al., 2019). For instance, treating candidiasis is challenging since there are few antifungal medications available, and those that are available are toxic to humans and have severe adverse effects. Antifungal medications can be added to different drug delivery methods to solve these problems (Raza et al., 2019). Based on a drug delivery strategy that includes immunization, liposomes represent intriguing prospects with widespread application. In mammalian cells, the efficiency of gene transfection has been reported to enhance by five to seven times by MEL-A, a glycolipid biosurfactant possessing cationic liposomes (Saikia et al., 2021). Cholesterol production may be induced by the application of liposomes which are made up of two hydrophobic tails, whereas noisomes are an ideal carrier for the application of drug delivery (Ge et al., 2019).

5. Conclusion and Future Prospects

Microbial surfactants are naturally synthesized metabolites with diverse commercial and biotechnological applications. Biosurfactants are released by microbes including yeast, bacteria and molds growing on hydrophobic sources. Different types of microbial surfactants are important ingredients in industries like food, cosmetic, detergent and pharmaceuticals. Surfactants are known for their low toxicity, high rates of biodegradation, and distinctive biological activities With stability at extreme temperatures, salt concentrations, and pH levels, they have become the preferred choice compared to their counterparts. Biosurfactants produced by *Bacillus* sp., possess tremendous applications in the pharmaceutical and biotechnological fields. To address new problems in the area of pharmaceuticals and therapeutics, application of efficient microbial surfactants is the need of the hour. Due to a multifold increase in drug resistance it is cumbersome to address drug resistant strains, hence the utility of effective microbial surfactants as drug molecules is an indication of their progressive applications in the field of pharmaceuticals. Microbial surfactants have shown anticancer, antimicrobial, antiviral and other activities which makes them attractive for researchers. However, more in-depth research is required to explore the action mechanism and efficacy of microbial surfactants in pharmaceuticals to solve multiple health problems.

References

Abdel-Shafy, H.I. & Mansour, M.S. (2018). Solid waste issue: Sources, composition, disposal, recycling, and valorization. *Egyptian Journal of Petroleum*, 27(4): 1275-1290.

Adu, S.A., Naughton, P.J., Marchant, R. & Banat, I.M. (2020). Microbial biosurfactants in cosmetic and personal skincare pharmaceutical formulations. *Pharmaceutics*, 12(11): 1099.

Anestopoulos, I., Kiousi, D.E., Klavaris, A., Maijo, M., Serpico, A., Suarez, A. et al. (2020). Marine-derived surface active agents: Health-promoting properties and blue biotechnology-based applications. *Biomolecules*, 10(6): 885.

Araújo, J., Rocha, J., Oliveira Filho, M., Matias, S., Júnior, S.O. & Padilha, C. (2018). Rhamnolipids biosurfactants from Pseudomonas aeruginosa – A review. *Biosciences Biotechnology Research Asia*, 15(4): 767-781.

Banat, I.M., Carboué, Q., Saucedo-Castaneda, G. & de Jesús Cázares-Marinero, J. (2021). Biosurfactants: The green generation of speciality chemicals and potential production using Solid-State Fermentation (SSF) technology. *Bioresource Technology*, 320: 124222.

Banjare, M.K., Behera, K., Kurrey, R., Banjare, R.K., Satnami, M.L., Pandey, S. et al. (2018). Self-aggregation of bio-surfactants within ionic liquid 1-ethyl-3-methylimidazolium bromide: A comparative study and potential application in antidepressants drug aggregation. *Spectrochimica Acta Part A: Molecular and Biomolecular Spectroscopy*, 199: 376-386.

Bhadani, A., Kafle, A., Ogura, T., Akamatsu, M., Sakai, K., Sakai, H. et al. (2020). Current perspective of sustainable surfactants based on renewable building blocks. *Current Opinion in Colloid & Interface Science*, 45: 124-135.

Bhadra, S., Chettri, D. & Kumar Verma, A. (2022). Biosurfactants: Secondary metabolites involved in the process of bioremediation and biofilm removal. *Applied Biochemistry and Biotechnology*, 1-27.

Bhattacharya, B., Ghosh, T.K. & Das, N. (2017). Application of bio-surfactants in cosmetics and pharmaceutical industry. *Scholars Academic Journal Pharmacy*, 6(7): 320-329.

Bjerk, T.R., Severino, P., Jain, S., Marques, C., Silva, A.M., Pashirova, T. et al. (2021). Biosurfactants: Properties and applications in drug delivery, biotechnology and ecotoxicology. *Bioengineering*, 8(8): 115.

Blount, P. & Iscla, I. (2020). Life with bacterial mechanosensitive channels, from discovery to physiology to pharmacological target. *Microbiology and Molecular Biology Reviews*, 84(1): e00055-19.

Borah, D., Chaubey, A., Sonowal, A., Gogoi, B. & Kumar, R. (2021). Microbial biosurfactants and their potential applications: An overview. *Microbial Biosurfactants*, 91-116.

Borsanyiova, M., Patil, A., Mukherji, R., Prabhune, A. & Bopegamage, S. (2016). Biological activity of sophorolipids and their possible use as antiviral agents. *Folia Microbiologica*, 61(1): 85-89.

Bradley, E., Forsberg, K., Betts, J.E., DeLuca, J.B., Kamitani, E., Porter, S.E. et al. (2019). Factors affecting pre-exposure prophylaxis implementation for women in the United States: A systematic review. *Journal of Women's Health*, 28(9): 1272-1285.

Bucci, A.R., Marcelino, L., Mendes, R.K. & Etchegaray, A. (2018). The antimicrobial and antiadhesion activities of micellar solutions of surfactin, CTAB and CPCl with terpinen-4-ol: Applications to control oral pathogens. *World Journal of Microbiology and Biotechnology*, 34(6): 1-9.

Calina, D., Sarkar, C., Arsene, A.L., Salehi, B., Docea, A.O., Mondal, M. et al. (2020). Recent advances, approaches and challenges in targeting pathways for potential COVID-19 vaccines development. *Immunologic Research*, 68(6): 315-324.

Ceresa, C., Fracchia, L., Fedeli, E., Porta, C. & Banat, I.M. (2021). Recent advances in biomedical, therapeutic and pharmaceutical applications of microbial surfactants. *Pharmaceutics*, 13(4): 466.

Chowdhury, S., Rakshit, A., Acharjee, A. & Saha, B. (2019). Novel amphiphiles and their applications for different purposes with special emphasis on polymeric surfactants. *ChemistrySelect*, 4(23): 6978-6995.

Claus, S., Jenkins Sánchez, L. & Van Bogaert, I.N.A. (2021). The role of transport proteins in the production of microbial glycolipid biosurfactants. *Applied Microbiology and Biotechnology*, 105: 1779-1793.

Coelho, A.L.S., Feuser, P.E., Carciofi, B.A.M., de Andrade, C.J. & de Oliveira, D. (2020). Mannosylerythritol lipids: Antimicrobial and biomedical properties. *Applied Microbiology and Biotechnology*, 104(6): 2297-2318.

Corazza, E., Abruzzo, A., Giordani, B., Cerchiara, T., Bigucci, F., Vitali, B. et al. (2022). Human lactobacillus biosurfactants as natural excipients for nasal drug delivery of hydrocortisone. *Pharmaceutics*, 14(3): 524.

Das, B., Kumar, B., Begum, W., Bhattarai, A., Mondal, M.H. & Saha, B. (2022). Comprehensive review on applications of surfactants in vaccine formulation, therapeutic and cosmetic pharmacy and prevention of pulmonary failure due to COVID-19. *Chemistry Africa*, 1-22.

Dhuldhaj, U.P. & Bora, C.R. (2021). Microbial surfactants: An overview. *Microbial Surfactants*, 1-26.

Domínguez Rivera, Á., Martínez Urbina, M.Á. & López y López, V.E. (2019). Advances on research in the use of agro-industrial waste in biosurfactant production. *World Journal of Microbiology and Biotechnology*, 35(10): 1-18.

dos Santos, J.R.A., Rocha, A.A., de Macedo, A.T., Santana, A.A., da Silva, J.B.S., de Souza, M.E.P. et al. (2022). Antifungal activity of biosurfactant against profound mycosis. *Green Sustainable Process for Chemical and Environmental Engineering and Science*, 257-287.

El-Khordagui, L., Badawey, S.E. & Heikal, L.A. (2021). Application of biosurfactants in the production of personal care products, and household detergents and industrial and institutional cleaners. *Green Sustainable Process for Chemical and Environmental Engineering and Science*, 49-96. Elsevier.

Evans, D.F. & Wennerström, H. (1999). The Colloidal Domain: Where Physics, Chemistry, Biology, and Technology Meet. 2nd Edition Wiley-VCH, USA. 632 pp. ISBN 0-471-24247-0, DOI: 10.1515/arh-2001-0032

Fenibo, E.O., Douglas, S.I. & Stanley, H.O. (2019). A review on microbial surfactants: Production, classifications, properties and characterization. *J. Adv. Microbiol*, 18(3): 1-22.

Gayathiri, E., Prakash, P., Karmegam, N., Varjani, S., Awasthi, M.K. & Ravindran, B. (2022). Biosurfactants: Potential and eco-friendly material for sustainable agriculture and environmental safety—A review. *Agronomy*, 12(3): 662.

Ge, X., Wei, M., He, S. & Yuan, W.E. (2019). Advances of non-ionic surfactant vesicles (niosomes) and their application in drug delivery. *Pharmaceutics*, 11(2): 55.

Gomes, A., Aguiar, L., Ferraz, R., Teixeira, C. & Gomes, P. (2021). The emerging role of ionic liquid-based approaches for enhanced skin permeation of bioactive molecules: A snapshot of the past couple of years. *International Journal of Molecular Sciences*, 22(21): 11991.

Gonçalves, R.F., Martins, J.T., Duarte, C.M., Vicente, A.A. & Pinheiro, A.C. (2018). Advances in nutraceutical delivery systems: From formulation design for bioavailability enhancement to efficacy and safety evaluation. *Trends in Food Science & Technology*, 78: 270-291.

Govindarajan, M. (2018). Amphiphilic glycoconjugates as potential anti-cancer chemotherapeutics. *European Journal of Medicinal Chemistry*, 143: 1208-1253.

Gupta, S., Raghuwanshi, N., Varshney, R., Banat, I.M., Srivastava, A.K., Pruthi, P.A. et al. (2017). Accelerated in vivo wound healing evaluation of microbial glycolipid containing ointment as a transdermal substitute. *Biomedicine & Pharmacotherapy*, 94: 1186-1196.

Handore, A.V., Khandelwal, S.R., Karmakar, R., Gupta, D.L. & Handore, D.V. (2022). Biosurfactants: The ecofriendly biomolecules of the upcoming era. *Microbial Surfactants*, 99-138. CRC Press.

Hazra, C., Kundu, D., Chatterjee, A., Chaudhari, A. & Mishra, S. (2014). Poly (methyl methacrylate) (core)–biosurfactant (shell) nanoparticles: Size controlled sub-100 nm synthesis, characterization, antibacterial activity, cytotoxicity and sustained drug release behavior. *Colloids and Surfaces A: Physicochemical and Engineering Aspects*, 449: 96-113.

He, Z., Zeng, W., Zhu, X., Zhao, H., Lu, Y. & Lu, Z. (2017). Influence of surfactin on physical and oxidative stability of microemulsions with docosahexaenoic acid. *Colloids and Surfaces B: Biointerfaces*, 151: 232-239.

Inamuddin, A.M., Asiri, A. & Suvardhan, K. (2019). Green Sustainable Process for Chemical and Environmental Engineering and Science. Elsevier.

Inès, M. & Dhouha, G. (2015). Lipopeptide surfactants: Production, recovery and pore forming capacity. *Peptides*, 71: 100-112.

Ismail, D.A., Youssif, M.A. & Negm, N.A. (2019). Biosurfactants: Properties and applications. *Surfactants in Tribology*, 25-56. CRC Press.

Israelachvili, J.N. (2011). Intermolecular and Surface Forces. Academic Press.

Jahan, R., Bodratti, A.M., Tsianou, M. & Alexandridis, P. (2020). Biosurfactants, natural alternatives to synthetic surfactants: Physicochemical properties and applications. *Advances in Colloid and Interface Science*, 275: 102061.

Jalali-Jivan, M., Rostamabadi, H., Assadpour, E., Tomas, M., Capanoglu, E., Alizadeh-Sani, M. et al. (2022). Recent progresses in the delivery of β-carotene: From nano/microencapsulation to bioaccessibility. *Advances in Colloid and Interface Science*, 102750.

Johnson, P., Trybala, A., Starov, V. & Pinfield, V.J. (2021). Effect of synthetic surfactants on the environment and the potential for substitution by biosurfactants. *Advances in Colloid and Interface Science*, 288: 102340.

Karlapudi, A.P., Venkateswarulu, T.C., Tammineedi, J., Kanumuri, L., Ravuru, B.K., Kodali, V.P. et al. (2018). Role of biosurfactants in bioremediation of oil pollution - A review. *Petroleum*, 4(3): 241-249.

Kashif, A., Rehman, R., Fuwad, A., Shahid, M.K., Dayarathne, H.N.P., Jamal, A. et al. (2022). Current advances in the classification, production, properties and applications of microbial biosurfactants – A critical review. *Advances in Colloid and Interface Science*, 102718.

Kiousi, D.E. (2022). Antiviral potential of marine and plant-derived bioactive molecules. Democritus University of Thrace, 1-79.

Kiran, G.S., Selvin, J., Manilal, A. & Sujith, S. (2011). Biosurfactants as green stabilizers for the biological synthesis of nanoparticles. *Critical Reviews in Biotechnology*, 31(4): 354-364.

Kiran, R.S., Pranitha, T., Shirisha, V., Meenakshi, V. & Rao, T.R. (2022). Review on pilot plant scale up techniques used in solid, liquid and semisolids. *Systematic Reviews in Pharmacy*, 13(6): 636-641.

Kumar, M. & Jain, C.P. (2020). Possible benefits of reformulating antiviral drugs with nanoemulsion system in the treatment of novel coronavirus infection. *Current Drug Therapy*, 2(3).

Kural, F.H. & Gürsoy, R.N. (2011). Formulation and characterization of surfactin-containing self-microemulsifying drug delivery systems SF-SMEDDS. *Hacettepe University Journal of the Faculty of Pharmacy*, 2: 171-186.

Liu, B., Wu, X., Liu, X. & Gong, M. (2022). An improved method of assessing marine utilization impact to describe the man-land relationship for coastal management: A case study of the Laizhou Bay, China. *Environmental Science and Pollution Research*, 1-13.

Luengo, G.S., Fameau, A.L., Leonforte, F. & Greaves, A.J. (2021). Surface science of cosmetic substrates, cleansing actives and formulations. *Advances in Colloid and Interface Science*, 290: 102383.

Magalhães, F.F., Nunes, J.C., Araújo, M.T., Ferreira, A.M., Almeida, M.R., Freire, M.G. & Tavares, A.P. (2021). Anti-cancer biosurfactants. *Microbial Biosurfactants*, 159-196. Springer, Singapore.

Malakar, C. & Deka, S. (2021). Biosurfactants against drug-resistant human and plant pathogens: Recent advances. *Biosurfactants for a Sustainable Future: Production and Applications in the Environment and Biomedicine*, 353-372.

Mallik, T. & Banerjee, D. (2022). Biosurfactants: The potential green surfactants in the 21st century. *Journal of Advanced Scientific Research*, 13(01): 97-106.

Maniglia, B.C., Laroque, D.A., de Andrade, L.M., Carciofi, B.A.M., Tenório, J.A.S. & de Andrade, C.J. (2019). Production of active cassava starch films; effect of adding a biosurfactant or synthetic surfactant. *Reactive and Functional Polymers*, 144: 104368.

Matos, R.P.C.B.C.D. (2019). Fermentation of white grape pomace by S. bombicola for the production of sophorolipids (Doctoral dissertation).

Meyer-Hoffert, U., Zimmermann, A., Czapp, M., Bartels, J., Koblyakova, Y., Gläser, R. et al. (2011). Flagellin delivery by Pseudomonas aeruginosa rhamnolipids induces the antimicrobial protein psoriasin in human skin. *PLoS One*, 6(1): e16433.

Mir, M.I., Quadriya, H., Kumar, B.K., Adeeb, S., Ali, M., Khan, M.Y. et al. (2022). 14 Biosurfactants of nitrogen fixers and their potential applications. *Microbial Surfactants: Applications in Food and Agriculture*. Vol. 2.

Mujumdar, S., Bashetti, S., Pardeshi, S. & Thombre, R.S. (2017). Industrial applications of biosurfactants. *Industrial Biotechnology*, 81-109. Apple Academic Press.

Mujumdar, S., Joshi, P. & Karve, N. (2019). Production, characterization, and applications of bioemulsifiers (BE) and biosurfactants (BS) produced by Acinetobacter spp.: A review. *Journal of Basic Microbiology*, 59(3): 277-287.

Mustafa, G., Zahid, M.T., Ihsan, S., Zia, I., Abbas, S.Z. & Rafatullah, M. (2022). Bacterial extracellular polymeric substances for degradation of textile dyes. *Polymer Technology in Dye-containing Wastewater*. 175-191. Springer, Singapore.

Naughton, P.J., Marchant, R., Naughton, V. & Banat, I.M. (2019). Microbial biosurfactants: Current trends and applications in agricultural and biomedical industries. *Journal of Applied Microbiology*, 127(1): 12-28.

Onaizi, S.A., Alsulaimani, M., Al-Sakkaf, M.K., Bahadi, S.A., Mahmoud, M. & Alshami, A. (2021). Crude oil/water nanoemulsions stabilized by biosurfactant: Stability and pH-switchability. *Journal of Petroleum Science and Engineering*, 198: 108173.

Oryan, A., Alemzadeh, E. & Moshiri, A. (2018). Potential role of propolis in wound healing: Biological properties and therapeutic activities. *Biomedicine & Pharmacotherapy*, 98: 469-483.

Patel, M., Siddiqui, A.J., Hamadou, W.S., Surti, M., Awadelkareem, A.M., Ashraf, S.A. et al. (2021). Inhibition of bacterial adhesion and antibiofilm activities of a glycolipid biosurfactant from Lactobacillus rhamnosus with its physicochemical and functional properties. *Antibiotics*, 10(12): 1546.

Patel, P., Bhatt, S. & Sayyed, R.Z. (2022). Biosurfactant: A biomolecule and its potential applications. *Microbial Surfactants: Applications in Food and Agriculture*, Vol. 2, 63-81.

Pirog, T.P., Geichenko, B.S. & Zvarych, A.O. (2019). Post-harvest treatment of vegetables with exometabolites of Nocardia vaccinii IMV B-7405, *Acinetobacter calcoaceticus* IMV B-7241 and *Rhodococcus erythropolis* IMV Ac-5017 to extend their shelf life. *Biotechnologia Acta*, 12(6): 46-55.

Qazi, M.J., Schlegel, S.J., Backus, E.H., Bonn, M., Bonn, D. & Shahidzadeh, N. (2020). Dynamic surface tension of surfactants in the presence of high salt concentrations. *Langmuir*, 36(27): 7956-7964.

Rathinam, P., Antony, S., Reshmy, R., Madhavan, A., Binod, P., Pandey, A. et al. (2022). Biosurfactant as an intervention for medical device associated infections. *Green Sustainable Process for Chemical and Environmental Engineering and Science*. 451-465. Academic Press.

Raza, A., Sime, F.B., Cabot, P.J., Maqbool, F., Roberts, J.A. & Falconer, J.R. (2019). Solid nanoparticles for oral antimicrobial drug delivery: A review. *Drug Discovery Today*, 24(3): 858-866.

Reeve, J.R., Thomas, R.K. & Penfold, J. (2021). Surface activity of ethoxylate surfactants with different hydrophobic architectures: The effect of layer substructure on surface tension and adsorption. *Langmuir*, 37(30): 9269-9280.

Ribeiro, B.G., Guerra, J.M. & Sarubbo, L.A. (2020). Biosurfactants: Production and application prospects in the food industry. *Biotechnology Progress*, 36(5): e3030.

Rocha e Silva, N.M.P., Meira, H.M., Almeida, F.C.G., Soares da Silva, R.D.C.F., Almeida, D.G., Luna, J.M. et al. (2019). Natural surfactants and their applications for heavy oil removal in industry. *Separation & Purification Reviews*, 48(4): 267-281.

Roque-Borda, C.A., Gualque, M.W.D.L., da Fonseca, F.H., Pavan, F.R. & Santos-Filho, N.A. (2022). Nanobiotechnology with therapeutically relevant macromolecules from animal venoms: Venoms, toxins, and antimicrobial peptides. *Pharmaceutics*, 14(5): 891.

Saborimanesh, N. & Mulligan, C.N. (2019). Biosurfactants and their applications in petroleum industry. *Microbial Biosurfactants and their Environmental and Industrial Applications*. 209-241. CRC Press.

Saikia, R.R., Deka, S. & Sarma, H. (2021). Biosurfactants from Bacteria and Fungi: Perspectives on Advanced Biomedical Applications. *Biosurfactants for a Sustainable Future: Production and Applications in the Environment and Biomedicine*, 293-315.

Saimmai, A., Riansa-Ngawong, W., Maneerat, S. & Dikit, P. (2020). Application of biosurfactants in the medical field. *Walailak Journal of Science and Technology*, 17(2): 154-166.

Sałek, K. & Euston, S.R. (2019). Sustainable microbial biosurfactants and bioemulsifiers for commercial exploitation. *Process Biochemistry*, 85: 143-155.

Santos, D.K.F., Rufino, R.D., Luna, J.M., Santos, V.A. & Sarubbo, L.A. (2016). Biosurfactants: Multifunctional biomolecules of the 21st century. *International Journal of Molecular Sciences*, 17(3): 401.

Saranraj, P., Sayyed, R.Z., Sivasakthivelan, P., Devi, M.D., Al Tawaha, A.R.M. & Sivasakthi, S. (2021). Microbial biosurfactants sources, classification, properties and mechanism of interaction. *Microbial Surfactants*, 243-265. CRC Press.

Shekhar, S., Sundaramanickam, A. & Balasubramanian, T. (2015). Biosurfactant producing microbes and their potential applications: A review. *Critical Reviews in Environmental Science and Technology*, 45(14): 1522-1554.

Silva, M.D.G.C.D., Durval, I.J.B., Silva, M.E.P.D. & Sarubbo, L.A. (2021). Potential applications of anti-adhesive biosurfactants. *Microbial Biosurfactants*. 213-225. Springer, Singapore.

Singh, P. & Cameotra, S.S. (2004). Potential applications of microbial surfactants in biomedical sciences. *TRENDS in Biotechnology*, 22(3): 142-146.

Srivastava, R.K., Bothra, N., Singh, R., Sai, M.C., Nedungadi, S.V. & Sarangi, P.K. (2022). Microbial originated surfactants with multiple applications: A comprehensive review. *Archives of Microbiology*, 204(8): 1-19.

Thakur, S., Singh, A., Sharma, R., Aurora, R. & Jain, S.K. (2020). Biosurfactants as a novel additive in pharmaceutical formulations: Current trends and future implications. *Current Drug Metabolism*, 21(11): 885-901.

Théatre, A., Hoste, A.C.R., Rigolet, A., Benneceur, I., Bechet, M., Ongena, M. et al. (2022). Bacillus sp.: A remarkable source of bioactive lipopeptides. *In:* Biosurfactants for the Biobased Economy (pp. 123-179). Cham: Springer International Publishing.

Vigneshwaran, C., Vasantharaj, K., Krishnanand, N. & Sivasubramanian, V. (2021). Production optimization, purification and characterization of lipopeptide biosurfactant obtained from Brevibacillus sp. AVN13. *Journal of Environmental Chemical Engineering*, 9(1): 104867.

Yang, X. & Yousef, A.E. (2018). Antimicrobial peptides produced by Brevibacillus spp.: Structure, classification and bioactivity: A mini review. *World Journal of Microbiology and Biotechnology*, 34(4): 1-10.

Yede, S. (2022). # 149 Improvement in solubility of poorly soluble Bosentan monohydrate through self-micro-emulsifying drug delivery system. *Journal of Pharmaceutical Chemistry*, 8(Supplement).

Zairi, A., Tangy, F., Bouassida, K. & Hani, K. (2009). Dermaseptins and magainins: Antimicrobial peptides from frogs' skin—New sources for a promising spermicides microbicide: A mini review. *Journal of Biomedicine and Biotechnology*, 2009.

Zaman, M. & Hamid, B. (2022). 7 Biosurfactants production and applications in food. *Microbial Surfactants: Applications in Food and Agriculture*, 2: 133-149.

Zouari, R., Moalla-Rekik, D., Sahnoun, Z., Rebai, T., Ellouze-Chaabouni, S. & Ghribi-Aydi, D. (2016). Evaluation of dermal wound healing and in vitro antioxidant efficiency of Bacillus subtilis SPB1 biosurfactant. *Biomedicine & Pharmacotherapy*, 84: 878-891.

CHAPTER

18

Nanobiosurfactants: Molecules of the 21st Century in Pharmacy and Cosmetics

Hulya Yilmaz[1], Meriam Bouri[2*], and Ebru Cubuk Demiralay[3]

[1] Sabanci University Nanotechnology Research and Application Center (SUNUM), Tuzla, Turkey

[2] Faculty of Engineering, Department of Genetics and Bioengineering, Yeditepe University, Istanbul, Turkey

[3] Suleyman Demirel University, Faculty of Pharmacy, Department of Basic Pharmaceutical Sciences, Isparta, Turkey

1. Introduction

Biosurfactants are surface-active molecules of microbial origins. In the past decades, these molecules have attracted significant interest mainly due to their low toxicity and economic sustainability, besides several additional advantages over chemical surfactants, such as biodegradability and biocompatibility. Therefore, biosurfactants have been useful in several applications, such as agri-foods, bioremediation, pharmacy, cosmetics, and oil industries (Jahan et al., 2020). Biosurfactants are produced by many micro-organisms as secondary metabolites. Bacteria, fungi, and yeast are the most reported biosurfactant producers either extra- or intra-cellularly. In microbial cells, biosurfactants are involved in various physiological processes such as motility, cell communications, gene and nutrient efflux regulation, pathogenesis, and biofilm formation (Sharma et al., 2021). In pharmaceutical industries, biosurfactants have emerged as potential drug delivery systems (Bjerk et al., 2021), antimicrobial agents, and therapeutic molecules. However, chemical synthesis based on structure-activity relationships is essential for the development of new amphiphiles with improved pharmacological properties, bioavailability, and biodegradability. With the recent advances in nanobiotechnology, biosurfactants have experienced new frontiers of research. Seeking new sustainable and less toxic molecules, green nanobiotechnology has explored the microbial kingdom as a biological system for nanoparticles (NPs).

* Corresponding author: mariem_bouri@hotmail.fr

Biogenesis of NPs using microbial metabolites (specifically biosurfactants) as well as their whole cells has offered specific advantages to nanomaterials. The use of microbes in the production of metal NPs could be considered as the most important application of these "bio-factories" in nanobiotechnology. Many microorganisms such as bacteria, fungi, yeasts, and viruses have been explored for their potential in the synthesis of metal NPs and/or stabilization. Microbial synthesis approaches offer several additional advantages over conventional methods, such as the ability to control the size and shape of NPs for specific applications in medicine and cosmetics (Dhanker et al., 2021). Nevertheless, the biosynthesis of NPs by microbes is influenced by various factors, such as environmental conditions and the availability of organic functional groups on the cell wall of the bacterium, (Koul et al., 2021). For example, the optimal synthesis of NPs such as silver NPs (AgNPs) and gold NPs (AuNPs) requires careful optimization of reaction conditions such as pH and temperature.

Although the most potential utilization of biosurfactants in nanotechnology is the production of metal NPs, microbial metabolites have been used either directly or indirectly in the synthesis and stabilization of various organic and inorganic nano-products. With the recent advances of nanobiotechnology, microbe-mediated nano-compounds are experiencing an outstanding progress in the context of product diversity, technology and application. Consequently, new glossary related to nanotechnological processing of biosurfactants is getting developed through literature. Terms such as "nano-biosurfactants", "nano-emulsions", "nano-carriers" and "nano-micelles" are being frequently used to describe nano-sized compounds derived from microbes and/or their biosurfactants. These nanomaterials have become a research hotspot in many areas, such as medicine, cosmetics, and agriculture. NPs issued from microbes could be described as the molecules of the century due to their low toxicity and sustainability besides several other advantages such as specific applications and biocompatibility. In this chapter, the term "nanobiosurfactants" is firstly introduced through a conceptual overview upon microbial nanotechnology based on biosurfactants. Then, there is a special focus on the synthesis and the classification of different nano-sized compounds that could be considered as nanobiosurfactants. Afterwards, several uses of these molecules in pharmacy and cosmetics are detailed to highlight their specific applications. Finally, potential perspectives related to nanobiosurfactant production, identification, characterization and safety are treated with regards to significant challenges of the sector.

2. Conceptual Overview

Biological systems such as microbes and their metabolites have been extensively used in nanotechnology, as green processes for the synthesis or the enhancement of NPs. Because of their rich diversity, these microbes are recognized as potential "bio-factories" for NPs synthesis, although the biochemical and molecular mechanisms involved in this biosynthesis have scope for more research to improve the yield and monodispersity of the produced particles (Płaza et al., 2014). Biosurfactants are surface active agents secreted by microorganisms that have been included in the synthesis of many bioactive products for specific applications in agri-food, chemistry, cosmetic, pharmaceutic, and even electronic industries. With the use of microbes and their products in biotechnology, various new research areas and new glossaries have emerged.

2.1 Biosurfactant Technology

Biosurfactants are biological surface-active compounds or microbial surface-active agents produced by living cells, mainly by microorganisms (Banat et al., 2010). They are surface-active biomolecules due to their amphiphilic nature characterized by the presence of both hydrophobic and hydrophilic moieties that offer them the potential to interfere between polar and nonpolar media (Franzetti et al., 2011). Their hydrophilic moiety is mostly either an alcohol or a carbohydrate, amino acid, cyclic peptide, phosphate and carboxylic acid. Their hydrophobic moiety can be a long-chain fatty acid, hydroxyl fatty acid or α-alkyl β-hydroxy fatty acid (Satpute et al., 2009). They can be on microbial cell surfaces (cell-bounds) or extracellularly secreted. Many microorganisms are known to be involved in the production of biosurfactants, mainly bacteria, fungi and yeasts. The extraction and purification of biosurfactants depend essentially on the microorganism and the secretion mode. In laboratory routines, the identification of biosurfactants relies essentially on the following tests: oil spreading (Morikawa et al., 2000), drop collapse method (Kuiper et al., 2004), emulsification capacity (%E24) (Al-Wahaibi et al., 2014) and surface tension. Given that the identification of efficient biosurfactant producers is time consuming and labor intensive, high throughput omics techniques have become very useful in biosurfactant research (Vivek et al., 2022). Biosurfactant are being used in a wide spectrum of industries, from agro-foods to pharmaceutical and cosmetic applications. Their features are mainly determined by their structures, molecular weights, and the location and size of their functional groups (Płaza et al., 2014). While the term biosurfactants specifically refers to low-molecular-weight surface active agents that efficiently lower surface and interfacial tension, the term bioemulsifer may also be used to describe high molecular-weight polymers which are more effective as emulsion-stabilizing agents (Rosenberg and Ron, 1999). Due to their important role in aggregation and stabilization of NP, biosurfactants have been recognized, recently, as a green alternative in nanobiotechnology applied in nanoparticle synthesis, stabilization and biological activity enhancement. Lately, there has been significant growth in biosurfactant research and market development. The advantageous characteristics of surfactants derived from microbial molecules in the 21st century make them attractive materials for various applications in nanobiotechnology (Fig. 1).

2.2 Nanobiotechnology and Microbial Nanotechnology

The word nanobiotechnology refers to the biological synthesis of NPs, in common usage. The areas of nanobiotechnology (also called bionanotechnology, or simply nanobiology) encompass biotechnology and nanotechnology. The emerging field of nanobiotechnology is exploring new methods to optimize the interconnection of various sustainable, cost-effective, and eco-friendly biological processes for nanoparticle production." NPs can be synthesized by various bio-processes such as plant extracts, industrial agro-wastes, microorganisms and standalone biomolecules (Płaza et al., 2014). Microorganisms and their products have been traditionally used in microbial biotechnology for many purposes, such as plant production and protection (Bouri et al., 2022), food industry (Graham and Ledesma-Amaro, 2023), cosmetic and pharmaceutical products (Abdel-Razek et al., 2020), and remediation technologies (Rabbani et al., 2014). Recently, microbial cells have been used for the synthesis of metal NPs as cleaners and eco-friendly alternatives for the synthesis of nano-sized

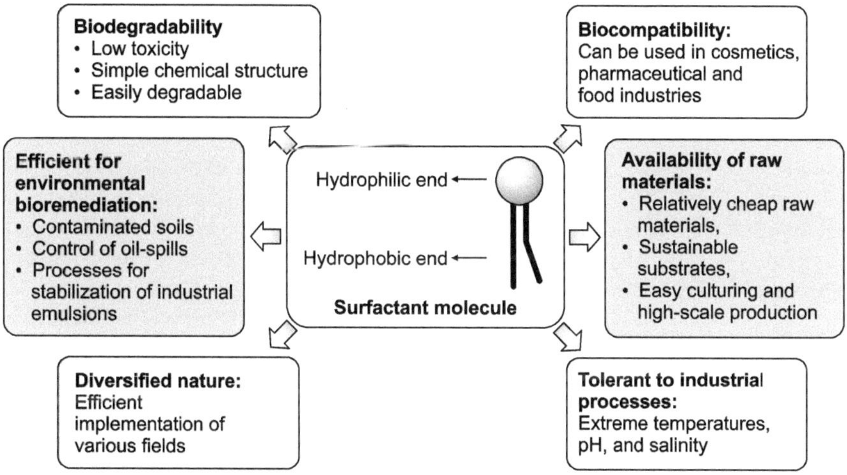

Fig. 1: Advantages of biosurfactants in green technology.

materials. The synthesis of NP by microorganisms is either intra- or extra-cellular, and it is thought to be made through cell wall reductive enzymes or soluble secreted enzymes (Rangarajan et al., 2014). Metal ions are leached from the solution by the microorganism and then reduced in their elemental form through enzymatic reactions during the metabolic activities of the microbial cells. Nevertheless, many unclear aspects of microbial synthesis of NPs are still to be deciphered especially with regards to the bioprocess and features of NPs (eg. size and shape, stability, etc.). The major disadvantage in the microbial synthesis of NPs is the complexity of the downstream process (Christopher et al., 2019). This can be tackled by the biosurfactant-mediated NP synthesis. Recent developments in nanotechnology have shown that NPs can also be synthesized or stabilized by biosurfactants from microorganisms, as capping agents. In this case, biosurfactant-mediated nanomaterials are not produced through the microbial cell "factory". Using biosurfactants from microbes for the synthesis of the metal NPs is easier than handling microbial biomasses. Biosurfactants from microorganisms can be used in the synthesis of NPs either through aggregation or stabilization processes. Biosurfactants act as adsorbents onto metallic NPs, then settle on the surface of NPs to avoid consequent aggregation. For example, rhamnolipid biosurfactants produced by *Pseudomonas aeruginosa* have been used as dispersants for NPs (Mulligan et al., 2001). The potential implication of biosurfactants in nanotechnology is essentially reported in the synthesis and stabilization process of Ag NPs (Christopher et al., 2019). Rhamnolipids and surfactins, produced respectively by *Pseudomonas aeruginosa* and *Bacillus subtilis*, are the most widely used biosurfactants for this purpose (Husseiny et al., 2007). Nevertheless, microorganisms and their biosurfactants have many applications in nanotechnology. Therefore, the term "nano-biosurfactant" could be used to describe the utilization of surfactants in nanotechnology.

2.3 Nanobiosurfactants

"Nanos" in Greek means "dwarf" and refers to things of one-billionth (10^{-9} m) in size. In the literature, the term "nano-biosurfactant" has been introduced by Hosein

Saeedi et al. (2014) to describe a biosurfactant (rhamnolipid) from an autochthonous *Pseudomonas aeruginosa*. More recently Debnath et al. (2021) have used "nano biosurfactants" to review the application of nanomaterials prepared from microorganisms or their metabolic products in nanobioremediation. Instead, expressions such as "nano emulsions" (Jafari et al., 2007), and "biosurfactant-based" nanomaterials (Chuo et al., 2021, Gupta, 2021) are more commonly used. The prefix "Nano" could be used with biosurfactants to designate surfactants from microorganisms that are used in the synthesis, the stabilization or even the enhancement of biological activities of nanomaterials. The resulting nano-products can be metallic or organic NPs, nano-emulsions, nanocarriers or simply microbial extracellular vesicles (for particulate nanobiosurfactants). Among the recently developed biosurfactant-based nano-formulations stand nano-structured biosurfactants such as nano-rod, lipid–polymer hybrid NPs, nano and self-nano-emulsifying drug delivery systems, nanoflower synthesis (Gupta, 2021) and nano micelles for the extraction of biomolecules (Chuo et al., 2021). However, the most considerable application of nanobiosurfactant materials is still relevant to the green synthesis of metal NPs. Their application in organic NP synthesis is still considered as an emerging field. While biosurfactants are essentially used as stabilizing and/or reducing agents in metal NP synthesis, they are usually applied to enhance the biological activity in organic NPs. For instance, biosurfactants have been used in nano-encapsulation systems, such as the co-encapsulation of rutinoside and β-carotene in liposomes to ameliorate nutrient delivery systems (Ji et al., 2023), the encapsulation of nisin into nano-vesicles using rhamnolipids as biosurfactants to increase the antibiotic efficacy (Niaz et al., 2019) and chitosan-based NPs as carriers for glycolipids as antimicrobial biosurfactants (Bettencourt et al., 2021).

3. Synthesis

3.1 Microbial Cell-mediated Synthesis of Metal NPs

NP synthesis is a crucial area of research in nanotechnology. While various methods have been developed for the synthesis of NPs, including biological, physical, and chemical methods, it is important to note that this chapter solely focuses on the biological synthesis of metal NPs by microbial cells. The significance of the biological synthesis of metal NPs cannot be overstated, and it is an area that requires further exploration and research (Ghosh et al., 2021). Biological synthesis has emerged as a sustainable, economical, and less harsh NP synthesis method than chemical or physical methods (Kim et al., 2018). Biological synthesis approaches, such as natural means, offer several advantages over conventional methods, such as the ability to control the size and shape of NPs for specific applications. Furthermore, biological synthesis provides high specificity and selectivity in synthesizing NPs. It is well-established that many organisms can produce inorganic materials intra- or extracellularly. During the intracellular process, the metal ions are actively taken up by the microbial cells through a complex series of mechanisms. Once inside the cell, the metal ions encounter a range of enzymes that play a role in transforming them into NPs. This transformation process is highly regulated and involves several distinct stages, including reduction and oxidation reactions that result in the formation of stable NPs. In contrast, the extracellular process involves the trapping of metal ions outside the cell surface, where they are exposed to a different set of enzymes and conditions. This process is also complex and involves multiple steps, including the reduction of metal ions by cellular enzymes to ultimately form NPs. Overall, the intracellular and extracellular processes for NP formation are fascinating examples of the complex interplay between microbial cells and their environment and highlight the remarkable ability of these microscopic organisms to transform and shape their surroundings (Koul ct al., 2021, Kalabegishvili ct al., 2012)

Researchers are investigating the use of various organisms such as bacteria, actinomycetes, fungi, yeasts, viruses, and algae as potential reducing or stabilizing agents for the synthesis of metal NPs. These NPs include gold (Au), silver (Ag), copper (Cu), cadmium (Cd), platinum(Pt), palladium (Pd), titanium (Ti), and (Zn) zinc. By utilizing the unique properties of these microorganisms, scientists hope to develop more efficient and sustainable methods for the production of metal NPs. Figure 2 provides a visual representation of some of these organisms being studied in this field (Chen & Mao, 2007, Ahmad et al., 2015, Khatoon et al., 2015). Over the past years, microbial-mediated synthesis of NPs has become increasingly popular, and for a good reason. This method offers a range of advantages over traditional methods, making it a clear choice for those seeking a more efficient and cost-effective approach to NPs synthesis. NPs are commonly coated with a lipid layer or biomolecules to improve their solubility and stability, which is essential for biomedical applications. The biosynthesis methods have proven to be highly beneficial in this regard, as they offer a natural and sustainable approach to produce these coatings. Unlike other synthetic methods, biosynthesis allows the incorporation of specific biomolecules that mimic the natural physiological environment, and thus enhance the biological functionality of NPs. This is particularly important, as the lack of physiological stability and solubility is often a bottleneck in the development of NPs for biomedical applications (Mazumder et al., 2016, Abdulla et al., 2021). Biogenic NPs present a range of challenges that must be addressed in order to enable large-scale applications. These challenges include lack of monodispersity, a time-intensive production process, low production rates, and batch-to-batch variations. To synthesize highly stable and well-characterized NPs, it is important to consider critical aspects such as the types of used organisms, their inheritable and genetic properties, optimal conditions for cell

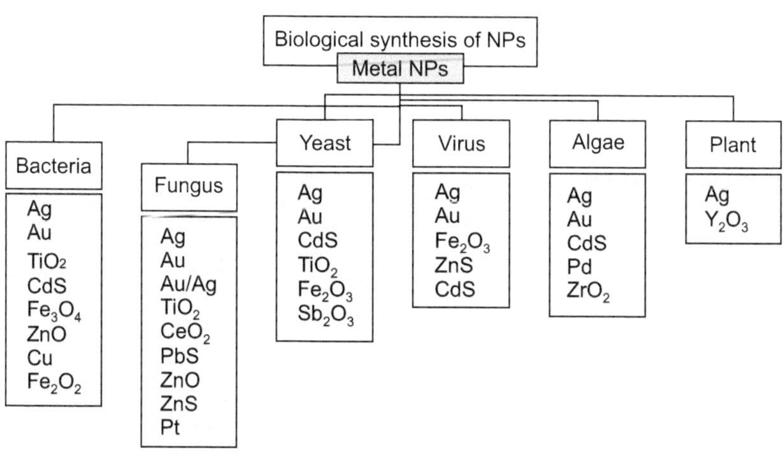

Fig. 2: Different metal NPs according to microbes used for their synthesis

growth and enzyme activity, and optimal reaction conditions. In addition, selecting the appropriate biocatalyst state is also crucial. By taking these factors into account, biological protocols can be optimized to facilitate the synthesis of biogenic NPs with high precision and reproducibility (Ghosh et al., 2021). It is important to keep in mind that most biomedical studies involving microbial NPs have only been conducted *in vitro*. In order to fully realize their potential *in vivo*, large-scale clinical trials and safety tests must be conducted. The potential benefits of microbial NPs in the field of medicine and healthcare are immense and cannot be ignored. This section provides a comprehensive overview of the latest research and developments in the field, emphasizing the potential of biological synthesis in producing NPs for various applications, particularly in the biomedical and pharmaceutical industries. With further in-depth studies, microbial NPs may prove to be a game-changer in improving the health and well-being of people across the globe.

Microorganisms' reduction of toxic metal species into metal NPs also plays a crucial role in detoxifying the immediate cell environment (Murray et al. 2017). In addition, biomolecules secreted by bacteria are utilized as capping and stabilizing agents during the synthesis of NPs. Microorganisms can uniquely manipulate the size and shape of biological NPs, setting them apart from plant-extract-based synthesis methods, which are often polydisperse and subject to seasonal variations in yield. Given these advantages, microorganisms are highly regarded as potential candidates for NPs' synthesis.

3.1.1 Bacteria-mediated Synthesis of Metal NPs

Bacteria that reduce metals are beneficial, as they function as catalysts for bioremediation and material synthesis. Through a biological process known as respiration, certain microorganisms are capable of synthesizing a diverse range of metal oxides. One prominent example is the genus *Shewanella*, which exhibits the unique ability to oxidize organic acids as electron donors and reduce inorganic metals as electron acceptors. This process, known as dissimilatory metal reduction, allows *Shewanella* to generate energy for growth and survival in anaerobic environments. The resulting metal oxides can have important implications in various fields, including environmental remediation and nanotechnology (Harris et al., 2018). Studies have revealed that bacterial strains that survive in harsh environments have the ability to produce metal NPs through biosynthesis. The process of biosynthesis of NPs from microbes is influenced by various factors. Environmental conditions such as reaction time, culture medium, temperature, pH, reactant concentrations and metallic salt to be reduced can govern the composition, size, symmetry, and shape of NPs (Hulkoti et al. 2014, Zhang et al., 2020). Lin et al. (2001) conducted research demonstrating that the reduction process of NPs involves several factors, such as pH, temperature, and the availability of organic functional groups on the cell wall of the bacterium. These factors initiate or induce the reduction of metal ions, which ultimately leads to the formation of NPs. In a recent study conducted by Koul et al. (2021), *Bacillus cereus* was found to be capable of extracellular biofabrication of Ag NPs at an ambient temperature in heavy metals-contaminated soil. The biosynthesized Ag NPs displayed exceptional surface plasmon polariton properties and were found to be useful in a variety of applications. This finding opens up new possibilities for the development of eco-friendly and cost-effective methods for the synthesis of Ag NPs. Another study of Yumei et al. (2017) has clearly demonstrated that pH and temperature have a profound impact on the biosynthesis of Ag NPs using *Arthrobacter* sp. The optimal synthesis of Ag NPs ranging in size from 9-72 nm was achieved at a pH range of 7-8 and temperature of 70°C, with a lower concentration of $AgNO_3$ required at 1 mM. The study also found that the synthesis of Ag NPs occurred only within a narrow pH range of 7-8, and any deviation from this range resulted in the cessation of NP synthesis. Furthermore, incubation at 90°C reduced fabrication time from 10 to 2 minutes. These results underscore the critical importance of optimizing reaction conditions to achieve the desired synthesis of Ag NPs. As an example, the size and site of biogenic Au NPs were reported to be influenced by the pH of the solution. Specifically, the size of the Au NPs produced by *Shewanella algae* was found to be 10-20 nm in periplasmic space and 50-500 nm extracellularly when the pH of the solution was 7.0 and 1, respectively (Konishi et al., 2004). From another side, *Rhodopseudomonas capsulata* produced spherical Au NPs of 10-20 nm diameters when the solution pH was 7.0, while Au nanoplates were formed when the pH of the solution was 4.0 (He et al., 2008).

3.1.2 Fungi and Yeast-mediated Synthesis of Metal NPs

Fungi and yeast have immense potential in the green synthesis of NPs for various applications. Studies have shown that Copper (II) oxide (CuO) NPs synthesized using *Aspergillus niger* (G3-1) have insecticidal properties and promote the growth of wheat grain germination (Badawy et al., 2021). Similarly, nanocomposites containing CuO and ZnO, synthesized using *Penicillium corylophilum* (As-1), have high photocatalytic activity (Fouda et al., 2020). Spherical titanium dioxide (TiO_2) NPs synthesized using Baker's yeast have good photocatalytic and antimicrobial activity, while TiO_2 NPs synthesized using *Trichoderma viride* have pesticidal activity against *Helicoverpa armigera* (Koul et al., 2021, Peiris et al., 2018, Chinnaperumal et al, 2018). Iron oxide NPs synthesized using fungi have a high adsorption capacity towards Cr(VI), and cobalt oxide NPs synthesized using *Aspergillus nidulans* have a promising capacitance in energy storage (Bhargava et al., 2013). These findings demonstrate the potential of fungi and yeast in developing eco-friendly and sustainable solutions for various fields.

3.1.3 Algae-mediated Synthesis of Metal NPs

Algae cells have different secondary metabolites and can synthesize various types of metal NPs. Several studies have reported the biosynthesis of NPs using different species of algae, including AgCl NPs using *Sargassum plagiophyllum* and *Chlorella vulgaris* (da Silva Ferreira et al., 2017), and Ag NPs using *Caulerpa racemosa*, *Laminaria japonica*, and *Portieria hornemannii* (Kim et al., 2018, Bhuyar et al., 2020). The NPs synthesized using the algae extract showed reasonable to strong antimicrobial activity against different pathogenic bacteria.

3.1.4 Virus-mediated Synthesis of Metal NPs

By allowing the templated assembly of millions of identical NPs within living cells, virus-based NPs (VNPs) have become a game changer in the field. Numerous studies have shown that viruses can be successfully used for the biosynthesis of NPs (Raja et al., 2020, Flenniken et al., 2009, Klem et al., 2005, Blum et al., 2004, Sirotkin et al., 2014) that can be applied in targeted drug delivery and cancer treatment (Slocik et al. 2005, Kobayashi et al. 2012). Once established in their hosts, viruses replicate quickly, making the cost-effective manufacturing of VNPs on an industrial scale possible.

VNPs have a high degree of structural control, providing an engineering scaffold that is superior to synthetic particles (Koudelka et al., 2015). They are composed mainly of protein and can be chemically modified by adding conjugates to specific amino acid side chains. VNPs are biocompatible, biodegradable, and efficiently deliver cargo to target cells, making them an ideal choice for various applications. Contrarily to VNPs from mammalian viruses, VNPs derived from bacteriophages and plant viruses are considered safe as they cannot infect humans. Therefore, the medical applications of VNPs based on bacteriophages and plant viruses have gained a lot of attention. Virus capsids, the outer protein shells of viruses, can be utilized as biological nanofactories for synthesizing nanomaterials. This is achieved through the repetitive arrangement of protein subunits to form highly symmetrical architectures that exhibit precise three-dimensional structures, incredible stability and uniformity in their size and shape (Sirotkin et al., 2014, Bothner et al 1998). Each protein subunit that makes up the virus capsid features repeating patterns of amino acid side chains that can direct the nucleation of inorganic materials. This unique feature could create novel hybrid molecules that can be displayed on the interior or exterior surfaces of the capsid (Flenniken et al., 2009, Klem et al., 2005). The approach provides a wide range of sizes, shapes, compositions, and physicochemical properties that can have different applications. For example, the cowpea chlorotic mottle viruses have been used as templates to synthesize AuNPs, which were found to be highly efficient in photothermal therapy (Slocik et al., 2005). Similarly, tobacco mosaic virus (TMV) was used to synthesize uniform-shaped AuNPs (size 5 nm), which were highly effective in cancer treatment (Kobayashi et al., 2012). Nanocarriers synthesized using the red clover necrotic mosaic virus (RCNMV) were successfully tested for the regulated delivery of the drug doxorubicin for the treatment of cancer (Cao et al., 2014). Overall, the utilization of virus capsids in nanotechnology has the potential to revolutionize the field, and further research and development in this area is warranted.

3.2 Biosurfactant-mediated Synthesis of Metal NPs

Biosurfactants are a class of compounds that exhibit immense potential in various industries due to their unique properties (see Table 1). Those with lower molecular weight, such as phospholipids, glycolipids, and low-molecular-weight lipopeptides, are of particular interest as they possess the ability to reduce surface and interfacial tension (Burilova et al., 2018, Pashirova, 2019). Among glycolipids, the most notable ones are rhamnolipids, trehalolipids, mannosylerythritol lipids, cellobiose lipids, (mostly derived from *Pseudomonas*), and sophorolipids derived from *Candida* and related species. In addition, larger biosurfactant molecules such as polysaccharides, lipoproteins, and lipopolysaccharides possess therapeutic potential due to their ability to adhere to surfaces and act as emulsifiers.

Recently, biosurfactants have been successfully used for the synthesis or the stabilization of metal NPs. In a study by Kiran et al. (2010), an unidentified glycolipid produced by a *Brevibacterium casei MSA19* bacterium, found in a marine sponge, was used as a stabilizer in the biosynthesis of Ag NPs under solid-state fermentation. The produced NPs were stable for two months, and the glycolipid acted as a stabilization agent, preventing the formation of aggregates. Another study conducted by Saikia et al. (2013), has shown that rhamnolipid, produced by a *P. aeruginosa* strain, could enhance the stability of colloidal Ag NPs at salt concentrations ranging from 2 to 60 mg NaCl/mL.

Table 1: Different types of biosurfactants used in nanobiotechnology

Biosurfactants		**Microbial strains**	**References**
Glycolipids	Rhamnolipids	*Pseudomonas aeruginosa, Pseudomonas* sp., *Nocardiodies* sp.	Sobrinho et al., 2013
	Sophorolipids	*Candida* spp., *Candida bombicola, Candida magnoliae, Candida apicola, Candida kuoi Candida bogoriensis, Torulopsis bombicola, T. apicola*	Ahmadi-Ashtiani et al., 2020, Sharma et al., 2021, Sobrinho et al., 2013
	Trehalolipids	*Rhodococcus erythropolis, Rhodococcus* sp., *Arthrobacter* spp., *Nocardia* sp., *Corynebacterium* sp., *Mycobacterium* sp., *Actinomycetales*	
Lipopeptides and lipoproteins	Peptide-lipid	*Bacillus licheniformis*	Ahmadi-Ashtiani et al., 2020, Sharma et al., 2021, Sobrinho et al., 2013
	Serrawettin	*Serratia marcescens*	
	Subtilisin	*Bacillus subtilis, Bacillus mojavensis*	
	Surfactin	*Bacillus subtilis*	
	Polymyxin	*Bacillus polymyxa*	
	Gramicidin	*Bacillus brevis*	
	Viscosin	*Pseudomonas fluorescens*	
Phospholipids, neutral lipids and fatty acids	Bile salts	*Myroides* sp.	Sobrinho et al., 2013
	Corynomycolic Acids	*Nocardia erythropolis, Corynebacterium lepus*	Ahmadi-Ashtiani et al., 2020
	Phosphatidyl ethanolamine	*Acinetobacter* spp.	Ahmadi-Ashtiani et al., 2020, Sharma et al., 2021
Particulate surfactant	Vesicles	*Serratia* sp., *Acinetobacter calcoaceticus*	Sobrinho et al., 2013,
	Whole cells	*Acinetobacter calcoaceticus*	Ahmadi-Ashtiani et al., 2020
Polymeric biosurfactants	Emulsan	*Acinetobacter calcoaceticus*	Ahmadi-Ashtiani et al., 2020, Sobrinho et al., 2013
	Liposan	*Candida lipolytica*	
	Mannan lipid protein	*Candida tropicalis Saccharomyces cerevisiae*	
	Biodispersan	*Acinetobacter calcoaceticus*	
	Carbohydrate lipid-protein	*Pseudomonas fluorescens*	Sobrinho et al., 2013
	Alasan	*Acineto radioresistens*	Ahmadi-Ashtiani et al., 2020

Rhamnolipids can be considered the most effectively used biosurfactants in NPs synthesis, providing a more eco-friendly and efficient approach. For example, purified rhamnolipids from the culture supernatant of *P. aeruginosa* were used for synthesizing Ag NPs (Kumar and Mamidyala, 2011) with remarkable antimicrobial effects on both Gram-negative and Gram-positive microbial pathogens (Ganesh et al., 2010). In a study conducted by Palanisamy (2008), commercial rhamnolipids were utilized as stabilizing agents to synthesize nickel oxide nanorods, which exhibited an approximate diameter of 22 nm and a length ranging between 150 and 250 nm. Furthermore, using a similar procedure, Palanisamy and Raichur (2009) obtained spherical nickel oxide NPs with uniform distribution. Similarly, Worakitsiri et al. (2011) used rhamnolipids as a template in the synthesis of polyaniline nano-fibers and nanotubes, resulting in the production of highly crystalline and electrically conductive NPs. A novel method of synthesizing Ag NPs in a water-in-oil microemulsion stabilized by commercial rhamnolipid was attempted by Xie et al. (2006).

Sophorolipids also figure among the most popular biosurfactants used in NPs' production. Sophorolipids derived from oleic acid, have been identified as promising capping and reducing agents in the synthesis of cobalt NPs (Płaza et al., 2014). This breakthrough in biomedical research has resulted in the production of NPs that possess superparamagnetic properties, excellent stability, and an average size of approximately 50 nm. Sophorolipids have demonstrated their potential in surface capping of NPs, forming particles with an average size below 100 nm, which makes them an effective tool for biomedical applications. The sophorose layer coating on the iron oxide NPs further enhances the potential of these NPs in the field of biomedical research. Sophorolipids produced by *Starmerella bombicola* yeast cells from oleic acid have been found to be highly effective as capping and reducing agents in the synthesis of cobalt NPs. The resulting NPs exhibit superparamagnetic properties and excellent stability, making them ideal materials in medicinal and diagnostic applications (Kasture et al., 2008). Larger ZnO particles have been produced by sophorolipids from linoleic acid compared to those produced from oleic acid. These observations highlight the importance of biosurfactant molecular size and distribution in determining the features of NPs formation (Kasture et al., 2007).

Surfactin is another biosurfactant that proved to be an excellent template and stabilizing agent for the synthesis of Ag NPs and Au NPs. Surfactin is the main biosurfactant produced by different *Bacillus* species, mainly *B. subtilis* and *B. amyloliquefaciens* (Ali et al., 2022a,b). For example, Reddy et al. (2009) have used commercial surfactin to stabilize Ag NPs and produce more uniformly shaped NPs, under different pH and temperature conditions. More recently, Singh et al. (2011) have utilized surfactin from *B. amyloliquefaciens*, to synthesize cadmium sulfide (CdS) NPs with remarkable stability. The crude surfactin extract was directly used to synthesize the CdS-NPs, and it played a significant rolc in stabilizing and protecting the NPs. The CdS-NPs coated with surfactin were observed to remain structurally intact for up to six months without any apparent changes. The approach being used has the potential to significantly transform the process of synthesizing stable NPs for various applications.

3.3 Synthesis of Microbial Nano-emulsion

Microbial nano-emulsions are self-aggregated colloidal systems that act as pools to control the properties of NPs. Most NP formulations are effectively based on nanometric-scaled emulsions, so-called nanoemulsions. Nanoemulsified NPs are mainly produced by three different methods, which are a high-energy method, a low-energy spontaneous emulsification method, and a low-energy phase inversion temperature method (Cid, 2018). The major feature of nanoemulsions that makes them prime candidates for NP engineering is the great stability of their droplet suspensions. The stability of an emulsion is hampered in two ways: flocculation by coalescence and Ostwald ripening. In nanoemulsion systems, steric stabilization naturally prevents flocculation, essentially due to the sub-micrometric droplet size. The second one is the reduction of the configurational entropy, which occurs when the inter-droplet distance becomes lower than the adsorbed layer thickness (López-Quintela et al., 1993). Hence, microbial nanoemulsions are prime NP engineering candidates due to their superior stability and unique properties. This method allows control of size, shape, and crystal structure. The same reagents can generate different shapes and sizes of seeds. Varying reaction time, temperature, and conditions can produce different geometries for various research fields (Jianguo et al., 2016). Nanoemulsion-based NPs can be used as delivery systems to enhance the antimicrobial activity of bioactive compounds against microorganisms. Wang et al. (2004) have developed a cost-effective method to synthesize magnetite NPs using a single microemulsion for the first time. They prepared a microemulsion by solubilizing NaOH into a DBS/ethanol/toluene system and added ferric and ferrous salts to the mixture. The magnetite NPs were extracted using magnetic separation and washed in ethanol/water solution. In another study, Wei et al. (2017) have developed a core-shell NP for dual-modal imaging in cancer diagnosis. The NP is activated by acidity and has a responsive MR imaging behavior. In an acidic aqueous solution, $CaCO_3$ melts to produce CO_2 bubbles that can be used to obtain a US signal. The NPs were confirmed to have dual-modal magnetic resonance/ultrasonic imaging capabilities *in vivo*. These findings may lead to NPs developing reactive dual-modal imaging skills. More recently, Tianimoghadam and Salabat (2018) have developed a novel micro-emulsion technique to synthesize thiol-functionalized Au NPs. The results indicated that the synthesized Au NPs were spherical and monodispersed, with 3 to 4 nm diameters. Overall, this novel micro-emulsion technique could offer a promising approach to synthesize thiol-functionalized Au NPs with high purity and monodispersity, which could potentially be applied in various fields such as catalysis, sensing, and biomedical applications.

3.4 Microbial Extracellular Vesicles

NPs such as liposomes, metal-based NPs, and polymers have been frequently utilized as drug carriers. However, the straightforward conjugation of synthetic nanomaterials is not an efficient strategy to simulate the intercellular interactions that are vital for successful NPs' trafficking and delivery (Liu et al., 2021). Conversely, microbial-derived extracellular vesicles (MEVs), which are nanosized particles, hold great promise as a new class of drug delivery carrier due to their capacity to encapsulate and deliver multiple bioactive molecules. Moreover, recent studies have revealed that MEVs can traverse distant organs via systemic circulation in the presence of microbial disorders and an inflammatory environment (Chronopoulos and Kalluri, 2020). MEVs are being used in nano-therapy as an antitumor vaccine or targeting vehicle. For example, extracellular vesicles from *E. coli* were reported to target cancer cells *in vivo* and subsequently induce anti-tumor responses through the production of antitumor cytokines CXCL10 and interferon-γ (Budamagunta et al., 2023, Kim et al., 2017).

Hybrid NPs created by combining MEVs with traditional synthetic materials have shown promise in cancer cell-targeted therapy and targeted delivery of chemotherapeutic agents. MEVs coated with Au NPs have demonstrated enhanced biostability and a faster immune response. Bioengineered MEVs with kinesin spindle protein (KSP) siRNA were able to target and kill cancer cells (Zhuang et al., 2021, Chronopoulos and Kalluri, 2020, Gao et al., 2019, Gao et al., 2014, Gardiner et al., 2013). MEVs can be loaded with bioactive molecules either *in vivo* or *in vitro*. *In vivo*, molecules can be loaded in generated extracellular vesicles by treating cells with those molecules. *In vitro*, genetic engineering has enabled the use of MEVs to deliver specific drugs and agents. Purified extracellular vesicles can be encapsulated with different types of compounds through electroporation (Gujrati et al., 2014). Nevertheless, safety issues related mainly to purification and identification of MEVs are still challenging analytical techniques, and assessing substantial risks.

4. Classification

As described above, the new raised term of nano-biosurfactant describes nano-materials directly produced or mediated by microorganisms with surfactant properties. They include metallic or organic NPs, nano-emulsions, nanocarriers or particulate nanobiosurfactants to designate extracellular vesicles and fimbriae such as in *Acinetobacter* sp. strains (Vecino et al., 2021), or even the whole microbial cell (Chauhan et al., 2022). Extracellular vesicles can be composed of proteins, phospholipids or lipopolysaccharides according to their microorganisms. Fimbriae are of a polymeric protein nature located at the surface of bacterial cells (Mol and Oudega, 1996).

Nanobiosurfactants could be classified according to their chemical compositions, molecular weight, ionic charges or secretion types.

4.1 Chemical Composition

Most commonly, nanobiosurfactants are classified according to the chemical nature of their biosurfactants (glycolipids, lipopeptides and lipoproteins, glycolipopeptides, glycopeptides and glycoproteins, fatty acids and lipids, polymerics) (Vecino et al., 2021) or their generated metal NPs (Iron oxide, Ag, Au, zinc oxide, etc.) (Chauhan et al., 2022). Rhamnolipids and surfactin are among the most used glycolipids and lipopeptides, respectively, in the production of metallic NPs such as Ag, Au and nickel oxide. For example, Ag and iron oxide NPs coated with rhamnolipids (from *Pseudomonas* species) were used for controlling antibiotic-resistant bacteria, thanks to their antiadhesive and antibacterial features (Khalid et al., 2019). Surfactin recovered from culture supernatant of *B. subtilis* was used to study the stabilization of Au NPs (Reddy et al., 2009).

4.2 Molecular Weight

Biosurfactants can be also classified according to their molecular weights. Low molecular weight biosurfactants (LMW) range from simple free fatty acids and phospholipids to amino acids linked to lipids, lipopeptides, and glycolipids (Kubicki et al., 2019). High molecular weight polymeric biosurfactants (HMW) are referred to as bioemulsifiers to distinguish them from low molecular weight metabolites. LMW biosurfactants are more effective in reducing the surface tension of water-air and water-oil interfaces, whereas HMW biosurfactants are more effective in stabilizing water-in-oil emulsions (Vijayakumar and Saravanan, 2015). Emulsan is regarded as the prototypical HMW bioemulsifier consisting of a heteropolysaccharide backbone with a repeating tri-saccharide unit (Zuckerberg et al., 1979). *Acinetobacter calcoaceticus* RAG-1 isolated from the Mediterranean Sea was reported to be an emulsan producer. Amphiphilic proteins such hydrophobins secreted by different filamentous fungi may also be considered as polymeric biosurfactants since they are surface-active (Linder, 2009). According to Cox et al. (2009), hydrophobins represent promising targets for biotechnology.

4.3 Ionic Charges

Based on the ionic charges of their hydrophilic head group, biosurfactants can also be grouped as anionic, cationic or non-ionic. For example, sophorolipids produced mainly by many species of yeasts are known as nonionic biosurfactants with emulsifying, foaming, wetting, and high-detergent capacities (Kim et al., 2002). The ionic nature of biosurfactants is very important in cosmetic safety, to prevent skin damage, eye irritation or adsorption onto hair (Moldes et al., 2020).

4.4 Secretion Type: Intracellular, Extracellular, or Cell-bounded

Biosurfactants used in nanotechnology can be intracellular, extracellular or cell-bound (associated to the plasmatic membrane biosurfactants) (Satpute et al., 2016). While most of biosurfactants are recovered from cell-free culture media and have often glycolipidic and lipopeptidic natures, cell-bound biosurfactants are reported to be usually of glycolipopeptidic and glycoproteinic nature (Moldes et al., 2019). Rhamnolipid, surfactin, cyclic lipopeptides, iturins, fengycins are known to be extracellular biosurfactants (Vecino et al., 2021). Extracellularly produced biosurfactants have occupied surfactant markets to a large extent probably because of their safety and easier extraction and purification.

5. Potential Applications of Nanobiosurfactants

Nanobiosurfactants have gained immense popularity in varied fields, including agriculture, food, environment, medicine and cosmetics as illustrated in Fig. 3. The pharmaceutical and cosmetic industries have benefited immensely from these applications. In this section, we have delved into how these advancements have transformed these industries, and how they continue to shape the future of healthcare and beauty.

5.1 Nanobiosurfactants in the Pharmaceutical Industry

Nanobiosurfactants are promising materials in the pharmaceutical and cosmetic fields due to their unique properties, such as controlled sizes, biocompatible nature, non-toxicity, and antimicrobial properties. These NPs have found various biomedical applications, including but not limited to anti-biofilm agents, drug delivery systems, and antioxidants, anti-cancer, imaging or diagnostic agents, skin care products (Fracchia et al., 2019). In the pharmaceutical industry, nanobiosurfactants have been used to improve the bioavailability of poorly soluble drugs, enhance drug delivery to

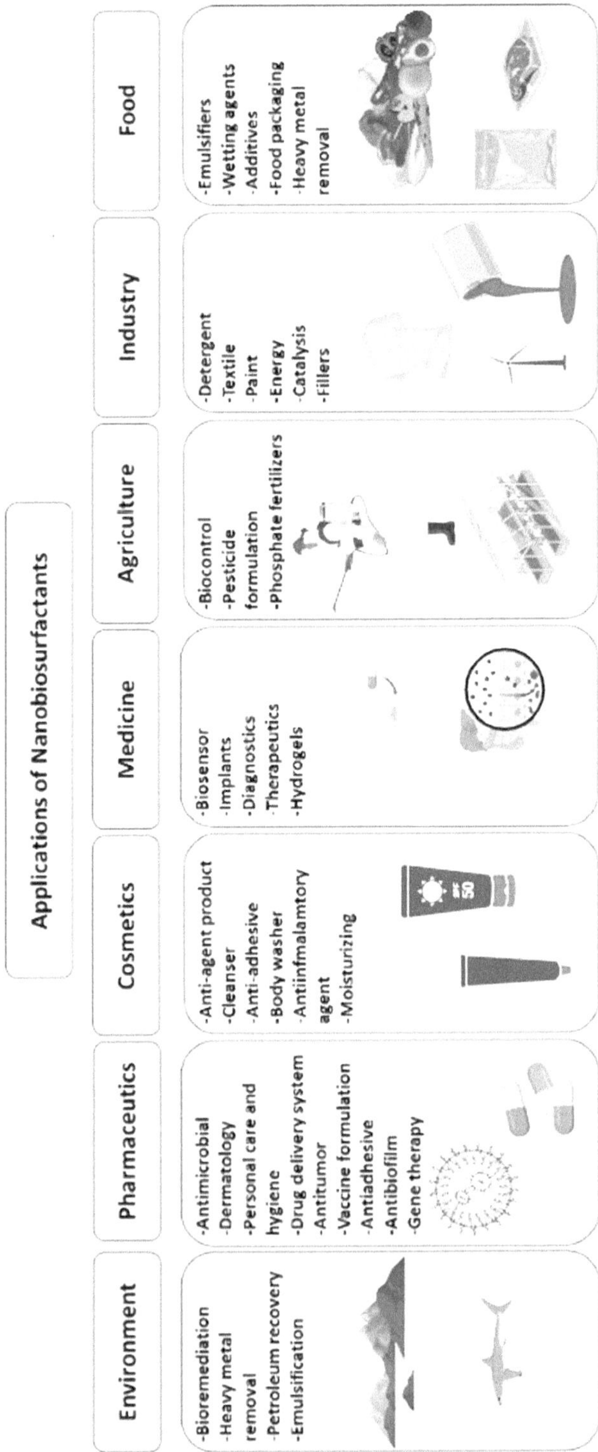

Fig. 3: Different applications of nanobiosurfactants.

targeted cells, prevent microbial infections, and even combat drug-resistant bacteria (Patra et al., 2018).

In the quest to combat drug-resistant bacteria, researchers have found a promising alternative to traditional antibiotics in the form of Aluminum oxide NPs (Al_2O_3 NPs). A recent study has demonstrated that these NPs can be synthesized using *Colletotrichum sp.* and then modified with plant oils from *Citrus medica* and *Eucalyptus globules* to create nano-activated oils (Gudkov et al., 2022). These oils exhibit powerful antimicrobial properties that can effectively eradicate food-spoiling pathogens. Another fascinating research has uncovered the biomedical properties of zinc oxide NPs (ZnO NPs) that can be bio fabricated using *Aspergillus terreus AF1*. The fungus acts as a reducing agent to convert ZnO into ZnO NPs, which are then capped by the proteins secreted by the fungus. The resulting spherical ZnO NPs (size 10–45 nm) exhibit potent antimicrobial activity against various bacteria and demonstrate a reasonable cytotoxic effect against Vero and Caco cell lines. Moreover, when incorporated into cotton fabric, these NPs exhibit moderate antibacterial activity against certain bacteria and are capable of blocking the passage of UV A and UV B rays (Anjum et al., 2021). In yet another study, ZnO NPs of different morphology were synthesized using two fungi, *A. niger* strain (G3-1) and *Fusarium keratoplasticum* strain (A1-3), which were isolated from soil. The nanorod ZnO NPs were found to be more effective against both Gram-positive and Gram-negative bacteria than hexagonal ZnO NPs. The bioactivity of NPs was influenced by their shape and size, which in turn were determined by variations in their physical and chemical properties (Sirelkhatim et al., 2015). Finally, ZnO NPs synthesized using *Xylaria acuta* isolated from the plant *Millingtonia hortensis* L.f. showed remarkable antimicrobial and anticancer properties against a range of bacteria, fungi, and human mammary gland carcinoma cells (Sumanth et al., 2020). Zinc oxide and copper oxide (CuO) NPs have emerged as promising antimicrobial nanomaterials and show efficacy against drug-resistant microbes.

Viruses can be modified via genetic engineering for the synthesis of nanomaterials, such as nanocomposites and nanoconjugates of metal NPs, for the treatment of cancer and targeted drug delivery. Various plant viruses, such as cowpea chlorotic mottle viruses and tobacco mosaic viruses, have been used as templates for the biosynthesis of NPs. These virus-mediated nanocarriers have shown great potential in targeted drug delivery and cancer treatment (Koul et al., 2021). Additionally, biosynthesized HEV NPs could be utilized as multifunctional delivery carriers for tumor imaging, tissue targeting, and therapeutic delivery (Kobayashi et al., 2012).

5.2 Nanobiosurfactants in the Cosmetic Industry

In the cosmetic industry, nanobiosurfactants have been used to create innovative products with unique functional properties, such as emulsification, dispersion, and stabilization of active ingredients. Overall, the potential applications of nanobiosurfactants are vast, and they have the potential to revolutionize the pharmaceutical and cosmetic industries (Adu et al., 2020).

The demand for natural and organic products has led to the use of nanobiosurfactants instead of synthetic compounds in cosmetic formulations. Studies have shown that they are less toxic than synthetic surfactants due to their features, including emulsification, wetting, foaming, cleansing, and surface activities, which make biosurfactants a promising alternative (Moldes et al., 2021). Glycolipids, such as sophorolipids, rhamnolipids and mannosylerythritol lipids, are the most commonly used in cosmetics. In particular, sophorolipids show good skin compatibility and excellent moisturizing properties (Adu et al., 2020). Rhamnolipids and glycolipids have often replaced chemical-based surfactants in cosmetic products as good stabilizing agents of oil-in-water emulsions (Vecino et al., 2015, Bai and McClements, 2016). For instance, Ferreira et al. (2017) have used biosurfactants obtained from *Lactobacillus paracasei* as a stabilizing agent in oil-in-water nano-emulsions containing essential oils and natural antioxidant extract for novel cosmetic formulations. In the last decade, the use of biosurfactants for the synthesis of NPs has experienced an increasing interest in cosmetic applications due to the unique features of the resulting nano-compounds (Shoeb et al., 2013). Sun protection products, such as emulsions and dispersions, are often unstable and can increase cutaneous permeability. To improve safety and efficacy, UV blocking minerals should be well dispersed into the aqueous phase, thus the need for surface-active compounds and emulsifiers. Solid lipid NPs (SLN), nanostructured lipid carriers (NLC), and microsponge delivery systems are currently used as carriers in cosmetic industry. The microsponge system is stable and cost-effective, while SLN and NLC systems act as UV blocking agents themselves and display a synergistic sun-protective effect with TiO_2 and ZnO NPs (Baroli et al, 2007, Rangarajan and Zatz, 2003). Overall, nanotechnology has enhanced the properties of biosurfactants to be used in the cosmetic industry, for better delivery.

6. Challenges and Future Perspectives

Although nanobiosurfactants have seen an astonishing increase in utilization in many areas since the last decade, biosurfactant nanotechnology is still facing some production and cost issues. The current enthusiasm for the use of nanobiosurfactants may be challenged by either general problems related to the biosurfactants industry or specific difficulties regarding the application of these nanomolecules. Below, we have summarized the most relevant issues in the context of nanobiosurfactant technology and discussed potential solutions.

6.1 Production

The drive for sustainable development has pushed industries to move towards biosurfactants as a green alternative. These compounds are still showing several residual problems related to the extraction and downstream process (over 60–80% of the total costs, according to Najmi et al. (2018) and low yield that prevent their imminent commercial exploitation (Ambaye et al., 2021, Prajapati et al., 2021). For example, strains of *P. aeruginosa* produce only 10–20 g/l of rhamnolipids (Saranraj et al., 2022a,b, Kitamoto et al., 2001).

First, the fermentative production of biosurfactants is affected by various factors such as availability of nutrients and their ratio, temperature, inoculum size, agitation speed, aeration, etc. Gurkok (2021) has reviewed factors necessary for the mass production of biosurfactants. The optimization of these parameters could enhance the production yield for a cost-effective product (Ambaye et al., 2021). Moreover, waste substrates could be used in biosurfactant production to promote the valorization of cheaper and renewable materials for better cost reduction (Makkar et al., 2011, Marchant and Banat, 2012). Nanobiosurfactant production presents extra difficulties since it takes technical know-how to set up new technology and conceptualize new products. To address this issue, more research on optimized cost-effective substrates, well-developed statistical models, and engineering solutions, such as microbial strain engineering, can create commercially viable fermentation processes. Although DNA recombinants can represent a potential alternative to optimize and enhance the mass-production of biosurfactants, knowledge about microbe cloning, functional and degradative characterization, and molecular pathways is still required (Satpute et al., 2016). Gaur et al. (2022) have underlined the opportunities in omics approaches for overcoming laborious and time-consuming biosurfactant screening methods and the feasibility of site remediation, for example. Furthermore, omics techniques, such as genome mining, could identify novel biosurfactant molecules.

Second, the extraction and purification of crude biosurfactants is another complex step to achieve cost-effective and large-scale industrial production. This is more difficult, especially for newly identified biosurfactants. The high diversity of microbes represents a potential source of biological materials for NP synthesis as much as it is challenging for their identification, extraction and purification. It is always difficult to find suitable organic solvents when the extraction of new biosurfactants is attempted. Moreover, the recovery of biosurfactants having the same polarity versus organic solvents is still difficult and costly, as mentioned by Gaur et al. (2022). The extraction of intracellularly produced metal NPs from microbes is another ambiguity in nano-biosurfactant technology. Although several methods have been described for microbial cell disintegration to recover metal NPs, this step can represent a challenging technique towards cost and ecological effectiveness of microbes mediated metal NPs' synthesis. For example, Sonkusre et al. (2014) have described a protocol of 4 steps to extract intracellular biogenic selenium NPs based mainly on lysis buffers, centrifugation, organic solvents and ultra-sonication. The same protocol has been also used more recently by Abdollahnia et al. (2020) for the extraction of Ag and selenium NPs from halophilic prokaryotes. Nevertheless, the harsh conditions experienced during the extraction process could induce changes by damaging or introducing exogenous molecules to the NPs that might alter their functionality.

6.2 Identification and Characterization

Several techniques have been described and optimized for the identification and characterization of microbial biosurfactants and their nanoproducts (Jimoh and Lin, 2019), essentially by coupling with mass spectrometry analysis. Thin-Layer Chromatography (TLC) and more recently High-Performance Thin-Layer Chromatography (HPTLC) are the most extensively used techniques for the identification of unknown ingredients in novel biosurfactants (Ambaye et al., 2021). Once the biosurfactant is well identified and characterized, the synthetic pathway could be deciphered and therefore genetic manipulation techniques could be used to produce specific biosurfactants such as the current sophorolipid production by *Candida bombicola* to produce surfactants tailored to meet specific needs (Saerens et al., 2011, Van Bogaert et al., 2009).

In NPs' synthesis with biosurfactants, biochemical and atomic components have to be well characterized to enhance the rate of biosynthesis and to control the features of the produced NPs such as the shape, the size and the crystallinity (Christopher et al., 2019). Although physical properties of NPs could be easily investigated through many techniques such as Electron Microscopy (EM), Dynamic Light Scattering (DLS) and flow cytometry, biochemical analyses are still a laborious task, especially in the identification and characterization of biological organic nanocompounds. To overcome this issue, instead of going through thorough characterization, only pertinent characteristics could be assessed. Buschmann et al. (2021) reviewed pertinent techniques of high relevance in drug delivery for the characterization of extracellular vesicles as particulate nanobiosurfactants. The shape of NPs could be a very important feature in the drug industry. For example, Ravichandran et al. (2016) have reported that nanorods exhibit sevenfold more buildup as compared to nanospheres. Therefore, additional characterization studies are required to create advanced properties in metallic nanobiosurfactants for more specific and distinguished applications in medicine and pharmacology. Functionalized NPs with biosurfactants should be more efficient to target particular diseases (Chauhan et al., 2022). For example, biosurfactants are being used in NPs synthesis to achieve better drug delivery and enhance the retention time in the body (Gupta & Sharma, 2019). After all, accurate standardization and rigorous quality control are required in any novel nanobiosurfactant engineering to reproducibly ensure the purity and the stability of the product.

6.3 Safety

Despite the fact that biosurfactants are known to be safer and even non-toxic compared to chemical surfactants, toxicity concerns of nanobiosurfactants can initially arise from the fact that they can be secreted by pathogenic or potentially pathogenic microbes. The possibility that these microbial compounds may retain harmful traits or induce undesirable effects when they are introduced in biomedical and pharmaceutical applications (Ceresa et al., 2023), cannot be discarded, at least theoretically. Although no major safety or health problem of the currently used or investigated biosurfactants has been reported, at least *in vitro* (Rana et al., 2021, Voulgaridou et al., 2021, Adu et al., 2023), the fact that some biosurfactants are extracted from potential pathogenic microorganisms such as Pseudomonas *aeruginosa* should be considered Cytotoxicity studies conducted on animal models to investigate nanobiosurfactants' *in vivo* toxicity and biological activities are still extremely rare which challenge their therapeutic applications (Smith et al., 2020, Ceresa et al., 2023). For instance, the utilization of rhamnolipids from *P. aeruginosa* was limited, at the time, by the fact that this bacterial species is classified as a class II pathogen in the United Kingdom. Nevertheless, rhamnolipids have been already produced in the USA on a commercial scale, with no reported problems (Marchant and Banat, 2012). However, precautions are needed to avoid allergic reactions and skin irritations (Markande et al. 2021). Although Class II pathogens are not highly infective, large-scale production should employ special measures towards the safety of employees and the environment during the downstream process. Environmental studies of biosurfactants and their products should be more extensive to determine ecological toxic doses of these nanocompounds. Bezerra de Souza Sobrinho et al. (2013) have demonstrated that high concentrations of biosurfactants from *Candida sphaerica* UCP 0995 can inhibit seed germination or root elongation of cabbage seeds.

Nanobiosurfactants may have more specific problems related to their nano-size. Although metallic nanobiosurfactants are reported to be safer as compared to conventional NPs, the reliability of these nanocompounds, on introduction to the human system, is not clearly understood until today. Hazra et al. (2013) didn't notice any change in cell viability of adult rat (*Rattus norvegicus*) hepatocytes for biosurfactant capped zinc sulphide NPs while conventional ones showed dose and time-dependent cytotoxicity. Furthermore, the risk that the nanobiosurfactant may lose its biological activity should be considered. Changes in the protein on encountering NPs can lead to its dysfunction (Chauhan et al., 2022) or target modification.

In any case, the utilization of nanobiosurfactants requires more intensive research for a better understanding about their biological and ecological safety before being considered as "eco-friendly nano-molecules".

7. Conclusion

It has been known for years that biosurfactants are diverse compounds derived from microorganisms with unique properties and are useful in various industries. These biosurfactants have shown remarkable potential in biomedical research specifically, due to their biocompatibility, low toxicity and specificity. With recent advances in nanobiotechnology, biosurfactant-based microbial engineering has experimented with new perspectives. These molecules have been used for the production of either metal or organic NPs, called nanobiosurfactants, with remarkable stability and several biological activities such as antimicrobial efficiency, anti-oxidant effects and anticancer abilities. The research hotspot of biosurfactant utilization in nanobiotechnology is still the synthesis of metal NPs which allows greater control of the size and shape of the produced NPs. Moreover, nanobiosurfactants are considered good alternatives for many chemical nano-sized products as they are low in toxicity and sustainable products making them the molecules of the 21st century for various applications, mainly targeting drug delivery and cancer treatment. Although nanobiosurfactants present various challenges that must be addressed to enable large-scale applications, enhance monodispersity and production rates, and decrease batch-to-batch variations, these nano-molecules continue to offer advantages over traditional methods in terms of efficiency, low toxicity, biocompatibility, and sustainability.

References

Abdel-Razek, A.S., El-Naggar, M.E., Allam, A., Morsy, O.M. & Othman, S.I. (2020). Microbial natural products in drug discovery. *Processes*, 8: 470. https://doi.org/10.3390/pr8040470

Abdollahnia, M., Makhdoumi, A., Mashreghi, M. & Eshghi, H. (2020) Exploring the potentials of halophilic prokaryotes from a solar saltern for synthesizing NPs: The case of silver and selenium. *PLoS ONE*, 15(3): e0229886. https://doi. org/10.1371/journal.pone.0229886

Abdulla, N.K., Siddiqui, S.I., Fatima, B., Sultana, R., Tara, N. & Hashmi, A.A. (2021). Silver-based hybrid nanocomposite: A novel antibacterial material for water cleansing. *J. Clean. Prod.*, 284: 124746. 10.1016/j.jclepro.2020.124746

Adu, S.A., Twigg, M.S., Naughton, P.J., Marchant, R. & Banat, I.M. (2023). Characterisation of cytotoxicity and immunomodulatory effects of glycolipid biosurfactants on human keratinocytes. *Appl. Microbiol. Biotechnol.*, 107: 137-152.

Adu, Simms A., Patrick J. Naughton, Roger Marchant & Ibrahim M. Banat. (2020). Microbial biosurfactants in cosmetic and personal skincare pharmaceutical formulations. *Pharmaceutics*, 12(11): 1099. https://doi.org/10.3390/pharmaceutics12111099

Ahmad, R., Mohsin, M., Ahmad, T. & Sardar, M. (2015). Alpha amylase assisted synthesis of TiO_2 NPs: Structural characterization and application as antibacterial agents. *J. Hazard. Mater.*, 283: 171-177. 10.1016/j.jhazmat.2014.08.073

Ahmadi-Ashtiani, H.R., Baldisserotto, A., Cesa, E., Manfredini, S., Sedghi Zadeh, H., Ghafori Gorab, M.G. et al. (2020). Microbial biosurfactants as key multifunction ingredients for sustainable cosmetics. *Cosmetics*, 7(2): 46. https://doi.org/10.3390/cosmetics7020046

Ali, S.A.M., Sayyed, R.Z., Mir, M.I., Hameeda, B., Khan, Y., Alkhanani, M.F. et al. (2022a). Induction of systemic resistance in maize and antibiofilm activity of surfactin from *Bacillus velezensis* MS20. *Front. Microbiol.*, 13: 879739. https://doi.org/10.3389/fmicb.2022.879739

Ali, S.A.M., Sayyed, R.Z., Reddy, M.S., Enshasy, H.E. & Hameeda, B. (2022b). Delving through quorum sensing and CRISPRi strategies for enhanced surfactin production. *In:* Sayyed, R.Z. (Ed.), *Biosurfactants: Production and Applications in Bioremediation/Reclamation*. 59-79. CRC Press, Taylor & Francis Group, USA.

Al-Wahaibi, Y., Joshi, S., Al-Bahry, S., Elshafie, A., Al-Bemani, A. & Shibulal, B. (2014). Biosurfactant production by Bacillus subtilis B30 and its application in enhancing oil recovery. *Colloids Surf. B Biointerfaces*, 114: 324-333. 10.1016/j.colsurfb.2013.09.022

Ambaye, T.G., Vaccari, M., Prasad, S. & Rtimi, S. (2021). Characterization and application of biosurfactant in various industries: A critical review on progress, challenges and perspectives. *Environ. Technol. Innov.*, 24. 10.1016/j.eti.2021.102090

Anjum, S., Hashim, M., Malik, S.A., Khan, M., Lorenzo, J.M., Abbasi, B.H. et al. (2021). Recent advances in zinc oxide NPs (ZnO NPs) for cancer diagnosis, target drug delivery, and treatment. *Cancers* (Basel), 13(18): 4570. doi: 10.3390/cancers13184570. PMID: 34572797; PMCID: PMC8468934

Badawy, A.A., Abdelfattah, N.A., Salem, S.S., Awad, M.F. & Fouda, A. (2021). Efficacy assessment of biosynthesized copper oxide NPs on stored grain insects and their impacts on morphological and physiological traits of wheat (*Triticum aestivum* l.) plant. *Biology*, 10: 233. doi: 10.3390/biology10030233

Bai, L. & McClements, D.J., 2016. Formation and stabilization of nanoemulsions using biosurfactants: Rhamnolipids. *J. Colloid Interfaces Sci.*, 479: 71.

Banat, I.M., Franzetti, A., Gandolfi, I., Bestetti, G., Martinotti, M.G., Fracchia, L. et al. (2010).. Microbial biosurfactants production, applications and future potential. *Appl. Microbiol. Biotechnol.*, 87: 427-444.

Baroli, B., Ennas, M.G., Loffredo, F., Isola, M., Pinna, R. & Lopez-Quintela, M.A. (2007). Penetration of metallic nanoparticles in human full-thickness skin. *J. Invest. Dermatol.*, 127: 1701-1712.

Bettencourt, A.F., Tomé, C. & Oliveira, T. (2021). Exploring the potential of chitosan-based particles as delivery-carriers for promising antimicrobial glycolipid biosurfactants. *Carbohydr Polym.*, 254: 1-12.

Bezerra de Souza Sobrinho, H., de Luna, J.M., Rufino, R.D., Figueiredo Porto, A.L. & Sarubbo, L.A. (2013). Assessment of toxicity of a biosurfactant from Candida sphaerica UCP 0995 cultivated with industrial residues in a bioreactor. *Electron. J. Biotechnol.*, 16: 4. 10.2225/vol16-issue4-fulltext-4

Bhargava, A., Jain, N., Barathi, M., Akhtar, M.S., Yun, Y.-S. & Panwar, J. (2013). Synthesis, characterization and mechanistic insights of mycogenic iron oxide nanoparticles. *J. Nanopart. Res.*, 15: 2031. doi: 10.1007/s11051-013-2031-5.

Bhuyar, P., Rahim, M.H.A., Sundararaju, S., Ramaraj, R., Maniam, G.P. & Govindan, N. (2020). Synthesis of silver NPs using marine macroalgae padina sp. and its antibacterial activity towards pathogenic bacteria. *Beni-Suef Univ. J. Basic Appl. Sci.*, 9: 3. doi: 10.1186/s43088-019-0031-y.

Bjerk, T.R., Severino, P., Jain, S., Marques, C., Silva, A.M., Pashirova, T. et al. (2021). Biosurfactants: Properties and applications in drug delivery, biotechnology and ecotoxicology. *Bioengineering* (Basel). 8(8): 115. doi: 10.3390/bioengineering8080115.

Blum, A.S., Soto, C.M., Wilson, C.D., Cole, J.D. & Kim, M. (2004). Cowpea mosaic virus as a scaffold for 3-D patterning of gold NPs. *Nano Lett.*, 4(5): 867-870.

Bothner, B., Dong, X.F., Bibbs, L., Johnson, J.E. & Siuzdak, G. 1998. Evidence of viral capsid dynamics using limited proteolysis and mass spectrometry. *J. Biol. Chem.*, 273(2): 673-676.

Bouri, M., Mehnaz, S. & Şahin, F. (2022). Extreme environments as potential sources for PGPR. *In:* Sayyed, R.Z., Uarrota, V.G. (Eds), *Secondary Metabolites and Volatiles of PGPR in Plant-Growth Promotion*. Springer, Cham. https://doi.org/10.1007/978-3-031-07559-9_12

Budamagunta, V., Shameem, N., Irusappan, S., Parray, J.A., Thomas, M., Marimuthu, S. et al. (2023). Microbial nanovesicle and extracellular polymeric substances mediate nickel tolerance and remediate heavy metal ions from soil. *Environmental Research*, 15: 114997, https://doi.org/10.1016/j.envres.2022.114997

Burilova, E.A., Pashirova, T.N., Lukashenko, S.S., Sapunova, A.S., Voloshina, A.D., Zhiltsova E.P. et al. (2018). Synthesis, biological evaluation and structure-activity relationships of self-assembled and solubilization properties of amphiphilic quaternary ammonium derivatives of quinuclidine. *J. Mol. Liq.*, 272: 722-730. doi: 10.1016/j.molliq.2018.10.008

Buschmann, D., Mussack, V. & Byrd, J.B. (2021). Separation, characterization, and standardization of extracellular vesicles for drug delivery applications. *Adv. Drug Deliv Rev.*, 174: 348-368. doi: 10.1016/j.addr.2021.04.027.

Cao, J., Guenther, R.H., Sit, T.L., Opperman, C.H., Lommel, S.A. & Willoughby, J.A. (2014). Loading and release mechanism of red clover necrotic mosaic virus derived plant viral NPs for drug delivery of doxorubicin. *Small.*, 10: 5126-5136. doi: 10.1002/smll.201400558.

Ceresa, C., Fracchia, L., Sansotera, A.C., De Rienzo, M.A.D. & Banat, I.M. (2023). Harnessing the potential of biosurfactants for biomedical and pharmaceutical applications. *Pharmaceutics*, 15: 2156. https://doi.org/10.3390/pharmaceutics15082156

Chauhan, A., Sangwan, N. & Avti, P.K. (2022). Biosurfactants role in nanotechnology for anticancer treatment, (Chapter 20). *In:* Inamuddin, Charles Oluwaseun Adetunji, Mohd Imran Ahamed (Eds), *Green Sustainable Process for Chemical and Environmental Engineering and Science*. 375-395. Academic Press. https://doi.org/10.1016/B978-0-323-85146-6.00011-5

Chen, X. & Mao, S.S. (2007). Titanium dioxide nanomaterials: Synthesis, properties, modifications, and applications. *Chem. Rev.*, 107: 2891-2959. 10.1021/cr0500535

Chinnaperumal, K., Govindasamy, B., Paramasivam, D., Dilipkumar, A., Dhayalan, A., Vadivel, A. et al. (2018). Bio-pesticidal effects of trichoderma viride formulated titanium dioxide nanoparticle and their physiological and biochemical changes on helicoverpa armigera (hub.). *Pestic Biochem. Physiol.*, 149: 26-36. doi: 10.1016/j.pestbp.2018.05.005

Christopher, F.C., Ponnusamy, S.K., Ganesan, J.J. & Ramamurthy, R. (2019). Investigating the prospects of bacterial biosurfactants for metal nanoparticle synthesis – A comprehensive review. *IET Nanobiotechnology*, 13: 243-249. https://doi.org/10.1049/iet-nbt.2018.5184

Chronopoulos, A. & Kalluri, R. (2020). Emerging role of bacterial extracellular vesicles in cancer. *Oncogene*, 39(46): 6951-6960.

Chuo, S.C., Mohd-Setapar, S.H., Ahmad, A. & Khatoon, A. (2021). Biosurfactants based nano micelles for extraction of biomolecules. *In:* Sarma, H., Joshi, S.J., Prasad, R., Jampilek, J. (Eds), *Biobased Nanotechnology for Green Applications. Nanotechnology in the Life Sciences*. Springer, Cham. https://doi.org/10.1007/978-3-030-61985-5_15

Cid, A. (2018). Synthesis of NPs by Microemulsion Method. IntechOpen. doi: 10.5772/intechopen.80633

Cox, P.W. & Hooley, P. (2009). Hydrophobins: New prospects for biotechnology. *Fungal Biol. Rev.*, 23: 40-47.

da Silva Ferreira, V., ConzFerreira, M.E., Lima, L.M.T.R., Frasés, S., de Souza, W. & Sant'Anna, C. (2017). Green production of microalgae-based silver chloride NPs with antimicrobial activity against pathogenic bacteria. *Enzym. Microb. Technol.*, 97: 114-121. doi: 10.1016/j.enzmictec.2016.10.018.

Debnath, M., Chauhan, N., Sharma, P. & Tomar, I. (2021). Potential of nano biosurfactants as an ecofriendly green technology for bioremediation. *Handbook of Nanomaterials for Wastewater Treatment*, 1039-1055. Jaipur: Elsevier, doi: 10.1016/B978-0-12-821496-1.00013-1

Dhanker, R., Hussain, T., Tyagi, P., Singh, K.J. & Kamble, S.S. (2021). The emerging trend of bio-engineering approaches for microbial nanomaterial synthesis and its applications. *Front Microbiol.*, 12: 638003. doi: 10.3389/fmicb.2021.638003

Ferreira, A., Vecino, X., Ferreira, D., Cruz, J.M., Moldes, A.B. & Rodrigues, L.R. (2017). Novel cosmetic formulations containing a biosurfactant from Lactobacillus paracasei. *Colloids and Surfaces B: Biointerfaces*, 155: 522-529. doi:10.1016/j.colsurfb.2017.04.026

Flenniken, M.L., Uchida, M., Lipold, L., Kang, S., Young, M.J. & Douglas, T. (2009). A library of protein cage architectures as nanomaterials. *Curr. Top. Microbiol. Immunol.*, 327: 71-73.

Fouda, A., Salem, S.S., Wassel, A.R., Hamza, M.F. & Shaheen, T.I. (2020;). Optimization of green biosynthesized visible light active cuo/zno nano-photocatalysts for the degradation of organic methylene blue dye. *Heliyon*, 6: e04896. doi: 10.1016/j.heliyon.2020.e04896

Fracchia, L., Ceresa, C. & Banat, I.M. (2019). Biosurfactants in cosmetic, biomedical and pharmaceutical industry. *In:* Banat, I.M., Thavasi R. (Eds), *Microbial Biosurfactants and Their Environmental and Industrial Applications*. 258-288. CRC Press, Boca Raton, FL, USA.

Franzetti, A., Gandolfi, I., Bestetti, G. & Banat, I.M. (2011). (Bio)surfactant and bioremediation, successes and failures. *In:* Płaza, G. (Ed.), *Trends in Bioremediation and Phytoremediation*. 145-156. Research Signpost: Kerala, India.

Ganesh, K.C., Mamidyala, S.K., Das, B., Sridhar, B., Devi, G.S. & Karuna, M.L. (2010). Synthesis of biosurfactant-based silver NPs with purified rhamnolipids isolated from Pseudomonas aeruginosa BS-161R. *J. Microbiol. Biotechnol.*, 20: 1061-1068.

Gao, F., Xu, L., Yang, B., Fan, F. & Yang, L. (2019). Kill the real with the fake: Eliminate intracellular Staphylococcus aureus using nanoparticle coated with its extracellular vesicle membrane as active-targeting drug carrier. *ACS Infect. Dis.*, 5(2): 218-227.

Gao, W., Fang, R.H., Thamphiwatana, S., Luk, B.T., Li, J., Angsantikul, P. et al. (2014). Modulating antibacterial immunity via bacterial membrane-coated nanoparticles. *Nano Lett.*, 15(2): 1403-1409.

Gardiner, C., Ferreira, Y.J., Dragovic, R.A., Redman, C.W. & Sargent, I.L. (2013). Extracellular vesicle sizing and enumeration by nanoparticle tracking analysis. *J. Extracell. Vesicles*, 2.

Gaur, Vivek K., Sharma, P., Gupta, S., Varjani, S., Srivastava, J.K. et al. 2022. Opportunities and challenges in omics approaches for biosurfactant production and feasibility of site remediation: Strategies and advancements. *Environ. Technol. Innov.*, 25, 102132, 10.1016/j.eti.2021.102132

Ghosh, S., Ahmad, R., Zeyaullah, M. and Khare, S.K. (2021). Microbial nano-factories: Synthesis and biomedical applications. *Front. Chem.*, 9: 626834. doi: 10.3389/fchem.2021.626834

Graham, A.E. & Ledesma-Amaro, R. (2023). The microbial food revolution. *Nat. Commun.*, 14: 2231. https://doi.org/10.1038/s41467-023-37891-1

Gudkov, S.V., Burmistrov, D.E., Smirnova, V.V., Semenova, A.A., Lisitsyn, A.B. et al. (2022). A mini review of antibacterial properties of Al_2O_3 NPs. *Nanomaterials* (Basel), 12(15): 2635. doi: 10.3390/nano12152635. PMID: 35957067; PMCID: PMC9370748.

Gujrati, V., Kim, S., Kim, S.H., Min, J.J., Choy, H.E. et al. (2014). Bioengineered bacterial outer membrane vesicles as cell-specific drug-delivery vehicles for cancer therapy. *ACS Nano*, 8(2): 1525-1537.

Gupta, R. & Sharma, D. (2019). Biofunctionalization of magnetite NPs with stevioside: Effect on the size and thermal behaviour for use in hyperthermia applications. *Int. J. Hyperther.*, 36(1): 301-311.

Gupta, S. (2021). Biosurfactant-based antibiofilm nano materials. *In:* Hemen Sarma, Majeti Narasimha and Vara Prasad (Eds), *Biosurfactants for a Sustainable Future*. 269-292. Wiley, USA. https://doi.org/10.1002/9781119671022.ch12

Gurkok, S. (2021). Important parameters necessary in the bioreactor for the mass production of biosurfactants. *Green Sustainable Process for Chemical and Environmental Engineering and Science*. Chapter 7, pp. 347-365. Elsevier. doi: 10.1016/B978-0-12-823380-1.00020-4

Harris, H.W., Sanchez-Andrea, I., Mclean, J.S., Salas, E.C., Tran, W. et al. (2018). Redox sensing within the genus Shewanella. *Front. Microbiol.*, 8: 2568. doi: 10.3389/fmicb.2017.02568

Hazra, C., Kundu, D., Chaudhari, A. & Jana, T. (2013). Biogenic synthesis, characterization, toxicity and photocatalysis of zinc sulfide NPs using rhamnolipids from Pseudomonas aeruginosa BS01 as capping and stabilizing agent. *Journal of Chemical Technology and Biotechnology*, 88: 1039-1048. doi: 10.1002/jctb.3934

He, S., Zhang, Y., Guo, Z. & Gu, N. (2008). Biological synthesis of gold nanowires using extract of rhodopseudomonas capsulata. *Biotechnol. Prog.*, 24: 476-480. doi: 10.1021/bp0703174.

Hosein Saeedi, L., Mazaheri Assadi, M., Mohammad Heydarian, S. & Jahangiri, M. (2014). The production and evaluation of a nano-biosurfactant. *Petroleum Science and Technology*, 32(2): 125-132, doi: 10.1080/10916466.2011.574176

Hulkoti, N.I. & Taranath, T. (2014). Biosynthesis of NPs using microbes—A review. *Colloids Surf. B Biointerfaces*, 121: 474-483. doi: 10.1016/j.colsurfb.2014.05.027.

Husseiny, M.I., El-Aziz, M.A., Badr, Y. & Mahmoud, M.A. (2007). Biosynthesis of gold NPs using Pseudomonas aeruginosa. *Spectrochimica Acta Part A: Molecular and Biomolecular Spectroscopy*, 67(3-4): 1003-1006.

Jafari, S.M., He, Y. & Bhandari, B. (2007). Optimization of nano-emulsions production by microfluidization. *Eur. Food Res. Technol.*, 225: 733-741. https://doi.org/10.1007/s00217-006-0476-9

Jahan, R., Bodratti, A.M., Tsianou, M. & Alexandridis, P. (2020). Biosurfactants, natural alternatives to synthetic surfactants: Physicochemical properties and applications. *Adv. Colloid Interface Sci.*, 275: 102061. doi: 10.1016/j.cis.2019.102061.

Ji, Y., Wang, Z., Ju, X., Deng, F. & Yang, F. (2023). Co-encapsulation of rutinoside and β-carotene in liposomes modified by rhamnolipid: Antioxidant activity, antibacterial activity, storage stability, and in vitro gastrointestinal digestion. *Journal of Food Science*, 88(5): 2064-2077.

Jianguo, F., Yali, S., Qianyao, Y., Chencheng, S. & Guantian, Y. (2016). Effect of emulsifying process on stability of pesticide nanoemulsions. *Colloid Surface A*, 497: 286e92.

Jimoh, A.A. & Lin, J. (2019). Biosurfactant: A new frontier for greener technology and environmental sustainability. *Ecotoxicol. Environ. Saf.*, 184: 10.1016/j.ecoenv.2019.109607

Kalabegishvili, T.L., Kirkesali, E.I., Rcheulishvili, A.N., Ginturi, E.N. & Murusidze, I.G. (2012). Synthesis of gold NPs by some strains of arthrobacter genera. *J. Mater. Sci. Eng. A Struct. Mater. Prop. Microstruct. Process.*, 2: 164-173.

Kasture, M., Singh, S., Patel, P., Joy, P.A. & Prabhune, A.A. (2007). Multiutility sophorolipids as nanoparticle capping agents: Synthesis of stable and water dispersible Co NPs. *Langmuir*, 23: 11409-11412.

Kasture, M.B., Patel, P., Prabhune, A.A., Ramana, C.V. & Kulkarni, A.A. (2008). Synthesis of NPs by sophorolipids: Effects of temperature and sophorolipid structure on size of particles. *J. Chem. Sci.*, 120: 515-520.

Khalid, H.F., Tehseen, B., Sarwar, Y., Hussain, S.Z., Khan, W.S. et al. (2019). Biosurfactant coated silver and iron oxide NPs with enhanced antibiofilm and antiadhesive properties. *J. Hazard. Mater.*, 364: 441-448.

Khatoon, N., Ahmad R. & Sardar M. (2015). Robust and fluorescent silver NPs using Artemisia annua: Biosynthesis, characterization and antibacterial activity. *Biochem. Eng. J.*, 102: 91-97. 10.1016/j.bej.2015.02.019

Kim, D.-Y., Saratale, R.G., Shinde, S., Syed, A., Ameen, F. & Ghodake, G. (2018). Green synthesis of silver NPs using Laminaria japonica extract: Characterization and seedling growth assessment. *J. Clean. Prod.*, 172: 2910-2918. doi: 10.1016/j.jclepro.2017.11.123.

Kim, K., Yoo, D., Kim, Y., Lee, B., Shin, D., & Kim, E.-K. (2002). Characteristics of sophorolipid as an antimicrobial agent. *J. Microbiol. Biotechnol.*, 12: 235-241.

Kim, O.Y., Park, H.T., Dinh, N.T.H., Choi, S.J., Lee, J., Kim, J.H. et al. (2017). Bacterial outer membrane vesicles suppress tumor by interferon-γ-mediated antitumor response. *Nat. Commun.*, 8(1): 626.

Kim, T.-Y., Kim, M.G., Lee, J.-H. & Hur, H.-G. (2018). Biosynthesis of nanomaterials by Shewanella species for application in lithium-ion batteries. *Front. Microbiol.*, 9: 2817. doi: 10.3389/fmicb.2018.02817

Kiran, G.S., Sabu, A. & Selvin, J. (2010). Synthesis of silver NPs by glicololid biosurfactant produced from marine Brevibacterium casei MSA 19. *J. Biotechnol.*, 148: 221-225.

Kitamoto, D., Ikegami, T. & Suzuki, G.T. (2001). Microbial conversion of n-alkanes into glycolipid biosurfactants, mannosylerythritol lipids, by Pseudozyma (Candida antarctica). *Biotechnology Letters*, 23: 1709-1714. https://doi.org/10.1023/A:1012464717259

Klem, M.T., Willits, D., Solis, D.J., Belcher, A.M., Young, M., & Douglas, T. (2005). Bio-inspired synthesis of protein-encapsulated CoPt NPs. *Adv. Funct. Mater.*, 15: 1489-1494.

Kobayashi, M., Tomita, S., Sawada, K., Shiba, K., Yanagi, H., Yamashita I. et al. (2012). Chiral meta-molecules consisting of gold NPs and genetically engineered tobacco mosaic virus. *Opt. Express.*, 20: 24856-24863. doi: 10.1364/OE.20.024856.

Konishi, Y., Ohno, K., Saitoh, N., Nomura, T. & Nagamine S. (2004). Microbial synthesis of gold NPs by metal reducing bacterium. *Trans. Mater. Res. Soc. Jpn.*, 29: 2341-2343.

Koudelka, K.J., Pitek, A.S., Manchester, M. & Steinmetz, N.F. (2015). Virus-based NPs as versatile nanomachines. https://doi.org/10.1146/annurev-virology-100114-055141

Koul, B., Poonia, A.K., Yadav, D. & Jin, J.O. (2021). Microbe-mediated biosynthesis of NPs: Applications and future prospects. *Biomolecules*, 11(6): 886. doi: 10.3390/biom11060886. PMID: 34203733; PMCID: PMC8246319.

Kubicki, S., Bollinger, A., Katzke, N., Jaeger, K.-E., Loeschcke, A. & Thies, S. (2019). Marine biosurfactants: Biosynthesis, structural diversity and biotechnological applications. *Mar. Drugs*, 17: 408. https://doi.org/10.3390/md17070408

Kuiper, I. (2004). Characterization of two Pseudomonas putida lipopeptide biosurfactants, putisolvin I and II, which inhibit biofilm formation and break down existing biofilms. *Molecular Microbiology*, 51: 97-113. doi: 10.1046/j.1365-2958.2003.03751.x.

Kumar, G.G. & Mamidyala, S.K. (2011). Extracellular synthesis of silver NPs using culture supernatant of Pseudomonas aeruginosa. *Colloids Surf. B Biointerfaces*, 84: 462-466.

Lin, Z.Y., Fu, J.K., Wu, J.M., Liu, Y.Y. & Cheng, H. (2001). Preliminary study on the mechanism of nonenzymatic bioreduction of precious metal ions. *Acta Physico-Chimica Sinica*, 17(5): 477-480. https://doi.org/10.3866/PKU.WHXB20010520

Linder, M.B. (2009). Hydrophobins: Proteins that self assemble at interfaces. *Curr. Opin. Colloid Interface Sci.*, 14: 356-363.

Liu, H., Zhang, Q., Wang, S., Weng, W., Jing, Y. & Su, J. (2021). Bacterial extracellular vesicles as bioactive nanocarriers for drug delivery: Advances and perspectives. *Bioact Mater.*, 14: 169-181. doi: 10.1016/j.bioactmat.2021.12.006. PMID: 35310361; PMCID: PMC8892084.

López-Quintela, M.A. & Rivas, J. (1993). Chemical reactions in microemulsions: A powerful method to obtain ultrafine particles. *Journal of Colloid and Interface Science*, 158: 446-451.

Makkar, R.S. (2011). Advances in utilization of renewable substrates for biosurfactant production. *AMB Express*, 1: 5.

Marchant, R. & Banat, I.M. (2012). Microbial biosurfactants: Challenges and opportunities for future exploitation. *Trends Biotechnol.*, 30: 558-565.

Markande, A.R., Patel, D. & Varjani, S. (2021). A review on biosurfactants: Properties, applications and current developments. *Bioresour. Technol.*, 330: 124963.

Mazumder, J.A., Ahmad, R. & Sardar, M. (2016). Reusable magnetic nanocatalyst for the synthesis of silver and gold NPs. *Int. J. Biol. Macromol.*, 93: 66-74. 10.1016/j.ijbiomac.2016.08.073

Microbial technologies for environmental remediation: Potential issues, challenges, and future prospects. Microbe Mediated Remediation of Environmental Contaminants. INC, Elsevier, Charlotte Cockle, Duxford, United Kingdom (2020), pp. 271-286, 10.1016/B978-0-12-821199-1.00022-5

Mol, O. & Oudega, B. (1996). Molecular and structural aspects of fimbriae biosynthesis and assembly in Escherichia coli. *FEMS Microbiology Reviews*, 19(1): 25-52, https://doi.org/10.1111/j.1574-6976.1996.tb00252.x

Moldes, A.B., Vecino, X., Rodríguez-López, L., Rincón-Fontán, M. & Cruz, J.M. (2019). Glycoprotein and lipopeptide biosurfactants production, properties and applications. In:

Banat, I.M., Thavasi, R. (Eds.), *Microbial Biosurfactants and Their Environmental and Industrial Applications*. 106-128. Taylor & Francis Group.

Moldes, A.B., Rodríguez-López, L., Rincón-Fontán, M., López-Prieto, A., Vecino, X. & Cruz, J.M. (2021). Synthetic and bio-derived surfactants versus microbial biosurfactants in the cosmetic industry: An overview. *Int J Mol Sci.*, 22(5): 2371. doi: 10.3390/ijms22052371.

Moldes, A., Vecino, X., Rodríguez-López, L., Rincón-Fontán, M. & Cruz, J.M. (2020). Biosurfactants: The use of biomolecules in cosmetics and detergents. *New and Future Developments in Microbial Biotechnology and Bioengineering*, 163-185. doi:10.1016/b978-0-444-64301-8.00008-1

Morikawa, M., Hirata, Y. & Imanaka, T. (2000). A study on the structure–function relationship of the lipopeptide biosurfactants. *Biochim. Biophys. Acta.*, 1488: 211-218.

Mulligan, C.N., Yong, R.N. & Gibbs, B.F. (2001). Heavy metal removal from sediments by biosurfactants. *J. Hazard Mater*, 85: 111-125.

Murray, A.J., Zhu, J., Wood, J. & Macaskie, L.E. (2017). A novel biorefinery: Recovery of precious metals from spent automotive catalyst leachates into new catalysts effective in metal reduction and hydrogenation of 2-pentyne. *Miner. Eng.*, 113: 102-108. doi: 10.1016/j.mineng.2017.08.011

Najmi, Z., Ebrahimipour, G., Franzetti, A. & Banat, I.M. (2018). In situ downstream strategies for cost-effective bio/surfactant recovery. *Biotechnol. Appl. Biochem.*, 65: 523-532.

Niaz, T., Shabbir, S. & Noor, T. (2019). Antimicrobial and antibiofilm potential of bacteriocin loaded nanovesicles functionalized with rhamnolipids against foodborne pathogens. *LWT Food Sci Technol.*, 116: 108513-108513.

Palanisamy, P. (2008). Biosurfactant mediated synthesis of NiO nanorods. *Mater. Lett.*, 62: 743-746.

Palanisamy, P. & Raichur, A.M. (2009). Synthesis of spherical NiO NPs through a novel biosurfactant mediated emulsion technique. *Mater. Sci. Eng.*, 29: 199-204.

Pashirova, T.N., Sapunova, A.S., Lukashenko, S.S., Burilova, E.A., Lubina, A.P., Shaihutdinova Z.M. et al. (2019). Synthesis, structure-activity relationship and biological evaluation of tetracationic gemini Dabco-surfactants for transdermal liposomal formulations. *Int. J. Pharm.*, 575: 118953. doi: 10.1016/j.ijpharm.2019.118953

Patra, J.K., Das, G. & Fraceto, L.F. (2018). Nano based drug delivery systems: Recent developments and future prospects. *J. Nanobiotechnol.*, 16: 71. https://doi.org/10.1186/s12951-018-0392-8

Peiris, M., Guansekera, T., Jayaweera, P. & Fernando, S. (2018). Tio 2 NPs from baker's yeast: A potent antimicrobial. *J. Microbiol. Biotechnol.*, 28: 1664-1670. doi: 10.4014/jmb.1807.07005.

Płaza, G.A., Chojniak, J. & Banat, I.M. (2014). Biosurfactant mediated biosynthesis of selected metallic NPs. *Int. J. Mol. Sci.*, 15: 13720-13737. https://doi.org/10.3390/ijms150813720

Prajapati, P., Varjani, S., Singhania, R.R., Patel, A.K., Awasthi, M.K., Sindhu, R. et al. (2021). Critical review on technological advancements for effective waste management of municipal solid waste – Updates and way forward. *Environ. Technol. Innov.*, 23: 101749. https://doi.org/10.1016/J.ETI.2021.101749

Rabbani, A., Zainith S., Deb, V.K., Das, P., Bharti, P., Rawat, D.S. et al. (2014). A stereospecific pathway diverts β-oxidation intermediates to the biosynthesis of rhamnolipid biosurfactants. *Chem. Biol.*, 21: 156-164. doi: 10.1016/j.chembiol.2013.11.010.

Raja Muthuramalingam Thangavelu, Rajendran Ganapathy, Pandian Ramasamy, Kathiravan Krishnan (2020). Fabrication of virus metal hybrid nanomaterials: An ideal reference for bio semiconductor. *Arabian Journal of Chemistry*, 13(1): 2750-2765, ISSN 1878-5352, https://doi.org/10.1016/j.arabjc.2018.07.006.

Rana, S., Singh, J., Wadhawan, A., Khanna, A., Singh, G. & Chatterjee, M. (2021). Evaluation of in vivo toxicity of novel biosurfactant from Candida parapsilosis loaded in PLA-PEG polymeric NPs. *J. Pharm. Sci.*, 110: 1727-1738.

Rangarajan, M. & Zatz, J.L. (2003). Effect of formulation on the topical delivery of alpha-tocopherol. *J. Cosmet Sci.*, 54: 161-174.

Rangarajan, V., Majumder, S. & Sen, R. (2014). Biosurfactant-mediated NPs synthesis: A green and sustainable approach. *In:* Mulligan, C.N., Sharma, S.K., Mudhoo, A. (Eds), *Biosurfactants: Research Trends and Applications*. 217-229. CRC Press, Taylor & Francis Group: Boca Raton, FL, USA.

Ravichandran, M., Jagadale, P. & Velumani, S. (2016). Inorganic nanoflotillas as engineered particles for drug and gene delivery. *Engineering of Nanobiomaterials*, 429-483. Elsevier.

Reddy, A.S., Chen, C.Y., Chen, C.C., Jean, J.S., Fan, C.W., Chen, H.R. et al. (2009). Synthesis of gold NPs via an environmentally benign route using a biosurfactant. *J. Nanosci. Nanotechnol.*, 9(11): 6693-6699.

Reddy, A.S., Chen, C.Y., Baker, S.C., Chen, C.C., Jean, J.C., Fan, C.W. et al. (2009). Synthesis of silver nanoparticles using surfactin: A biosurfactant as stabilizing agent. *Mater. Lett.*, 63: 1227-1230.

Rosenberg, E. & Ron, E.Z. (1999). High- and low-molecular mass microbial surfactants. *Appl. Microbiol. Biotechnol.*, 52: 154-162.

Saerens, K.M.J. (2011). Cloning and functional characterization of the UDP-glucosyl transferase UgtB1 involved in sophorolipid production by Candida bombicola and creation of a glucolipid producing yeast strain. *Yeast*, 28: 279-292.

Saikia, J.P., Bharali, P. & Konwar, B.K. (2013). Possible protection of silver NPs against salt by using rhamnolipid. *Colloids Surf. B Biointerfaces*, 104: 330-332.

Saranraj, P., Sayyed, R.Z., Sivasakthivelan, P., Hasan, M.S., Al-Tawaha, A.R.M.A & Amala, K. (2022a). Microbial biosurfactants: Methods of investigation, characterization, current market value and applications. *In:* Sayyed, R.Z. (Eds), *Biosurfatnats: Production and Applications in Bioremediation/Reclamation.* 19-34. CRC Press, Taylor & Francis Group, USA.

Saranraj, P., Sayyed, R.Z., Hamzah, K.J., Asokan, N., Sivasakthivelan, P. & Al-Tawaha, A.R.M.A. (2022b). *In:* Sayyed, R.Z. (Eds), *Biosurfactants: Production and Applications in Bioremediation/Reclamation.* 1-18. CRC Press, Taylor & Francis Group, USA.

Satpute, S.K., Kulkarni, G.R., Banpurkar, A.G., Banat, I.M., Mone, N.S., Patil, R.H. & Cameotra, S.S. (2016). Biosurfactant/s from Lactobacilli species: Properties, challenges and potential biomedical applications. *J. Basic Microbiol.*, 56: 1-19.

Satpute, S.K., Banpurkar, A.G., Dhakephalkar, P.K., Banat, I.M. & Chopade, B.A. (2009). Methods for investigating biosurfactants and bioemulsifiers: A review. *Crit. Rev. Biotechnol.*, 30: 127-144.

Sharma, J., Sundar, D. & Srivastava, P. (2021). Biosurfactants: Potential agents for controlling cellular communication, motility, and antagonism. *Front. Mol. Biosci.*, 8: 727070. doi: 10.3389/fmolb.2021.727070

Shoeb, E., Akhlaq, F., Badar, U., Akhter, J. & Imtiaz, S. (2013). Classification and industrial applications of biosurfactants. *Part - I: Natural and Applied Sciences*. Academic Research International, 4: 243-252.

Singh, B.R., Dwivedi, S., Al-Khedhairy, A.A. & Musarrat, J. (2011). Synthesis of stable cadmium sulfide nanoparticles using surfactin produced by Bacillus amyloliquifaciens strain KSU-109. *Colloids Surf. B Biointerfaces*, 85: 207-213.

Sirelkhatim, A., Mahmud, S., Seeni, A., Kaus, N.H.M., Ann, L.C., Bakhori, S.K.M. et al. (2015). Review on zinc oxide NPs: Antibacterial activity and toxicity mechanism. *Nanomicro Lett.*, 7(3): 219-242. doi: 10.1007/s40820-015-0040-x. Epub 2015 Apr 19. PMID: 30464967; PMCID: PMC6223899.

Sirotkin, S., Mermet, A., Bergoin, M., Ward, V. & Van Etten, J.L. (2014). Viruses as NPs: Structure versus collective dynamics. *Phys. Rev. E*, 90: 022718.

Slocik, J.M., Naik, R.R., Stone, M.O. & Wright, D.W. (2005). Viral templates for gold nanoparticle synthesis. *J. Mater. Chem.*, 15: 749-753. doi: 10.1039/b413074j.

Smith, M.L., Gandolfi, S., Coshall, P.M. & Rahman, P.K.S.M. (2013). Biosurfactants: A COVID-19 perspective. *Front. Microbiol.*, 11: 1341. doi:10.3389/fmicb.2020.01341

Sobrinho, H.B.S., Luna, J.M., Rufino, R.D., Porto, A.L.F. & Sarubbo, L.A. (2013). Biosurfactants: Classification, Properties and Environmental Applications, in Recent Developments in Biotechnology. 1-29. Govil, J.N. (ed.) Houston, TX: Studium Press LLC.

Sonkusre, P., Nanduri, R., Gupta, P. & Cameotra, S.S. (2014). Improved extraction of intracellular biogenic selenium NPs and their specificity for cancer chemoprevention. *Journal of Nanomedicine and Nanotechnology*, 5(2): 1-9.

Sumanth, B., Lakshmeesha, T.R., Ansari, M.A., Alzohairy, M.A., Udayashankar, A.C., Shobha, B. et al. (2020). Mycogenic synthesis of extracellular zinc oxide NPs from Xylaria acuta and its nanoantibiotic potential. *Int J Nanomedicine*, 15: 8519-8536. doi: 10.2147/IJN.S271743.

Tianimoghadam, S. & Salabat, A. (2018). A microemulsion method for preparation of thiol-functionalized gold nanoparticles. *Particuology*, 37: 33-36. doi: 10.1016/j.partic.2017.05.007

Van Bogaert, I.N.A. (2009). Knocking out the MFE-2 gene of Candida bombicola leads to improved medium-chain sophorolipid production. *FEMS Yeast Res.*, 9: 610-617.

Vecino, X., Barbosa-Pereira, L., Devesa-Rey, R., Cruz, J.M. & Moldes A.B. (2015). Optimization of extraction conditions and fatty acid characterization of Lactobacillus pentosus cell-bound biosurfactant/bioemulsifier. *J. Sci. Food Agric.*, 95: 313.

Vecino, X., Rodríguez-López, L., Rincón-Fontán, M., Cruz, J.M. & Moldes, A.B. (2021). Nanomaterials synthesized by biosurfactants. *Comprehensive Analytical Chemistry*, 267-301. doi:10.1016/bs.coac.2020.12.008

Vijayakumar, S. & Saravanan, V. (2015). Biosurfactants – types, sources and applications. *Res. J. Microbiol.*, 10(5): 181.

Vivek K. Gaur, Poonam Sharma, Shivangi Gupta, Sunita Varjani, J.K. Srivastava, Jonathan W.C. et al. (2022). Opportunities and challenges in omics approaches for biosurfactant production and feasibility of site remediation: Strategies and advancements. *Environmental Technology & Innovation*, 25. doi.org/10.1016/j.eti.2021.102132.

Voulgaridou, G.P., Mantso, T., Anestopoulos, I., Klavaris, A., Katzastra, C., Kiousi, D.E. et al. (2021). Toxicity profiling of biosurfactants produced by novel marine bacterial strains. *Int. J. Mol. Sci.*, 22: 2383.

Wang, J., Chen, Q., Zeng, C. & Hou, B. (2004). Magnetic-field-induced growth of single-crystalline Fe_3O_4 Nanowires. *Adv. Mater.*, 16: 137-140. https://doi.org/10.1002/adma.200306136

Wei, Z., Lin, X., Wu, M., Zhao, B., Lin, R., Zhang, Y. et al. (2017). Core-shell $NaGdF_4@CaCO_3$ nanoparticles for enhanced magnetic resonance/ultrasonic dual-modal imaging via tumor acidic micro-environment triggering. *Scientific Reports*, 7: 5370. doi: 10.1038/s41598-017-05395-w

Worakitsiri, P., Pornsunthorntawee, O., Thanpitcha, T., Chavadej, S., Weder, C., & Rujiravanit, R. (2011). Synthesis of polyaniline nanofibers and nanotubes via rhamnolipid biosurfactant templating. *Synth. Meth.*, 161: 298-306.

Xie, Y., Ye, R. & Liu, H. (2006). Synthesis of silver NPs in reverse micelles stabilized by natural biosurfactant. *Colloids Surf. A Physicochem. Eng. Asp.*, 2: 175-178.

Yumei, L., Yamei, L., Qiang, L. & Jie, B. (2017). Rapid biosynthesis of silver NPs based on flocculation and reduction of an exopolysaccharide from arthrobacter sp. B4: Its antimicrobial activity and phytotoxicity. *J. Nanomater.*, 2017: 9703614. doi: 10.1155/2017/9703614.

Zhang, D., Ma, X.-L., Gu, Y., Huang, H. & Zhang, G.-W. (2020). Green synthesis of metallic NPs and their potential applications to treat cancer. *Front. Chem.*, 8: 799. doi: 10.3389/fchem.2020.00799.

Zhuang, Q., Xu, J., Deng, D., Chao, T., Li, J., Zhang, R. et al. (2021). Bacteria-derived membrane vesicles to advance targeted photothermal tumor ablation. *Biomaterials*. 268: 120550.

Zuckerberg, A., Diver, A., Peeri, Z., Gutnick, D.L. & Rosenberg, E. (1979). Emulsifier of Arthrobacter RAG-1: Chemical and physical properties. *Appl. Environ. Microbiol.*, 37: 414-420.

CHAPTER

19

Applications of Sophorolipids in Medicine, Therapeutics and Agriculture: An Overview

Anamika Ghatak*

Department of Microbiology, THK Jain College, Cossipore, Kolkata, India

1. Introduction

A variety of microorganisms can produce surface-active amphiphilic biomolecules which have both hydrophilic and lipophilic groups in a single moiety. These are natural products and are coined as 'biosurfactants'. These can be classified as glycolipids, lipopeptides, phospholipids, lipopolysaccharides, fatty acids and polymers. Microorganisms such as bacteria, fungi, and yeasts produce these as extracellular secondary metabolites. Commercially available surfactants are basically emulsifiers containing petroleum as their main component and are produced by extensive chemical processes. Exertion of toxicity to the biosphere and generation of serious ecological hazards are the inevitable consequences of a prolonged use of these chemicals (Shafiei, 2014, Shah et al., 2016). So, bioaccumulation, toxicity and biodegradability of surfactants are increasingly becoming issues of concern now-a-days, thus, a promising alternative is biosurfactant (Olanya et al., 2018).

2. Structure and Properties of Sophorolipids

Sophorolipids (SLs) are mainly glycolipid biosurfactants. The features that make sophorolipids superior to synthetic surfactants include: stability in a wide range of pHs, temperatures, and salinity (Chandran and Das, 2012). Low-foaming and detergent properties (Hirata and Oda, 2009) or water hardness (high concentration of divalent cations) does not affect their interfacial properties significantly. Synergism between acidic and lactonic forms of SLs increases surfactant activities (Hirata et al., 2009). These are readily biodegradable and exhibit good surface activity (critical micelle concentration (CMC), surface tension and emulsification behavior) (Ma et al.,

* Corresponding author: anughatak@gmail.com

2012). SLs can be produced in large quantities from many renewable resources such as agro-industrial wastes and can be recovered with the aid of simplified methodologies (Pekin et al., 2005, Zhou and Kosaric, 1995, Ashby et al., 2005, Palme et al., 2010).

Sophorolipids are mainly glycolipids, structurally composed of a sophorose disaccharide (2′-O-β-D-glucopyranosyl-1-βD-glucose) linked by a β-glycosidic bond to a long-chain of fatty acids (Jiménez-Peñalver et al., 2018, Solaiman et al., 2017). They consist of a hydrophobic fatty acid tail of 16 or 18 carbon atoms and a hydrophilic carbohydrate head, sophorose (Fig. 1). Sophorose is a glucose disaccharide with an unusual β-1,2 bond and can—in the case of sophorolipids—be acetylated on the 6′- and/or 6″- positions. One terminal or subterminal hydroxylated fatty acid is β-glycosidically linked to the sophorose molecule. The carboxylic end of this fatty acid is either free (acidic or open form) or internally esterified at the 4″ or in some rare cases at the 6′- or 6″-position (lactonic form) (Figs 1 & 2). The hydroxy fatty acids generally have 16-18 carbon atoms and can have one or more unsaturated bonds. (Asmer et al., 1988). They are produced as a mix of structurally related molecules, reaching up to 40 different types and associated isomers.

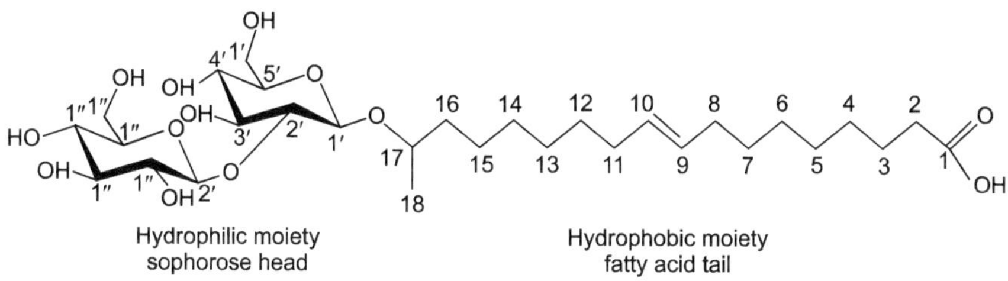

Fig. 1: General structure of sophorolipids.

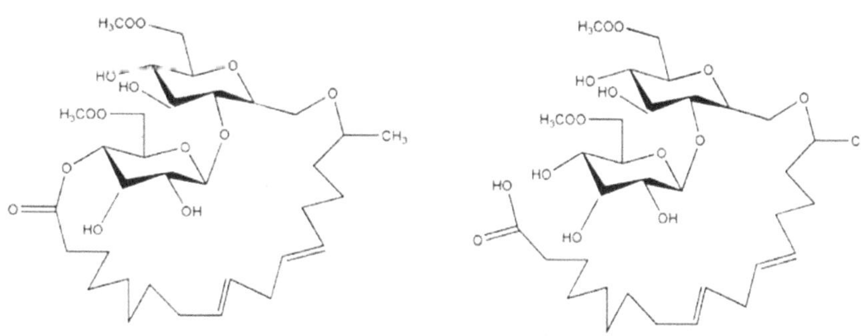

A Sophorolipid (Lactonic) C18:2 **B** Sophorolipid (Acidic) C18:2

Fig. 2: Chemical structure of (A) lactonic and (B) acidic forms of sophorolipids.

In general, acidic sophorolipids are more soluble and have a better foaming capacity, whereas the lactonic forms have better surface tension and antimicrobial properties (Zhang et al., 2017). Sophorolipids biosynthesis involves a complex pathway. At the initial phase, lipase enzymes promote the release of fatty acids, which may undergo β-oxidation or hydroxylation, resulting in a hydroxylated fatty acid chain. Hereafter, the enzymes glycosyltransferase I and II, that couple two molecules of glucose at the C1 and C2 positions to the hydroxyl group ω or ω-1 of the fatty acid forming the non-acetylated acidic sophorolipids (Van Bogaert et al., 2011). Two enzymes responsible for the lactonic and acetylated structures are lactonesterase and acetyltransferase, respectively (Bajaj et al., 2012). They are synthesized from a single hydrophilic carbon source (carbohydrate) or by association with hydrophobic sources (lipids, hydrocarbons, vegetable oils or animal fat). Studies have shown that the hydrophilic substrate is directed to produce the sophorose portion, while the hydrophobic component is responsible for making the lipid tail (Davila et al., 1994).

3. Microbial Sources and Production

In the early 1960s, Gorin et al. (1961) first described an extracellular glycolipid synthesized by the yeast *Torulopsis magnolia*. Various yeast strains can produce sophorolipids (SLs) and mannosylerythritol lipids (MELs) abundantly, which are representative glycolipid biosurfactants (Van et al., 2007). Sophorolipids are produced mainly by non-pathogenic yeasts strains, such as *Candida bombicola* (also known as *Torulopsis bombicola*), *Candida apicola* and *Rhodotorula bogoriensis*, while MELs are produced mainly by *Pseudozyma aphidis*, *Pseudozyma antarctica* and *Pseudozyma rugulosa* (Konishi et al., 2007). The extent of production of SLs has already been reported to be much higher as compared with production of other biosurfactants, which eventually broadens its commercialization (Sen et al., 2017).

Table 1: Microorganisms producing sophorolipids

Microorganisms	References
Candida bombicola ATCC 22214	Dolman et al., 2016
Rhodotorulababjevae YS3	Sen et al., 2017
Candida albicans	Kurtzman et al., 2010
Candida parapsilosis	Garg et al., 2018
Rhodotorula bogoriensis	Van Bogaert et al., 2011
Wickerhamiella domercqiae	Ma et al., 2014
Starmerella bombicola	Chen et al., 2019
Candida kuoi	Price et al., 2012
Cryptococcus sp.	Van Bogaert et al., 2011

It has already been discussed that the production of SLs increases when both the hydrophilic and the hydrophobic sources are available (Davila et al., 1994). The hydrophilic substrate governs the production of the sophorose portion, while the hydrophobic component is the source of the lipid tail. Glucose and oleic acid are considered the preferred sources of the metabolism for sophorolipid production (Kosaric and Vardar-Sukan, 2014, Jadhav et al., 2019), although the utilization and effects of other sources has already been documented. Use of agro-industrial residues is also expanding for lowering the production costs. Several hydrophobic sources have been tested in the production of sophorolipids and an increase in the yield was observed according to an increase in the number of carbon atoms in the fatty acids chain (C18 > C16 > C14 > C12) (Cavalero and Cooper, 2003).

Several organic and inorganic nitrogen sources, such as urea, peptone, $NaNO_3$, $(NH_4)_2SO_4$, malt extract and yeast extract were tested to determine their effects on SLs production (Chen et al., 2019). Yeast extract has been found to be the most efficient source as it also contains nutrients such as pantothenic acid, thiamine and pyridoxine and traces of zinc, iron and magnesium. At low concentration, sophorolipids production is stimulated, although, at high levels, it can stimulate the primary metabolism and deplete the glucose source. Yeast extract also interferes in the balance of acidic and lactonic forms. Casas et al. (1999) has studied that long fermentation time including low concentrations of yeast extract promoted the increase of lactonic forms (Casas and García-Ocha, 1999).

According to a literature source, the optimum temperature for sophorolipids production varies from 25 to 30°C (Dolman et al., 2017). The optimum pH for *C. bombicola* growth is between 5.0 to 6.0 (initial pH) and 3.5 during the stationary phase to produce sophorolipids. Usually, the decrease in pH during this phase can be correlated directly with the consumption of the nitrogen source and the generation of organic acids. A pH level 3.5 can promote up to 27.6% increase in sophorolipids production (Davila et al., 1997).

Agro-industrial wastes are of low-cost materials and can be used as alternative substrates to make the production process more viable economically and eco-friendly (Satpute et al., 2017). Several alternative substrates for the production of sophorolipids have already been studied: soy molasses, sugar beet and cane molasses, deproteinized whey, animal fat, fatty acid residues, used cooking oil, dairy waste water, by-products of biodiesel and of soybean oil refining, wheat and rice straw, potato peel, corn husk and sugarcane bagasse (Santos et al., 2017). Some pre-treatments are often needed to improve the production, for example sugarcane molasses must be clarified before use for removal of inhibitory components, and corn husks need to be hydrolyzed to facilitate absorption and increase final yield. Minucelli et al. (2017) studied the production of sophorolipids by *C. bombicola* with molasses and sugarcane juice, sucrose, glucose, poultry industry residual fat and sunflower oil as hydrophobic carbon source (Minucelli et al., 2017).

In one study, Hoa et al. (2017) used *C. Bombicola* to produce sophorolipids from sugarcane molasses and coconut oil (10%), using a temperature of 25°C, pH 6, 180 fermentor rpm for a duration of 168 hours (Hoa et al., 2017). The authors obtained a maximum yield of 10 g/L. Therefore, it is always important to select appropriate the type of substrate. Whether natural or semi-synthetic, the substrate(s) must have adequate nutritional value for microbial growth as well as both considerable and competitive production and yield (Satpute et al., 2017).

4. Applications

4.1.A Biomedicine and Therapeutics

SLs can find a diverse range of applications as biologically active compounds in biomedicine, particularly as antimicrobial, antitumor, antiviral and immune-modulator. The destabilization of cell membranes' structure by the action of SLs can modify membrane permeability and the cell surface which eventually destroys the cell (Joshi-Navare et al., 2013, Banat et al., 2010).

4.1.B Anticancer Activity

Several kinds of tumor cells are affected by the anticancer activity of SLs and this might be a potential bio-tool for cancer treatment. SLs of *Wickerhamiella domercqiae* show cytotoxic effects in several cancerous cell lines. SLs exhibited significant inhibition of cell proliferation on cancer cells of H7402 (liver cancer line). These molecules can induce apoptosis by blocking the cell cycle at the G1 phase and partly at the S phase, increase the activity of caspase-3 and increase intracellular concentration of Ca^{2+} (Chen et al., 2006). SLs produced by *Wickerhamiella domercqiae* exhibited different cytotoxic effects in cell lines KYSE 109 and KYSE 450 (human esophageal cancer). Diacetylated lactonic SLs promote a better inhibition on these two cell lines than that of monoacetylated lactonic SL. The monounsaturated SLs had the strongest cytotoxic effect whereas the unsaturated and saturated sophorolipids had a weaker and the weakest cytotoxic effect (only 20% of cells were inhibited at 60 mg/mL concentration), respectively. Monounsaturated or diunsaturated and monoacetylated or diacetylated acidic SL groups have little anticancer activity. In other words, there is a correlation between SL types (congeners) and anti-cancer activities (Shao et al., 2012). The anticancer responses are dose and derivative dependent. Natural mixtures of SLs (containing a combination of eight isoforms lactonic and acidic SLP) or derivatives (ethyl ester, methyl ester, ethyl ester monoacetate, ethyl ester diacetate, acidic SLs and diacetylated lactonic SLs) exhibit diverse responses against human pancreatic carcinoma cells. These distinct responses are produced due to different mechanisms for killing pancreatic cancer cells by necrosis. The cytotoxicity of SLs is specific to malignant cells since no cytotoxicity is observed against normal human cells, which minimizes the side effects commonly associated with current therapeutic regimes (Fu et al., 2008).

4.1.C Antimicrobial Activity

The chemical structure of the SLs and the cell wall structure of microorganisms are the two major determining factors of antimicrobial activity of SLs.Gram positive, gram negative bacteria and yeast cells are inhibited by SLs differently due to differences in their cell wall structure and osmolarities. Varying degrees of action of mono-, di- and deacetylated lactonic SLs plays a significant role in inhibiting microorganisms selectively. This make SLs potential components in healthcare, cosmetics and skincare products (De Oliveira et al., 2015).

The hydrophobic moiety of SLs produced by *Candida bombicola* derived from lauryl–myristyl alcohol exhibit antimicrobial activity against both gram negative and gram positive bacteria and yeast,e.g., gram negative bacteria such as *Escherichia coli* (in gastroenteritis, urinary tract infections and neonatal meningitis) and *Pseudomonas aeruginosa* (cross infections in hospitals and clinics) are completely inhibited. In gram negative bacteria, shrinking of the cell surface causes irregularities on the natural form whereas in gram positive bacteria cell lysis occurs (Shah et al., 2005).

SLs enhance the activity of many antibiotics such as tetracyclinc. Tetracycline alone could not promote total inhibition against *Staphylococcus aureus* until the end of 6 hours of exposure but when associated with SLs promote total inhibition before 4 hours of exposure (Shah et al., 2005).

4.1.D Antiviral and Spermicidal Activity

Acidic SLs are weak spermicides and exhibit good antiviral activity against HIV and are least cytotoxic of the structural analogs tested. Lactonic forms of SLs exhibit high spermicidal, cytotoxic, and proinflammatory activities with low virucidal activity (Shah et al., 2005). Fatty acids of shorter-carbon-chain have higher potency as virucidal agents and lower spermicidal activity. Longer-carbon-chain has lower potency as virucidal agents and higher spermicidal activity.

The effects of SLs against human HIV (RNA virus), Epstein-Barr virus (a Herpes DNA virus), and influenza virus (RNA virus) were also studied (Shah et al., 2005). The acidic SL (open-ring nonacetylated SL) was found to be more virucidal than a mixture of lactonic SLs.

4.1.E Immuno-modulatory Activity

People with septic shock often experience multiple organ failure owing to induction of cytokine cascades. Septic shock can activate the coagulation cascade and apoptosis, causing further organ damage and disseminated intravascular coagulation. Administration of SLs after induction of intraabdominal sepsis significantly decreases mortality in a rat model. This mortality decrease is assumed to be mediated by a decreased number of macrophages, decrease in nitric oxide and proinflammatory cytokines production and modulation of inflammatory responses (Bluth et al., 2006). SLs decrease IgE production in U266 cells (IgE producing myeloma cell line) by down-regulating important genes involved in the functioning of IgE in a synergistic manner (Hagler et al., 2007).

The entry of a virus into the host cell induces a cytokine storm that causes overproduction of early response proinflammatory cytokines like TNFα, IL-6, and IL1β. This cytokine storm activates coagulation pathways that eventually increase the risk of vascular hyper permeability and multi-organ failure, leading to death. The purified natural mixtures of SLs have immunomodulatory properties as they have been shown to down-regulate the inflammatory cytokines and up-regulate anti-inflammatory cytokines in a sepsis rat model (Bluth et al., 2006). They block lethal effects of septic shock disease by significantly reducing the IL-1β (42.5%, proinflammatory cytokine), TNF-α (50%, proinflammatory cytokine), and macrophage nitric oxide (NO: 28%), and by increasing the TGF-β1 (anti-inflammatory cytokine: 11.7%) (Bluth et al., 2006).

4.1.F Drug Delivery System

SLs can self-assemble into various structures due to their amphiphilic features and these structures form different shapes and sizes depending upon factors such as the temperature, pH and incubation-time of the reaction mixture.

Sodium salts combine with acidic SLs (SLsNa) to spontaneously form vesicles that act as skin penetration enhancers for active ingredients. The main active component of triterpene glycosides is mogrosides V and it exerts anti-cancer activity, anti-atherosclerotic effects, anti-allergy activity, and anti-diabetic effects in animal models. Those vesicles significantly increase the amount of mogroside V that penetrate through the skin (Imura et al., 2014).

SLs increase the transdermal absorption of lactoferrin (LCF). The effects of LCF on cell proliferation activities and levels of collagen IV, elastic fiber components and so on are not actually depressed by SLs. Ishii et al. (2012) experimentally proved that, SLs, in turn activated tropoelastin gene expression. In this way, a balance between the action of both LCF and SLs has been maintained synergistically (Ishii et al., 2012).

4.1.G Sophorolipids and Covid-19 Management

The antiviral activities of the SLs and their derivatives have been tested against Herpes virus (Ebsteir Barr virus), human immunodeficiency virus (HIV), and influenza virus in in-vitro cell-free virus inactivation assays. The possible mechanism of virucidal activity through viral membrane perturbation or disruption has been elucidated in detail by Shah et al. (2005). The acetyl groups in the SL structure play a crucial role in promoting its antiviral activity by imparting hydrophilicity to the SL (Borsanyiova et al., 2016). Moreover, the literature suggests that enveloped viruses are highly sensitive to the surfactants and the lipid contents of the enveloped viruses are easily solubilized by the surfactants (Strauss and Strauss, 2008; Conley et al., 2017). Chemical surfactants/detergents are toxic and unsafe for human health, Therefore natural compounds like SLs could be effective virucidal agents to kill the SARS-CoV-2 by disrupting and solubilising its lipid envelope. Among many essential viral proteins, spike (S) glycoprotein plays an essential role in viral attachment to the ACE2 (Angiotensin Converting Enzyme II) on host cells (like a human respiratory epithelial cell). This attachment is essential for the pathogenesis of a virus. The two possible mechanisms of SL are:

- The solubilisation of viral envelope, thus degrading the components of the virus, and
- Inhibition of the interaction between the virus and ACE2 (Fig. 3a); and the inhibition of the cytokine storm and activation of anti-apoptotic genes (Figs 3b and 3c).

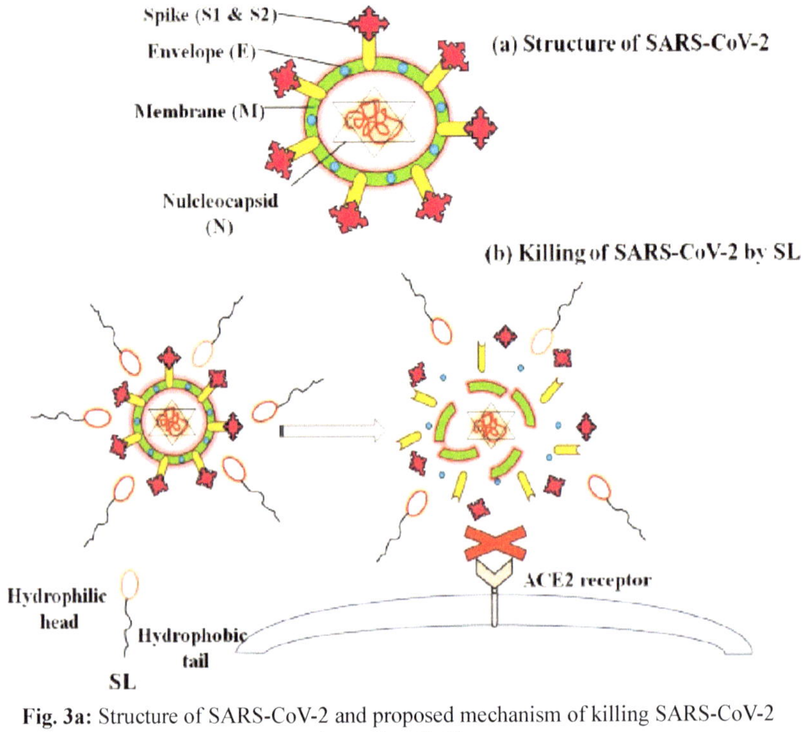

Fig. 3a: Structure of SARS-CoV-2 and proposed mechanism of killing SARS-CoV-2 by sophorolipids.

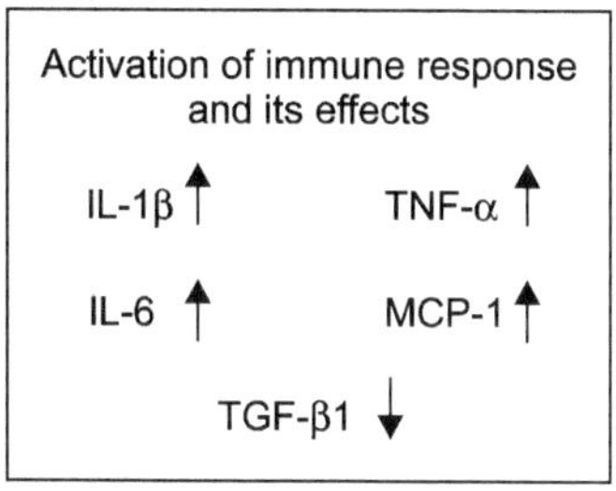

Fig. 3b

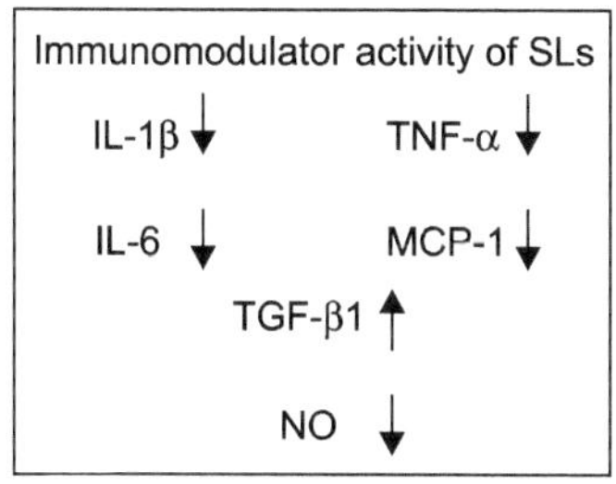

Fig. 3c

4.1.H Anticancer Activity

Chen et al. 2006, purified lactonic sophorolipid that can kill human liver cancer cells effectively. The cytotoxic effect varies with the degree of saturation of the fatty acid, acetylation of sophorose and lactonization or ring opening in SLs (Shao L. et al., 2012). Nawale et al. (2017), elucidated the growth regulatory function of two sophorolipids derivatives (SLB and SLC) in human cervical HeLa cancer cells via inducing apoptosis. SLAs stimulate the activation of caspase-3, 8 and 9 with modification of mitochondrial membrane potential. SLC was found to be more cytotoxic than SLB. Both SLs showed similar apoptotic effect in human cervical cancer cells in a time-dependent manner; arresting the cells at G1/S phase for 6, 12, 18 and 24 h. Thus, it has been proved that both SLs can act as anti-tumor agents.

4.2 Agriculture and Environment

They have the ability to form complexes with metals (micelles), thus increasing metal solubility and bioavailability in the soil solution (Seneviratne et al., 2017). They solubilize and emulsify hydrophobic contaminants like crude oil, diesel, kerosene, engine oil and motor oil etc., and remove heavy metals (cadmium, iron, mercury, chromium, lead, zinc) and pesticides in aqueous phases (Tang et al., 2018). Patowary et al. (2018) reported that sophorolipids and other biosurfactants can increase the apparent solubility of PAHs, increasing their mobility and biodegradability (Patowary et al. 2018). The acidic and lactonic forms of sophorolipids from *Rhodotorula babjevae* YS3 were experimented for their antifungal action against *Colletotrichum gloeosporioides*, *Fusarium verticillioides*, *Fusarium oxysporum*, *Corynespora cassiicola* and *Trichophyton rubrum* (Sen et al., 2017). Ali et al. (2022) demonstrated that one lipopeptide, surfactin, obtained from *Bacillus velezensis* MS20, a marine bacterium can induce systemic resistance in maize crop, making it a good candidate in agricultural practices. Microbial-enhanced oil recovery (MEOR) is among the major thrust areas of research in the oil industry (Saranraj et al., 2022). Sophorolipids addition was found to be effective in promoting 80% of biodegradation of crude oil (Minucelli et al., 2017).

Conclusion

SLs could be a potential sustainable alternative for synthetic surfactants. This article presented the properties of SLs and their applications in areas including agriculture, biomedicine and therapeutics. Currently, various companies are trying to produce SLs derived eco-friendly products such as cosmetics, cleaning and agro-based products. The reason behind the huge interest in SLs is due to their wide structural and functional diversity and a wide range of applications in various areas. Choice of substrates, microorganisms, optimization of production parameters and proper application are some of the major factors for successful implementation of biosurfactants conceptually.

References

Ali, S.A.M., Sayyed, R.Z., Mir, M.I., Hameeda, B., Khan, Y., Alkhanani, M.F. et al. (2022a). Induction of systemic resistance in maize and antibiofilm activity of surfactin from *Bacillus velezensis* MS20. *Front. Microbiol.*, 13: 879739.

Ashby, R.D., Nunez, A., Solaiman, D.K.Y. & Foglia, T.A. (2005). Sophorolipid biosynthesis from a biodiesel co-product stream. *Journal of the American Oil Chemists Society*, 82(9): 625–630. http://dx.doi.org/10.1007/s11746-005-1120-3

Asmer, H.J., Lang, S., Wagner, F. & Wray, V. (1988). Microbial production, structure elucidation and bioconversion of sophorose lipids. *J. Am. Oil Chem. Soc.*, 65(9): 1460-1466. https://doi.org/10.1007/BF0289830

Bajaj, V., Tilay, A. & Annapure, U. (2012). Enhanced production of bioactive sophorolipids by *Starmerella bombicola* NRRL Y-17069 by design of experiment approach with successive purification and characterization. *J. Oleo Sci.*, 61(7): 377-386. https://doi.org/10.5650/jos.61.377

Banat, I.M., Franzetti, A., Gandolfi, I., Bestetti, G., Martinotti, M.G., Fracchia, L. et al. (2010). Microbial biosurfactants production, applications and future potential. *Applied Microbiology and Biotechnology*, 87(2): 427-444. https://doi.org/10.1007/s00253-010-2589-0

Bluth, M.H., Kandil, E., Mueller, C.M., Shah, V. & Zhang, H. (2006). Sophorolipids block lethal effects of septic shock in rats in a cecal ligation and puncture model of experimental sepsis. *Critical Care Medicine*, 34(1): 188. doi: 10.1097/01.ccm.0000196212.56885.50

Borsanyiova, M., Patil, A., Mukherji, R. & Prabhune, A. (2016). Biological activity sophorolipids and their possible use as antiviral agents. *Folia Microbiol* (Praha). 61(1): 85-89. https://doi.org/10.1007/s12223-015-0413-z

Casas, J.A. & García-Ocha, F. (1999). Sophorolipid production by *Candida bombicola*: Medium composition and culture methods. *J. Biosci. Bioeng.*, 88(5): 488-494. https:// doi: 10.1016/s1389-1723(00)87664-1

Cavalero, D.A. &Cooper, D.G. (2003). The effect of medium composition on the structure and physical state of sophorolipids produced by *Candida bombicola* ATCC 22214. *J. Biotechnol.*, 103(1): 31-41. https://doi. org/10.1016/S0168-1656(03)00067-1

Chandran, P. & Das, N. (2012). Role of sophorolipid biosurfactant in degradation of diesel oil by *Candida tropicalis*. *Bioremediation Journal*, 16(1): 19-30. https://doi.org/10.1002/ejlt.200900153

Chen, J., Sing, X., Zhang, H., Qu, Y.B. & Miao, J.Y. (2006). Sophorolipid produced from the new yeast strain *Wickerhamiella domercqiae* induces apoptosis in H7402 human liver cancer cells. *Applied Microbiology and Biotechnology*, 72(1): 52-59. doi: 10.1007/s00253-005-0243-z.

Chen, Y., Lin, Y.T.X., Li. Q. & Chu, J. (2019). Real-time dynamic analysis with low-field nuclear magnetic resonance of residual oil and sophorolipids concentrations in the fermentation process of Starmerella bombicola. *J. Microbiol. Meth.*, 157: 9-15. https://doi.org/10.1016/j.mimet.2018.12.007

Conley, L., Tao, Y. & Henry, A. (2017). Evaluation of ecofriendly zwitterionic detergents for enveloped virus inactivation. *Biotechnol Bioeng.*, 114(4): 813-820. https://doi.org/10.1002/bit.26209

Davila, A.M., Marchal, R. & Vandecasteele, J.P.(1997). Sophorose lipid fermentation with differentiated substrate supply for growth and production phases. *Appl. Microbiol. Biotechnol.*, 47: 496-501. https://doi.org/10.1007/s002530050962

Davila. A.M., Marchal. R. &Vandecasteele, J.P. (1994). Sophorose lipid production from lipid precursors: Predictive evaluation of industrial substrates. *J. Ind. Microbiol. Biotechnol.*, 13(4): 249-257. https://doi.org/10.1007/BF01569757

De Oliveira, M.R., Baldo, C., Magri, A. & Municelli, T. (2015). Review: Sophorolipids a promising biosurfactant and its applications. *International Journal of Advanced Biotechnology and Research*, 6(2): 161-174.

Dolman, M.B., Kaisermann, C., Martin, J.P. & Winterburn, B.J. (2017). Integrated sophorolipid production and gravity separation. *Process Biochem.*, 54: 162-171. https://doi.org/10.1016/j.procbio.2016.12.021

Fu, S.L., Wallner, S.R., Bownw, W.B. & Hagler, M.D. (2008). Sophorolipids and their derivatives are lethal against human pancreatic cancer cells. *The Journal of Surgical Research*, 148(1): 77-82. doi:10.1016/j.jss.2008.03.005

Garg, M. & Chatterjee, M. (2018). Isolation, characterization and antibacterial effect of biosurfactant from *Candida parapsilosis*. *Biotechnol Rep.*, 18: e00251. https://doi.org/10.1016/j.btre.2018.e00251

Gorin, P.A.J., Spencer, J.F.T. & Tulloch, A.P. (1961). Hydroxy fatty acid glycosides of sophorose from Torulopsis magnoliae. *Canadian Journal of Chemistry*, 39(6199): 846-855.

Hagler, M.. Norowitz, T.S., Chice, S. & Wallner, S.R. (2007), Sophorolipids decrease IgE production in U266 cells by downregulation of BSAP (Pax5), TLR-2, STAT3 and IL-6. *Journal of Allergy and Clinical Immunology*, 119(1): 1030. doi:10.1016/j.jaci.2006.12.399

Hirata, Y., Ryu, M. & Igarashi, K. (2009). Natural synergism of acid and lactone type mixed sophorolipids in interfacial activities and cytotoxicities. *Journal of Oleo Science*, 58(11): 565-572. doi: 10.5650/jos.58.565

Hirata, Y., Ryu, M. & Oda, Y. (2009). Novel characteristics of sophorolipids, yeast glycolipid biosurfactants, as biodegradable low-foaming surfactants. *Journal of Bioscience and Bioengineering*, 108(2): 142-146. doi:10.1016/j.jbiosc.2009.03.012

Hoa et al. (2017). Production and characterization of sophorolipids produced by Candida bombicola using sugarcane molasses and coconut oil. *J. Sc. Technol.*, 22(2): 66-75. Article ID.: APST-22-02-06

Imura, T., Morita, T., Fukuoka, T., Ryu, M., Igarashi, K., Hirata, Y. & Kitamoto, D. (2014). Spontaneous vesicle formation from sodium salt of acidic sophorolipid and its application as a skin penetration enhancer. *Journal of Oleo Science*, 63(2): 141-147. doi:10.5650/jos.,ess13117

Ishii, N., Kobayashi, T., Matsumiya, K., Hirata, Y. & Ryu, M. (2012). Transdermal administration of lactoferrin with sophorolipid. *Biochemistry and Cell Biology*, 90(3): 504-512. https://doi.org/10.1139/o11-065

Jadhav, J.V., Pratap, A.P. & Kale, S.B. (2019). Evaluation of sunflower oil refinery waste as feedstock for production of sophorolipids. *Proc. Biochem.*, 78: 15-24. https://doi.org/10.1016/j.procbio.2019.01.015

Jiménez-Peñalver, P., Castellejos, M., Koh, A., Gross, R., Sanchez, A., Font, X. & Gea, T. (2018). Production and characterization of sophorolipids from stearic acid by solid-state fermentation, a cleaner alternative to chemical surfactants. *J. Clean Prod.*, 172: 2735-2747. https://doi. org/10.1016/j.jclepro.2017.11.138

Joshi-Navare, K., Khanvilkar, P. & Prabhune, A. (2013). Jatropha oil derived sophorolipids: Production and characterization as laundry detergent additive. *Biochemistry Research International*. 2013: 1-11. http://dx.doi.org/10.1155/2013/169797

Konishi, M., Morita, T., Fukuoka, T., Imura, T., Kakugawa, K. & Kitamoto, D. (2007). Production of different types of mannosylerythritol lipids as biosurfactants by the newly isolated yeast strains belonging to the genus *Pseudozyma*. *Appl. Microbiol. Biotechnol.*, 75: 521-531. https://doi.org/10.1007/s00253-007-0853-8

Kosaric, N. & Vardar-Sukan, F. (2014). Biosurfactants: Production and Utilization – Processes, Technologies, and Economics. CRC Press, USA, pp. 389. https://doi.org/10.1201/b17599

Kurtzman, C.P., Price, N.P.J., Ray, K.J. & Kuo, T. (2010). Production of sophorolipid biosurfactants by multiple species of the *Starmerella (Candida) bombicola* yeast clade. *FEMS MicrobiolLett.*, 311(2): 140-146. https://doi.org/10.1111/j.1574-6968.2010.02082.x

Ma, X., Li. H., Wang, D. & Song, X. (2014). Sophorolipid production from delignined corncob residue by *Wickerhamiella domercqiae* var. sophorolipid CGMCC 1576 and *Cryptococcus curvatus* ATCC 96219. *Environ. Biotechnol.*, 98: 475-483. https://doi.org/10.1007/s00253-013-4856-3

Ma, X.J., Li, H. & Song, X. (2012). Surface and biological activity of sophorolipid molecules produced by *Wickerhamiella domercqiae* var. sophorolipid CGMCC 1576. *Journal of Colloid and Interface Science*, 376(1): 165- 172. https://doi.org/10.1016/j.jcis.2012.03.007

Minucelli, T., Ribeiro-Viana, R.M., Borsato, D., Andrade, G., Cely, M.V.T., de Oliveira, M.R. et al. (2017). Sophorolipids production by *Candida bombicola* ATCC 22214 and its potential application in soil bioremediation. *Waste Biomass Valori*, 8(3): 743-753. https://doi.org/10.1007/s12649-016-9592-3

Nawale, L., Dubey, P. & Chaudhari, B. (2017). Anti-proliferative effect of novel primary cetyl alcohol derived sophorolipids against human cervical cancer cells HeLa. *PLoS ONE*, 12(4): 1-14. https://doi.org/10.1371/journal.pone.0174241

Olanya, O.M., Ukuku, D.O. & Mukhopadhyay, S. (2018). Reduction in *Listeria monocytogenes*, *Salmonella enterica* and *Escherichia coli* O157: H7 in vitro and on tomato by sophorolipid and sanitiser as affected by temperature and storage time. *Int. J. Food Sci. Technol.*, 53: 1303-1315. https://doi.org/10.1111/ijfs.13711

Palme, O., Comanescu, G., Stoineva, I., Radel, S., Benes, E. & Lang, S. (2010). Sophorolipids from *Candida bombicola*: Cell separation by ultrasonic particle manipulation. *European Journal of Lipid Science and Technology*, 112(6): 663-673. https://doi.org/10.1002/ejlt.200900163

Patowary, R., Patowary, K., Kalita, M.C. & Deka, S. (2018). Application of biosurfactant for enhancement of bioremediation process of crude oil contaminated soil. *Int. Biodeterior. Biodegrad.*, 129: 50-60. https://doi.org/10.1016/j.ibiod.2018.01.004

Pekin, G., Vardar-Sukan, F. & Kosaric, N. (2005). Production of sophorolipids from *Candida bombicola* ATCC 22214 using Turkish corn oil and honey. *Engineering in Life Sciences*, 5(4): 357-362. https://doi.org/10.1002/elsc.200520086

Price, N.P.J., Ray, K.J., Vermillion, K.E., Dunlap, C.A. & Kurtzman, C.P. (2012). Structural characterization of novel sophorolipid biosurfactants from a newly identified species of *Candida* yeast. *Carbohydrate Res.*, 348(1): 33-41. https://doi.org/10.1016/j.carres.2011.07.016

Santos, D.K., Meira, H. & M-Luna, J. (2017). Biosurfactant production from *Candida lipolytica* in bioreactor and evaluation of its toxicity for application as a bioremediation agent. *Process Biochem.*, 54: 20-27. https://doi.org/10.1016/j.procbio.2016.12.020

Saranraj, P., Sayyed, R.Z., Sivasakthivelan, P., Hasan, M.S., Al-Tawaha, A.R.M.A. & Amala, K. (2022). Microbial biosurfactants: Methods of investigation, characterization, current market value and applications. *In:* Sayyed, R.Z. (Eds), *Biosurfactants: Production and Applications in Bioremediation/Reclamation.* 19-34. CRC Press, Taylor & Francis Group, USA.

Satpute, S.K., Płaza, G.A. & Banpurkar, A.G. (2017). Biosurfactants' production from renewable natural resources: Example of innovative and smart technology in circular bioeconomy. *Manag. Syst. in Prod. Eng.*, 25(1): 46-54. https://doi.org/10.1515/mspe-2017-0007

Sen, S., Borah, S.N., Bora, A. & Deka, S. (2017). Production, characterization, and antifungal activity of a biosurfactant produced by *Rhodotorula babjevae* YS3. *Microb. Cell Fact.*, 16: 95. https://doi.org/10.1186%2Fs12934-017-0711-z

Seneviratne, M., Seneviratne, G., Madawala, H. & Vithanage, M. (2017). Role of rhizospheric microbes in heavy metal uptake by plants. *In:* Singh, J.S. et al. (Eds), *Agro-Environmental Sustainability, Vol. 2: Managing Environmental Pollution*. 147-163. Springer. https://doi.org/10.1007/978-3-319-49727-3_8

Shafiei, Z., Abdul Hamid, A, Fooladi, T. & Yusoff, W.M.W. (2014). Surface active components: Review. *Current Research Journal of Biological Sciences*, 6(2): 89-95, 2014. http://dx.doi.org/10.19026/crjbs.6.5503

Shah, N., Nikam, R., Gaikwad, S., Sapre, V. & Kaur, J. (2016). Biosurfactant: Types, detection methods, importance and applications. *Indian J Microbiol Res*, 3(1): 5-10. http://dx.doi.org/10.5958/2394-5478.2016.00002.9

Shah, V., Azim, A., Seyoum, T., Zalenskaya, I. & Hagver, R. (2005). Sophorolipids, microbial glycolipids with anti-human immunodeficiency virus and sperm-immobilizing activities. *Antimicrobial Agents and Chemotherapy*, 49(10): 4093-4100. https://doi.org/10.1128%2FAAC.49.10.4093-4100.2005

Shao, L., Song, X., Ma, X., Li, H. & Qu, Y. (2012). Bioactivities of sophorolipid with different structures against human esophageal cancer cells. *The Journal of Surgical Research*, 173(2): 286-291. https://doi.org/10.1016/j.jss.2010.09.013

Solaiman, D.K.Y., Ashby, R.D. & Uknalis, J. (2017). Characterization of growth inhibition of oral bacteria by sophorolipid using a microplate-format assay. *J. Microbiol. Meth.*, 136: 21-29. https://doi.org/10.1016/j.mimet.2017.02.012

Strauss, J.H. & Strauss, E.G. (2008). The structure of viruses. *Viruses and Human Disease*, 8: 35-62. Massachusetts, USA, Elsevier. https://doi.org/10.1016%2FB978-0-12-373741-0.50005-2

Tang, J. He, J., Xin, X., Hu, H. & Liu, T. (2018). Biosurfactants enhanced heavy metals removal from sludge in the electrokinetic treatment. *Chen. Eng. J.*, 334: 2579-2592. https://doi.org/10.1016/j.cej.2017.12.010

Van Bogaert, I.N.A., Saerens, K., De Muynck, C. & Develter, D. (2007). Microbial production and application of sophorolipids. *Appl. Microbiol. Biotechnol.*, 76: 23-34. https://doi.org/10.1007/s00253-007-0988-7

Van Bogaert, I.N.A., Zhang, J. & Soetaert, W. (2011). Microbial synthesis of sophorolipids. *Process Biochem.*, 46(4): 821-833. https://doi.org/10.1016/j.procbio.2011.01.010

Zhang, X., Ashby, R.D., Solaiman, D.K.Y., Liu, Y. & Fan, X. (2017). Antimicrobial activity and inactivation mechanism of lactonic and free acid sophorolipids against *Escherichia coli* O157:H7. *Biocatal. Agric. Biotechnol.*, 11: 176-182. https://doi.org/10.1016/j.bcab.2017.07.002

Zhou, Q.H. & Kosaric, N. (1995). Utilization of canola oil and lactose to produce biosurfactant with *Candida bombicola*. *Journal of the American Oil Chemists Society*, 72(1): 67-71. https://doi.org/10.1007/BF02635781

CHAPTER

20

Sophorolipids: In Medicine and Therapeutics

Rania N. Ghaleb[1,2], **and Bhosale Hemlata**[1*]

[1] DST-FIST and UGC-SAP Sponsored School of Life Sciences Swami Ramanand Teerth Marathwada University, Nanded, India

[2] Department of Microbiology, Faculty of Applied Sciences, Taiz University, Taiz, Yemen

1. Introduction

Biosurfactants (BSs) refer to amphiphilic compounds that are synthesized on biological surfaces, primarily on microbial cells. These compounds consist of hydrophobic and hydrophilic components that are either excreted extracellularly or present on the microbial cell fluid phases. Consequently, they effectively reduce the tension at the surface and interface, thereby impacting surface tension and interfacial tension, respectively (Desai & Banat, 1997, Diaz et al., 2015, Jahan et al., 2020, Santos et al., 2016, Bee et al., 2019, Ali et al., 2022a, Ali et al., 2022b, Sadiq et al., 2022, Shah et al., 2022, Saranraj et al., 2022a, Ravinder et al., 2022). Microbial production of BSs offers advantages over their chemical counterparts, including reduced toxicity, enhanced biodegradability, effectiveness across varying environmental conditions, and improved biocompatibility (Desai and Banat, 1997, Díaz De Rienzo et al., 2015). BSs are structurally diverse, environmentally friendly, and have functional qualities that benefit various industries, including medicine and the environment (Smyth et al., 2010, Jahan et al., 2020). BSs can be categorized based on their chemical composition and the microorganisms from which they originate (Desai and Banat, 1997, Díaz De Rienzo et al., 2015, Patel et al., 2022, Zaman et al., 2022), based on their molecular mass (Rosenberg and Ron, 1999), extracellularly or attachments to the cell wall (Hajfarajollah et al., 2018). BSs with low molecular mass are usually made up of glycolipids or lipopeptides. The most extensively studied glycolipid biosurfactants are rhamnolipids, trehalolipids, and sophorolipids (Rosenberg and Ron, 1999). These BSs consist of disaccharides that have been acylated with long-chain fatty acids or hydroxy fatty acids (Jahan et al., 2020).

* Corresponding author: bhoslehemlata@gmail.com

Sophorolipids (SLs) are low molecular mass glycolipid biosurfactants, which are produced by different types of yeast such as *Starmerella bombicola*, *Candida bastistaeic*, *C. floricola* and *C. apicola* (Jahan et al., 2020). SLs are composed of disaccharide sophorose chains that are connected via β-1,2 linkages. The hydroxyl groups at positions 6′- and 6″- are often subjected to acetylation. The lipid component is linked to the reducing end through a glycosidic bond. The carboxyl group located at the end of the fatty acid molecule has the potential to exist in two different forms: the lactonic form or the hydrolyzed form, which results in the production of an anionic surfactant (Rosenberg and Ron, 1999, Cavalero and Cooper, 2003, Shao et al., 2012, Callaghan et al., 2016, Santos et al., 2016). Various applications of SLs in food, agriculture, bioremediation, drug delivery systems, and biomedicine were reported (Saranraj et al., 2022b, Pal et al., 2023). SLs are microbially derived biosurfactants possessing bioactive properties such as antimicrobial, immunomodulatory and anticancer effects. This chapter summarizes the applications of SLs in medicine and therapeutics, as shown in Fig. 1.

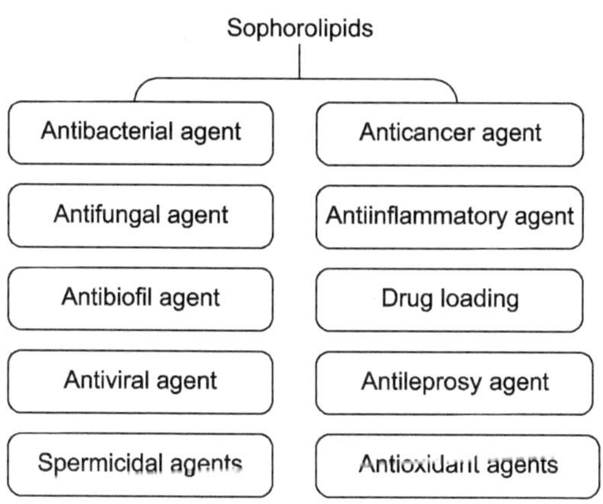

Fig. 1: Applications of sophorolipids in medicine and therapeutics.

2. Role of Sopholipids

2.1 As Antimicrobial Agents

BSs have the ability to make microbial cell membranes more fluid, which increases the permeability of the cytoplasmic membrane. BSs, in particular, have been found to exhibit antimicrobial properties against various microorganisms, including pathogenic and non-pathogenic strains of bacteria (both gram-negative and gram-positive), as well as yeast and fungi. The exact mechanism by which BSs achieve this remains uncertain, but there are several possibilities. For example, BSs can attach themselves to microbial cell surfaces, thus affecting their integrity and disrupting their nutrient cycle. The fatty acid molecules present in biosurfactants are also capable of penetrating the cytoplasmic membrane, leading to changes in the membrane's size and occurrence of undesired ultrastructural changes. In addition, they can modify the composition of membrane fatty acids, negatively impacting membrane permeability and interfering with the functioning of membrane proteins and phospholipids. In some cases, the use of BSs can cause disruptions in the endocrine system, deplete membrane lipid content, and interfere with plasma proteins (Sen et al., 2020, Glover et al., 1999, Silveira et al., 2020, Pal et al., 2023, Silveira et al., 2019).

2.2 Antibacterial Agent

Why BSs not antibiotics?, The answer to this question is very simple. Antibiotic overuse and abuse have been linked to the emergence and spread of antimicrobial-resistant strains in recent decades (Lin et al., 2020). *Enterococcus faecalis*, *Staphylococcus aureus*, *Klebsiella pneumoniae*, *Acinetobacter baumanii*, *Pseudomonas aeruginosa*, and *Enterobacter sp.* are the common multi-drug resistant oraganisms (MDROs) linked to serious hospital-acquired infections that can be fatal (Xiaoxiao Gu et al., 2023). MDROs are microorganisms that are either intrinsically resistant to antibiotics or have developed resistance to one or more classes of antibiotics. *Escherichia coli* has been found to be resistant to sulfamethoxazole-trimethoprim globally, leading to increased usage of fluoroquinolones and cephalosporins. Antibiotic resistance is increasing and Gram-negative members of Enterobacteriaceae that generate extended-spectrum beta-lactamases (ESBLs) are a potential source of serious infections that are difficult to treat (Cox and Wright, 2013, Tenney et al., 2018). Furthermore, Gram-positive bacteria resistance epidemiology has dramatically changed with global concern for MRSA and VRE (Huang et al., 2019).

There are two main types of resistance to antimicrobial drugs: intrinsic and acquired. Intrinsic resistance is a characteristic specific to the genus or species of bacteria. Acquired resistance, on the other hand, is unique to a particular strain of bacteria and can arise due to the acquisition of foreign resistance genes or through mutations in chromosomal target genes (Stratton, 2000).

The identification of new and effective compounds for fighting infections is a challenging task. The development of such compounds requires high investments, and as a result, the number of novel antimicrobials discovered has significantly decreased in recent decades. Since 1962, only two new antibiotic classes have been approved for use, highlighting the need for greater investment and research in this field (Gudiña et al., 2016). The rise of MDROs has made it more urgent to develop new and effective antibiotics. This is particularly important because aging, immunosuppression, and invasive surgeries increase the risk of serious infections, while globalization has helped MDROs spread. MDROs infections are causing higher rates of illness, death, and healthcare expenses. Therefore, finding new therapeutic leads is crucial (Monciardini et al., 2014). Hence, the microbiota appears to be a promising and abundant source for the development of new drugs, particularly novel antibiotics to combat diseases and MDROs, which pose a major public health concern (Romano et al., 2017), and to concerns over the long term impact of synthetic chemicals on the environment and human health (Zhang et al., 2017).

SLs are natural glycolipids that possess antimicrobial properties and show great potential in the pharmaceutical industry. Their antimicrobial actions occur between the exponential and stationary phases (Płaza and Achal, 2020), and its safety is examined when combined in topically applied creams for wound healing. The toxicity of SLs on mammalian cells was assessed by *in vitro* viability experiments. The results indicated that concentrations of SLs below 0.5 mg/ml did not have any detrimental effects on either endothelium or keratinocyte-derived cell lines. Furthermore, the results obtained from *in vivo* experiments employing a mouse skin wounding assay indicated that the application of creams containing SLs did not have any impact on the temporal progression of wound healing. Additionally, histological analysis of the regenerated skin tissue confirmed that the healing process observed in the experimental group was comparable to that observed in the control group, as there was no indication of inflammation (Lydon et al., 2017). Several studies have examined the effects of SLs on the growth of both Gram-negative and Gram-positive bacteria. According to the results, the bactericidal effects of sophorolipids vary depending on the concentration, type of bacteria, sophorolipid derivative, length of hydroxy fatty acid carbons, and purity.

SLs obtained using lauryl alcohol (SLLA) completely inhibited gram-negative bacteria *Escherichia coli* (ATCC 8739) and *Pseudomonas aeruginosa* (ATCC 9027) at 30 and 1 μg ml^{-1} for 2 and 4 hours, respectively. Gram-positive bacteria *Staphylococcus aureus* (ATCC 6358) and *Bacillus subtilis* (ATCC 6633) showed complete inhibition at 6 and 1 μg ml^{-1} after 4 hours (Dengle-Pulate et al., 2014). The study investigated the impact of SLs on the growth of Gram-negative *Cupriavidus necator* ATCC 17699 and Gram-positive *Bacillus subtilis* BBK006. The results indicate that SLs have bactericidal effects at concentrations of 5% v/v (Diaz et al., 2015). It exhibited effective antibacterial activities against foodborne pathogen *Staphylococcus aureus* with MICs of 32 μg ml^{-1} (Chen et al., 2020). The efficacy of different types of SLs in reducing populations of pathogenic *Escherichia coli* O157:H7 was studied. It was found that the effectiveness of antimicrobial properties of SLs depended on various factors such as their type, duration, treatment concentration, and the presence of ethanol. The lactonic forms of stearic and oleic SLs were more effective in reducing the populations of *E. coli* O157:H7 compared to their free-acid counterparts. The susceptibility of different strains of *E. coli* O157:H7 to SLs varied. The study revealed that SLs inactivate bacteria, including *E. coli* O157:H7, by damaging the cell membrane. Ethidium monoazide-PCR amplification was used to analyze the mechanism of inactivation. The combination of low amounts of ethanol with SLs has the potential to effectively inactivate pathogenic *E. coli* O157:H7 (Zhang et al., 2017). Excellent antibacterial activity against *B. subtilis* was observed for derivative SL-ester, with an inhibition efficiency exceeding 96% at a concentration of 1 g L^{-1}. This activity was superior to that of natural SLs and derivative SL-COOH (Tang et al., 2020). The derivative SL-ester has been found to have the lowest MIC and MBC values of 0.05 and 0.20mg ml^{-1}, respectively. It has been found to have the most effective inhibitory effect against *S. aureus*, followed by *P. aeruginosa*. However, it has no significant inhibitory effect on *E. coli* and *Lactobacillus* sp. The derivative SL-ester has a greater impact on the cell membrane of *S. aureus* by "leaking," while it exhibits a stronger effect on the cell wall of *P. aeruginosa* by "blasting" (Ma et al., 2022). The antibacterial properties of acidic sophorolipids have been seen against *Enterococcus faecalis* and *P. aeruginosa*, which are common nosocomial pathogens. Remarkably, even at concentrations as low as 5 mg/ml, there were substantial decreases in colony-forming units (CFU) (Lydon et al., 2017). Furthermore, not only do sophorolipid derivatives exhibit differences in their antibacterial effects, the length of hydroxy fatty acid carbons can also have a significant impact. For instance, production of SLs from *Rhodotorula bogoriensis* (C22-SL) exhibits higher antibacterial activity against *Propionibacterium acnes* compared to SLs from *C. bombicola* (C18-SL) (Solaiman et al., 2015).

SLs generated reactive oxygen species in both *B. subtilis* and *E. coli* and flow cytometric analysis indicated that 60 mg L^{-1} of biosurfactants killed 65.8% of *B. subtilis* and 4% of *E. coli.* (Gaur et al., 2019). Bacteria exposed to sophorolipids showed a reduction in viability at doses of 500 μg ml^{-1} and 2,000 μg ml^{-1} against Gram-positive bacteria (*E. faecium*, *S. aureus* and *Streptococcus mutans*). and Gram-negative bacteria (*Proteus mirabilis*, *E. coli*, *Salmonella enterica* subsp. *enterica*), respectively (Fontoura et al., 2020). Recently, an innovative protocol using antimicrobial coated cytocompatible chitosan-sophorolipids hydrogel mesh was produced by 3D-printing, showed antibacterial activity against *S. aureus* planktonic bacteria with 61% growth inhibition when compared to the control (Narciso et al., 2023). Another innovative protocol is antibacterial photodynamic therapies (APDT) using photosensitizers which are less effective due to inadequate photosensitizers bioaccumulation. Hence, SLs are used to enhance the bioaccumulation of photosensitizers inside bacterial cells which increase the bactericidal effect of photosensitizers. Gu et al., (2023) reported using conjugated sophorolipid to toluidine blue (SL-TB) as APDT against *P. aeruginosa* and *S. aureus*. They found that free photosensitizers toluidine blue (TB) after radiation reduce the $\log_{10}$ (CFU) of *P. aeruginosa* to 4.5 and *S. aureus* to 7.9. SL-TB conjugates reduced *P. aeruginosa* and *S. aureus* CFU by 6.3 and 9.7 $\log_{10}$ units, respectively. The fluorescence quantitative results showed that SL- TB could accumulate more than TB in bacterial cells leading to more fatal effects (Xiaoxiao Gu et al., 2023). Filipe and colleagues created biodegradable films from cassava starch, pullulan, and sophorolipids 2.5%, 5%, and 10%) to control skin pathogens, with or without citric acid as a crosslinking agent. All films that included sophorolipids at various concentrations exhibited complete inhibition of the growth of both *S. aureus* and *S. epidermidis* (Filipe et al., 2023).

2.3 As an Antifungal Agent

The most common cause of both superficial and systemic fungal infections is *Candida albicans*. Unfortunately, *C. albicans* have developed resistance to most antifungals, posing a challenge for treatment. Most studies of the effect of SLs were conducted on *C. albicans* strains focusing on inhibiting planktonic growth as well as biofilm formation. The MIC (minimum inhibitory concentration) for SLs was determined against *C. albicans* and non-*albicans Candida* (NAC). MIC_{80} of *C. albicans* was found to be 60 μg ml^{-1} and on non-*albicans Candida* (NAC) the MIC_{80} for *C. tropicalis* and *C. glabrata* was 60 μg ml^{-1} and 120 μg ml^{-1}, respectively. *C. lusitaniae* was found to be the most susceptible (MIC_{80} 30 μg ml^{-1}) among the tested NAC strains (Haque et al., 2016).

Lactonic and acidic sophorolipids mixtures displayed antifungal activity against an opportunistic yeast pathogen *C. albicans* by effectively reducing hyphal growth inhibition with SLs concentration and incubation time. At 64 μg ml^{-1} SLs treatment, hyphae were significantly shorter than untreated (no SLs) and entirely inhibited at 500 and 1000 μg ml^{-1} after 5 hours with reducing cell survival and remaining in yeast form (Alfian et al., 2022). Another study on the other hand showed that SLs obtained using lauryl alcohol (SLLA) inhibited *C. albicans* (ATCC 2091) with 0% cell survival at 50 μg ml^{-1} in an incubation period of 4 hours (Dengle et al., 2014).

Excellent antifungal activity of derivative SL-ester was observed against *Moesziomyces* sp. The inhibition efficiency exceeded 83% at a concentration of 1 g L^{-1}, which is significantly better than natural SLs and SL-COOH (Tang et al., 2020). In a study conducted by Sen et al. (2020), SLs were identified as a potential treatment for dermatophytosis of the skin caused by *Trichophyton mentagrophytes*.

The study assessed the efficacy of SL-YS3, a sophorolipid produced by *Rhodotorula babjevae* YS3, in the treatment of dermatophytosis caused by *T. mentagrophytes*. The topical application of SL-YS3 for a duration of 21 days was conducted on cutaneous dermatophytosis in a mouse model. The evaluation involved assessing the percentage of culture recovery from skin samples, doing macroscopic observations, and performing histological analyses. The findings indicated that the infected mice were effectively treated and cured (Sen et al., 2020).

2.4 As Antibiofilm Agents

Infections caused by biofilm formation on medical device surfaces are a serious concern. This is especially true for commonly used invasive devices such as silicone-based ones, which are becoming increasingly resistant to antibacterial drugs. As a result, there is a growing need for alternative antibiofilm surfaces. The study demonstrated the ability of SLs (5% v/v) in disrupting biofilms generated by both single and mixed cultures of *B. subtilis* BBK006 and *S. aureus* ATCC 9144 under static and flow conditions, as confirmed by scanning electron microscopy (Diaz et al., 2015). Mendes et al. (2021) reported using a mixture of acidic sophorolipids and purified lactonic sophorolipids in two methods to inhibit biofilm formation of *S. aureus* on catheter surfaces (medical grade silicone). The two methods are:

(a) An antiadhesive strategy through covalent bonding of a mixture of acidic sophorolipids to the surface which is a cytocompatible surface reduced biofilm formation by 90%.
(b) Release strategy using isolated purified lactonic sophorolipids which show a higher effect on biofilm formation (Mendes et al., 2021).

The most common cause of both superficial and systemic fungal infections is *C. albicans*. Unfortunately, *C. albicans* biofilms develop resistance to most antifungals and represent the source of dispersing cells to invade other sites and establish new infections, due to the previous biofilm formation posing a challenge for treatment. *C. albicans* biofilms demonstrated the presence of a complex structured biofilm having hyphae and yeast cells. Biofilms formed in the presence of low SL concentrations such as 60 μg ml^{-1} were devoid of hyphal organization and consisted mostly of yeast. However, at moderate concentrations (such as 120 μg ml^{-1}), biofilm cells were found to have perforated outer membranes with a swollen and deformed morphology. SLs at a high concentration such as 240 μg ml^{-1} and 480 μg ml^{-1} aggregated cell populations with wrinkled surfaces can be seen (Haque et al., 2016).

SLs were found to inhibit *C. albicans* and NAC biofilm formation as well as reduce the viability of preformed biofilms. It is found that BIC_{80} (biofilm inhibiting concentration) of *C. glabrata* was the highest (480 μg ml^{-1}) whereas, BIC_{80} for *C. albicans, C. tropicalis* and *C. lusitaniae* was 120 μg ml^{-1} (Haque et al., 2016). Another study examined the effect of sophorolipid against *C. albicans* biofilm at two different phases of biofilm formation (adherence phase and maturation) to reveal that 125 μg ml^{-1} of SLs can reduce *C. albicans* 90-min-old biofilm by approximately 50% while 500 μg ml^{-1} of SLs could reduce the biofilm of *C. albicans* in a mature biofilm by approximately 50% in 24 hours (Alfian et al., 2022).

SLs lead to a substantial increase in the generation of reactive oxygen species (ROS) in *C. albicans* and the expression of genes associated with oxidative stress, namely *SOD1* and *CAT1*. Elevated levels of ROS within cellular structures lead to endoplasmic reticulum (ER) stress, resulting in the release of calcium ions (Ca^{2+}) into the cytoplasm and the disruption of the mitochondrial membrane potential (MMP). The data obtained from quantitative real time-polymerase chain reaction (qRT-PCR) analysis indicated that SL treatment resulted in an upregulation of the ER stress marker *HAC1*. The results obtained from flow cytometric analysis (using Annexin V/PI staining) suggest that cell death may have been due to necrosis (Haque et al., 2019).

Recently, an antimicrobial chitosan-sophorolipids hydrogel mesh was produced by 3D-printing, to reduce the biofilm activity of *S. aureus* to 2 log units inhibition when compared to the control (Narciso et al., 2023).

2.5 Sophorolipid Acts in Synergy with Antibiotics as Adjuvant to Enhance their Efficiency

SLs have been found to enhance the effectiveness of antibiotics when used in combination with them. For instance, when SLs and tetracycline were used together against *S. aureus*, total inhibition was achieved within 4 hours, as opposed to a 25% inhibition rate observed when tetracycline was used alone for 6 hours. Similarly, the combination of SLs and cefaclor was found to be more effective against *E. coli*, with a 48% greater inhibition rate observed within 2 hours, compared to cefaclor alone. Scanning electron microscopy revealed that the combination caused destruction of bacterial cell membranes and generation of pores. This is because SLs have amphiphilic properties that allow them to easily pass through similar cell membranes, thereby facilitating the entry of medication molecules (Joshi-Navare and Prabhune, 2013). The use of acidic sophorolipids as an adjuvant at subinhibitory concentrations of 2 and 4 mg ml^{-1} can reduce the MIC of kanamycin and cefotaxime against *E. faecalis* and *P. aeruginosa*. Between the two concentrations, 4 mg ml^{-1} is the most effective in reducing the antibiotic MICs. It is worth noting that 4 mg ml^{-1} alone also has antibacterial activity against both strains (Lydon et al., 2017). More et al. (2019) used calcium alginate as a binding agent, with sericin and SLs as the primary ingredients in their experiment. The components of the mixture were biocompatible and biodegradable, making it ideal for evaluating the wound healing properties in Wistar rats. The study used povidone as a readily available ointment for comparison. The animal group that received treatment with sericin and SLs cream showed rapid contraction, closure, and healing, compared to the control and commercial ointment groups. Histological examinations verified the results, revealing increased rates of fibroblast proliferation, angiogenesis, and keratinization.

SLs and nisin increased *S. aureus* cell permeability, released intracellular contents and enhanced extracellular enzyme activities, while reducing respiratory-chain dehydrogenase activity and fractional inhibitory concentration (FIC) index for SLs and nisin combination was 1.5, indicating that the effect of SLs with nisin was additive (Chen et al., 2020). SLs and palmarosa essential oil are used in cosmetic formulations. Their association resulted in a self-preserving cosmetic formulation with great stability. SLs and palmarosa oil acted synergistically and additively to reduce the concentration required to kill *S. aureus* and *S. epidermidis* by 98.4% and 50%, respectively (Filipe et al., 2022).

2.6 Sophorolipids as Antiviral and Spermicidal Agents

The antiviral activity of SLs is believed to result from the disturbance or damage of the viral membrane. The presence of acetyl groups in the structure of SLs is crucial in enhancing its antiviral effectiveness by providing hydrophilic properties to the compound (Daverey et al., 2021). Shah et al. (2005), demonstrated that diacetate ethyl ester of sophorolipid has strong virucidal properties against HIV. Additionally, it was found that this compound also has potent spermicidal activity, with the ability to immobilize almost 100% of spermatozoa within 120 seconds. The study suggested that a longer chain length in the creation of micelles is positively correlated with the spermicidal impact, while an inverse relationship is observed in its virucidal characteristic. Research was conducted to investigate the impact of SLs on Epstein-Barr virus. The Epstein-Barr virus was absorbed on Daudi lymphoid cell lines and the viroid capsid antigen was evaluated after adding the SLs. All the examined SLs showed anti-herpes virus activity, with Ethyl 17-L-[(2′-O-β-D-glucopyranosyl-β-D-glucopyranosyl)-oxy]-cis-9-octadecenoate having a stronger antiviral effect with EC_{50} value of <0.03 μM (Gross and Shah, 2007, Borsanyiova et al., 2016). During the Covid-19 pandemic, researchers were tasked with finding effective treatment options for patients to help save lives. With the knowledge that both HIV and Epstein-Barr viruses are enveloped viruses, it is suggested that SLs could be a potential antiviral treatment for SARS-CoV-2 (Daverey et al., 2021).

2.7 Role of Sophorolipids in Immunity

Several natural and synthetic components of SLs have been demonstrated to exhibit distinct anti-inflammatory properties. The anti-inflammatory effects associated with sepsis involve various mechanisms, such as the decrease in nitric oxide levels, the regulation of inflammatory cytokines, and the manipulation of cell surface adhesion molecules. A research has provided evidence indicating that SLs have the ability to alter the expression of pro-inflammatory cytokines, such as Interleukin (IL)-1α, IL-1β, and IL-6, among other cytokines (Bluth et al., 2006a, Hardin et al., 2007). According to Hardin et al. (2007), administering SLs after inducing intra-abdominal sepsis can enhance the survival rate. The study provides empirical evidence that SLs have the potential to reduce sepsis-related mortality through various dose regimens and derivatives. This finding adds to the existing research that highlights the viability of SLs as a possible therapeutic option for sepsis. However, it's worth noting that toxicity becomes apparent when the dosage administered to septic animals exceeds 75 to 150 times the therapeutic dose. SLs showed treatment effect in a dextran sulfate sodium (DSS)-induced colitis mouse model when supplemented in their food. The growth performance was enhanced three days after DSS treatment with the administration of dietary SL. Additionally, the histopathological score was seen to be lower in the DSS-treated SLs group compared to the DSS-treated control group. The groups treated with SPL exhibited a notable increase in mucosal thickness and goblet cell counts when compared to the control group. In a similar manner, the supplementation of SLs resulted in an upregulation in gene expression levels for mucin-2, interleukin-10 (IL-10), and transforming growth factor-β, as well as an increase in the concentration of short-chain fatty acid as compared to the control groups (Kwak et al., 2022).

The U266 cells, when grown in cRPMI or sucrose vehicle, exhibited a significant increase in IgE production, with levels reaching 520 IU+/-32. The addition of SLs resulted in a significant reduction in IgE production, with the maximum effect observed at a concentration of 1.0 μg ml^{-1} (416+/-8SE) ($p<0.01$). Equivalent levels of soluble CD23 (sCD23) and cell surface expression of CD23 were observed across all experimental groups. The addition of higher concentrations (10 μg ml^{-1}) of SLs exhibited a correlation with a bimodal cell surface expression of CD38 on the cell surface, as well as an elevation in the proportions of plasma-like cells as compared to the control group (14% and 4% respectively) ($p<0.05$) (Bluth et al., 2006b). Recently, Xu et al. (2023), reported that SLs can be a useful treatment for histamine-dependent itch by decreasing PLC/IP3R signaling pathway activation and modulating TRPV1 activity when studied on mice with histamine-induced scratching behaviors.

The preliminary assessment of pro-inflammatory cytokine production HaCaT cells was conducted using a semi-quantitative array. The results indicated that lactonic SLs and di-RL led to an increase in the production of IL-8. The level of IL-8 produced was above the threshold required for detection through the use of ELISA (Adu et al., 2023).

2.8 Sophorolipids as Anticancer Agents

SLs possess distinct characteristics within the glycolipid family due to their high susceptibility to chemoenzymatic modifications. This feature grants them the potential for specific treatments targeting particular diseases. In this context, it has been demonstrated that SLs or certain derivatives exhibit anticancer properties, which suggest that SLs may have potential as a treatment option for cancer (Fu et al., 2008). The administration of SLs treatment resulted in inducing programmed cell death (apoptosis) in liver cancer cells (H7402). This was achieved by blocking the progression of the cell cycle at the G1 phase and partially at the S phase. Moreover, the treatment also activated caspase-3, an enzyme involved in apoptosis, and resulted in an increase in the concentration of calcium ions within the cytoplasm (Chen et al., 2006a). The SLs compound demonstrated a notable suppression of cellular growth across four distinct cell lines [H7402 (liver cancer line), A549 (lung cancer line), HL60 and K562 (leukemia lines)], with the degree of inhibition directly corresponding to the concentration of the drug, which varied from 0 to 62.5 μg ml^{-1}. When the drug concentration reached 62.5 μg ml^{-1}, there was an observed cell viability of approximately 0% (Chen et al., 2006b). It was observed that the cell viability in H7402 liver cancer cells decreased in a dose-dependent manner upon the administration of SLs. However, in the case of HL-7702 normal liver cells, the administration of SLs did not have any noticeable effect. Similarly, in the Chang liver cell line, a slight decline in cell viability was observed (Chen et al., 2006a). The same effects were observed by Rashad et al. (2014), on hepato- cellular carcinoma HepG2 and A549. In a study conducted by Fu et al. (2008), it was discovered that SLs had notable anticancer properties when tested against human pancreatic carcinoma cells. The cytotoxicity exhibited consistent results across all tested doses, with a natural combination serving as the mediator (20 ± 4%). In contrast, the cytotoxicity levels exhibited by the methyl ester derivative were significantly higher (63 ± 5%) when compared to the other derivatives (ethyl ester diacetate, 36 ± 6%, ethyl ester monoacetate, 18 ± 7%; $P < 0.05$). Acidic sophorolipids were observed to reduce the viability of colorectal cancer lines with dose-dependent reduction. However, they proved to be non-toxic to normal human colonic and lung cells. In addition, the administration of acidic sophorolipids at a dosage of 50 mg kg^{-1} orally for a duration of 70 days to $Apc^{min+/-}$ mice was found to be well-tolerated. This treatment led to an elevation in hematocrit levels, along with a reduction in both splenic size and red pulp area. The administration of oral food did not demonstrate any significant impact on the quantity or dimensions of tumors in this particular model (Callaghan et al., 2016, Callaghan et al., 2022). It was found that the cytotoxicity of SLs becomes stronger as the length of the carbon chain increases in different SL derivatives. Di-acetylated lactonic C18:1 sophorolipid inhibited the viability of human cervical cancer HeLa cells *in vitro* and *in vivo* without significant toxicity to tumor-bearing mice, the inhibition carried out by apoptosis and cell cycle was blocked at G0 phase and partly at G2 phase (Li et al., 2017).

Nawale et al. (2017), has reported that SLs synthesized using cetyl alcohol as the substrate (referred to as SLCA) significantly inhibit the survival of HeLa and HCT 116 cells without affecting the viability of normal human umbilical vein endothelial cells (HUVEC). The SLCA blocks the cell cycle progression of HeLa cells at the G1/S phase in a time-dependent manner, and induces apoptosis in HeLa cells through an increase in intracellular Ca^{2+} leading to the depolarization of mitochondrial membrane potential. Additionally, the SLCA increases the caspase-3, -8, and -9 activity. The human keratinocyte and malignant melanocyte cell lines, HaCaT and SK-Mel-28, were subjected to treatment with acidic sophorolipid and lactonic sophorolipid preparations. The concentrations of these preparations ranged from 0 to 100 $\mu g\ ml^{-1}$. The cytotoxic effect of acidic sophorolipids on both cell lines was found to be insignificant. The viability of SK-MEL-28 cells was considerably reduced ($p < 0.05$) when exposed to lactonic sophorolipids at concentrations beyond 40 $\mu g\ ml^{-1}$. Similarly, HaCaT cells showed a significant reduction in viability at concentrations above 60 $\mu g\ ml^{-1}$. Furthermore, the adverse effect of lactonic sophorolipids on the SK-MEL-28 cell line was significantly greater compared to HaCaT cells ($p \leq 0.05$ at 40 $\mu g\ ml^{-1}$; $p \leq 0.001$ at 60 $\mu g\ ml^{-1}$) (Adu et al., 2022). Breast cancer cell line (MDA-MB-321) also reported to be inhibited by lactonic SLs (C18:0, C18:1and C18:3) with higher cytotoxic effects compared to acidic SLs. Furthermore, SLs effectively suppress the migration of MDA-MB-231 cells while maintaining cell viability, as well as the ability to enhance the levels of intracellular ROS (Ribeiro et al., 2015).

A study conducted by Shao et al. (2012) indicates that the inhibitory effect of diacetylated lactonic sophorolipid on two types of human esophageal cancer cell lines KYSE 109 and KYSE 450 was more potent compared to monoacetylated lactonic sophorolipid. Specifically, complete inhibition of cell growth was observed at a concentration of 30 mg ml^{-1} for diacetylated lactonic sophorolipid, but a dosage of 60 mg ml^{-1} was required for entire inhibition with monoacetylated lactonic sophorolipid. The variation in the unsaturation degree of hydroxyl fatty acids in SL molecules significantly impacted their cytotoxicity towards esophageal cancer cells. The sophorolipid containing a single double bond in the fatty acid component had the most potent cytotoxic activity against two esophageal cancer cell lines, resulting in complete suppression at a dosage of 30 mg ml^{-1}. The acidic sophorolipid demonstrated minimal efficacy in inhibiting the growth of esophageal cancer cells (Shao et al., 2012). Significantly, a complex of curcumin with acidic sophorolipid enhanced the cytotoxicity effect of curcumin on breast cancer cell lines, MCF-7 and MDA-MB-231, due to the presence of the glucose moiety (Singh et al., 2014). To overcome the limitations of inadequate intracellular transport commonly observed with biosurfactants, poly (lactic-co-glycolic acid) (PLGA) nanocapsules encapsulating sophorolipids were synthesized_and employed for the targeted delivery of sophorolipids to tumor cell lines (CT26 murine colon cancer) in both *in vitro* and *in vivo* models. The impact of peripheral hydrophilicity was assessed. Formulations containing a 10% density of poly ethylene glycol have shown a significant reduction of over 80% in the viability of cancer cells over a 72 hour period. Additionally, these formulations exhibited an increased absorption by CT26 cells. The aforementioned formulations demonstrated increased tumor accumulation and a prolonged blood circulation profile in comparison to the nanocapsules lacking poly ethylene glycol. The experimental group of animals, which received sophorolipid-loaded nanocapsules, exhibited a 57% reduction in tumor growth compared to the control group. A comparative analysis of tumor mass within the identical research cohort revealed the most significant decrease when comparing the control group to the cohorts treated with free medication (Haggag et al., 2020).

2.9 Role of Sophorolipids in Improving the Drug Loading and Stability

Although SLs demonstrate promising applicability in biomedical fields, their translational applications for disease treatment are limited by their higher production cost, poor aqueous solubility, lower purification levels and substandard knowledge to understand their interactions with cells (Banat et al., 2010). Hence nanoparticle-based delivery systems are considered to improve bioavailability of SLs in cells (Shehzad et al., 2014, Klippstein et al., 2015). Recently Haggag et al. (2020), used glyceryl monocaprate based nanocapsules formulations to deliver SL to CT26 murine colon carcinoma and found more growth inhibition of CT26 cells compared to control nanocapsules.

Peng et al., (2018) developed a nano-delivery system to improve the aqueous solubility and oral bioavailability of curcumin. Curcumin is a beneficial nutraceutical that offers several health benefits to mankind. In this study SL-coated curcumin nanoparticles were formulated which showed 82% encapsulation efficiency and 14% loading capacity for curcumin.

The use of nanocarriers developed for delivery of oral therapeutic agents, is constrained due to the physical stability, mucus penetration, cellular uptake and transport of nanocarriers in the gastrointestinal tract. Liu et al. (2022), developed SL coated nanoparticles for oral delivery of therapeutic agents targeted for improved bioavailability and control of breast cancer metastasis. The formulated nanoparticles exhibited extended diffusion in mucus due to higher affinity of SL for mucin that offered protection of nanoparticles during mucus penetration. Recently Chen & Zhang (2022), reported improved stabilization and dispersion of liposomes in the presence of SLs due to lower particle sizes and PDIs and increased serum stability. Traditional anticancer agent etoposide is commercially available in non-aqueous oral and parenteral solutions. The aqueous solubility and comfort for drug administration can be improved with biosurfactant based nano-formulations. Ma et al. (2022), developed a novel sophorolipid based strategy for the preparation of etoposide-loaded emulsions. Acidic and lactonic sophorolipids were used instead of tween 80. The performance of acidic SLs was better than tween 80 and lactonic SLs. Acidic SLs- etoposide emulsions showed higher drug loading capacity, slow drug release rate and more antitumor activity against ovarian cancer cell line A2780 than tween-80-based emulsions and commercial etoposide injections. In the study made by Gu et al. (2023), a sophorolipid developed a sophorolipid-associated membrane-biomimetic choline phosphate-poly(lactic-co-glycolic) acid hybrid nanoparticles showing improved endocytosis due to more physical stability in gastrointestinal tract and rapid mucus diffusion. Poor availability and low solubility of hydrophobic drugs limits their body absorption and hence affects applicability. Hydrophobic drugs have poor bioavailability and low solubility, which makes it difficult for them to be absorbed by the body. To address this challenge, solid lipid nanoparticles (SLNs) have been suggested as a biocompatible vector for drug encapsulation. Kanwar and his colleagues reported the creation of SLNs using the solvent injection method and sophorolipids as the lipidic phase. The leprosy drugs rifampicin and dapsone (4 mg) were loaded into the SLNs. The SLNs with the drug-loaded lipidic phase demonstrated good drug concentration release (Kanwar et al., 2018).

3. Conclusion

The present review presented the applications of SLS in biomedical and pharmaceutical sectors focusing on their antibacterial, antifungal, antiviral, and anticancer properties. The use of SLs in nanocarrier formulations for improved drug delivery and stability has also been reviewed in detail. Based on their eco friendly nature, and applications free of side effects, SLs can act as a good alternative for their synthetic counterparts. We expect similar increasing interest in SLs based research activities among the scientific fraternity in the near future.

References

Adu, S.A., Twigg, M.S., Naughton, P.J., Marchant, R. & Banat, I.M. (2022). Biosurfactants as anticancer agents: Glycolipids affect skin cells in a differential manner dependent on chemical structure. *Pharmaceutics*, 14(2): 360. https://doi.org/10.3390/pharmaceutics14020360

Adu, S.A., Twigg, M.S., Naughton, P.J., Marchant, R. & Banat, I.M. (2023). Characterization of cytotoxicity and immunomodulatory effects of glycolipid biosurfactants on human keratinocytes. *Appl. Microbiol. Biotechnol.*, 107(1): 137-152. https://doi.org/10.1007/s00253-022-12302-5

Alfian, A.R., Watchaputi, K., Sooklim, C. & Soontorngun, N. (2022). Production of new antimicrobial palm oil-derived sophorolipids by the yeast Starmerella riodocensis sp. nov. against Candida albicans hyphal and biofilm formation. *Microb. Cell Fact.*, 21(1): 1-18. https://doi.org/10.1186/s12934-022-01852-y

Ali, S.A.M., Sayyed, R.Z., Mir, M.I., Hameeda, B., Khan, Y., Alkhanani, M.F. et al. (2022a). Induction of systemic resistance in maize and antibiofilm activity of surfactin from *Bacillus velezensis* MS20. *Front. Microbiol.*, 13: 879739. https://doi.org/10.3389/fmicb.2022.879739

Ali, S.A.M., Sayyed, R.Z., Reddy, M.S., Enshasy, H.E. & Hameeda, B. (2022b). Delving through quorum sensing and CRISPRi strategies for enhanced surfactin production. *In:* Sayyed, R.Z. 59-79. (Eds), *Biosurfactants: Production and Applications in Bioremediation/ Reclamation.* CRC Press, Taylor & Francis Group, USA.

Banat, I.M., Franzetti, A., Gandolfi, I., Bestetti, G., Martinotti, M.G., Fracchia, L. et al. (2010). Microbial biosurfactants production, applications and future potential. *Appl. Microbiol. Biotechnol.*, 87(2): 427-444. https://doi.org/10.1007/s00253-010-2589-0

Bee, H., Khan, M.Y. & Sayyed, R.Z. (2019). Microbial surfactants and their significance in agriculture. *In:* Sayyed Reddy Antonious (Eds), *PGPR: Prospects for Sustainable Agriculture.* 205-216. Springer-Nature, Singapore.

Bluth, Martin H., Kandil, E., Mueller, C.M., Shah, V., Lin, Y.-Y., Zhang, H. et al. (2006a). Sophorolipids block lethal effects of septic shock in rats in a cecal ligation and puncture model of experimental sepsis. *Crit Care Med.*, 34(1): E188. https://doi.org/10.1097/01.CCM.0000196212.56885.50

Bluth, M.H., Smith-Norowitz, T.A., Hagler, M., Beckford, R., Chice, S., Shah, V. et al. (2006b). Sophorolipids decrease IgE production in U266 cells. *J. Allergy Clin. Immunol.*, 117(2): S202. https://doi.org/10.1016/j.jaci.2005.12.797

Borsanyiova, M., Patil, A., Mukherji, R., Prabhune, A., Bopegamage, S. (2016). Biological activity of sophorolipids and their possible use as antiviral agents. *Folia Microbiol.* (Praha). 61(1): 85-89. https://doi.org/10.1007/s12223-015-0413-z

Callaghan, B., Lydon, H., Roelants, S.L.K.W., Van Bogaert, I.N.A., Marchant, R., Banat, I.M. et al. (2016). Lactonic sophorolipids increase tumor burden in Apcmin+/- mice. *PLoS One*. 11(6): 1-16. https://doi.org/10.1371/journal.pone.0156845

Callaghan, B., Twigg, M.S., Baccile, N., Van Bogaert, I.N.A., Marchant, R., Banat, I.M. et al. (2022). Microbial sophorolipids inhibit colorectal tumor cell growth in vitro and restore haematocrit in Apcmin+/− mice. *Appl. Microbiol. Biotechnol.*, 106(18): 6003-6016. https://doi.org/10.1007/s00253-022-12115-6

Cavalero, D.A. & Cooper, D.G. (2003). The effect of medium composition on the structure and physical state of sophorolipids produced by Candida bombicola ATCC 22214. *J. Biotechnol.*, 103(1): 31-41. https://doi.org/10.1016/S0168-1656(03)00067-1

Chen, H. & Zhang, Q. (2022). Surface functionalization of piperine-loaded liposomes with sophorolipids improves drug loading and stability. *J. Pharm. Innov.*, https://doi.org/10.1007/s12247-022-09687-1

Chen, J., Song, X., Zhang, H., Qu, Y. & Miao, J. (2006a). Sophorolipid produced from the new yeast strain Wickerhamiella domercqiae induces apoptosis in H7402 human liver cancer cells. *Appl. Microbiol. Biotechnol.*, 72(1): 52-59. https://doi.org/10.1007/s00253-005-0243-z

Chen, J., Song, X., Zhang, H. & Qu Y. (2006b). Production, structure elucidation and anticancer properties of sophorolipid from Wickerhamiella domercqiae. *Enzyme Microb Technol.*, 39(3): 501-506. https://doi.org/10.1016/j.enzmictec.2005.12.022

Chen, J., Zhifei, L.U., An, Z., Ji, P. & Liu, X. (2020). Antibacterial activities of sophorolipids and nisin and their combination against foodborne pathogen Staphylococcus aureus. *Eur. J. Lipid Sci. Technol.*, 122(3): 1-26. https://doi.org/10.1002/ejlt.201900333

Cox, G. & Wright, G.D. (2013). Intrinsic antibiotic resistance: Mechanisms, origins, challenges and solutions. *Int. J. Med. Microbiol.*, 303(6-7): 287-292. https://doi.org/10.1016/j.ijmm.2013.02.009

Daverey, Amita, Dutta, K., Joshi, S. & Daverey, Achlesh. (2021). Sophorolipid: A glycolipid biosurfactant as a potential therapeutic agent against COVID-19. *Bioengineered*, 12(2): 9550-9560. https://doi.org/10.1080/21655979.2021.1997261

Dengle-Pulate, V., Chandorkar, P., Bhagwat, S. & Prabhune, A.A. (2014). Antimicrobial and SEM studies of sophorolipids synthesized using lauryl alcohol. *J. Surfactants Deterg.*, 17(3): 543-552. https://doi.org/10.1007/s11743-013-1495-8

Desai, J.D. & Banat, I.M. (1997). Microbial production of surfactants and their commercial potential. *Microbiol. Mol. Biol. Rev.*, 61(1): 47-64. https://doi.org/10.1128/.61.1.47-64.1997

Díaz De Rienzo, M.A., Banat, I.M., Dolman, B., Winterburn, J. & Martin, P.J. (2015). Sophorolipid biosurfactants: Possible uses as antibacterial and antibiofilm agent. *N. Biotechnol.*, 32(6): 720-726. https://doi.org/10.1016/j.nbt.2015.02.009

Filipe, G.A., Bigotto, B.G., Baldo, C., Gonçalves, M.C., Kobayashi, R.K.T., Lonni, A.A.S.G. et al. (2022). Development of a multifunctional and self-preserving cosmetic formulation using sophorolipids and palmarosa essential oil against acne-causing bacteria. *J. Appl. Microbiol.*, 133(3): 1534-1542. https://doi.org/10.1111/jam.15659

Filipe, G.A., Silveira, V.A.I., Gonçalves, M.C., Beltrame Machado, R.R., Nakamura, C.V., Baldo C. et al. (2023). Bioactive films for the control of skin pathogens with sophorolipids from Starmerella bombicola. *Polym. Bull.*, 80(10): 10809-10823. https://doi.org/10.1007/s00289-022-04575-7

Fontoura, I.C.C., da, Saikawa, G.I.A., Silveira, V.A.I., Pan, N.C., Amador, I.R., Baldo, C. et al. (2020). Antibacterial activity of sophorolipids from Candida bombicola against human pathogens. *Brazilian Arch. Biol. Technol.*, 63: 1-10. https://doi.org/10.1590/1678-4324-2020180568

Fu, S.L., Wallner, S.R., Bowne, W.B., Hagler, M.D., Zenilman, M.E., Gross, R. et al. (2008). Sophorolipids and their derivatives are lethal against human pancreatic cancer cells. *J. Surg Res.*, 148(1): 77-82. https://doi.org/10.1016/j.jss.2008.03.005

Gaur, V.K., Regar, R.K., Dhiman, N., Gautam, K., Srivastava, J.K., Patnaik, S. et al. (2019). Biosynthesis and characterization of sophorolipid biosurfactant by Candida spp.: Application as food emulsifier and antibacterial agent. *Bioresour. Technol.*, 285: 121314. https://doi.org/10.1016/j.biortech.2019.121314

Glover, R.E., Smith, R.R., Jones, M.V., Jackson, S.K. & Rowlands, C.C. (1999). An EPR investigation of surfactant action on bacterial membranes. *FEMS Microbiol Lett.*, 177(1): 57-62. https://doi.org/10.1016/S0378-1097(99)00289-X

Gross, R.A. & Shah, V. (2007). Anti-herpes virus properties of various forms of sophorolipids. Polytech Univ assignee.

Gu, Xiaoxiao, Xu, L., Yuan, H., Li, C., Zhao, J., Li, S. & Yu, D. (2023). Sophorolipid-toluidine blue conjugates for improved antibacterial photodynamic therapy through high accumulation. *RSC Adv.*, 13(17): 11782-11793. https://doi.org/10.1039/d3ra01618h

Gu, Xiaoyan, Zhang, R., Sun, Y., Ai, X., Wang, Y., Lyu, Y. et al. (2023). Oral membrane-biomimetic nanoparticles for enhanced endocytosis and regulation of tumor-associated macrophage. *J. Nanobiotechnology.*, 21(1): 206. https://doi.org/10.1186/s12951-023-01949-5

Gudiña, E., Teixeira, J. & Rodrigues, L. (2016). Biosurfactants produced by marine microorganisms with therapeutic applications. *Mar. Drugs*, 14(2): 38. https://doi.org/10.3390/md14020038

Haggag, Y., Elshikh, M., El-Tanani, M., Bannat, I.M., McCarron, P. & Tambuwala, M.M. (2020). Nanoencapsulation of sophorolipids in PEGylated poly(lactide-co-glycolide) as a novel approach to target colon carcinoma in the murine model. *Drug Deliv Transl Res.*, 10(5): 1353-1366. https://doi.org/10.1007/s13346-020-00750-3

Hajfarajollah, H., Eslami, P., Mokhtarani, B. & Akbari Noghabi, K. (2018). Biosurfactants from probiotic bacteria: A review. *Biotechnol Appl Biochem.*, 1-40. https://doi.org/10.1002/bab.1686

Haque, F., Alfatah, M., Ganesan, K & Bhattacharyya, M.S. (2016). Inhibitory effect of sophorolipid on Candida albicans biofilm formation and hyphal growth. *Sci. Rep.*, 6(September 2015): 1-11. https://doi.org/10.1038/srep23575

Haque, F., Verma, N.K., Alfatah, M., Bijlani, S. & Bhattacharyya, M.S. (2019). Sophorolipid exhibits antifungal activity by ROS mediated endoplasmic reticulum stress and mitochondrial dysfunction pathways in Candida albicans. *RSC Adv.*, 9(71): 41639-41648. https://doi.org/10.1039/c9ra07599b

Hardin, R., Pierre, J., Schulze, R., Mueller, C.M., Fu, S.L., Wallner, S.R. et al. (2007). Sophorolipids improve sepsis survival: Effects of dosing and derivatives. *J. Surg. Res.*, 142(2): 314-319. https://doi.org/10.1016/j.jss.2007.04.025

Huang, L., Zhang, R., Hu, Y., Zhou, H., Cao, J., Lv, H. et al. (2019). Epidemiology and risk factors of methicillin-resistant Staphylococcus aureus and vancomycin-resistant enterococci infections in Zhejiang China from 2015 to 2017. *Antimicrob Resist Infect Control.*, 8(1): 90. https://doi.org/10.1186/s13756-019-0539-x

Jahan, R., Bodratti, A.M., Tsianou, M. & Alexandridis, P. (2020). Biosurfactants, natural alternatives to synthetic surfactants: Physicochemical properties and applications. *Adv Colloid Interface Sci.*, 275: 102061. https://doi.org/10.1016/j.cis.2019.102061

Joshi-Navare, K. & Prabhune, A. (2013). A biosurfactant-sophorolipid acts in synergy with antibiotics to enhance their efficiency. *Biomed Res Int.*, 2013: 1-8. https://doi.org/10.1155/2013/512495

Kanwar, R., Gradzielski, M. & Mehta, S.K. (2018). Biomimetic solid lipid nanoparticles of sophorolipids designed for antileprosy drugs. *J Phys Chem B.*, 122(26): 6837-6845. https://doi.org/10.1021/acs.jpcb.8b03081

Klippstein, R., Wang, J.T.-W., El-Gogary, R.I., Bai, J., Mustafa, F., Rubio, N. et al. (2015). Passively targeted curcumin-loaded PEGylated PLGA nanocapsules for colon cancer therapy in vivo. *Small.*, 11(36): 4704-4722. https://doi.org/10.1002/smll.201403799

Kwak, M.J., Ha, D.J., Choi, Y.S., Lee, H. & Whang, K.Y. (2022). Protective and restorative effects of sophorolipid on intestinal dystrophy in dextran sulfate sodium-induced colitis mouse model. *Food Funct.*, 13(1): 161-169. https://doi.org/10.1039/d1fo03109k

Li, H., Guo, W., Ma, X., Li, J. & Song, X. (2017). In vitro and in vivo anticancer activity of sophorolipids to human cervical cancer. *Appl Biochem Biotechnol.*, 181(4): 1372-1387. https://doi.org/10.1007/s12010-016-2290-6

Lin, L.Z., Zheng, Q.W., Wei, T., Zhang, Z.Q., Zhao, C.F., Zhong, H. et al. (2020). Isolation and characterization of fengycins produced by Bacillus amyloliquefaciens JFL21 and its broad-spectrum antimicrobial potential against multidrug-resistant foodborne pathogens. *Front Microbiol.*, 11(December). https://doi.org/10.3389/fmicb.2020.579621

Liu, Y., Shen, J., Shi, J., Gu, X., Chen, H., Wang, X. et al. (2022). Functional polymeric core–shell hybrid nanoparticles overcome intestinal barriers and inhibit breast cancer metastasis. *Chem Eng J.*, 427(August 2021): 131742. https://doi.org/10.1016/j.cej.2021.131742

Lydon, H.L., Baccile, N., Callaghan, B., Marchant, R., Mitchell, C.A. & Banat, I.M. (2017). Adjuvant antibiotic activity of acidic sophorolipids with potential for facilitating wound healing. *Antimicrob Agents Chemother.*, 61(5). https://doi.org/10.1128/AAC.02547-16

Ma, X., Wang, T., Yu, Z., Shao, J., Chu, J., Zhu, H. & Yao, R. (2022). Formulation and physicochemical and biological characterization of etoposide-loaded submicron emulsions with biosurfactant of sophorolipids. *AAPS Pharm Sci Tech.*, 23(6): 181. https://doi.org/10.1208/s12249-022-02329-2

Ma, X.J., Wang, T., Zhang, H.M., Shao, J.Q., Jiang, M., Wang, H. et al. (2022). Comparison of inhibitory effects and mechanisms of lactonic sophorolipid on different pathogenic bacteria. *Front Microbiol.*, 13(September): 1-10. https://doi.org/10.3389/fmicb.2022.929932

Mendes, R.M., Francisco, A.P., Carvalho, F.A., Dardouri, M., Costa, B., Bettencourt, A.F. et al. (2021). Fighting S. aureus catheter-related infections with sophorolipids: Electing an antiadhesive strategy or a release one? *Colloids Surfaces B Biointerfaces*, 208(August): 112057. https://doi.org/10.1016/j.colsurfb.2021.112057

Monciardini, P., Iorio, M., Maffioli, S., Sosio, M. & Donadio, S. (2014). Discovering new bioactive molecules from microbial sources. *Microb Biotechnol.*, 7(3): 209-220. https://doi.org/10.1111/1751-7915.12123

More, S.V., Koratkar, S.S., Kadam, N.R., Agawane, S.B. & Prabhune, A.A. (2019). Formulation and evaluation of wound healing activity of sophorolipid-sericin gel in wistar rats. *Pharmacogn Mag.*, 15: 123-127. https://doi.org/10.4103/PM.PM_392_18

Narciso, F., Cardoso, S., Monge, N., Lourenço, M., Martin, V., Duarte, N. et al. (2023). 3D-printed biosurfactant-chitosan antibacterial coating for the prevention of silicone-based associated infections. *Colloids Surfaces B Biointerfaces*. 230(August): 113486. https://doi.org/10.1016/j.colsurfb.2023.113486

Nawale, L., Dubey, P., Chaudhari, B., Sarkar, D. & Prabhune, A. (2017). Anti-proliferative effect of novel primary cetyl alcohol derived sophorolipids against human cervical cancer cells HeLa. *PLoS One*, 12(4): 1-14. https://doi.org/10.1371/journal.pone.0174241

Pal, S., Chatterjee, N., Das, A.K., McClements, D.J. & Dhar, P. (2023). Sophorolipids: A comprehensive review on properties and applications. *Adv Colloid Interface Sci.*, 313(August). https://doi.org/10.1016/j.cis.2023.102856

Patel, P., Bhatt, S., Patel, H., Marcelino, L.A. & Sayyed, R.Z. (2022). Biosurfactant - A biomolecule and its potential applications. *In:* Sayyed, R.Z. and Enshasy, H.E. (Eds), *Biosurfatnats: Production and Applications in Food and Agriculture*. Vol II, 133-149. CRC Press, Taylor & Francis Group, USA.

Peng, S., Li, Z., Zou, L., Liu, W., Liu, C. & McClements, D.J. (2018). Enhancement of curcumin bioavailability by encapsulation in sophorolipid-coated nanoparticles: An in vitro and in vivo study. *J Agric Food Chem.*, 66(6): 1488-1497. https://doi.org/10.1021/acs.jafc.7b05478

Płaza, G. & Achal, V. (2020). Biosurfactants: Eco-friendly and innovative biocides against biocorrosion. *Int J Mol Sci.*, 21(6): 2152. https://doi.org/10.3390/ijms21062152

Rashad, M.M., Nooman, M.U., Ali, M.M., Al-kashef, A.S. & Mahmoud, A.E. (2014). Production, characterization and anticancer activity of Candida bombicola/sophorolipids by means of solid state fermentation of sunflower oil cake and soybean oil. *Grasas y Aceites*, 65(2): e017. https://doi.org/10.3989/gya.098413

Ravinder, R., Manasa, M., Roopa, D., Bukhari, N.A., Hatamleh, A.A., Khan, M.Y. et al. (2022). Biosurfactant producing multifarious Streptomyces puniceus RHPR9 of Coscinium fenestratum rhizosphere promotes plant growth in chilli. *Plos One*, 17(3): e0264975. https://doi.org/10.1371/journal.pone.0264975

Ribeiro, I.A.C., Faustino, C.M.C., Guerreiro, P.S., Frade, R.F.M., Bronze, M.R., Castro, M.F. et al. (2015). Development of novel sophorolipids with improved cytotoxic activity toward MDA-MB-231 breast cancer cells. *J Mol Recognit.*, 28(3): 155-165. https://doi.org/10.1002/jmr.2403

Romano, G., Costantini, M., Sansone, C., Lauritano, C., Ruocco, N. & Ianora, A. (2017). Marine microorganisms as a promising and sustainable source of bioactive molecules. *Mar Environ Res.*, 128: 58-69. https://doi.org/10.1016/j.marenvres.2016.05.002

Rosenberg, E. & Ron, E.Z. (1999). High- and low-molecular-mass microbial surfactants. *Appl Microbiol Biotechnol.*, 52(2): 154-162. https://doi.org/10.1007/s002530051502

Sadiq, M.B., Khan, M.R. & Sayyed, R.Z. (2022). Biosurfactant mediated synthesis and stabilization of nanoparticles. *In:* Sayyed, R.Z. (Eds), *Biosurfactants: Production and Applications in Bioremediation/Reclamation.* 158-168. CRC Press, Taylor & Francis Group, USA.

Santos, D.K.F., Rufino, R.D., Luna, J.M., Santos, V.A. & Sarubbo, L.A. (2016). Biosurfactants: Multifunctional Biomolecules of the 21st Century. *Int J Mol Sci.*, 1-31. https://doi.org/10.3390/ijms17030401

Saranraj, P., Sayyed, R.Z., Sivasakthivelan, P., Hasan, M.S., Al-Tawaha, A.R.M.A & Amala, K. (2022a). Microbial biosurfactants: Methods of investigation, characterization, current market value and applications. *In:* Sayyed, R.Z. (Eds), *Biosurfactants: Production and Applications in Bioremediation/Reclamation.* 19-34. CRC Press, Taylor & Francis Group, USA.

Saranraj, P., Sayyed, R.Z., Hamzah, K.J., Asokan, N., Sivasakthivelan, P., & Al-Tawaha, A.R.M.A. (2022b). *In:* Sayyed, R.Z. (Eds), *Biosurfatnats: Production and Applications in Bioremediation/Reclamation.* 1-18. CRC Press, Taylor & Francis Group, USA.

Sen, S., Borah, S.N., Kandimalla, R., Bora, A. & Deka, S. (2020). Sophorolipid biosurfactant can control cutaneous dermatophytosis caused by Trichophyton mentagrophytes. *Front Microbiol.*, 11(March): 1-15. https://doi.org/10.3389/fmicb.2020.00329

Shah, I., Hamid, B., Zaman, M., Fatima, S., Farooq, S., Datta, R. et al. (2022). Microbial biosurfactants: An eco-friendly approach for bioremediation of contaminated environments. *In:* Sayyed, R.Z. (Eds), *Biosurfactants: Production and Applications in Bioremediation/Reclamation.* 197-207. CRC Press, Taylor & Francis Group, USA.

Shah, V., Doncel, G.F., Seyoum, T., Eaton, K.M., Zalenskaya, I., Hagver, R. et al. (2005). Sophorolipids, microbial glycolipids with anti-human immunodeficiency virus and sperm-immobilizing activities. *Antimicrob Agents Chemother.*, 49(10): 4093-4100. https://doi.org/10.1128/AAC.49.10.4093-4100.2005

Shao, L., Song, X., Ma, X., Li, H. & Qu, Y. (2012). Bioactivities of sophorolipid with different structures against human esophageal cancer cells. *J Surg Res.*, 173(2): 286-291. https://doi.org/10.1016/j.jss.2010.09.013

Shehzad, A., Ul-Islam, M., Wahid, F. & Lee, Y.S. (2014). Multifunctional polymeric nanocurcumin for cancer therapy. *J Nanosci Nanotechnol.*, 14(1): 803-814. https://doi.org/10.1166/jnn.2014.9103

Silveira, V.A.I., Marim, B.M., Hipólito, A., Gonçalves, M.C., Mali, S., Kobayashi, R.K.T. et al. (2020). Characterization and antimicrobial properties of bioactive packaging films based on polylactic acid-sophorolipid for the control of foodborne pathogens. *Food Packag Shelf Life*, 26(October): 1-7. https://doi.org/10.1016/j.fpsl.2020.100591

Silveira, V.A.I., Nishio, E.K., Freitas, C.A.U.Q., Amador, I.R., Kobayashi, R.K.T., Caretta, T. et al. (2019). Production and antimicrobial activity of sophorolipid against Clostridium perfringens and Campylobacter jejuni and their additive interaction with lactic acid. *Biocatal Agric Biotechnol.*, 21: 101287. https://doi.org/10.1016/j.bcab.2019.101287

Singh, P.K., Wani, K., Kaul-Ghanekar, R., Prabhune, A. & Ogale, S. (2014). From micron to nano-curcumin by sophorolipid co-processing: Highly enhanced bioavailability, fluorescence, and anti-cancer efficacy. *RSC Adv.*, 4(104): 60334-60341. https://doi.org/10.1039/c4ra07300b

Smyth, T.J.P., Perfumo, A., Marchant, R. & Banat, I.M. (2010). Handbook of hydrocarbon and lipid microbiology. Timmis, K.N. (Ed.) Berlin, Heidelberg: Springer Berlin Heidelberg. https://doi.org/10.1007/978-3-540-77587-4

Solaiman, D.K.Y., Ashby, R.D. & Crocker, N.V. (2015). High-titer production and strong antimicrobial activity of sophorolipids from Rhodotorula bogoriensis. *Biotechnol Prog.*, 31(4): 867-874. https://doi.org/10.1002/btpr.2101

Stratton, C.W. (2000). Mechanisms of bacterial resistance to antimicrobial agents. *J Med Liban.*, 48(4): 186-198. https://doi.org/10.1128/microbiolspec.arba-0019-2017

Tang, Y., Ma, Q., Du, Y., Ren, L., Van Zyl, L.J. & Long, X. (2020). Efficient purification of sophorolipids via chemical modifications coupled with extractions and their potential applications as antibacterial agents. *Sep Purif Technol.*, 245(February): 116897. https://doi.org/10.1016/j.seppur.2020.116897

Tenney, J., Hudson, N., Alnifaidy, H., Li, J.T.C. & Fung, K.H. (2018). Risk factors for acquiring multidrug-resistant organisms in urinary tract infections: A systematic literature review. *Saudi Pharm J.*, 26(5): 678-684. https://doi.org/10.1016/j.jsps.2018.02.023

Xu, R.Q., Ma, L., Chen, T., Zhang, W.X., Chang, K. & Wang, J. (2023). Sophorolipid inhibits histamine-induced itch by decreasing PLC/IP3R signaling pathway activation and modulating TRPV1 activity. *Sci Rep.*, 13(1): 1-13. https://doi.org/10.1038/s41598-023-35158-9

Zaman, M., Hassan, S., Fatima, S., Hamid, B., Farooq, S., Qayoom, I. et al. (2022). Biosurfactants production and applications in food. *In:* Sayyed, R.Z. and Enshasy, H.E. (Eds), *Biosurfactants: Production and Applications in Food and Agriculture.* Vol II, 225-241. CRC Press, Taylor & Francis Group, USA.

Zhang, X., Ashby, R.D., Solaiman, D.K.Y., Liu, Y. & Fan, X. (2017). Antimicrobial activity and inactivation mechanism of lactonic and free acid sophorolipids against Escherichia coli O157:H7. *Biocatal Agric Biotechnol.*, 11: 176-182. https://doi.org/10.1016/j.bcab.2017.07.002

CHAPTER

21

Sophorolipid: A Glycolipid Biosurfactant Applications in Therapeutics and Cosmetics

Nahida Fatima[1,3*], Susmitha B.[2], Anuradha B.S[3], Ashwitha Kodaparthi[4], M. Yahya Khan[5], and Bee Hameeda[*6]

[1] Department of Microbiology, Kasturba Gandhi Degree & PG College for Women, Marredpally, Hyderabad, India

[2] Department of Microbiology, St. Pious X Degree & PG College for Women, Nacharam, Hyderabad, India

[3] Department of Microbiology, Chaitanya (Deemed to be University) Hanamkonda, India

[4] Department of Microbiology, MNR Degree & PG College, Kukatpally, Hyderabad, Telangana

[5] Kalam's Institute of Science, Tilak Nagar Road, Nallakunta, Hyderabad, India

[6] Department of Microbiology, University College of Science, Osmania University, Hyderabad, India

1. Introduction

Biosurfactants are surface-active amphiphilic compounds produced by various microorganisms viz., yeast, fungi, or bacteria with varying molecular structures and surface activity (Campos et al., 2013). Biosurfactants, alternatively recognized as natural surfactants, emanate from microorganisms when fermented in laboratory conditions, through separation methodologies such as extraction, precipitation and purification. According to Subsanguan et al. (2022), biosurfactants are either negatively charged or neutral. The intricate structure of biosurfactants is influenced by a diverse microbial origin, the substrates used for their cultivation, and the specific growth conditions (Santos et al., 2016). The extensively studied types of microbial surfactants are rhamnolipids, sophorolipids, mannosyl-erythritol lipids (MELs), and surfactin (Yousef et al., 2022; Banat et al., 2010). Based on molecular weight,

* Corresponding authors: nahida.rafeeq@gmail.com; drhami2009@gmail.com

microbial surfactants are classified as low molecular weight and high molecular weight biosurfactants (Varvaresou & Iakovou, 2015). The high molecular weight microbial surfactants include two major classes , lipoproteins and lipopolysaccharides, whereas the low molecular weight microbial surfactants include phospholipids, glycolipids and lipopeptides (Grice et al., 2011, Yousef et al., 2022, Hospes et al., 2022). Glycolipids are an extensively investigated group of low molecular weight biosurfactants used in different sectors such as agro-food-industry (Bujak et al., 2015). Among them, sophorolipids are of great interest to researchers due to their unique properties, which make them suitable for applications in food processing, bioremediation, cosmetics, therapeutics, and health sector (Santos et al., 2016, Vecino et al., 2017, Naughton et al., 2019, Peyrat et al., 2019).

Cosmetics are blends of chemical substances produced from synthetic or natural sources and used for skin and personal care (Schneider et al., 2001). It is crucial that any cosmetic product should not have adverse effects on the skin, and contain safe ingredients (Ferreira et al., 2017, Adu et al., 2023). Personal skincare products must be formulated in such a way that they provide nutrients to the skin and its microflora, by the inhibition of pathogens. Therefore, the manufacturers of cosmetic products are in search of novel and improved components in their products which possess the above-mentioned characteristics (Ferreira et al., 2017, Nassan et al., 2017). Today, petrochemical or synthetic surfactants are used as surface-active agents in most of the personal skincare products, which are not good for the skin and its microflora (Fiume et al., 2017, Amani et al., 2021). Chemical surfactants, obtained from the non-renewable resources viz., petroleum have detrimental effects on the human skin and its microflora. Moreover, long term usage of these chemical surfactants on the skin causes the solubilization of the epidermis and the cytosol lipids, thus, affecting the impermeability of the skin (Babalola et al., 2020). Because of all these reasons, there is a dire need to switch from synthetic surfactants to renewable and natural compounds which minimize side-effects yet have a broad spectrum of action against pathogens (Kong et al., 2017). Sophorolipids are such compounds which have exceptional properties like antimicrobial, anticancer, and immunomodulatory activities. Due to their unique properties, sophorolipids are considered to be a better alternative for chemical surfactants in therapeutic sectors (Haque et al., 2019, Díaz-Rodríguez et al., 2018, Valotteau et al., 2017). This chapter highlights the properties of sophorolipids and their applications as an efficient substitute for chemical surfactants in formulations of cosmetics and pharmaceutical compounds.

2. Sophorolipids

The glycolipid class of biosurfactants, sophorolipids are extracellular compounds produced by *Torulopsis magnoliae* and were initially discovered by Gorin et al. (1961). Tulloch et al. (1968) reported that *Candida bogoriensis* which is renamed as *Rhodotorula bogoriensis* also produce extracellular glycolipids. Spencer et al. (1970) discovered the production of sophorolipids from *Candida bombicola* or *Starmerella bombicola* (Rosa & Lachance, 1998). Yeasts are the primary producers and also produce high yield of sophorolipids. *Saccharomyces* sp. when provided with carbohydrates and lipids in their media, synthesize sophorolipids (Christova et al., 2015). Chen et al. (2006) discovered another sophorolipid producing yeast strain, *Wickerhamiella domericqiae*.

The most efficient producer of sophorolipids is *C. bombicola* which was isolated from bumblebee honey (Davila et al., 1997, Gao et al., 2012). Originally, *C. bombicola* was known as *Torulopsis bombicola*, later it was named as *S. bombicola* which is a teleomorph of *C. bombicola*. Lachance et al. (2011) has reported the discovery of *S. bombicola* strains in south Africa, from concentrated grape juice. The natural habitats of *S. bombicola* are insects and plants, hence the sophorolipids produced by this yeast strain do not cause any harm to humans. It was reported by Solaiman et al. (2014) that low cost/waste substrates can be used to produce sophorolipids. Kurtzman et al. (2010) has reported that *S. bombicola* is the only organism from the genus *Starmerella* which produces sophorolipids.

3. Sophorolipids' Structure and Properties

Sophorolipids are amphiphilic surfactant molecules that interact with the phase boundary in heterogeneous systems. They consist of sophorose, a hydrophilic carbohydrate head, coupled with a hydrophobic fatty acid tail comprisingeither 16 or 18 carbon atoms. Regarding sophorolipids, sophorose, a glucose disaccharide featuring a typical β-1,2 bond, has the potential for acetylation at the 6′- and/or 6″-positions. A β-glycosidically linked terminal/subterminal hydroxylated fatty acid is attached to the sophorose molecule. These fatty acids' carboxylic ends can be free, internally esterified at the 4″ position, and in a few rare instances, esterified at the 6′/6″ position, lactonic form (Fig. 1). The hydroxy fatty acid itself has one or more unsaturated bonds and typically has 16 or 18 carbon atoms.

Hence, the sophorolipids produced by *C. bombicola* constitute a blend of interconnected molecules exhibiting distinctions in lactonization, acetylation patterns,

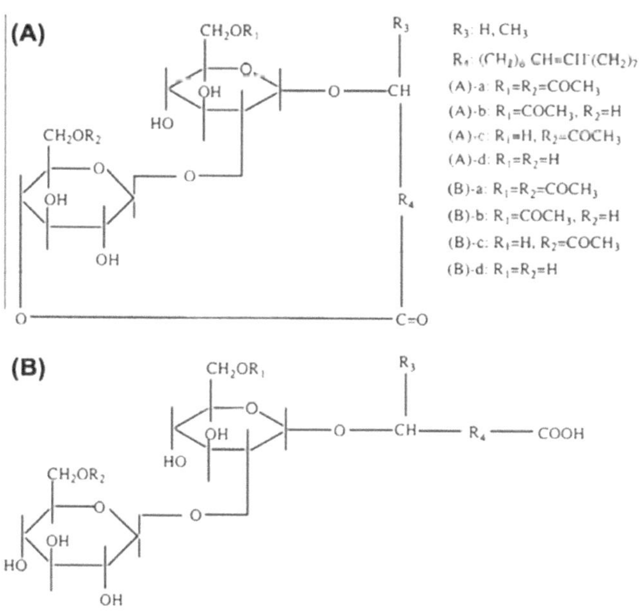

Fig. 1: Chemical Structure of Sophorolipid (a) Lactone and (b) Acid types (Ma et al., 2012)

and the composition of the fatty acid segment—including chain length, saturation, and hydroxylation positioning. Asmer et al. (1988) initially elucidated this structural diversity. They employed thin layer chromatography and medium pressure liquid chromatography techniques to fractionate the sophorolipid amalgam, identifying 14 principal components predominantly distinguished by their acetylation and lactonization configurations. But disparities in the length of the fatty acids and their hydroxylation pattern were not considered. High-performance liquid chromatography (HPLC) with gradient elution was utilised by Davila et al. (1993) to separate the sophorolipid mixture, and evaporative light scattering was employed to identify each sophorolipid separately. They identified over 20 components after paying close attention to the fatty acid chain analysis (Asmer et al., 1988).

When sophorolipids are dissolved in water, the surface tension undergoes a remarkable reduction from 72.8 mN/m to a range between 40 and 30 mN/m, reaching a critical micelle concentration of 40 to 100 mg/l. With a hydrophilic/lipophilic balance spanning between 10 to 13, sophorolipids serve as effective agents for detergent purposes or as stabilizers in oil-in-water emulsions. Notably, the diverse structural classifications engender significant disparities in physicochemical properties, as highlighted by Van Bogaert et al. (2007).

3.1 Lactonic Sophorolipids

Lactonic sophorolipids consist of a sophorose part linked with an extended chain of hydroxyl fatty acids. These carboxyl groups of fatty acids are always esterified internally at the 4′-position of the sophorose, in certain instances, at the 6′- or 6″-position, resulting in the formation of a closed ring structure (Van Bogaert et al., 2007). Although they contain an extra acetyl group, lactonic sophorolipids are less soluble in water than acidic sophorolipids. Lactonized sophorolipids typically exhibit good surface activity and exhibit a wide range of potential applications. In addition, it was shown that all the lactonic forms of sophorolipids exhibit greater antimicrobial activity than the acidic forms, making them more suitable for use in a variety of industries (Ciesielska et al., 2014). According to Shah et al., the potency and activity of lactonic sophorolipids were found to be significantly affected by the presence of acetyl groups in the compounds. According to this study, acetylation of sophorolipids has been demonstrated to have strong spermicidal properties along with anti-human immunodeficiency virus (HIV) properties. This reduces the hydrophilicity of the compounds (Shah et al., 2005).

3.2 Acidic Sophorolipids

Sophorolipids in acidic forms typically have free carboxyl groups made of fatty acid (FA) tails connected to the sophorose part via glycosidic bond. The acidic sophorolipid variants typically exhibit greater solubility as biosurfactants (Shu et al., 2021). To investigate the potential antimicrobial properties of acidic sophorolipids, Lydon et al. (2017) carried out a study to assess the antimicrobial activity of acidic sophorolipids in the process of creating pharmaceutical products intended for use in wound healing. With no hints of adverse effects or infections, the research demonstrated that the acidic sophorolipids have anti-microbial activity in both *in vitro* and *in vivo* experimental models. Acidic sophorolipids may be used as antimicrobial agents for the same reason that they speed up healing and reduce inflammation.

4. Sophorolipids as Alternatives

Synthetic surfactants are divided into four major classes, i.e., cationic, non-ionic, amphoteric, and anionic owing to the charges carried on the hydrophilic heads followed by the separation of the molecule in water (Corazza et al., 2010, Falk, 2019). The most commonly used chemical surfactants are SDS (Sodium dodecyl sulphate) cocamidopropyl betaine, SLS (Sodium lauryl sulphate) and cocamide (Jadhav et al., 2019, Corazza et al., 2010). Important functions carried out by chemical surfactants are cleaning the dirt from the body surface and hair, foaming, emulsification, dissolution of non-miscible solutions, conditioning, moisturizing and moistening.

The unique properties of chemical surfactants are their amphiphilic structure, absorption and surface tension reduction (Santos et al., 2016, Ananthapadmanabhan et al., 2004). Due to these properties, chemical surfactants are formulated in personal care products, but their use over a longer period of time has shown detrimental effects on the skin such as irritation, allergic reactions and change in the microflora of the skin (Bouslimani et al., 2019, Bujak et al., 2015). Physicochemical characteristics of synthetic surfactants, concentration, duration of contact with cutaneous layer may lead to adverse effects (Seweryn, 2018). Keratinocytes make up the epidermis of the skin and the uppermost epidermal layer is known as stratum corneum. The stratum corneum is made up of corneocytes that are formed due to the terminal differentiation of keratinocytes (Yanase & Hatta, 2017, Eckhart & Tschachler, 2018). Ceramides, cholesterol, cholesterol esters and fatty acids are the different types of lipids which fill the intracellular spaces of corneocytes. These intracellular corneocyte lipids control the amount of epidermal water loss, and therefore helps to retain the skin moisture. When the chemical surfactants interact with the skin they cause delipidation, i.e., a change in the equilibrium of intracellular lipids (Moore & Rawlings, 2017, Spada et al., 2018). Moreover, the other issues with chemical surfactants on the skin include protein denaturation, allergic reactions, skin irritation, swelling of the epidermal layer, affecting the immune cells in the skin such as keratinocytes and Langerhans cells thereby affecting the immunity (Bujak et al., 2015, Ananthapadmanabhan et al., 2004). Effects of synthetic surfactants depend upon their concentration and state (i.e., monomer/micelle) and the aggregates of surfactant monomers result in the development of micelles (Ananthapadmanabhan et al., 2004, Som et al., 2012). Micelle formation occurs at a particular temperature and concentration known as critical micelle concentration (CMC). It is reported that the chemical surfactant can penetrate through the skin in both the states, i.e., the micelle form and the monomer form. The micelles are highly unstable forms which break down into monomers once in contact with the skin (Morris et al., 2019, Seweryn, 2018)

Advanced research in the field of microbial biotechnology has resulted in the identification of potential microbial surfactants which can also improve the skin health (Lourith et al., 2009, Naughton et al., 2019, Seweryn 2018, Akbari et al., 2018, Marchant, 2012). On comparison of sophorolipids with the synthetic surfactants, sophorolipids have several advantages including biodegradability, less toxicity, biological compatibility, specificity, stability at extreme conditions of pH, temperature, and salt concentrations, cost effectiveness and environmental friendliness. Moreover, sophorolipids are produced from cheaper and renewable resources. Currently, sophorolipids are being examined for application as biosurfactants in lowering interfacial tension, enhancing disintegration of hydrocarbons and easy dissolution

(Bay et al., 2020) Sophorolipids derived from *S. bombicola* yeast have gained importance due to their application in the cosmetics and pharmaceutical industries (Schommer & Gallo, 2013). Thus, many researchers are interested in sophorolipids because they are natural compounds that seldom have undesirable side effects.

5. Effect of Sophorolipids on Skin and its Microflora

Bacteria, fungi, virus, and eukaryotic microorganisms form the predominant microflora associated with human skin (Weyrich et al., 2015). Colonization of the skin microflora occurs during and after birth. The Human Microbiome Project (HMP), 16S rRNA phenotyping the metagenomic sequencing has helped us to understand the microflora of skin (Cundell, 2016). Furthermore, the existence of microbial diversity of the skin has been reported (Lunjani et al., 2019). The skin microbial diversity is due to several factors such as exposure to environmental microbes, gender of host, region, diet or food, use of make-up and skin care products (Bouslimani et al., 2019, Cundell, 2016, Ohue & Nishikawa, 2019). Grice et al. (2009) based on 16S rRNA phenotyping has identified about 19 bacteria phyla in 20 various skin sites of many healthy individuals. The majority of the bacteria belong to the following four phylums: *Actinobacteria*, *Firmicutes*, *Proteobacteria* and *Bacteroidetes*. Among others, *Cutibacterium* spp., *Staphylococci* spp. and *Corynebacteria* spp. were the predominant bacteria (Cosseau et al., 2016). Human skin is considered as an ecosystem with an active immune system regulating the pathogenic and commensal microorganisms (Grice et al., 2011, Musthaq et al., 2018).

Sophorolipids possess physicochemical properties which are important for maintaining good skin health. For example, the hydrophobic ends of sophorolipids are good moisturizing agents for dry and rough skin. In addition, *Cutibacterium acnes*, earlier known as *Propionibacterium acnes* digest the triglycerides of biosurfactants and produce fatty acids making the sin pH acidic. Acidic pH of the skin, inhibits proliferation of disease-causing microbes and maintains the good skin microflora. In addition, the hydrophobic units of sophorolipids act like antioxidant and stops the formation of free radicals on exposure to UV light (Vecino et al., 2017, Rocha & Bagatin, 2018, Smith et al., 2020, Timm et al., 2020). The unique structure of biosurfactants increases their permeability through the skin membrane thereby enhancing skin protection and hair repair (Yanase & Hatta, 2017, Heinrich et al., 2014). Due to emulsification, foaming, wetting and solubilizing characteristics, sophorolipids are used as active compounds in body creams, lotions, powder, shampoo, and other self-care products (Gharaei-Fa, 2010). It is been reported that apart from sophorolipids, the other biosurfactants suitable for human skin are rhamnolipids, MELs and surfactin (Vecino et al., 2017, Peyrat et al., 2019, Sen et al., 2020). Certain other microbial metabolites, bacteriocins, enzymes, and defensins are the inhibitory substances present on the skin surface which act as the natural defence mechanism of the skin. Sophorolipids can be effective in skin care therapies and show inhibitory effects against the pathogenic microorganisms such as *Pseudomonas* sp., *S. pyogenes, C. acnes,* etc. (Vecino et al., 2017, Holland & Bojar, 2002). Commercial products containing sophorolipids produced by Kao Co. Ltd. are humectants which are skin moisturizers (Futara et al., 2002). Sophorolipids based make-up cosmetic brands include Sofina and Soliance produced for skin application with Sopholiance™- S as the key ingredient, Relipidium™ (2020). A combination of sophorolipids and an anionic surfactant was developed by Cox et al. (2013). They used this combination in shampoo and shower gel formulations. It has been reported by Kulkarni and Choudhary (2011) that a mixture of sophorolipids and cocoamidopropyl betaine are produced by *S. bombicola.* This mixture can be used in making body washes. Sophorolipids derived from *T. bombicola* were modified to form long chain alkyl sophorolipids when treated with alkylene oxides. These modified structures act as good moisturising agents (Kulkarni and Choudhary, 2011). Muthuswamy et al. (2008) suggested the application of sophorolipids for topical use due to their antibacterial and antiviral activity. Thus, biosurfactants are considered as the best alternative to regular drugs, even though the microbicidal effect of biosurfactants varies from one type to another (Gharaei-Fa, 2010).

6. Effectiveness of Sophorolipids as Antimicrobial Agents

In the mid-nineteenth century pharmaceutical companies world-wide produced about 20 classes of new antibiotics which was a major breakthrough in the reduction of the mortality rate due to bacterial infections (Yang et al., 2019). However, in the past five decades there has been a decrease in the discovery and production of new drugs. Antibiotics are used to treat various infections and diseases, such as acne, eczema and atopic dermatitis (Lim et al., 2018, Sohn, 2018, Dempster et al., 2019). But many pathogenic bacteria have become antibiotic resistant; thus, there is an urgent need for an alternative with effective antimicrobial properties (Lim et al., 2018, Silveira et al., 2018). Sophorolipids have been reported as an alternative with effective antimicrobial properties such as antibacterial, antifungal, antiviral and antibiofilm effects (Lydon et al., 2017, Juma et al., 2020, Anestopoulos et al., 2020, Fernandes et al., 2007). The structure, concentration, composition of the fatty acid chain, and type of microorganism involved, determines the effectiveness of a biosurfactant as antimicrobial agent. Sophorolipids have shown bactericidal effect against pathogenic Gram-positive bacteria like *S. aureus*, the causative organism of atopic dermatitis, and *C. acnes* which causes acne vulgaris (Sohn, 2018, Silveira et al., 2018). The mechanism by which sophorolipids execute bactericidal activity involves rupturing the cell membrane of pathogenic bacteria, leading to the release of their cytoplasmic contents, such as intracellular enzymes like malate dehydrogenase. Presence of the malate dehydrogenase enzyme outside the bacteria cell indicates the action of sophorolipids on the cellular membrane (Akbari et al., 2018, Holland & Bojar, 2002). The antimicrobial properties of sophorolipids have made them highly sought-after substances for use in the formulation of cosmetics and pharmaceuticals for skincare (Varvaresou & Iakovou 2015). According to the studies conducted by Lydon et al. (2017) acidic sophorolipids at a low concentration of 5 mg/mL kill the Gram-negative pathogens viz., *Escherichia coli*, *P. aeruginosa*, etc. De Rienzo et al. (2015), reported the antibacterial activity of sophorolipids at a low concentration of 5% (v/v) against Gram positive bacteria. Mostly, sophorolipids show a huge microbicidal effect on Gram +ve bacteria than Gram –ve bacteria (Fernandes et al., 2007), because the outer membrane of Gram negative bacteria gives them protection (Ferreira et al., 2017). However, Gram positive bacteria are the major skin inhabiting bacteria (Grice & Segre, 2011, Cogen et al., 2008). Studies suggest that due to lack of moisture and increased osmotic pressure on skin, very less colonization and multiplication of Gram negative bacteria is observed (Tarale et al., 2015).

Researchers assumed that high antimicrobial activity of sophorolipids might be because of its association with antibiotics. This association causes the increase in the permeability of antibiotics through bacterial cell membranes (Borsanyiova et al., 2015). To prove this hypothesis, Juma et al. (2020) conducted an experiment in which he treated the methicillin resistant *S. aureus* (MRSA) with the minimum concentrations of sophorolipids in combination with tetracycline. This study, explained swollen bacterial cell under SEM (scanning electron microscopy) as well as cell surface destruction.

7. Sophorolipids as Antibiofilm Agents

Biofilm inhibition is another important property of sophorolipids (Vaishnavi et al., 2019). Biofilms are nothing but bacterial cells enclosed in a matrix which is made up of substances such as extracellular polymer compounds like carbohydrates, lipids, proteins, and extracellular DNA. They carry out important functions such as increasing the survival period of bacteria, their propagation, protection from predation, resistance to antimicrobial agents and increasing pathogenicity (Lopez et al., 2010, Brandwein et al., 2016, Flemming et al., 2007). According to Romling and Balsalobre (2012), biofilm development is responsible for up to 80% of human skin infections such as atopic dermatitis, acne vulgaris and chronic wounds which are caused by both Gram negative and Gram positive bacteria. Sophorolipids, show antibiofilm properties and are the most promising choice in wound healing and skincare creams (Naughton et al., 2019, Valotteau et al., 2019).

The growth of pathogenic fungal strain *C. albicans* is also inhibited by sophorolipids. In addition, sophorolipids also cause the inhibition of biofilm formation by these pathogenic fungi (Haque et al., 2016, 2017). The mechanisms by which sophorolipids exhibit antifungal activity includes the alteration in cell permeability and the release of the ROS (Reactive oxygen species) which may result in the augmented oxidative stress leading to fungal necrosis and apoptosis (Haque et al., 2019). Furthermore, sophorolipids antifungal activity against *Trichophyton* helps in controlling infections viz., tineapedis along with the deadly dermatophytosis. Thus, sophorolipids remain promising alternative compounds which can be used in various cosmetics and pharmaceuticals (Morais et al., 2017).

8. Low Toxicity Effect of Sophorolipids

Biosurfactants consist of a hydrophilic and hydrophobic moiety in its structure (Kim et al., 2011, Lee et al., 2004). Due to their distinct chemical and biological qualities biosurfactants are becoming significantly crucial in fields like environmental cleaning, food, biomedical, textile, and cosmetic sciences (Jin et al., 2015). Compared to the available products in the market, sophorolipids prove that they cause less skin surface irritation (Araújo et al., 2022, Brakemeier et al., 1998). Sophorolipid is the only biosurfactant that can be produced in quantities higher than 300 g/l among those documented in the literature (Lee et al., 2010). Furthermore, it can be made using renewable resources like whey. The adoption of sophorolipids as antimicrobial agents in human health care products is encouraged by their distinct qualities, which include high water-retention capacity, water solubility, and low skin toxicity, in addition to their production from renewable low cost substrates resulting in economic benefits (Cho et al., 2006).

Previous research studies have indicated that the use of chemical synthetic constituents in cosmetic preparations can cause allergic reactions, especially on the skin. Statistical analysis conducted in the U.K on the adverse effects of cosmetics used on the skin revealed that approximately 23% of women and 14% of men are affected in a year (Orton & Wilkinson, 2004). As reported by Bouslimani et al. (2019) petroleum-based compounds like propylene glycol found in antiperspirants remains on the skin for about 2 weeks, which could change the microbiota of skin. These issues can be resolved by using sophorolipids in skin care cosmetic products due to their low cytotoxicity. For this reason they are widely accepted for use in therapeutic and cosmetic productions (Santos et al., 2016, Sewcryn, 2018). According to Stipcevic et al. (2005), the cytotoxic effect of sophorolipids merely depends on their concentration. It has been reported that sophorolipids produced from *C. bombicola* ATCC 22214 (Subsanguan et al., 2022) possess antimicrobial activity toward *P. acnes*, and show lower toxic effects on the skin surface.

9. Sophorolipids with Wound Healing Properties

Chronic wounds are a persistent challenge for wound care professionals and consume a significant amount of healthcare resources globally. Wound care and biofilm prevention are regarded as critical priorities in the healing process by the scientific community and medical professionals. Furthermore, wound care needs to be done effectively and accurately. Any modifications that impede the healing process run the risk of damaging tissue and delaying the healing process. Wound healing can be slowed down by a number of variables, including inflammation, oxidation, and microbial infection. Biofilms in chronic wounds retard the healing process by contributing to infections. The bacteria *P. aeruginosa* and *S. aureus* are most frequently detected in chronic wounds. Therefore, it has been claimed that use of sophorolipids, can help address these issues (Hillman et al., 2022). Sophorolipids have unique biological properties such as antimicrobial activity, stimulation of skin dermal fibroblasts and collagen production. They are highly compatible with human skin and show less inflammation as wound-healing agents (Araújo et al., 2022, Ceresa et al., 2021). Additionally, these bioproducts expedite the process of re-epithelialization and foster collagen deposition, consequently accelerating the pace of wound healing (Ohadi et al., 2017, Zouari et al., 2016).

In a recent experimental inquiry aimed at showcasing the wound healing characteristic of sophorolipids, a cell culture replica utilizing human dermal fibroblasts stands as a surrogate for human skin. Studies have revealed that sophorolipids influence the expression of elastase inhibition collagen I mRNA within human skin fibroblasts, @ IC_{50} (38.5 μg/mL). Moreover, an assessment of *in vitro* wound healing in the human colorectal adenocarcinoma (HT-29) cell line demonstrated a noteworthy escalation in collagenase-1 expression in HT-29 cells following a 48-hour treatment regimen involving sophorolipids (Kwak et al., 2021), which explains the wound healing property of sophorolipids.

10. Sophorolipids as an Antiviral Agent

Sophorolipids, because of their detergent, therapeutic, and viricidal properties are now being recommended to showcase a vital role in the control the COVID-19 pandemic along with other viral diseases. Subramaniam *et al.* (2020) have emphasized the anti-inflammatory properties of sophorolipids to control the COVID-19 pathogenesis. The virucidal property of sophorolipids was examined against Herpes, Influenza and HIV (Human Immunodeficiency Virus in the laboratory by a technique called virus inactivation assay (Azim et al., 2006, Felse et al., 2007, Daverey et al., 2021). Destruction of the cell membrane inside the virus has been reported as the probable mode of action of the antiviral activity of sophorolipids (Shah et al., 2005). The antiviral property of sophorolipids is mainly due to the presence of alkyl groups in their structure which imparts the deliquescent property to sophorolipids.

The properties of sophorolipids which are helpful in controlling COVID-19 are listed below. As reported by Conley et al. (2017), sophorolipids solubilize the lipid membrane of viruses, especially the enveloped viruses. Felse et al. (2007) reported the antiviral property of sophorolipids. They also exhibit immunomodulatory activity i.e they inhibit the overproduction of cytokines and reduce inflammations, thus they can be used in the treatment of lung injuries (Bluth et al., 2006). Hence, sophorolipids can provide protection to patients suffering from COVID-19 and other viral diseases, however, further studies in this direction are required to confirm their application in countering emerging or re-emerging viral diseases in the near future.

11. Sophorolipids' Anticancer Activity

The main drawback of the current cancer therapeutic agents is the major side effects it causes which is due to the effect of drugs on normal cells. Researchers are looking for natural compounds with effective anticancer properties and zero cytotoxicity against normal cells. It has been reported that sophorolipids target only cancer cells. They do not target normal cells (Nawale et al., 2017). Sophorolipids' structure plays a major role in exhibiting its anticancer properties. In an assay experiment conducted using an *ex-vivo* rat aorta ring (Kithur et al., 2019), it has been observed that a natural sophorolipid which consists of both, the acidic and lactonic forms, acts as an angiogenesis inhibitor (cancer blood vessel growth inhibitor). Sophorolipids with only the lactonic form showed anticancer properties when used in an experiment conducted on tumor cell lines (Haggag et al., 2020). Researchers have reported that diacetylated lactonic sophorolipids with 18C monounsaturated fatty acid have shown the highest anticancer activity. These sophorolipids inhibit the growth of human breast cancer cells, oesophageal cancer cells and human cervical cancer cells (Li et al., 2016, Nawale et al., 2017). Recent studies showed the effectiveness of the role played by apoptosis on cancer cells reduced, as the degree of unsaturation (number of double bonds) of the sophorolipids increased. Sophorolipids show anticancer activity against human cells (cervical cancer) only when they bind to the cetyl alcohol group. Liang et al. (2020) reported that this combined structure of sophorolipids bound to cetyl alcohol increases the calcium levels in the cells, causing depolarisation of the mitochondrial membrane, which induces apoptosis. Effective anticancer activities like induction of angiogenesis and growth inhibition of liver cancer cells were shown by lactonic sophorolipids (Kithur et al., 2019). Therefore, sophorolipids are emerging as molecules for cancer treatment requiring further research.

12. Sophorolipids Nanoparticles

Nanoparticles synthesized via chemical or physical methods are toxic not only to human health but also to the environment. Sophorolipids-mediated metallic nanoparticles have shown antibacterial or antifungal (antimicrobial) activities (Singh et al., 2009). Sophorolipids due to their sustainable properties have seized the attention of researchers. They can be used both as reducing and the capping agents in the production of valuable nanoparticles. Sophorolipid capped gold nanoparticles (Au-NPs-SL) showed high antimicrobial potential against Gram negative bacteria. In addition, they also show a bactericidal effect against non-dividing cells which are the major cause of health problems, as most of the drugs available can only kill metabolically active cells.These sophorolipid based nanoparticles also exhibit the bactericidal effect against multidrug resistance bacteria. Due to the characteristic features like biocompatibility, nontoxicity to cells, gold nanoparticles are used in drug delivery systems. Gellan gum reduced and sophorolipids conjugated capped gold nanoparticles are highly efficient anticancer agents that effectively kill the human glioma cell line LN-229. Baccile et al. (2013) stated that iron oxide nanoparticles could be developed from acidic sophorolipids obtained from *C. bombicola*, due to the high stability of sophorolipids. Similarly, zinc oxide nanoparticles can be formed from sophorolipids due to their excellent stability. These nanoparticles exhibit antimicrobial activity against the test microorganisms, *S. enterica* and *C. albicans* (Basak et al., 2014). Diacetate acidic sophorolipids produced by *Cryptococcus* sp. also exhibit antibacterial and antifungal activities. Hence, it is observed that sophorolipids exhibit antimicrobial activity but there has been an enhancement in its antimicrobial effects when complemented with iron oxide. This concept of combined antimicrobial effect of sophorolipids and iron oxide can be researched further to produce a functional nanoparticle (Delbeke et al., 2016). Furthermore, it was observed that sophorolipids with quaternary ammonium compounds hold bactericidal activity. SLNs (Solid lipid nanoparticles) are made up of those lipids which are stabilised by a surfactant solution. These SLNs are used as a carrier for the delivery of hydrophobic compounds such as drugs, additives or cosmetics (Muller, Mader et al., 2000, Muller, Radtke et al., 2002). They protect the encapsulated drug from chemical degradation and allow the slow release of the drug. Rifampicin and dapsone are the broad-spectrum hydrophobic antibiotics used for the treatment of *Mycobacterium leprae*. Due to their hydrophobic nature these drugs are relatively less soluble thereby showing a low therapeutic effect (Singh et al., 2013, Borges et al., 2013). Therefore, SLNs are used to overcome problems like solubility, stability and improving the bioavailability of these drugs. The sophorolipids based solid nanoparticles serve as promising drug carriers. Due to the inherent properties they possess, sophorolipids can be used as a lipid source for the development of sophorolipids based SLNs. Cortes-Sanchez et al. (2013) reported that lactonic sophorolipids are preferred as compared to acidic sophorolipids in the development of nanoparticles. Sophorolipids, with their dual roles as reducing agents and micelle-forming structures, play a vital part in the field of nanotechnology. The use of sophorolipid nanoparticles (NPs) for targeted drug-delivery is an even more promising way to revolutionize the treatment for various diseases.

13. Sophorolipids and Biomedical Applications

The capping and metal-reducing abilities of sophorolipids demonstrate high control over reaction parameters. This characteristic feature can be utilized in biomedical

Table 1: Details of sophorolipids in cosmetics and therapeutics

Application	Microbial source of sophorolipids	Biocompatibility	Mode of action	References
Cosmetic and wound healing	*Starmerella bombicola*	Emulsifier	Face creams and body lotions, skin ointments	Zerhusen et al. (2019)
		Microbicidal	Prevents acne	Ashby et al. (2011)
		Increase absorption of lactoferrin	Multiplication of fibroblast. Used in cosmetics	Ishii et al. (2012)
		Percutaneous penetration	Transferosome hydrogel used in cosmetics	Naik et al. (2019)
		Low interfacial surface tension	Skin penetration enhancer	Imura et al. (2014)
		Microbicidal	In combination with drugs such as kanamycin used in wound healing	Lydon et al. (2017)
		Antibacterial	Wound healing when used in rats	Hirata et al. (2021)
	Rhodotorula bogoriensis	Microbicidal	Inhibits the growth of C. acne Used in skin care products	Solaiman et al. (2020)
	Pseudohyphozyma Bogoriensis	Bactericidal	Inhibits the growth of P. acne used in skin care products	Solaiman et al. (2020)
	Starmerella bombicola	Immunomodulatory properties	Used in coatings to enhance the terminating process of inflammatory response. wound healing	Diaz et al. (2018)
Microbicidal	*Starmerella bombicola*	Bactericidal	Induces cell death in planktonic cells	De Rienzo et al. (2015)
		Biofilm inhibition	Prevents the growth of bacteria	Archana et al. (2019)
		Antibacterial	Inhibits gram-+ve bacteria	Valotteau et al. (2019)
		Microbicidal	In combination with antibiotics, inhibits the growing bacteria	Joshi Navare & Prabhune (2013)

Contd...

Contd...

Application	Microbial source of sophorolipids	Biocompatibility	Mode of action	References
Antimicrobial	*Starmerella bombicola*	Antimicrob al Antifungal	Inhibit bacterial and Mycotic infections	Dengle Pulate et al. (2013)
		Fungicidal	Prevent the growth of Candida	Haque et al. (2016)
		Antibiofilm Fungicidal	Prevent the infections caused by Candida sps.	Haque et al. (2017)
		Antibiofilm Fungicidal	Inhibit Mycotic infections	Haque et al. (2019)
	Candida tropicalis	Bactericidal	Stops the growth of gram-positive bacteria	Ankulkar & Chavan (2019)
	Rhodotorula babjevae	Fungicidal	Inhibits the contagious fungal infections to skin or scalp	Sen et al. (2020)
Anticancer	*Starmerella bombicola*	Anticance	Inhibits growth of human colon, cervical and breast cancer cells	Wang et al. (2021)
	Wickerhamiella domercqiae	Anticancer	Inhibits human esophageal cancer cell growth	Li et al. (2016)

applications such as biosensing agents, MRI enhancers etc (Yamamoto et al., 2012, Paulino et al., 2016, Bonte, 2011). Vesicles formed due to sodium salts of acidic sophorolipids help the active ingredients penetrate through the skin. Mogrosides V possess characteristic properties such as anti-cancer, anti-allergy, and anti-diabetic effects in animal models. It is the active component of triterpene glycosides. The vesicles formed by sodium salts of acidic sophorolipids increase the penetration of mogroside V through the skin surface (Ikarashi et al., 2017). It is observed that sophorolipids enhance the absorption of lactoferrin through the skin without affecting it or its functions. In addition, sophorolipids enhance the expression of the tropoelastin gene which indicates the synergistic association between sophorolipids and lactoferrin which results in enhanced effects on cell proliferation and collagen synthesis (Takahashi et al., 2012).

14. Sophorolipids Global Market

The global sophorolipids market stood at USD 510.29 million in 2022 and is projected to burgeon to USD 776.21 million by 2031, indicating a compound annual growth rate (CAGR) of 5.8%. Many fast-moving consumer goods are increasingly using sophorolipids as active ingredients. There is a rising demand for sophorolipids in the agriculture, cancer medications, and other therapeutic segments which contributes to an increase in revenue growth from their sales. The growing demand for sustainable natural products, the unique properties of sophorolipids and their ability to grow on cheap renewable sources are the main criteria increasing their sales. The main challenge in the large scale production of sophorolipids is product recovery. However, researchers have put an effort into standardizing the media composition and optimizing various product recovery processes to improve sophorolipid production (Ashby et al., 2013). The shift in consumer preferences is a crucial factor contributing to the enhancement in sophorolipids' utilization.

15. Challenges in Application of Sophorolipids

An efficient cosmetic and personal care product should provide moisturization, protection, cleansing and prevention (Rodan et al., 2016). The criteria to prefer sophorolipids over chemical surfactants in the beauty and healthcare industries depends on their ability to produce constant or enhanced results with their formulations at reasonable market prices. The realms encompassing genome sequencing, metabolic engineering, and the utilization of microbial enzymes must be thoroughly examined and evaluated as these are promising optional methods used to improve the production and structural variance of sophorolipids (Saika et al., 2018). Furthermore, microorganisms usually produce biosurfactants as a mixture of relative compounds instead of one compound and these compounds exhibit different bioactivities. To precisely determine the effectiveness of microbial surfactants and their suitable concentration, pure preparations of the specific compound must be examined (Rahimi et al., 2019).

16. Conclusion

Sophorolipids are derived from natural sources, they are more efficient when compared to chemical surfactants. Sophorolipids possess all the characteristic features such as biodegradability, low toxicity and environment friendly and thus are considered as the best alternative to chemical surfactants. Sophorolipids have gained attention over the years due to their diversified structure and production from cheap renewable sources. Hence, sophorolipids are natural and efficient compounds applied in agriculture, cosmetics, nanotechnology, therapeutics, and drug delivery. Further, exploration is imperative to isolate novel microbial strains adept at expeditiously generating sophorolipids across varied cultural environments. Moreover, augmenting sophorolipid productivity necessitates delving into strain enhancement via methodologies like genetic recombination and mutation. Future investigations may harness nanotechnology to deploy sophorolipids within the realms of the food and therapeutic industries. Advancing research demands through integration of safety and cutting-edge technologies will broaden the use of sophorolipids in the beauty and health sectors.

References

Adu, S.A., Twigg, M.S., Naughton, P.J., Marchant, R. & Banat, I.M. (2023). Characterisation of cytotoxicity and immunomodulatory effects of glycolipid biosurfactants on human keratinocytes. *Applied Microbiology and Biotechnology*, 107(1): 137–152. https://doi.org/10.1007/s00253-022-12302-5

Akbari, S., Abdurahman, N.H., Yunus, R.M., Fayaz, F. & Alara, O.R. (2018). Biosurfactants—A new frontier for social and environmental safety: A mini review. *Biotechnology Research and Innovation*, 2(1): 81–90. https://doi.org/10.1016/j.biori.2018.09.001

Amani, P., Karakashev, S.I., Grozev, N.A., Simeonova, S., Miller, R., Rudolph, V. et al. (2021). Effect of selected monovalent salts on surfactant stabilized foams. *Advances in Colloid and Interface Science*, 295: 102490. https://doi.org/10.1016/j.cis.2021.102490

Ananthapadmanabhan, K.P., Moore, D.J., Subramanyan, K., Misra, M. & Meyer, F. (2004). Cleansing without compromise: The impact of cleansers on the skin barrier and the technology of mild cleansing. *Dermatologic Therapy*, 17(Suppl 1): 16–25. https://doi.org/10.1111/j.1396-0296.2004.04s1002.x

Anestopoulos, I., Kiousi, D.E., Klavaris, A., Galanis, A., Salek, K., Euston, S.R. et al. (2020). Surface active agents and their health-promoting properties: Molecules of multifunctional significance. *Pharmaceutics*, 12(7): 688. https://doi.org/10.3390/pharmaceutics12070688

Ankulkar, R. & Chavan, M. (2019). Characterisation and application studies of sophorolipid biosurfactant by Candida tropicalis RA1. *Journal of Pure and Applied Microbiology*, 13(3): 1653–1665. https://doi.org/10.22207/jpam.13.3.39

Araújo, J.M., Monteiro, J., Silva, D.H.D.S., Alencar, A., Silva, K., Coelho, L.E. et al. (2022). Surface-active compounds produced by microorganisms: Promising molecules for the development of antimicrobial, anti-inflammatory, and healing agents. *Antibiotics*, 11(8): 1106. https://doi.org/10.3390/antibiotics11081106

Archana, K., Sathi Reddy, K. & Parameshwar, J. (2019). Isolation and characterization of sophorolipid producing yeast from fruit waste for application as antibacterial agent. *Environmental Sustainability*, 2: 107–115. https://doi.org/10.1007/s42398-019-00069

Ashby, R.D., McAloon, A.J., Solaiman, D.K.Y., Yee, W. & Reed, M.J. (2013). A process model for approximating the production costs of the fermentative synthesis of sophorolipids. *Journal of Surfactants and Detergents*, 16(5): 683–691. https://doi.org/10.1007/s11743-013-1466-0

Ashby, R.D., Zerkowski, J.A., Solaiman, D.K.Y. & Liu, L. (2011). Biopolymer scaffolds for use in delivering antimicrobial sophorolipids to the acne-causing bacterium Propionibacterium acnes. *New Biotechnology*, 28(1): 24–30. https://doi.org/10.1016/j.nbt.2010.08.001

Asmer, H., Lang, S., Wagner, F. & Wray, V. (1988). Microbial production, structure elucidation and bioconversion of sophorose lipids. *Journal of the American Oil Chemists' Society*, 65(9): 1460–1466. https://doi.org/10.1007/bf02898308

Azim, A., Shah, V., Doncel, G.F., Peterson, N., Gao, W. & Gross, R. (2006). Amino acid conjugated sophorolipids: A new family of biologically active functionalized glycolipids. *Bioconjugate Chemistry*, 17(6): 1523–1529. https://doi.org/10.1021/bc060094n

Babalola, M.O., Ayodeji, A.O., Bamidele, O.S. & Ajele, J.O. (2020). Biochemical characterization of a surfactant-stable keratinase purified from Proteus vulgaris EMB-14 grown on low-cost feather meal. *Biotechnology Letters*, 42(12): 2673–2683. https://doi.org/10.1007/s10529-020-02976-0

Baccile, N., Noiville, R., Stievano, L. & Van Bogaert, I. (2013). Sophorolipids-functionalized iron oxide nanoparticles. *Physical Chemistry Chemical Physics*, 15(5): 1606–1620. https://doi.org/10.1039/c2cp41977g

Banat, I.M., Franzetti, A., Gandolfi, I., Bestetti, G., Martinotti, M.G., Fracchia, L. et al. (2010). Microbial biosurfactants production, applications and future potential. *Applied Microbiology and Biotechnology*, 87(2): 427–444. https://doi.org/10.1007/s00253-010-2589-0

Basak, G., Das, D. & Das, N. (2014). Dual role of acidic diacetate sophorolipid as biostabilizer for ZnO nanoparticle synthesis and biofunctionalizing agent against Salmonella enterica and Candida albicans. *Journal of Microbiology and Biotechnology*, 24(1): 87–96. https://doi.org/10.4014/jmb.1307.07081

Bay, L., Barnes, C.J., Fritz, B.G., Thorsen, J., Restrup, M.E.M. & Rasmussen, L. (2020). Universal Dermal Microbiome in Human Skin. *mBio*, 11(1): e02945-19. https://doi.org/10.1128/mBio.02945-19

Bluth, M.H., Kandil, E., Mueller, C.M., Shah, V., Lin, Y.Y., Zhang, H. et al. (2006). Sophorolipids block lethal effects of septic shock in rats in a cecal ligation and puncture model of experimental sepsis. *Critical Care Medicine*, 34(1): 188–195. https://doi.org/10.1097/01.ccm.0000196212.56885.50

Bonté, F. (2011). Skin moisturization mechanisms: New data. *Annales Pharmaceutiques Francaises*, 69(3): 135–141. https://doi.org/10.1016/j.pharma.2011.01.004

Borges, V.R., Simon, A., Sena, A.R., Cabral, L.M. & de Sousa, V.P. (2013). Nanoemulsion containing dapsone for topical administration: A study of in vitro release and epidermal permeation. *International Journal of Nanomedicine*, 8: 535–544. https://doi.org/10.2147/IJN.S39383

Borsanyiova, M., Patil, A., Mukherji, R., Prabhune, A. & Bopegamage, S. (2015). Biological activity of sophorolipids and their possible use as antiviral agents. *Folia Microbiologica*, 61(1): 85–89. https://doi.org/10.1007/s12223-015-0413-z

Bouslimani, A., da Silva, R., Kosciolek, T., Janssen, S., Callewaert, C., Amir, A. et al. (2019). The impact of skin care products on skin chemistry and microbiome dynamics. *BMC Biology*, 17(1): 47. https://doi.org/10.1186/s12915-019-0660-6

Brakemeier, A., Wullbrandt, D. & Lang, S. (1998). Candida bombicola: Production of novel alkyl glycosides based on glucose/2-dodecanol. *Applied Microbiology and Biotechnology*, 50(2): 161–166. https://doi.org/10.1007/s002530051271

Brandwein, M., Steinberg, D. & Meshner, S. (2016). Microbial biofilms and the human skin microbiome. *NPJ Biofilms and Microbiomes*, 2: 3. https://doi.org/10.1038/s41522-016-0004-z

Bujak, T., Wasilewski, T. & Nizioł-Łukaszewska, Z. (2015). Role of macromolecules in the safety of use of body wash cosmetics. *Colloids and Surfaces B: Biointerfaces*, 135: 497–503. https://doi.org/10.1016/j.colsurfb.2015.07.051

Campos, J.M., Stamford, T.L., Sarubbo, L.A., de Luna, J.M., Rufino, R.D. & Banat, I.M. (2013). Microbial biosurfactants as additives for food industries. *Biotechnology Progress*, 29(5): 1097–1108. https://doi.org/10.1002/btpr.1796

Ceresa, C., Fracchia, L., Fedeli, E., Porta, C. & Banat, I.M (2021). Recent advances in biomedical, therapeutic and pharmaceutical applications of microbial surfactants. *Pharmaceutics*, 13(4): 466. https://doi.org/10.3390/pharmaceutics13040466

Chen, J., Song, X., Zhang, H. & Qu, Y. (2006). Production, structure elucidation and anticancer properties of sophorolipid from Wickerhamiella domercqiae. *Enzyme and Microbial Technology*, 39(3): 501–506. https://doi.org/10.1016/j.enzmictec.2005.12.022

Cho, Z.H., Hwang, S.C., Wong, E.K., Son, Y.D., Kang, C.K., Park, T.S. et al. (2006). Neural substrates, experimental evidences and functional hypothesis of acupuncture mechanisms. *Acta Neurologica Scandinavica*, 113(6): 370–377. https://doi.org/10.1111/j.1600-0404.2006.00600.x

Christova, N., Lang, S., Wray, V., Kaloyanov, K., Konstantinov, S. & Stoineva, I. (2015). Production, structural elucidation, and in vitro antitumor activity of trehalose lipid biosurfactant from Nocardia farcinica strain. *Journal of Microbiology and Biotechnology*, 25(4): 439–447. https://doi.org/10.4014/jmb.1406.06025

Ciesielska, K., Van Bogaert, I., Chevineau, S., Li, B., Groeneboer, S., Soetaert, W. et al. (2014). Exoproteome analysis of Starmerella bombicola results in the discovery of an esterase required for lactonization of sophorolipids. *Journal of Proteomics*, 98: 159–174. https://doi.org/10.1016/j.jprot.2013.12.026

Cogen, A.L., Nizet, V. & Gallo, R.L. (2008). Skin microbiota: A source of disease or defence? *British Journal of Dermatology*, 158(3): 442–455. https://doi.org/10.1111/j.1365-2133.2008.08437.x

Conley, L., Tao, Y., Henry, A., Koepf, E., Cecchini, D., Pieracci, J. et al. (2017). Evaluation of eco-friendly zwitterionic detergents for enveloped virus inactivation. *Biotechnology and Bioengineering*, 114(4): 813–820. https://doi.org/10.1002/bit.26209

Corazza, M., Lauriola, M.M., Zappaterra, M., Bianchi, A. & Virgili, A. (2010). Surfactants, skin cleansing protagonists. *Journal of the European Academy of Dermatology and Venereology*, 24(1): 1–6. https://doi.org/10.1111/j.1468-3083.2009.03349.x

Cortés-Sánchez, A. de J., Hernández-Sánchez, H. & Jaramillo-Flores, M.E. (2013). Biological activity of glycolipids produced by microorganisms: New trends and possible therapeutic alternatives. *Microbiological Research*, 168(1): 22–32. https://doi.org/10.1016/j.micres.2012.07.002

Cosseau, C., Romano-Bertrand, S., Duplan, H., Lucas, O., Ingrassia, I., Pigasse, C. et al. (2016). Proteobacteria from the human skin microbiota: Species-level diversity and hypotheses. *One Health*, 2: 33–41. https://doi.org/10.1016/j.onehlt.2016.02.002

Cox, T.F., Crawford, R.J., Gregory, L.G., Hosking, S.L. & Kotsakis, P. (2013). U.S. Patent No. 8,563,490. Washington, DC: U.S. Patent and Trademark Office.

Cundell, A.M. (2016). Microbial ecology of the human skin. *Microbial Ecology*, 76(1): 113–120. https://doi.org/10.1007/s00248-016-0789-6

Daverey, A., Dutta, K., Joshi, S. & Daverey, A. (2021). Sophorolipid: A glycolipid biosurfactant as a potential therapeutic agent against COVID. *Bioengineered*, 12(2): 9550–9560. https://doi.org/10.1080/21655979.2021.1997261

Davila, A., Marchal, R. & Vandecasteele, J.P. (1997). Sophorose lipid fermentation with differentiated substrate supply for growth and production phases. *Applied Microbiology and Biotechnology*, 47(5): 496–501. https://doi.org/10.1007/s002530050962

Davila, A.M., Marchal, R., Monin, N. & Vandecasteele, J.P. (1993). Identification and determination of individual sophorolipids in fermentation products by gradient elution high-performance liquid chromatography with evaporative light-scattering detection. *Journal of Chromatography*, 648(1): 139–149. https://doi.org/10.1016/0021-9673(93)83295-4

De Freitas Ferreira, J., Vieira, E.A. & Nitschke, M. (2019). The antibacterial activity of rhamnolipid biosurfactant is pH dependent. *Food Research International*, 116: 737–744. https://doi.org/10.1016/j.foodres.2018.09.005

Delbeke, E.I.P., Movsisyan, M., Van Geem, K.M. & Stevens, C.V. (2016). Chemical and enzymatic modification of sophorolipids. *Green Chemistry*, 18(1): 76–104. https://doi.org/10.1039/c5gc02187a

Dempster, C., Marchant, R. & Banat, İ.M. (2019). Antimicrobial and antibiofilm potential of biosurfactants as novel combination therapy against bacterium that cause skin infections. *Access Microbiology*, 1(1A). https://doi.org/10.1099/acmi.ac2019.po0566

Dengle-Pulate, V., Chandorkar, P., Bhagwat, S.S. & Prabhune, A. (2013). Antimicrobial and SEM studies of sophorolipids synthesized using lauryl alcohol. *Journal of Surfactants and Detergents*, 17(3): 543–552. https://doi.org/10.1007/s11743-013-1495-8

De Rienzo, M.A.D., Banat, I.M., Dolman, B., Winterburn, J. & Martin, P. (2015). Sophorolipid biosurfactants: Possible uses as antibacterial and antibiofilm agent. *New Biotechnology*, 32(6): 720–726. https://doi.org/10.1016/j.nbt.2015.02.009

Díaz-Rodríguez, P., Chen, H., Erndt-Marino, J., Liu, F., Totsingan, F., Gross, R.A. et al. (2018). Impact of select sophorolipid derivatives on macrophage polarization and viability. *ACS Applied Bio Materials*, 2(1): 601–612. https://doi.org/10.1021/acsabm.8b00799

Eckhart, L. & Tschachler, E. (2018). Control of cell death-associated danger signals during cornification prevents autoinflammation of the skin. *Experimental Dermatology*, 27(8): 884–891. https://doi.org/10.1111/exd.13700

Falk, N.A. (2019). Surfactants as antimicrobials: A brief overview of microbial interfacial chemistry and surfactant antimicrobial activity. *Journal of Surfactants and Detergents*, 22(5): 1119–1127. https://doi.org/10.1002/jsde.12293

Felse, P.A., Shah, V., Chan, J., Rao, K.J. & Gross, R.A. (2007). Sophorolipid biosynthesis by Candida bombicola from industrial fatty acid residues. *Enzyme and Microbial Technology*, 40(2): 316–323. https://doi.org/10.1016/j.enzmictec.2006.04.013

Fernandes, P.A.V., De Souza Arruda, I.R., Santos, A.F.A.B.D., De Araújo, A.A., Maior, A.M.S. & Ximenes, E.A. (2007). Antimicrobial activity of surfactants produced by Bacillus subtilis R14 against multidrug-resistant bacteria. *Brazilian Journal of Microbiology*, 38(4): 704–709. https://doi.org/10.1590/s1517-83822007000400022

Ferreira, A., Vecino, X., Ferreira, D., Cruz, J.M., Moldes, A.B. & Rodrigues, L.R. (2017). Novel cosmetic formulations containing a biosurfactant from Lactobacillus paracasei. *Colloids and Surfaces B: Biointerfaces*, 155: 522–529. https://doi.org/10.1016/j.colsurfb.2017.04.026

Fiume, M.M., Heldreth, B., Bergfeld, W.F., Belsito, D.V., Hill, R.A., Klaassen, C.D. et al. (2017). Safety assessment of diethanolamine and its salts as used in cosmetics. *International Journal of Toxicology*, 36(5_suppl2): 89S–110S. https://doi.org/10.1177/1091581817707179

Flemming, H.C., Neu, T.R. & Wozniak, D.J. (2007). The EPS matrix: The "house of biofilm cells". *Journal of Bacteriology*, 189(22): 7945–7947. https://doi.org/10.1128/JB.00858-07

Futara, T., Igarashi, K. & Hirata, Y. (2002). Low-foaming detergent compositions (World Patent 2003/002700). Saraya Co., Ltd., Osaka.

Gao, R., Falkeborg, M., Xu, X. & Guo, Z. (2012). Production of sophorolipids with enhanced volumetric productivity by means of high cell density fermentation. *Applied Microbiology and Biotechnology*, 97(3): 1103–1111. https://doi.org/10.1007/s00253-012-4399-z

Gharaei-Fa, E. (2010). Biosurfactants in pharmaceutical industry: A mini-review. *American Journal of Drug Discovery and Development*, 1(1): 58–69. https://doi.org/10.3923/ajdd.2011.58.69

Gorin, P.A., Spencer, J.F.T. & Tulloch, A.P. (1961). Hydroxy fatty acid glycosides of sophorose from Torulopsis magnoliae. *Canadian Journal of Chemistry*, 39(4): 846–855. Https://Doi.Org/10.1139/V61-104

Grice, E.A., Kong, H.H., Conlan, S., Deming, C., Davis, J., Young, A. et al. (2009). Topographical and temporal diversity of the human skin microbiome. *Science*, 324(5931): 1190–1192. https://doi.org/10.1126/science.1171700

Grice, E.A. & Segre, J.A. (2011). The skin microbiome. *Nature Reviews Microbiology*, 9(4): 244–253. https://doi.org/10.1038/nrmicro2537

Haggag, Y., Elshikh, M., El-Tanani, M., Bannat, I.M., McCarron, P. & Tambuwala, M.M. (2020).

Nanoencapsulation of sophorolipids in PEGylated poly(lactide-co-glycolide) as a novel approach to target colon carcinoma in the murine model. *Drug Delivery and Translational Research*, 10(5): 1353–1366. https://doi.org/10.1007/s13346-020-00750-3

Haque, F., Alfatah, M. & Ganesan, K. (2016). Inhibitory Effect of sophorolipid on Candida albicans biofilm formation and hyphal growth. *Sci. Rep.*, 6: 23575 https://doi.org/10.1038/srep23575

Haque, F., Sajid, M., Cameotra, S.S. & Battacharyya, M.S. (2017). Anti-biofilm activity of a sophorolipid-amphotericin B niosomal formulation against Candida albicans. *Biofouling*, 33(9): 768–779. https://doi.org/10.1080/08927014.2017.1363191

Haque, F., Verma, N.K., Alfatah, M., Bijlani, S. & Bhattacharyya, M.S. (2019b). Sophorolipid exhibits antifungal activity by ROS mediated endoplasmic reticulum stress and mitochondrial dysfunction pathways in *Candida albicans*. *RSC Advances*, 9(71): 41639–41648. https://doi.org/10.1039/c9ra07599b

Heinrich, K., Heinrich, U. & Tronnier, H. (2014). Influence of different cosmetic formulations on the human skin barrier. *Skin Pharmacology and Physiology*, 27(3): 141–147. https://doi.org/10.1159/000354919

Hillman, P.F., Lee, C. & Nam, S. (2022). Microbial natural products with wound-healing properties. *Processes*, 11(1): 30. https://doi.org/10.3390/pr11010030

Hirata, Y., Igarashi, K., Ueda, A. & Quan, G.L. (2021). Enhanced sophorolipid production and effective conversion of waste frying oil using dual lipophilic substrates. *Bioscience, Biotechnology, and Biochemistry*, 85(7): 1763–1771. https://doi.org/10.1093/bbb/zbab075

Holland, K.T. & Bojar, R. (2002). Cosmetics. *American Journal of Clinical Dermatology*, 3(7): 445–449. https://doi.org/10.2165/00128071-200203070-00001

Hospes, R., Hospes, B.I., Reiss, I., Bostedt, H. & Gortner, L. (2002). Molecular biological characterization of equine surfactant protein A. *Journal of Veterinary Medicine. A, Physiology, Pathology, Clinical Medicine*, 49(10): 497–498. https://doi.org/10.1046/j.1439-0442.2002.00489.x

Ikarashi, N., Kon, R., Kaneko, M., Mizukami, N., Kusunoki, Y. & Sugiyama, K. (2017). Relationship between aging-related skin dryness and aquaporins. *International Journal of Molecular Sciences*, 18(7): 1559. https://doi.org/10.3390/ijms18071559

Imura, T., Morita, T., Fukuoka, T., Ryu, M., Igarashi, K., Hirata, Y. et al. (2014). Spontaneous vesicle formation from sodium salt of acidic sophorolipid and its application as a skin penetration enhancer. *Journal of Oleo Science*, 63(2): 141–147. https://doi.org/10.5650/jos.ess13117

Ishii, N., Kobayashi, T., Matsumiya, K., Ryu, M., Hirata, Y., Matsumura, Y. et al. (2012). Transdermal administration of lactoferrin with sophorolipid. This article is part of a Special Issue entitled Lactoferrin and has undergone the Journal's usual peer review process. *Biochemistry and Cell Biology*, 90(3): 504–512. https://doi.org/10.1139/o11-065

Jadhav, J., Pratap, A.P. & Kale, S. (2019). Evaluation of sunflower oil refinery waste as feedstock for production of sophorolipid. *Process Biochemistry*, 78: 15–24. https://doi.org/10.1016/j.procbio.2019.01.015

Jin, S.H., Lee, J.E., Yun, J.H., Kim, I., Ko, Y. & Park, J.B. (2015). Isolation and characterization of human mesenchymal stem cells from gingival connective tissue. *Journal of Periodontal Research*, 50(4): 461–467. https://doi.org/10.1111/jre.12228

Joshi-Navare, K. & Prabhune, A. (2013). A biosurfactant-sophorolipid acts in synergy with antibiotics to enhance their efficiency. *BioMed Research International*, 2013: 512495. https://doi.org/10.1155/2013/512495

Juma, A., Lemoine, P., Simpson, A.B.J., Murray, J., O'Hagan, B., Naughton, P. et al. (2020). Microscopic investigation of the combined use of antibiotics and biosurfactants on methicillin resistant Staphylococcus aureus. *Frontiers in Microbiology*, 11. https://doi.org/10.3389/fmicb.2020.01477

Kim, S.H., Lee, S.O., Park, J.B., Park, I.A., Park, S.J., Yun, S.C. et al. (2011). A prospective longitudinal study evaluating the usefulness of a T-cell-based assay for latent tuberculosis infection in kidney transplant recipients. *American Journal of Transplantation: Official*

Journal of the American Society of Transplantation and the American Society of Transplant Surgeons, 11(9): 1927–1935. https://doi.org/10.1111/j.1600-6143.2011.03625.x

Kithur Mohamed, S., Asif, M., Nazari, M.V., Baharetha, H.M., Mahmood, S., Yatim, A.R.M. et al. (2019). Antiangiogenic activity of sophorolipids extracted from refined bleached deodorized palm olein. *Indian Journal of Pharmacology*, 51(1): 45–54. https://doi.org/10.4103/ijp.IJP_312_18

Kong, L., Saar, K.L., Jacquat, R., Hong, L., Levin, A., Gang, H. et al. (2017). Mechanism of biosurfactant adsorption to oil/water interfaces from millisecond scale tensiometry measurements. *Interface Focus*, 7(6): 20170013. https://doi.org/10.1098/rsfs.2017.0013

Kulkarni, S. & Choudhary, P. (2011). Production and isolation of biosurfactant-sophorolipid and its application in body wash formulation. *Asian J. Microbiol. Biotechnol. Environ. Sci.*, 13: 217–221. https://www.researchgate.net/publication/283925681

Kurtzman, C.P., Price, N.P., Ray, K.J. & Kuo, T.M. (2010). Production of sophorolipid biosurfactants by multiple species of the Starmerella (Candida) bombicola yeast clade. *FEMS Microbiology Letters*, 311(2): 140–146. https://doi.org/10.1111/j.1574-6968.2010.02082.x

Kwak, M.J., Park, M.Y., Kim, J., Lee, H. & Whang, K.Y. (2021). Curative effects of sophorolipid on physical wounds: In vitro and in vivo studies. *Veterinary Medicine and Science*, 7(4): 1400–1408. https://doi.org/10.1002/vms3.481

Lachance, M.A., Wijayanayaka, T.M., Bundus, J.D. & Wijayanayaka, D.N. (2011). Ribosomal DNA sequence polymorphism and the delineation of two ascosporic yeast species: Metschnikowia agaves and Starmerella bombicola. *FEMS Yeast Research*, 11(4): 324–333. https://doi.org/10.1111/j.1567-1364.2011.00718.x

Lee, H.S., Oh, J.Y., Lee, J.H., Yoo, C.G., Lee, C.T., Kim, Y.W. et al. (2004). Response of pulmonary tuberculomas to anti-tuberculous treatment. *The European Respiratory Journal*, 23(3): 452–455. https://doi.org/10.1183/09031936.04.00087304

Lee, S., Kim, S., Park, I., Chung, S., Chandra, M.S. & Choi, Y. (2010). Isolation, purification, and characterization of novel fengycin S from Bacillus amyloliquefaciens LSC04 degrading-crude oil. *Biotechnology and Bioprocess Engineering*, 15(2): 246–253. https://doi.org/10.1007/s12257-009-0037-8

Li, H., Guo, W., Ma, X., Li, J. & Song, X. (2016). In vitro and in vivo anticancer activity of sophorolipids to human cervical cancer. *Applied Biochemistry and Biotechnology*, 181(4): 1372–1387. https://doi.org/10.1007/s12010-016-2290-6

Liang, W., Guan, W., Chen, R., Wang, W., Li, J., Xu, K. et al. (2020). Cancer patients in SARS-CoV-2 infection: A nationwide analysis in China. *The Lancet Oncology*, 21(3): 335–337. https://doi.org/10.1016/S1470-2045(20)30096-6

Lim, J.S., Park, H.S., Cho, S. & Yoon, H.S. (2018). Antibiotic susceptibility and treatment response in bacterial skin infection. *Annals of Dermatology*, 30(2): 186. https://doi.org/10.5021/ad.2018.30.2.186

López, D., Vlamakis, H. & Kolter, R. (2010). Biofilms. *Cold Spring Harbor Perspectives in Biology*, 2(7): a000398. https://doi.org/10.1101/cshperspect.a000398

Lourith, N. & Kanlayavattanakul, M. (2009). Natural surfactants used in cosmetics: Glycolipids. *International Journal of Cosmetic Science*, 31(4): 255–261. https://doi.org/10.1111/j.1468-2494.2009.00493.x

Lunjani, N., Hlela, C. & O'Mahony, L. (2019). Microbiome and skin biology. *Current Opinion in Allergy and Clinical Immunology*, 19(4): 328–333. https://doi.org/10.1097/aci.0000000000000542

Lydon, H.L., Baccile, N., Callaghan, B., Marchant, R., Mitchell, C.A. & Banat, I.M. (2017). Adjuvant antibiotic activity of acidic sophorolipids with potential for facilitating wound healing. *Antimicrobial Agents and Chemotherapy*, 61(5): e02547-16. https://doi.org/10.1128/AAC.02547-16

Ma, X., Li, H. & Song, X. (2012). Surface and biological activity of sophorolipid molecules produced by Wickerhamiella domercqiae var. sophorolipid CGMCC 1576. *Journal of Colloid and Interface Science*, 376(1): 165–172.

Marchant, R. & Banat, I.M. (2012). Biosurfactants: A sustainable replacement for chemical surfactants. *Biotechnology Letters*, 34(9): 1597–1605. https://doi.org/10.1007/s10529-012-0956-x

Moore, D.J. & Rawlings, A.V. (2017). The chemistry, function and (patho)physiology of stratum corneum barrier ceramides. *International Journal of Cosmetic Science*, 39(4): 366–372. https://doi.org/10.1111/ics.12399

Morais, I.M.C., Cordeiro, A.L., Teixeira, G.S., Domingues, V.S., Nardi, R.M.D., Monteiro, A.S. et al. (2017). Biological and physicochemical properties of biosurfactants produced by Lactobacillus jensenii P_{6A} and Lactobacillus gasseri P_{65}. *Microbial Cell Factories*, 16(1): 155.

Morris, S.A.V., Thompson, R.T., Glenn, R.W., Ananthapadmanabhan, K.P. & Kasting, G.B. (2019). Mechanisms of anionic surfactant penetration into human skin: Investigating monomer, micelle and sub micellar aggregate penetration theories. *International Journal of Cosmetic Science*, 41(1): 55–66. https://doi.org/10.1111/ics.12511

Müller, R.H., Mäder, K. & Gohla, S. (2000). Solid lipid nanoparticles (SLN) for controlled drug delivery: A review of the state of the art. *European Journal of Pharmaceutics and Biopharmaceutics: Official Journal of Arbeitsgemeinschaft fur Pharmazeutische Verfahrenstechnik e.V*, 50(1): 161–177. https://doi.org/10.1016/s0939-6411(00)00087-4

Müller, R.H., Radtke, M. & Wissing, S.A. (2002). Solid lipid nanoparticles (SLN) and nanostructured lipid carriers (NLC) in cosmetic and dermatological preparations. *Advanced Drug Delivery Reviews*, 54(Suppl 1): S131–S155. https://doi.org/10.1016/s0169-409x(02)00118-7

Musthaq, S., Mazuy, A. & Jakus, J. (2018). The microbiome in dermatology. *Clinics in Dermatology*, 36(3): 390–398. https://doi.org/10.1016/j.clindermatol.2018.03.012

Muthusamy, K., Gopalakrishnan, S., Ravi, T.K. & Sivachidambaram, P. (2008). Biosurfactants: Properties, commercial production and application. *Current Science*, 94(6): 736–747. http://www.jstor.org/stable/24100627

Naik, N., Abhyankar, I., Darne, P., Prabhune, A. & Madhusudhan, B. (2019). Sustained transdermal release of lignans facilitated by sophorolipid based transferosomal hydrogel for cosmetic application. *International Journal of Current Microbiology and Applied Sciences*, 8(02): 1783–1791. https://doi.org/10.20546/ijcmas.2019.802.210

Nassan, F.L., Coull, B.A., Gaskins, A.J., Williams, M.A., Skakkebaek, N.E., Ford, J.B. et al. (2017). Personal care product use in men and urinary concentrations of select phthalate metabolites and parabens: Results from the environment and reproductive health (EARTH) study. *Environmental Health Perspectives*, 125(8): 087012. https://doi.org/10.1289/EHP1374

Naughton, P.J., Marchant, R., Naughton, V. & Banat, I.M. (2019). Microbial biosurfactants: Current trends and applications in agricultural and biomedical industries. *Journal of Applied Microbiology*, 127(1): 12–28. https://doi.org/10.1111/jam.14243

Nawale, L., Dubey, P., Chaudhari, B., Sarkar, D. & Prabhune, A. (2017). Anti-proliferative effect of novel primary cetyl alcohol derived sophorolipids against human cervical cancer cells HeLa. *PloS One*, 12(4): e0174241. https://doi.org/10.1371/journal.pone.0174241

Ohadi, M., Forootanfar, H., Rahimi, H.R., Jafari, E., Shakibaie, M., Eslaminejad, T. et al. (2017). Antioxidant potential and wound healing activity of biosurfactant produced by Acinetobacter junii B6. *Current Pharmaceutical Biotechnology*, 18(11): 900–908. https://doi.org/10.2174/1389201018666171122121350

Ohue, Y. & Nishikawa, H. (2019). Regulatory T (Treg) cells in cancer: Can Treg cells be a new therapeutic target? *Cancer Science*, 110(7): 2080–2089. https://doi.org/10.1111/cas.14069

Orton, D. & Wilkinson, J.D. (2004). Cosmetic allergy. *American Journal of Clinical Dermatology*, 5(5): 327–337. https://doi.org/10.2165/00128071-200405050-00006

Paulino, B.N., Pessôa, M.G., Mano, M.C., Molina, G., Neri-Numa, I.A. & Pastore, G.M. (2016). Current status in biotechnological production and applications of glycolipid biosurfactants. *Applied Microbiology and Biotechnology*, 100(24): 10265–10293. https://doi.org/10.1007/s00253-016-7980-z

Peyrat, L., Tsafantakis, N., Georgousaki, K., Ouazzani, J., Genilloud, O., Trougakos, I.P. et al. (2019). Terrestrial microorganisms: Cell factories of bioactive molecules with skin protecting applications. *Molecules*, 24(9): 1836. https://doi.org/10.3390/molecules24091836

Rahimi, K., Lotfabad, T.B., Jabeen, F. & Ganji, S.M. (2019). Cytotoxic effects of mono- and di-rhamnolipids from Pseudomonas aeruginosa MR01 on MCF-7 human breast cancer cells. *Colloids and Surfaces B: Biointerfaces*, 181: 943–952. https://doi.org/10.1016/j.colsurfb.2019.06.058

Relipidium BC 10096 https://www.carecreations.basf.com/product-formulations/products/products-detail/RELIPIDIUMBC10096/307139450 (accessed on 28 June 2020) referred (2023)

Rocha, M.A. & Bagatin, E. (2018). Skin barrier and microbiome in acne. *Archives of Dermatological Research*, 310(3): 181–185. https://doi.org/10.1007/s00403-017-1795-3

Rodan, K., Fields, K., Majewski, G. & Falla, T.J. (2016). Skincare bootcamp. Plastic and reconstructive surgery. *Global Open*, 4: e1152. https://doi.org/10.1097/gox.0000000000001152

Römling, U. & Balsalobre, C. (2012). Biofilm infections, their resilience to therapy and innovative treatment strategies. *Journal of Internal Medicine*, 272(6): 541–561. https://doi.org/10.1111/joim.12004

Rosa, C.A. & Lachance, M.A. (1998). The yeast genus Starmerella gen. nov. and Starmerella bombicola sp. nov., the teleomorph of Candida bombicola (Spencer, Gorin & Tullock) Meyer & Yarrow. *International Journal of Systematic Bacteriology*, 48(Pt 4): 1413–1417. https://doi.org/10.1099/00207713-48-4-1413

Santos, D.K.F.D., Rufino, R.D., De Luna, J.M., Santos, V.L.D.A. & Sarubbo, L.A. (2016). Biosurfactants: Multifunctional biomolecules of the 21st century. *International Journal of Molecular Sciences*, 17(3): 401. https://doi.org/10.3390/ijms17030401

Saika, A., Koike, H., Fukuoka, T. & Morita, T. (2018). Tailor-made mannosylerythritol lipids: Current state and perspectives. *Applied Microbiology and Biotechnology*, 102(16): 6877–6884. https://doi.org/10.1007/s00253-018-9160-9

Schneider, G., Gohla, S., Schreiber, J., Kaden, W., Schönrock, U., Diembeck, W. et al. (2001). Skin cosmetics. *Ullmann's Encyclopedia of Industrial Chemistry*. https://doi.org/10.1002/14356007.a24_219

Schommer, N.N. & Gallo, R.L. (2013). Structure and function of the human skin microbiome. *Trends in Microbiology*, 21(12): 660–668. https://doi.org/10.1016/j.tim.2013.10.001

Sen, S., Borah, S.N., Kandimalla, R., Bora, A. & Deka, S. (2020). Sophorolipid biosurfactant can control cutaneous dermatophytosis caused by Trichophyton mentagrophytes. *Frontiers in Microbiology*, 11: 329. https://doi.org/10.3389/fmicb.2020.00329

Seweryn, A. (2018). Interactions between surfactants and the skin – Theory and practice. *Advances in Colloid and Interface Science*, 256: 242–255. https://doi.org/10.1016/j.cis.2018.04.002

Shah, V., Doncel, G.F., Seyoum, T., Eaton, K.M., Zalenskaya, I., Hagver, R. et al. (2005). Sophorolipids, microbial glycolipids with anti-human immunodeficiency virus and sperm-immobilizing activities. *Antimicrobial Agents and Chemotherapy*, 49(10): 4093–4100. https://doi.org/10.1128/AAC.49.10.4093-4100.2005

Shu, Q., Lou, H., Wei, T., Liu, X. & Chen, Q. (2021). Contributions of glycolipid biosurfactants and glycolipid-modified materials to antimicrobial strategy: A review. *Pharmaceutics*, 13(2): 227. https://doi.org/10.3390/pharmaceutics13020227

Silveira, V.A.I., Freitas, C.A.U.Q. & Celligoi, M.A.P.C. (2018). Antimicrobial applications of sophorolipid from Candida bombicola: A promising alternative to conventional drugs. *Journal of Applied Biology and Biotechnology*, 6(6): 87–90. https://doi.org/10.7324/jabb.2018.60614

Singh, H., Bhandari, R. & Kaur, I.P. (2013). Encapsulation of Rifampicin in a solid lipid nanoparticulate system to limit its degradation and interaction with Isoniazid at acidic pH. *International Journal of Pharmaceutics*, 446(1-2): 106–111. https://doi.org/10.1016/j.ijpharm.2013.02.012

Singh, S., Patel, P., Jaiswal, S., Prabhune, A., Ramana, C.V. & Prasad, B.L.V. (2009). A direct method for the preparation of glycolipid–metal nanoparticle conjugates: Sophorolipids as reducing and capping agents for the synthesis of water re-dispersible silver nanoparticles and their antibacterial activity. *New Journal of Chemistry*, 33(3): 646–652. https://doi.org/10.1039/b811829a

Smith, R., Russo, J., Fiegel, J. & Brogden, N. (2020). Antibiotic delivery strategies to treat skin infections when innate antimicrobial defense fails. *Antibiotics* (Basel, Switzerland), 9(2): 56. https://doi.org/10.3390/antibiotics9020056

Sohn E. (2018). Skin microbiota's community effort. *Nature*, 563(7732): S91–S93. https://doi.org/10.1038/d41586-018-07432-8

Solaiman, D.K.Y., Ashby, R.D., Núñez, A. & Crocker, N.V. (2020). Low-temperature crystallization for separating monoacetylated long-chain sophorolipids: Characterization of their surface-active and antimicrobial properties. *Journal of Surfactants and Detergents*, 23(3): 553–563. https://doi.org/10.1002/jsde.12396

Solaiman, D.K., Liu, Y., Moreau, R.A. & Zerkowski, J.A. (2014). Cloning, characterization, and heterologous expression of a novel glucosyltransferase gene from sophorolipid-producing Candida bombicola. *Gene*, 540(1): 46–53. https://doi.org/10.1016/j.gene.2014.02.029

Som, I., Bhatia, K. & Yasir, M. (2012). Status of surfactants as penetration enhancers in transdermal drug delivery. *Journal of Pharmacy & Bioallied Sciences*, 4(1): 2–9. https://doi.org/10.4103/0975-7406.92724

Spada, F., Barnes, T.M. & Greive, K.A. (2018). Skin hydration is significantly increased by a cream formulated to mimic the skin & rsquo's own natural moisturizing systems. *Clinical, Cosmetic and Investigational Dermatology*, 11: 491–497. https://doi.org/10.2147/ccid.s177697

Spencer, J.F., Gorin, P.A. & Tulloch, A.P. (1970). Torulopsis bombicola sp.n. *Antonie van Leeuwenhoek*, 36(1): 129–133. https://doi.org/10.1007/BF02069014

Stipčević, T., Piljac, T. & Isseroff, R.R. (2005). Di-rhamnolipid from displays differential effects on human keratinocyte and fibroblast cultures. *Journal of Dermatological Science*, 40(2): 141–143. https://doi.org/10.1016/j.jdermsci.2005.08.005

Subramaniam, M.D., Venkatesan, D., Iyer, M., Subbarayan, S., Vivekanandhan, G., Roy, A. et al. (2020). Biosurfactants and anti-inflammatory activity: A potential new approach towards COVID-19. *Current Opinion in Environmental Science & Health*, 17: 72–81. https://doi.org/10.1016/j.coesh.2020.09.002

Subsanguan, T., Khondee, N., Rongsayamanont, W. & Luepromchai, E. (2022). Formulation of a glycolipid:lipopeptide mixture as biosurfactant-based dispersant and development of a low-cost glycolipid production process. *Scientific Reports*, 12(1). https://doi.org/10.1038/s41598-022-20795-3

Takahashi, M., Morita, T., Fukuoka, T., Imura, T. & Kitamoto, D. (2012). Glycolipid biosurfactants, mannosylerythritol lipids, show antioxidant and protective effects against H_2O_2-induced oxidative stress in cultured human skin fibroblasts. *Journal of Oleo Science*, 61(8): 457–464. https://doi.org/10.5650/jos.61.457

Tarale, P., Gawande, S. & Jambhulkar, V. (2015b). Antibiotic susceptibility profile of bacilli isolated from the skin of healthy humans. *Brazilian Journal of Microbiology*, 46(4): 1111–1118. https://doi.org/10.1590/s1517-838246420131366

Timm, C.M., Loomis, K., Stone, W., Mehoke, T., Brensinger, B., Pellicore, M. et al. (2020). Isolation and characterization of diverse microbial representatives from the human skin microbiome. *Microbiome*, 8(1): 58. https://doi.org/10.1186/s40168-020-00831-y

Tulloch, A.P., Spencer, J.F.T. & Deinema, M. (1968). A new hydroxy fatty acid sophoroside from Candida bogoriensis. *Canadian Journal of Chemistry*, 46(3): 345–348. https://doi.org/10.1139/v68-057

Vaishnavi, K.V., Safar, L. & Devi, K. (2019). Biofilm in dermatology. *Journal of Skin and Sexually Transmitted Diseases*, 1: 3–7. https://doi.org/10.25259/jsstd_14_2019

Valotteau, C., Baccile, N., Humblot, V., Roelants, S., Soetaert, W., Stevens, C.V. et al. (2019). Nanoscale antiadhesion properties of sophorolipid-coated surfaces against pathogenic bacteria. *Nanoscale Horizons*, 4(4): 975–982. https://doi.org/10.1039/c9nh00006b

Valotteau, C., Banat, I.M., Mitchell, C.A., Lydon, H., Marchant, R., Babonneau, F. et al. (2017). Antibacterial properties of sophorolipid-modified gold surfaces against gram positive and gram-negative pathogens. *Colloids and Surfaces B: Biointerfaces*, 157: 325–334. https://doi.org/10.1016/j.colsurfb.2017.05.072

Van Bogaert, I.N., Saerens, K., De Muynck, C., Develter, D., Soetaert, W. & Vandamme, E.J. (2007). Microbial production and application of sophorolipids. *Applied Microbiology and Biotechnology*, 76(1): 23–34. https://doi.org/10.1007/s00253-007-0988-7

Varvaresou, A. & Iakovou, K. (2015). Biosurfactants in cosmetics and biopharmaceuticals. *Letters in Applied Microbiology*, 61(3): 214–223. https://doi.org/10.1111/lam.12440

Vecino, X., Cruz, J., Moldes, A.B. & Rodrigues, L.R. (2017). Biosurfactants in cosmetic formulations: Trends and challenges. *Critical Reviews in Biotechnology*, 37(7): 911–923. https://doi.org/10.1080/07388551.2016.1269053

Wang, X., Xu, N., Li, Q., Chen, S., Cheng, H., Yang, M. et al. (2021). Lactonic sophorolipid-induced apoptosis in human HepG2 cells through the Caspase-3 pathway. *Applied Microbiology and Biotechnology*, 105(5): 2033–2042. https://doi.org/10.1007/s00253-020-11045-5

Weyrich, L.S., Dixit, S., Farrer, A.G. & Cooper, A. (2015). The skin microbiome: Associations between altered microbial communities and disease. *Australasian Journal of Dermatology*, 56(4): 268–274. https://doi.org/10.1111/ajd.12253

Yamamoto, S., Morita, T., Fukuoka, T., Imura, T., Yanagidani, S., Sogabe, A. et al. (2012). The moisturizing effects of glycolipid biosurfactants, mannosylerythritol lipids, on human skin. *Journal of Oleo Science*, 61(7): 407–412. https://doi.org/10.5650/jos.61.407

Yanase, K. & Hatta, I. (2017). Disruption of human stratum corneum lipid structure by sodium dodecyl sulphate. *International Journal of Cosmetic Science*, 40(1): 44–49. https://doi.org/10.1111/ics.12430

Yang, Y., Ashworth, A.J., Willett, C.D., Cook, K.J., Upadhyay, A., Owens, P.R. et al. (2019). Review of antibiotic resistance, ecology, dissemination, and mitigation in U.S. broiler poultry systems. *Frontiers in Microbiology*, 10. https://doi.org/10.3389/fmicb.2019.02639

Yousef, H., Alhajj, M. & Sharma, S. (2023). Anatomy, Skin (Integument), Epidermis. [Updated 2022 Nov 14]. In: StatPearls [Internet]. Treasure Island (FL): StatPearls Publishing. Available from: https://www.ncbi.nlm.nih.gov/books/NBK470464/

Zerhusen, C., Bollmann, T., Gödderz, A., Fleischer, P., Glüsen, B. & Schörken, U. (2019). Microbial synthesis of nonionic long-chain sophorolipid emulsifiers obtained from fatty alcohol and mixed lipid feeding. *European Journal of Lipid Science and Technology*, 122(1). https://doi.org/10.1002/ejlt.201900110

Zouari, R., Moalla-Rekik, D., Sahnoun, Z., Rebai, T., Ellouze-Chaabouni, S. & Ghribi-Aydi, D. (2016). Evaluation of dermal wound healing and in vitro antioxidant efficiency of Bacillus subtilis SPB1 biosurfactant. *Biomedicine & Pharmacotherapy = Biomedecine & Pharmacotherapie*, 84: 878–891. https://doi.org/10.1016/j.biopha.2016.09.084

CHAPTER

22

Revolutionizing Sustainable Cosmetic Formulations with Endophytic Bacterial Biosurfactants

Anita V. Handore[1*], Sharad R. Khandelwal[1], Arpita M. Gupte[2], Pravin G. Morankar[3], Sharmila S. Ghangale[4], Amol N. More[5], Soham V. Kale[1], and Dilip V. Handore[1]

[1] Research and Development Department, Phytoelixir Pvt. Ltd. Nashik, India
[2] School of Biotechnology and Bioinformatics, DY Patil (Deemed to be University), Navi Mumbai, India
[3] St. John Institute of Pharmacy and Research, Palghar, India
[4] Department of Biotechnology, Changu Kana Thakur Arts, Commerce and Science College, New Panvel, India
[5] Research and Development Department, Biotox Laboratories, Nashik, India

1. Introduction

Cosmetic products are an integral part of our daily lives, encompassing a wide range of items including toothpaste, soap, shampoo, deodorant, skincare products, perfume, and makeup, among others. Cosmetic products are an integral part of our daily lives, encompassing a wide range of items including toothpaste, soap, shampoo, deodorant, skincare products, perfume, and makeup, among others. The cosmetic industry holds significant environmental, social, and economic implications, prompting a quest for natural, safe, ecofriendly and sustainable products (Saranraj et al., 2022, Handore et al., 2022).

Contemporary consumer preferences lean toward the incorporation of natural ingredients in cosmetics, driven by their potential to match or even surpass the benefits offered by chemical-based counterparts. Within this context, biosurfactants, the surface-active substances produced by bacteria have emerged as a natural solution with considerable promise for cosmetic formulations. Owing to their biodegradability, potential health impacts, low toxicity, and compatibility with various skin types,

* Corresponding author: avhandore@gmail.com

application of such eco friendly safe products could be efficiently used for sustainable cosmetic formulations (Saranraj et al., 2022, Patel et al., 2022).

2. Microbial Biosurfactants in Cosmetic Formulations

The microbial biosurfactants derived from bacteria have diverse properties and are being explored for a wide range of cosmetic applications. Based on their chemical composition and molecular weight, these are typically categorized as low molecular weight biosurfactants (e.g. Glycolipids, lipopeptides, lipoproteins, fatty acids, and phospholipids etc.) and high molecular weight biosurfactants encompassing polymeric biosurfactants, characterized by more complex molecular structures with higher molecular weights.

Glycolipids are the most studied biosurfactants in cosmetic and personal care formulations. For example, patented cosmetics containing rhamnolipids have been used as anti-wrinkle and anti-aging products.

A patented cosmetic product, incorporated with few rhamnolipid biosurfactants in concentrations of 0.001% to 5% was designed for the treatment of aging signs (Piljac et al., 1999, Lourith et al., 2009, Vecino et al., 2017). In 2013, Owen and Fan carried out an assessment of oligomeric biosurfactants in various topically applied formulations, such as anti-aging creams and facial gels and conditioning hair masks. Their research was focused on the incorporation of a polymeric acylated biosurfactant, as an ingredient with a concentration range of 10 to 1000 ppm in these formulations (Vecino et al., 2017).

The diversity of bacterial biosurfactants (BB) and their sources underscores the potential for their use in various cosmetic applications. In this regard, such biosurfactants have been the focus of numerous patents of which, rhamnolipids account for 50% of the registered patents. This highlights the significant interest and research in these biosurfactants and their potential commercial applications Rhamnolipids could be produced by *Pseudomonas aeruginosa* (Janek et al., 2010, Belgacem et al., 2015). Trehalolipids could be produced by *Rhodococcus erythropolis* (Kundu et al., 2013). Besides, Surfactin is also one of the most extensively studied lipopeptide biosurfactants which is produced by various *Bacillus spp.* including *Bacillus subtilis.* (Coronel-Leon et al., 2015, Ma Z et al., 2014, Vecino et al., 2017).

In the cosmetic industry, numerous compounds for cosmetic applications undergo enzymatic conversion facilitated by various lipases and whole cells. Surfactants with a minimum shelf life of 3 years are highly sought after in this industry, leading to a preference for saturated acyl groups over unsaturated compounds. In cosmetics, Monoglycerides are one of the widely used surfactants, which are obtained from glycerol-tallow (1.5:2) with an impressive 90% yield, achieved by using *P. fluorescens* lipase treatment. This method underscores the industry's commitment to efficient and sustainable processes in crafting stable and long-lasting cosmetic formulations.

3. Biosurfactant Based Cosmetic Products

Biosurfactants are increasingly being explored for use in cosmetic products due to their unique properties, including their biodegradability, low toxicity, and potential benefits as shown in Table 1.

Table 1: Biosurfactant based products

Category of cosmetics	Biosurfactants based Products	References
Cleansing Products	**Facial Cleansers: Biosurfactants** are commonly used in facial cleansers, such as face washes and makeup removers, to effectively remove dirt, makeup, and impurities while being gentle on the skin. **Body Washes and Shower Gels:** Biosurfactants can create a rich and foamy lather in body washes, providing a pleasant and cleansing experience.	Adu, et al., 2020
Hair Care Products	**Shampoos and Conditioners:** Biosurfactants are used in shampoos for their ability to clean hair effectively while being mild on the scalp. They can also be included in conditioners to improve hair manageability and softness.	Adu, et al., 2020
Emulsified Products	**Moisturizers:** In moisturizing creams and lotions, biosurfactants act as emulsifiers, helping to combine oil and water-based ingredients for a stable and well-textured product. **Sunscreen:** Some sunscreens incorporate biosurfactants to ensure even distribution of active sun-blocking ingredients, resulting in a smooth and non-greasy application.	Adu, et al., 2020
Makeup and Skincare	**Foundation:** Some foundation formulations use biosurfactants to create a smooth, even coverage and improve the wearability of makeup products. **Skincare Serums:** Biosurfactants may be included in serums to enhance the absorption of active ingredients into the skin.	Heinrich et al., 2014, Adu, et al., 2020
Fragrance Products	**Perfumes and Body Sprays:** biosurfactants can be used to improve the even dispersion of fragrances in these products, ensuring a consistent scent application.	Gupta et al.2015
Oral Care	**Toothpaste:** Some toothpaste formulations utilize biosurfactants for their foaming and cleaning properties, promoting oral hygiene.	Elshikh et al., 2017
Deodorants	**Roll-Ons and Sprays:** Biosurfactants can be found in deodorant formulations to improve the spread ability and consistency of the product on the skin.	Zirwas et al., 2008
Antimicrobial Products Acne Treatments	Biosurfactants with antimicrobial properties can be used in acne treatments to combat acne-causing bacteria and promote clearer skin.	Fernandes et al., 2007, Gomaa et al., 2013, Urzedo et al., 2018, Adu, et al., 2020

Table 2: Application of microbial biosurfactants in cosmetic industries

Type of microbial biosurfactant	Application	References
Sophorolipid	Cleanser in shower gel and shampoo, Body washer, Anti-inflammatory agent, Cleanser for anti dandruff shampoo, Moisturizer, Skin cleanser, Body cleanser	Cox et al., 2013, Kulkarni et al., 2011, Hagler et al., 2007
Glycolipid	Antifungal shampoo, Hair-care conditioning polymers, Hair care formulation, Cleanser in shampoo	Parry et al., 2018, Mimee et al., 2005, Fernández-Peña et al., 2020
Lipopeptide	Hair care formulation, Rosemary oil/ water emulsions, Dried hair care formulation, Stabilizing agent for antidandruff formulations based on Zn pyrithione powder, Sunscreen formulations based on mica powder, Pickering emulsions containing vitamin E, Stabilizing agent of vitamin C, Antiacne formulation, Antimicrobial agent for dermal application, Preservative and irritant agent	Rincón-Fontán et al., 2016, Rodríguez-López et al., 2019 Gómez-Graña et al., 2017, Knoth et al., 2019, Das et al., 2008
MELs	Anti-ageing product, Skin roughness preventing agent, Makeup product, Antimicrobial agent	Takahashi et al., 2012, Kitagawa et al. 2010, Kitagawa et al., 2015, Kitamoto et al., 1993
Glycolipopeptide	Rosemary oil/water emulsions, Cosmetic formulation with antioxidants, Bioactivity against skin pathogens, Antimicrobial and anti-adhesive agent, Preservative and irritant agent	Vecino et al., 2015, Vecino, et al., 2017, Ferreira et al., 2017, Vecino et al., 2018, Rodríguez-López et al., 2019
Rhamnolipid	Anti-ageing product, Cleanser in shampoos, Anti-adhesive products, Cleanser for anti dandruff shampoo, Moisturizing skin cleanser, Body cleanser	Desanto 2011, Nickzad 2014
Rhamnolipid/ Sophorolipid	Cleanser for anti dandruff shampoo, Moisturizing skin cleanser, Body cleanser	Allef et al., 2016
Oligomeric biosurfactant	Conditioning agent for hair products	Owen et al., 2013

4. Biosurfactant Production by Bacteria

Bacteria produce biosurfactants as versatile tools that enable them to adapt to various environmental conditions, enhance their survival and competitiveness, and improve their interactions with both abiotic and biotic components of their surroundings. These molecules play a crucial role in the ecological and physiological strategies of bacterial species. Biosurfactants (BB) for a variety of reasons serve several important functions in the Bacterial world due to the following key reasons:

- **Surface Tension Reduction:** Bacteria can access nutrients, move through the liquid, and form biofilms more easily by reducing surface tension.
- **Nutrient Acquisition:** Bacteria use biosurfactants to aid in the acquisition of nutrients. They produces biosurfactants to emulsify and break down complex organic compounds, facilitating their utilization.
- **Cell Mobility:** Biosurfactants can help bacteria move efficiently in their environment by reducing the surface tension at cell-substrate interface. They also enable processes like swarming, gliding motility, and movement on solid surfaces, which are essential for bacterial survival and resource exploration.
- **Biofilm Formation:** Biosurfactants can play a crucial role in the early stages of biofilm formation by promoting initial attachment of bacterial cells to surfaces. Additionally, they contribute to the structural stability of biofilms.
- **Interactions with Other Microbes:** In complex microbial communities, bacteria may use biosurfactants to interact with other microorganisms. They can influence the behavior of neighboring bacteria, fungi, or other microorganisms by modifying the local environment or aiding in the formation of multispecies biofilms.
- **Competitive Advantage:** Bacteria may produce biosurfactants to gain a competitive advantage over other microorganisms. For example, they can inhibit the growth of competing species by producing biosurfactants that disrupt their cell membranes or biofilm formation.
- **Adaptation to Stressful Environments:** In harsh or stressful environments, such as those with high salinity or the presence of toxic compounds, bacteria may produce biosurfactants as a protective mechanism. These molecules can help mitigate the effects of stressors by forming protective layers or reducing the impact of harmful substances.

4.1 Cosmetic Formulations with Bacterial Biosurfactants (BB)

Cosmetic formulations with Bacterial biosurfactants offer various functional benefits. They contribute to the development of cosmetics that are gentle on the skin, environmentally sustainable, and aligned with the growing trend toward natural, clean and safe beauty. The use of specific biosurfactants and their concentrations could vary depending on the desired properties and objectives of the cosmetic products.

4.2 Relevant Key Properties of Bacterial Biosurfactants (BB) for Cosmetic Formulations

During the formulation of cosmetics, it's crucial to consider the specific characteristics of biosurfactants along with their compatibility with other ingredients. Following properties of BB could collectively demonstrate the reason for gaining traction in the cosmetic industry as a more eco-friendly and skin-friendly alternative to traditional surfactants:

- **Low Toxicity:** These biosurfactants have minimal toxicity, making them safer for both consumers and the environment.
- **Gentle Cleansing Property:** These biosurfactants offer gentle and effective cleansing, making them suitable for sensitive or dry skin. Therefore, they can be

used in facial cleansers, body washes, and shampoos to create a rich and stable foam while effectively removing dirt, oil, and impurities from the skin and hair, even suitable for sensitive or dry skin.

- **Emulsifying Power:** They serve as effective emulsifiers, facilitating the blending of oil and water-based ingredients for stable cosmetic formulations in products like creams, lotions, and serums. Such stable and well-textured formulations could be easy to apply and spread on the skin.
- **Foaming Property:** Biosurfactants can enhance the foaming properties of cosmetic products, such as foaming facial cleansers and shaving creams.
- **Stabilization Property:** Biosurfactants can help to stabilize emulsions and suspensions in products like sunscreens, moisturizers, and some other beauty products. Use of these Stabilizers could ensure consistent texture and performance by inseparable distribution throughout the product over time.
- **Moisturizing and Conditioning Property:** Some biosurfactants have moisturizing and skin-conditioning properties. They can help improve the hydration of the skin, making them suitable for products like creams and lotions.
- **Antimicrobial Property:** Some biosurfactants exhibit antimicrobial properties, which can be beneficial in products designed to control acne, prevent infections, or promote overall skin health.
- **Biodegradability:** Biosurfactants are environmentally friendly as they are naturally biodegradable, reducing the ecological impact of cosmetic products.
- **Sustainability:** Being biodegradable, such biosurfactants produced from natural renewable resources could contribute to the sustainability of cosmetic products.

5. Multifaceted Interplay between Bacterial Biosurfactants and Human Skin Microbiome

The skin functions as a self-contained ecosystem. Over time, there has been a concept of improving the health of the skin microbiome by incorporating bioactive compounds, like microbial biosurfactants, into beauty products. (Kumari et al., 2021, Karnwal et al., 2023). Bacteria are the most prevalent microorganisms found on the skin, and the bacterial varieties found on and within individuals are often quite similar. It is reported that approximately one billion diverse microorganisms are found to inhabit 1 sq.cm of human skin (Peng and Biswas, 2020, Bianchi et al., 2016, Valsecchi et al., 2014, Karnwal et al., 2023). Typically, the skin's microbial diversity expands due to continuous exposure to atmospheric microorganisms. This expansion is regulated by factors such as T lymphocytes and related host traits, including factors like sexuality, environment, nutrition, location, and the use of personal and cosmetic products. (Vanlay et al., 2022, Wei et al., 2022). It's evident that bacteria are the most common microorganisms found on the skin, with the interpersonal and intrapersonal varieties of bacteria being nearly identical. The exploration of the interplay between biosurfactants, the skin microbiome, and human skin health is a multifaceted and evolving field of research. This intersection offers insights into how these natural compounds could impact the skin's ecology, health, and overall well-being. The skin microbiome plays a crucial role in maintaining skin health by influencing factors such as pH, moisture, and the immune response. Biosurfactants could interact with the skin micro biomes and affect their composition and activity. Bacterial biosurfactants possess important physicochemical properties that are beneficial for maintaining

Table 3: Bacterial biosurfactant for cosmetic industry

Bacteria	Biosurfactant	References
A. calcoaceticus	Emulsan	Handore et al., 2022, Choi et al.,1996
A. radioresistens	Alasan	Handore et al., 2022, Navonvenezia et al.,1995
Agrobacterium sp. *Pseudomonas* sp., *T. thiooxidan*	Ornithine lipids	Handore et al., 2022, Desai et al., 1997
Acinetobacter sp.	Phospholipids	Handore et al., 2022, Kosaric 2001
A. calcoaceticus, P. marginilis, P. Maltophila	Vesicles & fimbriae	Handore et al., 2022, Desai et al.,1997
B. subtilis, B. licheniformis, B. magiterium, B. amyloliquefaciens ST34	Surfactin/Iturin	Handore et al., 2022, Dey et al., 2015, Dey,et al., 2017, Ndlovu et al., 2017
Bacillus sp.	Amino acids lipids	Handore et al., 2022, Cotter et al., 2005
Bacillus licheniformis, B. subtilis	Lichenysin	Handore et al., 2022, Yakimov et al., 1997
B. glumae, B. plantarii, B. thailandensis, P. aeruginosa, Pseudomonas sp., *P. chlororaphis, S. rubidea*	Rhamnolipids	Handore et al., 2022, Ndlovu et al., 2017
B. subtilis	Subtilisin	Handore et al., 2022, Sutyak et al., 2008
B. subtilis SPB1	Lipopeptide	Handore et al., 2022, Zouari et al., 2016
B. licheniformis	Peptide lipids	Handore et al., 2022, Begley et al., 2009
Cyanobacteria	Whole cell	Handore et al., 2022, Levy et al., 1990
C. michiganensis sub sp. *insidiosus*	Fatty acids/neutral lipids	Handore et al., 2022, Herman et al., 2002
Corynebacterium sp., *R. erythropolis, Arthrobacter* sp., *N. erythropolis, Mycobacterium* sp	Trehalose lipids	Handore et al., 2022, Muthusamy et al., 2008
L. fermentum	Diglycosyl diglycerides	Handore et al., 2022, Mulligan et al., 2001
P. fluorescens, L. mesenteriods	Viscosin	Handore et al., 2022, Banat et al., 2010
P. fluorescens, D. polmorphus	Carbohydrate-lipid	Handore et al., 2022, Nerurkar et al., 2011
P. aeruginosa	Protein PA	Handore et al., 2022, Hisatsuka et al., 1971
S. marcescens	Serrawettin	Handore et al., 2022, Lai et al., 2009

healthy skin. Their fatty acid components are effective in moisturizing and hydrating rough and dry skin surfaces. Besides, they show antioxidant potential and support to reduce the generation of free radicals induced by UV light exposure (Thakur et al., 2021, Phetcharat et al., 2019).

Moreover, the degradation of triglycerides in microbial biosurfactants into fatty acid chains by *Cutibacterium acnes* plays a key role in regulating the skin's pH to maintain an acidic environment and promotes adhesion of indigenous skin micro flora and prohibits development of pathogenic bacteria. This mechanism contributes to the preservation of a healthier skin microbiota essential for maintaining the skin health. (Ahle et al., 2022).

The interplay between biosurfactants and skin microbiome can be relevant in managing skin conditions such as acne and eczema. Balancing the microbiome and reducing the presence of harmful bacteria may contribute to skin health. Some biosurfactants may serve as prebiotics, providing nourishment for beneficial skin bacteria. This can support the growth and activity of these beneficial microorganisms. Therefore, formulating the skin products with biosurfactants could promote and support the and balance the skin microbiome and maintain skin health (Karnwal et al., 2023).

6. Significance of Endophytic Bacterial Bio-surfactants in Cosmetic Industry

Endophytes are microorganisms that inhabit the plant tissue without causing any harmful effects on the host plant. Such microorganisms are able to grow under environmentally stressed conditions like environments with low pH, deficiency in nutrients, and low water availability media which favor their growth.

Although both endophytes and epiphytes can produce biosurfactants, the key difference lies in their ecological niches, adaptation strategies, and the roles of their synthesized biosurfactants in their respective interactions with plants.

Bio-surfactants produced by Endophytes can aid in various processes within the plant, such as nutrient uptake and defense against pathogens. Whereas, biosurfactants produced by Epiphytes can aid in their attachment to the plant surface. These biosurfactants help in reducing surface tension, allowing them to adhere to plant structures. Understanding these differences is crucial for harnessing their unique properties for various cosmetic industrial applications. Bacterial endophytes often influence the host plant's ability to take up hydrocarbons from the soil and translocate them into various plant tissues. Biosurfactants produced by endophytic bacteria facilitate their biofilm development and cell signaling. This enhances plant-endophyte interaction and the adoption of endophytic bacteria. Biosurfactants in cosmetic formulations align with the industry's need to transition towards more sustainable and environmentally friendly practices. It will help to address the concerns related to biodegradability, resource utilization, and consumer preferences for eco-conscious products. However, it's essential for researchers, cosmetic formulators, and industry stakeholders to continue exploring and developing such sustainable alternatives to further improve their efficacy and applicability in cosmetic products.

The potential scope of endophyte-derived bacterial biosurfactants in the cosmetic industry, particularly concerning sustainability and eco-friendliness, is closely tied to the growing emphasis on environmentally responsible practices. As consumer awareness continues to drive demand for sustainable and clean beauty products, such biosurfactants could have the opportunity to play a pivotal role in meeting these demands could help to reduce the industry's ecological footprint, and support the broader goals of environmental conservation and sustainability. Following are the key aspects highlighting their scope:

6.1 Sustainability and Eco-Friendliness of Endophytic Bacterial Bio-surfactants

These endophytic bacterial biosurfactants could offer a sustainable sourcing option, as they are derived from bacteria residing within the plants. Such biosurfactants are often biodegradable, which could minimize the potential for long-lasting pollution. Their production typically involves fewer chemical processes compared to traditional surfactants, resulting in fewer chemical residues and byproducts in the manufacturing process. Their sourcing and production may result in a lower ecological footprint compared to traditional surfactants, which are often energy-intensive to produce. Besides, it could support plant biodiversity conservation. Many regulatory bodies are increasingly encouraging the use of sustainable and eco-friendly ingredients in cosmetics. The use of endophyte-derived biosurfactants could align well with these regulatory trends. Besides, there is a growing demand for cosmetic products which are not only effective but also eco-friendly. Above and beyond, unlike some synthetic surfactants that break down into micro plastics, biodegradable biosurfactants do not pose a threat to marine life and ecosystems. This addresses growing concerns about micro plastic pollution.

In this context, these biosurfactants could meet this demand, attracting environmentally conscious consumers. Moreover, embracing eco-friendly ingredients and sustainable practices could improve a company's corporate social responsibility image and strengthen its commitment to environmental stewardship. Moreover, companies adopting such endophyte-derived biosurfactants as part of their sustainability initiatives can gain a competitive advantage by offering products with a lower environmental impact.

6.2 Natural and Clean Beauty Trends

As consumers become increasingly conscious of the ingredients they apply to their skin and the environmental impact of beauty products, the demand for natural, non-toxic, and plant-derived ingredients along with clean formulations is on the rise. Endophyte-derived biosurfactants align well with these trends. Thereby, the potential scope of such biosurfactants in the cosmetic industry, with regard to natural and clean beauty trends, is substantial. As clean beauty continues to gain momentum, cosmetic companies that embrace such biosurfactants can offer products that cater to this growing segment of the market, while also reducing their acceptance with respect to the environmental footprint and thereby contributing to promotion of natural and eco-friendly practices. Such biosurfactants are derived from plant associated microorganisms having good potential of biodegradation, which is a key attribute for clean beauty products. They break down naturally without harming the environment. Such biosurfactants tend to be milder and less likely to cause skin or scalp irritation compared to some synthetic surfactants. This is essential for clean beauty products that prioritize non-toxic ingredients. Besides, the cosmetic companies can emphasize the natural sourcing of biosurfactants on product labels, increasing transparency for consumers who seek natural and clean formulations. Moreover, the cosmetic companies can tailor endophyte-derived biosurfactants to meet specific clean beauty criteria, such as avoiding certain synthetic chemicals or allergens.

These biosurfactants could contribute to obtaining clean beauty certifications and labels for enhancing their product credibility. Such cosmetic products that incorporate biodegradable biosurfactants may qualify for eco-certifications or labels that highlight their environmentally responsible attributes. Further development of new strains of promising endophytes biosynthesizing biosurfactants is the ongoing area of research. This ongoing innovation expands the scope for these biodegradable ingredients.

6.3 Customization and Innovation

The potential scope of endophyte-derived biosurfactants in the cosmetic industry, focusing on customization and innovation, is driven by the industry's need to create products that are tailored to consumer preferences and evolving market trends. Cosmetic companies can harness the versatility of such valuable biosurfactants to develop innovative and unique formulations that cater to a wide range of skin and hair care needs, enhancing consumer satisfaction and brand competitiveness.

Following are the reasons for utilization of these biosurfactants for driving innovation and customization:

- **Versatile Ingredient:** The bacterial endophytic biosurfactants are versatile and can be customized to meet specific formulation requirements. They can serve multiple functions, such as emulsifying, foaming, and stabilizing, providing a wide range of applications.
- **Addressing Diverse Skin and Hair Concerns:** Cosmetic companies can tailor formulations to address various skin and hair concerns, such as dryness, sensitivity, acne, or aging. The adaptability of biosurfactants allows for the creation of products that cater to different needs.
- **Sensory Experience Enhancement:** The versatility of these biosurfactants enables the enhancement of the sensory experience of using cosmetic products. They can contribute to the texture, feel, and fragrance of products, making them more appealing to consumers.
- **Innovative Delivery Systems:** These biosurfactants can be utilized for innovative delivery systems, including encapsulation of active ingredients. This allows for controlled release and targeted delivery of specific benefits, such as anti-aging or skin brightening.
- **Clean Beauty Formulations:** The clean beauty movement emphasizes the use of natural, non-toxic ingredients. Cosmetic companies can innovate by formulating products that meet clean beauty criteria while incorporating these biosurfactants as multifunctional, natural ingredients.
- **Green Product Development:** The customization of formulations with these biosurfactants enables the development of eco-friendly and sustainable products, catering to environmentally conscious consumers.
- **Texture and Consistency Control:** Cosmetic companies can use these biosurfactants to control the texture and consistency of their products, ensuring that they meet consumer expectations in terms of feel and application.
- **Enhanced Color Cosmetics:** In color cosmetics, such as foundations and eyeshadows, these biosurfactants can be employed to improve pigment dispersion, leading to better color payoff and longer-lasting wear.

- **Microbiome-Friendly Formulations:** The ability to customize formulations with these biosurfactants can lead to the development of products that support a healthy skin microbiome, which is increasingly important in cosmetic innovation.
- **Consumer-Centric Solutions:** Customized formulations with these biosurfactants can provide consumer-centric solutions for unique needs, such as products for specific age groups, skin types, or ethnicities.
- **Competitive Advantage:** Brands that use these biosurfactants in innovative and customized formulations can differentiate themselves in the market, offering products with unique features and benefits.
- **Research and Development Opportunities:** Collaborations between cosmetic companies and biotechnology organizations can lead to the discovery of new strains of bacterial endophyte-derived biosurfactants with novel properties and applications, further driving innovation in the industry.

This way, the cosmetic companies that embrace the biodegradability of these biosurfactants can differentiate themselves from competitors by offering products with a lower environmental impact. Also, cosmetic companies adopting such endophyte-derived biosurfactants demonstrate their commitment to eco-friendly and ethical practices, resonating with consumers who seek responsible brands.

6.4 Collaborative Research

Collaborations between cosmetic companies and biotechnology organizations can drive research and development efforts, leading to the discovery of new applications and improved formulations. The potential scope of bacterial endophyte-derived biosurfactants in the cosmetic industry, with a focus on collaborative research, is promising. Collaborations between cosmetic companies and biotechnology organizations, as well as academic institutions, can drive innovation and the development of novel cosmetic formulations. This leads to the discovery of new applications Collaborative research could expand the scope of biosurfactants in the industry due to the following reasons:

- **Strain Discovery and Optimization:** Collaborative research can lead to the discovery of new endophyte strains with unique bio-surfactant properties. These strains can be optimized for specific cosmetic applications, providing a wider array of ingredients for formulators to work with.
- **Improved Formulations:** Researchers can work with cosmetic companies to develop formulations that maximize the benefits of such biosurfactants. This includes optimizing concentration, stability, and sensory properties.
- **New Applications:** Collaborative efforts can uncover innovative applications for biosurfactants in cosmetics. These may include advanced delivery systems, color cosmetics, and products targeting specific skin and hair concerns.
- **Tailored Solutions:** Based on these biosurfactants, research partnerships allow for the creation of tailored solutions for different consumer segments, such as products for sensitive skin, anti-aging, or eco-friendly formulations.
- **Safety and Efficacy Testing:** Collaborative research can facilitate extensive safety and efficacy testing, ensuring such bio-surfactant-based products meet regulatory standards and consumer expectations.
- **Consumer-Driven Products:** Research partnerships allow for the development of products that are directly influenced by consumer feedback and demands.

- **Market Expansion:** Collaborative research can open new market opportunities by discovering novel applications and benefits of biosurfactants in cosmetics.
- **Enhanced Product Performance:** With the input of research partners, cosmetic companies can improve the overall performance and quality of their products, leading to greater consumer satisfaction.
- **Competitive Advantage:** Brands that actively engage in collaborative research can gain a competitive edge by being at the forefront of innovation and offering unique products.
- **Regulatory Compliance:** Collaborative research can assist in navigating the complex landscape of cosmetic regulations, ensuring that these bio-surfactant-based products meet industry standards and customer safety expectations.

The potential scope of endophyte-derived biosurfactants in the cosmetic industry, with the support of collaborative research, is vast. This collaborative approach could enable the development of cutting-edge cosmetic products that align with the evolving needs and preferences of consumers, while also contributing to the industry's sustainability and eco-friendliness goals. By combining the expertise of biotechnology organizations and cosmetic companies, the potential for innovation and growth in the industry is significant.

6.5 Regulatory Support

The potential scope of endophyte-derived biosurfactants in the cosmetic industry, supported by regulatory agencies, is substantial. Regulatory compliance not only ensures consumer safety but also fosters innovation, market access, and the growth of eco-friendly and sustainable cosmetic products. As regulatory bodies continue to support natural and environmentally responsible ingredients, the cosmetic industry can further explore and expand the applications of these biosurfactants, meeting both regulatory requirements and consumer demands.

Regulatory support could enhance the scope of biosurfactants in the industry due to the following reasons:

- **Natural and Sustainable**: Regulatory bodies are increasingly supportive of natural and sustainable ingredients in cosmetics. Endophyte-derived biosurfactants, being sourced from plant-associated microorganisms, align with these preferences, making it easier for cosmetic companies to gain regulatory approval.
- **Biodegradability and Environmental Compliance:** biosurfactants are often biodegradable, addressing concerns related to environmental impact. Regulatory support for eco-friendly ingredients encourages the use of such biosurfactants in cosmetic formulations.
- **Safety Assessments:** Regulatory agencies require thorough safety assessments for cosmetic ingredients. biosurfactants, especially those derived from well-studied bacterial endophytes, can undergo rigorous safety evaluations to meet regulatory standards, ensuring consumer safety.
- **Reduced Chemical Additives:** Such biosurfactants can reduce the need for certain chemical additives and preservatives in cosmetic formulations. Regulatory bodies often favor simpler ingredient lists with fewer synthetic chemicals, making biosurfactants an attractive option.
- **Transparency and Ingredient Disclosure:** Cosmetic regulations increasingly

emphasize transparency in ingredient labeling. Such biosurfactants offer a natural and clear labeling option, promoting transparency and compliance with regulatory guidelines.

- **Collaborative Research and Data Sharing:** Regulatory support can facilitate collaborative research between industry and regulatory bodies, leading to a better understanding of the safety and efficacy of such biosurfactants. Shared data can expedite the approval process.
- **Harmonization of Standards:** Regulatory support can encourage the harmonization of standards related to such biosurfactants, ensuring consistency in regulations across different regions. This simplifies the market entry process for cosmetic companies.
- **Innovation Incentives:** Regulatory agencies may provide incentives or fast-track approvals for innovative and sustainable ingredients. Such biosurfactants, being innovative and eco-friendly, can benefit from such incentives, encouraging their adoption in cosmetic formulations.
- **Support for Green Initiatives:** Many governments and regulatory bodies support green and sustainable initiatives. Cosmetic companies using these biosurfactants can leverage this support, aligning their products with national and international environmental goals.
- **Consumer Safety and Trust:** Regulatory support ensures that products on the market are safe for consumer use. Cosmetic products formulated with these biosurfactants can instill trust in consumers, knowing that these ingredients have met regulatory standards.
- **Global Market Access:** Regulatory approval in one region often facilitates market access in other regions through mutual recognition agreements. Such biosurfactants with regulatory approval can thus enter multiple markets more easily.
- **Compliance with Emerging Regulations:** Such biosurfactants align with emerging regulations related to sustainability, natural ingredients, and eco-friendly practices. Regulatory support ensures that these products comply with future regulations as well.

6.6 Positive Brand Image

The use of sustainable and natural ingredients like bacterial endophyte-derived biosurfactants can enhance a brand's image and appeal to conscious consumers. The potential scope of such biosurfactants in the cosmetic industry, with regard to positive brand image, is substantial. A positive brand image is crucial for the success of any cosmetic company as it influences consumer perceptions, trust, and loyalty. Contribution of endophyte-derived biosurfactants to create a positive brand image:

- **Sustainability:** Brands incorporating such biosurfactants demonstrate a commitment to sustainability. These eco-friendly alternatives to traditional surfactants align with environmentally conscious consumers, enhancing the brand's image as socially responsible and environmentally friendly.
- **Innovation:** Embracing innovative solutions such as bacterial endophyte-derived biosurfactants showcases a brand's commitment to staying at the forefront of the industry. This positions the brand as forward-thinking, creative, and willing to invest in research and development.

- **Transparency:** Such biosurfactants, being natural and derived from specific microorganisms, allow for clear and transparent communication with consumers. Brands can openly share information about the sourcing, production, and benefits of these ingredients, fostering trust and transparency.
- **Clean Beauty Credentials:** The clean beauty movement emphasizes using non-toxic, natural ingredients. Incorporating such biosurfactants in formulations aligns with clean beauty standards, appealing to consumers seeking products with minimal and natural ingredients.
- **Consumer Education:** Brands can educate consumers about the eco-friendly nature of these biosurfactants, highlighting their positive impact on the environment. Informed consumers are more likely to develop a positive perception of the brand.
- **Non-Irritating Formulations:** Such biosurfactants are often milder on the skin, making them suitable for sensitive skin types. Brands offering non-irritating formulations enhance their reputation for producing products that prioritize consumer well-being and comfort.
- **Product Efficacy:** These biosurfactants contribute to the overall efficacy of cosmetic products. Brands can highlight the effectiveness of their formulations, leading to positive consumer experiences and word-of-mouth recommendations.
- **Positive Reviews and Testimonials:** Satisfied consumers are likely to share their positive experiences with such bio-surfactant-based products. Genuine testimonials and positive reviews enhance the brand's image and credibility.
- **Partnerships and Collaborations:** Brands collaborating with biotechnology organizations or research institutions for the development of such bio-surfactant-based products demonstrate their dedication to innovation and scientific advancement, bolstering their reputation.
- **Corporate Social Responsibility:** Brands using sustainable ingredients like such biosurfactants contribute to social and environmental well-being. Communicating these efforts, such as supporting local communities or conservation initiatives, enhances the brand's image as socially responsible.
- **Regulatory Compliance:** Ensuring that such bio-surfactant-based products meet regulatory standards reinforces the brand's commitment to quality, safety, and compliance, enhancing consumer trust.
- **Premium Brand Perception:** These Bio-surfactant-based formulations can enhance the premium image of a brand. Consumers often associate natural and innovative ingredients with higher quality, allowing brands to position themselves in the premium segment.
- This way, the potential scope of bacterial endophyte-derived biosurfactants in the cosmetic industry, concerning positive brand image, is vast. By aligning with consumer values, demonstrating innovation, cnsuring transparency, and prioritizing sustainability and safety, brands can build a strong, positive, and enduring image in the minds of consumers, leading to brand loyalty and market success. The combination of sustainability, functionality, and alignment with consumer preferences positions bacterial endophyte-derived biosurfactants as a promising ingredient in the cosmetic industry of the future. Their eco-friendly nature, potential for innovation, and suitability for clean beauty formulations make them a valuable asset for cosmetic companies looking to meet evolving market demands.

7. Screening of Endophytes for Bio-surfactant Production

7.1 Inoculation of Endophytes and Collection of Bio-surfactant

Isolate the pure colonies of endophytes and inoculate them into the biosurfactant production media. This media should be designed to promote biosurfactant production. You can add a carbon source like glucose or glycerol for growth and biosurfactant production. Prepare a mineral salt medium by autoclaving at 121 °C (30 min). Then inoculate the endophytic microbes in the mineral salt medium (gL^{-1}: 0.5 g NH_4Cl, 4 g NaCl, 0.5 g KH_2PO_4, 1 g Na_2HPO_4 and 0.5 g $MgSO_4.7H_2O$) modified with 10 mL/L of oil (Olive oil/kerosene) in a conical flask. Allow the mixture to incubate at room temperature for 10 days. Thereafter, centrifuge the medium (10,000 rpm, 20 min, 4 °C) for harvesting the biosurfactant. Collect the cell free supernatant of endophytes for further biosurfactant screening tests (Chander Suresh et al., 2012).

7.2 Drop Collapse Assay

Take a polystyrene 96-well microplate and add 2 μL of oil to each well and keep it aside for 24 h at 22 °C. Thereafter, add 5 μL of filtered cell-free supernatant to the center of the oil-coated well. Observe the plates after 2 min. Flat oil drop and rounded oil drops indicate positive and negative results respectively (Jain et al., 1991).

7.3 Oil Spread Assay

Take distilled water (20 mL) with 1 mL of oil on the water surface in a petri plate. Then add cell-free supernatant (20 μL) to the oil surface. Observe the plate. The clear zone diameter indicates oil displacement activity (Morikawa et al., 2000).

7.4 Emulsification Index (E24)

Take cell-free culture extracts (3 mL) in test tubes and add oil (2 mL) to them. Vortex the mixture (2 min, 70 rpm) and allow it to incubate for 24 hours. Then compare the emulsified oil height (cm) with the total volume and calculate the emulsification using the formula:

$$E - 24 \text{ index}\% = \text{Height of emulsion/Total height of solution}) \times 100$$

(Ali et al., 2013)

7.5 Characterization

Further characterize the biosurfactants produced, including their chemical composition, stability, and specific applications in cosmetic formulations.

7.6 Molecular Identification of the Bacterial Endophyte-Producing Bio-surfactant

The bacterial endophyte showing the highest biosurfactant production potential should be identified using microscopic morphology as well as molecular analysis based on 16S rRNA gene sequence. Then, construct the phylogenetic tree by the maximum parsimony method. The maximum parsimony tree will represent the branching patterns and relationships among the organisms based on the 16S rRNA gene sequences.

8. Systematic Approach for Developing Novel Cosmetic Formulations Using Endophytic Bacterial Bio-surfactants (EBB)

Formulating cosmetic products with EBB involves a systematic approach to ensure the efficacy, safety, and stability of the final products. A general approach involves the following steps.

8.1 Define the Product Concept

This crucial initial phase sets the foundation for the entire development process and ensures that the final product aligns with the goals and meets the needs of consumers. Following are the steps to define the product concept effectively:

- Market research
- Identifying target consumers
- Define the Product Category
- Specify the Key Benefits
- Formulation Considerations
- Sustainability and Ethical Considerations
- Claims and Marketing Strategy
- Packaging and Branding
- Budget and Pricing Strategy
- Regulatory Compliance
- Competitive Analysis

8.2 Select Suitable Bio-surfactants

The choice of biosurfactant can significantly impact the performance of the product, market positioning, and consumer appeal. The pivotal stage of developing novel cosmetic formulations involves the careful selection of appropriate biosurfactants to ensure the creation of products that are not only effective and safe but also marketable. The following factors should be taken into account while choosing biosurfactants:

- Identify the Specific Needs of Your Formulation
- Assess the Source and Origin
- Analyze the Surfactant's Chemical Structure
- Emulsifying Capacity
- Compatibility with Other Ingredients
- Skin Compatibility
- Desired Texture and Sensory Properties
- Preservation and Microbial Stability
- Regulatory Approval
- Sustainability and Ethical Considerations
- Cost and Availability
- Research and Development
- Performance Testing

8.3 Identify Supporting Ingredients

The selection of supporting ingredients is a very important process. In this regard, ensure that selection of complementary ingredients aligns seamlessly with the envisioned product concept, formulation prerequisites, and specific benefits of consumers. Moreover, it is imperative to conduct thorough stability and compatibility tests to ascertain the harmonious interaction of the chosen ingredients in the final formulation. Collaborating with cosmetic chemists or experienced formulators could be advantageous in fine-tuning the selection of supporting ingredients for the innovative cosmetic formulation. Determination of concentration of endophytic bacterial biosurfactants in cosmetic formulations is also an essential step in developing novel products. Therefore this parameter should also be considered while making cosmetic formulations. Some important parameters should be taken into account while choosing biosurfactants, understand the formulation needs, evaluate the role of biosurfactants, consider targeted effects, incorporate emollients and oils,humectants, antioxidants,preservatives, thickeners and stabilizers, fragrance and essential oils,pH adjusters, skin conditioners, sunscreen actives, colorants and pigments, specialty actives, ethical and sustainable ingredients.

8.4 Conduct Compatibility Testing

Conducting comprehensive compatibility tests is essential, especially while assessing the interaction between biosurfactants and other formulation ingredients, to detect any potential issues. Consistent monitoring and testing throughout the developmental stages are imperative to guarantee the stability and safety of the cosmetic product for consumer use. The results of these tests could serve as a basis for making necessary adjustments to the formulation, ensuring that the product maintains its stability and performs optimally along with preventing the issues such as phase separation, texture alterations, or diminished product performance.

During Endophytic Bacterial Biosurfactants (EBB) based cosmetic formulations, some common compatibility tests such as, physical compatibility, centrifugation test, heat stability test, freeze-thaw test, rheological testing, solubility and dispersibility testing, pH testing, chemical compatibility testing, microbiological testing, accelerated stability testing, long-term stability testing, interfacial tension measurement are needed.

8.5 Conduct Stability Testing

Stability testing is an ongoing process, and it should be repeated as and when needed. This crucial step ensures that the cosmetic products maintain their intended quality, efficacy, and appearance over time. To conduct stability testing for cosmetic formulations with biosurfactants some important steps should be considered as: selection of suitable containers and conditions, preparation of test samples, accelerated stability testing like, real-time stability testing, evaluation of physical changes, pH, viscosity and microbiological testing, chemical analysis, compatibility testing, optimization and reformulation, final product assessment etc.

8.6 Assess Skin Compatibility

Assessing skin compatibility for cosmetic formulations is a critical ongoing process in the development process and it should be repeated to ensure that the product is safe and gentle on the skin. For novel cosmetic formulations,select suitable biosurfactants, ethical considerations, patch testing, repeat insult patch testing (RIPT), Hypoallergenic testing, objective measurements, dermatologist evaluation, pH testing, consumer sensory testing, regulatory compliance, reformulation and improvement etc.

8.7 Optimize Texture and Sensory Properties

Optimizing the texture and sensory properties of a cosmetic formulation is an iterative process that requires attention to detail and a deep understanding of the biosurfactants and other ingredients used. By focusing on user experience and continuous improvement, the product can stand out in the market.

The optimization related to Texture and Sensory Properties should be carried out using the following steps:

- Define the Product Goals
- Select Appropriate biosurfactantsConduct Preliminary Formulation
- Texture Assessment
- Adjust Bio-Surfactant Concentration
- Incorporate Co-surfactants and Emollients
- Microstructure Analysis
- Fragrance Selection
- Color and Appearance
- pH Adjustment
- Consumer Sensory Testing
- Iterative Formulation Adjustments
- Stability Testing
- Quality Control and Validation
- Packaging and Labeling
- Market Testing

8.8 Packaging and Presentation to Enhance the Products Marketability

Packaging and presentation are essential aspects of developing novel cosmetic formulations using biosurfactants. This is because packaging and presentation can create a strong brand image to attract consumers, and effectively communicate the value of the cosmetic formulation. The packaging should not only protect the product but also convey its qualities and appeal to potential customers. Following are the steps to enhance the product's marketability to contribute to its success in the competitive cosmetics industry.

- Understand the Target Market
- Create Brand Identity
- Eco-Friendly Packaging with respect to (w.r.t) protection and preservation
- User-Friendly Design
- Aesthetic Appeal
- Clear Product Information
- Labeling Compliance
- Transparency and Clarity
- Sensory Experience
- Sample Sizes and Travel-Friendly Options
- Fragrance Preservation
- Innovative Design
- Market Research and Testing
- Quality Control

- Cost Considerations
- Sustainability Claims
- Legal and Regulatory Compliance
- Promotion and Marketing

8.9 Regulatory Compliance

Compliance with cosmetic regulations is non-negotiable in the cosmetics industry. Failing to meet regulatory requirements can result in serious consequences, including product recalls, fines, and damage to the brand's reputation. Therefore, it's essential to prioritize regulatory compliance from the early stages of product development and throughout the product's lifecycle.

The following steps are essential for Regulatory Compliance:

- Understand Cosmetic Regulations
- Ingredient Safety Assessment
- Labeling Compliance
- Notifying or Registering the Product
- Cosmetic Ingredient Review (CIR)
- Preservatives and Stabilizers
- Good Manufacturing Practices (GMP)
- Microbiological Safety
- Safety Assessment for Finished Product
- Post-Market Surveillance
- Legal and Liability Considerations
- Third-Party Testing and Certification

8.10 Efficacy Testing

Efficacy testing for novel cosmetic formulations containing biosurfactants is essential to ensure that the product delivers the intended benefits and performs as expected. It helps to build trust with consumers and can give a competitive advantage in the cosmetics market.

Some important aspects should be considered for performing efficacy testing such as, Define Efficacy Parameters, Select Relevant Testing Methods, Create a Test Plan, Ethical Considerations, Test Subjects Selection, Baseline Measurements, Product Application, Assessment Period, Data Collection and Analysis, Repeat Testing as Needed, Consumer Feedback, Regulatory Compliance, Quality Control.

8.11 Production and Quality Control

Production and quality control are critical aspects for developing novel cosmetic formulations using biosurfactants. Ensuring consistent product quality and safety is essential. Effective production and quality control can help build trust in your brand and foster customer loyalty.

The following are the steps for production and quality control of such formulations:

A. Production

- Establish Standard Operating Procedures (SOPs)
- Ingredient Sourcing and Quality Assurance
- Quality Control of Raw Materials

- Determining appropriate Batch Size and Scaling
- GMP Compliance
- Manufacturing Equipment and Cleanrooms
- Quality Assurance and Documentation
- Stability Testing of Bulk Formulation

B. Quality Control

- Quality Control Testing
- Microbiological Testing
- Packaging Quality Control
- Batch Release Testing
- Labeling Verification
- In-Process Quality Control
- Compliance with Regulatory Standards
- Post-Market Surveillance
- Documentation and Records
- Personnel Training
- Continuous Improvement

8.12 Consumer Testing

Consumer testing is a valuable part of the product development process, it helps to create a cosmetic formulation that not only meets regulatory and safety requirements but also resonates with the target consumers. It provides valuable feedback from potential users and helps to evaluate the product's effectiveness, safety, and overall satisfaction. Consumer testing should be carried out with some steps as, define objectives and parameters, select test groups, informed consent and ethical considerations, survey questionnaires, focus groups, blind or double-blind testing, product application and usage instructions, testing duration, data collection and analysis, iterative formulation adjustments, comparative testing, packaging and sensory testing, consumer perception studies, safety and allergenicity checks, regulatory compliance, documentation and reporting, product refinement, product launch.

8.13 Launching and Marketing a Novel Cosmetic Formulation

It is an ongoing process that requires flexibility and adaptability to the changing needs and preferences of the target consumers. Effective marketing can help to build brand awareness and establish a loyal customer base for the cosmetic product containing EBB. Following steps should be introduced for successful product launching and marketing:

- Market Research and Competitive Analysis
- Product Positioning
- Branding and Packaging
- Regulatory Compliance
- Production and Quality Assurance
- Marketing Materials
- Online Presence
- Social Media Marketing
- Influencer Collaborations

- Content Marketing
- Email Marketing
- Product Launch Event
- Sales Channels
- Customer Support and Feedback
- Post-launch Evaluation
- Compliance with Feedback

8.14 Post-market Monitoring

Post-market monitoring is a crucial step in the life cycle of novel cosmetic formulations. It involves ongoing surveillance and assessment of the product's safety, efficacy, and consumer satisfaction after it has been launched. It allows continuous improvement and adaptation to changing market conditions and consumer preferences. Regular feedback and vigilance are key for maintaining successful and safe cosmetic formulations. The following steps should be considered for post-market monitoring:

- Establish a System for Feedback Collection
- Monitor Adverse Events
- Track Consumer Satisfaction
- Regulatory Compliance
- Product Stability Testing
- Quality Control and Assurance
- Investigate and Address Complaints
- Periodic Formulation Reviews
- Stay Informed
- Maintain Product Documentation
- Transparency and Communication
- Product Improvement and Innovation
- Continuous Compliance
- Customer Education
- Product Recalls and Safety Alerts
- Regulatory Reporting etc.

9. Limitations and Challenges

Although, use of endophytic bacterial biosurfactants (EBB) in cosmetics offers numerous advantages, they comes with the following limitations and challenges associated with their industrial use:

- **Production Costs:** The production of such biosurfactants can be more expensive than synthetic surfactants, which can affect the cost of cosmetic formulations.
- **Scalability:** Scaling up the production of these biosurfactants to meet the demand of the cosmetics industry can be a challenge. Large-scale production requires optimized fermentation processes and cost-effective production methods.
- **Consistency:** Maintaining consistent quality and composition of biosurfactants can be challenging due to variations in endophytic bacterial strains and fermentation conditions.
- **Allergenic Potential:** Some individuals may be sensitive or allergic to specific biosurfactants, just as with any cosmetic ingredient. This requires careful formulation to ensure product safety.

- **Limited Supply:** Sometime sourcing of specific endophytic bacterial strains for biosurfactant production may be limited, leading to supply chain challenges.
- **Formulation Compatibility:** The compatibility of such biosurfactants with other ingredients in cosmetic formulations can be complex and may require formulation adjustments.
- **Research and Development:** Continued research is required to explore new sources, optimize production processes, and identify applications in cosmetics, which demands sufficient investment and resources.

10. Recommendations and Future Perspectives

- Continued exploration and development of sustainable alternatives, aiming to enhance their efficacy and suitability for cosmetic products, is crucial for researchers, cosmetic formulators, and industry stakeholders.
- Essentially, biosurfactants produced by microorganisms typically exist as mixtures rather than in isolation as single compounds. Within these mixtures, various bioactive compounds may manifest distinct bioactivities. Consequently, it becomes imperative to analyze the individual bioactive compounds in their pure form to ascertain efficacy and pinpoint optimal dosages for application.
- Devising diverse strategies to address the mentioned challenges is a crucial measure in guaranteeing the safety and effectiveness of biosurfactants in cosmetic formulations.
- Cosmetic manufacturers may need to work closely with experts in microbiology and formulation chemistry to develop effective, safe and stable products.
- Thorough research and evaluations in emerging areas such as molecular sequencing, metabolic process engineering, and the application of microbial enzymes are essential to fully harness the maximum potential of these biosurfactants.
- Utilizing prebiotic and probiotic bacteria in cosmetic formulations presents a promising alternative to biosurfactants produced by pathogenic organisms. To delve into this innovative approach, a profound comprehension of the genetic makeup of organisms producing prebiotic and probiotic biosurfactants is essential.
- Research focused on addressing the lack of toxicology data and enhancing production yields is crucial. Improvements in these aspects can pave the way for the necessary enhancements to fully leverage the use of biosurfactants in large-scale sustainable cosmetic production
- The exact mechanisms by which biosurfactants interact with the human body remain incompletely understood. Ongoing research in microbial biotechnology, pharmaceutical studies, and cosmetic science is anticipated to provide further insights into this realm.
- Addressing various challenges related to biosurfactants, such as cost-effective production, scalability, and their adoption on a large industrial scale, is imperative.

11. Conclusion

The potential of endophytic bacterial biosurfactants to enhance the effectiveness of cosmetics holds considerable implications for human health. Embracing sustainable cosmetic formulations utilizing endophytic bacterial biosurfactants represents an innovative approach aligned with the increasing preference for safe, eco-friendly and natural cosmetic products. Pertinent research in this area holds promising potential for the development of new, personalized formulations. By harnessing the biosynthetic capabilities of endophytes for biosurfactant production, we contribute to the creation of sustainable cosmetic formulations that not only benefit our well-being but also support the health of our planet. This could resonate with the rising demand for eco-friendly and skin-friendly cosmetic products, marking a positive stride towards a healthier future.

Acknowledgement

The authors are grateful to the Research and Development Department, Phytoelixir Pvt. Ltd. Nashik. Special thanks to Mr. Abhijeet Jagtap, Mr. Sunil Ghayal, Mr. V.C. Handore and Mrs. Hira V. Handore, for their valuable assistance.

References

Adu, Simms A., Naughton, Patrick J., Marchant, Roger & Banat, Ibrahim M. (2020). Microbial biosurfactants in cosmetic and personal skincare pharmaceutical formulations. *Pharmaceutics*, 12(11): 1099.

Ahle, C.M., Stødkilde, K., Poehlein, A., Bömeke, M., Streit, W.R., Wenck, H. et al. (2022). Interference and co-existence of *Staphylococci* and *Cutibacterium* acnes within the healthy human skin microbiome. *Communications Biology*, 5(1): 923.

Ali, S.R., Chowdhury, B.R., Mondal, P. & Rajak, S. (2013). Screening and characterization of biosurfactants producing microorganism from natural environment (whey spilled soil). *J. Nat. Sci. Res.*, 3: 53-59.

Allef, P., Hartung, C. & Schilling, M. (March 2016). Aqueous hair and skin cleaning compositions comprising biosurfactants. Patent US 9,271,908B2.

Banat, I.M., Franzetti, A., Gandolfi, I., Bestetti, G., Martinotti, M.G., Fracchia, L. et al. (2010). Microbial biosurfactants production, applications and future potential. *Appl. Microbiol. Biotechnol.*, 87: 427-444.

Begley, M., Cotter, P.D., Hill, C. & Ross, R.P. (2009). Identification of a novel two-peptide lantibiotic, lichenicidin, following rational genome mining for LanM proteins. *Applied Environmental Microbiology*, 75: 5451-5460.

Belgacem, Z.B., Bijttebier, S., Verreth, C., Voorspoels, S., Van de Voorde, I., Aerts, G. et al. (2015). Biosurfactant production by *Pseudomonas* strains isolated from floral nectar. *Journal of Applied Microbiology*, 118(6): 1370-1384.

Bianchi, P., Theunis, J., Casas, C., Villeneuve, C., Patrizi, A., Phulpin, C. et al. (2016). Effects of a new emollient-based treatment on skin microflora balance and barrier function in children with mild atopic dermatitis. *Pediatric Dermatology*, 33(2): 165-171.

Choi, W.J., Choi, H.G. & Lee, W.H. (1996). Effects of ethanol and phosphate on emulsan production by Acinetobacter calcoaceticus RAG-1. *Journal of Biotechnology*, 45(3): 217-225.

Coronel-León, J., de Grau, G., Grau-Campistany, A., Farfan, M., Rabanal, F., Manresa, A. et al. (2015). Biosurfactant production by AL 1.1, a *Bacillus licheniformis* strain isolated from Antarctica: Production, chemical characterization and properties. *Ann Microbiol.*, 65: 2065-2078.

Cotter, P.D., Hill, C. & Ross, R.P. (2005). Bacteriocins: Developing innate immunity for food. *Nature Review Microbiology*, 3: 777-788.

Cox, T.F., Crawford, R.J., Gregory, L.G., Hosking, S.L. & Kotsakis, P. (October 2013). Mild to the skin, foaming detergent composition. Patent US8,563,490B2.

Das, P., Mukherjee, S. & Sen, R. (2008). Antimicrobial potential of a lipopeptide biosurfactant derived from marine *Bacillus Circulans*. *J. Appl. Microbiol.*, 104: 1675-1684.

Desai, J.D. & Banat, I.M. (1997). Microbial production of surfactants and their commercial potential. *Microbiol Mol. Biol. Rev.*, 61: 47-64.

Desanto, K. (July 2011). Rhamnolipid-based formulations. Patent US 7,985,722B2.

Dey, G., Bharti, R., Dhanarajan, G., Das, S., Dey, K.K., Kumar, B.N.P. et al. (2015). Marine lipopeptide Iturin A inhibits Akt mediated GSK3β and FoxO3a signaling and triggers apoptosis in breast cancer. *Sci. Rep.*, 5: 1-14.

Dey, G., Bharti, R., Das, A.K., Sen, R. & Mandal, M. (2017). Resensitization of Akt induced Docetaxel resistance in breast cancer by 'Iturin A': A lipopeptide molecule from marine bacteria *Bacillus megaterium*. *Sci. Rep.*, 7: 1-11.

Elshikh, M., Moya-Ramírez, I., Moens, H., Roelants, S., Soetaert, W., Marchant, R. et al. (2017). Rhamnolipids and lactonic sophorolipids: Natural antimicrobial surfactants for oral hygiene. *J. Appl. Microbiol.*, 123: 1111-1123.

Fernandes, P.A.V., De Arruda, I.R., Dos Santos, A.F.A.B., De Araújo, A.A., Maior, A.M.S. & Ximenes, E.A. (2007). Antimicrobial activity of surfactants produced by *Bacillus subtilis* R14 against multidrug-resistant bacteria. *Braz. J. Microbiol.*, 38: 704-709.

Fernández-Peña, L., Guzmán, E., Leonforte, F., Serrano-Pueyo, A., Regulski, K., Tournier-Couturier, L. et al. (2020). Effect of molecular structure of eco-friendly glycolipid biosurfactants on the adsorption of hair-care conditioning polymers. *Colloids Surf. B: Biointerfaces*, 185: 110578.

Gupta, S., Gupta, C. & Garg, A. (2015). A biotechnological approach to microbial based perfumes and flavours. *J Microbiol Exp.*, 2(1): 11-18.

Gómez-Graña, S., Perez-Ameneiro, M., Vecino, X., Pastoriza-Santos, I., Perez-Juste, J., Cruz, J.M. et al. (2017). Biogenic synthesis of metal nanoparticles using a biosurfactant extracted from corn and their antimicrobial properties. *Nanomaterials*, 7: 139.

Gomaa, E.Z. (2013). Antimicrobial activity of a biosurfactant produced by *Bacillus licheniformis* strain m104 grown on whey. *Braz. Arch. Biol. Technol.*, 56: 259-268.

Hagler, M., Smith-Norowitz, T.A., Chice, S., Wallner, S.R., Viterbo, D., Mueller, C.M. et al. (2007). Sophorolipids decrease IgE production in U266 cells by downregulation of BSAP (Pax5), TLR-2, STAT3 and IL-6. J. *Allergy Clin. Immunol.*, 119: S263.

Handore, A.V., Khandelwal, S.R., Karmakar, R., Gupta, D.L. & Handore, D.V. (2022). Biosurfactants: The ecofriendly biomolecules of the upcoming era. *Microbial Surfactants* (1st Edition). eBook ISBN9781003260165.

Heinrich, K., Heinrich, U. & Tronnier, H. (2014). Influence of different cosmetic formulations on the human skin barrier. *Skin Pharmacol. Physiol.*, 27: 141-147.

Herman, D. & Maier, R. (2002). Biosynthesis and applications of glycolipid and lipopeptide biosurfactants. *Lipid Biotechnology* (1st Edition). CRC Press. 623-648.

Hisatsuka, K., Nakahara, T., Sano, N. & Yamada, K. (1971). Formation of rhamnolipid by *Pseudomonas aeruginosa* and its function in hydrocarbon fermentation. *Agricultural Biology and Chemistry*, 35(5): 686-692.

Jain, D.K., Collins-Thompson, D.L., Lee, H. & Trevors, J.T. (1991). A drop-collapsing test for screening surfactant-producing microorganisms. *J. Microbiol Methods*, 13: 271-279.

Janek, T., Łukaszewicz, M., Rezanka, T. & Krasowska, A. (2010). Isolation and characterization of two new lipopeptide biosurfactants produced by *Pseudomonas fluorescens* BD5 isolated from water from the Arctic Archipelago of Svalbard. *Bioresource Technol.*, 101: 6118-6123.

Karnwal, A., Shrivastava, S., Al-Tawaha, A.R.M.S., Kumar, G., Singh, R., Kumar, A. et al. (2023). Microbial biosurfactant as an alternate to chemical surfactants for application in cosmetics industries in personal and skin care products: A critical review. *BioMed Research International*, 1-22.

Kitagawa, M., Suzuki, M., Yamamoto, S., Sogabe, A., Kitamoto, D., Imura, T. et al. (January 2010). Biosurfactant-containing skin care cosmetic and skin roughness-improving agent. Patent US 20,100,004,472.

Kitagawa, M., Nishimoto, K. & Tanaka, T. (November 2015). Cosmetic pigments, their production method, and cosmetics containing the cosmetic pigments. Patent US 9,181,436B2.

Kitamoto, D., Yanagishita, H., Shinbo, T., Nakane, T., Kamisawa, C. & Nakahara, T. (1993). Surface active properties and antimicrobial activities of mannosylerythritol lipids as biosurfactants produced by *Candida Antarctica*. *J. Biotechnol*., 29: 91-96.

Kosaric, N. (2001). Biosurfactants and their application for soil bioremediation. *Food Technology and Biotechnology*, 39: 295-304.

Knoth, D., Rincón-Fontán, M., Stahr, P.L., Pelikh, O., Eckert, R.W., Dietrich, H. et al. (2019). Evaluation of a biosurfactant extract obtained from corn for dermal application. *Int. J. Pharm*., 564: 225-236.

Kulkarni, S. & Choudhary, P. (2011). Production and isolation of biosurfactant-sophorolipid and its application in body wash formulation. *Asian J. Microb. Biotechnol. Environ. Sci*., 13: 217-221.

Kumari, A., Kumari, S., Prasad, G.S. & Pinnaka, A.K. (2021). Production of sophorolipid biosurfactant by insect derived novel yeast *Metschnikowia churdharensis* f.a., sp. nov. and its antifungal activity against plant and human pathogens. *Frontiers in Microbiology*, 12: 678668.

Kundu, D., Hazra, C., Dandi, N. & Chaudhari, A. (2013). Biodegradation of 4-nitrotoluene with biosurfactant production by *Rhodococcus pyridinivorans* NT2: Metabolic pathway, cell surface properties and toxicological characterization. *Biodegradation*. 24: 775-793.

Lai, C.C., Huang, Y.C., Wei, Y.H. & Chang, J.S. (2009). Biosurfactant-enhanced removal of total petroleum hydrocarbons from contaminated soil. *Journal of Hazardous Materials*, 167: 609-614.

Levy, N., Bar-Or, Y. & Magdassi, S. (1990). Flocculation of bentonite particles by a cyanobacterial bioflocculant. *Colloids and Surfaces*, 48: 337-349.

Lourith, N. & Kanlayavattanakul, M. (2009). Natural surfactants used in cosmetics: Glycolipids. *International Journal of Cosmetic Science*, 31(4): 255-261.

Ma, Z. & Hu, J. (2014). Production and characterization of iturinic lipopeptides as antifungal agents and biosurfactants produced by a marine *Pinctada martensii* derived *Bacillus mojavensis* B0621A. *Appl Biochem Biotechnol*. 173: 705-715.

Mimee, B., Labbé, C., Pelletier, R. & Bélanger, R.R. (2005). Antifungal activity of flocculosin, a novel glycolipid isolated from *Pseudozyma Flocculosa*. *Antimicrob. Agents Chemother*., 49: 1597-1599.

Morikawa, M., Hirata, Y. & Imanaka, T. (2000). A study on the structure – Function relationship of lipopeptide biosurfactants. *Biochim. Biophys. Acta*, 1488: 211-218.

Mulligan, C.N., Yong, R.N. & Gibbs, B.F. (2001). Heavy metal removal from sediments by biosurfactants. *J Hazard Mater*., 85: 111-125.

Muthusamy, K.K., Gopalakrishnan, S., Ravi, T. & Sivachidambaram, P. (2008). Biosurfactants: Properties, commercial production and application. *Current Science*, 94: 736-747.

Navonvenezia, S., Zosim, Z., Gottieb, A., Legmann, R., Carmeli, S., Ron, E.Z. & Rosenberg, E. (1995). Alasan, a new bioemulsifier from *Acinetobacter radioresistens*. *Appl Environ Microbiol*., 61: 3240-3244.

Ndlovu, T., Rautenbach, M., Vosloo, J.A., Khan, S. & Khan, W. (2017). Characterization and antimicrobial activity of biosurfactant extracts produced by *Bacillus amyloliquefaciens* and *Pseudomonas aeruginosa* isolated from a wastewater treatment plant. *AMB Express*, 7(1).

Nerurkar, A.S. (2010). Structural and molecular characteristics of lichenysin and its relationship with surface activity. *Adv Exp Med Biol*., 672: 304-315.

Nickzad, A. & Déziel, E. (2014). The involvement of rhamnolipids in microbial cell adhesion and biofilm development—An approach for control. *Lett. Appl. Microbiol*., 58: 447-453.

Owen, D. & Fan, L. (2013). Oligomeric biosurfactants in dermato cosmetic compositions. Patent US 8431523 B2.

Parry, N.J. & Stevenson, P.S. (April 2018). Personal care compositions. Patent EP 2,931,237B1

Patel, P., Bhatt, S., Patel, H., Sayyed, R.Z. & Aguilar-Marcelino, L. (2022). Biosurfactant: A biomolecule and its potential applications. *Microbial Surfactants*, 63-81.

Peng, M. & Biswas, D. (2020). Environmental influences of high-density agricultural animal operation on human forearm skin microflora. *Microorganisms*, 8(10).

Phetcharat, T., Dawkrajai, P., Chitov, T., Mhuantong, W., Champreda, V. & Bovonsombut, S. (2019). Biosurfactant-producing capability and prediction of functional genes potentially beneficial to microbial enhanced oil recovery in indigenous bacterial communities of an onshore oil reservoir. *Curr Microbiol.*, 76: 382-391.

Piljac, T. & Piljac, G. (1999). Use of rhamnolipids in wound healing, treating burn shock, atherosclerosis, organ transplants, depression, schizophrenia and cosmetics. (European Patent 1 889 623). Paradigm Biomedical Inc., New York.

Rincón-Fontán, M., Rodríguez-López, L., Vecino, X., Cruz, J.M. & Moldes, A.B. (2016). Adsorption of natural surface-active compounds obtained from corn on human hair. *RSC Advances*, 6(67).

Rodríguez-López, L., Rincón-Fontán, M., Vecino, X., Cruz, J.M. & Moldes, A.B. (2019). Preservative and irritant capacity of biosurfactants from different sources: A comparative study. *Journal of Pharmaceutical Sciences*, 108(7): 2296-2304.

Saranraj, P., Sayyed, R.Z., Sivasakthivelan, P., Hasan, M.S., Al-Tawaha, A.R.M.A. & Amala, K. (2022). Microbial biosurfactants: Methods of investigation, characterization, current market value and applications. *In:* Sayyed, R.Z. (Eds), *Biosurfactants: Production and Applications in Bioremediation/Reclamation*. 19-34. CRC Press, Taylor & Francis Group, USA.

Suresh Chander, C.R., Lohitnath, T., Kumar, D.J.M. & Kalaichelvan, P.T. (2012). Production and characterization of biosurfactant from bacillus subtilis MTCC441 and its evaluation to use as bioemulsifier for food bio-preservative. *Adv. Appl. Sci. Res. Res. Libr.*, 3: 1827-1831.

Sutyak, K.E., Wirawan, R.E., Aroutcheva, A.A. & Chikindas, M.L. (2008). Isolation of the *Bacillus subtilis* antimicrobial peptide subtilosin from the dairy product derived *Bacillus amyloliquefaciens*. *Journal of Applied Microbiology*, 104: 1067-1074.

Takahashi, M., Morita, T., Fukuoka, T., Imura, T. & Kitamoto, D. (2012). Glycolipid biosurfactants, mannosylerythritol lipids, show antioxidant and protective effects against H(2)O(2)-induced oxidative stress in cultured human skin fibroblasts. *J. Oleo Sci.*, 61: 457-464.

Thakur, P., Saini, N.K., Thakur, V.K., Gupta, V.K., Saini, R.V. & Saini, A.K. (2021). Rhamnolipid the glycolipid biosurfactant: Emerging trends and promising strategies in the field of biotechnology and biomedicine. *Microbial Cell Factories*, 20(1): 1-15.

Urzedo, C.A., Freitas, Q., Akemi, V., Silveira, I., Pedrine, M.A. & Celligoi, C. (2018). Antimicrobial applications of sophorolipid from *Candida bombicola*: A promising alternative to conventional drugs. *Adv. Biotechnol. Microbiol.*, 9: 555753

Valsecchi, C., Marseglia, A., Montagna, L., Tagliacarne, S.C., Elli, M., Licari, A. et al. (2014). Evaluation of the effects of a probiotic supplementation with respect to placebo on intestinal microflora and secretory IgA production, during antibiotic therapy, in children affected by recurrent airway infections and skin symptoms. *Journal of Biological Regulators and Homeostatic Agents*, 28(1): 117-124.

Vanlay, M., Samnang, S., Jung, H.-J., Choe, P., Kang, K.K. & Nou, I.-S. (2022). Interspecific and intraspecific hybrid rootstocks to improve horticultural traits and soil-borne disease resistance in tomato. *Genes*, 13(8): 1468.

Vecino, X., Barbosa-Pereira, L., Devesa-Rey, R., Cruz, J.M. & Moldes, A.B. (2015). Optimization of extraction conditions and fatty acid characterization of *Lactobacillus Pentosus* cell-bound biosurfactant/bioemulsifier. *Journal of the Science of Food and Agriculture*, 95: 313-320.

Vecino, X., Cruz, J.M., Moldes, A.B. & Rodrigues, L.R. (2017). Biosurfactants in cosmetic formulations: Trends and challenges. *Critical Reviews in Biotechnology*, 37(7): 911-923.

Vecino, X., Rodríguez-López, L., Ferreira, D., Cruz, J.M., Moldes, A.B. & Rodrigues, L.R. (2018). Bioactivity of glycolipopeptide cell-bound biosurfactants against skin pathogens. *International Journal of Biological Macromolecules*, 109: 971-979.

Wei, Y., Li, Z., Wedegaertner, T.C., Jaconis, S., Wan, S., Zhao, Z. et al. (2022). Conservation and divergence of phosphoenolpyruvate carboxylase gene family in cotton. *Plants* (Basel, Switzerland). 11(11): 1482.

Yakimov, M.M., Amro, M.M., Bock, M., Boseker, K., Fredrickson, H.L. & Lessel, D.G. (1997). The potential of *Bacillus licheniformis* strains for in situ enhanced oil recovery. *J. Petrol. Sci. Eng.*, 18: 147-160.

Zirwas, M.J. & Moennich, J. (2008). Antiperspirant and deodorant allergy: Diagnosis and management. *J. Clin. Aesthet. Dermatol.*, 1: 38-43.

Zouari, R., Moalla-Rekik, D., Sahnoun, Z., Rebai, T., Ellouze-Chaabouni, S. & Ghribi-Aydi, D. (2016). Evaluation of dermal wound healing and in vitro antioxidant efficiency of *Bacillus subtilis* SPB1 biosurfactant. *Biomed. Pharmacother.*, 84: 878-891.

CHAPTER

23

Therapeutic Applications of Biosurfactant and Bioemulsifiers

Vipul P. Patel[1*], Sharav A. Desai[1] and Kirti C. Thombare[1]

[1] Department of Pharmaceutical Biotechnology, Sanjivani College of Pharmaceutical Education & Research, Savitribai Phule Pune University, Kopargaon - 423603, Maharashtra, India

1. Introduction

Surfactants are amphoteric molecules.They may originate from microorganisms or may be synthesized chemically, depending on whether they are produced by microbial fermentation or through chemical synthesis. Biosurfactants are surfactants that can be produced by microorganisms (Van Hamme et al., 2006). By selecting biosurfactants over chemical surfactants, industries can make more environmentally friendly choices which can have a positive impact on the environment, and human health from sustenance and ecological perspectives. These environmental reasons are driving the increasing adoption of biosurfactants in various applications including therapeutics, biomedical uses, pharmaceuticals, and cosmetics and more. Biosurfactants are generally surface-active amphiphilic molecules that are produced through microbes. These microorganisms produce biosurfactants on their microbial cell surfaces or are excreted extracellularly. For the production of biosurfactant bacteria, algae, filamentous fungi, yeast, actinomycetes, and many more microorganisms are commonly used. Glycolipids, lipopeptides, fatty acids, phospholipids, saponins, polysaccharides, and alkyl polyglycosides are the biosurfactants produced by microorganisms (Singh et al., 2019). Biosurfactants have a microbial origin, high biodegradability, low ecotoxicity, multifunctionality, long-term physicochemical stability, resistance to high pH and temperature changes, and efficiency of production from renewable energy sources. Biosurfactants have hydrophilicity and hydrophobicity shown by groups in their structure and tend to assemble in the spaces between the fluid phases and thus minimize the interfacial and surface tensions, respectively (Naughton et al., 2019). Biosurfactants are commonly used as potential therapeutics. Apart from this they find application in bioremediation, increased oil recovery, and the food and cosmetics

* Corresponding author: v_pharmacy@yahoo.co.in

markets (Gudiña et al., 2013). Biosurfactants have potential use in vaccine development and gene therapy, as well as potentially important immuno-modulatory molecules and antibacterial, antifungal, and antiviral medicines. The utilization of biosurfactants as suitable anti-adhesive coating agents for medical insertional materials that lower hospital infections is demonstrated by their drastic effectiveness against a number of diseases without requiring the use of synthetic chemicals and medications. In addition to being adjuvants for antigens, ligands for binding immunoglobulins, inhibitors of fibrin clot development, and activators of fibrin clot lyses, biosurfactants have been applied to gene transfection (Rodrigues and Teixeira, 2010).

2. Classification of Biosurfactants

According to the Polarity, chemical composition, and origin of microorganisms of biosurfactants, they can be categorized as:

(a) Glycolipids,
(b) Lipopeptides or lipoproteins,
(c) Phospholipids and fatty acids (mycolic acids),
(d) Polysaccharide-lipid complexes,
(e) Complete cell surface type,
(f) Polymeric surfactants,
(g) The particulate type.

3. Therapeutic Applications of Biosurfactants

Due to their immunomodulatory, hemolytic, antiviral, antibacterial, anti-adhesive, and anticancer activities, biosurfactants provide a wide range of potential therapeutic uses. Therapeutic applications of Biosurfactants are discussed below.

3.1 Antimicrobial Activity of Lipopeptides and Glycolipids

The frequent use of antimicrobials has accelerated the emergence of an increasing number of bacterial strains that are resistant to drugs, which could have major potential impacts on global healthcare systems. The World Health Organisation estimates that

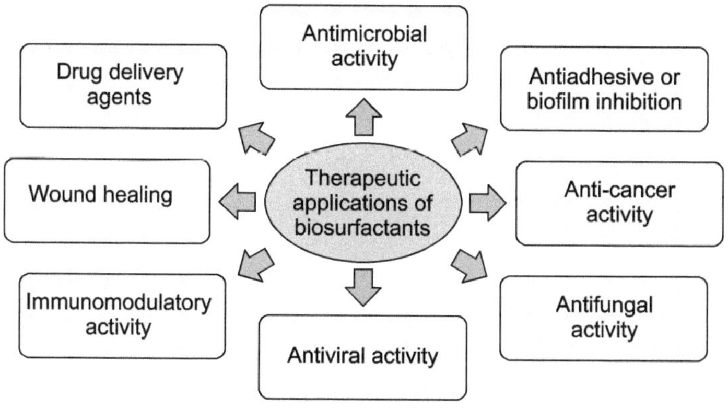

Fig. 1: Therapeutic applications of biosurfactants.

antibiotic resistance causes 700,000 accidental deaths annually and 25,000 deaths in Europe alone, with an estimated 1.5 billion euros in costs (Ceresa et al., 2021).

Bacteria, fungi, and yeast produce biosurfactant compounds and exhibit a broad-spectrum of antimicrobial properties. Glycolipids and lipopeptides show a broad range of antimicrobial activities and are currently applied in several fields (Ndlovu et al., 2017). Antimicrobial lipopeptides include the following: cyclic lipopeptides like Daptomycin from *Streptomyces roseosporus*, Polymyxin B, Pumilacidin and Lichenysin from *Bacillus polymyxa*, *Bacillus pumilus, and Bacillus licheniformis*, and finally viscosin from *Pseudomonas*. Antimicrobial lipopeptides include Surfactin, Iturin, Fengycin, Mycosubtilins, and Bacillomycins produced by *Bacillus subtilis*. Antimicrobial features of glycolipids, such as rhamnolipids from *Pseudomonas aeruginosa*, mannosylerythritol lipids (MEL-A and MEL-B) from *Candida antarctica*, and sophorolipids from *Candida bombicola*, have also been identified (Fracchia et al., 2015).

The mechanism of action of lipopeptides, such as fengycin and surfactin, involves their antibacterial properties because of their capacity to self-associate and form pore-bearing channels or micellar aggregates in the lipid membrane. Lipopeptides typically result in membrane rupture, increased membrane permeability, metabolite leaks, and cell lysis because of these characteristics. Furthermore, modifications to membrane composition and disruptions to protein structures affect the production and transit of energy inside membranes (Carrillo et al., 2003, Fracchia et al., 2015). Lipopeptides that have a lipid tail length of 10–12 carbon atoms have higher bactericidal action, while those that have a lipid tail length of 14–16 carbon atoms have higher antifungal action (Mandal et al., 2013). An environmental strain of *Brevibacillus laterosporus* has been found to produce a novel cationic lipopeptide that exhibits antimicrobial properties against Gram-positive bacteria, such as methicillin-resistant *Staphylococcus aureus* (MRSA), vancomycin-resistant *Lactobacillus plantarum*, and *Enterococcus faecalis* (Yang et al., 2016).

A biological assay was carried out to investigate the antibacterial activity of *Rhodococcus opacus* R7 through the production of a novel bioactive molecule that is a member of the biosurfactant class. The biological assay demonstrated the antibacterial activity of *R. opacus* R7 biosurfactant peptide against *Escherichia coli* ATCC 29522 and *Staphylococcus aureus* ATCC 6538, with MIC values of 2.6 mg mL^{-1} and 1.7 mg mL^{-1}, respectively (Zampolli et al., 2022).

The biosurfactant produced by *Bacillus subtilis* C19 demonstrates a specific antibiotic action against six human pathogens: *Escherichia coli, Pseudomonas aeruginosa, Salmonella enterica Typhi, Listeria monocytogenes, Candida albicans,* and *Staphylococcus aureus*. It functions as an antifungal agent and can stop the growth of *Candida albicans* without preventing the growth of gram-positive or gram-negative bacteria (Yuliani et al., 2018).

In the most recent study, a lipopeptide generated by *Bacillus licheniformis* M104 was investigated as an antimicrobial agent against *Candida albicans, Pseudomonas aeruginosa, Escherichia coli, Salmonella typhimurium, Proteus vulgaris,* and *Klebsiella pneumoniae*, as well as Gram-positive bacteria (*Bacillus subtilis, Bacillus thuringiensis, Bacillus cereus, Staphylococcus aureus,* and *Bacillus monocytogenes*). The microorganisms under investigation exhibit potent antimicrobial activity through the use of biosurfactant lipopeptide with *Staphylococcus aureus* showing increased susceptibility (Gomaa, 2013).

Production of biosurfactants from lactic acid bacteria (LAB), including *Lactobacillus plantarum* and *Pediococcus acidilactici* exhibit significant antibacterial activity against *Staphylococcus aureus* CMCC 26003 in vitro (Yan et al., 2019). Glycolipid derived from *Lactobacillus acidophilus* NCIM 2903, has been identified as one of the most potent antimicrobials, reducing the surface tension to 27 Mn/m (Satpute et al., 2018). The antimicrobial mechanism of action of biosurfactants has been studied,revealing that their disruptive activity on cell membranes is attributed to their amphiphilic nature. But there is growing evidence that biosurfactants play an integral role in quorum sensing signaling, and this has become a current area of research in biosurfactants (Khan et al., 2019, Yan et al., 2019).

Table 1: List of Biosurfactants having antibacterial activity

Biosurfactants	Source	References
Polymyxin A and Polymyxin B	*Bacillus polymyxa*	Landman et al., 2008
Polymyxin B		
Surfactin	*Bacillus subtilis*	Vater et al., 2002
Iturin		
Fengycin		
Mycosubtilins		
Bacillomycins		
Pumilacidin	*Bacillus pumilus*	Naruse et al., 1990a
Lichenysin	*Bacillus licheniformis*	Grangemard et al., 2001
WH1 Fungin	*Bacillus amyloliquefaciens*	Barrantes et al., 2021
Daptomycin	*Streptomyces roseosporus*	Barrantes et al., 2021
Viscosin	*Pseudomonas fluorescens*	Saini et al., 2008a

3.2 Antiadhesive or Biofilm Inhibition Properties

Biosurfactants are getting expanded attention as potential therapeutic agents, predominantly in the development of antiadhesive agents. The development of biosurfactant-based antiadhesive agents holds promise for fighting infections and decreasing the spread of antibiotic resistance. The ability of biosurfactants to prevent pathogenic microbes from adhering to solid surfaces or infection sites suggests that their adhesion to implant material solid surfaces may provide a new and efficient way to prevent the colonization of harmful microbes (Rodrigues and Teixeira, 2010).

Biofilms denote a massive scope in the biomedical field because they are strongly related to chronic healthcare-associated infections (HAI) and antimicrobial resistance (Ceresa et al., 2021, Sharma et al., 2019). Important research was performed for the estimation of antibiofilm activity for the two biosurfactants derived against a few clinical strains of *Salmonella Typhimurium, Enterococcus faecali, Streptococcus mutans, Staphylococcus aureus, Escherichia coli,* and *Pseudomonas aeruginosa* using *Lactobacillus brevis* and *Bacillus* spp. Thus, these biosurfactants could prevent biofilm formation and remove biofilms previously formed by infective bacteria (Haddaji et al., 2022).

The anti-adhesive or anti-biofilm activity of biosurfactants derived from bacteria such as *Bacillus subtilis* VSG4 and *Bacillus licheniformis* VS16 against *Staphylococcus aureus* ATCC 29523, *Salmonella typhimurium* ATCC 19430, and *Bacillus cereus* ATCC 11778 was assessed, based on research outcomes. At doses of 3-5 mg/mL, these biosurfactants exhibited anti-adhesive action. Additionally, both biosurfactants had strong anti-biofilm capabilities, with the VSG4 biosurfactant exhibiting biofilm eradication from 63.9 to 80.03% and the VS16 biosurfactant exhibited biofilm eradication from 61.1 to 68.4% (Giri et al., 2019).

In the biofilm activity evaluation study, the co-incubation of TIM96, a mixture of surfactin, iturin, and fengicin, with *Trichosporon* spp. has been shown to inhibit the formation of biofilms by reducing thickness and cell viability, up to 99.2%, and preventing cell adhesion (up to 96.89%). According to estimations from earlier investigations, under co-incubation circumstances, a *Bacillus subtilis* strain's surfactin considerably facilitated the dislodging of biofilms and strongly inhibited *Staphylococcus aureus* adherence on surfaces such as glass, polystyrene, and stainless steel (Liu et al., 2019).

The antibiofilm activity of different types of rhamnolipids and sophorolipids against oral bacterial infections, including *Streptococcus oralis, Actinomyces naeslundii, Neisseria mucosa,* and *Streptococcus sanguinis*, was investigated. Biosurfactants were able to remove pre-existing 12-hour-old biofilms in the range of 50–100% for all the experimentally examined strains. They also prevented biofilm development in the above-mentioned strains in a range of 60–90% in co-incubation and pre-coating conditions (Elshikh et al., 2017).

A mixture of three sophorolipids, the antibiofilm activity of SLA (acidic congeners), SL18 (lactonic congeners), and SLV (combination of both congeners) against *Candida albicans, Pseudomonas aeruginosa, and Staphylococcus aureus* has been investigated. These biosurfactants prevented up to 90–95% of the growth of microbial biofilms in co-incubation conditions (Ceresa et al., 2020).

Antiadhesive properties of biosurfactants produced by *Lactobacilli* are frequently evaluated. As they are able to disrupt energy-generating processes and membrane functions in order to cause cell membrane rupture, they lower the hydrophobicity of the cell surface, and decrease microbial adhesion to surfaces (Walencka et al., 2008). The potential biofilm inhibition activity of biosurfactants produced from *Lactobacillus acidophilus* was evaluated. In reality, approximately 80% inhibitions were observed for *Bacillus subtilis* and *Staphylococcus aureus*, and 59–65% for *Proteus vulgaris*, *Pseudomonas aeruginosa, Pseudomonas putida,* and *Escherichia coli* (Ceresa et al., 2021).

3.3 Anticancer

The capacity of biosurfactants to regulate a various roles played by mammalian cells and, consequently, their potential to act as antitumor agents by disrupting specific processes involved in the genesis of cancer, is one of the most captivating discoveries published about biosurfactants (Gudiña et al., 2013). It has been demonstrated that these molecules take part in a number of intercellular molecular recognition processes, including signal transduction, cell differentiation, and immunological responses (Rodrigues et al., 2006).

Surfactin is one of the lipopeptides that have been extensively studied for possible anticancer action against several cancer cell lines. Surfactin powerfully suppressed down PI3K or phosphoinositide 3 kinases, and Akt protein kinase B (PKB), a serine/threonine-specific protein kinase involved in a number of biological processes including cell proliferation and apoptosis, shown in Fig 2. It is well recognized that this pathway is essential for controlling proapoptotic incidents, such as cell cycle arrest. These findings generally imply that surfactin has the ability to inhibit cancer cell survival and downregulate the cell cycle (Gudiña et al., 2013). Human malignant melanocyte (SK-MEL-28) spontaneously transformed human keratinocyte (HaCaT) cell lines treated with glycolipids like rhamnolipids and sophorolipids. These biosurfactants cause necrosis, which kill cells. Furthermore, it was demonstrated that sophorolipids considerably prevented the SK-MEL-28 cells' migration (Adu et al., 2022).

A new biosurfactant derived from *Candida parapsilosis* has been investigated for cytotoxicity against breast cancer cells (MDA-MB-231) in polymeric nanoparticle form. For the study, copolymers of polylactic acid with poly (ethylene glycol) (PLA–PEG) biosurfactant-encapsulated polymeric nanoparticles were used. Encapsulated biosurfactants enabled the breast cancer cell line to undergo apoptosis, which destroyed the cancer cells. Thus, to regulate breast cancer cells, PLA–PEG polymeric nanoparticles can be used as an appropriate vehicle for the controlled release of a new biosurfactant that was derived from *Candida parapsilosis* (Wadhawan et al., 2022).

When mono- and di-Rhamnolipids (RL-1 and RL-2) produced by *Pseudomonas aeruginosa* were tested in vitro against human cell lines (MCF-10A, MCF-7, and MDA-MB-231) for breast cancer, they demonstrated both anticancer and autophagy inhibitory activities (Semkova et al., 2021).

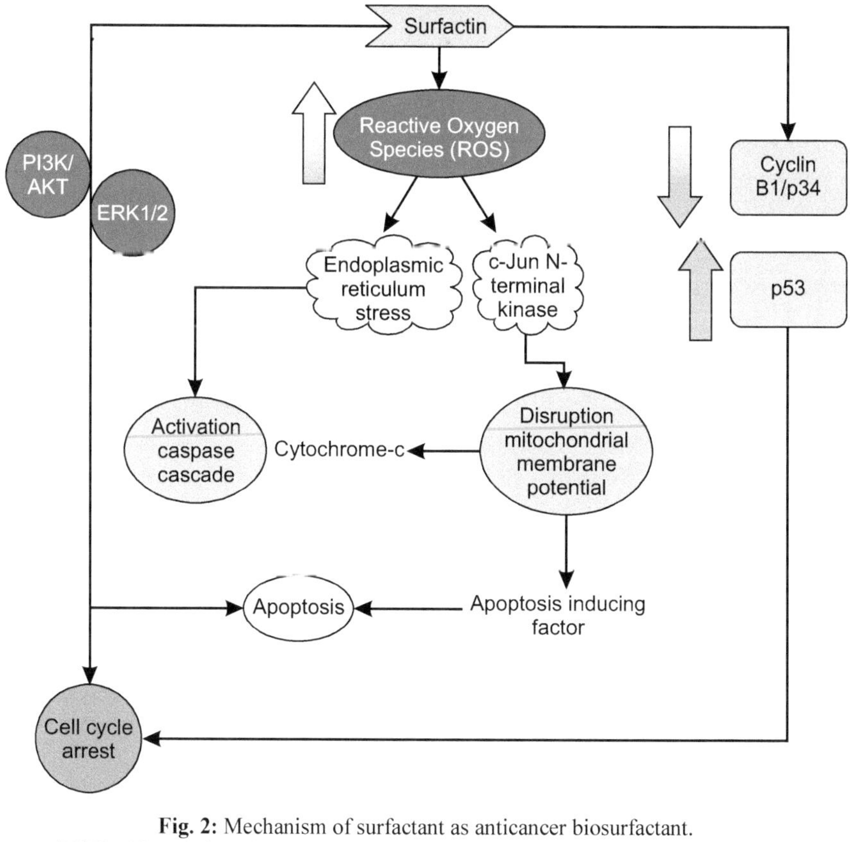

Fig. 2: Mechanism of surfactant as anticancer biosurfactant.
(PI3K – Phosphoinositide-3 Kinase, ERK – Extracellular Regulated Protein Kinase)

Table 2: Biosurfactants having anticancer activity

Biosurfactants	Cancer type	Activity	References
Mannosylerythritol lipids	Myelogenous leukemia	Differentiation, Growth inhibition	Fan et al., 2016
Sophorolipids	Liver cancer, Lung cancer, Esophageal cancer	Apoptosis induction, cell cycle arrest, and growth inhibition	Haque et al., 2021
Rhamnolipids	Breast cancer	Autophagy inhibition	Semkova et al., 2021
Viscosin	Prostate cancer	Migration inhibition	Saini et al., 2008b
Surfactins	Hepatocellular carcinoma, Breast cancer	Apoptosis induction, cell cycle arrest, and growth inhibition	Cao et al. 2010, Zhou et al., 2020
Serratamolide	B-chronic lymphocytic leukemia	Apoptosis induction	Escobar-Díaz et al., 2005
Monoolein	Cervical cancer, Leukemia	Growth inhibition	Gudiña et al., 2013
Glycolipids	Melanoma	Inhibit cell migration	Adu et al., 2022

Subsequently, it has also been proposed that the most likely mechanism of biosurfactant anticancer activity is their capacity to rupture cell membranes, leading to a number of processes such as increased membrane permeability and lysis, and metabolite leakage (Janek et al., 2013).

3.4 Antifungal

Antifungal biosurfactants have established potential in therapeutic applications because of their unique properties with potential biomedicine benefits. *Bacillus* spp. make a wide range of lipopeptide biosurfactants. Lipopeptides demonstrated efficacious antifungal activity against a range of phytopathogens and their associated ailments, including *Trichoderma atroviride* (ear rot and root rot), *Fusarium moniliforme* (rice bakanae disease), *Fusarium oxysporum* (root rot), and *Fusarium solani* (root rot) (Sarwar et al., 2018). *Bacillus amylofaciens* strain AR2 produced biosurfactant growing on sucrose and dextrose-based Minimal Salt Medium (MSM) demonstrating potential antifungal activity (Singh et al., 2014).

The researchers assessed the antifungal activity of a biosurfactant produced by *Rhodotorula babjevae* YS3 against a number of plant and human pathogens, including *Trichophyton rubrum* (MTCC 8477), *Fusarium verticilliodes* (MTCC 10556), *Fusarium oxysporum* f. sp. *pisi* (ITCC 4814), *Corynespora cassiicola* (ITCC 6748), and *Fusarium gloeosporioides* (ITCC 6434). Biosurfactants exhibit good antifungal properties with low minimum inhibitory concentrations against *Candida gloeosporioides, F. oxysporum* f. sp. *pisi,* and *F. verticillioides*. However, no inhibitory activity against *C. cassiicola* was detected at the minimum inhibitory concentrations (Sen et al., 2017).

Biosurfactants possessing antifungal activity exhibit great potential in the therapeutic domain, providing substitutes for conventional antifungal drugs, decreasing the risk of resistance formation, and promoting prospects for novel approaches in medication delivery and wound healing. They will probably play a more significant role in clinical settings and provide better treatment results for fungal infections in the future due to research and development.

3.5 Antivirals

Additionally, compared to non-enveloped viruses, biosurfactants have been shown to have antiviral activity, primarily against enveloped viruses including retroviruses and herpes viruses. The physicochemical interactions and inhibitory effect of the surfactants on the virus envelope are believed to be the real cause. The antiviral efficacy of the sophorolipids and rhamnolipids alginate complex was demonstrated against both the herpes simplex virus and HIV (Fracchia et al., 2015). A global economic crisis is being caused by the coronavirus illness (2019 (COVID-19), which is infecting and killing millions of people. Biosurfactants (BSs) can be used in pharmacological, therapeutic, hygienic, and environmental settings to prevent, control, and manage pandemics. The biopeptide Cyclosporine A (CsA), produced by the fungus *Tolypocladium inflatum*, stopped the influenza virus from proliferating by blocking protein synthesis (budding or assembly) without interfering with RNA replication or adsorption (Sarangi et al., 2022). A research investigation on herpes simplex virus type 1 (HSV-1) in 1990 demonstrated the antiviral potential of pumilacidin, a biosurfactant produced by *Bacillus subtilis*. The pumilacidin substance, a complex of biosurfactant molecules linked to surfactin, was shown in the study to have high antiviral activity (Barrantes et al., 2021, Barrantes et al. 2021).

3.6 Immuno-modulatory Activities

Biosurfactant molecules have immunomodulatory properties through immune system activation or repression. It has been discovered that glycolipid biosurfactants (GB) influence the cellular and humoral immune systems. *Pseudomonas aeruginosa* primarily secretes a rhamnolipid which is a bacterial exotoxin. Previous research demonstrated the immunomodulatory properties of rhamnolipid; yet, recent investigations have revealed that rhamnolipid is a molecule that acts like a soldier to ensure *Pseudomonas aeruginosa* survival inside the host. According to the research, rhamnolipid significantly contributes to the inhibition of human beta defensin-2 (hBD-2) synthesis, an antimicrobial peptide. Rhamnolipids prevent protein kinase-C (PKC) and diacylglycerol (DAG) from interacting. Subsequently, the reduction of hBD-2 synthesis results from the blocking of PKC interaction with DAG, which in turn blocks the activity of Mitogen-Activated Protein Kinase (MAPK) (Sajid et al., 2020).

By down-regulating Toll-Like Receptor-2 (TLR-2), Paired Box Protein (PAX5), Signal Transducer, and Activation of Transcription (STAT3), and Interleukin-6 (IL-6) gene expression, sophorolipid dramatically reduces the synthesis of IgE (Immunoglobulin E) in U266 cells, a myeloma cell line that produces IgE. These potentially demonstrate the anti-inflammatory properties of sophorolipid and may be a novel treatment option for treating diseases with altered IgE production (Hagler et al., 2007, Sajid et al., 2020).

Nuclear Factor Kappa B (NF-KB) translocation from the cytoplasm to the nucleus is decreased as a result of decreased IκB kinase (IκK) and IκB phosphorylation, which is mediated by the lipopeptide surfactin. Surfactins' inhibitory actions on macrophage antigen presentation are linked to a reduction in the activation of several signalling molecules, including NF-κB, phosphoinositide 3-kinase (PI-3 K), and MAPK. The features listed above make it clear that surfactin is a strong immunosuppressive agent and that it may be used in place of immunosuppressive medications for organ transplantation and autoimmune illnesses (Park and Kim, 2009, Sajid et al., 2020).

3.7 Wound Healing

Hemostasis, inflammation, proliferation, and remodeling are the four main stages of the natural biological process of wound healing. While the innate immune response has been the driving force for wound healing (Guo and DiPietro, 2010), in an evaluation study, the potential activity of the biosurfactant produced by Acinetobacter junii B6 for wound healing was discovered (Ohadi et al., 2018). Rhamnolipids show antibacterial activity against some gram-positive as well as gram-negative bacteria like *Bacillus subtilis, Staphylococcus aureus, Streptococcus faecalis, Micrococcus luteus* and *Salmonella typhimurium*. This antibacterial characteristic of rhamnolipids may be effective in wound-healing and help with overcoming the healing delay caused by bacterial infection. Because of its hemolytic property, rhamnolipid is toxic at higher concentrations and can even cause cell necrosis. However, at lower concentrations, rhamnolipid inhibits fibroblast proliferation and promotes keratinocyte proliferation in existing serum, indicating that it may be a useful agent for wound healing (Sana et al., 2018).

3.8 As Drug Delivery Agents

The use of biosurfactants as a drug delivery system (DDS) in order to improve the oral bioavailability of some medicines that show poor water solubility has proven to be a significant problem in the field of pharmaceutical sciences. These compounds have huge potential for emulsification and auto assembly. Thus, various techniques have been used to create delivery systems that can increase the oral bioavailability of hydrophobic medications, like MDDS (microemulsion drug delivery systems). They are typically globular in shape and include cosolvents, surfactants, cosurfactants, and lipids (Bjerk et al., 2021)

Daptomycin and polymyxin B, lipopeptides generated from antibiotics by microorganisms utilized in the manufacturing of medication delivery systems, are the best-characterized lipopeptides (Oliveira et al., 2020, Wang et al., 2020).

Table 3: Biosurfactant and biosurfactant derived compounds as therapeutics

Biosurfactants	Activity	References
Mupirocin (Pseudomonic acid)	Antibacterial	Sutherland et al., 1985
Oxazolidine linezolid	Antibacterial	Kaskatepe and Ozturk, 2023
Surfactin	Antibacterial	Vater et al., 2002
Viscosin	Antibacterial	Saini et al., 2008a
Rapeseed sophorolipids	Antibacterial, Cleansing	Cho et al., 2022
WH1 Fungin	Antibacterial	Barrantes et al., 2021
Daptomycin	Antibacterial	Barrantes et al., 2021
Amphotericin B	Antifungal and Biofilm inhibition	Haque et al., 2017
Micafungin	Antifungal and prevent Biofilm formation	Hijazi et al., 2023
Rufisan	Antiadhesive	Barrantes et al., 2021
Dispersin B	Inhibit biofilm formation	Barrantes et al., 2021
Rhamnolipids	Antibacterial, Antiadhesive, Emulsifier	Thakur et al., 2021

4. Bioemulsifiers

Emulsifiers are a broad range of substances called surface-active agents or surfactants. An emulsifier is a substance that functions by stabilizing chemical reactions and lowering their rate of occurrence. Bioemulsifiers are also known as surface active biomolecular compounds because of their various benefits over chemical surfactants, including their non-toxicity, biodegradability, foaming, biocompatibility, efficiency at low concentrations, and good selectivity in a range of pH, temperatures, and salinities. Bioemulsifiers are made of yeast, fungi, and bacteria and can be found in a variety of natural resources. Compared to biosurfactants, bioemulsifiers have a higher molecular weight (Alizadeh-Sani et al., 2018). Complex combinations of lipopolysaccharides, lipoproteins, heteropolysaccharides, and proteins are called bioemulsifiers (Uzoigwe et al., 2015a).

Bioemulsifiers are multifunctional substances that are essential in many industrial applications because they can stabilize oil-in-water emulsions (Alvarez et al., 2018). These compounds are efficient at emulsifying two immiscible liquids, even at low concentrations, but are not as effective at lowering surface tension. Protein, fatty acid, and polysaccharide components together improve the emulsifying efficiency of bioemulsifiers (Uzoigwe et al., 2015a). Microbially produced bioemulsifiers are advantageous and might constitute a key alternative to synthetic emulsifiers. The physicochemical properties of bioemulsifiers that are used in therapeutic development are listed in Table 4.

Table 4: Physicochemical properties of bioemulsifiers

Bioemulsifiers	Physicochemical properties	References
Alasan	Emulsification	Walzer et al., 2006
Amyloid	Cell aggregation, adhesion, biofilm formation	Markande et al., 2018
Uronic acid	Emulsification	Jain et al., 2013
Glycolipid	Surface active agent, stability of emulsion, Rheological properties	Adetunji and Olaniran, 2019
Proteoglycan	Emulsification of heating oils	Marin et al., 1996
Lipopolysaccharides	Emulsification of polycyclic aromatic hydrocarbon	Ortega-de la Rosa et al., 2018

4.1 Therapeutic Applications of Bioemulsifiers

Many useful compounds, including insecticides, medications, and other substances, are soluble in organic solvents like alcohol to produce solutions. Most of these solvents are expensive and dangerous, therefore using them as solutions is discouraged, especially when utilizing them on humans. Emulsifiers are beneficial because of their increased stability, reduced toxicity, ecological sustainability, and human friendliness, particularly bioemulsifiers. They have many applications in every sector of the economy and all aspects of daily life, including the oil and gas sector, medicines, hygiene and textiles, paint, detergents, supplies for cleaning, food processing, cement, beer, and drinks (Gharaei-Fa, 2010).

It is widely recognized that high molecular weight biosurfactants, such as emulsifiers, have many exposed reactive groups, which gives them a high surface

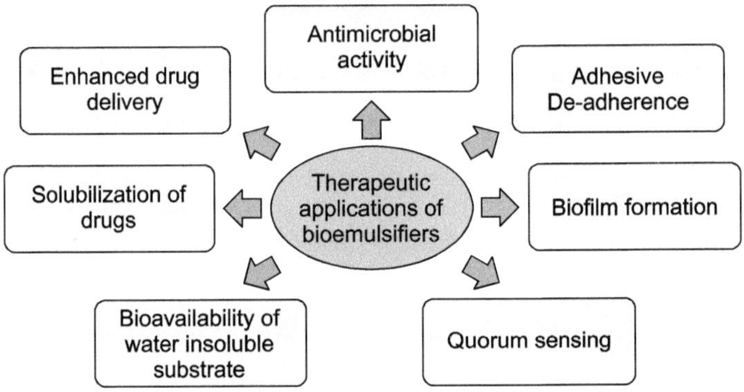

Fig. 3: Therapeutic applications of bioemulsifiers.

adhesion capability and the ability to develop persistent monolayers. When fresh water is added with mixing, bioemulsifiers are known to generate stable emulsions and dispersions that barely agglomerate and can re-emulsify because they remain associated with the droplets (Uzoigwe et al., 2015a). Bioemulsifiers are in use as drug carriers from solubilizing drugs to controlled release. Emulsions are used for oral, topical, and parenteral routes of drug delivery. For the oral administration of medications, emulsions of water in oil (W/O) are said to work effectively. While both W/O and O/W emulsions have been examined for parenteral drug delivery, mainly O/W emulsions are used. For cancer, Alzheimer's disease, thrombosis therapy, marketed emulsion formulations are used. For the encapsulation of bioactive compounds, double emulsions are considered to be outstanding systems (Masuda et al., 2003). Bioemulsifiers are produced from different sources like bacteria, fungi, yeast, and algae as shown in Table 5.

Table 5: Bioemulsifiers produced by different sources

Bioemulsifier	**Source**	**References**
Lauryl fructose	*Pseudomonas* spp.	Alizadeh-Sani et al., 2018
Alasan or EKA53	*Acinetobacter radioresistant*	
Emulsan	*Acinetobacter calcoaceticus* RAG 1	
Lipopeptide	*Bacillus pseudomycoides* BS6	
Rhamnolipid	*Pseudomonas cepacia* CCT6659	
Lichenynsin	*Bacillus subtilis*	
Subtilisin	*Bacillus subtilis*	
Surfactin	*Bacillus amyloliquefaciens*	
Mannoprotein	*Saccharomyces* spp.	Cameron et al., 1988
Liposan	*Candida lipolytica*	Alizadeh-Sani et al., 2018
Rhodotorula yeast	*Rhodotorula glutinis*	
Sophorolipids	*Torulopsis petrophilum*, *Torulopsis apicola*	
Exopolysaccharide	*Dunaliella salina*	Laroche, 2022
Sugar + Fatty acids + Proteins	*Phormidium* sp. Strain J-1	

4.2 Antimicrobial Activity

Microorganism-produced bioemulsifiers have received increasing attention and are more likely to be applied in many industries. The antibacterial activity of bioemulsifiers works by rupturing cell membrane integrity, which results in cell lysis. Iturin A, a lipopeptide, breaks the plasma membranes of yeast cells by increasing the electrical conductance of the membrane and causing intramembranous particles to accumulate within the cells, however, the exact mechanism by which the bioemulsifier impacts membrane integrity varies. It has been demonstrated that the lipopeptide surfactin increases membrane permeability by interacting with the phospholipids in cell membranes. For a rhamnolipid, a glycolipid is expected to function on the lipid or outer proteins of cell membranes. Therefore, a similar mechanism has been proposed, resulting in membrane structural fluctuations. The antifungal activity of rhamnolipids has been linked to zoospore lysis (Hyder, 2015).

The most effective bioemulsifier production has been demonstrated by *Serratia* spp. and *Serratia marcescens* (S10) from insects. The bioemulsifier has demonstrated antifungal action against *Aspergillus niger, Geotricum* spp., and *Candida albicans* in addition to antibacterial activity against *Salmonella* spp., *Staphylococcus aureus,* and *Lesteria* spp. Compared to the antibacterial activity, the stated antifungal activity was greater. The study also undoubtedly confirmed the antimicrobial effect of the bioemulsifier against different bacteria like *Escherichia coli, Staphylococcus aureus, Pseudomonas aeruginosa* pathogen, *Bacillus subtilis, Bacillus cereus, Staphylococcus epdermidis* and it provides high inhibition. The rhamnolipid mixture derived from *Pseudomonas aeruginosa* was shown by the researcher to exhibit inhibitory activity against a variety of bacteria and fungi, including *Aspergillus niger, Chaetonium globosum, Enicillium crysogenum, Aureobasidium pullulans, Botrytis cinerea,* and *Rhizoctonia solani*, at different concentrations. Thus, the research study suggested the use of this bioemulsifier compound with pharmaceutical and cosmetics for dermal and other applications and observed to have a lethal effect on pathogenic bacteria (Kadhem et al., 2019). Another study demonstrates the bioemulsifying ability of *Levilactobacillus brevis* S4 and *Lactiplantibacillus plantarum* S5. They exhibited antibacterial action against microorganisms classified as Gram-positive or Gram-negative (Tchakouani et al., 2023). Thus, bioemulsifiers have potent activity as antimicrobials and have more future potential.

4.3 Adherence and De-adherence to the Surface

When biosurfactants, such as bioemulsifiers, come into contact with an interface, they create a conditioning layer that modifies the original surface's wet ability and influences bacterial adhesion and de-adhesion. Regardless of the characteristics of the bacterial surface, the presence of bioemulsifiers influences bacterial adherence. The type of bacteria that can adhere to a surface that has been wetted with a bioemulsifier depends on how the bioemulsifier is oriented on the surface (Perfumo et al., 2010). Bioemulsifiers have been reported extensively to change cell surface hydrophobicity in *Serratia marcescens, Alcanivorax borkumensis, Pseudomonas aeruginosa*, and *Acinetobacter calcoaceticus.* Because of their wettability, bioemulsifiers promote the formation of a conditioning film at the interface, converting a hydrophobic to a hydrophilic condition and vice versa. The suitable bacteria can adhere better due to this alteration in interface qualities. It is well known that bacteria swarm and glide to explore their local environment. This process is thought to be a constant desorption assisted by a bioemulsifier along a two-dimensional interface where the complementary roles of flagellins and flagellum are utilized (Markande and Anuradha S. Nerurkar, 2013).

4.4 Biofilm Formation

The wetting of the surface by a bioemulsifier leads to the preparation of a conducive environment for bacterial attachment, facilitating reversible adhesion of the bacteria resulting in the formation of high-density, attached microbial communities frequently embedded in extracellular matrices, also called biofilms (Ortega-Morales et al., 2010).

After initial conditioning of the surfaces, the exopolymeric bioemulsifiers help bacteria in forming a biofilm and are also present as an integral part of the matrix which shields its inhabitants from predators, biocides, dehydration, and other dangerous environmental conditions. Adhesion of bacteria to a solid surface can happen in both turbulent and stationary aqueous phases. Bacterial aggregation, motility, and film development at this interface are influenced by sedimentation and the capillary (or drainage) forces brought on by the pressure of the liquid flowing between the solid surface and the bacterial surface (Sadowska et al., 2010). The surface active compound from *Brevibacterium casei* MSA19 shows activity against pathogenic biofilms in vitro (Kiran et al., 2010).

4.5 Quorum Sensing

Bacteria use a widely used communication mechanism called quorum sensing to control the expression of certain genes based on the presence or lack of a small signal molecule called an auto-inducer. In another context, bacteria make and release auto-inducers, and the entire bacterial population changes its gene expression when the concentration achieves a particular value. This can result in pathogenesis, modulation of association with higher organisms, or community assembly (biofilm formation). In nature, quorum sensing mechanisms are numerous (Rutherford and Bassler, 2012).

Recent research has shown that both high- and low-molecular-weight biosurfactants belong to the biosurfactant category, and can affect swarming behaviour by functioning as chemotactic-like stimuli. Rhamnolipids have a well-established function in both surface motility in general and swarming motility specifically. There is a selective benefit to producing bioemulsifiers at high cell densities. It has been proposed that pathogen-produced emulsifiers are virulence factors that are generated when the cell density reaches a certain level, allowing for a targeted attack on the host (Glick et al., 2010).

4.6 Bioavailability of Water-insoluble Substrates

It has been noted that the interfacial surface area (an interface) between water and substrate/air can restrict the speed of bacterial growth. When the interfacial surface area becomes limited for bacteria growing on any interface, the biomass increases arithmetically rather than exponentially(Ron and Rosenberg, (2002). Emulsifying substances naturally contribute to bioremediation by expanding the surface area of insoluble substrates. Each cell forms its microenvironment and emulsification takes place when the cell is capable of micellar interaction and emulsification takes place near the cell surface without microscopic-level mixing. The consequence of this process may not be visible macroscopically. Surface active substances, such as bioemulsifiers, can increase the growth of bound or inaccessible substrates by making them more soluble in water or by desorbing them from surfaces (Ron and Rosenberg, 2001).

4.7 Solubilization of Drugs

Drugs that are not highly soluble in water can be made readily available for absorption and metabolism by using bioemulsifiers. This can enhance a drug's absorption through the skin and digestive system, which can be especially helpful for medications that are taken orally or topically. Studies have indicated that the anti-cancer medication paclitaxel, which is poorly soluble in water, may be dissolved using the bioemulsifier rhamnolipid. It has been demonstrated that the combination of rhamnolipid with paclitaxel inhibits cancer cell proliferation more effectively than paclitaxel alone (Tran et al., 2013, Ma, 2013).

4.8 Enhanced Drug Delivery

Medications can be effectively delivered to particular tissues or organs by using bioemulsifiers. This can be achieved through the design of drug-delivery systems that are based on bioemulsifiers that target particular transport pathways or receptors. Research indicated that the anti-cancer medication doxorubicin might be administered to the brain via the bioemulsifier sophorolipid. It has been demonstrated that the combination of sophorolipid and doxorubicin is more effective than doxorubicin by itself at slowing the growth of brain tumors (Gudiña et al., 2013).

5. Conclusion

Biosurfactants and bioemulsifiers are versatile bioactive agents with promising therapeutic applications. Their antimicrobial properties, ability to inhibit biofilm formation, and immunomodulatory properties make them valuable tools in the fight against infections. Bioemulsifiers also play an important role in microbe control and drug delivery. Future research in this field holds the promise of revealing even more therapeutic applications.

References

Adetunji, Adegoke Isiaka & Ademola Olufolahan Olaniran. (2019). Production and characterization of bioemulsifiers from acinetobacter strains isolated from lipid-rich wastewater. *3 Biotech*, 9(4): 151. doi: 10.1007/s13(205-019-1683-y.

Adu, Simms A., Matthew S. Twigg, Patrick J. Naughton, Roger Marchant & Ibrahim M. Banat. (2022). Biosurfactants as anticancer agents: Glycolipids affect skin cells in a differential manner dependent on chemical structure. *Pharmaceutics*, 14(2): 360. doi: 10.3390/pharmaceutics140(20360.

Alizadeh-Sani, Mahmood, Hamed Hamishehkar, Arezou Khezerlou, Maryam Azizi-Lalabadi, Yaghob Azadi, Elyas Nattagh-Eshtivani et al. (2018). Bioemulsifiers derived from microorganisms: Applications in the drug and food industry. *Advanced Pharmaceutical Bulletin*, 8(2): 191–199. doi: 10.15171/apb.(2018.023.

Alvarez, Vanessa Marques, Diogo Jurelevicius, Rodrigo Vassoler Serrato, Eliana Barreto-Bergter, & Lucy Seldin. (2018). Chemical characterization and potential application of exopolysaccharides produced by Ensifer Adhaerens JHT2 as a bioemulsifier of edible oils. *International Journal of Biological Macromolecules*, 114: 18–25. doi: 10.1016/j.ijbiomac.(2018.03.063.

Barrantes, A., Schujman, G., de Mendoza, D. & Altabe, S. (2021). Antiviral potential of biosurfactant pumilacidin against herpes simplex virus type 1. *Journal of Microbiological Methods*, 184: 106219. doi:10.1016/j.mimet.2021.106219

Barrantes, Kenia, Juan José Araya, Luz Chacón, Rolando Procupez-Schtirbu, Fernanda Lugo, Gabriel Ibarra et al. (2021). Antiviral, antimicrobial, and antibiofilm properties of biosurfactants. *Biosurfactants for a Sustainable Future*, 245–268. Wiley.

Bjerk, Thiago R., Patricia Severino, Sona Jain, Conrado Marques, Amélia M. Silva, Tatiana Pashirova et al. (2021). Biosurfactants: Properties and applications in drug delivery, biotechnology and ecotoxicology. *Bioengineering*, 8(8): 115. doi: 10.3390/bioengineering8080115.

Cameron, D.R., Cooper, D.G. & Neufeld, R.J. (1988). The mannoprotein of *Saccharomyces cerevisiae* is an effective bioemulsifier. *Applied and Environmental Microbiology*, 54(6): 20–25. doi: 10.1128/aem.54.6.14(20-1425.1988.

Cao, Xiao-Hong, Ai-Hua Wang, Chun-Ling Wang, De-Zhi Mao, Mei-Fang Lu, Yun-Qian Cui et al. (2010). Surfactin induces apoptosis in human breast cancer MCF-7 cells through a ROS/JNK-mediated mitochondrial/caspase pathway. *Chemico-Biological Interactions*, 183(3): 357–362. doi: 10.1016/j.cbi.(2009.11.027.

Carrillo, Carmen, José A. Teruel, Francisco J. Aranda & Antonio Ortiz. (2003). Molecular mechanism of membrane permeabilization by the peptide antibiotic surfactin. *Biochimica et Biophysica Acta (BBA) – Biomembranes*, 1611(1–2): 91–97. doi: 10.1016/S0005-2736(03)00029-4.

Ceresa, C., Fracchia, L., Williams, M., Banat, I.M. & Díaz De Rienzo, M.A. (2020). The effect of sophorolipids against microbial biofilms on medical-grade silicone. *Journal of Biotechnology*, 309: 34–43. doi: 10.1016/j.jbiotec.(2019.12.019.

Ceresa, Chiara, Letizia Fracchia, Emanuele Fedeli, Chiara Porta & Ibrahim M. Banat. (2021). Recent advances in biomedical, therapeutic and pharmaceutical applications of microbial surfactants. *Pharmaceutics*, 13(4): 466. doi: 10.3390/pharmaceutics13040466.

Cho, Wei Yan, Jeck Fei Ng, Wei Hsum Yap, & Bey Hing Goh. (2022). Sophorolipids—Bio-based antimicrobial formulating agents for applications in food and health. *Molecules*, 27(17): 5556. doi: 10.3390/molecules27175556.

Elshikh, M., Moya-Ramírez, I., Moens, H., Roelants, S., Soetaert, W., Marchant, R. et al. (2017). Rhamnolipids and lactonic sophorolipids: Natural antimicrobial surfactants for oral hygiene. *Journal of Applied Microbiology*, 123(5): 1111–1123. doi: 10.1111/jam.13550.

Escobar-Díaz, E., López-Martín, E.M., Hernández del Cerro, M., Puig-Kroger, A., Soto-Cerrato, V., Montaner, B. et al. (2005). AT514, a cyclic depsipeptide from *Serratia marcescens*, induces apoptosis of B-chronic lymphocytic leukemia cells: Interference with the Akt/NF-KB survival pathway. *Leukemia*, 19(4): 572–579. doi: 10.1038/sj.leu.2403679.

Fan, Linlin, Hongji Li, Yongwu Niu & Qihe Chen. (2016). Characterization and inducing melanoma cell apoptosis activity of mannosylerythritol lipids-A produced from *Pseudozyma aphidis*. *PLOS ONE*, 11(2): e0148198. doi: 10.1371/journal.pone.0148198.

Fracchia, Letizia, Jareer J. Banat, Massimo Cavallo, Chiara Ceresa & Ibrahim M. Banat. (2015). Potential therapeutic applications of microbial surface-active compounds. *AIMS Bioengineering*, 2(3): 144–162. doi: 10.3934/bioeng.2015.3.144.

Gharaei-Fa, Eshrat. (2010). Biosurfactants in pharmaceutical industry: A mini-review. *American Journal of Drug Discovery and Development*, 1(1): 58–69. doi: 10.3923/ajdd.(2011.58.69.

Giri, S.S., Ryu, E.C., Sukumaran, V. & Park, S.C. (2019). Antioxidant, antibacterial, and anti-adhesive activities of biosurfactants isolated from Bacillus strains. *Microbial Pathogenesis*, 132: 66–72. doi: 10.1016/j.micpath.(2019.04.035.

Glick, Rivka, Christie Gilmour, Julien Tremblay, Shirley Satanower, Ofir Avidan, Eric Déziel et al. (2010). Increase in rhamnolipid synthesis under iron-limiting conditions influences surface motility and biofilm formation in *Pseudomonas aeruginosa*. *Journal of Bacteriology*, 192(12): 2973–2980. doi: 10.1128/JB.01601-09.

Gomaa, Eman Zakaria. (2013). Antimicrobial activity of a biosurfactant produced by *Bacillus licheniformis* strain M104 grown on whey. *Brazilian Archives of Biology and Technology*, 56(2): 259–268. doi: 10.1590/S1516-8913(2013000(200011.

Grangemard, Isabelle, Jean Wallach, Regine Maget-Dana & Françoise Peypoux. (2001). Lichenysin: A more efficient cation chelator than surfactin. *Applied Biochemistry and Biotechnology*, 90(3): 199–210. doi: 10.1385/ABAB:90:3:199.

Gudiña, Eduardo J., Vivek Rangarajan, Ramkrishna Sen, & Lígia R. Rodrigues. (2013). Potential therapeutic applications of biosurfactants. *Trends in Pharmacological Sciences*, 34(12): 667–675. doi: 10.1016/j.tips.(2013.10.002.

Guo, S. & DiPietro, L.A. (2010). Factors affecting wound healing. *Journal of Dental Research*, 89(3): 219–229. doi: 10.1177/002(2034509359125.

Haddaji, Najla, Abdelkarim Mahdhi, Nouha Bouali, Mouna Ghorbel, Olfa Bechambi, Nadia Leban et al. (2022). Biosurfactants as inhibitors of the adhesion of pathogenic bacteria. *Emirates Journal of Food and Agriculture*. doi: 10.9755/ejfa.(2022.v34.i1.2803.

Hagler, M., Smith-Norowitz, T.A., Chice, S., Wallner, S.R., Viterbo, D., Mueller, C.M. et al. (2007). Sophorolipids decrease IgE production in U266 cells by downregulation of BSAP (Pax5), TLR-2, STAT3 and IL-6. *Journal of Allergy and Clinical Immunology*, 119(1): S263. doi: 10.1016/j.jaci.(2006.12.399.

Haque, Farazul, Mohd Sajjad, Ahmad Khan & Naif AlQurashi. (2021). ROS-mediated necrosis by glycolipid biosurfactants on lung, breast, and skin melanoma cells. *Frontiers in Oncology*, 11. doi: 10.3389/fonc.(2021.622470.

Haque, Farazul, Mohammad Sajid, Swaranjit Singh Cameotra & Mani Shankar Battacharyya. (2017). Anti-biofilm activity of a sophorolipid-amphotericin B Niosomal formulation against *Candida Albicans*. *Biofouling*, 33(9): 768–779. doi: 10.1080/08927014.(2017.1363191.

Hijazi, Duaa M., Lina A. Dahabiyeh, Salah Abdelrazig, Dana A. Alqudah, & Amal G. Al-Bakri. (2023). Micafungin effect on Pseudomonas aeruginosa metabolome, virulence and biofilm: Potential quorum sensing inhibitor. *AMB Express*, 13(1): 20. doi: 10.1186/s13568-023-01523-0.

Hyder, Nadhem H. (2015). Production, characterization and antimicrobial activity of a bioemulsifier produced by *Acinetobacter baumanii* AC5 utilizing edible oils. *Iraqi Journal of Biotechnology*, 14(2).

Jain, Rakeshkumar M., Kalpana Mody, Nidhi Joshi, Avinash Mishra & Bhavanath Jha. (2013). Production and structural characterization of biosurfactant produced by an alkaliphilic bacterium, Klebsiella sp.: Evaluation of different carbon sources. *Colloids and Surfaces B: Biointerfaces*, 108: 199–204. doi: 10.1016/j.colsurfb.2013.03.002.

Janek, Tomasz, Anna Krasowska, Agata Radwańska & Marcin Łukaszewicz. (2013). Lipopeptide Biosurfactant Pseudofactin II induced apoptosis of melanoma A 375 cells by specific interaction with the plasma membrane. *PLoS ONE*, 8(3): e57991. doi: 10.1371/journal.pone.0057991.

Kadhem, Bashar, Rajwa Essa & Nibras Mahmood. (2019). Antimicrobial activity of a bioemulsifier produced by *Saccharomyces cerevisiae*. *Journal of Garmian University*, 6(1): 546–554. doi: 10.24271/garmian.1031.

Kaskatepe, Banu, & Sukran Ozturk. (2023). Assessment of synergistic activity of rhamnolipid and linezolid against methicillin-resistant *Staphylococcus aureus* in-vitro and in-vivo with Galleria mellonella larvae model. *Microbial Pathogenesis*, 174: 105945. doi: 10.1016/j.micpath.(2022.105945.

Khan, Fazlurrahman, Sandra Folarin Oloketuyi & Young-Mog Kim. (2019). Diversity of bacteria and bacterial products as antibiofilm and antiquorum sensing drugs against pathogenic bacteria. *Current Drug Targets*, 20(11): 1156–1179. doi: 10.2174/13894501(206661904 23161249.

Kiran, George Seghal, Balu Sabarathnam & Joseph Selvin. (2010). Biofilm disruption potential of a glycolipid biosurfactant from marine *Brevibacterium casei*. *FEMS Immunology & Medical Microbiology*, 59(3): 432–438. doi: 10.1111/j.1574-695X.(2010.00698.x.

Landman, David, Claudiu Georgescu, Don Antonio Martin & John Quale. (2008). Polymyxins Revisited. *Clinical Microbiology Reviews*, 21(3): 449–465. doi: 10.1128/CMR.00006-08.

Laroche, Céline. (2022). Exopolysaccharides from microalgae and cyanobacteria: Diversity of strains, production strategies, and applications. *Marine Drugs*, 20(5): 336. doi: 10.3390/md(20050336.

Liu, Jin, Wei Li, Xiaoyu Zhu, Haizhen Zhao, Yingjian Lu, Chong Zhang et al. (2019). Surfactin effectively inhibits Staphylococcus aureus adhesion and biofilm formation on surfaces. *Applied Microbiology and Biotechnology*, 103(11): 4565–4574. doi: 10.1007/s00253-019-09808-w.

Ma, Ping. (2013). Paclitaxel nano-delivery systems: A comprehensive review. *Journal of Nanomedicine & Nanotechnology*, 04(02). doi: 10.4172/2157-7439.1000164.

Mandal, Santi M., Aulus E.A.D. Barbosa & Octavio L. Franco. (2013). Lipopeptides in microbial infection control: Scope and reality for industry. *Biotechnology Advances*, 31(2): 338–345. doi: 10.1016/j.biotechadv.(2013.01.004.

Marin, M., Pedregosa, A. & Laborda, F. (1996). Emulsifier production and microscopical study of emulsions and biofilms formed by the hydrocarbon-utilizing bacteria *Acinetobacter calcoaceticus* MM5. *Applied Microbiology and Biotechnology*, 44(5): 660–667. doi: 10.1007/BF00172500.

Markande, Anoop R. & Anuradha S. Nerurkar. (2013). Biochemical diversity of microbial bioemulsifiers and their roles in the natural environment. *Journal of Chemical Information and Modeling*, 53(9).

Markande, Anoop R., Venkata R. Vemuluri, Yogesh S. Shouche & Anuradha S. Nerurkar. (2018). Characterization of *Solibacillus silvestris* strain AM1 that produces amyloid bioemulsifier. *Journal of Basic Microbiology*, 58(6): 523–531. doi: 10.1002/jobm.201700685.

Masuda, Kazuyoshi, Kazutoshi Horie, Ryuji Suzuki, Takayoshi Yoshikawa & Koichiro Hirano. (2003). Bioemulsifiers. *Pharmaceutical Research*, 20(1): 130–134. doi: 10.1023/A:1022267312869.

Naruse, Nobauki, Osamu Tenmyo, Seikichi Kobaru, Hideo Kamei et al. (1990a). Pumilacidin, a complex of new antiviral antibiotics. Production, isolation, chemical properties, structure and biological activity. *The Journal of Antibiotics*, 43(3): 267–280. doi: 10.7164/antibiotics.43.267.

Naughton, P.J., Marchant, R., Naughton, V. & Banat, I.M. (2019). Microbial biosurfactants: Current trends and applications in agricultural and biomedical industries. *Journal of Applied Microbiology*, 127(1): 12–28. doi: 10.1111/jam.14243.

Ndlovu, Thando, Marina Rautenbach, Johann Arnold Vosloo, Sehaam Khan & Wesaal Khan. (2017). Characterization and antimicrobial activity of biosurfactant extracts produced by *Bacillus amyloliquefaciens* and *Pseudomonas aeruginosa* isolated from a wastewater treatment plant. *AMB Express*, 7(1): 108. doi: 10.1186/s13568-017-0363-8.

Ohadi, Mandana, Hamid Forootanfar, Hamid Reza Rahimi, Elham Jafari, Mojtaba Shakibaie, Touba Eslaminejad et al. (2018). Antioxidant potential and wound healing activity of biosurfactant produced by *Acinetobacter junii* B6. *Current Pharmaceutical Biotechnology*, 18(11): 900–908. doi: 10.2174/1389(201018666171122121350.

Oliveira, D.M.L., Rezende, P.S., Barbosa, T.C., Andrade, L.N., Bani, C., Tavares, D.S. et al. (2020). Double membrane based on Lidocaine-coated polymyxin-alginate nanoparticles for wound healing: In vitro characterization and in vivo tissue repair. *International Journal of Pharmaceutics*, 591: 120001. doi: 10.1016/j.ijpharm.2020.120001.

Ortega-de la Rosa, Nestor, D., Jose, L., Vázquez-Vázquez, Sergio Huerta-Ochoa, Miquel Gimeno et al. (2018). Stable bioemulsifiers are produced by *Acinetobacter bouvetii*

UAM25 growing in different carbon sources. *Bioprocess and Biosystems Engineering*, 41(6): 859–869. doi: 10.1007/s00449-018-19(20-5.

Ortega-Morales, Benjamín Otto, Manuel Jesús Chan-Bacab, Susana del Carmen, De la Rosa-García & Juan Carlos Camacho-Chab (2010). Valuable processes and products from marine intertidal microbial communities. *Current Opinion in Biotechnology*, 21(3): 346–352. doi: 10.1016/j.copbio.(2010.02.007.

Park, Sun Young & Young Hee Kim. (2009). Surfactin inhibits immunostimulatory function of macrophages through blocking NK-KB, MAPK and Akt pathway. *International Immunopharmacology*, 9(7–8): 886–893. doi: 10.1016/j.intimp.(2009.03.013.

Perfumo, A., Smyth, T.J.P., Marchant, R. & Banat, I.M. (2010). Production and roles of biosurfactants and bioemulsifiers in accessing hydrophobic substrates. *In: Handbook of Hydrocarbon and Lipid Microbiology*, 1501–1512. Berlin, Heidelberg: Springer Berlin Heidelberg.

Rodrigues, Lígia, Ibrahim M. Banat, José Teixeira & Rosário Oliveira. (2006). Biosurfactants: Potential applications in medicine. *Journal of Antimicrobial Chemotherapy*, 57(4): 609–618. doi: 10.1093/jac/dkl024.

Rodrigues, Lígia R. & José A. Teixeira. (2010). Biomedical and Therapeutic Applications of Biosurfactants. pp. 75-87. *In:* Biosurfactants. Springer. doi:10.1007/978-3-642-14490-5_4

Ron, Eliora Z. & Eugene Rosenberg. (2001). Natural roles of biosurfactants. *Environmental Microbiology* 3(4): 229–36. doi: 10.1046/j.1462-29(20.(2001.00190.x.

Ron, Eliora Z. & Eugene Rosenberg. (2002). Biosurfactants and oil bioremediation. *Current Opinion in Biotechnology*, 13(3): 249–52. doi: 10.1016/S0958-1669(02)00316-6.

Rutherford, S.T. & Bassler, B.L. (2012). Bacterial quorum sensing: its role in virulence and possibilities for its control. *Cold Spring Harbor Perspectives in Medicine*, 2(11): a012427–a012427. doi: 10.1101/cshperspect.a012427.

Sadowska, B., Walencka, E., Wieckowska-Szakiel, M. & Różalska, B. (2010). Bacteria competing with the adhesion and biofilm formation by *Staphylococcus aureus*. *Folia Microbiologica*, 55(5): 497–501. doi: 10.1007/s12223-010-0082-x.

Saini, Harvinder S., Blanca E. Barragán-Huerta, Ariel Lebrón-Paler, Jeanne E. Pemberton, Refugio R. Vázquez et al. (2008a). Efficient purification of the biosurfactant viscosin from *Pseudomonas Libanensis* strain M9-3 and its physicochemical and biological properties. *Journal of Natural Products*, 71(6): 10111015. doi. 10.1021/np800069u.

Sajid, Mohammad, Mohd Sajjad Ahmad Khan, Swaranjit Singh Cameotra & Abdullah Safar Al-Thubiani. (2020). Biosurfactants: Potential applications as immunomodulator drugs. *Immunology Letters*, 223: 71–77. doi: 10.1016/j.imlet.2020.04.003.

Sana, Santanu, Sriparna Datta, Dipa Biswas, Biswajit Auddy, Mradu Gupta et al. (2018). Excision wound healing activity of a common biosurfactant produced by *Pseudomonas* sp. *Wound Medicine*, 23: 47–52. doi: 10.1016/j.wndm.(2018.09.006.

Sarangi, Manoj Kumar, Sasmita Padhi, L.D. Patel, Goutam Rath, Sitansu Sekhar Nanda et al. (2022). Theranostic efficiency of biosurfactants against COVID-19 and similar viruses – A review. *Journal of Drug Delivery Science and Technology*, 76: 103764. doi: 10.1016/j.jddst.(2022.103764.

Sarwar, Ambrin, Günter Brader, Erika Corretto, Gajendar Aleti, Muhammad Abaidullah et al. (2018). Qualitative analysis of biosurfactants from *Bacillus* species exhibiting antifungal activity. *PLOS ONE* 13(6): e0198107. doi: 10.1371/journal.pone.0198107.

Satpute, Surekha, Nishigandha Mone, Parijat Das, Arun Banpurkar & Ibrahim Banat. (2018). Lactobacillus acidophilus derived biosurfactant as a biofilm inhibitor: A promising investigation using microfluidic approach. *Applied Sciences*, 8(9): 1555. doi: 10.3390/app8091555.

Semkova, Severina, Georgi Antov, Ivan Iliev, Iana Tsoneva, Pavel Lefterov, Nelly Christova et al. (2021). Rhamnolipid biosurfactants—Possible natural anticancer agents and autophagy inhibitors. *Separations*, 8(7): 92. doi: 10.3390/separations8070092.

Sen, Suparna, Siddhartha Narayan Borah, Arijit Bora & Suresh Deka. (2017). Production, characterization, and antifungal activity of a biosurfactant produced by *Rhodotorula babjevae* YS3. *Microbial Cell Factories*, 16(1): 95. doi: 10.1186/s12934-017-0711-z.

Sharma, Divakar, Lama Misba & Asad U. Khan. (2019). Antibiotics versus biofilm: An emerging battleground in microbial communities. *Antimicrobial Resistance & Infection Control,* 8(1): 76. doi: 10.1186/s13756-019-0533-3.

Singh, Anil Kumar, Ria Rautela & Swaranjit Singh Cameotra. (2014). Substrate dependent in vitro antifungal activity of *Bacillus* sp strain AR2. *Microbial Cell Factories,* 13(1): 67. doi: 10.1186/1475-2859-13-67.

Singh, P., Patil, Y. & Rale, V. (2019). Biosurfactant production: Emerging trends and promising strategies. *Journal of Applied Microbiology,* 126(1): 2–13. doi: 10.1111/jam.14057.

Sutherland, R., Boon, R.J., Griffin, K.E., Masters, P.J., Slocombe, B. & White, A.R. (1985). Antibacterial activity of Mupirocin (Pseudomonic acid), a new antibiotic for topical use. *Antimicrobial Agents and Chemotherapy,* 27(4): 495–498. doi: 10.1128/AAC.27.4.495.

Tchakouani, Galvany Franck Yamagueu, Hippolyte Tene Mouafo, Richard Marcel Nguimbou, Nadège Donkeng Nganou & Augustin Mbawala. (2023). Antibacterial activity of bioemulsifiers/biosurfactants produced by *Levilactobacillus brevis* <scp>S4</Scp> and *Lactiplantibacillus plantarum* <scp>S5</Scp> and their utilization to enhance the stability of cold emulsions of milk chocolate drinks. *Food Science & Nutrition.* doi: 10.1002/fsn3.3740.

Thakur, Priyanka, Neeraj K. Saini, Vijay Kumar Thakur, Vijai Kumar Gupta, Reena V. Saini & Adesh K. Saini. (2021). Rhamnolipid the glycolipid biosurfactant: Emerging trends and promising strategies in the field of biotechnology and biomedicine. *Microbial Cell Factories,* 20(1): 1. doi: 10.1186/s12934-0(20-01497-9.

Tran, Thao T.D., Phuong H.L. Tran, Tran N. Khanh, Toi V. Van & Beom-Jin Lee. (2013). Solubilization of poorly water-soluble drugs using solid dispersions. *Recent Patents on Drug Delivery & Formulation,* 7(2): 122–133. doi: 10.2174/1872211311307020004.

Uzoigwe, Chibuzo, J. Grant Burgess, Christopher J. Ennis & Pattanathu K.S.M. Rahman. (2015a). Bioemulsifiers are not biosurfactants and require different screening approaches. *Frontiers in Microbiology* 6. doi: 10.3389/fmicb.2015.00245.

Van Hamme, Jonathan, D., Ajay Singh & Owen P. Ward. (2006). Physiological aspects. *Biotechnology Advances*, 24(6): 604–620. doi: 10.1016/j.biotechadv.(2006.08.001.

Vater, Joachim, Bärbel Kablitz, Christopher Wilde, Peter Franke, Neena Mehta & Swaranjit Singh Cameotra. (2002). Matrix-assisted laser desorption ionization-time of flight mass spectrometry of lipopeptide biosurfactants in whole cells and culture filtrates of *Bacillus Subtilis* C-1 isolated from petroleum sludge. *Applied and Environmental Microbiology,* 68(12): 6210–6219. doi: 10.1128/AEM.68.12.6210-6219.(2002.

Wadhawan, Aishani, Joga Singh, Himani Sharma, Shristi Handa, Gurpal Singh, Ravinder Kumar et al. (2022). Anticancer biosurfactant-loaded PLA–PEG nanoparticles induce apoptosis in human MDA-MB-231 breast cancer cells. *ACS Omega,* 7(6): 5231–5241. doi: 10.1021/acsomega.1c06338.

Walencka, E., Różalska, S., Sadowska, B. & Różalska, B. (2008). The influence of *Lactobacillus acidophilus*-derived surfactants on staphylococcal adhesion and biofilm formation. *Folia Microbiologica,* 53(1): 61–66. doi: 10.1007/s12223-008-0009-y.

Walzer, Gil, Eugene Rosenberg & Eliora Z. Ron. (2006). The Acinetobacter Outer Membrane Protein A (OmpA) is a secreted emulsifier. *Environmental Microbiology,* 8(6): 1026–1032. doi: 10.1111/j.1462-29(20.(2006.00994.x.

Wang, Jing, Jin Zhang, Kai Liu, Jinfeng He, Yongqiang Zhang, Shengfu Chen et al. (2020). Synthesis of gold nanoflowers stabilized with Amphiphilic Daptomycin for enhanced photothermal antitumor and antibacterial effects. *International Journal of Pharmaceutics* 580: 119231. doi: 10.1016/j.ijpharm.2020.119231.

Yan, Xin, Shanshan Gu, Xingyang Cui, Yunjia Shi, Shanshan Wen, Hongyan Chen et al. (2019). Antimicrobial, anti-adhesive and anti-biofilm potential of biosurfactants isolated from *Pediococcus acidilactici* and *Lactobacillus plantarum* against *Staphylococcus aureus* CMCC26003. *Microbial Pathogenesis,* 127: 12–20. doi: 10.1016/j.micpath.2018.11.039.

Yang, Xu, En Huang, Chunhua Yuan, Liwen Zhang & Ahmed E. Yousef. (2016). Isolation and structural elucidation of Brevibacillin, an antimicrobial lipopeptide from *Brevibacillus laterosporus* that combats drug-resistant gram-positive bacteria. *Applied and Environmental Microbiology,* 82(9): 2763–2772. doi: 10.1128/AEM.00315-16.

Yuliani, Hanif, Meka Saima Perdani, Imelda Savitri, Meilani Manurung, Muhamad Sahlan, Anondho Wijanarko et al. (2018). Antimicrobial activity of biosurfactant derived from *Bacillus subtilis* C19. *Energy Procedia,* 153: 274–278. doi: 10.1016/j.egypro.2018.10.043.

Zampolli, Jessica, Alessandra De Giani, Alessandra Di Canito, Guido Sello & Patrizia Di Gennaro. (2022). Identification of a novel biosurfactant with antimicrobial activity produced by *Rhodococcus opacus* R7. *Microorganisms,* 10(2): 475. doi: 10.3390/microorganisms100(20475.

Zhou, Shengnan, Ge Liu & Shimei Wu. (2020). Marine bacterial surfactin CS30-2 induced necrosis-like cell death in Huh7.5 liver cancer cells. *Journal of Oceanology and Limnology,* 38(3): 826–833. doi: 10.1007/s00343-019-9129-2.